I0041904

INTRODUCTION TO
THE THEORY OF THE
EARLY UNIVERSE
Hot Big Bang Theory

SECOND EDITION

INTRODUCTION TO
THE THEORY OF THE
EARLY UNIVERSE
Hot Big Bang Theory

SECOND EDITION

VALERY A RUBAKOV

Russian Academy of Sciences, Russia & Moscow State University, Russia

DMITRY S GORBUNOV

Russian Academy of Sciences, Russia

World Scientific

NEW JERSEY · LONDON · SINGAPORE · BEIJING · SHANGHAI · HONG KONG · TAIPEI · CHENNAI · TOKYO

Published by

World Scientific Publishing Co. Pte. Ltd.

5 Toh Tuck Link, Singapore 596224

USA office: 27 Warren Street, Suite 401-402, Hackensack, NJ 07601

UK office: 57 Shelton Street, Covent Garden, London WC2H 9HE

Library of Congress Cataloging-in-Publication Data

Names: Gorbunov, D. S. (Dmitriĭ Sergeevich) | Rubakov, V. A.

Title: Introduction to the theory of the early universe : hot big bang theory / Valery A. Rubakov
   (Russian Academy of Sciences, Russia & Moscow State University, Russia),
   Dmitry S. Gorbunov (Russian Academy of Sciences, Russia).

Other titles: Vvedenie v teoriiu ranneĭ Vselennoĭ. English | Hot big bang theory

Description: 2nd edition. | New Jersey : World Scientific, 2017. |
   Includes bibliographical references and index.

Identifiers: LCCN 2017014302| ISBN 9789813209879 (hardcover : alk. paper) |
   ISBN 9813209879 (hardcover : alk. paper) | ISBN 9789813209886 (pbk : alk. paper) |
   ISBN 9813209887 (pbk : alk. paper)

Subjects: LCSH: Big bang theory. | Expanding universe.

Classification: LCC QB991.B54 G6713 2017 | DDC 523.1/8--dc23

LC record available at https://lccn.loc.gov/2017014302

British Library Cataloguing-in-Publication Data

A catalogue record for this book is available from the British Library.

Copyright © 2018 by World Scientific Publishing Co. Pte. Ltd.

All rights reserved. This book, or parts thereof, may not be reproduced in any form or by any means, electronic or mechanical, including photocopying, recording or any information storage and retrieval system now known or to be invented, without written permission from the publisher.

For photocopying of material in this volume, please pay a copying fee through the Copyright Clearance Center, Inc., 222 Rosewood Drive, Danvers, MA 01923, USA. In this case permission to photocopy is not required from the publisher.

Desk Editor: Ng Kah Fee

Typeset by Stallion Press

Email: enquiries@stallionpress.com

To Olesya and Elvira

Preface to the 2nd Edition

Particle physics and cosmology enjoyed rapid development between the first and the second editions of this book. Experimental research resulted in a wealth of new data, which forced us to correct and in some places rewrite chapters on dark matter, phase transitions and baryon asymmetry of the Universe. We have also made substantial revisions in other chapters. The numerical values of particle physics and cosmological parameters have been updated in accordance to contemporary data.

Besides that, we have corrected numerous misprints and drawbacks that existed in the first edition. We are indebted to our numerous colleagues for their input, and also students at the Department of Particle Physics and Cosmology, which has been recently created at the Physics Faculty of the Lomonosov Moscow State University (http://ppc.inr.ac.ru).

Preface to the 1st Edition

It is clear by now that there is a deep interconnection between cosmology and particle physics, and between macro- and micro-worlds. This book is written precisely from this perspective. We present here the results on the homogeneous and isotropic Universe at the hot stage of its evolution and at subsequent stages. This part of cosmology is often dubbed as the Hot Big Bang theory. In the accompanying book we study the theory of cosmological perturbations (inhomogeneities in the Universe), inflationary theory and the theory of post-inflationary reheating.

This book grew from the lecture course which had been taught for a number of years at the Department of Quantum Statistics and Field Theory of the Physics Faculty of the Lomonosov Moscow State University. This course is aimed at undergraduate students specializing in theoretical physics. We decided, however, to add a number of more advanced Chapters and Sections which we mark by asterisks. The reason is that there are problems in cosmology (nature of dark matter and dark energy, mechanism of the matter-antimatter asymmetry generation, etc.) which have not found their compelling solutions yet. Most of the additional material deals with hypotheses on these problems that at the moment compete with each other.

Knowledge of material taught in general physics courses is in principle sufficient for reading the main Chapters of this book. So, the main Chapters must be understandable by undergraduate students. The necessary material on General Relativity and particle physics is collected in the Appendices which, of course, do not pretend to give comprehensive account of these areas of physics. On the other hand, some parts labeled by asterisks make use of the methods of classical and quantum field theory as well as nonequilibrium statistical mechanics, so basic knowledge of these methods is required for reading these parts.

Literature on cosmology is huge, and presenting systematic and comprehensive bibliography would be way out of the scope of this book. To orient the reader, at the end of this book we give a list of monographs and reviews where the issues we touch upon are considered in detail. Certainly, this list is by no means complete. We occasionally refer to original literature, especially in those places where we present concrete results without detailed derivation.

Both observational cosmology and experimental particle physics develop very fast. Observational and experimental data we quote, the results of their compilations and fits (values of the cosmological parameters, limits on masses and couplings of hypothetical particles, etc.) will most probably get more precise even before this book is published. This drawback can be corrected, e.g., by using the regularly updated material of Particle Data Group at http://pdg.lbl.gov/.

We would like to thank our colleagues from the Institute for Nuclear Research of the Russian Academy of Sciences — F. L. Bezrukov, S. V. Demidov, V. A. Kuzmin, D. G. Levkov, M. V. Libanov, E. Y. Nugaev, G. I. Rubtsov, D. V. Semikoz, P. G. Tinyakov, I. I. Tkachev and S. V. Troitsky for participation in the preparation of the lecture course and numerous helpful discussions and comments. Our special thanks are to S. L. Dubovsky who participated in writing this book at an early stage. We are deeply indebted to V. S. Berezinsky, A. Boyarsky, A. D. Dolgov, D. I. Kazakov, S. Y. Khlebnikov, V. F. Mukhanov, I. D. Novikov, K. A. Postnov, M. V. Sazhin, M. E. Shaposhnikov, A. Y. Smirnov, A. A. Starobinsky, R. A. Sunyaev, A. N. Tavkhelidze, O. V. Verkhodanov, A. Vilenkin, M. B. Voloshin and M. I. Vysotsky for many useful comments and criticism.

Contents

Chapter 1

Cosmology: A Preview

The purpose of this Chapter is to give a preview of the field which we consider in this and the accompanying book. The presentation here is at the qualitative level, and is by no means complete. Our purpose is to show the place of one or another topic within the entire area of cosmology.

Before proceeding, let us introduce units and conventions that we use throughout this book.

1.1. Units

We mostly use the "natural" system of units in which the Planck constant, speed of light and the Boltzmann constant are equal to 1,

$$\hbar = c = k_B = 1.$$

Then the mass M, energy E and temperature T have the same dimension (since $[E] = [mc^2]$, $[E] = [k_B T]$). A convenient unit of mass and energy is $1\,\text{eV}$ or $1\,\text{GeV} = 10^9\,\text{eV}$; the proton mass is then equal to $m_p = 0.938\,\text{GeV}$, and $1\,\text{K}$ is approximately $10^{-13}\,\text{GeV}$. Time t and length l in the natural system have dimension M^{-1} (since $[E] = [\hbar\omega]$, $[\omega] = [t^{-1}]$ and $[l] = [ct]$), with $1\,\text{GeV}^{-1} \sim 10^{-14}\,\text{cm}$ and $1\,\text{GeV}^{-1} \sim 10^{-24}\,\text{s}$. We give the coefficients relating various units in Tables 1.1 and 1.2.

Problem 1.1. *Check the relations given in Tables 1.1 and 1.2. What are 1 Volt (V), 1 Gauss (G), 1 Hertz (Hz) and 1 Angström (Å) in natural system of units?*

In natural system of units, the Newton gravity constant G has dimension M^{-2}. This follows from the formula for the gravitational potential energy $V = -G\frac{m_1 m_2}{r}$, since $[V] = M$, $[r^{-1}] = M$. It is convenient to introduce the Planck mass M_{Pl} in the following way,

$$G = M_{Pl}^{-2}.$$

Numerically

$$M_{Pl} = 1.2 \cdot 10^{19}\,\text{GeV}, \tag{1.1}$$

1

Table 1.1.   Conversion of natural units into CGS units.

Energy	$1\,\mathrm{GeV} = 1.6 \cdot 10^{-3}\,\mathrm{erg}$
Mass	$1\,\mathrm{GeV} = 1.8 \cdot 10^{-24}\,\mathrm{g}$
Temperature	$1\,\mathrm{GeV} = 1.2 \cdot 10^{13}\,\mathrm{K}$
Length	$1\,\mathrm{GeV}^{-1} = 2.0 \cdot 10^{-14}\,\mathrm{cm}$
Time	$1\,\mathrm{GeV}^{-1} = 6.6 \cdot 10^{-25}\,\mathrm{s}$
Particle number density	$1\,\mathrm{GeV}^3 = 1.3 \cdot 10^{41}\,\mathrm{cm}^{-3}$
Energy density	$1\,\mathrm{GeV}^4 = 2.1 \cdot 10^{38}\,\mathrm{erg} \cdot \mathrm{cm}^{-3}$
Mass density	$1\,\mathrm{GeV}^4 = 2.3 \cdot 10^{17}\,\mathrm{g} \cdot \mathrm{cm}^{-3}$

Table 1.2.   Conversion of CGS units into natural units.

Energy	$1\,\mathrm{erg} = 6.2 \cdot 10^2\,\mathrm{GeV}$
Mass	$1\,\mathrm{g} = 5.6 \cdot 10^{23}\,\mathrm{GeV}$
Temperature	$1\,\mathrm{K} = 8.6 \cdot 10^{-14}\,\mathrm{GeV}$
Length	$1\,\mathrm{cm} = 5.1 \cdot 10^{13}\,\mathrm{GeV}^{-1}$
Time	$1\,\mathrm{s} = 1.5 \cdot 10^{24}\,\mathrm{GeV}^{-1}$
Particle number density	$1\,\mathrm{cm}^{-3} = 7.7 \cdot 10^{-42}\,\mathrm{GeV}^3$
Energy density	$1\,\mathrm{erg} \cdot \mathrm{cm}^{-3} = 4.8 \cdot 10^{-39}\,\mathrm{GeV}^4$
Mass density	$1\,\mathrm{g} \cdot \mathrm{cm}^{-3} = 4.3 \cdot 10^{-18}\,\mathrm{GeV}^4$

and the Planck length, time and mass are

$$l_{Pl} = \frac{1}{M_{Pl}} = 1.6 \cdot 10^{-33}\,\mathrm{cm},$$

$$t_{Pl} = \frac{1}{M_{Pl}} = 5.4 \cdot 10^{-44}\,\mathrm{s}, \tag{1.2}$$

$$M_{Pl} = 2.2 \cdot 10^{-5}\,\mathrm{g}.$$

The gravitational interactions are weak precisely because M_{Pl} is large.

Problem 1.2. *Check the relations* (1.1) *and* (1.2).

Problem 1.3. *What is the ratio of gravitational interaction energy to Coulomb energy for two protons?*

The traditional unit of length in cosmology is megaparsec,

$$1\,\mathrm{Mpc} = 3.1 \cdot 10^{24}\,\mathrm{cm}.$$

Let us also introduce a convention which we use in this book. The subscript 0 denotes present values of quantities which can depend on time. As an example, $\rho(t)$ denotes the energy density in the Universe as a function of time, while $\rho_0 \equiv \rho(t_0)$ is always its present value.

There are several units of length that are used in astronomy, depending on sizes of objects and length scales considered. Besides the metric system, in use are astronomical unit (a.u.), which is the average distance from the Earth to the Sun,

$$1\,\text{a.u.} = 1.5 \cdot 10^{13}\,\text{cm};$$

light year, the distance that a photon travels in one year,

$$1\,\text{year} = 3.16 \cdot 10^7\,\text{s}, \quad 1\,\text{light year} = 3 \cdot 10^{10}\,\frac{\text{cm}}{\text{s}} \cdot 3.16 \cdot 10^7\,\text{s} = 0.95 \cdot 10^{18}\,\text{cm};$$

and parsec (pc) — distance from which an object of size 1 a.u. is seen at angle 1 arc second,

$$1\,\text{pc} = 2.1 \cdot 10^5\,\text{a.u.} = 3.3\,\text{light year} = 3.1 \cdot 10^{18}\,\text{cm}.$$

To illustrate the hierarchy of spatial scales in the Universe, let us give the distances to various objects expressed in the above units.

10 a.u. is the average distance to Saturn, 30 a.u. is the same for Pluto, 100 a.u. is the estimate of maximum distance which can be reached by solar wind (particles emitted by the Sun). 100 a.u. is also the estimate of the maximum distance to cosmic probes (Pioneer 10, Voyager 1, Voyager 2). Further out is the Oort cloud, the source of the most distant comets, which is at the distance of 10^4–10^5 a.u. $\sim$0.1–1 pc.

The nearest stars — Proxima and Alpha Centauri — are at 1.3 pc from the Sun. The distance to Arcturus and Capella is more than 10 pc, the distances to Canopus and Betelgeuse are about 100 pc and 200 pc, respectively; Crab Nebula — the remnant of supernova seen by naked eye — is 2 kpc away from us.

The next point on the scale of distances is 8 kpc. This is the distance from the Sun to the center of our Galaxy. Our Galaxy is of spiral type, the diameter of its disc is about 30 kpc and the thickness of the disc is about 250 pc. The distance to the nearest dwarf galaxies, satellites of our Galaxy, is about 30 kpc. Fourteen of these satellites are known; the largest of them — Large and Small Magellanic Clouds — are 50 kpc away. Search for new, dimmer satellite dwarfs is underway; we note in this regard that only eight of Milky Way satellites were known by 1994.

The mass density of the usual matter in usual (not dwarf) galaxies is about 10^5 higher than the average over the Universe.

The nearest usual galaxy–the spiral galaxy M31 in Andromeda constellation — is 800 kpc away from the Milky Way. Despite the large distance, it occupies a sizeable area on the celestial sphere: its angular size is larger than that of the Moon! Another nearby galaxy is in the Triangulum constellation. Our Galaxy together with the Andromeda and Triangulum galaxies, their satellites and other 35 smaller galaxies constitute the Local Group, the gravitationally bound object consisting of more than 50 galaxies.

The next scale in this ladder is the size of clusters of galaxies, which is 1–3 Mpc. Rich clusters contain thousands of galaxies. The mass density in clusters exceeds the average density over the Universe by a factor of a hundred and even sometimes a thousand. The distance to the center of the nearest cluster, which is in the Virgo constellation, is about 15 Mpc. Its angular size is about 5 degrees. Clusters of galaxies are the largest gravitationally bound systems in the Universe.

1.2. The Universe Today

We begin our preview with the brief discussion of the properties of the present Universe (more precisely, of its observable part).

1.2.1.   *Homogeneity and isotropy*

The Universe is homogeneous and isotropic at large spatial scales. The sizes of the largest structures in the Universe — superclusters of galaxies and gigantic voids — reach[1] tens of megaparsec. At scales exceeding 200 Mpc, all parts of the Universe look the same (homogeneity). Likewise, there are no special directions in the Universe (isotropy). These facts are well established by deep galaxy surveys which collected data on more than a million galaxies.

About 20 superclusters are known by now. The Local Group belongs to a supercluster with the center in the direction of Virgo constellation. The size of this supercluster is about 30 Mpc, and besides the Virgo cluster and Local Group it contains about a hundred groups and clusters of galaxies. Superclusters are rather loose: the density of galaxies in them is only twice higher than the average in the Universe. The nearest to Virgo is the supercluster in the Hydra and Centaurus constellations; its distance to the Virgo supercluster is about half a hundred megaparsec.

The largest catalog of galaxies and quasars up to date is the freely available catalog of SDSS [2] (Sloan Digital Sky Survey). This catalog is the result of the analysis of the data collected during almost 8 years of operation of a dedicated telescope, 2.5 meters in diameter, which is capable of measuring simultaneously spectra of 640 astrophysical objects in 5 optical bandpasses (photon wavelengths $\lambda = 3800\text{–}9200$ Å). The catalog includes millions celestial objects. Most of the data has been processed by now; measurements of spectra of more than 1.8 million galaxies and more than 300 thousand quasars resulted in the creation of a 3-dimensional map covering a large part of the visible Universe. Its area exceeds a quarter of the sky. There are other catalogs which cover smaller parts of the Universe (see, e.g., Ref. [3] for the next-to-largest catalog based on the 2dF Galaxy Redshift Survey).

The early SDSS results are illustrated in Fig. 13.1 in color pages, where positions of 40 thousand galaxies and 4 thousand quasars are presented. The covered part of the celestial sphere has the area of 500 squared degrees. Recognizable are clusters of galaxies and voids. Isotropy and homogeneity of the Universe are restored at spatial scales of order 100 Mpc and larger. Color of each dot refers to the type of the astrophysical object. The domination of one type over others is, generally speaking, caused by peculiarities of structure formation and evolution. Thus, what one observes is partially pictured in *time* rather than in space.

Indeed, from the distance of 1.5 Gpc, where the distribution of bright red elliptical galaxies (red dots in Fig. 13.1) is at maximum, light travels to the Earth for about 5 billion years. At that epoch, the Universe was different (for instance, there was no Solar system yet). One more reason for choosing objects of a certain type is the finite sensitivity of a telescope. In particular, only highly luminous objects can be detected at the largest distances, while the highest-luminosity, continuously shining objects in the Universe are quasars.

1.2.2.   *Expansion*

The Universe expands: the distances between galaxies increase.[2] Loosely speaking, the space, being always homogeneous and isotropic, stretches out.

---

[1]This is a somewhat loose statement: most accurate estimates are obtained from the galaxy correlation function, which falls off as power-law at large separations.

[2]Of course, this does not apply to galaxies that are gravitationally bound to each other in clusters.

To describe this expansion, one introduces the scale factor $a(t)$ which grows in time. The distance between two far away objects in the Universe is proportional to $a(t)$ and the number density of particles decreases as $a^{-3}(t)$. The rate of the cosmological expansion, i.e., the relative growth of distances in unit time, is characterized by the Hubble parameter,

$$H(t) = \frac{\dot{a}(t)}{a(t)}.$$

Hereafter, the dot denotes the derivative with respect to the cosmic time t. The Hubble parameter depends on time; its present value, according to our convention, is denoted by H_0.

The expansion of the Universe gives rise also to the growth of the wavelength of a photon emitted in distant past. Like other distances, the photon wavelength increases proportionally to $a(t)$; the photon experiences redshift. This redshift z is determined by the ratio of photon wavelengths at absorption and emission,

$$\frac{\lambda_{ab}}{\lambda_{em}} \equiv 1 + z. \tag{1.3}$$

Clearly, this ratio depends on the moment of the emission (assuming that the photon is detected today on the Earth), i.e., on the distance to the source. Redshift is a directly measurable quantity: the wavelength at emission is determined by the physics of the emission process (e.g., by energies of an excited and the ground state of an atom), while λ_{ab} is the measured wavelength. Thus, one identifies the system of emission (or absorption) lines and determines how much they are shifted to the red spectral region, and in this way one measures the redshift.

In reality, the identification of lines makes use of patterns which are characteristic of particular objects, see Fig. 1.1, Ref. [5]. If the spectrum contains absorption dips, as in Fig. 1.1, then the object whose redshift is being measured is between the emitter and observer.[3] The peaks in the spectrum — emission lines — mean that the object is an emitter itself.

For $z \ll 1$, the distance to the source r and the redshift are related by the Hubble law

$$z = H_0 r, \quad z \ll 1. \tag{1.4}$$

At larger z the redshift-distance relation is more complicated, which we will discuss in detail in this book.

The determination of absolute distances to far away sources is a complicated problem. One of the methods is to measure the photon flux from a source whose absolute luminosity is assumed to be known. These sources are sometimes called standard candles.

---

[3] Photons of definite wavelengths experience resonant absorption by atoms and ions, with subsequent isotropic emission. This leads to the loss of photons reaching the observer.

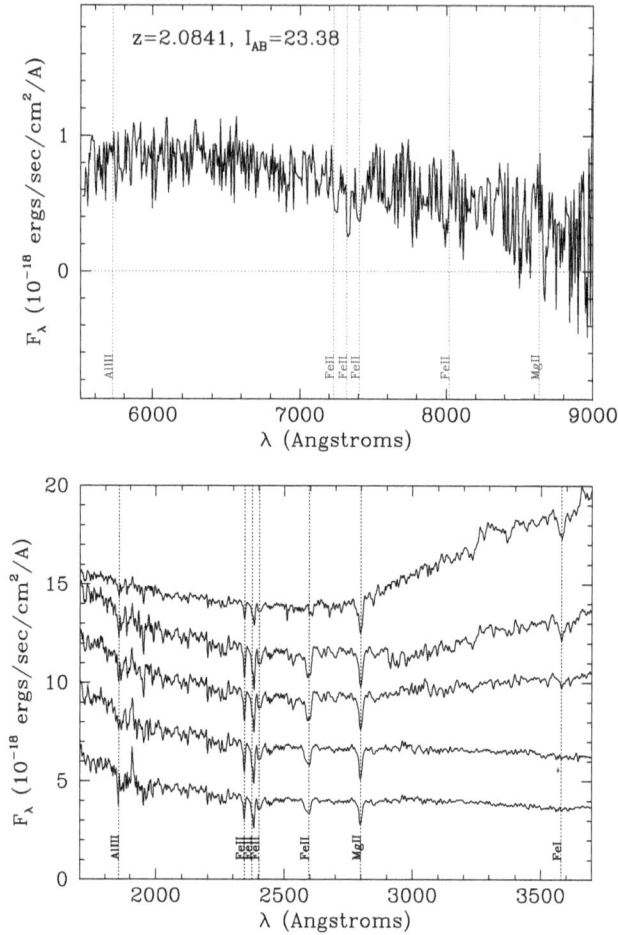

Fig. 1.1.   Absorption lines of distant galaxies [5]. The upper panel shows the measurement of the differential energy flux from a far away galaxy ($z = 2.0841$). The vertical lines show the position of atomic lines whose identification has been used to measure redshift. The spectra of nearer galaxies have more pronounced dips. The plot with the spectra of these galaxies, shifted to comoving frame, is shown in the lower panel.

Systematic uncertainties in the determination of H_0 were not particularly well known until recently and they are still fairly large. We note in this regard that the value of the Hubble constant as determined by Hubble in 1929 was $550\,\mathrm{km/(s \cdot Mpc)}$. The contemporary determinations give [1]

$$H_0 = (67.3 \pm 1.2)\,\frac{\mathrm{km}}{\mathrm{s \cdot Mpc}}. \tag{1.5}$$

Problem 1.4. *Relate the dimensionless redshift to distance expressed in Mpc.*

Let us comment on the traditional unit for the Hubble parameter used in (1.5). A naive interpretation of the Hubble law (1.4) is that the redshift is caused by the radial motion of galaxies from the Earth with velocities proportional to the distances,

$$\mathbf{v} = H_0 \mathbf{r}, \quad v \ll 1. \tag{1.6}$$

Then the redshift (1.4) is interpreted as the longitudinal Doppler effect (at $v \ll c$, i.e., $v \ll 1$ in natural units, the Doppler shift equals to $z = v$). According to this interpretation, the dimension of the Hubble parameter H_0 is [velocity/distance]. We stress, however, that the interpretation of the cosmological redshift in terms of the Doppler effect is unnecessary, and often inadequate. The right way is to use the relation (1.4) as it is.

Problem 1.5. *Consider a system of many particles in Newtonian mechanics. Show that it is spatially homogeneous and isotropic if and only if the density of the particles is constant over space, and the relative velocity of each pair of particles i and j is related to the distance between them by the "Hubble law"*

$$\mathbf{v}_{ij} = H_0 \mathbf{r}_{ij},$$

where H_0 is independent of spatial coordinates. Hereafter, bold face letters denote three-vectors, $\mathbf{v} = (v_1, v_2, v_3)$.

The quantity H_0 is usually parameterized in the following way,

$$H_0 = h \cdot 100 \, \frac{\text{km}}{\text{s} \cdot \text{Mpc}}, \tag{1.7}$$

where h is a dimensionless parameter of order one (see (1.5)),

$$h = 0.673 \pm 0.012.$$

We use the value $h = 0.7$ in numerical estimates throughout this book.

One type of objects used for measuring the Hubble parameter are Cepheids, stars of variable brightness whose variability is related to absolute luminosity in a known way. This relationship is measured by observing Cepheids in compact systems like Magellanic Clouds. Since Cepheids in one and the same system are, to good approximation, at the same distance from us, the ratio of their visible brightness to absolute luminosity is the same for every star. The periods of Cepheid pulsations range from a day to tens of days, and during this period the brightness varies within an order of magnitude. The results of observations show that there is indeed a well-defined relation between the period and luminosity: the longer the period, the brighter the star. Hence, Cepheids serve as standard candles.

Cepheids are giants and super-giants, so they are visible at large distances from our Galaxy. By measuring their spectra, one finds redshift of each of them, and by measuring the period of pulsations one obtains the absolute luminosity and hence the distance. Using these data, one measures the Hubble constant in (1.4). Figure 1.2 shows the Hubble diagram — redshift-distance relation — obtained in this way [10].

Besides Cepheids, there are other objects used as standard candles. These include, in particular, supernovae of type Ia. The determination of the Hubble parameter from the observations of remote standard candles is shown in Fig. 1.3.

Cosmology: A Preview

Hubble Diagram for Cepheids (flow–corrected)

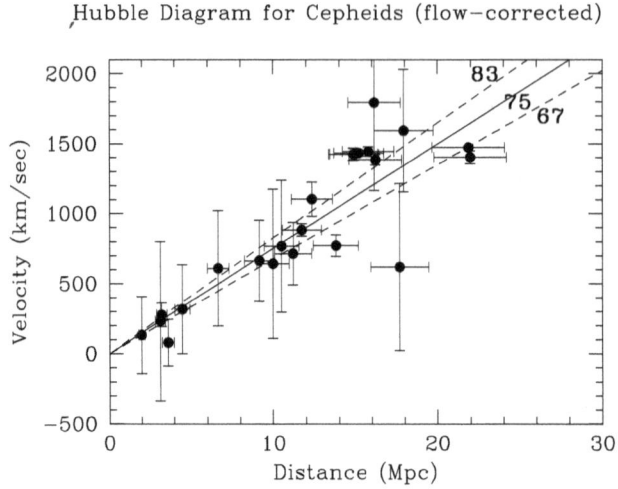

Fig. 1.2.   Hubble diagram for Cepheids [10]. The solid line shows the Hubble law with the Hubble constant $H_0 = 75\,\mathrm{km}/(\mathrm{s}\cdot\mathrm{Mpc})$, as determined from these observations. The dashed lines show the uncertainty in the determination of the Hubble parameter.

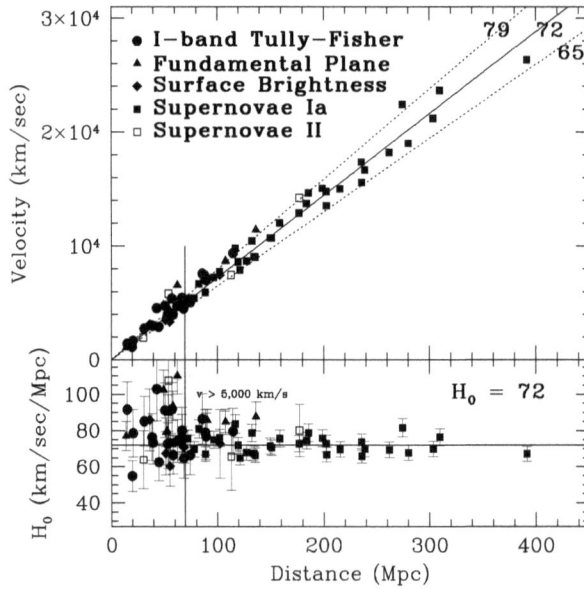

Fig. 1.3.   Hubble diagram for remote standard candles including supernovae of type Ia [10]. The solid line shows the Hubble law with the value of the Hubble parameter $H_0 = 72\ \mathrm{km}/(\mathrm{s}\cdot\mathrm{Mpc})$ as determined from these data. Dashed lines correspond to experimental uncertainty in the Hubble parameter.

1.2.3. *Age of the Universe and size of its observable part*

The Hubble parameter in fact has dimension $[t^{-1}]$, so the present Universe is characterized by the time scale

$$H_0^{-1} = \frac{1}{h} \cdot \frac{1}{100} \frac{\mathrm{s} \cdot \mathrm{Mpc}}{\mathrm{km}} = h^{-1} \cdot 3 \cdot 10^{17}\,\mathrm{s} = h^{-1} \cdot 10^{10}\,\mathrm{yrs}$$

and the scale of distances

$$H_0^{-1} = h^{-1} \cdot 3000\,\mathrm{Mpc}\,, \tag{1.8}$$

which gives, for $h = 0.7$,

$$H_0^{-1} = 1.4 \cdot 10^{10}\,\mathrm{yrs} \tag{1.9a}$$
$$= 4.3 \cdot 10^3\,\mathrm{Mpc}. \tag{1.9b}$$

Crudely speaking, all distances in the Universe will become twice larger in about 10 billion years; galaxies at distance of order 3 Gpc from us move away with velocities comparable to the speed of light. We will see that the time scale H_0^{-1} gives the order of magnitude estimate for the age of the Universe, and the distance scale H_0^{-1} is roughly the size of its observable part. We will discuss the notions of the age and size of the observable part in the course of presentation, and here we point out that bold extrapolation of the cosmological evolution back in time (made according to the equations of classical General Relativity) leads to the notion of the Big Bang, the moment at which the classical evolution begins. Then the age of the Universe is the time passed from the Big Bang, and the size of the observable part (horizon size) is the distance travelled by signals emitted at the Big Bang and moving at the speed of light (more accurate estimate of the horizon size is 15 Mpc). We note in passing that the actual size of our Universe is larger, and most probably much larger than the horizon size; the spatial size of the Universe may be infinite in General Relativity.

Irrespective of the cosmological data, there exist observational lower bounds on the age of the Universe t_0. Various independent methods give similar bounds at the level

$$t_0 \gtrsim 13\,\mathrm{billion\ years} = 1.3 \cdot 10^{10}\,\mathrm{yrs}. \tag{1.10}$$

One of these methods makes use of the distribution of luminosities of white dwarfs. White dwarfs are compact stars of high density, whose masses are similar to the solar mass. They slowly cool down and get dimmer. There are white dwarfs of various luminosities in the Galaxy, but the number of them sharply drops off below a certain luminosity. This means that there is a maximum age of white dwarfs, which, of course, should be smaller than the age of the Universe. This maximum age is found from the energy balance of a white dwarf (see, e.g., Ref. [12]). In this way the bound like (1.10) is obtained.

Other methods include the studies of the radioactive element abundances in the Earth core, in meteorites (see, e.g., Ref. [13]), and in the metal-poor[4] stars (e.g., Ref. [14]), the comparison (e.g., Ref. [15]) of the stellar evolution curve for main-sequence stars

---

[4]The term "metals" in astrophysics is used for all elements heavier than helium.

on the Herzsprung–Russel diagram (luminosity-color or brightness-temperature) with the abundance of the oldest stars in metal-poor globular clusters,[5] the analysis of relaxation processes in stellar clusters, measurement of the abundance of hot gas in clusters of galaxies, etc.

1.2.4. *Spatial flatness*

Homogeneity and isotropy of the Universe do not imply, generally speaking, that at each moment of time the 3-dimensional space is Euclidean, i.e., that the Universe has zero spatial curvature. Besides the 3-plane (3-dimensional Euclidean space), there are two homogeneous and isotropic spaces, 3-sphere (positive spatial curvature) and 3-hyperboloid (negative curvature). A fundamental observational result of recent years is the fact that the spatial curvature of our Universe is very small, if not exactly zero. Our 3-dimensional space is thus Euclidean to a very good approximation. We will repeatedly get back to this statement, both for quantifying it and for explaining which observational data set bounds on the spatial curvature. We only note here that the main source of these bounds is the study of the temperature anisotropy of the Cosmic Microwave Background (CMB), and that at the qualitative level, these bounds mean that the radius of spatial curvature is much greater than the size of the observable part of the Universe, i.e., much greater than H_0^{-1}.

We note here that CMB data are also consistent with the trivial spatial topology. If our Universe had compact topology (e.g., topology of 3-torus) and its size were of the order of the Hubble length, CMB temperature anisotropy would show a certain regular pattern. Such a pattern is absent in measured anisotropy, see Ref. [11].

1.2.5. *"Warm" Universe*

The present Universe is filled with Cosmic Microwave Background (CMB), gas of non-interacting photons, which was predicted by the Hot Big Bang theory and discovered in 1964. The number density of CMB photons is about 400 per cubic centimeter. The energy distribution of these photons has thermal, Planckian spectrum. This is shown in Fig. 1.4 [16]. The CMB temperature is [1]

$$T_0 = 2.7255 \pm 0.0006 \text{ K}. \tag{1.11}$$

The temperature of photons coming from different directions on celestial sphere is the same at the level of better than 10^{-4} (modulo dipole component, see below); this is yet another evidence for homogeneity and isotropy of the Universe.

Still, the temperature does depend on the direction in the sky. The angular anisotropy of the CMB temperature has been measured, as shown in Fig. 1.5 [38] (see Fig. 13.2 on color pages). It is of order $\delta T/T_0 \sim 10^{-4}$–$10^{-5}$.

---

[5]Globular clusters are structures of sizes of order 30 pc inside galaxies; they can contain hundreds of thousand and even millions of stars.

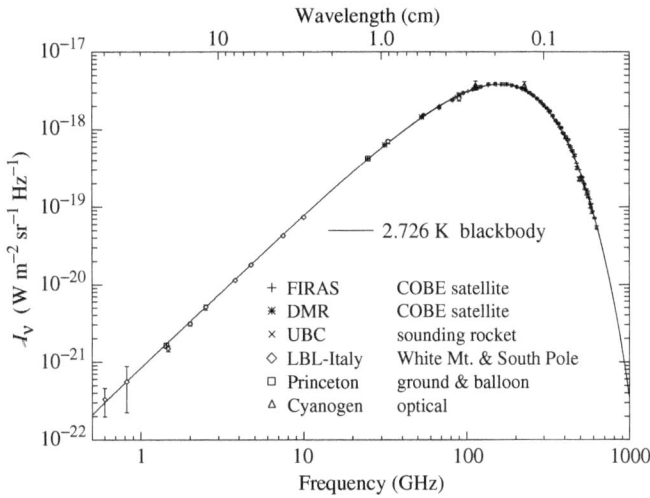

Fig. 1.4.   CMB spectrum. The compilation of the data is made in Ref. [16]. The solid line shows the Planckian (black body) spectrum.

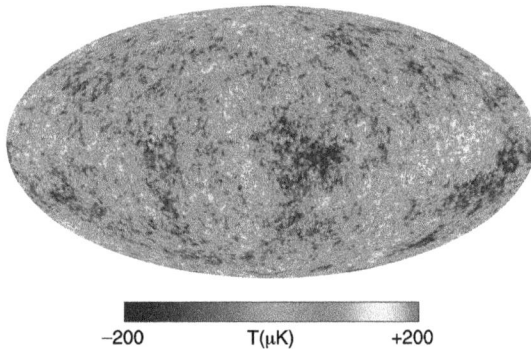

Fig. 1.5.   WMAP data [38]: angular anisotropy of CMB temperature, i.e., variation of the temperature of photons coming from different directions in the sky; see Fig. 13.2 for color version. The average temperature and dipole component are subtracted. The observed variation of temperature is at the level of $\delta T \sim 100 \ \mu$K, i.e., $\delta T/T_0 \sim 10^{-4}-10^{-5}$.

We will repeatedly come back to CMB anisotropy and polarization, since, on the one hand, they encode a lot of information about the present and early Universe and, on the other hand, they can be measured with high precision.

Let us note that the existence of CMB means that there is a special reference frame in our Universe: this is the frame in which the gas of photons is at rest. The solar system moves with respect to this frame towards Hydra constellation. The velocity of this motion determines the dipole component of the measured CMB anisotropy [18],

$$\delta T_{\text{dipole}} = 3.346 \ \text{mK}. \tag{1.12}$$

Problem 1.6. *Making use of the value of the dipole component, estimate the velocity of motion of the Solar system with respect to CMB.*

Problem 1.7. *Estimate the seasonal modulation of the CMB anisotropy caused by the motion of the Earth around the Sun.*

The present Universe is transparent to the CMB photons:[6] their mean free path well exceeds the horizon size H_0^{-1}. This was not the case in the early Universe, when photons actively interacted with matter.

Problem 1.8. Greisen–Zatsepin–Kuzmin effect [20, 21]. *Interaction of photon with proton at sufficiently high energies may lead to the absorption of photon and creation of π-meson. Let the cross-section of the latter process in the center-of-mass frame be (in fact, this is a pretty reasonable approximation for this problem)*

$$\sigma = \begin{cases} 0 & at \ \sqrt{s} < m_\Delta \\ 0.5 \ \mathrm{mb} & at \ \sqrt{s} > m_\Delta \end{cases},$$

where $\sqrt{s}$ is the total energy of photon and proton, $m_\Delta = 1200 \, \mathrm{MeV}$ (Δ is the resonance mass), $1 \, \mathrm{mb} = 10^{-27} \, \mathrm{cm}^2$.

Find the mean free path of a proton in the present Universe with respect to this process as a function of proton energy. At what distance from the source does proton lose 2/3 of its energy? Ignore all photons (e.g., emitted by stars), except for CMB.

Since the CMB temperature T depends on the direction $\mathbf{n}$ of celestial sphere, it is convenient to perform its decomposition over spherical harmonics $Y_{lm}(\mathbf{n})$. The latter form the basis of functions on a sphere, and the decomposition is the closest analog of the Fourier decomposition. The temperature fluctuation δT in the direction $\mathbf{n}$ is conveniently defined as

$$\delta T(\mathbf{n}) = T(\mathbf{n}) - T_0 - \delta T_{\mathrm{dipole}}$$

and its decomposition is

$$\delta T(\mathbf{n}) = \sum_{l,m} a_{l,m} Y_{lm}(\mathbf{n}),$$

where the coefficients $a_{l,m}$ obey $a_{l,m}^* = (-1)^m a_{l,-m}$, so that temperature is real. The multipoles l correspond to fluctuations of characteristic angular size π/l. The current measurements are capable of studying angular scales ranging from the largest ones to less than $0.1°$ ($l \sim 1000$, see Fig. 1.6 [39]).

---

[6]This is not completely true in some regions of the Universe. As an example, photons scatter off hot gas ($T \sim 10 \, \mathrm{keV}$) in clusters of galaxies and gain some energy. Thus, CMB is warmer in the directions towards clusters. This is called the Sunyaev–Zeldovich effect [19]. It is small but is measured in observations.

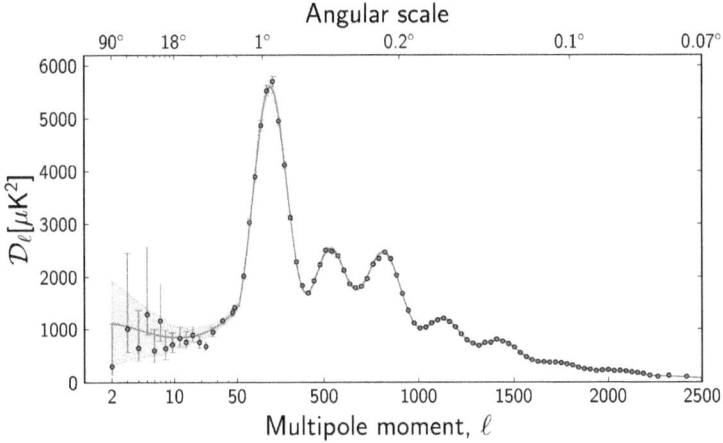

Fig. 1.6.   CMB angular anisotropy as measured by Planck experiment [39]. The theoretical curve is obtained within the ΛCDM model (see Chapter 4); finite width of this curve (shadow) illustrates cosmic variance, which is due to the fact that only one (our) Universe is observed.

The observational data are consistent with the property that temperature fluctuations $\delta T(\mathbf{n})$ are Gaussian random field, i.e., that the coefficients $a_{l,m}$ are statistically independent for different l and m,

$$\langle a_{l,m} a^*_{l',m'} \rangle = C_{lm}\delta_{ll'}\delta_{mm'}, \tag{1.13}$$

where brackets mean averaging over an ensemble of Universes like ours. The coefficients C_{lm} do not depend on m in isotropic Universe, $C_{lm} = C_l$. They determine the correlation of temperature fluctuations in different directions,

$$\langle \delta T(\mathbf{n_1})\delta T(\mathbf{n_2}) \rangle = \sum_l \frac{2l+1}{4\pi} C_l P_l(\cos\theta),$$

where P_l are the Legendre polynomials, functions of the angle θ between the vectors $\mathbf{n_1}$ and $\mathbf{n_2}$. In particular, the temperature fluctuation is

$$\langle \delta T^2 \rangle = \sum_l \frac{2l+1}{4\pi} C_l \approx \int \frac{l(l+1)}{2\pi} C_l\, d\ln l.$$

Thus, the quantity $\mathcal{D}_l \equiv \frac{l(l+1)C_l}{2\pi}$ determines the contribution to the fluctuation of a decimal interval of multipoles. It is this quantity that is shown in Fig. 1.6.

It is important that the measurement of the CMB anisotropy gives not just a number, but a large set of data, the values of C_l for different l. This set is determined by numerous parameters of the present and early Universe, hence its measurement provides a lot of cosmological information. Additional information comes from the measurement of CMB polarization.

1.3.   Energy Balance in the Present Universe

A dimensional estimate of the energy density in the Universe may be obtained in the following way. Given the energy density ρ_0, the density of "gravitational charge" is of order $G\rho_0$. Since the dynamics of the Universe is governed by gravity, the "charge" $G\rho_0$ must somehow be related to the present expansion rate. The "charge" has dimension of M^2; the same dimension as H_0^2. This suggests that $\rho_0 \sim H_0^2 G^{-1} = M_{Pl}^2 H_0^2$. Indeed, we will see that the present energy density in a *spatially flat* Universe is given by

$$\rho_c = \frac{3}{8\pi} H_0^2 M_{Pl}^2.$$

With precision better than 1%, this is the energy density in our Universe today.[7] Numerically

$$\rho_c = 1.88 \cdot 10^{-29} h^2 \frac{\text{g}}{\text{cm}^3} = 0.53 \cdot 10^{-5} \frac{\text{GeV}}{\text{cm}^3} \quad \text{at} \quad h = 0.7 . \tag{1.14}$$

According to the data of cosmological observations which we will discuss in due course, the contribution of baryons (protons, nuclei) into the total present energy density is[8] about 4.6%,

$$\Omega_B \equiv \frac{\rho_B}{\rho_c} = 0.046.$$

The contribution of relic neutrinos of all types is even smaller; the cosmological bound is

$$\Omega_\nu \equiv \frac{\sum \rho_{\nu_i}}{\rho_c} < 0.0055, \tag{1.15}$$

where the sum runs over the three species of neutrinos ν_e, ν_μ, ν_τ and anti-neutrinos $\bar\nu_e$, $\bar\nu_\mu$, $\bar\nu_\tau$. We emphasize that there is still no cosmological evidence for the neutrino mass; it is rather likely that the neutrino contribution is quite a bit smaller than the right-hand side of Eq. (1.15). Other known stable particles give negligible contribution to the present total energy density. Thus, the dominating material in the present Universe is something unknown.

This "something unknown" in fact consists of two fractions, one of which is capable of clustering, and another is not. The former component is called "dark matter". Its contribution to the energy density is about 25%.

We will discuss the results (Big Bang Nucleosynthesis, CMB anisotropy, structure formation) which show that dark matter cannot consist of known particles. Most probably it is made of new stable particles which were non-relativistic in very distant past and remain non-relativistic today (cold, or possibly warm dark matter). This is one of a few experimental evidences for New Physics beyond the Standard

---

[7] This 1% has to do with observationally allowed effect of spatial curvature.
[8] Note that only 10% of baryons are in stars. Most of baryons are in hot gas.

Model of particle physics. Direct detection of dark matter particles is an extremely important, and yet unsolved problem of particle physics.

According to current viewpoint, the rest of energy in the present Universe, about 70%, is homogeneously spread over space. This is not matter consisting of some unknown particles, but rather a unconventional form of energy of vacuum type. It is called by different names: dark energy, vacuum-like matter, quintessence, cosmological constant, Λ-term. We will use the term "dark energy" and will use the terms "quintessence" and "cosmological constant" for dark energy with specific properties: in the case of cosmological constant the energy density does not depend on time, while for quintessence weak dependence, instead, exists.

It is not excluded that observational data which are quoted as showing the presence of dark energy can be explained in an alternative way. One possibility is that gravity deviates from General Relativity at cosmological distance and time scales. There is theoretical activity in the latter direction indeed, but it is out of the scope of this book to discuss it in any detail. We will assume throughout that gravitational interactions are described by General Relativity.

We will further discuss dark energy and observations leading to this notion in due course. Here we mention the property that unlike the energy (mass) density of non-relativistic particles which decays as $a^{-3}(t)$ as the Universe expands, dark energy density either does not depend on time at all, or depends on time very weakly. Hence, at some stage of the cosmological evolution dark energy starts to dominate. The transition from matter dominated to dark energy dominated expansion occurred in our Universe at $z \simeq 0.5$.

Density of baryons and dark matter in clusters of galaxies is determined by various methods of measurement of the gravitational potential, i.e., total mass distribution. As an example, the left panel of Fig. 1.7 (see Fig. 13.4 on color pages) shows mass distribution in a

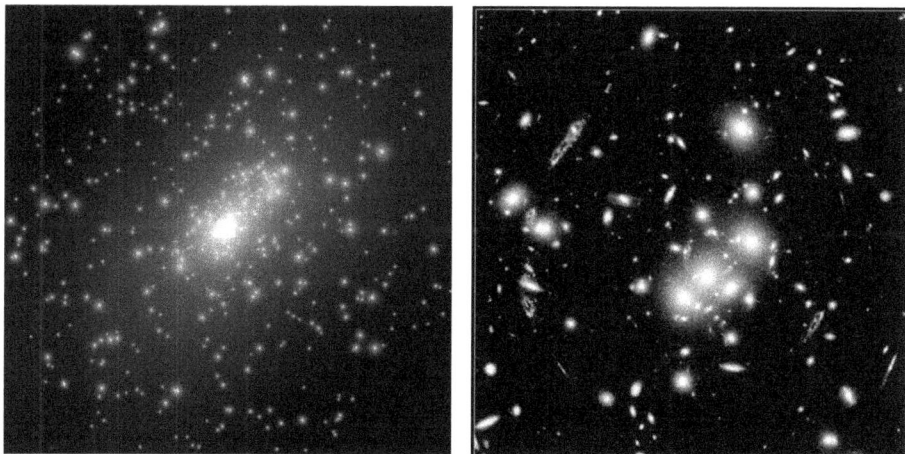

Fig. 1.7.   Cluster of galaxies CL0024 + 1654 [22]; see Fig. 13.4 for color version.

cluster, obtained by the method of gravitational lensing [22]. The idea is that light rays from galaxies residing behind the cluster get bent by the gravitational field of the cluster. This gives rise to multiple, distorted images of the distant galaxies[9] (see the right panel of Fig. 1.7). Hence, this method enables one to measure the gravitational potential in a cluster irrespectively of the sort of matter producing it. The result is that visible matter, whose density can be determined independently, makes rather small fraction of total mass; most of the mass is due to dark matter. The latter is clustered; its density is inhomogeneous over the Universe. Assuming that the ratio of dark and visible mass in the Universe is the same as in clusters of galaxies,[10] one finds that the mass density of baryons and dark matter together constitutes about 30% of the total energy,

$$\rho_M \approx 0.3\rho_c. \tag{1.16}$$

Gravitational lensing is by no means the only way to measure the gravitational potential in clusters of galaxies. In particular, X-ray observations show that most of baryons in clusters are in hot, ionized intergalactic gas. The total mass of baryons in the gas exceeds the mass of baryons in luminous matter by an order of magnitude. As a matter of fact, X-rays are produced by electrons. So, the observations give the spatial distributions of their number density $n_e(\mathbf{r})$ and temperature $T(\mathbf{r})$. Assuming spherical symmetry, one obtains the spatial distribution of total mass density $\rho(\mathbf{r})$ from the equation of hydrostatic equilibrium,

$$\frac{dP}{dR} = -\mu n_e(R) m_p \frac{GM(R)}{R^2}, \quad M(R) = 4\pi \int_0^R \rho(r) r^2 dr, \tag{1.17}$$

where $\mu n_e(R)$ is the number density of baryons in the gas, and μ is determined by its chemical composition (the gas is electrically neutral, so the numbers of baryons and electrons are the same up to factor μ). The pressure P is mostly due to electrons, and it is related to temperature in the usual way, $P(R) = n_e(R) T_e(R)$. All quantities in (1.17) except for $M(R)$ are obtained from observations, so the mass $M(R)$ is uniquely determined. Again, this method gives the results consistent with (1.16); see, e.g., Ref. [23].

The same conclusion follows from the study of the motion of galaxies and their groups in clusters. Assuming that the relaxation processes for galaxies have been over, one makes use of the virial theorem to infer the mass of a cluster,

$$3M\langle v_r^2 \rangle = G \frac{M^2}{R}. \tag{1.18}$$

Here M and R are the mass and size of a cluster, and $\langle v_r^2 \rangle^{1/2}$ is the dispersion of projections of velocities on the line of sight. The latter is determined by the analysis of the Doppler effect in the spectra of either entire cluster or individual galaxies. The masses obtained from the virial theorem (1.18) by far exceed the sum of masses of individual galaxies in clusters (even including dark galactic haloes); this means that most of the mass is due to dark matter which is distributed smoothly over the cluster.

It is known for a long time that nearby galaxy groups and clusters move towards Virgo constellation. Assuming that this motion is due to the gravitational attraction by the central cluster of galaxies, one estimates its mass. This estimate again shows that the mass of galaxies is too small, so this Virgo cluster contains dark matter.

---

[9]This is called strong lensing, as opposed to weak lensing which only affects the intensity of light.
[10]This is not an innocent assumption, since most of galaxies are *not* in clusters; conversely, clusters host only about 10% of galaxies and probably about 10% of dark matter.

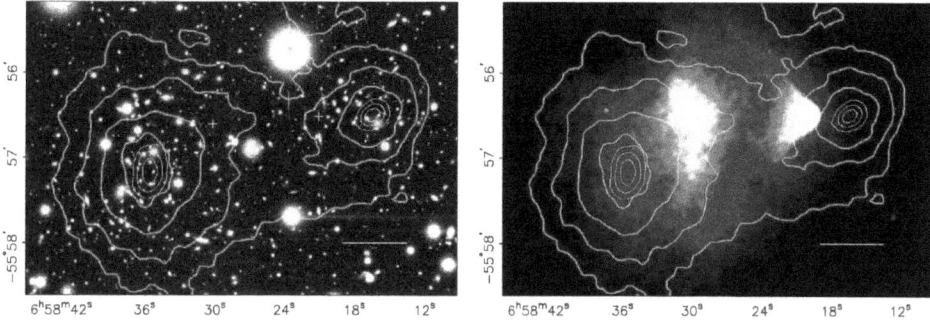

Fig. 1.8. Observation of "Bullet cluster" 1E0657-558, two collided clusters of galaxies [24]; see Fig. 13.5 for color version. Lines show gravitational equipotential surfaces, the bright regions in the right panel are regions of hot baryon gas.

A particularly convincing argument for dark matter in clusters of galaxies comes from the observation of clusters just after their collision. The result [24] is shown in Fig. 1.8 (see Fig. 13.5 on color pages). Bright colors in the right panel show the distribution of hot gas whose X-ray emission has been observed by Chandra telescope. This gas contains about 90% of all baryons in both clusters, while galaxies add the remaining 10%. The gravitational potential is measured via gravitational lensing; the distribution of galaxies follows the gravitational potential. It is clear that the gravitational potential is not at all produced by baryons; rather, its source is dark matter. Dark matter and galaxies here are collisionless: they passed through each other and move away from each other with the original speed. Hot gas loses its velocity due to collisions, and it lags behind dark matter.

Dark matter also explains the rotational velocities of stars, gas clouds, globular clusters and satellite dwarf galaxies at the periphery of *galaxies*. Under the assumption of circular motion, the dependence of velocity $v(R)$ on distance R from the galactic center follows from Newton's law,

$$v(R) = \sqrt{G\frac{M(R)}{R}}, \quad M(R) = 4\pi \int_0^R \rho(r)r^2dr,$$

where $\rho(r)$ is the mass density. Observationally, $v(R) = $ constant sufficiently far away from the center, see Fig. 1.9 [25], while the contribution of luminous matter to ρ would lead to $v(R) \propto 1/\sqrt{R}$. This is explained by assuming that luminous matter is embedded into dark cloud of larger size — galactic halo.

Clustered dark matter could in principle consist of the usual particles, baryons and electrons, if they were contained in dark objects like neutron stars and brown dwarfs. These are dim, dense objects of small sizes. To cope with observations, the latter should be present not only in the disc of our Galaxy but also in the halo. Similar distribution should be characteristic to other galaxies. The density of these objects can be found observationally. Moving across the line of sight between the Earth and a star (say, from a nearby dwarf galaxy) these objects would serve as gravitational lenses. Some evidence of lensing of this type (microlensing) has indeed been observed [26], but it is by far not as frequent as necessary for explaining the whole of dark matter in the halo [26, 27]. Furthermore, many of the candidate compact objects are not suitable for other reasons. As an example, neutron stars are remnants of supernovae. The latter are main sources of oxygen, silicon and other heavy elements. The abundances of these elements in galaxies are well known.

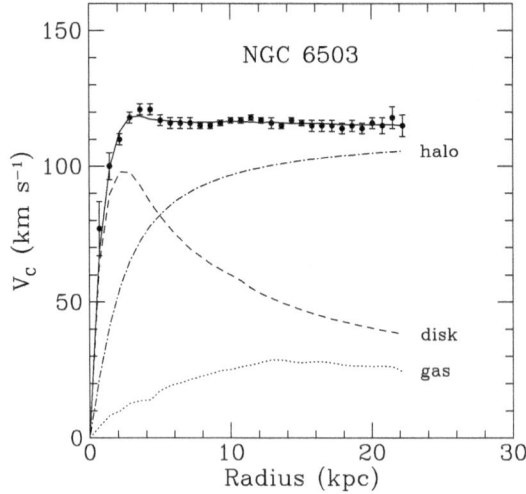

Fig. 1.9. Rotation curve of a galaxy NGS 6503 [25]. Different curves show the contributions of three major components of matter to the gravitational potential.

In this way the number of supernova explosions, and hence neutron stars, is estimated, and this number is definitely too small.

There are several observational results showing the existence of *dark energy*. We already mentioned that the Universe is spatially flat to high precision: the present total energy density coincides with the critical density ρ_c to better than 1%. On the other hand, the estimate for the density of clustered energy is given by (1.16), which is considerably smaller than ρ_c. The rest of the energy density is attributed to dark energy.

An independent argument for dark energy is as follows. We will see that the expansion rate of the Universe now and in the past depends on its energy content. The expansion rate determines in turn the relationship between redshift and visible brightness for distant "standard candles". This relationship has been measured by making use of supernovae Ia (SNe Ia) which have been argued to have the properties of standard candles.[11] Already the first data [28, 29] have shown that distant supernovae are relatively dimmer than nearby ones. This is interpreted as evidence for the accelerated expansion of the Universe today and in recent past. In General Relativity, accelerated expansion is possible only if there is dark energy whose density weakly depends on time (or does not depend on time at all).

There are several other, independent arguments based, in particular, on the estimate of the age of the Universe, structure formation, cluster abundance, CMB anisotropy. All of them point to the existence of dark energy whose density today is at the level of $0.7\rho_c$. The hope is that future observations will shed light on the nature and properties of this energy component in our Universe.

One of the candidates to dark energy is vacuum. Particle physics theories often ignore vacuum energy, as it serves merely as a reference point for the energy, while

---

[11] Nearby SNe Ia show empirical relation between the absolute luminosity at maximum brightness and the time behavior of the light emission: brighter supernovae shine longer. Assuming the same relation for distant supernovae, their absolute luminosities are inferred from the durations of their bursts.

one is interested in masses and energies of particles — excitations about vacuum. Completely different situation occurs in General Relativity: vacuum energy, like any other form of energy, gravitates. If gravitational fields are not extremely strong, vacuum is the same everywhere anytime, so its energy density is constant in space and time. Hence, vacuum energy does not cluster, so it is indeed an excellent candidate for dark energy. The problem, though, is that the energy density has dimension M^4, and one would expect offhand that the value of vacuum energy density would be of the order of the fourth power of the mass scale of fundamental interactions. These scales are 1 GeV for strong interactions, 100 GeV for electroweak interactions and $M_{Pl} \sim 10^{19}$ GeV for gravitational interactions themselves. Thus, one would estimate the corresponding contributions to vacuum energy as follows,

$$\rho_{vac} \sim 1\,\mathrm{GeV}^4, \quad \text{for strong interactions}$$

$$\sim 10^8\,\mathrm{GeV}^4, \quad \text{for electroweak interactions}$$

$$\sim 10^{76}\,\mathrm{GeV}^4, \quad \text{for gravitational interactions.} \tag{1.19}$$

Any of these estimates exceeds by many orders of magnitude the actual dark energy density

$$\rho_\Lambda \sim \rho_c \sim 10^{-5}\,\frac{\mathrm{GeV}}{\mathrm{cm}^3} \sim 10^{-46}\,\mathrm{GeV}^4. \tag{1.20}$$

This is a serious problem for theoretical physics which is dubbed *cosmological constant problem*: it is a mystery that the vacuum energy density is so small compared to the estimates (1.19), and the even greater mystery is that it is different from zero (if dark energy is vacuum energy indeed). Without exaggeration, this is one of the major problems, if not *the* major problem, of fundamental physics. There are many puzzles and coincidences here, which require fine tuning of parameters of different nature. Too large vacuum energy density (but still many orders of magnitude below the particle physics scales) would be incompatible with our existence: large and positive vacuum energy would lead to very fast cosmological expansion totally suppressing formation of galaxies and stars, while the Universe with large and negative vacuum energy would recollapse long before any structure would form. A coincidence calling for explanation is that the three different energy components — dark energy, dark matter and baryons — are of the same order of magnitude in the present Universe ("Why now?"). These components have different origins, so *a priori* they would give contributions of different orders of magnitude.

Let us stress that vacuum is by no means the only dark energy candidate discussed in literature; we will consider some other candidates later on in this book.

1.4. Future of the Universe

The future of the Universe is mostly determined by its geometry and the properties of dark energy.

We will see that the contribution of the spatial curvature into effective energy density is inversely proportional to the scale factor *squared*, a^{-2}. So, if spatial curvature is non-zero, it will sooner or later start to dominate over energy density of non-relativistic matter which decays like a^{-3}.

Hence, in the long run, the competition will be between the spatial curvature and dark energy. If the latter is time-dependent and will relax to *zero* sufficiently rapidly in future, then the expansion of Universe with positive curvature (closed model) will slow down, then terminate, and eventually the Universe will recollapse to singularity. The Universe with negative spatial curvature will expand forever, though its expansion will slow down. Clusters of galaxies will move away from each other to larger and larger distances. The same would happen to galaxies not bound to clusters. All systems that are not gravitationally bound will disappear. The same properties hold for spatially flat Universe with dark energy relaxed to zero (but the expansion in that case will be even slower).

If dark energy density does not depend on time, like in the case of vacuum energy,[12] or depends on time weakly, then the dominant player will be dark energy. Positive dark energy will lead to *exponential* expansion; the Universe will expand forever with (almost) constant acceleration.

One cannot exclude also the possibility that the dark energy will become negative in distant future. In that case the dark energy will slow the expansion down, and in the end the Universe will recollapse to singularity.

We stress that it is *in principle* impossible to predict the ultimate fate of the Universe on the basis of cosmological observations only. These observations enable one, generally speaking, to figure out the dependence (or independence) of dark energy density on time in the past, but the behavior of dark energy in the future can only be hypothesized. To predict the distant future of our Universe, one would need to know the nature of dark energy (or, more generally, the precise reason for the accelerated expansion of the Universe at present). Whether and how such a knowledge can be obtained is hard to tell. Nevertheless, one can extrapolate, with reasonable confidence, the future evolution of the Universe in 10–20 billion years. During this time, the Universe will expand at rate comparable to the present Hubble rate.

Yet another possibility discussed in literature is that the dark energy density will *grow* in future. If this growth is sufficiently fast, the Universe will end up in Big Rip: in finite time the scale factor will become infinite, interactions between particles will be insufficient to keep them in bound states, and all bound systems, including atoms and nuclei, will disintegrate and point-like particles will move away from each other to infinite distance.

---

[12]We do not discuss here the possibility of the *cosmological phase transition* which would change the energy balance and thus have major effect on the evolution of the Universe.

1.5. Universe in the Past

The very fact that the Universe expands clearly implies that it was denser and warmer in the past. On the basis of General Relativity and standard thermodynamics we will see that matter had higher and higher temperature and density at earlier and earlier epochs, and that at most stages it was in thermal equilibrium. Hot Big Bang theory is precisely the theory of such a Universe. Going back in time, and, accordingly, up in temperature, we find a number of particularly important "moments" (better to say, more or less lengthy periods) in the cosmological evolution; see Fig. 1.10. Let us briefly discuss some of them.

1.5.1. *Recombination*

At relatively low temperatures the usual matter in the Universe was in the state of neutral gas (mainly hydrogen). At an earlier stage, i.e., at higher temperatures, the binding energy was insufficient for keeping electrons in atoms, and the matter was in the state of baryon–electron–photon plasma. The temperature of recombination — the transition from plasma to gas — is determined, very crudely speaking, by the binding energy in hydrogen atom, 13.6 eV. We will see, however, that recombination

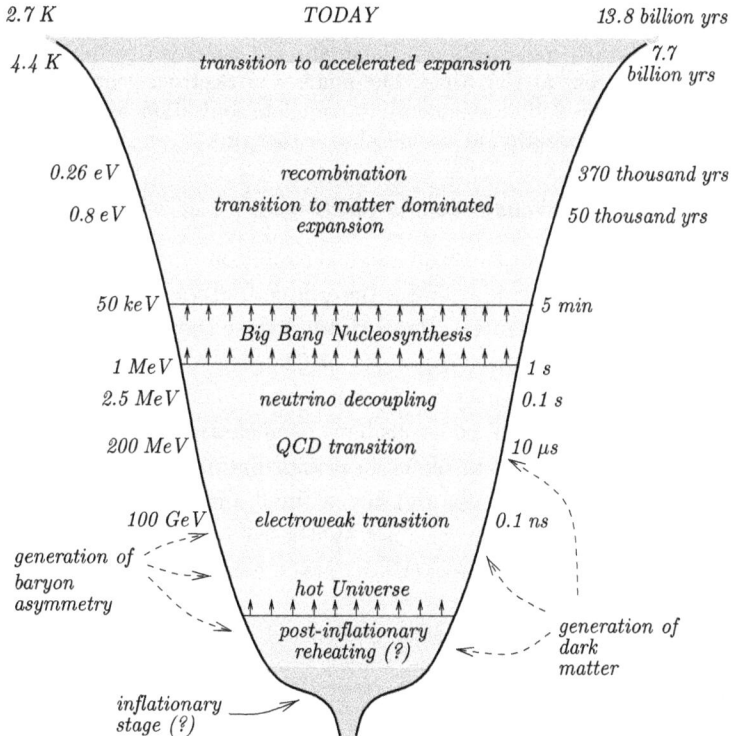

Fig. 1.10.   Stages of the evolution of the Universe.

occurred at a somewhat lower temperature, $T \sim 0.26\,\text{eV}$. This is a very important epoch: before it photons actively scattered off electrons in the plasma (and at even earlier times they were emitted and absorbed), while after recombination the neutral gas was transparent to photons. Hence, the CMB that we see today comes from the recombination epoch, and it carries information about the properties of the Universe at the epoch when its temperature was about $0.26\,\text{eV} \approx 3000\,\text{K}$ and age about 370 thousand years.

We have already pointed out that the high degree of CMB isotropy shows that the Universe was highly homogeneous at recombination, the density perturbations $\delta\rho/\rho$ were comparable to temperature fluctuations and were roughly of order 10^{-4}–10^{-5}. Nevertheless, these perturbations in the end have given rise to structures: first stars, then galaxies, then clusters of galaxies.

In fact, the optical depth (scattering probability) for photons after recombination is different from zero and is equal to $\tau \simeq 0.08$–0.10. The reason is the reionization in the Universe which begins at the time of active formation of the first stars, $z \lesssim 20$.

The fact that hydrogen in the Universe is almost completely ionized ($n_H/n_p < 10^{-5}$) at $z \leq 6$, has been known for a long time from the observation of hydrogen emission lines of quasars: if the light from quasars traveled through hydrogen, it would get completely absorbed. Indeed, even though the light from quasars gets redshifted, there would be enough hydrogen atoms with appropriate velocities to absorb it.

Evidence for the early reionization, $z \sim 10$, comes from CMB anisotropy and polarization. In particular, the polarization at large angular scales is due to scattering of relic photons off free electrons at that time. The number of electrons required to explain the optical depth $\tau \simeq 0.08$–0.10 corresponds to complete ionization of cosmic hydrogen at $z \simeq 10$–12 (or partial ionization at a somewhat earlier time).

1.5.2. *Big Bang Nucleosynthesis (BBN)*

Another important epoch in the cosmological evolution occurs at much higher temperatures, whose order of magnitude is determined, crudely speaking, by the scale of binding energies in nuclei, i.e., 1–10 MeV. Again, the actual temperatures are somewhat smaller; the reason is discussed in Chapter 8. In any case, at high temperatures protons and neutrons were free in cosmic plasma, but after the Universe cooled down due to expansion, neutrons have been captured into nuclei. As a result, besides hydrogen, there are light nuclei in primordial plasma: mostly helium-4 (the most tightly-bound light nucleus) and also a small amount of deuterium, helium-3 and lithium-7; heavier elements were not synthesized in the early Universe.[13] This epoch of Big Bang Nucleosynthesis (BBN) is the earliest epoch studied directly so

---

[13]Heavy elements are produced during stellar evolution. In particular, one of the important elements in the nucleosynthesis chain, carbon, is synthesized in the fusion of three ^{4}He-nuclei. This process is possible only at very high densities reached in stellar interiors after hydrogen has been burned out. All other elements are synthesized from carbon. Relatively light elements, including iron, are produced in thermonuclear reactions in stars; heavier elements are synthesized as a result of neutron capture in stars and supernova explosions, and some elements are produced, presumably, in the processes of proton or positron capture.

far: the calculation of the light element abundances makes use of General Relativity and known microscopic physics (physics of nuclei and weak interactions), while measurements of these primordial abundances, though difficult, are quite precise by now.

Good agreement between BBN theory and observations is one of the cornerstones of the theory of the early Universe. Let us stress that BBN epoch lasted from about 1 to 300 seconds after the Big Bang, the relevant temperatures range from 1 MeV to 50 keV.

The difficulty in measuring primordial element abundances is that most of the baryonic material in the Universe has been reprocessed in stars, and its composition is different from that of primordial plasma. Nevertheless, it is possible to find places in the Universe where matter can be claimed, with good confidence, to have not been reprocessed, and its composition coincides with the primordial composition.

1.5.3. *Neutrino decoupling*

While photons last scattered at temperature 0.26 eV, neutrino interactions with cosmic plasma, as we will see, terminated at temperature 2–3 MeV. Before that, neutrinos were in thermal equilibrium with the rest of matter, and after that, they freely propagate through the Universe. We will calculate the temperature and number density of neutrinos, and here we note that they are of the same order of magnitude as the temperature and number density of photons. Unfortunately, direct detection of relic neutrinos is a very difficult, and may even be an unsolvable problem.

Problem 1.9. *If there exist neutrinos of ultra-high energies in Nature, they may scatter off relic neutrinos. The neutrino–neutrino cross-section in the Standard Model of particle physics is very small. Its maximum value, $\sigma_{\nu\nu} = 0.15\,\mu b = 1.5 \cdot 10^{-31}$ cm^2, is reached at center-of-mass energy $\sqrt{s} \approx M_Z \approx 90\,GeV$ when neutrinos pair-annihilate through the resonant Z-boson production. The observation of photons produced in Z-decay chains would be indirect evidence for the existence of relic neutrinos.*

Find the mean free path of an ultra-high energy neutrino in the present Universe with respect to the above process. Does one expect a cutoff in the spectrum of ultra-high energy neutrinos, similar to the GZK cutoff in the spectrum of ultra-high energy protons (see Problem 1.8)?

The role of neutrinos in the present Universe is not particularly important. However, the neutrino density in the early Universe is an important parameter of BBN theory. Primordial nucleosynthesis occurs in expanding Universe, and the neutrino component affects the expansion rate and hence cooling rate of plasma at the time of BBN. The latter is important for inequilibrium processes of the light element synthesis. The success of BBN theory gives decisive evidence for the existence of relic neutrinos.

Relic neutrinos play a role in structure formation; they affect CMB angular spectrum too. A combination of cosmological data yields the limits on the sum of masses of all neutrino cpecies which are in the range [94, 99]

$$\sum m_\nu < 0.2 \div 0.5 \text{ eV}.$$

Note that these limits are stronger than direct experimental bound. It is not unlikely that cosmological methods will actually provide the determination of the neutrino masses.

In this and the accompanying book, we will study relic neutrinos and their effects in some details.

1.5.4.   *Cosmological phase transitions*

Going further back in time, we come to the epochs which have not been directly probed by observations so far. Hence, we have to make more or less reasonable extrapolations. It is likely that the history of the hot Universe goes back to temperatures of the order of hundreds GeV and quite possibly to even higher temperatures. At so high temperatures there are epochs of interest, at least from the theoretical viewpoint. Some of them can be loosely called epochs of phase transitions.

— Transition[14] from quark-gluon matter to hadronic matter. Its temperature; is determined by the energy scale of strong interactions and is about 200 MeV. At much higher temperatures quarks and gluons behave as individual particles (rather strongly interacting towards the transition epoch), while at lower temperatures they are confined in colorless bound states, hadrons. At the same time (or almost the same time) there was the transition associated with chiral symmetry breaking.

— Electroweak transition [30–33]. Simplifying the situation we can say that at temperatures above 100 GeV (energy scale of electroweak interactions), the Englert–Brout–Higgs condensate is absent, and W- and Z-bosons have zero masses. The present phase with broken electroweak symmetry, Higgs condensate and massive W- and Z-bosons is the result of the electroweak transition[15] that occurred at temperature of order 100 GeV.

— Grand Unified transition. There are hints towards Grand Unification, the hypothesis that at energies and temperatures above 10^{16} GeV there is no distinction between strong, weak and electromagnetic interactions: these interactions are unified into a single force. If so, and if these temperatures existed in the Universe, then at the temperature of Grand Unification $T_{GUT} \sim 10^{16}$ GeV there was the corresponding phase transition. We note, however, that maximum temperature in the Universe may well be below T_{GUT}, so the phase of Grand Unification may not exist in the early

---

[14]This is likely a smooth crossover rather than phase transition proper.
[15]In fact, the situation is somewhat more complicated: an order parameter is absent in electroweak theory (at least within the Standard Model of particle physics), so the phase transition can be absent. Indeed, given the experimental bound on the Higgs boson mass, there is smooth crossover in the Standard Model instead of the phase transition.

Universe. This is the case in many models of inflation: there, the reheat temperature is lower than T_{GUT}.

1.5.5. *Generation of baryon asymmetry*

The present Universe contains baryons (protons, neutrons) and practically no antibaryons. Quantitative measure of the baryon abundance is the ratio of baryon and photon number densities. The studies of BBN and CMB give

$$\eta_B \equiv \frac{n_B}{n_\gamma} = 6.05 \cdot 10^{-10} \tag{1.21}$$

with precision of about 1%. The baryon number is conserved at sufficiently low energies and temperatures, and we will see that in the early Universe n_B/n_γ is of the same order of magnitude as given in (1.21). Thus, baryon asymmetry η_B is one of the most important parameters of cosmology.

At temperatures of hundreds of MeV and higher, there were a lot of quarks and antiquarks in the cosmic plasma, which were continuously pair-created and annihilated. Thus, unlike in the present Universe, there were almost as many particles with negative baryon number (antiquarks) as those with positive baryon number (quarks). Simple thermodynamical arguments, to be given in this book, show that the number of quark–antiquark pairs at high temperatures is about the same as the number of photons, so the baryon asymmetry can be understood as determining the following ratio

$$\frac{n_q - n_{\bar{q}}}{n_q + n_{\bar{q}}} \sim \eta_B \sim 10^{-10}. \tag{1.22}$$

Here n_q and $n_{\bar{q}}$ are number densities of quarks and antiquarks, respectively. We see that in the early Universe, there existed one uncompensated quark per 10 billion of quark–antiquark pairs. It is this tiny excess that is responsible for the existence of baryonic matter in the present Universe: as the Universe expanded and cooled down, antiquarks annihilated with quarks, while uncompensated quarks remained there and in the end formed protons and neutrons.

One of the problems of cosmology is to explain the very existence of the baryon asymmetry [34, 35], as well as to understand its value (1.21). It is extremely implausible that the small excess of quarks over antiquarks (1.22) existed in the Universe from the very beginning, i.e., that it is one of the initial data of the cosmological evolution; it is much more reasonable to think that the Universe "in the beginning" was baryon-symmetric. The same conclusion comes from inflationary theory. Asymmetry (1.22) *was generated* in the course of the cosmological evolution due to processes with baryon number non-conservation. We will discuss possible mechanisms of generation of this asymmetry, but we stress right away that today there is no unique answer to the question of its origin. Here we note only that baryon asymmetry was generated most probably at very high temperatures, at least 100 GeV and maybe much higher, although its generation at lower temperatures is not completely excluded.

The problem of the baryon asymmetry cannot be solved within the Standard Model of particle physics. This is another cosmological hint towards New Physics beyond the Standard Model.

1.5.6.   *Generation of dark matter*

Which particles make non-baryonic clustered dark matter is not known experimentally. One expects that these are stable or almost stable particles that do not exist in the Standard Model of particle physics. Hence, the very existence of dark matter is a very strong argument for incompleteness of the Standard Model. This makes the detection and experimental study of the dark matter particle extremely interesting and important. On the other hand, the lack of experimental information on the properties of these particles makes it impossible to give a unique answer to the question of the mechanism of the dark matter generation in the early Universe. We will discuss various dark matter particle candidates in this book, and here we make one remark. We will see that hypothetical stable particles of mass in the GeV–TeV range, whose annihilation cross-section is comparable to weak cross-sections, do not completely annihilate in the course of the cosmological evolution, and that their resulting mass density in the present Universe is naturally of the order of the critical density ρ_c. Therefore, these particles are *natural* dark matter candidates, especially because they exist in some extensions of the Standard Model, including supersymmetric extensions.

Particles which we have briefly described are called WIMPs (weakly interacting massive particles). Their freeze-out, i.e., termination of annihilation, occurred at a temperature somewhat below their mass, i.e., $T \sim 1-100 \, \text{GeV}$.

Of course, there are many other dark matter particle candidates besides WIMPs. These include axion, gravitino, sterile neutrino, etc. Some of them are discussed in this book.

1.6.   Structure Formation in the Universe

We discussed in previous Sections the most important stages of the cosmological evolution. Each of them, be it nucleosynthesis or recombination, has finite time duration. There is, however, a process in the Universe that began at a very early epoch and continues at present. This process is formation of structures — first stars, galaxies, clusters of galaxies, superclusters. The order we put them here is not random: smaller objects get formed earlier.

The theory describing structure formation is based on the Jeans instability, the gravitational instability of matter density perturbations. It should be assumed, of course, that the perturbations have already existed at the very early stage of the cosmological evolution, even though they were very small in amplitude.

The source of primordial perturbations is not particularly relevant to the theory of structure formation. Rapid growth of density perturbations occurs at that stage of the cosmological expansion when the dominating energy component is mass of non-relativistic matter. Transition to this stage happened 50 thousand years after Big Bang; before that the Universe was so hot that the dominant component was relativistic ("radiation"). The epoch of this transition is called radiation–matter equality. At that time the density perturbations had small amplitude, $\delta\rho/\rho \sim 10^{-4}$–$10^{-5}$. Regions of higher density are sources of gravitational potential, they attract surrounding matter, and the density in these regions becomes even higher. This is precisely the physical reason for gravitational instability.[16] Once the overdensity is large enough, the overdense region becomes gravitationally bound and starts living its own life; in particular, the size of this region does not grow despite the expansion of the Universe. Instead, gravitational interaction within this region leads to its collapse to an object of much smaller size. In this way protostars and protogalaxies get formed. First stars were formed at $z \sim 10$ and somewhat earlier. Active galaxy formation occured at redshifts $2-4$, though some galaxies were formed much earlier: there is evidence for galaxies at redshifts of order 10.

The mass of an object — galaxy, cluster of galaxies — is determined by the size of the primordial overdense region. So, the number densities of galaxies and clusters and their mass distribution reflect the spectrum of primordial perturbations. Existing observational data on structures are consistent with the simplest "flat" primordial spectrum, called Harrison–Zeldovich spectrum. The defining property of the latter is its scale invariance: in a certain sense, perturbations of different sizes have one and the same amplitude.

Perturbations existing in cosmic medium at recombination give rise to CMB temperature anisotropy and polarization. Hence, the primordial spectrum can be determined also from the CMB observations. Notably, the spectra found from CMB and structures are in good agreement with each other.

Structure formation gives yet another argument for the presence of dark matter: without dark matter, density perturbations would start to grow after recombination only, and by now they would not have developed into structures yet. Furthermore, it follows from the theory of structure formation that the major part of dark matter must be *cold* (or warm), i.e., it must consist of particles that became non-relativistic at a very early epoch. If a large fraction of dark matter were *hot*, i.e., consisted of particles remaining relativistic until late times, then the formation of objects of relatively small size would be suppressed, in contradiction to observations. An important example of hot dark matter is the neutrino of mass $m_\nu \sim 1$–10^{-3} eV. This suggests that cosmology is capable of setting bounds on neutrino masses.

We discuss in the accompanying book the structure formation and the role played in it by different components of cosmic matter.

---

[16]This mechanism does not work for relativistic particles, since weak gravitational field cannot keep them inside an overdense region.

1.7.   Before the Hot Epoch

1.7.1.   *Argument from observations*

Cosmological observations unambiguously suggest that the hot epoch of the cosmological evolution discussed in Sec. 1.5 was preceded by some other epoch with quite unusual properties. We discuss this issue in detail in the accompanying book, and here we point out that this remarkable conclusion follows from the analysis of matter perturbations — the perturbations which in the end have become galaxies and other structures. The perturbations in baryon–electron–photon component before recombination were nothing but acoustic waves[17]; we will see that the sound speed in this medium $u_s \approx 1/\sqrt{3}$. The wavelength of each of the acoustic wave increases due to the expansion of the Universe, $\lambda(t) \propto a(t)$, while the wave vector (momentum) $q(t) = 2\pi/\lambda(t)$ decreases,

$$q(t) = \frac{k}{a(t)},$$

where k is a time-independent quantity called conformal momentum. The frequency of each wave also decreases,

$$\omega(t) = u_s q(t).$$

The time-dependence of the density contrast of the baryon–electron–photon component for a mode of a definite wave vector $\mathbf{k}$ has oscillatory behavior (with qualifications which we do not specify here),

$$\frac{\delta\rho_{B\gamma}}{\rho_{B\gamma}}(t) = A_{\mathbf{k}} \cos\left[\int_0^t u_s \frac{k}{a(t)}\, dt + \varphi_{\mathbf{k}}\right], \tag{1.23}$$

where $A_{\mathbf{k}}$ is the amplitude of the wave, $\varphi_{\mathbf{k}}$ is a time-independent phase, and notation $B\gamma$ refers to baryon–electron–photon component.

   This oscillatory behavior was not characteristic of the acoustic waves at all times. At the radiation domination epoch, the scale factor increases as $a(t) \propto \sqrt{t}$, whereas the Hubble paramter equals $H(t) = (2t)^{-1}$. Therefore, at early times the Hubble parameter was larger than frequency, $H(t) \gg \omega(t)$, i.e., the expansion was faster than the oscillations, and the oscillations simply had no time to occur. This property can be also phrased in the following way. Suppose that the cosmological evolution began right from the hot epoch. The signals emitted at the Big Bang and propagating with the speed of light would travel by time t to distance $l_H(t) \sim t \sim H^{-1}(t)$. This would be a maximum size of causally connected region at time t, i.e., the cosmologcal horizon size. Since for given k the wavelength $\lambda(t)$ grows like

---

[17]There were density perturbations in dark matter too, but this is not particularly important here.

$a(t) \propto \sqrt{t}$, we have for small t

$$\lambda(t) \gg l_H(t).$$

This means that for each wave there was an epoch when its wavelength exceeded the horizon size! In such a situation the perturbation is called superhorizon. We note that the perturbations responsible for galaxies, and even more so for clusters of galaxies, were superhorizon until rather late times: their horizon entrance (time at which $\lambda(t) \sim l_H(t)$) occured at fairly low temperatures, $T \ll 100$ keV, i.e., after the nucleosynthesis epoch. So, there is no guesswork at this point.

It is highly implausible that the primordial perturbations were built in as an initial condition of the cosmological evolution. It is natural to think, instead, that they were *generated* by some physical processes. Then causality guarantees that this generation cannot happen at early hot epoch when the perturbation wavelength exceeds the cosmological horizon size. Thus, there are two options: either the perturbations were generated relatively late at the hot epoch (when their wavelengths were smaller than the horizon size) or the hot epoch was preceded by some other epoch — the epoch responsible for the generation of perturbations.

It is remarkable that the cosmological data discriminate between the two options: the data tell that the perturbations were actually generated at some pre-hot epoch. The point is that if the perturbations existed already at early hot epoch when they were superhorizon (and hence they were generated before the hot epoch), the phase of oscillations in (1.23) is uniquely determined.[18] For example, this phase equals zero for not very long waves, $\varphi_{\mathbf{k}} = 0$; this is the property of solutions to the wave equations in the Universe that expands as $a(t) \propto \sqrt{t}$. On the other hand, if the perturbations were generated at the late hot epoch, when they were subhorizon, there is no reason for the phase to be fixed: different waves can have different phases. One expects, instead, that the phase is a random function of momentum $\mathbf{k}$. The latter property is indeed characteristic of concrete models of the generation of density perturbations at the hot epoch, which will be briefly discussed in Chapter 12.

Once the phase in (1.23) is fixed (let us take for definiteness $\varphi_{\mathbf{k}} = 0$, which is correct for most of perturbations, except for the those with the largest wavelengths), the waves at the recombination time t_r have a definite phase which depends on the wave vector k:

$$\frac{\delta \rho_{B\gamma}}{\rho_{B\gamma}}(t_r; k) = A_{\mathbf{k}} \cos \left(\int_0^{t_r} u_s \frac{k}{a(t)}\, dt \right). \qquad (1.24)$$

As a result, the density contrast at recombination, $\delta\rho_{B\gamma}(k)$, is an oscillating function of k; for some momenta, the density (and hence temperature) is at maximum while for other momenta it is, roughly speaking, equal to zero. This shows up in the oscillatory behavior of the CMB temperature angular spectrum, see Fig. 1.6. The

---

[18]Under an assumption that the perturbations are adiabatic; this assumption is confirmed by the data.

perturbations for which cosine in (1.24) equals ± 1 are the largest in amplitude. They are seen at definite angles, which correspond to the peaks in Fig. 1.6. Notably, the value of the initial phase $\varphi_{\mathbf{k}}$ in (1.23) inferred from observations agrees with the theoretical prediction.

The oscillatory dependence of $\delta \rho_{B\gamma}(t_r; k)$ on momentum is reflected also in the property of the galaxy distribution in the late Universe, called baryon acoustic oscillations (BAO). This feature has been also observed experimentally.

On the other hand, if the phase $\varphi_{\mathbf{k}}$ in (1.23) took random values for various momenta, the oscillatory behavior of the CMB angular spectrum and BAO would be absent, in contradiction to the observations.

Thus, observational cosmology definitely requires that the phase $\varphi_{\mathbf{k}}$ in (1.23) is uniquely defined, and this means that the density perturbations existed at the very beginning of the hot epoch and hence were generated before that epoch.

1.7.2. *Drawbacks of the Hot Big Bang theory*

Besides the problem of primordial perturbations, the Hot Big Bang theory briefly discussed in Sec. 1.5 has other drawbacks. These are due to the fact that this theory needs very special — and unnatural — initial conditions, otherwise it would grossly fail to describe the early and present Universe. We discuss in detail the problem of initial conditions in the accompanying book, and here we give two examples showing what sort of problems we are talking about.

The first of these examples has to do with the properties of the cosmological horizon in the Hot Big Bang theory. As we pointed out, the horizon size $l_H(t)$ at time t in this theory equals t modulo a numerical coefficient of order one. In particular, the horizon size at recombination is $l_H(t_r) \approx 10^6$ light yrs $\simeq 300$ kpc. Since then the size of this region has increased due to the expansion of the Universe, and today it is equal to $l_H(t_r)a_0/a(t_r) \approx 300$ Mpc. This is substantially smaller than the size of the present horizon $l_H(t_0) \approx 15$ Gpc. Thus, the observable part of the Universe contains of order $50^3 \simeq 10^5$ regions which were causally disconnected at recombination epoch (within the Hot Big Bang theory). One would think that these regions should be quite different from each other. However, their properties both today and at recombination are exactly the same; in particular, the temperature of the CMB photons coming from different regions is the same to better than 0.01%. This inconsistency is called the horizon problem: the homogeneity of the Universe at large scales does not have natural explanation in the Hot Big Bang theory and can only be obtained within this theory by imposing very special initial conditions.

Another example is somewhat different. As we already know, the Universe is warm, so it can be characterized by its entropy. Entropy density is of order of photon number density; in the present Universe

$$s \sim 10^3 \, \text{cm}^{-3}.$$

Thus, the estimate for the entropy in the observable part of the Universe, whose size is $R_0 \sim 10^4 \, \text{Mpc} \sim 10^{28} \, \text{cm}$, is

$$S \sim sR_0^3 \sim 10^{88}.$$

This huge dimensionless number is one of the properties of our Universe. Why does the Universe have such a large entropy? The Hot Big Bang theory does not answer this question, since entropy is (almost) conserved during the hot stage. The huge entropy has to be introduced into the Hot Big Bang theory "by hands", as one of the initial data. This uncomfortable situation is called the entropy problem.

1.8. Inflationary Theory

Both the problem of initial conditions and the problem of primordial perturbations find elegant solutions in inflationary theory [40–44]. According to this theory, the hot cosmological epoch was preceded by the epoch of exponential expansion (inflation). During the inflationary epoch, the initially small region of the Universe (whose spatial size was comparable, say, to the Planck length l_{Pl}) inflated to a very large size, typically many order of magnitudes larger than the size of the part of the Universe we see today. This in the end explains flatness, homogeneity and isotropy of the observable part of the Universe. Due to the exponential character of expansion, the duration of the inflationary epoch may be short: the initial condition problems of the Hot Big Bang theory is solved provided that the duration is greater than $(50-70)H_{infl}^{-1}$ where H_{infl} is the Hubble parameter at inflation, say $H_{infl} \sim 10^{-6}M_{Pl}$ (H_{infl} may be quite a bit smaller). In that case, the minimum duration of inflation is of order $10^8 t_{Pl} \sim 10^{-35} \, \text{s}$. It is likely that inflation lasted much longer, but in any case it is plausible that we deal with microscopic time scale.

The inflationary regime occurs if the energy density in the Universe depends on time very weakly. Energy density of conventional matter — gases of particles — does not have this property. Therefore, most models of inflation make use of hypothetical new field(s).[19] Under certain conditions this new field — inflaton — is spatially homogeneous and changes slowly in time in inflating region. Its potential energy changes slowly too, and this gives rise to the inflationary expansion regime.

At some moment of time the conditions for inflationary expansion get violated, and inflation terminates. A new epoch — post-inflationary reheating — sets in, at which the inflaton energy is tranferred to the energy of conventional matter. As a result, the Universe heats up to very high temperature, and the Hot Big Bang epoch begins. Reheating occurs with entropy generation, which provides the solution to the entropy problem.

---

[19]The reservation here alludes to the possibility of employing the Englert–Brout–Higgs field of the Standard Model of particle physics [316]. Yet another possibility is to have extra terms in the action for gravitational field. This case is often equivalent to the introduction of new field(s).

Inflationary theory was originally proposed as a solution to the initial data problems, but it soon became clear that it also solves the problem of primordial perturbations [45–49]. The original source of perturbations are vacuum fluctuations of quantum field(s), in the simplest version, vacuum fluctuations of the inflaton field itself. These fluctuations get strongly enhanced at inflationary stage due to strong time-dependence of the gravitational field in the Universe. At the end of inflation they are reprocessed into density perturbations of conventional matter. The amplitude of primordial perturbations depends on unknown parameters of a model, but their power spectrum (dependence on wavelength) is uniquely calculable within a given model of inflation. Notably, most models predict almost flat (almost Harrison–Zeldovich [50–52]) power spectrum, in gross agreement with observational data. However, a typical prediction is a slight tilt in the spectrum, which makes it different from the Harrison–Zeldovich one. Small red (negative) tilt of the power spectrum has been recently discovered by the analysis of the CMB data.

Another prediction of inflationary models is the existence of primordial gravitational waves [40]. They are also generated at inflationary stage from vacuum fluctuations, now of the gravitational field. In some models the amplitudes of gravity waves of wavelengths comparable to the present Hubble size are of order $10^{-5} – 10^{-6}$. These gravity waves would affect CMB temperature anisotropy [53–56] and polarization [57–62]. These effects have not been discovered yet, but they may be observable in future. Discovery of primordial gravity waves will not only be a strong argument in favor of inflation, but also will enable us to determine its fundamental parameter, the inflationary Hubble scale H_{infl}.

Presently, inflationary theory is well developed. We study various aspects of this theory in the accompanying book.

1.8.1. *Alternatives to inflation*

Cosmological inflation is by far the most popular and best understood scenario for the epoch that preceeded the Hot Big Bang epoch. Yet there are alternatives, which have not been ruled out. One alternative is the bouncing Universe which starts from contraction, experiences the bounce and then enters the expansion regime (a version is pulsating Universe). We briefly discuss the bouncing Universe scenario in the accmpanying book. Another option is the Genesis scenario [63] which assumes that the Universe was initially empty and had Minkowski geometry, then the energy density grew and the Universe expanded at a higher and higher rate until this regime terminated and the Universe entered the hot epoch. These scenarios require very exotic fields, more exotic than the inflaton.

As opposed to inflation, its alternatives predict unobservably small amplitude of primordial gravity waves. There are other, more subtle predictions which discriminate these scenarios from inflation and which can be discovered observationally, at least in principle. It is quite plausible that with the development of theory and observations, we will soon or late learn which regime preceded the Hot Big Bang epoch. Isn't it fascinating that the observations of the Universe at large are capable of sheding light on the first moments of the cosmological history characterized by extremely high energy densities and expansion rates and extremely small distance and time scales?

To conclude our preview we note that it misses some of the topics which we study in this book. We hope, nevertheless, that this preview makes the content of the book reasonably clear.

Chapter 2

Homogeneous Isotropic Universe

In this book we use the basic notations and equations of General Relativity; see Appendix A. The notations and conventions are summarized in Sec. A.11.

2.1. Homogeneous Isotropic Spaces

To a very good approximation, our Universe is homogeneous and isotropic at sufficiently large scales. This means that at given moment of time, the geometry of space is the geometry of homogeneous and isotropic manifold. There are only three such manifolds[1] (up to overall scale): 3-dimensional sphere, 3-dimensional Euclidean space (3-plane) and 3-dimensional hyperboloid. The geometry of 3-dimensional sphere is best understood by imagining that it is embedded into (fictitious) 4-dimensional Euclidean space and writing the equation of 3-sphere in the standard form,

$$(y^1)^2 + (y^2)^2 + (y^3)^2 + (y^4)^2 = R^2,$$

where y^a ($a = 1, \ldots, 4$) are coordinates of the 4-dimensional Euclidean space and R is the radius of the 3-sphere. Let us introduce three angles χ, θ and ϕ, so that

$$
\begin{aligned}
y^1 &= R\cos\chi, \\
y^2 &= R\sin\chi\cos\theta, \\
y^3 &= R\sin\chi\sin\theta\cos\phi, \\
y^4 &= R\sin\chi\sin\theta\sin\phi.
\end{aligned}
\tag{2.1}
$$

Then the distance between two points on the 3-sphere with coordinates (χ, θ, ϕ) and $(\chi + d\chi, \theta + d\theta, \phi + d\phi)$ is

$$dl^2 = (dy^1)^2 + (dy^2)^2 + (dy^3)^2 + (dy^4)^2$$

---

[1] It is important that the 3-dimensional metric has Euclidean signature, i.e., it can locally be cast into the standard form $dl^2 = (dx^1)^2 + (dx^2)^2 + (dx^3)^2$.

$$= R^2\{[d(\cos\chi)]^2 + [d(\sin\chi\cos\theta)]^2 + [d(\sin\chi\sin\theta\cos\phi)]^2 \qquad (2.2)$$
$$+ [d(\sin\chi\sin\theta\sin\phi)]^2\}.$$

After simple calculation we find that the metric of the 3-sphere has the following form,

$$\text{3-sphere:} \qquad dl^2 = R^2[d\chi^2 + \sin^2\chi(d\theta^2 + \sin^2\theta d\phi^2)]. \qquad (2.3)$$

Note that no trace of the fictitious 4-dimensional Euclidean space is left in this formula, as should be the case.

In analogy to 3-sphere, 3-dimensional hyperboloid is conveniently described by its embedding into fictitious 4-dimensional Minkowski space with metric

$$ds^2 = -(dy^1)^2 + (dy^2)^2 + (dy^3)^2 + (dy^4)^2,$$

while the equation for the hyperboloid is

$$(y^1)^2 - (y^2)^2 - (y^3)^2 - (y^4)^2 = R^2. \qquad (2.4)$$

We are interested in the one connected component $y^1 > 0$.

Problem 2.1. *Show that the hyperboloid is indeed a homogeneous and isotropic space. Hint: Begin with defining what precisely is a homogeneous and isotropic space.*

Coordinates on 3-hyperboloid can be introduced in analogy to (2.1):

$$y^1 = R\cosh\chi,$$
$$y^2 = R\sinh\chi\cos\theta,$$
$$y^3 = R\sinh\chi\sin\theta\cos\phi,$$
$$y^4 = R\sinh\chi\sin\theta\sin\phi.$$

The calculation of the distance between two points on the hyperboloid is analogous to (2.2). It gives

$$\text{3-hyperboloid:} \qquad dl^2 = R^2[d\chi^2 + \sinh^2\chi(d\theta^2 + \sin^2\theta d\phi^2)]. \qquad (2.5)$$

For completeness, let us also write the metric of 3-plane (3-dimensional Euclidean space),

$$\text{3-plane:} \qquad dl^2 = (dx^1)^2 + (dx^2)^2 + (dx^3)^2. \qquad (2.6)$$

One of the properties of homogeneous and isotropic spaces is that all covariant geometrical quantities are expressed through the metric tensor γ_{ij} and, possibly, tensors δ^j_i and E_{ijk}, the latter existing in any Riemannian space; see Appendix A (here $i, j = 1, 2, 3$; we denote the metric tensor of 3-dimensional space by γ_{ij} to distinguish it from the metric tensor $g_{\mu\nu}$ of 4-dimensional space-time). Furthermore, the coefficients do not depend on coordinates. In particular, the Riemann tensor is equal to

$$^{(3)}R_{ijkl} = \frac{\varkappa}{R^2}(\gamma_{ik}\gamma_{jl} - \gamma_{il}\gamma_{jk}), \tag{2.7}$$

where we introduced the parameter $\varkappa = 0, \pm 1$, which distinguishes 3-plane, 3-sphere and 3-hyperboloid,

$$\varkappa = \begin{cases} +1, & \text{3-sphere} \\ 0, & \text{3-plane} \\ -1, & \text{3-hyperboloid} \end{cases}. \tag{2.8}$$

It follows from (2.7) that the Ricci tensor is

$$^{(3)}R_{ij} = 2\frac{\varkappa}{R^2}\gamma_{ij}, \tag{2.9}$$

and the curvature scalar is constant in space and equal to $6\varkappa R^{-2}$.

Problem 2.2. *Obtain the relations (2.7) and (2.9) by direct calculation.*

Metrics of 3-sphere, 3-hyperboloid and 3-plane can be written in a unified form. To this end, let us first note that by introducing the coordinate $\rho = R\chi$ on the sphere and hyperboloid, and spherical coordinates (ρ, θ, ϕ) on 3-plane, the metrics (2.3), (2.5) and (2.6) can be written as follows,

$$dl^2 = d\rho^2 + r^2(\rho)(d\theta^2 + \sin^2\theta d\phi^2), \tag{2.10}$$

where

$$r(\rho) = \begin{cases} R\sin(\rho/R), & \text{3-sphere} \\ \rho, & \text{3-plane} \\ R\sinh(\rho/R), & \text{3-hyperboloid} \end{cases}. \tag{2.11}$$

The interpretation of quantities entering (2.10) is obvious: ρ is the geodesic (shortest) distance from the origin to a point with coordinates (ρ, θ, ϕ), while $r(\rho)$ determines the area of two-dimensional sphere at distance ρ from the origin, $S = 4\pi r^2(\rho)$. It is also clear that an interval of length l at distance ρ from the origin is seen from the origin at angle

$$\Delta\theta = \frac{l}{r(\rho)}.$$

Instead of ρ, one can choose r as the radial coordinate. With this choice we obtain, e.g., for hyperboloid,

$$d\rho^2 = \frac{dr^2}{\cosh^2(\frac{\rho}{R})} = \frac{dr^2}{1 + \frac{r^2}{R^2}},$$

so that the metrics of the three spaces take the following form,

$$dl^2 = \frac{dr^2}{1 - \varkappa \frac{r^2}{R^2}} + r^2(d\theta^2 + \sin^2\theta d\phi^2), \tag{2.12}$$

where the parameter $\varkappa$ is the same as in (2.8). Note that the coordinates (r, θ, ϕ) cover only half of the 3-sphere: the region $0 \leq r < R$ is part of the 3-sphere extending from the origin (pole) to the 2-dimensional surface of maximum area (equator of the 3-sphere). This is the reason for the coordinate singularity in the metric (2.12) at $r = R$ for $\varkappa = 1$.

2.2.   Friedmann–Lemaître–Robertson–Walker Metric

Expanding homogeneous isotropic Universe is described by the metric

$$ds^2 = dt^2 - a^2(t)\gamma_{ij}dx^i dx^j, \tag{2.13}$$

where $\gamma_{ij}(x)$ is the metric of unit 3-sphere (metric (2.3) with $R = 1$), unit 3-hyperboloid or 3-plane. Metric (2.13) is called the Friedmann–Lemaître–Robertson–Walker metric (FLRW). One distinguishes closed Universe (the space is 3-sphere, $\varkappa = +1$), and open and flat Universe (the space is 3-hyperboloid and 3-plane, $\varkappa = -1$ and $\varkappa = 0$, respectively). For closed and open Universe, the scale factor $a(t)$ at given moment of time is the radius of spatial curvature. On the other hand, for spatially flat Universe, the scale factor itself does not have physical significance, since at a particular moment of time it may be set equal to any number (say, unity) by rescaling the spatial coordinates. What has the physical meaning in the flat Universe is the ratio of scale factors at different times, $a(t_1)/a(t_2)$, and, in particular, the Hubble parameter

$$H(t) = \frac{\dot{a}(t)}{a(t)}.$$

Hereafter, the dot denotes derivative with respect to time.

Note that in the cases of closed and open Universe the spatial coordinates x^i entering (2.13) are dimensionless, while the scale factor has dimension of length. On the other hand, in the case of spatially flat Universe it is convenient to assign the dimension of length to the coordinates x^i, while treating the scale factor as a dimensionless quantity.

The metric of homogeneous and isotropic Universe has the form (2.13) in a certain reference frame. This frame is singled out by the fact that space is the same everywhere at each moment of time. Furthermore, this frame is comoving: world lines of particles which are at rest with respect to this frame are geodesic, i.e., these particles are free. Before showing that, let us point out that for these particles one has $ds^2 = dt^2$, i.e., the time coordinate t has the meaning of proper time of particles at rest. In the present Universe, these particles may be thought of as galaxies, if one disregards their local (peculiar) motion caused by gravitational potentials (which are due to, e.g., nearby galaxies).

Let us show that the world line $x^i = $ const obeys the geodesic equation in metric (2.13),

$$\frac{du^\mu}{ds} + \Gamma^\mu_{\nu\lambda} u^\nu u^\lambda = 0, \tag{2.14}$$

where u^μ is 4-velocity (see Appendix A). To this end, let us calculate the Christoffel symbols,

$$\Gamma^\mu_{\nu\lambda} = \frac{1}{2} g^{\mu\sigma} (\partial_\nu g_{\lambda\sigma} + \partial_\lambda g_{\nu\sigma} - \partial_\sigma g_{\nu\lambda}). \tag{2.15}$$

Nonzero components of the metric tensor are

$$g_{00} = 1, \quad g_{ij} = -a^2(t)\gamma_{ij}(x),$$

while for the inverse tensor we have

$$g^{00} = 1, \quad g^{ij} = -\frac{1}{a^2(t)}\gamma^{ij}(x).$$

It is clear that

$$\Gamma^0_{00} = 0, \quad \Gamma^0_{0i} = 0, \quad \Gamma^i_{00} = 0$$

(in the expression in parenthesis in (2.15), at least two indices are equal to zero, but $g_{0i} = 0$, $\partial_\mu g_{00} = 0$). For Γ^i_{0j} we have

$$\Gamma^i_{0j} = \frac{1}{2} g^{ik} \partial_0 g_{jk} = \frac{\dot{a}}{a}\delta^i_j. \tag{2.16}$$

The remaining Christoffel symbols are also straightforwardly calculated,

$$\Gamma^0_{ij} = a\dot{a}\gamma_{ij}, \tag{2.17}$$

$$\Gamma^i_{jk} = {}^{(3)}\Gamma^i_{jk}, \tag{2.18}$$

where ${}^{(3)}\Gamma^i_{jk}$ are the Christoffel symbols for metric γ_{ij}.

Let us now turn to Eq. (2.14). The only non-vanishing component of the 4-velocity $u^\mu = dx^\mu/ds$ of a particle at rest is

$$u^0 = \frac{dx^0}{ds} = \frac{dt}{dt} = 1.$$

Equation (2.14) is obviously satisfied, since $du^\mu/ds = 0$ and $\Gamma^\mu_{00} = 0$ for any μ. Thus, the world lines of particles which are at rest in our reference frame are indeed geodesic.

Problem 2.3. *If one chooses a time coordinate which does not coincide with proper time of particles at rest, FLRW metric takes the form*

$$ds^2 = N^2(t)dt^2 - a^2(t)\gamma_{ij}dx^i dx^j.$$

Show by direct calculation that in this metric too, world lines of particles at rest are geodesic (this is of course obvious, since these lines are the same lines as in the text).

To end this Section, we note that for both closed and open models, the spatial curvature can often be neglected, so that one can use the spatially flat metric

$$\gamma_{ij} = \delta_{ij}. \tag{2.19}$$

This is certainly possible for processes at spatial scales much smaller than the curvature radius $a(t)$. We already mentioned in Chapter 1 that our Universe is spatially flat to a very good approximation, both in the past and at present. Hence, the approximation (2.19) is actually very good for all scales. We will quantify this statement in what follows.

2.3.   Redshift. Hubble Law

The scale factor $a(t)$ grows in time, and the distances between points of fixed spatial coordinates x^i grow too — the Universe expands. Because of that, the wavelength of a photon emitted long ago by a distant source increases as the photon moves towards an observer, i.e., the photon wavelength gets redshifted. To descibe this phenomenon, let us write the action for free electromagnetic field in space–time with metric $g_{\mu\nu}(x)$:

$$S = -\frac{1}{4} \int d^4x \sqrt{-g} g^{\mu\nu} g^{\lambda\rho} F_{\mu\lambda} F_{\nu\rho}, \tag{2.20}$$

where, as usual,

$$F_{\mu\nu} = \nabla_\mu A_\nu - \nabla_\nu A_\mu = \partial_\mu A_\nu - \partial_\nu A_\mu,$$

and A_μ is the electromagnetic vector-potential. The action (2.20) is the simplest covariant generalization of the action of Maxwell's electrodynamics; the factor $\sqrt{-g}$ ensures the invariance of 4-volume (see Appendix A). One could in principle add to the action (2.20) other invariant terms vanishing in the Minkowski limit, like

$$\delta S = \frac{\alpha}{M_{Pl}^2} \int d^4x \sqrt{-g} g^{\mu\nu} R^{\lambda\rho} F_{\mu\lambda} F_{\nu\rho} \tag{2.21}$$

where $R^{\lambda\rho}$ is the Ricci tensor, α is a dimensionless constant, and the factor M_{Pl}^{-2} is included on dimensional grounds. However, terms like (2.21) are negligibly small compared to (2.20) if the space–time curvature is small compared to the Planck value, $|R_{\mu\nu}| \ll M_{Pl}^2$, and the constant α is not extremely large. Therefore, at all classical stages of the evolution of the Universe, when its parameters are far from Planckian, freely propagating photons are described by the action (2.20).

Consider now the propagation of a photon in the homogeneous isotropic Universe. Realistically, the photon wavelength is small compared to the spatial curvature radius, even if the Universe is open or closed. Therefore, the Universe can be considered spatially flat, and one can use the metric

$$ds^2 = dt^2 - a^2(t)\delta_{ij}dx^i dx^j. \tag{2.22}$$

It is convenient to use conformal time η instead of cosmic time t. The former is defined by

$$dt = a d\eta, \tag{2.23}$$

i.e.,

$$\eta = \int \frac{dt}{a(t)}.$$

In terms of this new time coordinate the metric is

$$ds^2 = a^2(\eta)[d\eta^2 - \delta_{ij}dx^i dx^j]. \tag{2.24}$$

In other words,

$$g_{\mu\nu} = a^2(\eta)\eta_{\mu\nu}, \tag{2.25}$$

where $\eta_{\mu\nu}$ is the Minkowski metric. The metric (2.25) differs from the Minkowski metric by overall time-dependent rescaling, i.e., the metric of homogeneous isotropic Universe has conformally flat form in coordinates (η, x^i). In these coordinates one has

$$g^{\mu\nu} = a^{-2}\eta^{\mu\nu}, \quad \sqrt{-g} = a^4.$$

By substituting these expressions into (2.20) we obtain that in coordinates (η, x^i) the action of electromagnetic field coincides with

$$S = -\frac{1}{4} \int d^4x\, \eta^{\mu\nu}\eta^{\lambda\rho} F_{\mu\lambda} F_{\nu\rho}. \tag{2.26}$$

This property is inherent in a theory of massless *vector* field; in theories of other fields, the action in conformal coordinates (η, x^i) in general does not reduce to the flat space–time action. In this regard, free electromagnetic fields is called conformal, while other fields, generally speaking, are not conformal.

It follows from (2.26) that solutions to equations for the free electromagnetic field in the Universe with metric (2.22) (or, equivalently, with metric (2.24)) are superpositions of plane waves

$$A_\mu^{(\alpha)} = e_\mu^{(\alpha)} e^{ik\eta - i\mathbf{kx}},$$

where $\mathbf{k}$ is a constant vector (called coordinate momentum, or conformal momentum), $k = |\mathbf{k}|$, and $e_\mu^{(\alpha)}$ are the standard polarization vectors of a photon, $\alpha = 1, 2$.

We stress that $\mathbf{k}$ is not the physical momentum of a photon, and k is not the physical frequency, since $d\mathbf{x}$ and $d\eta$ are not physical distance and physical time interval. The quantity $\Delta x = 2\pi/k$ is the coordinate wavelength of a photon, while the physical wavelength at time t according to (2.22) is

$$\lambda(t) = a(t)\Delta x = 2\pi \frac{a(t)}{k}. \tag{2.27}$$

Similarly, $\Delta\eta = 2\pi/k$ is the period of electromagnetic wave in conformal time, while according to (2.23), the period in physical time[2] t is

$$T = a(t)\Delta\eta = 2\pi \frac{a(t)}{k}. \tag{2.28}$$

Hence, the physical momentum $\mathbf{p}$ and physical frequency of a photon at time t are

$$\mathbf{p}(t) = \frac{\mathbf{k}}{a(t)}, \quad \omega(t) = \frac{k}{a(t)}. \tag{2.29}$$

In expanding Universe, the scale factor $a(t)$ increases in time, the physical wavelength (2.27) grows, while the physical momentum and frequency decrease: they get redshifted. If a photon was emitted at time t_i with given wavelength λ_i (e.g., in the process of transition of hydrogen atom from excited to ground state), then its wavelength at the Earth is equal to

$$\lambda_0 = \lambda_i \frac{a_0}{a(t_i)} \equiv \lambda_i[1 + z(t_i)]. \tag{2.30}$$

As usual, the index 0 refers to the present value of the relevant quantity.

The quantity

$$z(t) = \frac{a_0}{a(t)} - 1 \tag{2.31}$$

is called redshift. The further is the object emitting photons, the longer these photons travel through the Universe to the Earth, the smaller is $a(t_i)$: objects at larger distances have larger redshifts. The redshift is a directly measurable quantity; its measurement boils down to the indentification of an emission (or absorption) lines and determination of their actual wavelengths; see Chapter 1.

Let us stress that formulas (2.30) and (2.31) are general; they are valid at all z.

For not so distant objects, the difference $(t_0 - t_i)$ — the photon travel time — is not very large, and we can expand in $(t_0 - t_i)$,

$$a(t_i) = a_0 - \dot{a}(t_0)(t_0 - t_i).$$

---

[2]Note that we assume here that the period T is much smaller than the time scale of the evolution of the scale factor. This is of course true at all cosmological epochs of interest.

In terms of the present value of the Hubble parameter $H_0 = H(t_0)$ one obtains

$$a(t_i) = a_0[1 - H_0(t_0 - t_i)].$$

Hence, to the linear order in $(t_0 - t_i)$ one has

$$z(t_i) = H_0(t_0 - t_i).$$

Finally, the travel time is equal to the distance to the emitter, up to corrections of order $(t_0 - t_i)^2$,

$$r = t_0 - t_i.$$

In this way one obtains the Hubble law,

$$z = H_0 r, \quad z \ll 1. \tag{2.32}$$

When deriving it, we assumed that $(t_0 - t_i)$ is not large; this corresponds to small redshift, as we indicated in Eq. (2.32).

The Hubble parameter H_0 is one of the fundamental cosmological parameters of the present Universe. We have already discussed its importance in Chapter 1. Here we recall that the measured value is given by (1.5); for the definition of the related parameter h see (1.7), while the associated time and distance scales are presented in (1.9a) and (1.9b).

To end this Section, let us make the following comment. Our derivation of Eqs. (2.27)–(2.29) and, accordingly, Eq. (2.30) was based on the fact that electromagnetic field is conformal, i.e., in coordinates (η, x^i) its action reduces to the action in Minkowski space–time. However, these formulas are of general character; they are valid for all massless particles.

Consider as an example a massless scalar field with action

$$S = \frac{1}{2} \int d^4x \sqrt{-g} g^{\mu\nu} \partial_\mu \phi \partial_\nu \phi. \tag{2.33}$$

In conformal coordinates (η, x^i), the explicit form of this action in space–time with metric (2.24) is

$$S = \frac{1}{2} \int d^3x d\eta \, a^2(\eta) \eta^{\mu\nu} \partial_\mu \phi \partial_\nu \phi. \tag{2.34}$$

The scalar field with action (2.33) is not conformal: the action (2.34) does not reduce to the flat space–time action by any change of variables. By varying (2.34) with respect to ϕ we obtain the equation for the scalar field in metric (2.24),

$$\frac{1}{a^2} \partial_\eta(a^2 \partial_\eta \phi) - \partial_i \partial_i \phi = 0 \tag{2.35}$$

(summation over $i = 1, 2, 3$ is assumed; $\partial_i \partial_i$ is Laplacian in flat 3-dimensional space). In the first place, let us notice that the operator in the left-hand side of this equation — covariant d'Alembertian — does not depend on x^i, so the solution can be written as a linear combination of 3-dimensional plane waves,

$$\phi = \frac{1}{a(\eta)} f(\eta) e^{-i\mathbf{kx}}.$$

The factor $a^{-1}(\eta)$ is introduced for convenience; due to this factor, Eq. (2.35) becomes the equation for $f(\eta)$ without first-order time derivative,

$$\partial_\eta^2 f - \frac{\partial_\eta^2 a}{a} f + \mathbf{k}^2 f = 0. \tag{2.36}$$

If the expansion rate of the Universe and its time derivative are small compared to the frequency, the second term in Eq. (2.36) can be neglected, and the solutions to the scalar field equation are superpositions of plane waves

$$\phi = \frac{1}{a(\eta)} e^{ik\eta - i\mathbf{kx}}. \tag{2.37}$$

Coordinate frequency and momentum, k and $\mathbf{k}$, are again independent of time, which again leads to Eqs. (2.27)–(2.29), as promised. Note that the factor $a^{-1}(\eta)$ in the solution (2.37) compensates for the factor $a^2(\eta)$ in the Lagrangian in (2.33); in other words, the action for the field $f(\mathbf{x}, \eta) = a(\eta)\phi(\mathbf{x}, \eta)$ reduces to the flat space–time action of massless scalar field up to corrections containing time derivatives of the scale factor.

Problem 2.4. *Give quantitative formulation of conditions under which the second term in Eq. (2.36) is small compared to the third term. Express these conditions in terms of the physical frequency, Hubble parameter and its derivative with respect to physical time t.*

Problem 2.5. *Consider photons and massless scalar particles (actions (2.20) and (2.33), respectively). Show that the relation (2.31) between redshift and scale factor remains valid in open and closed Universes, provided that the wavelength $\lambda(t)$ is small compared to the radius of spatial curvature $a(t)$ at all times between emission and absorption.*

2.4. Slowing Down of Relative Motion

Physical momenta of *massive* free particles also decrease as

$$\mathbf{p} = \frac{\mathbf{k}}{a(t)}, \tag{2.38}$$

where $\mathbf{k}$ is time-independent coordinate momentum. To see this, let us consider the geodesic equation

$$\frac{du^\mu}{ds} + \Gamma^\mu_{\nu\lambda} u^\nu u^\lambda = 0, \tag{2.39}$$

for 4-velocity of a free particle,

$$u^\mu = \frac{dx^\mu}{ds}$$

(see Appendix A). Let us note right away that u^i are not the physical values of spatial components of 4-velocity: since the physical distances dX^i are related to coordinate distances by $dX^i = a(t)dx^i$, the physical components are

$$U^i = \frac{dX^i}{ds} = a(t)u^i.$$

The physical momenta are expressed through the physical velocities in the usual way,

$$p^i = mU^i,$$

while the usual 3-velocities

$$v^i = \frac{dX^i}{dt}$$

are related to U^i by

$$U^i = \frac{v^i}{\sqrt{1-\mathbf{v}^2}}. \tag{2.40}$$

Indeed, the 4-velocities u^μ obey the relation $g_{\mu\nu}u^\mu u^\nu = 1$. In metric (2.25) the latter relation is

$$(u^0)^2 - a^2 u^i u^i = 1$$

(summation over i is assumed), or

$$\left(\frac{dt}{ds}\right)^2 - (U^i)^2 = 1.$$

Therefore,

$$v^i = \frac{dX^i}{dt} = \frac{dX^i}{ds}\frac{ds}{dt} = \frac{U^i}{\sqrt{1+\mathbf{U}^2}},$$

which gives precisely Eq. (2.40).

Let us come back to the geodesic equation (2.39). For metric (2.22), the only non-vanishing components of connection are given by (2.16) and (2.17). Thus, spatial components of Eq. (2.39) ($\mu = i = 1,2,3$) read

$$\frac{du^i}{ds} + \Gamma^i_{0j}u^0 u^j + \Gamma^i_{j0}u^j u^0 = 0,$$

or

$$\frac{du^i}{ds} + 2\frac{\dot{a}}{a}\frac{dt}{ds}u^i = 0.$$

In terms of the physical components, the latter equation is

$$\frac{dU^i}{dt} = -\frac{\dot{a}}{a}U^i.$$

Hence, velocities of free particles decrease in time as

$$U^i = \frac{\text{const}}{a(t)},$$

i.e., the law (2.38) is indeed valid.

Thus, the velocities of free particles with respect to comoving frame decrease in expanding Universe; particles gradually "freeze in". In particular, if at early cosmological stages massive particles were relativistic, they become non-relativistic at later stages. This behavior is characteristic of neutrino, if its mass is well above 10^{-4} eV.

To conclude this Section, let us give another derivation of Eq. (2.38). This derivation has to do with solutions to massive field equations in expanding Universe. Consider as an example the theory of massive scalar field with action

$$S = \int d^4x \sqrt{-g} \left(\frac{1}{2}g^{\mu\nu}\partial_\mu\phi\partial_\nu\phi - \frac{m^2}{2}\phi^2 \right).$$

In metric (2.24) this action has the form

$$S = \int d^3x d\eta \left(\frac{1}{2}a^2\eta^{\mu\nu}\partial_\mu\phi\partial_\nu\phi - \frac{m^2a^4}{2}\phi^2 \right).$$

An equation obtained from this action by varying with respect to ϕ is the Klein–Gordon equation in expanding Universe (in conformal coordinates)

$$\frac{1}{a^2}\partial_\eta(a^2\partial_\eta\phi) - \partial_i\partial_i\phi + m^2a^2\phi = 0. \tag{2.41}$$

This equation again does not contain spatial coordinates explicitly, so its solutions are again superpositions of plane waves

$$\phi = \frac{1}{a(\eta)}f(\eta)e^{-i\mathbf{kx}},$$

where $\mathbf{k}$ is independent of time. The physical wavelength of a particle is thus given by (2.27), so its physical momentum indeed redshifts according to (2.38).

Problem 2.6. *Show that for slowly varying scale factor, solutions to Eq. (2.41) are superpositions of*

$$\phi = \frac{1}{a(\eta)\sqrt{\Omega(\eta)}} e^{i\int^{\eta} \Omega(\eta)d\eta} e^{-i\mathbf{k}\mathbf{x}} \cdot (1 + \mathcal{O}(\partial_{\eta}a))$$

where $\Omega(\eta) = \sqrt{k^2 + m^2 a^2(\eta)}$. *Thus, coordinate frequency (derivative of the exponent with respect to conformal time) equals* $\Omega(\eta)$, *while the physical frequency is*

$$\omega(\eta) = \frac{\Omega(\eta)}{a(\eta)} = \sqrt{\mathbf{p}^2 + m^2},$$

as should be the case in relativistic physics.

2.5.   Gases of Free Particles in Expanding Universe

Let us consider the behavior of gases of stable non-interacting particles in a homogeneous isotropic Universe. We will discuss local properties like number density, so we again neglect the spatial curvature and use metric (2.22). The gas of particles is characterized by the distribution function $f(\mathbf{X}, \mathbf{p})$, so that $f(\mathbf{X}, \mathbf{p})d^3\mathbf{X}d^3\mathbf{p}$ is the number of particles in physical volume element $d^3\mathbf{X}$ in interval of physical momenta $d^3\mathbf{p}$. The distribution in general depends on time and is not one of the equilibrium distribution functions. We will consider homogeneous gases, whose distribution functions are independent of coordinates and depend only on momenta (and time).

We have seen in previous Sections that coordinate momenta of free particles $\mathbf{k}$ are independent of time. The coordinate volume $d^3\mathbf{x}$ is also constant in time, so the distribution function written in terms of coordinate momentum is time-independent,

$$f(k) = \text{const.}$$

The number of particles in an element of comoving phase space is also independent of time,

$$f(k)d^3\mathbf{x}d^3\mathbf{k} = \text{const.}$$

The comoving phase space volume coincides with the physical one,

$$d^3\mathbf{x}d^3\mathbf{k} = d^3(a\mathbf{x})d^3\left(\frac{\mathbf{k}}{a}\right) = d^3\mathbf{X}d^3\mathbf{p}.$$

Hence, the time dependence of the distribution function of free particles is entirely determined by the redshift of momenta,

$$f(\mathbf{p}, t) = f(\mathbf{k}) = f[a(t) \cdot \mathbf{p}].$$

If the distribution function is known at some moment of time and equal to $f_i(\mathbf{p})$, then at later times it is equal to

$$f(\mathbf{p}, t) = f_i \left(\frac{a(t)}{a_i} \mathbf{p} \right). \tag{2.42}$$

We stress that, generally speaking, this formula is valid for free particles only.

   We will see that the known particles, as well as (most likely) dark matter particles actively interacted with each other in the early Universe, so that they were in thermal equilibrium. At some moment of time (different moment for different types of particles) the Universe expanded to the extent that the density and temperature became sufficiently small, and interactions between particles switched off. At that moment the particles had thermal distribution function, and after that the distribution function evolved as given by[3] (2.42).

   Let us consider in more detail two limiting cases. We begin with massless particles; the most interesting among them are photons after they decouple from other particles (i.e., after recombination). At the time t_i when their interaction with matter switches off, they have Planckian distribution function. The latter depends only on the ratio of momentum and temperature at that time, $\frac{|\mathbf{p}|}{T_i}$:

$$f_i(\mathbf{p}) = f_{\mathrm{Pl}} \left(\frac{|\mathbf{p}|}{T_i} \right) = \frac{1}{(2\pi)^3} \frac{1}{e^{|\mathbf{p}|/T_i} - 1}. \tag{2.43}$$

For massless fermions this would be the Fermi–Dirac distribution at zero mass (see Chapter 5), which also depends only on the ratio $\frac{|\mathbf{p}|}{T_i}$. According to (2.42), the distribution function at later times is

$$f(\mathbf{p}, t) = f_{\mathrm{Pl}} \left(\frac{a(t)|\mathbf{p}|}{a_i T_i} \right) = f_{\mathrm{Pl}} \left(\frac{|\mathbf{p}|}{T_{\mathit{eff}}(t)} \right),$$

where

$$T_{\mathit{eff}}(t) = \frac{a_i}{a(t)} T_i. \tag{2.44}$$

Hence, the distribution function always has the equilibrium shape despite the fact that photons are not in thermal equilibrium. It follows from (2.44) that relic photons have Planckian spectrum with effective temperature decreasing in time according to

$$T_{\mathit{eff}}(t) \propto \frac{1}{a(t)}. \tag{2.45}$$

We will see in what follows that the same behavior (with reservations) is characteristic also of the distribution function of photons in the early Universe, when photons are in thermal equilibrium with matter.

---

[3]This statement is not exact for various reasons. One is that there are gravitational potentials in our Universe which are produced by galaxies, clusters of galaxies, etc. This fact is to good approximation irrelevant for relic photons, but it is very relevant for massive particles. The latter obeyed Eq. (2.42) at sufficiently early cosmological stages only, when the inhomogeneities in the Universe were small.

It goes without saying that the relation (2.45) is valid for other massless parti-cles, including fermions (if there are massless particles in Nature besides photons; gravitons are an entirely different story). Importantly, the effective temperature decreases according to (2.45) from the time at which these particles get out of thermal equilibrium with cosmic medium (time of decoupling). At later times their effective temperature may be different from the photon temperature. We encounter this situation in Chapter 7. If particles have small but non-zero mass m and at the time of decoupling were relativistic, their spectrum remains thermal, and effective temperature falls as $a^{-1}(t)$ as long as $T_{eff} \gg m$. At later times the spectrum is no longer thermal. Indeed, the distribution in momenta is always Planckian (if the particles are bosons), i.e., the distribution function has the form (2.43) with tem-perature $T_{eff}(t) \propto a^{-1}(t)$, while at $T_{eff} \ll m$ the equilibrium distribution is the Maxwell–Boltzmann distribution. Relic particles of this sort are called "hot dark matter"; it is hot in the sense that the particles decouple being relativistic and hence they always have massless distribution function in momenta.[4] An important example of hot dark matter is the relic neutrino.

Problem 2.7. *Consider particles of mass m that decouple at $T \gg m$. Show that their distribution in momenta is given by (2.43) at all times afterwards, including late times when $T_{eff} \ll m$.*

Let us now consider another limiting case of particles which decouple being non-relativistic. At the time of decoupling they have the Maxwell–Boltzmann distribu-tion function (see Chapter 5)

$$f(\mathbf{p}) = \frac{1}{(2\pi)^3} \exp\left(-\frac{m - \mu_i}{T_i}\right) \exp\left(-\frac{\mathbf{p}^2}{2mT_i}\right),$$

where T_i and μ_i are the temperature and chemical potential at that time. According to (2.42), the distribution function at later times is equal to

$$f(\mathbf{p}) = \frac{1}{(2\pi)^3} \exp\left(-\frac{m - \mu_i}{T_i}\right) \exp\left(-\frac{a^2(t)\mathbf{p}^2}{2ma_i^2 T_i}\right).$$

It can again be written in the Maxwell–Boltzmann form,

$$f(\mathbf{p}, t) = \frac{1}{(2\pi)^3} \exp\left(-\frac{m - \mu_{eff}}{T_{eff}}\right) \exp\left(-\frac{\mathbf{p}^2}{2mT_{eff}}\right), \qquad (2.46)$$

where

$$T_{eff}(t) = \left(\frac{a_i}{a(t)}\right)^2 T_i,$$

---

[4]The same property holds for "warm" dark matter; the distinction between hot and warm dark matter is discussed in Sec. 9.1.

and effective chemical potentail μ_{eff} is determined by the relation

$$\frac{m - \mu_{eff}(t)}{T_{eff}} = \frac{m - \mu_i}{T_i}.$$

Equation (2.46) means that the distribution function still has the equilibrium form[5] (although particles are no longer in thermal equilibrium), and the effective temperature changes as

$$T_{eff}(t) \propto \frac{1}{a^2(t)}. \tag{2.47}$$

It decreases faster than in the case of massless particles. Dark matter particles which were non-relativistic at their decoupling are called cold dark matter. As we pointed out in Chapter 1, cold (or possibly warm) dark matter makes about 25% of total energy density today.

---

[5]See footnote 3 in this Chapter.

Chapter 3

Dynamics of Cosmological Expansion

3.1. Friedmann Equation

The law of the cosmological expansion, i.e., the dependence of the scale factor a on time, is determined by the Einstein equations (see Appendix A),

$$R_{\mu\nu} - \frac{1}{2}g_{\mu\nu}R = 8\pi G T_{\mu\nu}.$$

Let us find their explicit form for homogeneous and isotropic metric (2.13). We begin with the calculation of the Ricci tensor,

$$R_{\mu\nu} = \partial_\lambda \Gamma^\lambda_{\mu\nu} - \partial_\mu \Gamma^\lambda_{\nu\lambda} + \Gamma^\lambda_{\mu\nu}\Gamma^\sigma_{\lambda\sigma} - \Gamma^\lambda_{\mu\sigma}\Gamma^\sigma_{\lambda\nu}. \tag{3.1}$$

The non-vanishing components of the Christoffel symbols are given by (2.16)–(2.18). These formulas imply, in particular, that

$$\Gamma^\mu_{0\mu} = \frac{\dot{a}}{a}\delta^i_i = 3\frac{\dot{a}}{a}, \quad \Gamma^\mu_{i\mu} = {}^{(3)}\Gamma^j_{ij}.$$

Let us first calculate R_{00}. Since $\Gamma^\mu_{00} = 0$, the only contributions come from the second and fourth terms in (3.1), and we obtain

$$R_{00} = -\partial_0 \Gamma^\lambda_{0\lambda} - \Gamma^\lambda_{0\sigma}\Gamma^\sigma_{0\lambda} = -\partial_0 \Gamma^\lambda_{0\lambda} - \Gamma^i_{0j}\Gamma^j_{0i}$$

$$= -\partial_0 \left(3\frac{\dot{a}}{a}\right) - \left(\frac{\dot{a}}{a}\right)^2 \delta^i_j \delta^j_i = -3\partial_0 \left(\frac{\dot{a}}{a}\right) - 3\left(\frac{\dot{a}}{a}\right)^2.$$

Finally,

$$R_{00} = -3\frac{\ddot{a}}{a}. \tag{3.2}$$

Let us now turn to the mixed components R_{0i}. Keeping only non-vanishing Christoffel symbols in (3.1), we write

$$R_{0i} = \partial_j \Gamma^j_{0i} - \partial_0 \Gamma^\lambda_{i\lambda} + \Gamma^j_{0i}\Gamma^\lambda_{j\lambda} - \Gamma^k_{0j}\Gamma^j_{ik}. \tag{3.3}$$

This expression in fact is equal to zero, since Γ^j_{0i} are independent of spatial coordinates, $\Gamma^\lambda_{i\lambda} = {}^{(3)}\Gamma^j_{ij}$ are calculated with static metric γ_{ij} and hence are independent of time and $\Gamma_{0k}^j \propto \delta_k^j$, which leads to the cancellation of the last two terms in (3.3). Hence,

$$R_{0i} = 0. \tag{3.4}$$

This should have been expected, since R_{0i} transform as components of a 3-vector under spatial rotations, and there is no special direction in isotropic space.

Let us finally calculate the spatial components R_{ij}. We again keep non-vanishing Christoffel symbols in (3.1) and write

$$R_{ij} = (\partial_0 \Gamma_{ij}^0 + \partial_k \Gamma_{ij}^k) - \partial_i \Gamma_{j\lambda}^\lambda$$
$$+ (\Gamma_{ij}^0 \Gamma_{0\sigma}^\sigma + \Gamma_{ij}^k \Gamma_{k\sigma}^\sigma) - (\Gamma_{ik}^0 \Gamma_{j0}^k + \Gamma_{i0}^k \Gamma_{jk}^0 + \Gamma_{il}^k \Gamma_{jk}^l), \tag{3.5}$$

where we collected in parentheses the terms that come from each of the four terms in (3.1). Taking into account (2.18), we combine into the Ricci tensor $^{(3)}R_{ij}$ those terms in (3.5) which contain spatial derivatives only; the tensor $^{(3)}R_{ij}$ is calculated with the 3-dimensional metric γ_{ij}. Four other terms are calculated directly, and we obtain

$$R_{ij} = \partial_0(\dot a a)\gamma_{ij} + \dot a a \gamma_{ij} \cdot 3\frac{\dot a}{a} - \dot a a \gamma_{ik} \cdot \frac{\dot a}{a}\delta_j^k - \frac{\dot a}{a}\delta_i^k \dot a a \gamma_{jk} + {}^{(3)}R_{ij}.$$

Finally,

$$R_{ij} = (\ddot a a + 2\dot a^2 + 2\varkappa)\gamma_{ij}, \tag{3.6}$$

where we made use of relation (2.9).

Let us now use Eqs. (3.2), (3.4) and (3.6), to find the scalar curvature,

$$R = g^{\mu\nu}R_{\mu\nu} = g^{00}R_{00} + g^{ij}R_{ij} = R_{00} - \frac{1}{a^2}\gamma^{ij}R_{ij}.$$

Since $\gamma^{ij}\gamma_{ij} = 3$, we have

$$R = -6\left(\frac{\ddot a}{a} + \frac{\dot a^2}{a^2} + \frac{\varkappa}{a^2}\right).$$

As a result, the 00-component of the left-hand side of the Einstein equations takes the simple form,

$$R_{00} - \frac{1}{2}g_{00}R = 3\left(\frac{\dot a^2}{a^2} + \frac{\varkappa}{a^2}\right). \tag{3.7}$$

The other components of the Einstein tensor $G_{\mu\nu} \equiv R_{\mu\nu} - \frac{1}{2}g_{\mu\nu}R$ are calculated in a similar way.

Let us now turn to the right-hand side of the Einstein equations. At cosmological epochs of interest to us here, matter in the Universe can be described macroscopically: it can be considered as homogeneous fluid with energy density $\rho(t)$ and pressure $p(t)$. This fluid as a whole is at rest with respect to comoving reference frame, so the only non-zero component of its 4-velocity is u^0. From the relation $g_{\mu\nu}u^\mu u^\nu = 1$ we have

$$u^0 = 1, \quad u_0 = 1.$$

Hence, the 00-component of energy-momentum tensor is (see Appendix A)

$$T_{00} = (p + \rho)u_0 u_0 - g_{00}p = \rho. \tag{3.8}$$

Combining (3.7) and (3.8) we conclude that the 00-component of the Einstein equations has the following form for homogeneous and isotropic Universe,

$$\left(\frac{\dot{a}}{a}\right)^2 = \frac{8\pi}{3}G\rho - \frac{\varkappa}{a^2}. \tag{3.9}$$

This is called Friedmann equation; it relates the rate of the cosmological expansion (the Hubble parameter $H = \dot{a}/a$) to the total energy density ρ and spatial curvature.

The Friedmann equation has to be supplemented with one more equation, since Eq. (3.9) contains two unknown functions of time, $a(t)$ and $\rho(t)$. To obtain the additional equation, it is convenient to consider the covariant conservation of the energy-momentum tensor (see Appendix A)

$$\nabla_\mu T^{\mu\nu} = 0.$$

Setting $\nu = 0$, we write

$$\nabla_\mu T^{\mu 0} \equiv \partial_\mu T^{\mu 0} + \Gamma^\mu_{\mu\sigma} T^{\sigma 0} + \Gamma^0_{\mu\sigma} T^{\mu\sigma} = 0. \tag{3.10}$$

Non-vanishing components of the energy-momentum tensor in the comoving frame are

$$T^{00} = g^{00} g^{00} T_{00} = \rho, \tag{3.11}$$

$$T^{ij} = g^{ik} g^{jl} T_{kl} = g^{ik} g^{jl} (-g_{kl} p) = \frac{1}{a^2} \gamma^{ij} p. \tag{3.12}$$

Here we made use of the fact that the spatial components of 4-velocity are zero in the comoving frame, so that $T_{ij} = (p + \rho)u_i u_j - p g_{ij} = -p g_{ij}$. Note that γ^{ij} entering (3.12) are time-independent components of the matrix inverse to γ_{ij}. We again use the expressions for non-zero Christoffel symbols, Eqs. (2.16)–(2.18), and write Eq. (3.10) in the following explicit form,

$$\dot{\rho} + 3\frac{\dot{a}}{a}(\rho + p) = 0. \tag{3.13}$$

To close the system of equations determining the dynamics of homogeneous and isotropic Universe, one has to specify the equation of state of matter,

$$p = p(\rho). \tag{3.14}$$

The latter equation is not a consequence of equations of General Relativity. The equation of state is determined by matter content in the Universe. For non-relativistic particles one has $p = 0$, while $p = \frac{1}{3}\rho$ for relativistic particles and $p = -\rho$ for vacuum; see Sec. 3.2.

Equations (3.9), (3.13) and (3.14) completely determine dynamics of the cosmological expansion. Let us make two remarks concerning Eqs. (3.13) and (3.14). First, if there exist several types of matter in the Universe which do not interact with each other, then the energy–momentum tensor of each type of matter obeys the covariant conservation equation independently. Therefore, Eqs. (3.13) and (3.14) must be satisfied by every type of matter separately. On the other hand, the energy density ρ in the Friedmann equation (3.9) is the sum of energy densities of all types of matter. Second, if matter in the Universe is in thermal equilibrium at zero chemical potentials, then Eq. (3.13) has simple interpretation. It can be written in the following form,

$$\frac{d\rho}{p + \rho} = -3d(\ln a). \tag{3.15}$$

The left-hand side of this equation coincides with $d(\ln s)$ where s is the entropy density, see Chapter 5. Hence, Eq. (3.13) reduces to the relation

$$sa^3 = \text{const},$$

which means the entropy conservation in comoving volume. In other words, due to the expansion of the Universe, entropy density decreases as the inverse element of the comoving volume, $s \propto a^{-3}$. We see in Sec. 5.2 that the same interpretation holds for systems at non-vanishing chemical potentials too.

To conclude this Section, let us make two comments. First, we note that when obtaining the equations governing the dynamics of the cosmological evolution we made use of only one Einstein equation, $R_{00} - \frac{1}{2}g_{00}R = 8\pi G T_{00}$, and only one of the covariant conservation equations, $\nabla_\mu T^{\mu 0} = 0$. One can show that other Einstein equations and covariant conservation equations are identically satisfied for solutions to Eqs. (3.9) and (3.13). Nevertheless, let us write for future reference the equation that follows from the ij-components of the Einstein equations. It is clear from (3.6) that ij-components of the Einstein tensor are proportional to γ_{ij}. Furthermore, the ij-components of the energy-momentum tensor $T_{ij} = -p\,g_{ij}$ are also proportional to γ_{ij}. So, all ij-components of the Einstein equations are reduced to a single Raychaudhuri equation,

$$2\frac{\ddot{a}}{a} + \frac{\dot{a}^2}{a^2} = -8\pi G p - \frac{\varkappa}{a^2}. \tag{3.16}$$

We do not use this equation in this book; it can be viewed as a consequence of the Friedmann equation (3.9) and covariant conservation of energy.

The second comment is that the assumption that matter filling the Universe is ideal fluid is in fact unnecessary for deriving Eqs. (3.9), (3.13) as well as Eq. (3.16).

The cosmological model we discuss is self-consistent if matter is homogeneous and isotropic. The isotropy means that the energy–momentum tensor necessarily has the diagonal structure, $T_{00} = \rho, T_{ij} = -p\,g_{ij}$, where the parameters ρ and p are energy density and *effective* pressure. Homogeneity requires that these parameters are functions of time only, $\rho = \rho(t), p = p(t)$. With this understanding, the analysis of this Section applies to the general case.

Problem 3.1. *Prove that equations*

$$R_{0i} - \frac{1}{2}g_{0i}R = 8\pi GT_{0i}, \quad R_{ij} - \frac{1}{2}g_{ij}R = 8\pi GT_{ij}, \quad \nabla_\mu T^{\mu i} = 0$$

are identically satisfied in the case of homogeneous and isotropic Universe, provided that the following equations are satisfied,

$$R_{00} - \frac{1}{2}g_{00}R = 8\pi GT_{00}, \quad \nabla_\mu T^{\mu 0} = 0.$$

Do not assume in the proof that the Universe is filled with ideal fluid; it is important only that matter is homogeneous and isotropic.

Problem 3.2. *Find equation of state in the model of a scalar field with the Lagrangian*

$$\mathcal{L} = -V_0\sqrt{1 - \partial_\mu\phi\,\partial^\mu\phi}, \quad V_0 > 0,$$

assuming that the field ϕ is classical and spatially homogeneous (i.e., it depends on time only).

3.2.   Sample Cosmological Solutions. Age of the Universe. Cosmological Horizon

Before discussing the realistic model of our Universe, let us present several examples of cosmological solutions. In this Section, we mostly consider a spatially flat model,

$$\varkappa = 0.$$

This is a very good approximation to the real Universe; we will see in what follows that the term $\varkappa/a^2$ in the Friedmann equation (3.9), if any, is small compared to the first term in the right-hand side at both present and early epochs.

In the spatially flat model, the Friedmann equation takes the following form,

$$\left(\frac{\dot{a}}{a}\right)^2 = \frac{8\pi}{3}G\rho. \tag{3.17}$$

Simple solutions described in this Section are obtained for the Universe filled with one type of matter. Then Eqs. (3.13) and (3.14) (or, equivalently, (3.14) and (3.15)) determine the energy density as a function of the scale factor, $\rho = \rho(a)$, and then the dependence of the scale factor on time is found from Eq. (3.17). Let us recall

(see Sec. 2.2) that in the case of a spatially flat Universe, the physically meaning-
ful quantity is the ratio of scale factors taken at different times, rather than the
scale factor itself. Therefore, one expects that a solution $a(t)$ will contain an arbitrary multiplicative constant. Also, Eqs. (3.13) and (3.17) are invariant under time
translation, so a solution will contain another arbitrary constant, "the beginning
of time".

3.2.1.   *Non-relativistic matter ("dust")*

We begin with the model of the Universe filled with non-relativistic matter whose
equation of state is

$$p = 0.$$

We obtain from Eq. (3.15)

$$\rho = \frac{\text{const}}{a^3}. \tag{3.18}$$

Everywhere in this Section, "const" denotes an arbitrary positive constant (which is
generally different in different formulas). Relation (3.18) has a simple interpretation:
number density of particles n decreases as the comoving volume increases, and
$a^3 n = \text{const}$, meaning the conservation of the total number of particles. Since the
energy density is given by $\rho = mn$, where m is mass of a particle, the energy density
behaves in the same way, i.e., $\rho a^3 = \text{const}$.

With account of (3.18), Eq. (3.17) takes the following form,

$$\left(\frac{\dot{a}}{a}\right)^2 = \frac{\text{const}}{a^3}$$

and has the solution

$$a(t) = \text{const} \cdot (t - t_s)^{2/3}, \tag{3.19}$$

where t_s is an arbitrary constant. The Universe expands, and its expansion decelerates, $\ddot{a} < 0$. The energy density is

$$\rho(t) = \frac{\text{const}}{(t - t_s)^2}. \tag{3.20}$$

The solution (3.19), (3.20) is singular at $t = t_s$: at that moment the scale factor is
equal to zero (all distances are vanishingly small), and the energy density is infinite.
This is an example of the cosmological singularity, the moment of Big Bang. We
will see that many other cosmological solutions start from the singularity. Clearly,
it does not make sense to extrapolate the classical evolution back to the Big Bang
singularity. What we learn is that it is quite possible that the Universe started off its
evolution from a state in which the energy density was very high (say, comparable

to the Planck energy density $\rho_{Pl} \sim M_{Pl}^4 \sim 10^{76} \, \mathrm{GeV}^4$), and for which classical laws of physics were not applicable.

In what follows, we will count time from the cosmological singularity (if a solution has this singularity). For the solution (3.19) and (3.20), this means that

$$t_s = 0.$$

Then t is the age of the Universe. Making use of (3.19), we relate it to the Hubble parameter,

$$H(t) = \frac{\dot{a}}{a}(t) = \frac{2}{3t}. \tag{3.21}$$

Making use of Eq. (3.17) once again, we find

$$\rho = \frac{3}{8\pi G} H^2 = \frac{1}{6\pi G} \frac{1}{t^2}. \tag{3.22}$$

Equalities (3.21) and (3.22) relate the physical quantities; they do not contain arbitrary parameters.

If most of the evolution of our Universe occurred in the regime of domination of non-relativistic matter, the age of the Universe would be given by (3.21), i.e.,

$$t_0 = \frac{2}{3H_0}. \tag{3.23}$$

Making use of (1.8), we would obtain

$$t_0 = h^{-1} \cdot 0.65 \cdot 10^{10} \ \mathrm{yrs} = 0.93 \cdot 10^{10} \ \mathrm{yrs} \quad (h = 0.7). \tag{3.24}$$

This value, even with an account of the uncertainty in the determination of H_0 (see (1.5)), would contradict independent bounds on the age of the Universe, $t_0 \geq 1.3 \cdot 10^{10}$ yrs, which we mentioned in Chapter 1. We will see that the situation becomes comfortable if the expansion of the Universe today is determined mostly by dark energy.

Let us make use of the solution (3.19) to introduce another important notion, the cosmological horizon. Imagine signals emitted at the moment of Big Bang and since then are traveling at the speed of light. We are interested in the distance $l_H(t)$ that such a signal travels from the point of its emission by the time t. The physical meaning of $l_H(t)$ is that it is equal to the size of the region (in infinite Universe!) causally connected by the time t: an observer living at the time t cannot know in principle what has happened outside the sphere of radius $l_H(t)$. This sphere is called the cosmological horizon, and $l_H(t)$ is the size of the observable part of the Universe at the time t. Clearly l_H increases in time; the horizon opens up.

Let us point out that another term for the cosmological horizon introduced here is "particle horizon". This is to be distinguished from "event horizon" which is discussed in Sec. 3.2.3.

For calculating $l_H(t)$, it is convenient to make use of conformal time η, see Sec. 2.3. Light-like geodesics obeying $ds^2 = 0$ in metric (2.24) are described by the equation

$$|d\mathbf{x}| = d\eta.$$

Therefore, the coordinate size of the horizon at time t is equal to $\eta(t)$, and its physical size is

$$l_H(t) = a(t)\eta(t) = a(t) \int_0^t \frac{dt'}{a(t')}. \tag{3.25}$$

For solution (3.19) we have

$$l_H(t) = 3t = \frac{2}{H(t)}. \tag{3.26}$$

If our Universe were matter dominated, the size of the horizon today would be equal to

$$l_{H,0} = \frac{2}{H_0}.$$

We find numerically from (1.8) that

$$l_{H,0} = h^{-1} \cdot 6000 \text{ Mpc}$$

$$= 0.86 \cdot 10^4 \text{ Mpc} = 2.6 \cdot 10^{28} \text{ cm} \quad (h = 0.7). \tag{3.27}$$

Yet another property of the horizon is the subject of the following problem.

Problem 3.3. *Show that signals emitted by sources which at time t are away from an observer at the distance $l_H(t)$, come to the observer at time t with infinite redshift.*

To conclude, in models with the cosmological horizon, the region in the Universe which is observable in principle, has a finite size even if the Universe itself is infinite.

3.2.2.   Relativistic matter ("radiation")

If the energy density in the Universe is due to relativistic matter, the equation of state is (see Chapter 5)

$$p = \frac{1}{3}\rho.$$

In this case Eq. (3.15) gives

$$\rho = \frac{\text{const}}{a^4}. \tag{3.28}$$

This behavior differs from (3.18) due to the fact that the expansion of the Universe leads not only to the dilution of the number density of particles, $n \propto a^{-3}$, but also to the redshift of the energy of each particle, $w \propto a^{-1}$ (the latter properties are valid for both non-interacting relativistic particles, cf. (2.29) and particles in thermal equilibrium; see Sec. 5.1).

Equation (3.17) becomes

$$\left(\frac{\dot{a}}{a}\right)^2 = \frac{\text{const}}{a^4}$$

and has the solution

$$a(t) = \text{const} \cdot t^{1/2}$$

(we again set the parameter t_s equal to zero). The properties of this solution are analogous to those of the solution (3.19) and (3.20): the Universe undergoes decelerated expansion; the moment $t = 0$ is the moment of the Big Bang singularity; the age is inversely proportional to the Hubble parameter (cf. (3.21))

$$H \equiv \frac{\dot{a}}{a} = \frac{1}{2t},$$

and the energy density is inversely proportional to the age squared (cf. (3.22)),

$$\rho = \frac{3}{8\pi G} H^2 = \frac{3}{32\pi G} \frac{1}{t^2}.$$

The horizon size is again finite,

$$l_H = a(t) \int_0^t \frac{dt'}{a(t')} = 2t = \frac{1}{H(t)}. \tag{3.29}$$

It is useful to relate the expansion rate (the Hubble parameter) to temperature, assuming thermal equilibrium between all types of particles and neglecting chemical potentials. At temperature T the energy density is (see Chapter 5)

$$\rho = \frac{\pi^2}{30} g_* T^4, \tag{3.30}$$

where

$$g_* = \sum_b g_b + \frac{7}{8} \sum_f g_f$$

is the effective number of degrees of freedom. Here summation runs over bosonic (b) and fermionic (f) particle species of masses smaller than T, g_b and g_f are the numbers of spin degrees of freedom of the boson b and fermion f. Using $G = M_{Pl}^{-2}$

(see Sec. 1.1) we write the relation (3.17) in the following form,

$$H = \frac{T^2}{M_{Pl}^*},\tag{3.31}$$

where

$$M_{Pl}^* = \sqrt{\frac{90}{8\pi^3 g_*}} M_{Pl} = \frac{1}{1.66\sqrt{g_*}} M_{Pl}\tag{3.32}$$

is the reduced Planck mass. We will repeatedly use the relation (3.31) keeping in mind that the parameter M_{Pl}^* depends on the effective number of degrees of freedom g_* and hence on temperature (since a particle of mass m contributes to g_* only at $T \gtrsim m$). This dependence, however, is rather weak, and when discussing the early Universe at certain stages of its evolution one can often treat M_{Pl}^* as a constant.

By comparing (3.28) with (3.30), we observe that the temperature of relativistic matter in thermal equilibrium is inversely proportional to the scale factor (up to slightly varying factor depending on g_*),

$$T(t) \approx \frac{\mathrm{const}}{a(t)}.\tag{3.33}$$

Let us recall that the same relation (which in that case is exact) holds for the effective temperature of a gas of non-interacting relativistic particles, see Sec. 2.5. Finally, it is useful to note that Eqs. (3.31) and (3.33) imply

$$\frac{\dot{T}}{T} \approx -H = -\frac{T^2}{M_{Pl}^*}.\tag{3.34}$$

The latter two relations, (3.33) and (3.34), are exact at those periods of the cosmological evolution in which the effective number of degrees of freedom g_* does not change.

3.2.3. *Vacuum*

In flat space–time, vacuum is the same in all inertial reference frames. It may have non-zero energy density, and the Lorentz-invariance dictates the form of its energy–momentum tensor,

$$T_{\mu\nu} = \rho_{vac}\eta_{\mu\nu}.\tag{3.35}$$

The vacuum energy density is equal to $T_{00} = \rho_{vac}$, while effective pressure, defined by $T_{ij} = -p\eta_{ij}$, is

$$p = -\rho_{vac}.$$

Thus, vacuum has an exotic equation of state $p = -\rho$; for positive energy density, the vacuum pressure is negative.

In the curved space–time of not very high curvature, the expression (3.35) remains valid in any locally-Lorentz frame, and in an arbitrary frame one has

$$T_{\mu\nu} = \rho_{vac} g_{\mu\nu}. \tag{3.36}$$

Here ρ_{vac} is a constant in space and time; in principle it should be calculable in a complete theory of elementary particles and their interactions. Up to now, no compelling calculation of the vacuum energy density has been made, and this is one of the major problems of fundamental physics.

Time-independent ρ_{vac} is consistent with Eq. (3.13), which for $p = -\rho$ gives $\dot{\rho} = 0$. This is in fact obvious: Eq. (3.13) is a consequence of the covariant conservation of the energy–momentum tensor, while the tensor (3.36) obeys covariant conservation law for $\rho_{vac} = $ const due to the fact that $\nabla_\mu g^{\lambda\rho} = 0$ (see Appendix A).

One can look at the right-hand side of the Einstein equations with $T_{\mu\nu} = $ const $\cdot$ $g_{\mu\nu}$ from a slightly different prospective. General covariance allows us to add to the Einstein–Hilbert action S_G of General Relativity, the term

$$S_\Lambda = -\Lambda \int \sqrt{-g} d^4 x.$$

By varying the action $(S_G + S_\Lambda)$ with respect to metric one obtains, in the absence of matter, the following equations (see Appendix A),

$$R_{\mu\nu} - \frac{1}{2} g_{\mu\nu} R - 8\pi G \Lambda g_{\mu\nu} = 0.$$

They are exactly the same as the Einstein equations with energy–momentum tensor (3.36), with identification $\Lambda = \rho_{vac}$. Historically this route was the first, and the parameter Λ is often called the cosmological constant for historical reasons. The difference between the cosmological constant and vacuum energy density is purely philological, at least at the current level of understanding of this issue.

The solution to the Friedmann equation (3.17) with $\rho = $ const $= \rho_{vac}$ is

$$a = \text{const} \cdot e^{H_{dS} t}, \tag{3.37}$$

where the Hubble parameter

$$H_{dS} = \sqrt{\frac{8\pi}{3} G \rho_{vac}}$$

is independent of time. The space–time with metric

$$ds^2 = dt^2 - e^{2 H_{dS} t} d\mathbf{x}^2 \tag{3.38}$$

is called de Sitter space.[1] It is the space–time of constant curvature.

---

[1] The coordinates $(t, \mathbf{x})$ cover only half of the de Sitter space, see [36] and the problem in this Section.

Problem 3.4. *Show that de Sitter metric obeys* (*cf.* (2.7))

$$R_{\mu\nu\lambda\rho} = -H_{ds}^2(g_{\mu\lambda}g_{\nu\rho} - g_{\mu\rho}g_{\nu\lambda}).$$

Unlike in previous examples of cosmological solutions, the Universe undergoes accelerated expansion, $\ddot{a} > 0$. Furthermore, de Sitter space does not have the initial singularity: even though $a(t)$ tends to zero as $t \to -\infty$, the metric can be made non-singular at any t by a coordinate transformation.

Problem 3.5. *Consider a fictitious flat five-dimensional space with metric*

$$ds^2 = (dy^0)^2 - (dy^1)^2 - (dy^2)^2 - (dy^3)^2 - (dy^4)^2.$$

Consider a hyperboloid embedded into this space according to equation

$$(y^0)^2 - (y^1)^2 - (y^2)^2 - (y^3)^2 - (y^4)^2 = -H^{-2} = \text{const.}$$

It is clear that this hyperboloid does not have singularities. Let us choose the coordinates (t, x^i), $i = 1, 2, 3$ on this hyperboloid, such that

$$
\begin{aligned}
y^0 &= -H^{-1}\sinh Ht - \frac{H}{2}\mathbf{x}^2 e^{Ht}, \\
y^i &= x^i e^{Ht}, \\
y^4 &= H^{-1}\cosh Ht - \frac{H}{2}\mathbf{x}^2 e^{Ht}.
\end{aligned}
\tag{3.39}
$$

Show that with this choice of coordinates, the metric induced on the hyperboloid from the five-dimensional space coincides with (3.38). *What part of the hyperboloid is covered by the coordinates $(t, \mathbf{x})$?*

In the case of de Sitter space, the cosmological (particle) horizon is absent. The "beginning of time" is shifted to $t = -\infty$. So, instead of Eq. (3.25) we have for the horizon size,

$$l_H(t) = a(t)\int_{-\infty}^t \frac{dt'}{a(t')} = e^{H_{ds}t}\int_{-\infty}^t dt' e^{-H_{ds}t'} = \infty,$$

which precisely means that the particle horizon is absent.

For spaces like de Sitter, another notion of horizon is introduced. It is quite different from the particle horizon discussed in Sec. 3.2.1. Namely, let there be an observer at the point $\mathbf{x} = 0$ at the moment of time t. Let us ask what is the size of the region from which signals emitted at *that* moment of time will never reach the observer (which will stay at the point $\mathbf{x} = 0$) in arbitrary distant *future*. Since the time-like geodesics obey $|d\mathbf{x}| = d\eta$, the coordinate size of this region is

$$\eta(t \to \infty) - \eta(t) = \int_t^\infty \frac{dt'}{a(t')},$$

and its physical size at time t equals

$$l_{dS} = a(t) \int_t^\infty \frac{dt'}{a(t')} = \frac{1}{H_{dS}}. \qquad (3.40)$$

The observer will never know about events that happen at a given moment of time at distances exceeding $l_{dS} = H_{dS}^{-1}$; this is the meaning of the de Sitter horizon. It is also called event horizon.

Problem 3.6. *Show that the solutions of Secs. 3.2.1 and 3.2.2 do not have event horizons.*

3.2.4. General barotropic equation of state $p = w\rho$

Let us continue with the brief discussion of a model with the matter equation of state

$$p = w\rho,$$

where w is a constant larger than -1. Non-relativistic and relativistic matter correspond to $w = 0$ and $w = 1/3$, respectively. Models with negative (and, generally, time-dependent) w attract considerable attention in recent years; matter with this effective equation of state has different names: quintessence, time-dependent Λ-term, etc.

At $w > -1$ the solution to Eq. (3.13) is

$$\rho = \frac{\text{const}}{a^{3(1+w)}}. \qquad (3.41)$$

We find from Eq. (3.17) that

$$a = \text{const} \cdot t^\alpha,$$

where

$$\alpha = \frac{2}{3} \frac{1}{1 + w}.$$

The parameter α is positive; there is the cosmological singularity at $t = 0$. The energy density behaves as

$$\rho = \frac{\text{const}}{t^2};$$

it tends to infinity as $t \to 0$. Since

$$\ddot{a} = \text{const} \cdot \alpha(\alpha - 1)t^{\alpha-2},$$

the expansion of the Universe decelerates ($\ddot{a} < 0$) for $\alpha < 1$ and accelerates at $\alpha > 1$. In terms of the equation-of-state parameter w we have

$$\text{(a)} \qquad w > -\frac{1}{3} : \text{deceleration},$$

$$\text{(b)} \qquad w < -\frac{1}{3} : \text{acceleration}.$$

Note that in the open Universe dominated by negative spatial curvature, the effective energy density is equal to $\rho = \text{const}/a^2$ (see Eq. (3.9) with $\varkappa = -1$). This would mean that $w = -1/3$ (see (3.41)). In that case the expansion would neither accelerate nor decelerate, $\ddot{a} = 0$.

The cases (a) and (b) differ also in the following respect. The models from the class (a) have cosmological (particle) horizon and do not have event horizon, while the situation is reverse for models from the class (b). Indeed, the particle horizon exists if the integral

$$\int_0^t \frac{dt'}{a(t')}$$

converges (see (3.25)). For $\alpha < 1$ (i.e. $w > -\frac{1}{3}$) this integral converges indeed, while for $\alpha > 1$ (i.e. $w < -\frac{1}{3}$) the integral diverges at the lower limit of integration. In the latter case the particle horizon gets shifted to spatial infinity. The existence of event horizon is determined by the convergence properties of the integral (see (3.40))

$$\int_t^\infty \frac{dt'}{a(t')}.$$

It diverges at the upper limit of integration for $\alpha < 1$ (event horizon is absent) and converges for $\alpha > 1$ (event horizon exists).

Problem 3.7. *Study the cosmological evolution in a model with equation of state $p = w\rho$, where w is time-independent and $w < -1$.*

Problem 3.8. *Is it possible in expanding Universe with equation of state $p = p(\rho)$ to evolve from the regime $(p+\rho) > 0$ to the regime $(p+\rho) < 0$ without violating the real-valuedness of the sound speed u_s defined as $u_s^2 = \partial p/\partial \rho$?*

3.3.   Solutions with Recollapse

For completeness, we briefly discuss in this Section the homogeneous and isotropic solutions, in which expansion of the Universe is followed by contraction (recollapse). This situation occurs if the right-hand side of the Friedmann equation (3.9) contains both positive and negative terms, and if positive terms decrease with growing scale factor faster than the absolute values of negative terms. Examples of possibly interesting negative contributions are the contribution of curvature in the closed model ($\varkappa = +1$) and the dark energy contribution. Regarding the latter, we know that it is positive at the present epoch, but we cannot exclude that it depends on time and will become negative in distant future.

As an example, let us consider the closed cosmological model with non-relativistic matter. Making use of (3.18) we write the Friedmann equation

$$\left(\frac{\dot{a}}{a}\right)^2 = \frac{a_m}{a^3} - \frac{1}{a^2},$$

(3.42)

where the parameter a_m is determined by the total mass of matter in the Universe. At $a \ll a_m$ the Universe expands in the same way as in the spatially flat case (Sec. 3.2.1). The expansion terminates when

$$a = a_m;$$

at that time the right-hand side of (3.42) vanishes.

Problem 3.9. *Find the relation between a_m and the total mass in the closed Universe. What would be the maximum size of the Universe having 1 kg of non-relativistic matter?*

The explicit solution has a simple form in terms of the conformal time η defined by $dt = a d\eta$ (see (2.23)). The Friedmann Eq. (3.42) then takes the form

$$\frac{1}{a^4}\left(\frac{da}{d\eta}\right)^2 = \frac{a_m}{a^3} - \frac{1}{a^2}$$

and its solution is

$$a = a_m \sin^2 \frac{\eta}{2}.$$

(3.43)

The expansion begins from singularity at $\eta = 0$, the Universe has maximum size at $\eta = \pi$, and at $\eta = 2\pi$ it collapses back to the singularity. The relation between the physical and conformal time is

$$t = \int a(\eta) d\eta = \frac{a_m}{2}(\eta - \sin \eta).$$

(3.44)

Thus, the total lifetime and the maximum size are related by $t_{tot} = \pi \cdot a_m$.

Problem 3.10. *Show that at $a \ll a_m$ the solution (3.43), (3.44) indeed coincides with the spatially flat solution (3.19).*

Similar situation occurs in the case when the expansion of the Universe terminates due to negative Λ-term. The Universe exists for finite time between its emergence from and recollapse to the singularity.

Problem 3.11. *Find the law of evolution $a = a(t)$ of the spatially flat Universe ($\varkappa = 0$) with negative, time-independent cosmological constant, assuming that* *matter in the Universe has equation of state $p = 0$. Find the total lifetime. Hint: Make use of the Friedmann equation in physical time.*

Problem 3.12. *Consider the Universe filled with matter whose equation of state is that of Chaplygin gas [37],*

$$p = -\frac{A}{\rho}. \tag{3.45}$$

(1) *Find the dependence of the Hubble parameter on the scale factor.*
(2) *Find the law of evolution $a = a(t)$ at small and large scale factors in all three cases, $\varkappa = 0, \pm 1$.*
(3) *Find the complete evolution $a = a(t)$ in the case of spatially flat Universe.*
(4) *What values of $\varkappa$ admit static solutions to the Einstein equations?*
(5) *What can be said about the future of the Universe, if it is known that at some moment of time the expansion of the Universe accelerates? Consider all three cases, $\varkappa = 0, \pm 1$.*
(6) *Consider a theory of a scalar field with action*

$$S_\phi = \int d^4x \sqrt{-g} \left[\frac{1}{2} g^{\mu\nu} \partial_\mu \phi \partial_\nu \phi - V(\phi) \right].$$

In the case of spatially flat Universe, find the scalar potential $V(\phi)$ for which the spatially homogeneous solution describes the same evolution as in the case (3), and the relation between the energy density and pressure has the form (3.45).

Chapter 4

ΛCDM: Cosmological Model with Dark Matter and Dark Energy

4.1. Composition of the Present Universe

Cosmological solutions studied in Sec. 3.2 are not realistic. Energy density in the present Universe is due to *non-relativistic matter* (baryons and dark matter, as well as those neutrino species whose mass is considerably larger than $5\,\mathrm{K} \sim 10^{-3}$ eV), *relativistic matter* (photons and light neutrino of mass less than 10^{-4} eV, if such a neutrino exists, see Appendix C) and *dark energy*. In general, the Universe may have non-vanishing *spatial curvature*. Therefore, all these contributions have to be included into the right-hand side of the Friedmann equation (3.9), and this equation takes the form

$$H^2 \equiv \left(\frac{\dot{a}}{a}\right)^2 = \frac{8\pi}{3}G(\rho_M + \rho_{rad} + \rho_\Lambda + \rho_{curv}), \tag{4.1}$$

where $\rho_M, \rho_{rad}, \rho_\Lambda$ are energy densities of non-relativistic matter, relativistic matter ("radiation") and dark energy, respectively, and by definition

$$\frac{8\pi}{3}G\rho_{curv} = -\frac{\varkappa}{a^2} \tag{4.2}$$

is the contribution due to spatial curvature. Let us introduce the critical density ρ_c by

$$\rho_c \equiv \frac{3}{8\pi G}H_0^2. \tag{4.3}$$

We stress that we always use the notion of critical density only in application to the present Universe; for us ρ_c is the time-independent quantity. Its meaning is that if the total energy density in the present Universe, $\rho_{M,0} + \rho_{rad,0} + \rho_{\Lambda,0}$, equals precisely ρ_c, then the Universe is spatially flat (since in this case $\rho_{curv} = 0$ and $\varkappa = 0$). The numerical value of the critical energy density is given in (1.14). We note that the average energy density in the present Universe is fairly small: it is equivalent to 5 proton masses per cubic meter.

Let us introduce the parameters,

$$\Omega_M = \frac{\rho_{M,0}}{\rho_c}, \quad \Omega_{rad} = \frac{\rho_{rad,0}}{\rho_c}, \quad \Omega_\Lambda = \frac{\rho_{\Lambda,0}}{\rho_c}, \quad \Omega_{curv} = \frac{\rho_{curv,0}}{\rho_c}. \quad (4.4)$$

Again, these parameters refer to the present Universe only, and by definition they do not depend on time. It follows from (4.1) and (4.3) that

$$\sum_i \Omega_i \equiv \Omega_M + \Omega_{rad} + \Omega_\Lambda + \Omega_{curv} = 1. \quad (4.5)$$

The parameters Ω_i are equal to relative contributions of different sorts of energy, and also of spatial curvature, into the right-hand side of the Friedmann equation (4.1) at the present epoch. Ωs make one of the most important sets of cosmological parameters.

It is rather straightforward to estimate the relative contribution of relativistic particles Ω_{rad}. It comes mainly from relic photons of temperature $T_0 = 2.726\,\mathrm{K}$. According to the Stefan–Boltzmann law, their energy density is (see also Chapter 5)

$$\rho_{\gamma,0} = 2\frac{\pi^2}{30}T_0^4,$$

where the factor 2 is due to two photon polarizations. Numerically,

$$\rho_{\gamma,0} = 2.6 \cdot 10^{-10}\frac{\mathrm{GeV}}{\mathrm{cm}^3},$$

and

$$\Omega_\gamma = 2.5 \cdot 10^{-5}h^{-2} = 5.0 \cdot 10^{-5}, \quad h = 0.7. \quad (4.6)$$

If there exist massless or light neutrinos (with $m_\nu \lesssim 5\,\mathrm{K} \sim 10^{-3}$ eV), their contribution is somewhat smaller than Ω_γ. Hence

$$\Omega_{rad} \lesssim 10^{-4}. \quad (4.7)$$

Because of this, the effect of relativistic particles on cosmological expansion is negligible today and in the late Universe.

We mentioned in Chapter 1 that observations of CMB anisotropy imply that the spatial curvature of the Universe is either zero or very small. Quantitatively, there is the bound on Ω_{curv},

$$|\Omega_{curv}| < 0.01. \quad (4.8)$$

We will discuss in what follows how this bound is obtained, and will simply use it for the time being.

There are several independent observational methods to determine Ω_M and Ω_Λ. We discuss some of them in this and the accompanying book, while others are just mentioned (see also Chapter 1). We quote here the current values,

$$\Omega_M = 0.315, \quad \Omega_\Lambda = 0.685 \tag{4.9}$$

with precision of about 5%. Hence, the present expansion of the Universe is determined to large extent by dark energy, and to less extent by non-relativistic matter.

The present energy density of non-relativistic matter is the sum of mass densities of baryons and dark matter,

$$\Omega_M = \Omega_B + \Omega_{DM},$$

with

$$\Omega_B = 0.050, \quad \Omega_{DM} = 0.265.$$

At least two types of neutrinos have masses exceeding the temperature in the present Universe, see Appendix C. Neutrinos of these types are non-relativistic today. We discuss neutrino contribution Ω_ν in Chapter 7, and here we assume for simplicity that it is negligible as compared to Ω_M. The number of electrons equals the number of protons by electric neutrality, so that[1]

$$\Omega_e \approx \frac{m_e}{m_p} \cdot \Omega_B \simeq 3 \cdot 10^{-5}.$$

Thus, the main contribution into Ω_M comes from dark matter. As we discuss in Chapter 9, dark matter is very likely *cold*.

Spatially flat cosmological model with non-relativistic cold dark matter and dark energy with parameters close to (4.9) will be called[2] ΛCDM model. In what follows we will specify this model further, by adding more cosmological parameters. Let us make one qualification right now: unless we state the opposite, we will assume, in the framework of ΛCDM that ρ_Λ is independent of time (cosmological constant $\equiv$ vacuum energy density).

ΛCDM model is consistent with the entire set of cosmological data. This does not necessarily mean, of course, that this model is exact, or that no alternative models are possible. In any case, ΛCDM serves as an important reference point among numerous cosmological models.

The relative contributions Ω_i to the right-hand side of the Friedmann equation (4.1) are characteristic of the present epoch only, since $\rho_{rad}, \rho_M, \rho_\Lambda$ and ρ_{curv} behave in time in different ways, namely $\rho_{rad} \propto a^{-4}$ (see (3.28)), $\rho_M \propto a^{-3}$ (see (3.18)), $\rho_{curv} \propto a^{-2}$ (see (4.2)), and, according to our assumption, ρ_Λ is

---

[1]The first equality here is approximate, since there exist neutrons bound into nuclei and contributing to Ω_B. This is not important for our purposes here.
[2]In general, ΛCDM often denotes wider class of models. We will use this term in a narrow sense given in the text. This model is also dubbed in literature as "concordance model".

independent of time. Thus, the Friedmann equation in the ΛCDM model can be written as follows,

$$\left(\frac{\dot{a}}{a}\right)^2 = \frac{8\pi}{3}G\rho_c\left[\Omega_M\left(\frac{a_0}{a}\right)^3 + \Omega_{rad}\left(\frac{a_0}{a}\right)^4 + \Omega_\Lambda + \Omega_{curv}\left(\frac{a_0}{a}\right)^2\right], \qquad (4.10)$$

or

$$H^2 = H_0^2[\Omega_M(1+z)^3 + \Omega_{rad}(1+z)^4 + \Omega_\Lambda + \Omega_{curv}(1+z)^2]. \qquad (4.11)$$

A subtlety here is that the number of relativistic species, as well as the number of non-relativistic ones, is different at different cosmological epochs. In particular, neutrinos of mass $m_\nu \gtrsim 5\,\mathrm{K}$ are non-relativistic today, but they were relativistic at early stages of the evolution. This subtlety is not very important for this Chapter, but one should keep in mind that the Friedmann equation in the form (4.10) must be used with care.

Another remark concerns dark energy. It cannot be excluded that ρ_Λ in fact depends on time. In particular, one can consider dark energy with equation of state $p_\Lambda = w_\Lambda \rho_\Lambda$ with $w_\Lambda \neq -1$; in that case its energy density evolves with scale factor according to power law (3.41). Observational data show that w_Λ belongs to the interval $-1.25 \lesssim w_\Lambda \lesssim -0.95$. Most of the conclusions in this Chapter remain valid for such a dark energy, although formulas get more cumbersome for $w_\Lambda \neq -1$. It is worth stressing that the question about time-dependence of ρ_Λ is extremely important both for cosmology and particle physics, since it is directly related to the nature of dark energy: if $\rho_\Lambda = $ const, then the dark energy is cosmological constant, while time-dependence of ρ_Λ would imply the existence of a new form of matter in Nature (e.g. fairly exotic scalar field dubbed quintessence).

4.2.   General Properties of Cosmological Evolution

Let us discuss, first at the qualitative level, which contributions to the right-hand side of the Friedmann equation are most relevant at different cosmological epochs. In the first place, the contribution due to curvature has never dominated. Indeed, it follows from (4.8) and (4.9) that this contribution, if any, is presently small compared to contributions of both non-relativistic matter and dark energy. In the past, the contribution of non-relativistic matter was enhanced with respect to that of curvature by a factor $a_0/a = 1 + z$, so that curvature was even less important. If the dark energy does not depend on time, the curvature contribution will be small in future as well: the spatial curvature will decrease as $1/a^2$, while ρ_Λ stays constant.

Talking about the future, let us note that all terms in the right-hand side of Eq. (4.10), except for Ω_Λ, decrease as a grows. Thus, the expansion rate of the Universe will be determined by dark energy, and the behavior of the scale factor will tend to the exponential law, (3.37) with $\rho_{vac} \equiv \rho_\Lambda = \rho_c \Omega_\Lambda$. Of course, this conclusion holds for time-independent ρ_Λ only. It is not known whether the latter assumption is valid, so distant future of the Universe cannot be reliably predicted, cf. Sec. 1.4.

Problem 4.1. *Give examples of the evolution of dark energy leading to scenarios for the future of the Universe outlined in Sec. 1.4. In particular, find the bounds on maximum size and lifetime of the closed Universe assuming that dark energy instantaneously switches off right after the present epoch. Hint: Use the bound (4.8) and the values (4.9).*

We will be interested in the past of our Universe. Equation (4.10) shows that the major contribution to the present energy density comes from dark energy. This contribution, ρ_Λ, has become important only recently. Before that, there was a long period of domination of non-relativistic matter ("matter domination"). At even earlier times, at sufficiently small a, relativistic matter was prevailing ("radiation domination"). Within the concepts developed so far in this book, the radiation dominated epoch began right from the cosmological singularity. This is precisely the Hot Big Bang picture. As we discussed in Sec. 1.7.1, this picture is certainly incomplete; nevertheless, we concentrate for the time being on the Hot Big Bang theory and study scenarios for the preceding epoch (inflation and alternatives) in the accompanying book.

"Moments" when the regimes of evolution change are of considerable interest. Let us discuss them in some details.

4.3.   Transition from Deceleration to Acceleration

We neglect the contributions due to relativistic matter and curvature and write Eq. (4.10) as follows,

$$\dot{a}^2 = \frac{8\pi}{3}G\rho_c\left(\frac{\Omega_M a_0^3}{a} + \Omega_\Lambda a^2\right).$$

This gives for acceleration

$$\ddot{a} = a\frac{4\pi}{3}G\rho_c\left(2\Omega_\Lambda - \Omega_M\left(\frac{a_0}{a}\right)^3\right).$$

Since $2\Omega_\Lambda > \Omega_M$ and hence $\ddot{a} > 0$, the expansion accelerates at the present epoch. In the past, at sufficiently large $z \equiv a_0/a - 1$, the expansion was decelerating, $\ddot{a} < 0$. The transition from deceleration to acceleration occurred at

$$\left(\frac{a_0}{a_{ac}}\right)^3 = \frac{2\Omega_\Lambda}{\Omega_M},$$

i.e., at

$$z_{ac} = \left(\frac{2\Omega_\Lambda}{\Omega_M}\right)^{1/3} - 1.$$

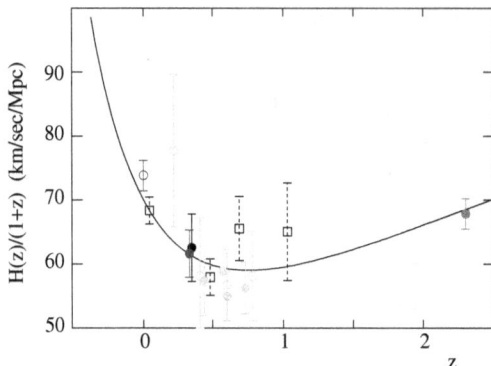

Fig. 4.1.   Evolution of the time derivative of the scale factor inferred from the observation of standard candles and baryon acoustic oscillations [74]. Transition from decrease to increase as z decreases is precisely the transition from deceleration to acceleration. The theoretical curve is obtained for spatially flat Universe with $h = 0.7$ and $\Omega_\Lambda = 0.73$.

For $\Omega_M = 0.315$, $\Omega_\Lambda = 0.685$, we find numerically

$$z_{ac} \approx 0.63.$$

Thus, transition from deceleration to acceleration occurred in the Universe pretty late. This transition is illustrated in Fig. 4.1, which shows the measured values of $H(z)/(1 + z) = \dot{a}(t)/a_0$.

Problem 4.2. *At what z did the contributions to energy density due to non-relativistic matter and dark energy become equal to each other?*

Since the dependence $\rho_M \propto a^{-3}$ is quite strong, while ρ_Λ depends on a weakly or does not depend on a at all, the transition occurred rather abruptly. Before the transition, the Universe expanded as $a \propto t^{2/3}$ (matter dominated epoch, see Sec. 3.2.1).

Problem 4.3. *At what z does the transition from deceleration to acceleration occur for dark energy with equation of state $p = w\rho$, $w = const$? For what value of the parameter w this transition would occur now? Give numerical estimate using the values (4.9).*

4.4.   Transition from Radiation Domination to Matter Domination

As we have already pointed out, in the Hot Big Bang theory the earliest epoch is radiation domination. The "moment" of the transition from radiation domination to matter domination ("equality") is very important from the viewpoint of the growth of density perturbations: we show in the accompanying book that perturbations behave in a very different way at these two epochs.

A crude estimate for the moment of equality is obtained from Eq. (4.10) and estimates (4.6) and (4.9). Neglecting dark energy and curvature at this moment, we find that the contributions of radiation and non-relativistic matter were equal at

$$z_{eq} + 1 = \frac{a_0}{a_{eq}} \sim \frac{\Omega_M}{\Omega_{rad}} \sim 10^4. \tag{4.12}$$

At that time the temperature in the Universe had the following order of magnitude,

$$T_{eq} = T_0(1 + z_{eq}) \sim 10^4 \text{ K} \sim 1 \text{ eV}. \tag{4.13}$$

Thus, equality occurred at rather distant past.

The estimates (4.12) and (4.13) have to be refined. At temperature of order 1 eV, not only photons, but also all three species of neutrinos are relativistic (see Appendix C). We see in Chapter 7 that neutrinos do not interact between themselves and with the rest of cosmic matter at this temperature. According to Sec. 2.5, their distribution functions are thermal nevertheless. We show in Chapter 7 that the effective temperature of neutrinos is

$$T_\nu = \left(\frac{4}{11}\right)^{1/3} T_\gamma, \tag{4.14}$$

where T_γ is the photon temperature. Everywhere in this book we identify the temperature in the Universe with the photon temperature, so that

$$T_\gamma \equiv T.$$

The energy density of relativistic neutrinos is given by the appropriately modified Stefan–Boltzmann law (see Chapter 5),

$$\rho_\nu = 3 \cdot 2 \cdot \frac{7}{8} \frac{\pi^2}{30} T_\nu^4, \tag{4.15}$$

where the factor 3 corresponds to three neutrino species, factor 2 is due to the existence of both neutrino and anti-neutrino, one polarization each, and factor 7/8 accounts for the Fermi statistics. Thus, the energy density of relativistic matter at the epoch of interest is

$$\rho_{rad} = \rho_\gamma + \rho_\nu = \left[2 + \frac{21}{4}\left(\frac{4}{11}\right)^{4/3}\right] \frac{\pi^2}{30} T^4, \tag{4.16}$$

where the first and second terms in square brackets come from photons and three neutrino species, respectively. Hence,

$$\rho_{rad} = 1.68\rho_\gamma = 1.68 \left(\frac{a_0}{a}\right)^4 \Omega_\gamma \rho_c. \tag{4.17}$$

The energy density of non-relativistic matter is still given by

$$\rho_M = \left(\frac{a_0}{a}\right)^3 \Omega_M \rho_c. \tag{4.18}$$

We find that equality between relativistic and non-relativistic components happens at

$$1 + z_{eq} = \frac{a_0}{a_{eq}} = 0.6 \frac{\Omega_M}{\Omega_\gamma},$$

and making use of (4.6) we obtain

$$1 + z_{eq} = 2.4 \cdot 10^4 \, \Omega_M h^2, \tag{4.19}$$

which for $\Omega_M = 0.31$ and $h = 0.7$ gives

$$1 + z_{eq} = 3.5 \cdot 10^3. \tag{4.20}$$

At that time, temperature is

$$T_{eq} = (1 + z_{eq}) T_0 = 5.6 \, \Omega_M h^2 \text{ eV} \tag{4.21a}$$

$$T_{eq} = 0.8 \text{ eV} \quad \text{for } h = 0.7, \quad \Omega_M = 0.31. \tag{4.21b}$$

The expressions (4.19)–(4.21b) refine the estimates (4.12), (4.13) with account of three light neutrino species.

Let us estimate the age of the Universe at equality. To simplify the calculation, we neglect non-relativistic matter before equality and note that during most of the evolution only photons and neutrinos were relativistic (the lightest of other particles, electrons and positrons, are relativistic at temperatures $T \gtrsim m_e = 0.5$ MeV only). Therefore, we use formulas of Sec. 3.2.2 with effective number of degrees of freedom[3] $\hat{g}_*$ obtained by comparing (3.30) with (4.16):

$$\hat{g}_* = 2 + \frac{21}{4} \left(\frac{4}{11} \right)^{4/3} = 3.36. \tag{4.22}$$

Making use of (3.29) and (3.31) we estimate the age as

$$t_{eq} \sim \frac{1}{2 H_{eq}} = \frac{M_{Pl}^*}{2 T_{eq}^2}, \tag{4.23}$$

where, as before, $M_{Pl}^* = M_{Pl}/(1.66 \sqrt{\hat{g}_*})$. Using (4.21a) and (4.22) we obtain for $h = 0.7$, $\Omega_M = 0.31$

$$t_{eq} \sim 3.2 \cdot 10^{36} \text{ GeV}^{-1} = 2.1 \cdot 10^{12} \text{s} = 70 \text{ thousand yrs.} \tag{4.24}$$

This time is of course much shorter than the present age of the Universe, $t_0 \approx 14$ billion years. We refine this estimate below, see Eq. (4.28).

To conclude this Section we note that the definition of equality $\rho_M = \rho_{rad}$ is unambiguous. However, from the viewpoint of the cosmological evolution, the transition from radiation domination to matter domination is actually not a well

---

[3] We reserve the notation g_* for the parameter that determines the entropy density, see Sec. 5.2. The parameters g_* and $\hat{g}_*$ coincide at $T \gtrsim 1$ MeV, so we use the notation g_* when discussing the Universe at so high temperatures.

defined moment in the history of the Universe, but the process whose duration is comparable to the Hubble time at that epoch, H_{eq}^{-1} (in other words, comparable to the age t_{eq}). The ratio between energy densities of non-relativistic and relativistic matter depends on scale factor rather weakly, $\rho_M/\rho_{rad} \propto a$, so this ratio does not change very much in Hubble time. Thus, the jump from the expansion law $a \propto t^{1/2}$ to $a \propto t^{2/3}$ at $t = t_{eq}$ would be only a crude approximation. The relation (4.23) between the temperature T_{eq} and age t_{eq} is approximate as well, as in our derivation of (4.23) we neglected non-relativistic matter at $t < t_{eq}$.

Let us calculate the age of the Universe at temperatures higher than the neutrino masses but well below 100 keV. In view of the cosmological bound on the neutrino masses given in Sec. 7.2, Eq. (7.15), we are talking here about temperatures exceeding 0.2 eV. The relevant range of temperatures includes T_{eq} as well as the temperature of recombination which we discuss in Chapter 6.

Let us first write the relation between the age and scale factor:

$$t(a) = \int_0^a \frac{da}{\dot{a}} = \int_0^a \frac{da}{aH}.$$

Starting from temperature of order 100 keV (a few minutes after Big Bang which we ignore), the composition of cosmic medium remains constant. Thus, the relation $T \propto a^{-1}$ is exact, and we have

$$t(T) = \int_T^\infty \frac{dT}{TH(T)}. \tag{4.25}$$

We neglect the dark energy contribution (this is a very good approximation at the epoch we consider) and write

$$H^2(T) = H_{eq}^2 \left[\frac{1}{2} \left(\frac{T}{T_{eq}} \right)^3 + \frac{1}{2} \left(\frac{T}{T_{eq}} \right)^4 \right], \tag{4.26}$$

where $H_{eq} \equiv H(T_{eq})$, the first and second terms in square barckets are contributions of non-relativistic and relativistic matter to the Friedmann equation, respectively; we make use of the fact that these contributions are equal to each other at $T = T_{eq}$ and that their sum at T_{eq} is $H^2(T_{eq})$. Now the integral (4.25) is straightforwardly calculated, and we get

$$t(T) = \frac{\sqrt{2}}{H_{eq}} \mathcal{F}_t(T/T_{eq}),$$

where

$$\mathcal{F}_t(T/T_{eq}) = \frac{2}{3} \left(1 + \frac{T_{eq}}{T} \right)^{3/2} - 2 \left(1 + \frac{T_{eq}}{T} \right)^{1/2} + \frac{4}{3}.$$

The Hubble parameter at equality can be written as follows:

$$H_{eq} = \left[2 \cdot \frac{8\pi}{3} G \rho_M(T_{eq}) \right]^{1/2} = \sqrt{2\Omega_M} H_0 (1 + z_{eq})^{3/2},$$

which gives finally

$$t(T) = \frac{\mathcal{F}_t(T/T_{eq})}{H_0 \sqrt{\Omega_M} (1 + z_{eq})^{3/2}}. \tag{4.27}$$

We insert here $\mathcal{F}_t(1) = (4 - 2\sqrt{2})/3$, $h = 0{,}7$, $\Omega_M = 0{,}31$, and obtain numerically

$$t_{eq} = 52 \text{ thousand yrs.} \qquad (4.28)$$

This refines the estimate (4.24).

Comoving horizon size at these temperatures is calculated in a similar way. We write

$$\eta = \int_0^t \frac{dt}{a(t)} = \int_0^a \frac{da}{a\dot{a}} = \int_0^a \frac{da}{a^2 H} = \frac{1}{a_{eq}} \int_0^{a/a_{eq}} \frac{d(a/a_{eq})}{(a/a_{eq})^2 H} = \frac{1}{a_{eq}} \int_T^\infty \frac{d(T_{eq}/T)}{(T_{eq}/T)^2 H(T)}.$$

We again make use of the relation (4.26) and obtain, upon calculating the integral,

$$\eta(T) = \frac{\sqrt{2}}{a_{eq} H_{eq}} \mathcal{F}_\eta(T/T_{eq}) = \frac{\mathcal{F}_\eta(T/T_{eq})}{a_0 H_0 \sqrt{\Omega_M}(1 + z_{eq})^{1/2}}, \qquad (4.29)$$

where

$$\mathcal{F}_\eta(T/T_{eq}) = 2\left(1 + \frac{T_{eq}}{T}\right)^{1/2} - 2.$$

We use this result in Sec. 6.4.

Problem 4.4. *Show that the expressions (4.27) and (4.29) reduce at $T \gg T_{eq}$ and $T \ll T_{eq}$ to known expressions for the age and horizon size at radiation domination and matter domination, respectively.*

4.5.   Present Age of the Universe and Horizon Size

The fact that the cosmological expansion for fairly long time has been affected by dark energy leads to different values of the present age of the Universe and present horizon size as compared to the estimates (3.24) and (3.27). To calculate them, we neglect spatial curvature and relativistic matter: as we discussed in Sec. 4.1, the curvature is negligible at all epochs, while the relativistic matter is relevant at early times only, $t \lesssim t_{eq}$. Then Eq. (4.10) becomes

$$\left(\frac{\dot{a}}{a}\right)^2 = H_0^2\left[\Omega_M\left(\frac{a_0}{a}\right)^3 + \Omega_\Lambda\right], \qquad (4.30)$$

where we made use of (4.3). Here

$$\Omega_M + \Omega_\Lambda = 1, \quad \Omega_\Lambda > 0. \qquad (4.31)$$

The solution to Eq. (4.30) is

$$a(t) = a_0\left(\frac{\Omega_M}{\Omega_\Lambda}\right)^{1/3}\left[\sinh\left(\frac{3}{2}\sqrt{\Omega_\Lambda}H_0 t\right)\right]^{2/3}. \qquad (4.32)$$

It is clear from this formula that the law of matter dominated expansion, $a \propto t^{2/3}$, is restored at early times, while at late times the scale factor grows exponentially, as should be the case.

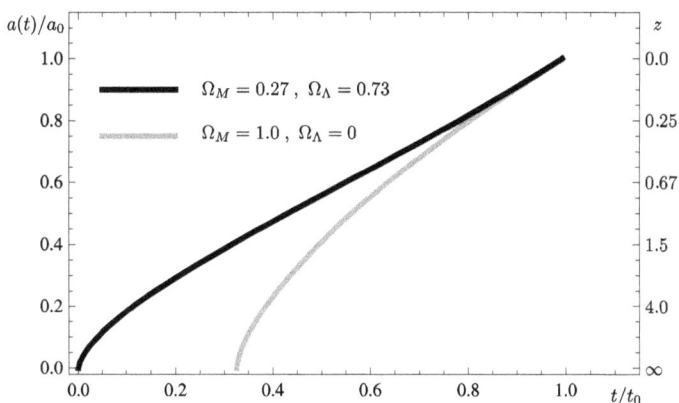

Fig. 4.2.   Evolution of $a = a(t)$ for spatially flat models.

The present age of the Universe is now found from the following equation,

$$\left(\frac{\Omega_M}{\Omega_\Lambda}\right)^{1/3}\left[\sinh\left(\frac{3}{2}\sqrt{\Omega_\Lambda}H_0 t_0\right)\right]^{2/3} = 1.$$

It is given by

$$t_0 = \frac{2}{3\sqrt{\Omega_\Lambda}}\frac{1}{H_0}\mathrm{Arsinh}\sqrt{\frac{\Omega_\Lambda}{\Omega_M}}. \tag{4.33}$$

At $\Omega_\Lambda \to 0$ and $\Omega_M \to 1$ we restore formula (3.24). For positive Ω_Λ the age exceeds $2/(3H_0)$. To see this, we plot the scale factor as function of time for matter dominated model ($\Omega_\Lambda = 0$) and ΛCDM model ($\Omega_\Lambda > 0$) in such a way that they are tangential to each other (values of both a and $\dot{a}$ coincide) at the present time, i.e., at $a = a_0$ (equality of derivatives corresponds to one and the same value of the present Hubble parameter $H_0 = (\dot{a}/a)_0$). Since the Friedmann equation for realistic model is given by (4.30) and the right-hand side for matter dominated model equals $H_0^2(a_0/a)^3$, the time derivative $\dot{a}$ is greater for matter dominated model at every $a < a_0$, so we come to the plot shown in Fig. 4.2.

The distance along the horizontal axis from the point of singularity $a = 0$ to the point $a = a_0$ is precisely the present age; it is clear that the age is larger for the Universe with $\Omega_\Lambda > 0$. We obtain from (4.33)

$$t_0 = 1.38 \cdot 10^{10} \text{ yrs} \quad \text{for } \Omega_M = 0.315, \quad \Omega_\Lambda = 0.685, \quad h = 0.673.$$

This age does not contradict the independent bounds discussed in Chapter 1. It is the presence of dark energy that removes the contradiction between the age of the Universe, calculated by using the measured value of the present Hubble parameter, and the bounds on this aged obtained by other methods.

Problem 4.5. *Consider an open model without dark energy (this model in fact is excluded by CMB data), in which $\Omega_M \neq 0$, $\Omega_{curv} \neq 0$, $\Omega_\Lambda = 0$ and $\Omega_M + \Omega_{curv} = 1$.*

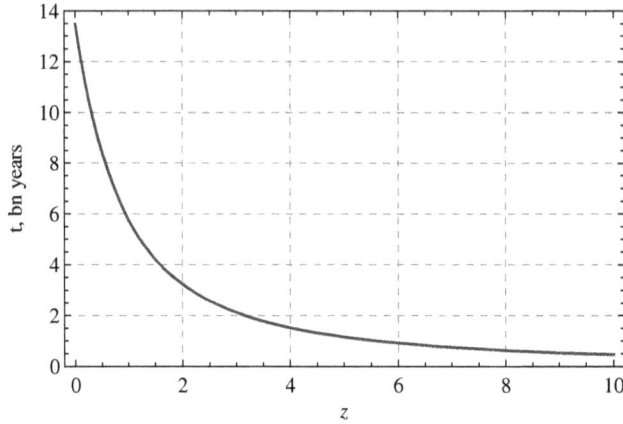

Fig. 4.3.   Age of the Universe t as function of redshift z in ΛCDM model with $\Omega_\Lambda = 0.7$ $h = 0.7$.

Find the present age of the Universe at given value of H_0. Calculate the numerical value for $\Omega_M \approx 0.3$ (estimated from the studies of clusters of galaxies) and $h = 0.7$.

Problem 4.6. *Find the present age of the Universe for dark energy with equation of state $p = w\rho$, $w = const$. Give numerical estimates for $w = -1.25$ and $w = -0.85$ with $\Omega_M = 0.3$, $\Omega_\Lambda = 0.7$.*

It is of interest to find the age of the Universe at different redshifts. We recall that $1 + z(t) = a_0/a(t)$ and obtain from (4.33)

$$t(z) = \frac{2}{3\sqrt{\Omega_\Lambda}} \frac{1}{H_0} \mathrm{Arsh} \sqrt{\frac{\Omega_\Lambda}{\Omega_M(1+z)^3}}. \tag{4.34}$$

The function $t(z)$ is plotted in Fig. 4.3. We emphasize that the formula (4.34) is valid only for $z \ll 1000$, when the energy density of radiation is negligible.

The discussion of the cosmological horizon in ΛCDM model is not so instructive. Nevertheless, let us make the estimate. According to general formula (3.25), the present horizon size is

$$l_{H,0} = a_0 \int_0^{t_0} \frac{dt}{a(t)}.$$

Since $a(t) \propto t^{2/3}$ at small t (see (4.32)), this integral is convergent at the lower limit, i.e., the cosmological horizon size is finite. For given H_0 it is larger than the value $2/H_0$ obtained for the flat model with matter but without dark energy. Numerically, for $\Omega_M = 0.315$, $\Omega_\Lambda = 0.685$, the estimate is

$$l_{H,0} = \frac{2}{H_0} \cdot 1.6 = 14.2 \text{ Gpc}, \quad h = 0.673. \tag{4.35}$$

Problem 4.7. *Show that the horizon size is greater than $2/H_0$ in the model with matter and positive dark energy. Confirm the numerical value (4.35).*

To end this Section, let us make the following remark. The bound (4.8) for Ω_{curv} together with the estimate (4.35) can be used to show that there are many regions of size $l_{H,0}$ outside our horizon. Let us recall in this regard that in the Hot Big Bang theory these regions are causally disconnected. In any case, we cannot obtain any information on what is going on in these regions; as an example, relic photons traveled the distance slightly smaller than $l_{H,0}$ since the recombination epoch.

Clearly, there is an infinite number of regions in question in open and flat models, so we are talking about closed Universe, 3-sphere.[4] It follows from the definitions (4.2), (4.3) and (4.4) that the radius of this sphere a_0 is related to Ω_{curv} as follows,

$$\frac{1}{a_0^2} = H_0^2|\Omega_{curv}|. \tag{4.36}$$

Comparing this with (4.35) and using the bound (4.8) we find

$$\frac{a_0}{l_{H,0}} = \frac{1}{3.2\sqrt{|\Omega_{curv}|}} > 3.1.$$

Thus, the radius of the Universe is larger than the horizon size. This becomes even more striking if we calculate the number of regions like ours. The latter is equal to the ratio of the volume of the 3-sphere $2\pi^2 a_0^3$ and the volume of a region of radius $l_{H,0}$:

$$N \approx \frac{2\pi^2 a_0^3}{(4\pi/3)l_{H,0}^3} = 4.7\left(\frac{a_0}{l_{H,0}}\right)^3 > 150. \tag{4.37}$$

Hence, observational data show that we see less than 1% of the whole Universe. In the accompanying book, we give theoretical arguments suggesting that Ω_{curv} is many orders of magnitude smaller than the observational bound (4.8), i.e., the number of regions outside our horizon is by many orders of magnitude larger than the bound (4.37).

4.6. Brightness–Redshift Relation for Distant Standard Candles

Let us discuss one of the important methods of determination of cosmological parameters, such as H_0, Ω_M, Ω_Λ and Ω_{curv}. This method is also capable of discriminating between the cosmological constant (vacuum energy) and time-dependent forms of dark energy. The method makes use of the simultaneous measurement of redshift z and visible brightness of "standard candles" which are at distances comparable to the horizon size. These standard candles are very luminous objects whose

---

[4]We assume here that the Universe is homogeneous and isotropic both inside and outside our horizon.

absolute luminosity is assumed to be known. Among the objects used as standard candles[5] are supernovae of type Ia (SNe Ia).

Let us find the relation between redshift and visible brightness of a source with absolute luminosity (energy emitted in unit time) L. Although the analysis that follows (but not concrete results!) is straightforwardly generalized to time-dependent dark energy, let us assume, for the time being, that dark energy density ρ_Λ does not depend on time. It is useful to keep spatial curvature not equal to zero and ignore the bound (4.8). Let us choose for definiteness open cosmological model with $\varkappa = -1$ and $\Omega_{curv} > 0$. The flat model is obtained in the limit $\Omega_{curv} \to 0$, or, equivalently, $a_0 \to \infty$, see (4.36).

Let us use the form (2.10) of the metric,

$$ds^2 = dt^2 - a^2(t)[d\chi^2 + \sinh^2 \chi(d\theta^2 + \sin^2 \theta d\phi^2)]. \tag{4.38}$$

As usual, the coordinate distance betwen the source emitting light at time t_i and observer at the Earth at time t_0 equals

$$\chi = \int_{t_i}^{t_0} \frac{dt}{a(t)}. \tag{4.39}$$

Let us first find the relation between the coordinate distance and redshift z of the source. To this end, we use the Friedmann equation in the form (4.10), and neglect radiation. Changing the integration variable in (4.39) from t to

$$z(t) = \frac{a_0}{a(t)} - 1,$$

we obtain

$$\chi = \int_0^z \frac{dz'}{a_0(\dot{a}/a)(z')}.$$

Using Eq. (4.10) we cast this integral into

$$\chi(z) = \int_0^z \frac{dz'}{a_0 H_0} \frac{1}{\sqrt{\Omega_M(z'+1)^3 + \Omega_\Lambda + \Omega_{curv}(z'+1)^2}}. \tag{4.40}$$

This integral cannot be evaluated analytically, but its numerical calculation is easy.

---

[5] As everywhere in this book, we do not discuss astrophysical and observational aspects of this problem. In particular, we leave aside the issue about the nature of SNe Ia, the question of why they are candidates for standard candles, etc.

According to (4.38), the physical area of the sphere crossed by photons today is

$$S(z) = 4\pi r^2(z),\tag{4.41}$$

where

$$r(z) = a_0 \sinh \chi(z).\tag{4.42}$$

The number of photons crossing unit surface at the observer's position is inversely proportional to S, while the energy of each photon differs from the energy at emission by the redshift factor $(1+z)^{-1}$. The same factor additionally enters the expression for the number of photons crossing unit surface *in unit time*, since the time intervals for the source and observer differ by factor $(1+z)^{-1}$. The latter point can be understood as follows. In conformal coordinates $(\eta, \mathbf{x})$ photons behave in the same way as in Minkowski space; see Sec. 2.3. Therefore, in these coordinates the time intervals between emission and absorption of two photons are equal to each other, $d\eta_e = d\eta_0$. This gives the relation between the physical time intervals, $dt_0 = (1+z)dt_e$.

Thus, the visible brightness (energy flux at observer's position) equals

$$J = \frac{L}{(1+z)^2 S(z)}.\tag{4.43}$$

This is the desired expression for the brightness–redshift relation for a source whose absolute luminosity L is assumed to be known.

Let us introduce *photometric distance* r_{ph} in such a way that the relation between L and J is formally the same as in Minkowski space–time,

$$J = \frac{L}{4\pi r_{ph}^2}.$$

We find from (4.43)

$$r_{ph} = (1+z) \cdot r(z),\tag{4.44}$$

where $r(z)$ is given by (4.42).

At first sight, the relation (4.43) involves five cosmological parameters: H_0, a_0, Ω_M, Ω_Λ and Ω_{curv}. In fact, the number of independent parameters is three in view of the relations (see (4.2) and (4.4), (4.5))

$$\Omega_M + \Omega_\Lambda + \Omega_{curv} = 1\tag{4.45}$$

and

$$|\Omega_{curv}| = \frac{1}{a_0^2 H_0^2}.\tag{4.46}$$

Note that at $z \ll 1$, one can neglect z' in the integrand in (4.40), then $\chi(z) = z/(a_0 H_0)$ and $r(z) = a_0\chi(z)$, so we get back to the Hubble law $r(z) = H_0^{-1}z$.

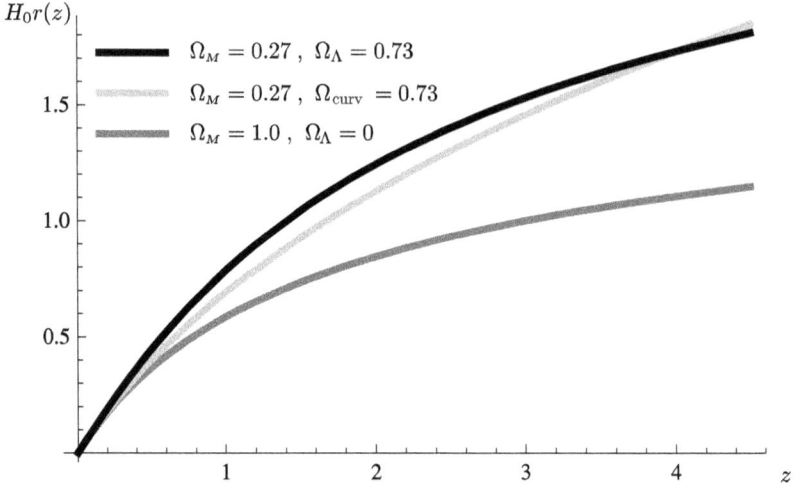

Fig. 4.4.   Dependence of $H_0 r(z)$ on redshift z for various cosmological models.

In that case, the leading order expression for brightness coincides with the usual formula

$$J = \frac{L}{4\pi r^2(z)}, \quad z \ll 1.$$

Let us now discuss the general case. It is clear from (4.40)–(4.46) that the three independent parameters enter the brightness–redshift relation in a non-trivial way. In principle, all of them may be determined by measurements in a wide range of z. This is illustrated in Fig. 4.4.

To understand what is shown in Fig. 4.4 we note that the dependence on cosmological parameters enters the formula (4.43) through the function $r(z)$. If we measure $r(z)$ in Hubble units H_0^{-1}, then

$$H_0 r(z) = \frac{1}{\sqrt{\Omega_{curv}}} \sinh \chi(z), \tag{4.47}$$

$$\chi(z) = \int_0^z \frac{\sqrt{\Omega_{curv}} dz'}{\sqrt{\Omega_M(1+z')^3 + \Omega_\Lambda + \Omega_{curv}(1+z')^2}}.$$

Hence, the right-hand side of (4.47) does not explicitly depend on H_0; it is this quantity that is shown in Fig. 4.4.

Let us first discuss black and dark-gray curves in Fig. 4.4, which correspond to spatially flat Universes with: $\Omega_M = 0.27$, $\Omega_\Lambda = 0.73$ (black curve) and $\Omega_M = 1$, $\Omega_\Lambda = 0$ (dark-gray curve). We obtain formulas for the flat model from the general expression (4.42) by taking the limit $a_0 \to \infty$, $\Omega_{curv} \to 0$. In this way we get

$$r(z) = \frac{1}{H_0} \int_0^z \frac{dz'}{\sqrt{\Omega_M(z'+1)^3 + \Omega_\Lambda}}, \quad \Omega_{curv} = 0, \tag{4.48}$$

$H_0 r(z)$

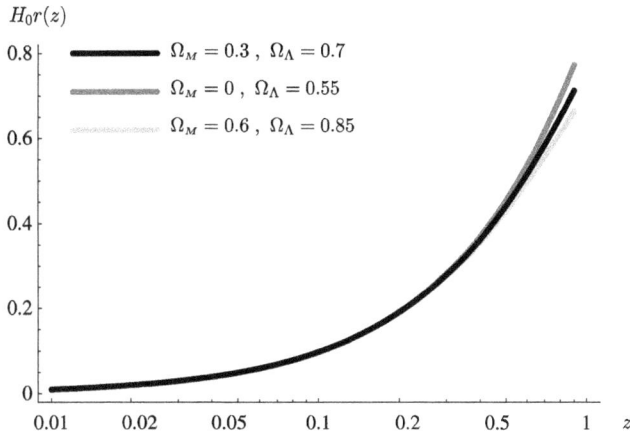

Fig. 4.5.   Degeneracy in the parameter space $(\Omega_M, \Omega_\Lambda)$. The cases $(\Omega_M = 0, \Omega_\Lambda = 0.55)$ and $(\Omega_M = 0.6, \Omega_\Lambda = 0.85)$ correspond to open and closed models, respectively, with $\Omega_{curv} = 1 - \Omega_M - \Omega_\Lambda$. Note that the horizontal scale here is logarithmic, unlike in Fig. 4.4.

with $\Omega_M + \Omega_\Lambda = 1$. It is clear that the larger is Ω_Λ (and, accordingly, the smaller is Ω_M), the faster the function $r(z)$ grows with z; distant supernovae are dimmer in ΛCDM model as compared to the flat model without dark energy. It is this property that have been observed [28, 29]. Figure 4.6 shows one of the early sets of data [64], while Fig. 4.7 illustrates uniqueness of the interpretation in terms of dark energy with enlarged dataset [65].

Now, black and light gray lines in Fig. 4.4 correspond to ΛCDM model and open model without the cosmological constant. They differ already at moderate z. Therefore, the model with $\Omega_\Lambda = 0$, $\Omega_{curv} = 0.7$ is also inconsistent with data. Overall, data on SNe 1a reject models without dark energy, see Figs. 4.6–4.9.

This is probably the strongest argument for dark energy. We stress, however, that there are other, independent arguments. Namely, we mentioned already the argument based on the extrapolation of the mass estimates of clusters of galaxies to the whole Universe (giving $\Omega_M \approx 0.3$), together with CMB bound on spatial curvature. We also presented the argument based on the age of the Universe. Other arguments come from the analysis of CMB and large scale structure; some of them are discussed in the accompanying book.

Let us now turn to Fig. 4.5. We present it here to illustrate the degeneracy in parameters: models with very different parameters give very similar results at moderate z; this range of z is of particular interest, since objects at large z are dim, and hence difficult to observe. To see what is going on, let us find the first correction to the Hubble law at small z. We make use of the relation $\Omega_\Lambda = 1 - \Omega_M - \Omega_{curv}$ and write to quadratic order in z

$$\chi(z) = \frac{1}{a_0 H_0} \left[z - \frac{z^2}{4} (3\Omega_M + 2\Omega_{curv}) \right].$$

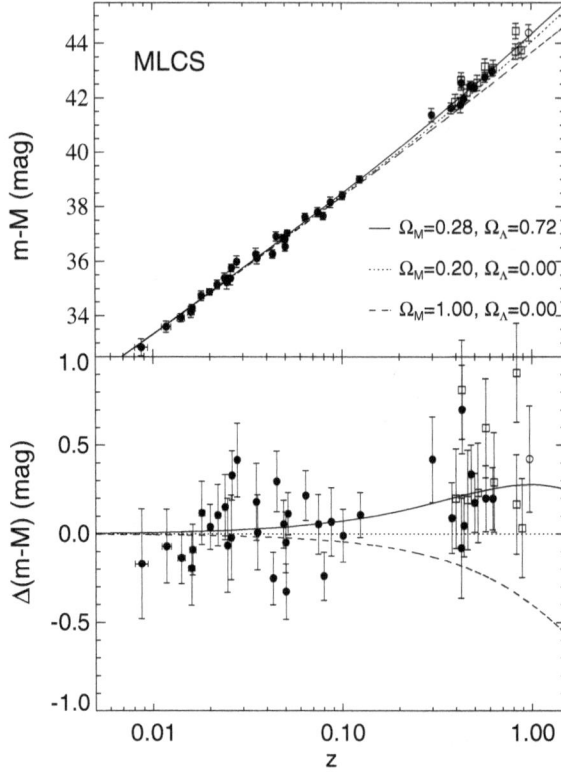

Fig. 4.6.   Hubble diagram for SNe Ia: early data [64]. The upper panel shows the brightness distribution of supernovae (appropriately corrected). The lower panel illustrates the incompatibility of observations to the CDM model with spatial curvature ($\Omega_M = 0.2, \Omega_\Lambda = 0, \Omega_{curv} = 0.8$, dotted line) and flat CDM model ($\Omega_M = 1, \Omega_\Lambda = 0, \Omega_{curv} = 0$, dashed line). The black line is the prediction of the ΛCDM model with $\Omega_M = 0.28, \Omega_\Lambda = 0.72, \Omega_{curv} = 0.0$ which is consistent with the data. The notation on vertical axis is related to brightness measure in astronomy, apparent magnitude. The difference $(m-M)$ is related to photometric distance by $m-M = 5\log_{10}(r_{ph}/\mathrm{Mpc})+25$. The larger $(m-M)$ the dimmer the object.

To this order $r(z) = a_0\chi(z) + \mathcal{O}(z^3)$, so

$$r(z) = \frac{1}{H_0}\left[z - \frac{z^2}{4}(3\Omega_M + 2\Omega_{curv})\right]. \tag{4.49}$$

The second term in the right-hand side is precisely the correction we are concerned about. It is clear from (4.49) that this correction depends only on the combination $(3\Omega_M + 2\Omega_{curv})$ or, in terms of Ω_M and Ω_Λ, on the combination $(\Omega_M - 2\Omega_\Lambda)$, but not on Ω_M and Ω_Λ separately. It is this property that is responsible for the degeneracy at small z. To study the degeneracy at moderate z one has to use higher order terms in z in Eq. (4.49). It turns out that the terms depending on the combination $(\Omega_M - 2\Omega_\Lambda)$ partially cancel at the orders z^2 and z^3 in the interesting range of parameters. Uncompensated contribution of order z^3 depends on another linear combination,

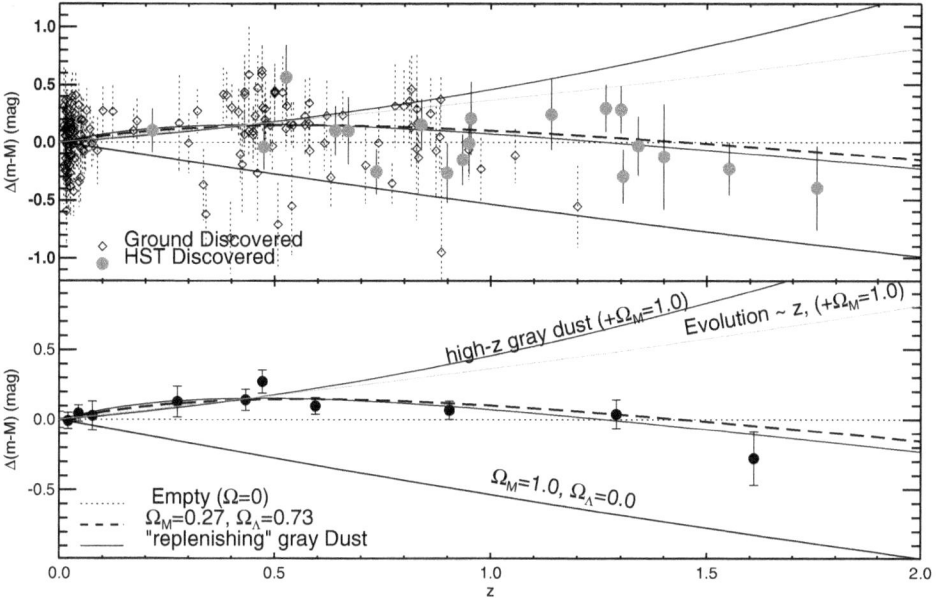

Fig. 4.7. Plots [65] (2004) illustrating different possible interpretations of observations of SNe Ia. They show the deviation from the prediction of a model of curved empty Universe ($\Omega_M = \Omega_\Lambda = 0, \Omega_{curv} = 1$; recall that such a Universe expands at constant speed, $\ddot{a} = 0$). Among cosmological models, considered are flat ΛCDM ($\Omega_M = 0.27, \Omega_\Lambda = 0.73$) and flat CDM model without dark energy ($\Omega_M = 1, \Omega_\Lambda = 0$). Model with evolution is the model in which absolute luminosity of SNe 1a decreases like z^{-1} (in flat CDM cosmology). Models with non-standard intergalactic medium are models with "dust" absorbing the supernovae light; the "dust" densities are $\rho(z) = \rho_0(1 + z)^\alpha$, where $\alpha = 3$ (dash-dotted line) and $\alpha = 3$ at $z < 0.5$ while $\alpha = 0$ at other z (thin black line). Data at upper panel are from the Hubble telescope (bullets) and terrestrial telescopes (diamonds). At the lower panel these data are combined and binned in z. We note that independent observations of SNe Ia [66] (2005) give results consistent with those shown here. More recent supernovae catalogs are also consistent with the data shown here, see Fig. 4.8.

($2\Omega_M - \Omega_\Lambda$). Hence, the observations studying the photometric distances at moderate z are sensitive to the latter combination. The data show that this combination is close to zero, while there is approximate degeneracy along the orthogonal linear combination. Therefore, the allowed region of parameters is stretched along the line $2\Omega_M - \Omega_\Lambda = 0$. This is seen in Fig 4.9 [67].

Problem 4.8. *Show that to the order z^3 the degeneracy in parameters is removed, i.e., $r(z)$ given by (4.42) depends in a non-trivial way on all three parameters H_0, Ω_M, Ω_Λ. Show, nevertheless, that in the interesting region of cosmological parameters there remains approximate degeneracy at moderate z along the line $2\Omega_M - \Omega_\Lambda = 0$.*

There is nothing surprising in the degeneracy in parameters. It is clear that the lowest in z correction to the Hubble law can depend only on the present values of

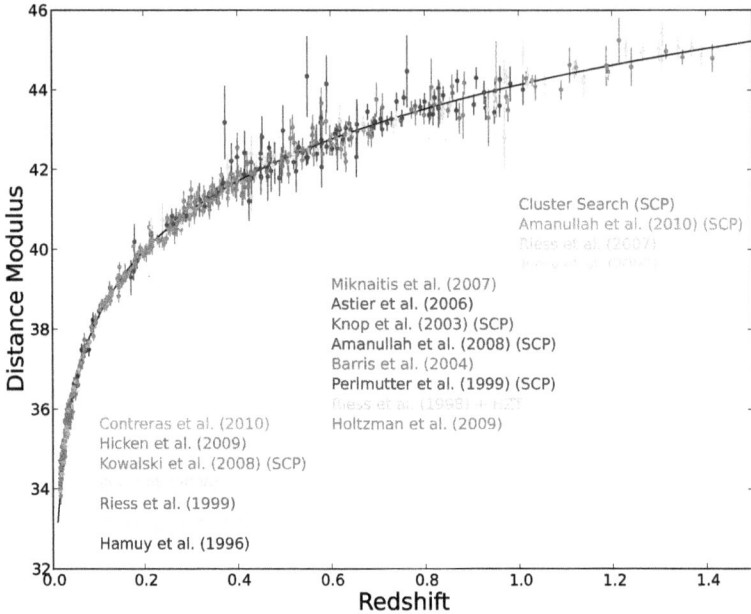

Fig. 4.8. Hubble diagram for distant objects [75] (2011). Solid curve is the best fit for spatially flat ΛCDM model. See Fig. 13.13 for color version.

the Hubble parameter and acceleration parameter. We define the latter as follows,[6]

$$q_0 = \frac{1}{H_0^2}\left(\frac{\ddot{a}}{a}\right)_0. \qquad (4.50)$$

By measuring one parameter q_0, one can determine only one combination of Ω_M and Ω_Λ, hence the degeneracy.

Regarding Fig. 4.9, let us make a general comment. We often show regions in a parameter plane allowed by one or another data. If we do not state the opposite, three regions embedded into each other correspond to regions allowed at 1σ, 2σ and 3σ confidence level (assuming Gaussian distribution for relevant quantity), i.e., at the confidence level of 68.3%, 95.4% and 99.7%.

Problem 4.9. *Find the lowest in z correction to the Hubble law, i.e., the function $r(z)$ at quadratic level in z, in terms of H_0 and q_0. Do not use the Friedmann equation. Show that after using the Friedmann equation, this expression coincides with (4.49).*

---

[6]It is traditional to use *deceleration* parameter, which differs by sign from the acceleration parameter (4.50). We think that using *deceleration* parameter for *accelerating* Universe does not make much sense.

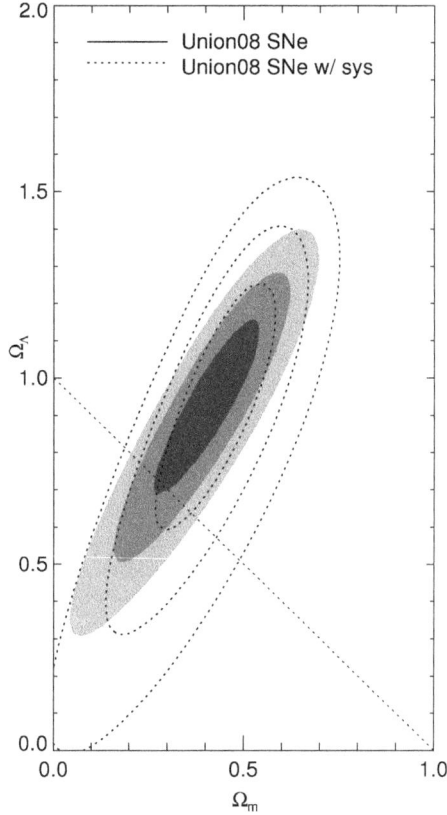

Fig. 4.9. Regions in the plane of cosmological parameters $(\Omega_M, \Omega_\Lambda)$, consistent with observational data on SNe Ia [67]. Dotted lines show contours obtained with proper estimates of possible systematics.

The approximate degeneracy in parameters makes it difficult to determine Ω_M, Ω_Λ and Ω_{curv} by studying standard candles only. On the other hand, CMB temperature anisotropy gives strong bound on Ω_{curv}: $|\Omega_{curv}| < 0.01$. Making use of this bound, one can determine Ω_M and Ω_Λ from the observations of SNe Ia. This is shown in Fig. 4.9, where the line $\Omega_{curv} = 0$ is denoted as $\Omega_{tot} = 1$; it is seen that supernovae observations give $(0.23 < \Omega_M < 0.39, 0.77 > \Omega_\Lambda > 0.61)$ at 95%CL.

To conclude, the observations of SNe Ia together with CMB and BAO measurements are one of the main sources of information on dark energy. The fit to existing cosmological data gives the following determination,

$$\Omega_M = 0.315^{+0.016}_{-0.017}, \quad \Omega_\Lambda = 0.685^{+0.017}_{-0.016}$$

at 68% CL.

Similar observations with better precision and statistics and at larger z will most likely make it possible to find the dependence of dark energy density on time (or establish strong bounds on this dependence). The existing data are consistent with

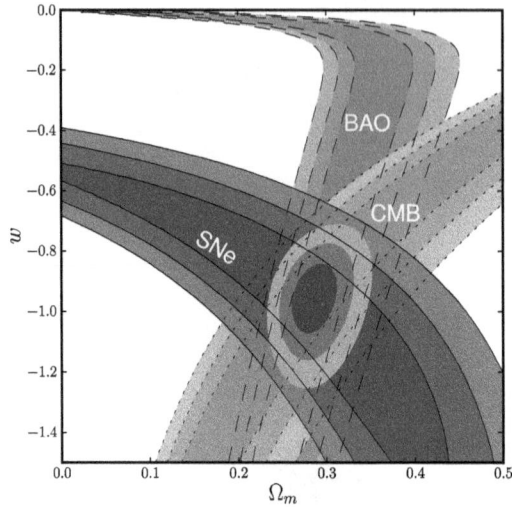

Fig. 4.10. Regions in the plane of parameters (Ω_M, w) allowed by observations of CMB anisotropy (orange, dotted lines), SNe Ia (blue, solid lines) and baryon acoustic oscillations (green, dashed lines). Shaded regions correspond to the combined analysis of all these data [75]. Small, middle size and large regions are allowed at confidence levels of 68.3%, 95.4% and 99.7%, including both statistical and systematic uncertainties.

time-independent dark energy (i.e., dark energy equation of state $p = -\rho$), and under the assumption of time-independence of the parameter w of the dark energy equation of state $p = w\rho$, these data give (see Fig. 4.10 and Fig. 13.6 on color pages)

$$-1.2 < w < -0.8. \qquad (4.51)$$

Refining this result is a very important task for future observations.

Problem 4.10. *Generalize formulas of this Section to the case of dark energy with equation of state $p = w\rho$, $w = const$. Taking $\Omega_{curv} = 0$ and $\Omega_M = 0.27$, draw plots of $r(z)$ for $w = -2$, $w = -1.5$, $w = -1$, $w = -0.75$ and $w = -0.5$. Making use of Fig. 4.9 show that existing data are indeed capable of determining w at the level of accuracy given in (4.51).*

The existing observations are consistent with the assumption that dark energy is the cosmological constant. Figure 4.11 shows the present physical distance r (4.48) as function of redshift in ΛCDM model.

4.7. Angular Sizes of Distant Objects

An important characteristic of an extended object (e.g. galaxy) is its angular size. In this regard, a useful concept is angular diameter distance D_a that relates the

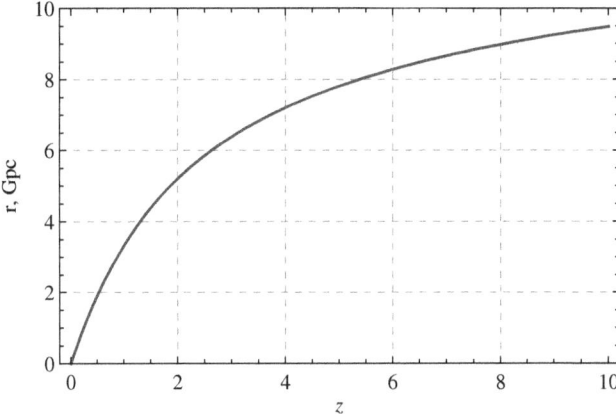

Fig. 4.11. Present physical distance (4.48) to object with redshift z in ΛCDM model with $\Omega_\Lambda = 0.7$, $h = 0.7$.

absolute size of an object d to angle $\Delta\theta$ at which this object is seen today,

$$d = D_a(z) \cdot \Delta\theta,$$

where z is the redshift of the object. To find the expression for $D_a(z)$ let us recall again that in conformal coordinates photons behave in the same way as in Minkowski space–time. Therefore, the coordinate size of an object is related to its coordinate distance from us χ and angular size $\Delta\theta$ by the relation

$$d_{conf} = \sinh\chi \cdot \Delta\theta.$$

The physical size of the object emitting photons at time t_i is equal to

$$d = a(t_i)d_{conf} = \frac{a(t_i)}{a_0} \cdot a_0 \sinh\chi \cdot \Delta\theta.$$

Making use of (4.42) we obtain

$$D_a(z) = \frac{1}{1+z}r(z),$$

where $r(z)$ is given by formulas of Sec. 4.6.

Angular diameter distance increases with z relatively slowly at moderate z, see Fig. 4.12. On the other hand, photometric distance (4.44) rapidly increases with z, so galaxies become more and more dim. At large z, the large distance from the Earth to the galaxies shows up not in the smallness of their angular sizes but in their low surface brightness (visible brightness of a region of unit angular size).

One way to measure the angular diameter distance $D_a(z)$ at various z (more precisely, a combination of $D_a(z)$ and the Hubble parameter $H(z)$) is to make use of the baryon acoustic oscillations, BAO. We consider them in detail in the accompanying book, and here we give qualitative treatment. To begin with, we point out that at

$H_0 D_a(z)$

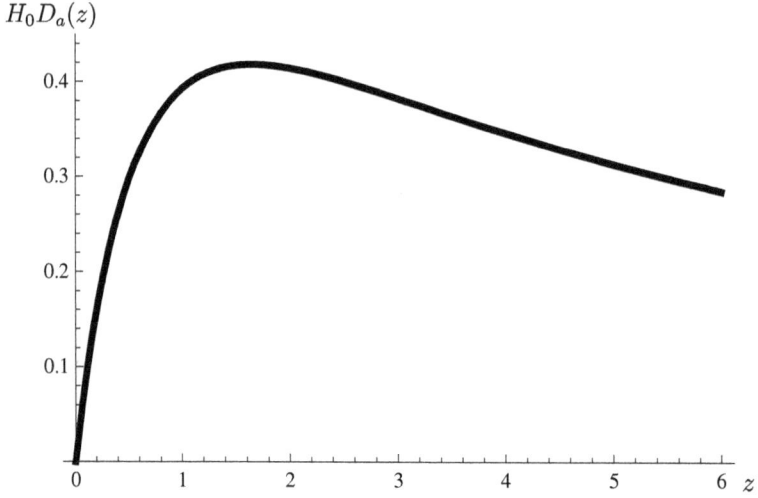

Fig. 4.12.   Angular diameter distance as a function of redshift in ΛCDM model ($\Omega_M = 0.27$, $\Omega_\Lambda = 0.73$, $\Omega_{curv} = 0$).

very early hot epoch, the perturbations in dark matter and baryon-electron-photon component coincide in space and, modulo a numerical factor, in amplitude (adiabatic perturbations):

$$\frac{\delta\rho_{CDM}(\mathbf{x})}{\rho_{CDM}} = \frac{3}{4}\frac{\delta\rho_{B\gamma}(\mathbf{x})}{\rho_{B\gamma}}.$$

Later on, but still before recombination, the baryon–photon perturbations propagate with the sound speed $u_s = \sqrt{\partial p/\partial\rho} \approx 1/\sqrt{3}$ (see Chapter 5), while the sound speed in dark matter vanishes, since its pressure equals zero. Let us imagine that initially there is overdensity of dark matter and baryons in some place in space. By the time of recombination t_r, the excess in dark matter remains at the same place, wile the excess of baryons (and photons) moves from this place (as a spherical sound wave, if the initial excess is spherically symmetric) to distance

$$r_s = a(t_r)\int_0^{t_r} u_s \frac{dt}{a(t)}. \tag{4.52}$$

To obtain this formula, we recall that the coordinate distance dx that a sound wave travels in time interval dt is given by

$$a(t)dx = u_s dt.$$

The quantity $r_s(t_r)$ is called acoustic horizon at recombination; it is reliably calculable once Ω_{CDM} and Ω_B are known. The corresponding present size is

$$r_s\frac{a_0}{a(t_r)} = 147 \text{ Mpc}.$$

After recombination, baryons no longer interact with photons, pressure in the baryon component and hence sound speed in it fall practically to zero, and the baryon perturbations get frozen out in space. As a result, the overdensities in baryons and in dark matter are separated by the distance r_s. The corresponding physical distance increases in time due to the cosmological expansion, $r_s(t) \propto a(t)$. Both dark matter and baryons participate in galaxy formation, so our discussion shows that there must be an additional correlation in the density of galaxies at distance $r_s(t)$. This feature in the galaxy correlation function has been discovered indeed [76].[7]

Thus, there is a standard ruler in the Universe at every z, whose size is

$$r_s(z) = r_s(t_r) \frac{1 + z_r}{1 + z}.$$

The measurement of an angle at which it is seen enables one to determine the angular diameter distance $D_a(z)$. In practice one measures the combination $[D_a^2(z) H^{-1}(z)]^{1/3}$, which is the geometric mean of distance scales in directions normal to the line of sight (D_a) and along the line of sight (H^{-1}). These measurements cover today a fairly large region $z < 1$; see Fig. 4.13.

Presently, BAO is an important instrument for studying the dynamics of the cosmological evolution at late epoch, and in particular for analyzing the dark energy properties.

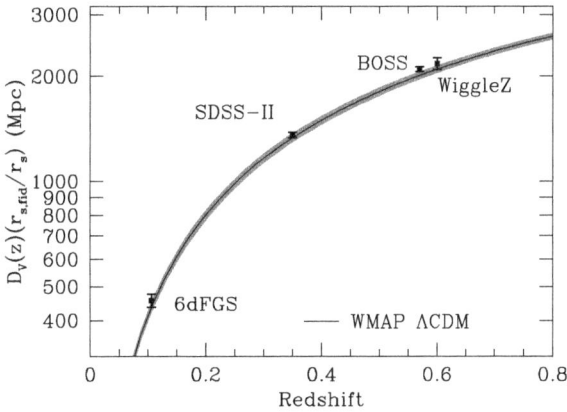

Fig. 4.13. The measured values of $D_V \equiv [z(1+z)^2 D_a^2(z)/H(z)]^{1/3}$, according to the data on structures [77]. Theoretical prediction is made in the ΛCDM model, the gray band corresponds to the cosmological parameters inferred from CMB data; r_s here denotes the present size of the sound horizon at recombination and $r_{s,fid}$ is its value obatined from CMB data.

---

[7]In Fourier space, it yields oscillations in the power spectrum, hence the name.

4.8. *Quintessence

Cosmological constant (vacuum energy density) is not the only possible reason of the accelerated expansion of the Universe at the present epoch. Nature of dark energy is one of the major problems of contemporary natural science. No wonder, numerous hypotheses have been put forward in this regard. One of these hypotheses is the existence of "quintessence", spatially homogeneous field whose energy plays the role of dark energy [68–70]. Unlike the cosmological constant, quintessence is a dynamical field, and its energy density depends on time. In terms of the effective equation of state $p = w\rho$ this means that $w \neq -1$ with, generally speaking, time-dependent w.

Quintessence is often (but not always) associated with a *scalar* field. As a concrete example, we consider one class of models of this sort later in this Section, but before that we study general properties of scalar field dynamics in expanding Universe. These results will be useful also in our discussions of other topics. Understanding the dynamics of scalar field is particularly important in the context of inflationary theory.

4.8.1. *Evolution of scalar field in expanding Universe*

Let us consider a theory of real scalar field with action

$$S = \int d^4x \sqrt{-g}\mathcal{L} = \int d^4x \sqrt{-g} \left[\frac{1}{2} g^{\mu\nu} \partial_\mu \varphi \partial_\nu \varphi - V(\varphi) \right], \qquad (4.53)$$

where $V(\varphi)$ is the scalar potential. Let us stick to spatially flat Universe whose metric has standard Friedmann–Lemaître–Robertson–Walker form (2.22) with scale factor $a(t)$ being a known function of time.

The equation for the scalar field is obtained, as usual, by varying the action (4.53) with respect to φ, and has the form

$$\frac{1}{\sqrt{-g}} \partial_\mu (\sqrt{-g} g^{\mu\nu} \partial_\nu \varphi) = -\frac{\partial V}{\partial \varphi}. \qquad (4.54)$$

Let us specify to spatially homogeneous (independent of coordinates) scalar field $\varphi(t)$ in the background metric (2.22). In that case, Eq. (4.54) reduces to

$$\ddot{\varphi} + 3H\dot{\varphi} = -\frac{\partial V}{\partial \varphi}. \qquad (4.55)$$

Equation (4.55) formally coincides with the equation for the classical mechanics of a "particle" with coordinate φ in the "potential" $V(\varphi)$, which experiences friction with time-dependent friction coefficient H. Depending on the strengths of the driving force and friction, two regimes are possible: (1) fast roll regime when $H\dot{\varphi} \ll V'$ (prime denotes the derivative with respect to φ), friction is weak, and the "particle" rapidly rolls down to the minimum of the potential $V(\varphi)$; (2) slow roll regime when friction is strong, and the "particle" barely moves. Let us begin with the second regime. In that case

$$H\dot{\varphi} \sim V'. \tag{4.56}$$

During the Hubble time H^{-1} the field changes by

$$\delta\varphi \sim \dot{\varphi}H^{-1} \sim \frac{V'}{H^2}.$$

This change is small compared to the field itself, $\delta\varphi \ll \varphi$, provided that

$$\frac{V'}{\varphi} \ll H^2. \tag{4.57}$$

For power-law potentials like $m^2\varphi^2$ or $\lambda\varphi^4$ one has $V' \sim V\varphi^{-1}$, so the condition for slow roll is

$$\frac{V}{\varphi^2} \ll H^2. \tag{4.58}$$

Thus, in the case of power-law potential, the condition (4.58) is necessary for slow roll, at which the value of the field remains practically constant during the cosmological evolution. Once this condition is violated, the field rapidly rolls down to the minimum of $V(\varphi)$.

Problem 4.11. *The condition (4.58) is not sufficient for slow roll. The second necessary condition is that the first term on the left-hand side of Eq. (4.55) does not exceed the second term, $\ddot{\varphi} \lesssim H\dot{\varphi}$. The latter condition ensures that velocity does not change much in one Hubble time, $\delta\dot{\varphi} \sim \ddot{\varphi}H^{-1} \lesssim \dot{\varphi}$, so the velocity remains small during the evolution. Find the second slow roll condition in terms of the potential $V(\varphi)$ and its derivatives, and $H(t)$ and its derivatives. Simplify this condition in the case of power-law dependence of $a(t)$ on time (like $a \propto t^{2/3}$, matter domination).*

Problem 4.12. *Show that for the power-law dependence of the scale factor on time, $a(t) = t^\alpha$, $\alpha > 1/3$, the general solution to (4.55) in the approximation $V' = const$ is*

$$\varphi = \varphi_i + C \cdot (t^2 - t_i^2) + d \cdot \left[1 - \left(\frac{t_i}{t}\right)^{3\alpha-1}\right], \tag{4.59}$$

where φ_i is the initial value of the field, $\varphi_i = \varphi(t_i)$. In the case $\dot{\varphi}(t_i) \ll H(t_i)\varphi(t_i)$, find the constants C and d in terms of φ_i and $\dot{\varphi}(t_i)$. Show that in this case the third term in (4.59) is always small. Find the values of t at which the second term in (4.59) is small compared to the initial value φ_i and show that for a power-law potential this time is in accordance to (4.58), and in general case to (4.57). Show that in this time interval the relation (4.56), as well as $\ddot{\varphi} \sim H\dot{\varphi}$, hold.

Although we will not need in this Section to know the behavior of the scalar field near the minimum of the scalar potential, let us consider how $\varphi(t)$ approaches this minimum. Let the minimum of $V(\varphi)$ be at $\varphi = 0$, and near the minimum the potential be

$$V(\varphi) = \frac{m^2}{2}\varphi^2. \tag{4.60}$$

Then Eq. (4.55) near the minimum has the form

$$\ddot{\varphi} + 3\frac{\dot{a}}{a}\dot{\varphi} + m^2\varphi = 0. \tag{4.61}$$

To analyze equations of this sort, it is convenient to get rid of the friction term proportional to $\dot{\varphi}$ and cast the equation into the form of the oscillator equation with time-dependent frequency (cf. end of Sec. 2.3). In our case the change of variables is given by

$$\varphi(t) = \frac{1}{a^{3/2}(t)} \cdot \chi(t),$$

where χ is a new unknown function which obeys

$$\ddot{\chi} + \left(m^2 - \frac{3}{2}\frac{\ddot{a}}{a} - \frac{3}{4}\frac{\dot{a}^2}{a^2}\right)\chi = 0. \tag{4.62}$$

Note that

$$\frac{\ddot{a}}{a} \sim \frac{\dot{a}^2}{a^2} = H^2.$$

It then follows from (4.62) that at $m^2 \ll H^2$ the mass term is negligible. In that case the potential (4.60) obeys the condition (4.58), and we come back to the slow roll regime.

Problem 4.13. *Obtain the results of the previous problem starting from Eq. (4.62).*

We are now interested in the opposite regime, $m^2 \gg H^2$. In this case, we can neglect terms of order H^2 in parenthesis in (4.62), so the solution has the form $\chi = \mathrm{const} \cdot \cos(mt + \beta)$ where β is an arbitrary phase. Hence, at $m^2 \gg H^2$ the field relaxes to the minimum in the following way,

$$\varphi(t) = \varphi_* \cdot \frac{\cos(mt + \beta)}{a^{3/2}(t)}, \tag{4.63}$$

where φ_* is some constant. The field oscillates near the minimum with decreasing amplitude. Note that the relative change of the amplitude in one period of oscillations is of order

$$\frac{\dot{a}}{a} \cdot m^{-1} \sim \frac{H}{m},$$

so it is small in the case under study, $m \gg H$.

The solution (4.63) can be found also from energy considerations. In the case of free massive scalar field, the action (4.53) in the Universe with metric (2.22) takes the form

$$S = \int d^4x \cdot a^3 \cdot \left(\frac{1}{2}\dot{\varphi}^2 - \frac{1}{2a^2}(\partial_i\varphi)^2 - \frac{m^2}{2}\varphi^2 \right).$$

If the scale factor were independent of time, the homogeneous field would have conserved energy,

$$E = \int d^3\mathbf{x} \cdot a^3 \cdot \left(\frac{1}{2}\dot{\varphi}^2 + \frac{m^2}{2}\varphi^2 \right).$$

The solution for the field equation would oscillate with frequency m, i.e., $\varphi \propto \cos(mt + \beta)$. If the scale factor slowly (adiabatically) grows in time, the oscillatory behavior of solutions persists, and energy is still (approximately) conserved, as we know from classical mechanics. The latter property means that

$$a^3(t) \left(\frac{1}{2}\dot{\varphi}^2 + \frac{m^2}{2}\varphi^2 \right) = \text{const.} \tag{4.64}$$

This tells that the amplitude of oscillations decreases as $a^{-3/2}$, i.e., we come back to the solution (4.63). Note that the approximate conservation law (4.64) can be interpreted as (approximate) energy conservation in comoving volume. Note also that for fast changing $a(t)$, there is no conservation of energy at all; energy is not conserved in time-dependent backgrounds (in this case $a(t)$). It does not make sense to talk about energy (including energy of the gravitational field) of the entire Universe; there is no such an integral of motion in General Relativity.[8]

To end this Section, let us find the energy–momentum tensor of the scalar field in different regimes. The general expression for energy–momentum tensor in the theory with action (4.53) is

$$T_{\mu\nu} = \frac{2}{\sqrt{-g}} \frac{\delta S}{\delta g^{\mu\nu}} = \partial_\mu\varphi \partial_\nu\varphi - g_{\mu\nu}\mathcal{L}. \tag{4.65}$$

For homogeneous scalar field and potential $V(\varphi)$ we have for non-zero components in the locally-Lorentz frame (i.e., setting the local metric $g_{\mu\nu}$ in (4.65) equal to $\eta_{\mu\nu}$)

$$T_{00} = \frac{1}{2}\dot{\varphi}^2 + V(\varphi) \equiv \rho_\varphi, \quad T_{ij} = \left(\frac{1}{2}\dot{\varphi}^2 - V(\varphi) \right)\delta_{ij} \equiv p_\varphi\delta_{ij}. \tag{4.66}$$

The estimate (4.56) gives *in the slow roll regime*

$$\dot{\varphi}^2 \sim \frac{V'^2}{H^2} \ll V' \cdot \varphi, \tag{4.67}$$

---

[8]There are certain cases in which one can define total energy (including energy of the gravitational field) in the framework of General Relativity. One special case is asymptotically flat space–time. So, the notion of energy (mass) of a gravitating body, away from which space–time is Minkowskian, is well-defined.

where we used (4.57) to obtain the inequality. Power-law potentials obey $V'\varphi \sim V$, so (4.67) gives

$$\dot{\varphi}^2 \ll V.$$

Hence, in the slow roll regime

$$\rho_\varphi \approx -p_\varphi \approx V(\varphi), \tag{4.68}$$

i.e., the equation of state is approximately the same as for vacuum, $p \approx -\rho$ (although it is clear from (4.66) that one always has $p > -\rho$).

To discuss *the regime of fast oscillations* about the minimum of the scalar potential, we use (4.60) and (4.63) and obtain

$$T_{00} = \frac{m^2\varphi_*^2}{2} \cdot \frac{1}{a^3(t)}, \quad T_{ij} = -\frac{m^2\varphi_*^2}{2}\frac{1}{a^3}\cos(2mt + 2\beta) \cdot \delta_{ij},$$

where we again used inequality $H \ll m$. Thus, energy density and pressure, averaged over a period of oscillations, are

$$T_{00} \equiv \rho_\varphi = \frac{m^2\varphi_*^2}{2} \cdot \frac{1}{a^3(t)}, \quad T_{ij} \equiv p_\varphi \cdot \delta_{ij} = 0. \tag{4.69}$$

We conclude that average energy–momentum tensor of the coherent scalar field oscillations coincides with the energy–momentum tensor of non-relativistic matter: pressure is zero, while energy density decreases as a^{-3}. As we already mentioned, the latter property reflects the energy conservation in comoving volume; see (4.64).

From the quantum field theory viewpoint, homogeneous oscillating scalar field (4.63) is to be viewed as a collection of free particles of mass m, all at zero spatial momentum (at rest). The number density of these particles is given by

$$n = \frac{\rho}{m} = \frac{1}{2}m\varphi_*^2 \cdot \frac{1}{a^3(t)}.$$

At any number density, $n(t)$ decreases as a^{-3}. The fact that pressure vanishes has natural interpretation in this picture: particles at rest do not produce any pressure.

4.8.2. *Accelerated cosmological expansion due to scalar field*

Accelerated expansion of the Universe at the present epoch may be explained by introducing scalar field φ (quintessence) with action (4.53) and choosing the potential $V(\varphi)$ and the present value of the field φ in such a way that the evolution of the field occurs in the slow roll regime. Furthermore, one should assume that the field φ is spatially homogeneous; initial data of the latter type are natural in the framework of inflationary theory, so the latter assumption is not particularly strong.

The effective equation of state in the slow roll regime is $p_\varphi \approx -\rho_\varphi$, see (4.68). Thus, the Universe indeed undergoes accelerated expansion, if the scalar field gives dominant contribution into the energy density. Let us find the conditions under which the slow roll regime indeed holds. Let us assume for definiteness that near the present value of φ the scalar potential may be approximated by power law, $V(\varphi) \propto \varphi^k$ where $|k|$ is not very large. Then the slow roll condition takes the form (4.58). Since energy density in the Universe is mostly due to the scalar field, the Friedmann equation is

$$H^2 = \frac{8\pi}{3M_{Pl}^2}\rho_\varphi = \frac{8\pi}{3M_{Pl}^2}V(\varphi), \tag{4.70}$$

where we used (4.68). By combining (4.58) and (4.70) we obtain

$$\varphi \gg M_{Pl}. \tag{4.71}$$

This is the form of the slow roll condition for power-law potentials and the Universe dominated by the scalar field itself.

Despite the fact that the value of the scalar field is extremely large, the value of the potential $V(\varphi)$ must be very small,

$$V(\varphi) \sim \rho_c$$

(more precisely, $V(\varphi) = 0.7\rho_c$). This means for power-law potential (and also for more general potentials) that it must be extremely flat. As an example, in the case $V(\varphi) = \frac{m^2}{2}\varphi^2$ the mass must be extremely small,

$$m \lesssim \frac{\sqrt{\rho_c}}{M_{Pl}} \sim 10^{-33}\,\text{eV},$$

while in the case $V(\varphi) = \lambda\varphi^4$ the coupling must be tiny,

$$\lambda \lesssim 10^{-122}.$$

Thus, the quintessence idea works for very exotic scalar potentials only.

Problem 4.14. *Show that for power-law potentials the condition (4.71) ensures $\ddot{\varphi} \ll H\dot{\varphi}$. Thus, the second slow roll condition (see Problem 4.11) is satisfied automatically.*

Quintessence models can, at least in principle, be tested by observations. In these models, the relation $p = -\rho$ for dark energy *is approximate*. Dark energy density depends on time, so the cosmological expansion is different from that in the model with cosmological constant.

Problem 4.15. *In quintessence model with potential $V(\varphi) = \frac{m^2}{2}\varphi^2$, find the present value of the dark energy equation of state parameter w as function of the present value $\varphi(t_0) = \varphi_0$. Choose the present value φ_0 in such a way that $w_0 = 0.95$, and find $w(z)$ as function of redshift at $2 > z > 0$. Take the values $\Omega_\varphi \equiv \Omega_\Lambda = 0.7$, $\Omega_M = 0.3$ and make use of the fact that the scalar field changes slowly at the present epoch.*

Problem 4.16. *Under conditions of the previous problem, and with $w_0 = 0.95$ find numerically the photometric distance $r_{ph}(z)$ as function of redshift (see Sec. 4.6). Draw a plot analogous to Fig. 4.4 and compare it to the plot with $\Omega_M = 0.3$, $\Omega_\Lambda = 0.7$ and time-independent ρ_Λ.*

Finally, let us note that the future of the Universe in quintessence models is, generally speaking, different from the future of the Universe with cosmological constant, see also discussion in Sec. 1.4. This difference is particularly strong in models where the field $\varphi(t)$ rolls down to negative values of the potential $V(\varphi)$: in that case expansion will eventually terminate, and the Universe will recollapse to singularity.

Problem 4.17. *Assuming that today $\varphi = \varphi_0 \gg M_{Pl}$, study the future of the Universe in a model with potential $V(\varphi) = \frac{m^2}{2}\varphi^2 + \epsilon\varphi$, where $|\epsilon| \ll m^2 M_{Pl}$ is a small parameter, and m is such that $V(\varphi_0) \sim \rho_c$. Hint: Use the results of Sec. 4.8.1.*

We stress that quintessence models do not, generally speaking, solve the cosmological constant problem. They only divide this problem into two parts: one is the question of why the "true" cosmological constant (vacuum energy) is zero, and another is the question of why the energy density of quintessence is so small. The first question cannot *in principle* be answered in quintessence models; the second has to do with *naturalness* of the scalar potential. Furthermore, there is a new question in quintessence models: what mechanism ensures the "correct" present value of the scalar field? A possible answer to the latter question is given in models with suitable scalar potentials [69–73]. A subclass of these are "tracker" models [70, 73] which we now discuss.

4.8.3. *Tracker field*

Let us consider a subclass of quintessence models, where the potential has the form[9]

$$V(\varphi) = \frac{M^{n+4}}{n\varphi^n},$$

where M is a parameter of dimension of mass, and $n > 2$ is a numerical parameter. At radiation or matter domination, when $a(t) \propto t^\alpha$, the field equation (4.55) takes the form

$$\ddot{\varphi} + \frac{3\alpha}{t}\dot{\varphi} - \frac{M^{n+4}}{\varphi^{n+1}} = 0. \tag{4.72}$$

It has special, "tracker" solution

$$\varphi^{(tr)}(t) = CM^{1+\nu}t^\nu, \tag{4.73}$$

---

[9]This potential is rather exotic from particle physics viewpoint.

where

$$\nu = \frac{2}{n+2},\qquad(4.74)$$

and $C = C(n, \alpha)$ is determined from Eq. (4.72).

Problem 4.18. *Show that* (4.73) *is indeed a solution to Eq.* (4.72). *Find* $C(n,\alpha)$ *entering* (4.73).

Solution (4.73) is a power-law attractor: if at initial time t_i the field φ_i is smaller than solution (4.73) taken at the same time, $\varphi_i < \varphi^{(tr)}(t_i)$, then the driving force

$$\mathcal{F}(\varphi) \equiv -V'(\varphi) = \frac{M^{n+4}}{\varphi^{n+1}}$$

is larger than the driving force for the solution (4.73),

$$\mathcal{F}(\varphi(t)) > \mathcal{F}(\varphi^{(tr)}(t)).$$

Hence, the solution with $\varphi_i < \varphi^{(tr)}(t_i)$ catches up the tracker solution (4.73). Conversely, a solution with $\varphi_i > \varphi^{(tr)}(t_i)$ rolls slower than the tracker solution, so the former also approaches the latter in the course of evolution. This precisely means that the solution (4.73) is an attractor: in sufficiently wide range of initial data, solutions tend to (4.73) at late times; the evolution of the field φ is basically independent of initial data and is described by the tracker solution.

The solution (4.73) obeys (we omit the superscript (tr) in what follows)

$$\dot{\varphi}^2 \sim V(\varphi) \propto \frac{1}{t^{2-2\nu}}.$$

It is clear, first, that the regime of evolution of the field is *not* the slow roll regime. Second, energy density ρ_φ (see (4.66)) decreases in time slower than the energy density of the dominant matter (relativistic or non-relativistic): the latter decreases as $H^2 \propto t^{-2}$. The relative contribution of the tracker field into total energy density increases in time as $t^{2\nu}$. Third, since $a \propto t^\alpha$, we have

$$\rho_\varphi \propto \frac{1}{a^{\frac{2-2\nu}{\alpha}}}.$$

Making use of (3.41) we find the following expression for the equation of state parameter w_φ entering $p_\varphi = w_\varphi \rho_\varphi$,

$$w_\varphi = -1 + \frac{2}{3}\frac{1-\nu}{\alpha}.$$

This gives, with account of (4.74) and the results of Sec. 3.2.4,

$$w_\varphi = w\frac{n}{n+2} - \frac{2}{n+2},\qquad(4.75)$$

where w is the parameter of equation of state of the dominant matter ($w = \frac{1}{3}$ and $w = 0$ for relativistic and non-relativistic matter, respectively). Thus, equation of state for tracker field depends on the equation of state of dominant matter. This is the origin of the term "tracker field". Note that at large n the equation of state of the tracker field is close to that of dominant matter, $w_\varphi \approx w$.

Problem 4.19. *By calculating pressure and energy density according to* (4.66), *derive* (4.75) *in an alternative way.*

Problem 4.20. *Show that solution* (4.73) *obeys* $V'' \propto H^2$ *at both radiation and matter domination. This is another reason to use the term "tracker field".*

Solution (4.73) is valid at times when energy density of the tracker field is smaller than matter energy density. The relative contribution of tracker energy into total energy density grows in time, and tracker field starts to dominate at some point. After that, the solution (4.73) is no longer valid. Let us find the value of the field at the time when it just starts to dominate. At that time

$$V(\varphi) \sim \dot{\varphi}^2 \sim \frac{\varphi^2}{t^2},$$

i.e.,

$$\rho_\varphi \sim \frac{\varphi^2}{t^2}.$$

Matter energy density at that time is

$$\rho_M = \frac{3}{8\pi} M_{Pl}^2 H^2(t) \sim \frac{M_{Pl}^2}{t^2}.$$

Requiring $\rho_\varphi \sim \rho_M$ we find that the tracker field starts to dominate when

$$\varphi \sim M_{Pl}.$$

Let us note in parenthesis that the latter condition also implies that the initial value of the tracker field may be fairly arbitrary, but should obey $\varphi_i \ll M_{Pl}$.

At later times the field $\varphi(t)$ increases, and fairly soon the relation (4.71) starts to hold. The evolution of φ enters the slow roll regime, and the expansion of the Universe enters the regime of acceleration.

Clearly, the cosmological evolution has rather special character at the transition period from matter domination to tracker field domination. Therefore, cosmological observations may well be capable to confirm or falsify the model. Existing data exclude neither cosmological constant nor tracker field; they are consistent with many other quintessence models too.

Problem 4.21. *Estimate, for various n, at what values of M the tracker field model is consistent with known facts about the accelerated cosmological expansion.*

Chapter 5

Thermodynamics in Expanding Universe

5.1. Distribution Functions for Bosons and Fermions

We consider in most Chapters of this book processes that occur in expanding Universe filled with various species of interacting particles. As we will see, the rates of interactions between these particles are often much higher than the expansion rate of the Universe, so the cosmic medium is in thermal equilibrium at any moment of time. So, we will need a number of basic relations and formulas from equilibrium thermodynamics. We collect them in this Chapter. We note here that as a rule, the most interesting periods in the cosmological evolution are those when one or another reaction *goes out* of equilibrium ("freezes out"). The laws of equilibrium thermodynamics are still useful in such a situation for semi-quantitative analysis, as they enable us to estimate the time of departure from equilibrium and determine the direction of inequilibrium processes.

The thermodynamical description of a system with various particle species is made in terms of chemical potential μ for each type of particles. If there is a reaction

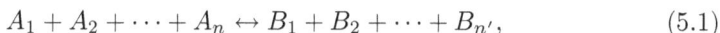

$$A_1 + A_2 + \cdots + A_n \leftrightarrow B_1 + B_2 + \cdots + B_{n'}, \tag{5.1}$$

where A_i, B_j denote types of particles, then the chemical potentials in thermal equilibrium[1] obey

$$\mu_{A_1} + \mu_{A_2} + \cdots + \mu_{A_n} = \mu_{B_1} + \mu_{B_2} + \cdots + \mu_{B_{n'}}. \tag{5.2}$$

In particular, any reaction involving charged particles may occur with emission of a photon (e.g., there is an inelastic scattering reaction $e^- e^- \to e^- e^- \gamma$). Therefore,

---

[1]More precisely, equality (5.2) holds when the system is in *chemical* equilibrium with respect to the reaction (5.1), i.e., when the rate of this reaction is higher than the rate at which external parameters evolve (higher than the expansion rate in our case).

it follows from Eq. (5.2) that photon has zero chemical potential,

$$\mu_\gamma = 0.$$

As another application of (5.2), let us consider the annihilation process

$$e^+ + e^- \leftrightarrow 2\gamma.$$

Since $\mu_\gamma = 0$, Eq. (5.2) gives

$$\mu_{e^-} + \mu_{e^+} = 0.$$

Clearly, the same relation is valid for any other particles and their antiparticles, since they can also annihilate into photons. Thus, chemical potential of an antiparticle equals to that of a particle taken with opposite sign.

A convenient way to deal with all relevant reactions in thermalized system with various particle species is to introduce chemical potentials μ_i to *conserved* quantum numbers $Q^{(i)}$ only. $Q^{(i)}$ should be independent of each other and should form a complete set of conserved quantum numbers. The chemical potential for particle of type A is then

$$\mu_A = \sum_i \mu_i Q_A^{(i)}, \qquad (5.3)$$

where $Q_A^{(i)}$ are quantum numbers carried by particle A. Say, at temperatures $200\,\mathrm{MeV} \ll T \ll 100\,\mathrm{GeV}$, conserved quantum numbers are baryon number B, lepton numbers L_e, L_μ, L_τ and electric charge Q (see Appendix B, color of quarks and gluons is irrelevant for us here), i.e., the complete set is $Q^{(i)} = B, L_e, L_\mu, L_\tau, Q$. In this case the chemical potential of, say, u-quark (baryon number $1/3$, electric charge $2/3$) is

$$\mu_u = \frac{1}{3}\mu_B + \frac{2}{3}\mu_Q,$$

while for d-quark, electron and electron neutrino we have

$$\mu_d = \frac{1}{3}\mu_B - \frac{1}{3}\mu_Q, \quad \mu_e = \mu_{L_e} - \mu_Q, \quad \mu_{\nu_e} = \mu_{L_e}.$$

Relations of the type (5.2) hold automatically for all reactions; an example is a weak process

$$u + e^- \to d + \nu_e.$$

Problem 5.1. *Show that the relations (5.2) hold automatically if chemical potentials are given by (5.3), and charges $Q^{(i)}$ are conserved in all reactions (5.1), i.e.,*

$$Q_{A_1}^{(i)} + \cdots Q_{A_n}^{(i)} = Q_{B_1}^{(i)} + \cdots Q_{B_{n'}}^{(i)}.$$

Now, the chemical potentials to the charges $Q^{(i)}$ can be found if charge densities $n_{Q^{(i)}}$ are known. Indeed, the number density of each particle species n_A is a function of μ_A, so the system of equations

$$\sum_A Q_A^{(i)} n_A = n_{Q^{(i)}}$$

together with (5.3) determines[2] all μ_i in terms of $n_{Q^{(i)}}$.

Particle interactions in cosmic plasma are often fairly weak; we will quantify this statement in appropriate places. In this case the equilibrium distributions in spatial momenta **p** in locally Lorentz frame are given by distribution functions of Bose- and Fermi-gases,

$$f(\mathbf{p}) = \frac{1}{(2\pi)^3} \frac{1}{e^{(E(\mathbf{p})-\mu)/T} \mp 1}.$$ (5.4)

Here

$$E(\mathbf{p}) = \sqrt{\mathbf{p}^2 + m^2}$$ (5.5)

is energy of a particle of mass m, T is the temperature. The minus sign in (5.4) refers to bosons, plus sign to fermions.

Sometimes it is possible to disregard ± 1 in the denominator in (5.4), then the distribution has the universal Maxwell–Boltzmann form,

$$f(\mathbf{p}) = \frac{1}{(2\pi)^3} e^{-(E(\mathbf{p})-\mu)/T}.$$ (5.6)

This form describes, in particular, low density gas of non-relativistic particles, for which $m \gg T$, $(m - \mu) \gg T$. In the latter case

$$f(\mathbf{p}) = \frac{1}{(2\pi)^3} e^{\frac{\mu-m}{T}} \cdot e^{-\frac{\mathbf{p}^2}{2mT}}.$$ (5.7)

Upon integrating the distribution function over momenta one obtains the number density of particle species i,

$$n_i = g_i \int f(\mathbf{p}) d^3\mathbf{p} = 4\pi g_i \int f(E) \sqrt{E^2 - m_i^2} E dE.$$ (5.8)

When obtaining the second relation, we integrated over angles and used the fact that

$$E dE = |\mathbf{p}| d|\mathbf{p}|,$$ (5.9)

following from (5.5). The factor g_i in (5.8) is the number of spin states (degrees of freedom). As an example, photons, electrons and positrons have $g_\gamma = g_{e^-} = g_{e^+} = 2$, for neutrinos and antineutrinos $g_\nu = g_{\bar\nu} = 1$, while the overall number of degrees of freedom of W^+-, W^--, Z^0-bosons and the Higgs boson in the Standard Model

---

[2]It is important here that quantum numbers $Q^{(i)}$ are independent and form a complete set.

at $T \gtrsim 100\,\text{GeV}$ is equal to 10: massive W^+-, W^--, Z^0-bosons have 3 polarizations each, while the Higgs boson adds one degree of freedom.[3]

It follows from the form of the distribution function (5.4) that the difference of number densities of particles and their antiparticles depends on chemical potential. These differences are typically very small at high enough temperatures as compared to number densities themselves. As an example, at temperatures $T \gtrsim 1\,\text{GeV}$ the asymmetry between quarks and antiquarks is of order 10^{-10}; see Sec. 1.5.5. The relative difference of electron and positron numbers is of the same order: the Universe is electrically neutral, so the net positive charge of quarks is compensated by the negative charge of electrons. Hence, chemical potentials are indeed very small in the early Universe and can often be neglected.

Energy density ρ_i for particle species i is given by the following integral

$$\rho_i = g_i \int f(\mathbf{p}) E(\mathbf{p}) d^3\mathbf{p} = 4\pi g_i \int f(E)\sqrt{E^2 - m_i^2}\, E^2 dE, \qquad (5.10)$$

where we again integrated over angles and used (5.9).

To find the expression for pressure, let us consider surface element ΔS orthogonal to the third axis. The number of particles of momenta between $\mathbf{p}$ and $\mathbf{p}+d\mathbf{p}$ hitting this surface during time Δt is

$$\Delta n = v_z f(\mathbf{p}) d^3\mathbf{p}\, \Delta S \Delta t,$$

where

$$v_z = \frac{p_z}{E} > 0$$

is the third component of velocity. Upon elastic scattering off the surface, particle whose third component of momentum is p_z transfers to the surface momentum

$$\Delta p_z = 2p_z.$$

Pressure is equal to the ratio of the total transfered momentum to the time Δt and area ΔS, so we have

$$
\begin{aligned}
p_i &= g_i \int_{p_z>0} 2\frac{p_z^2}{E} f(\mathbf{p}) d^3\mathbf{p} \\
&= \frac{4\pi g_i}{3} \int_0^\infty \frac{|\mathbf{p}|^4 d|\mathbf{p}|}{E(\mathbf{p})} f(\mathbf{p}) = \frac{4\pi g_i}{3} \int_0^\infty f(E)(E^2 - m_i^2)^{3/2} dE,
\end{aligned}
\qquad (5.11)
$$

---

[3] We show in Chapter 10 that electroweak symmetry is restored at high temperatures, and W- and Z-bosons are massless. So, another way of counting the degrees of freedom is more adequate: two degrees of freedom due to each of the W^+-, W^--, Z^0-bosons and four due to the complex Higgs doublet.

where we made use of the fact that only half of particles move towards the surface, and that on average

$$\langle p_z^2 \rangle = \frac{1}{3} \langle \mathbf{p}^2 \rangle = \frac{1}{3} \langle E^2 - m^2 \rangle,$$

due to spatial isotropy.

Problem 5.2. *Derive the expression* (5.11) *by calculating the energy–momentum tensor for gas of non-interacting relativistic particles whose distribution function is* $f(E)$.

Let us now study the expressions for number density, energy density and pressure in physically interesting limiting cases. We begin with gas of relativistic particles,

$$T \gg m_i$$

at zero chemical potential,

$$\mu_i = 0.$$

Expression (5.10) gives then the Stefan–Boltzmann law,

$$\rho_i = \frac{g_i}{2\pi^2} \int \frac{E^3}{e^{E/T} \mp 1} dE = \begin{cases} g_i \dfrac{\pi^2}{30} T^4 - \text{Bose} \\[2mm] \dfrac{7}{8} g_i \dfrac{\pi^2}{30} T^4 - \text{Fermi} \end{cases}. \tag{5.12}$$

The calculation of the integrals used in this Chapter is given, e.g., in the book [78]; we collect the relevant formulas in the end of this Section. As one should have expected on dimensional grounds, the energy density is equal to T^4 up to a numerical coefficient; the contribution of fermions differs from that of bosons by a factor 7/8. Note that relativistic particles are conveniently characterized by their helicity — spin projection onto the momentum direction; the parameter g_i is then the number of helicity states.

If there are various relativistic particles of one and the same temperature T, and chemical potentials are negligible, the energy density of the relativistic component is

$$\rho = g_* \frac{\pi^2}{30} T^4, \tag{5.13}$$

where

$$g_* = \sum_{\substack{\text{bosons} \\ \text{with } m \ll T}} g_i + \frac{7}{8} \sum_{\substack{\text{fermions} \\ \text{with } m \ll T}} g_i \tag{5.14}$$

is the effective number of degrees of freedom.

The expression (5.11) for pressure has a simple form in the relativistic case,

$$p_i = \frac{g_i}{6\pi^2} \int \frac{E^3}{e^{E/T} \mp 1} dE = \frac{\rho_i}{3}. \tag{5.15}$$

Thus, equation of state of relativistic matter is

$$p = \frac{1}{3}\rho.$$

This implies, in particular, that the sound speed squared is

$$u_s^2 = \frac{\partial p}{\partial \rho} = \frac{1}{3}.$$

Problem 5.3. *Show that the relation $p = \rho/3$ is valid for gas of relativistic particles which is not necessarily in thermal equilibrium.*

Finally, the expression (5.8) for number density gives in the relativistic case

$$n_i = \frac{g_i}{2\pi^2} \int \frac{E^2}{e^{E/T} \mp 1} dE = \begin{cases} g_i \dfrac{\zeta(3)}{\pi^2} T^3 & - \text{ Bose} & (5.16a) \\[2mm] \dfrac{3}{4} g_i \dfrac{\zeta(3)}{\pi^2} T^3 & - \text{ Fermi} & (5.16b) \end{cases}.$$

Numerical value for zeta-function here is

$$\zeta(3) \approx 1.2.$$

Making use of (5.12), (5.16) we find the average energy per particle,

$$\langle E \rangle = \frac{\rho_i}{n_i} \approx \begin{cases} 2.70\, T & - \text{ Bose} \\ 3.15\, T & - \text{ Fermi} \end{cases}. \tag{5.17}$$

As a simple example of application of these formulas, let us estimate the maximum temperature at which relativistic cosmic matter is in thermal equilibrium with respect to electromagnetic interactions. We consider here the high temperature case, $T \gg 1\,\mathrm{MeV}$, when electrons are relativistic. Relevant processes are Compton scattering, e^+e^--annihilation (Fig. 5.1), etc. Amplitudes of these processes are proportional to $\alpha \equiv e^2/(4\pi)$, and the cross-sections to α^2. Together with dimensional

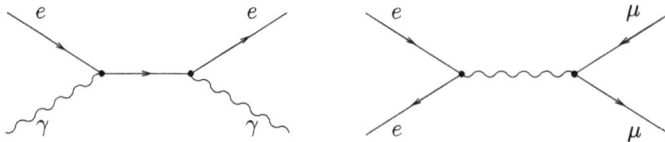

Fig. 5.1.   Feynman diagrams for Compton scattering and e^+e^--annihilation.

arguments, this gives for the rate (inverse of time between scatterings of a single particle with other particles in the medium)

$$\Gamma \sim \alpha^2 T.$$

The processes are in thermal equilibrium if this rate exceeds the cosmological expansion rate,

$$\Gamma \gg H(T).$$

We know from Sec. 3.2.2 that

$$H(T) = \frac{T^2}{M_{Pl}^*}.$$

Hence, the cosmic plasma is in thermal equilibrium at

$$T \ll \alpha^2 M_{Pl}^* \sim 10^{14}\,\text{GeV}.$$

Very similar but not exactly the same estimate is valid for weak and strong (QCD) interactions. Thus, the cosmic medium is indeed in thermal equilibrium during long epoch of the cosmological evolution. As we mentioned above, most interesting are in fact *inequilibrium* phenomena; we discuss many of them in this book.

Let us now consider the dilute gas of non-relativistic particles, whose distribution function is given by (5.6). In this case the expression for particle number density is

$$n_i = g_i \left(\frac{m_i T}{2\pi}\right)^{3/2} e^{\frac{\mu_i - m_i}{T}}, \tag{5.18}$$

while energy and pressure are

$$\rho_i = m_i n_i + \frac{3}{2} n_i T_i \tag{5.19}$$

and

$$p_i = T n_i \ll \rho_i. \tag{5.20}$$

As anticipated, non-relativistic gas has equation of state $p = 0$ up to corrections of order $\mathcal{O}(T/m)$.

Problem 5.4. *Show that at temperature exceeding the mass and chemical potential, $T \gg m, \mu$, the difference of the number densities of particles and their antiparticles of a given helicity is*

$$\Delta n = \mu \frac{T^2}{3} - Bose \tag{5.21}$$

and

$$\Delta n = \mu \frac{T^2}{6} - Fermi. \tag{5.22}$$

Hints: (1) *Recall that chemical potentials of particles and their antiparticles differ by sign only;* (2) *Take linear in μ term in the integral* (5.8) *with $m = 0$, integrate by parts and use formulas presented in the end of this Section.*

Problem 5.5. *Find the differences of the number densities of all relativistic Standard Model particles and their antiparticles in the cosmic medium at temperature $T = 400\,MeV$ for given densities of baryon and lepton numbers n_B, n_{L_e}, n_{L_μ}, n_{L_τ} in the realistic case $n_B, n_{L_e}, n_{L_\mu}, n_{L_\tau} \ll T^3$. Hints:* (1) *At this temperature, relativistic are u-, d-, s-quarks, gluons, photons, electrons, muons and all neutrino species; other Standard Model particles are non-relativistic;* (2) *neutrinos have one helicity, while other fermions have two;* (3) *quarks have three color states, gluons have eight.*

Let us give numerical values of useful integrals.

1.

$$\int_0^\infty \frac{dz}{e^z + 1} = \log 2.$$

2. For positive integer n

$$\int_0^\infty \frac{z^{2n-1}}{e^z - 1} dz = \frac{(2\pi)^{2n}}{4n} B_n,$$

$$\int_0^\infty \frac{z^{2n-1}}{e^z + 1} dz = \frac{2^{2n-1} - 1}{2n} \pi^{2n} B_n,$$

where B_n are Bernoulli numbers,

$$B_1 = \frac{1}{6}, \quad B_2 = \frac{1}{30}, \quad B_3 = \frac{1}{42}, \dots$$

3. For arbitrary x

$$\int_0^\infty \frac{z^{x-1}}{e^z - 1} dz = \Gamma(x)\zeta(x),$$

$$\int_0^\infty \frac{z^{x-1}}{e^z + 1} dz = (1 - 2^{1-x})\Gamma(x)\zeta(x),$$

where $\zeta(x)$ is the Riemann zeta-function, some of whose values are

$$\zeta(3) = 1.202, \quad \zeta(5) = 1.037, \quad \zeta(3/2) = 2.612, \quad \zeta(5/2) = 1.341.$$

Recall that for positive integer n

$$\Gamma(n) = (n-1)!$$

The values of gamma-function $\Gamma(x)$ at half-integer x can be found by making use of the value

$$\Gamma(1/2) = \sqrt{\pi}$$

and the relation $\Gamma(1 + x) = x\Gamma(x)$.

5.2. Entropy in Expanding Universe. Baryon-to-Photon Ratio

One of the main thermodynamical characteristics of a system is its entropy. Hence, it is useful to discuss entropy of cosmic medium in the expanding Universe. Recall that entropy in thermodynamics enters the first law. In the general case of variable number of particles the latter has the form

$$dE = TdS - pdV + \sum_i \mu_i dN_i, \qquad (5.23)$$

where S is entropy of the system and subscript i labels particle species. This subscript and the corresponding summation will be omitted wherever possible.

Energy E and number of particles are extensive quantities (proportional to volume of the system), while temperature and pressure are local characteristics independent of the volume. In accordance to the first law (5.23) entropy is extensive quantity. It is convenient to introduce densities

$$\rho \equiv \frac{E}{V}, \quad n \equiv \frac{N}{V}, \quad s \equiv \frac{S}{V}. \qquad (5.24)$$

Then the first law (5.23) is written as follows:

$$(Ts - p - \rho + \mu n)\, dV + (Tds - d\rho + \mu dn)\, V = 0. \qquad (5.25)$$

This relation is valid for both the entire system and any of its part. Using it for a region of constant volume inside the system we obtain

$$Tds = d\rho - \mu dn.$$

We now use (5.25) for the entire system and find for entropy density

$$s = \frac{p + \rho - \mu n}{T}.$$

An important example is relativistic matter at vanishing chemical potentials, for which

$$s = \frac{p + \rho}{T}. \qquad (5.26)$$

Making use of the latter relation and expressions of the previous Section for ρ and p we obtain the following expression for the contribution of ith particle species,

$$s_i = \frac{4}{3}\frac{\rho_i}{T} = \begin{cases} g_i \dfrac{2\pi^2}{45} T^3 & - \text{ Bose} \\[2ex] \dfrac{7}{8} g_i \dfrac{2\pi^2}{45} T^3 & - \text{ Fermi} \end{cases}. \qquad (5.27)$$

By comparing these expressions to formulas (5.16) for number density, we see that entropy and number of particles differ in relativistic case by a numerical factor of order 1 only.

To obtain the expression for entropy in non-relativistic case, we make use of (5.19) and (5.20) and write

$$s_i = \frac{5}{2} n_i + \frac{m_i - \mu_i}{T} n_i.$$

The chemical potential is related to the number density by Eq. (5.18), so we find finally

$$s_i = n_i \left\{ \frac{5}{2} + \log \left[\frac{g_i}{n_i} \left(\frac{m_i T}{2\pi} \right)^{3/2} \right] \right\}.$$

Number density of non-relativistic particles is always small in cosmic medium; at temperatures $T \lesssim 100\,\text{MeV}$, when protons and neutrons are non-relativistic, their number density is estimated as $n_B \sim 10^{-9} n_\gamma$; see (1.21). The same estimate applies to electrons which are non-relativistic at $T \lesssim 0.5\,\text{MeV}$. Thus, non-relativistic contribution to entropy is negligible, so the total entropy is given by

$$s = g_* \frac{2\pi^2}{45} T^3, \tag{5.28}$$

where the effective number of relativistic degrees of freedom is defined by (5.14).

One of the key properties of entropy making it a very useful quantity is the second law of thermodynamics. According to this law, entropy of any closed system can only increase, and it stays constant for equilibrium evolution, i.e., slow evolution during which the system always remains in thermal equilibrium. Let us see how this law works in expanding Universe, assuming that the evolution of cosmic medium is close to equilibrium. To this end, we come back to the relation (5.23) and recall that it is adequate to introduce chemical potentials to conserved quantum numbers, rather than to individual particle species; thus, dN_i in (5.23) should be understood as differentials of conserved quantum numbers. With this qualification, we apply the relation (5.23) to the comoving volume, $V = a^3$. Then the conservation of quantum numbers in comoving volume gives $dN_i = 0$, so that

$$T\frac{dS}{dt} \equiv T\frac{d(a^3 s)}{dt} = (p + \rho)dV + Vd\rho = a^3 \left[(p + \rho) \cdot 3\frac{\dot{a}}{a} + \dot{\rho} \right] = 0, \tag{5.29}$$

where we used the covariant conservation of energy in expanding Universe, Eq. (3.13). We see that total entropy in comoving volume is conserved,

$$sa^3 = \text{const.} \tag{5.30}$$

As we have already mentioned in Sec. 3.1, the covariant conservation of energy has simple interpretation in terms of the entropy conservation in comoving volume.

Entropy conservation is used for introducing quantitative, time-independent characteristics of asymmetries of conserved quantum numbers. In particular, as we discussed in Chapter 1, there is baryon asymmetry in the Universe, as there are protons and neutrons but none of their antiparticles today. At temperatures below

$T \sim 100 \, \text{GeV}$, there are no processes violating baryon number, so it is conserved in comoving volume,[4]

$$(n_B - n_{\bar{B}})a^3 = \text{const,} \qquad (5.31)$$

where n_B and $n_{\bar{B}}$ are number densities of baryons and antibaryons. By comparing (5.30) and (5.31) we see that ratio

$$\Delta_B = \frac{n_B - n_{\bar{B}}}{s} \qquad (5.32)$$

is a time-independent characteristic of the baryon asymmetry.

Let us give an estimate for the baryon asymmetry in the early Universe. When discussing the Universe at relatively low temperatures ($T \lesssim 100 \, \text{keV}$), one traditionally uses *baryon-to-photon ratio*

$$\eta_B = \frac{n_B}{n_\gamma},$$

where n_γ is the photon number density. At these temperatures η_B is independent of time, and it differs from Δ_B by a numerical factor of order 1. The latter factor is due to both the difference between n_γ and photon entropy s_γ (see (5.16) and (5.28)), and the neutrino contribution to entropy (see Chapter 7). At $T \lesssim 100 \, \text{keV}$ (but, strictly speaking, at temperatures exceeding the neutrino masses) entropy density is

$$s = \frac{2\pi^2}{45} g_{*,0} T^3, \qquad (5.33)$$

where

$$g_{*,0} = 2 + \frac{7}{8} \cdot 2 \cdot 3 \cdot \frac{4}{11} = \frac{43}{11}$$

is the effective number of relativistic degrees of freedom after neutrino decoupling.[5] The first and second terms here are due to photons and neutrinos, respectively (effective temperature of neutrinos is related to photon temperature $T_\gamma \equiv T$ by (4.14); the origin of the factor $\frac{7}{8} \cdot 2 \cdot 3$ is the same as in (4.15)).

Photons and neutrinos are non-interacting in the present Universe, and neutrinos have masses. Therefore, the formula (5.33) is not valid at late times. However, the combination aT remains constant starting from temperatures of order 100 keV (below electron mass) and up until now. Therefore, the relation (5.30) is valid for late Universe, if one *defines* the quantity s by Eq. (5.33) for all temperatures below

---

[4]In some scenarios, baryon asymmetry is *generated* at temperatures below 100 GeV, see, e.g., Sec. 11.6. We do not consider this possibility here.
[5]Note the difference between the parameters $g_{*,0}$ and $\hat{g}_*$, the latter entering (4.22).

100 keV. With this definition, the present value of entropy density is

$$s_0 = \frac{2\pi^2}{45} g_{*,0} T_0^3 = 2.9 \cdot 10^3 \text{ cm}^{-3}. \tag{5.34}$$

Thus, we have

$$\Delta_B = \frac{n_\gamma}{s} \eta_B = \frac{\frac{2\zeta(3)}{\pi^2}}{\frac{2\pi^2}{45} g_{*,0}} \cdot \eta_B = 0.14 \cdot \eta_B. \tag{5.35}$$

The value of baryon-to-photon ratio is obtained from studies of BBN and CMB. These two independent methods agree with each other and give [1]

$$\eta_B = (6.05 \pm 0.07) \cdot 10^{-10}. \tag{5.36}$$

In terms of the baryon asymmetry

$$\Delta_B = 0.86 \cdot 10^{-10}.$$

As we discussed in the previous Section, this means, in particular, that baryon chemical potential is very small at $T \gtrsim 200$ MeV.

Problem 5.6. *Estimate the value of the chemical potential for u-quark at temperature 1 GeV.*

To conclude this Section we make two comments. One is that the baryon-to-photon ratio is directly related to the relative energy density of baryons $\Omega_B = \rho_B/\rho_c = m_B n_B/\rho_c$, see Chapter 4. The present number of photons calculated according to (5.16a) is

$$n_\gamma = 2\frac{\zeta(3)}{\pi^2} T_0^3 = 411 \text{ cm}^{-3}.$$

Making use of the value (1.14) of critical density, we find from (1.21)

$$\Omega_B h^2 = 0.0221 \pm 0.0003 , \tag{5.37}$$

and hence $\Omega_B \simeq 0.05$, the value quoted in Sec. 4.1.

Problem 5.7. *Estimate the contribution of electrons and protons into total energy at $T = 50$ keV and $T = 1$ eV. Hint: Recall that electrons and protons are non-relativistic at these temperatures. Neglect the presence of light nuclei in the plasma.*

The second comment concerns the baryon–electron–photon plasma before recombination. As we discuss in Section 6.3, baryons in it are in thermal equilibrium with photons. The sound speed squared in this plasma is

$$u_s^2 = \frac{dp}{d\rho} = \frac{dp_\gamma}{d\rho_\gamma + d\rho_B} = \frac{4p_\gamma}{4\rho_\gamma + 3\rho_B} = \frac{1}{3(1 + R_B)},$$
(5.38)

where we use $p_\gamma = \rho_\gamma/3 \propto T^4$, $\rho_B \propto T^3$ and introduce the time-dependent parameter

$$R_B = \frac{3\rho_B}{4\rho_\gamma}.$$

At the recombination epoch[6] one has $T_r = 2970$ K and

$$R_B = \frac{3}{4} \cdot 0.75 \cdot \frac{m_p \eta_B n_\gamma}{\rho_\gamma} \simeq 0.46.$$
(5.39)

It follows from (5.38) and (5.39) that the sound speed before recombination is somewhat different from the relativistic value $u_s = 1/\sqrt{3}$. This fact is, of course, accounted for in the data analyis, see Section 1.7.1 in this regard.

Problem 5.8. *Estimate the contributions of electrons and protons to entropy density in the Universe at* $100\,keV \gg T \gg 1\,eV$. *Hint: at these temperatures electrons and protons are non-relativistic and do not combine into atoms (recombination has not happened yet); the number densities of electrons and protons are equal; neglect the helium and other nuclei and atomes.*

5.3.   *Models with Intermediate Matter Dominated Stage: Entropy Generation

Let us consider a possibility that at some early epoch, before relatively late radiation domination, the expansion of the Universe was matter dominated. There are at least two scenarios for how that can happen. One of them assumes that there exist new heavy particles $\tilde{G}$ in Nature whose lifetime τ is large enough, and that in the beginning of the Hot Big Bang epoch the expansion is radiation dominated, but the cosmic medium contains an admixture of $\tilde{G}$-particles. $\tilde{G}$-particles are non-relativistic at temperature below their mass. As the Universe expands, energy density of the relativistic component decreases as a^{-4}, while energy density of non-relativistic $\tilde{G}$-particles decreases as a^{-3}. Therefore, at a certain moment of time t_* the expansion becomes dominated by non-relativistic $\tilde{G}$-particles, and the intermediate matter

---

[6]We take into account the fact that recombination of helium occurs at much higher temperature, so helium atoms decouple from photons and do not contribute to R_B. The ratio of the number of baryons in helium to the total number of baryons (helium mass fraction) is about 25%, see Chapter 8, hence the factor 0.75 in (5.39).

dominated stage begins. The main assumption here is that

$$t_* \ll \tau \equiv \Gamma^{-1}, \tag{5.40}$$

where Γ is the total decay width of $\tilde{G}$-particle. During later evolution of the Universe, an important process is $\tilde{G}$-particle decay into light particles, say, quarks, leptons and other Standard Model particles. These rapidly (faster than in one Hubble time) thermalize and become part of the relativistic component. We will see that the epoch of $\tilde{G}$-particle domination ends up soon after the expansion rate slows down to $H \sim \Gamma$: at about that time, all $\tilde{G}$-particles decay away, the Universe becomes filled with hot relativistic matter and the radiation domination resumes. We note that thermal equilibrium is strongly violated in the time interval $t_* < t < \Gamma^{-1}$: the concentration of $\tilde{G}$-particles is very far from equilibrium. This property is of course a consequence of (5.40); it gives rise to strong entropy generation in the Universe.

This scenario is realized in some models with heavy gravitino, see Sec. 9.6.3, hence our notation for the new heavy particles. Another example is unstable sterile neutrino.

The second scenario makes use of the observation that the role of non-relativistic matter in cosmology may be played by massive spatially homogeneous scalar field oscillating near the minimum of its scalar potential, see Sec. 4.8.1. As we discussed in Sec. 4.8.1, this field can be viewed as a collection of large number of particles at rest. It is assumed that before the Hot Big Bang epoch, all energy is due to this field, and its oscillations begin at time t_* (before that the field evolves in the slow roll regime, see Sec. 4.8.1). Hence, there is no relativistic matter at all at time t_*. One assumes further that the decay width of the scalar particles again obeys $t_* \ll \Gamma^{-1}$, and that these decays is the main mechanism of the decay of the scalar field oscillations. In fact, the latter assumption is pretty strong: there are alternative mechanisms leading to the decay of coherent oscillations of scalar fields, which are often more efficient. We discuss some of these mechanisms in the accompanying book, and here we will proceed under the above assumption. Then, like in the first scenario, the Universe becomes radiation dominated when $H(t) \sim \Gamma$.

The analysis of the two scenarios goes in parallel up to some point. Let ρ_M and ρ_{rad} be energy densities of non-relativistic and relativistic component, respectively. Let us assume for simplicity that the effective number of relativistic degrees of freedom g_* does not change in time (the opposite case is the subject of the problem in the end of this Section). The number of heavy particles in comoving volume $(n_M a^3)$ decreases only due to their decays, hence

$$\frac{d}{dt}(n_M a^3) = -\Gamma n_M a^3.$$

Since $\rho_M \propto n_M$, we find

$$\dot{\rho}_M + 3H\rho_M = -\Gamma \rho_M. \tag{5.41}$$

Energy density of relativistic particles decreases due to the cosmological expansion, since their number density gets diluted and energy of each particle gets redshifted. However, energy is injected into this component due to decays of heavy particles. Hence, equation for ρ_{rad} has the form

$$\dot\rho_{rad} + 4H\rho_{rad} = \Gamma\rho_M. \tag{5.42}$$

The third equation that closes the system of evolution equations is the Friedmann equation

$$H^2 = \frac{8\pi}{3}G(\rho_M + \rho_{rad}). \tag{5.43}$$

Our purpose is to study the solutions to the system (5.41)–(5.43).

Problem 5.9. *Making use of Eqs. (5.41) and (5.42) show that the total energy and pressure (sums of energies and pressures of the two components) obey the covariant conservation equation $\dot\rho_{tot} = -3H(\rho_{tot} + p_{tot})$.*

In the first place, the solution to (5.41) is

$$\rho_M(t) = \frac{\text{const}}{a^3(T)}e^{-\Gamma t}. \tag{5.44}$$

At both radiation and matter domination one has $t \sim H^{-1}$, so the energy density of non-relativistic component is small at late times, when $\Gamma \gg H(t)$. In this regime the right-hand side of Eq. (5.42) can be neglected, and the energy density of relativistic matter falls as a^{-4}. This is power-law rather than exponential decrease, hence at $\Gamma \gg H(t)$ one has $\rho_{rad} \gg \rho_M$. Thus, the expansion is indeed radiation dominated at late times. The time of transition from matter domination $t_{MD\to RD}$ is roughly estimated as

$$\Gamma \sim H(t_{MD\to RD}). \tag{5.45}$$

The above reasoning does not exclude the possibility that the right-hand side of (5.45) contains a logarithmically large factor, so that the transition occurs somewhat later; neither it excludes earlier transition, at time when $\Gamma \ll H$. We will see that neither of this possibilities is realized.

Let us consider solutions to Eqs. (5.41)–(5.43) after the beginning of matter dominated stage, $t > t_*$, but at sufficiently early times when $\Gamma \ll H(t)$. In other words, we are interested in the time interval

$$t_* < t \ll \Gamma^{-1}.$$

Let us assume that the expansion is matter dominated in the whole interval; we will justify this assumption *a posteriori*. At these times the exponential factor in (5.44) is nearly equal to 1, i.e., the loss of particles due to their decays is negligible. Hence, the evolution of the Universe is described by formulas of Sec. 3.2.1; in particular, $a \propto t^{2/3}$ and

$$H = \frac{2}{3t}, \quad \rho_M = \frac{3}{8\pi G}H^2 = \frac{1}{6\pi G t^2}.$$

By substituting the latter expressions into (5.42), we obtain the equation

$$\dot{\rho}_{rad} + \frac{8}{3t}\rho_{rad} = \frac{\Gamma}{6\pi G}\frac{1}{t^2}.$$

Its general solution is

$$\rho_{rad} = \frac{\Gamma}{10\pi G}\frac{1}{t} + \frac{C}{t^{8/3}}, \tag{5.46}$$

where C is an arbitrary constant. This constant is determined by initial data and it is different in the two scenarios outlined above.

Let us begin with the first of them. We will see that at the time t_* of the transition to matter domination, the first term in (5.46) is small, and the property that $\rho_{rad}(t_*) \simeq \rho_M(t_*)$ gives

$$\frac{C}{t_*^{8/3}} \simeq \frac{1}{6\pi G t_*^2}.$$

Hence, the behavior of ρ_{rad} at $t \gg t_*$ is

$$\rho_{rad} = \frac{\Gamma}{10\pi G}\frac{1}{t} + \frac{1}{6\pi G}\frac{t_*^{2/3}}{t^{8/3}} \equiv \rho_{gen} + \rho_{init}. \tag{5.47}$$

The first term here is due to injection of energy through decays of heavy particles, while the second is the energy density of the relativistic matter existing in the Universe initially. Note that the first term decreases in time relatively slowly, while the second term decays as a^{-4}, as should be the case. Our basic assumption $\Gamma \ll t_*^{-1}$ implies that the first, induced term is indeed small at $t = t_*$; it starts to dominate at $t \simeq t_*^{2/5}\Gamma^{-3/5} \ll \Gamma^{-1}$, i.e., long before the transition back to radiation domination. At the latter time, efficient generation of entropy begins.

Let us point out that ρ_{rad}, and hence the temperature of relativistic component, monotonically decreases in time. This may be a surprise: one might naively expect that maximum temperature in the Universe is reached at $H(t) \sim \Gamma$, when most of heavy particles decay. The above analysis shows that the smallness of the relative number of heavy particle decays at early times is compensated for by high number density of these particles, so the contribution of the decays into ρ_{rad} is not small at small t.

Problem 5.10. *Show that temperature of the hot component monotonically decreases at $H(t) \lesssim \Gamma$ as well.*

In the second scenario, $\rho_{rad} = 0$ at $t = t_*$, which gives

$$\rho_{rad} = \frac{\Gamma}{10\pi G}\left(\frac{1}{t} - \frac{t_*^{5/3}}{t^{8/3}}\right). \tag{5.48}$$

Energy density of hot matter rapidly (during time of order t_*) increases from zero to its maximum value[7]

$$\rho_{rad,max} \sim \frac{\Gamma}{10\pi G t_*},$$

and then decreases as t^{-1}. The maximum temperature is reached soon after the beginning of matter dominated stage.

In both scenarios one has $\rho_{rad} \ll \rho_M$ at times $t \ll \Gamma^{-1}$. The energy density of hot component approaches that of heavy particles as t approaches $t_{MD\rightarrow RD} \sim \Gamma^{-1}$, i.e., the time when $H \sim \Gamma$. This means that the relation (5.45) is valid without large logarithms. Right after the transition to radiation domination one can use formulas of Sec. 3.2.2 for estimates, so the relation (5.45) can be written as

$$\Gamma \sim \frac{T_{MD\rightarrow RD}^2}{M_{Pl}^*}.$$

It is known from the analysis of Big Bang Nucleosynthesis (see Chapter 8) that the expansion was radiation dominated at temperatures of at least $T_{BBN} \sim 1\,\text{MeV}$, so there is a bound $T_{MD\rightarrow RD} > T_{BBN}$. This bound gives the bound on the lifetime of heavy particles,

$$\tau = \Gamma^{-1} \lesssim \frac{M_{PL}^*(T_{BBN})}{T_{BBN}^2} \simeq 1\,\text{s}. \tag{5.49}$$

This bound is important, e.g., for some models with heavy gravitino, see Sec. 9.6.3.

Decays of heavy particles is a strong source of entropy in the above scenarios. In the first of them, entropy density at the time of the transition back to radiation domination is mostly due to these decays, and its value is $s_{gen} \sim \rho_{gen}^{3/4}(t = \Gamma^{-1})$. If there were no heavy particles, the entropy density would be equal to $s_{init} \sim \rho_{init}^{3/4}(t = \Gamma^{-1})$, see (5.47). This gives the entropy growth factor

$$\frac{s_{gen}}{s_{init}} \sim \frac{1}{\sqrt{\Gamma t_*}}.$$

Clearly, this factor is large for $t_* \ll \Gamma^{-1}$ i.e., entropy generation is very effective. In the second scenario, there is no entropy in the beginning at all, so all entropy comes from decay processes.

---

[7]Since the reheating is fast, the approximation of instantaneous beginning of matter dominated stage is not adequate. For this reason, the numerical coefficient in (5.48) depends on the details of evolution at the beginning of that stage.

Problem 5.11. *Generalize the analysis of this Section to the case when effective number of relativistic degrees of freedom g_* changes in time. Hint: Instead of Eq. (5.42), derive and use similar equation for entropy of the hot component.*

5.4. *Inequilibrium Processes

We often encounter in this book a situation when most processes in cosmic medium are fast, but there is one or several slow processes whose low rate keeps the medium out of complete thermal equilibrium.

Let us give a typical (but by no means unique) example of such a situation. Let there exist new long living heavy particle X (say, of mass $m_X \gtrsim 100\,\text{GeV}$), and the only processes by which the number of X-particles and their antiparticles $\bar{X}$ in comoving volume can change are their pair creation and annihilation. Let us assume that scattering of X-particles off the Standard Model particles (quarks, leptons, photons, etc.) occurs at sufficiently high rate. At temperature T exceeding the mass of X-particles m_X, creation and annihilation of X-particles are typically also rapid, so they have equilibrium abundance. Thus, the system is in complete thermal equilibrium at $T \gtrsim m_X$. Let us consider symmetric medium with equal number of particles X and their antiparticles $\bar{X}$. At $T \ll m_X$, the number density of X-particles exponentially decays as temperature decreases, and the annihilation of X-particles occurs less and less often simply because the probability for an X-particle to meet its antiparticle becomes smaller and smaller. This is an example of the situation we are going to consider: at $T \ll m_X$ all processes except for annihilation and creation of X-$\bar{X}$-pairs are fast, while the latter processes are slow, and the abundances n_X and $n_{\bar{X}} = n_X$ may differ from the equilibrium abundance. In this case the medium still has well-defined temperature and numbers of particles (including numbers of X and $\bar{X}$ themselves). The distribution functions are given by formulas of Sec. 5.1; the fact that the system is out of thermal equilibrium is reflected by the property that $\mu_X \neq 0$, and $\mu_X = \mu_{\bar{X}}$ for symmetric matter (recall that $\mu_X = 0$ for symmetric matter in equilibrium).

Leaving aside for a while the cosmological expansion, let us discuss how the system relaxes to thermal equilibrium in these cases. In general, as the system relaxes, its effective temperature and number densities of all particles evolve in time. Temperature changes, if the relaxation occurs with heat release; this is the case in our example if the X-particle abundance is higher than in equilibrium, $n_X > n_X^{eq}$, so the relaxation proceeds via annihilation (rather than creation) of X-$\bar{X}$-pairs. Still, at any moment of time, temperature and chemical potentials take well-defined values. So, at any time the system has a certain free energy $F(T, N_i)$.[8] Let us recall

---

[8]It is often more convenient to use Grand potential rather than free energy, considering chemical potentials rather than particle numbers as the thermodynamical parameters. In this Section, however, we will discuss states with fixed number of particles. The formula (5.50) is valid precisely for these states.

that free energy is related to the number of states (partition function) by

$$Z = e^{-F/T}, \tag{5.50}$$

where we assume that the system has finite, albeit large, volume. As the system approaches thermal equilibrium, free energy tends to its minimum, and partition function to its maximum.

The system relaxes to thermal equilibrium due to "direct" slow process (in our example this is X-$\bar{X}$-annihilation; we assume for definiteness that $n_X > n_X^{eq}$). Let us denote the rate of these processes (number of their occurrences per unit time per unit volume) by Γ_+.[9] There are of course inverse processes whose rate is denoted by Γ_- (these are processes of X-$\bar{X}$-pair creation in our example). Let us find the relation between Γ_+ and Γ_-.

As a result of a single direct process the system makes a transition from a state with smaller partition function Z_-, and hence larger free energy F_-, to the state with larger partition function Z_+ and smaller free energy F_+. Let i_- be one of *microscopic* states[10] before the direct process, and j_+ be one of microscopic states that can appear after the direct process. The probability that the system is in the initial state i_- is

$$P(i_-) = \frac{e^{-E(i_-)/T}}{Z_-},$$

where $E(i_-)$ is the energy of the state i_- (It is important here that the system is in thermal equilibrium with respect to all processes except for the one we consider.) Let $\gamma(i_- \to j_+)$ be the probability of the process $i_- \to j_+$. Then the total probability of the direct process in the system is

$$\Gamma_+ = \sum_{i_-,j_+} P(i_-) \cdot \gamma(i_- \to j_+) = \frac{1}{Z_-} \sum_{i_-,j_+} e^{-E(i_-)/T} \gamma(i_- \to j_+).$$

Similarly, the total probability of an inverse process is

$$\Gamma_- = \frac{1}{Z_+} \sum_{i_-,j_+} e^{-E(j_+)/T} \gamma(j_+ \to i_-).$$

The probabilities of the process $i_- \to j_+$ and the inverse process $j_+ \to i_-$ are equal to each other, $\gamma(j_+ \to i_-) = \gamma(i_- \to j_+)$. Making use of this fact, and also the

[9]The notation has to do with the fact that the direct process increases the value of the partition function; this process is thermodynamically favored.
[10]As an example, in classical mechanics this is a state in which position and velocity of every particle are well-defined.

energy conservation, $E(i_-) = E(j_+)$, we obtain

$$\frac{\Gamma_+}{\Gamma_-} = \frac{Z_+}{Z_-} = e^{(F_+ - F_-)/T},$$

and finally

$$\frac{\Gamma_+}{\Gamma_-} = e^{-\Delta F/T}, \qquad (5.51)$$

where ΔF is the difference of free energies after and before the direct process; according to our definitions, $\Delta F < 0$. In what follows, Γ_+ and Γ_- denote the number of direct and inverse processes occurring *per unit time per unit volume*. Clearly, the formula (5.51) is valid for these quantities too. The relation (5.51) is known as the general formula of detailed balance.

In our example, we have

$$\Delta F = \frac{\partial F}{\partial N_X} \Delta N_X + \frac{\partial F}{\partial N_{\bar{X}}} \Delta N_{\bar{X}},$$

where $\Delta N_X = \Delta N_{\bar{X}} = -1$ is the change in the number of X- and $\bar{X}$-particles in a single direct process (annihilation). Hence,

$$\Delta F = -2\mu_X, \qquad (5.52)$$

where we made use of the facts that $\partial F/\partial N_X = \mu_X$ and $\mu_{\bar{X}} = \mu_X$.

The relation (5.51) shows, in the first place, that free energy indeed decreases during relaxation: $\Gamma_+ > \Gamma_-$ for $\Delta F < 0$. Now, if $|\Delta F| \gg T$ then $\Gamma_+ \gg \Gamma_-$, so the inverse processes are negligible. Conversely, if $|\Delta F| \ll T$ then the deviation from thermal equilibrium is small, $(\Gamma_+ - \Gamma_-) \ll \Gamma_\pm$, so direct and inverse processes occur at almost equal rates. In the latter case, the rates are approximately the same as in equilibrium $\Gamma_+ \approx \Gamma_- \approx \Gamma^{eq}$, so the relaxation rate is given by

$$\Gamma_+ - \Gamma_- = -\frac{\Delta F}{T}\Gamma^{eq}, \quad |\Delta F| \ll T. \qquad (5.53)$$

In the situations we discuss, it is often possible to introduce local (independent of volume V) parameter characterizing the medium, $n = N/V$. In our example this is the number density of X-particles, while N is the total number of X-particles in the system. The quantity N does not change in fast processes and changes by ΔN and $(-\Delta N)$ in each direct and inverse process. Still leaving aside the expansion of the Universe, let us write down the equation for $n(t)$,

$$\frac{dn}{dt} = \Delta N \cdot (\Gamma_+ - \Gamma_-) \qquad (5.54)$$

(recall that $\Gamma_\pm$ are the rates per unit volume). The free energy as function of N is at minimum in thermal equilibrium, so

$$(\Delta F)^{eq} = \left(\frac{\partial F}{\partial N}\right)^{eq} \cdot \Delta N = 0.$$

Near thermal equilibrium, ΔF is proportional to the deviation of n from its equilibrium value n^{eq}, hence

$$\Delta F = -\alpha \left(n - n^{eq}\right),$$

where α is a positive parameter. Equation (5.54) then takes the form

$$\frac{dn}{dt} = \frac{\alpha}{T} \Delta N \Gamma^{eq} \left(n - n^{eq}\right). \tag{5.55}$$

Since n does not depend on time in thermal equilibrium,

$$\frac{dn^{eq}}{dt} = 0.$$

Equation (5.55) can be written as

$$\frac{d\left(n - n^{eq}\right)}{dt} = \frac{\alpha}{T} \Delta N \Gamma^{eq} \left(n - n^{eq}\right). \tag{5.56}$$

It is now clear that the relaxation to thermal equilibrium has exponential character.[11]

In our example of symmetric medium, the abundance of particles X and $\bar{X}$ at $T \ll m_X$ is given by (5.18) with non-vanishing $\mu_X = \mu_{\bar{X}}$, hence

$$n_X = n_{\bar{X}} = e^{\mu_X/T} n_X^{eq}.$$

Thus, near thermal equilibrium, i.e., at $\frac{\mu_X}{T} \ll 1$, Eq. (5.52) gives

$$\Delta F = -2\mu_X = -2T \frac{n_X - n_X^{eq}}{n_X^{eq}}.$$

We see that $\alpha = \frac{2T}{n_X^{eq}}$ and, since $\Delta N = -1$, the relaxation to thermal equilibrium is described by the equation

$$\frac{d\left(n_X - n_X^{eq}\right)}{dt} = -2 \frac{\Gamma^{eq}}{n_X^{eq}} \left(n_X - n_X^{eq}\right). \tag{5.57}$$

Finally, let us note that the equilibrium rate of annihilation processes per unit volume is

$$\Gamma^{eq} = \Gamma_X^{eq} \cdot n_X^{eq},$$

where $\Gamma_{X,eq} = \tau_X^{-1}$ is the lifetime of X-particle in the equilibrated medium. We get

$$\frac{d\left(n_X - n_X^{eq}\right)}{dt} = -2\Gamma_X^{eq} \left(n_X - n_X^{eq}\right),$$

which is the *Boltzmann equation* for the number density in our example.

---

[11] Note that the right-hand side of Eq. (5.56) is always negative: for $n > n^{eq}$ the direct process decreases n, i.e., $\Delta N < 0$, and vice versa.

The latter result can be obtained also in an elementary way. Let n_X and $n_{\bar{X}}$ be the number densities of particles and antiparticles. Then the probability of annihilation of a given particle X with some antiparticle $\bar{X}$ per unit time is

$$\Gamma_X = \sigma_{\text{ann}} \cdot n_{\bar{X}} \cdot v_X,$$

where v_X is X-particle velocity, and σ_{ann} is the annihilation cross-section (during time $\tau_X = \Gamma_X^{-1}$ a particle of cross-section σ_{ann} meets precisely one antiparticle $\bar{X}$). Hence, the annihilation probability per unit time per unit volume is

$$\Gamma_+ = \Gamma_X \cdot n_X = \sigma_{\text{ann}} \cdot n_{\bar{X}} \cdot v_X \cdot n_X = \Gamma^{eq} \frac{n_{\bar{X}}}{n_{\bar{X}}^{eq}} \frac{n_X}{n_X^{eq}}. \tag{5.58}$$

The probability of the inverse process of X-$\bar{X}$-pair creation is independent of the actual number of X-particles in the medium, so it is given by

$$\Gamma_- = \Gamma^{eq}.$$

Hence

$$\frac{d\left(n_X - n_X^{eq}\right)}{dt} = \Gamma^{eq}\left(1 - \frac{n_{\bar{X}} n_X}{n_{\bar{X}}^{eq} n_X^{eq}}\right).$$

In symmetric medium, $n_{\bar{X}} = n_X$ and $n_{\bar{X}}^{eq} = n_X^{eq}$, so for $|n_X - n_X^{eq}| \ll n_X^{eq}$ we recover Eq. (5.57).

Equation (5.54) is no longer valid in expanding Universe. As the Universe expands, the parameters like number density decrease as a^{-3} even if slow processes are switched off. So, it makes sense to consider the quantity N in comoving volume, $N \propto na^3$. Instead of (5.54), we now have the following equation,

$$\frac{d\left(na^3\right)}{dt} = \Delta N \cdot (\Gamma_+ - \Gamma_-) \cdot a^3, \tag{5.59}$$

where Γ_+ and Γ_- are still the rates of slow processes per unit physical volume, which in general obey the relation (5.51). Clearly, Γ_+ and Γ_- depend on time, since temperature and number densities decrease in time. Near thermal equilibrium, Eq. (5.55) is generalized to

$$\frac{d\left(na^3\right)}{dt} = \frac{\alpha}{T}\Delta N \cdot \Gamma^{eq}\left(n - n^{eq}\right) \cdot a^3. \tag{5.60}$$

Equation (5.59) is the Boltzmann equation for the number density in expanding Universe. We will repeatedly encounter equations like this in the course of our study.

Chapter 6

Recombination

6.1.   Recombination Temperature in Equilibrium Approximation

At temperatures exceeding the binding energy of electrons in atoms, the matter in the Universe was in the phase of plasma consisting of electrons, photons and baryons. As we discuss in Chapter 8, at temperatures below a few dozen keV the baryon component was predominantly protons (about 75% of total mass) and ^{4}He nuclei (about 25% of total mass). As the Universe cools down, it becomes thermodynamically favorable for baryons and electrons to combine into atoms. This cosmological epoch is called recombination. Before recombination, the plasma was opaque because of photon scattering off free electrons, while after recombination relic photons propagate freely. These are the CMB photons we see today.

We are interested in recombination of hydrogen, since recombination of helium occurs earlier and has very little effect on properties of photon–electron–proton plasma afterwards. The recombination of helium is the subject of Problem 6.5.

We begin our analysis by determining the temperature at which formation of atoms becomes thermodynamically favorable. We will assume temporarily that the expansion of the Universe is adiabatically slow. Under this assumption, the medium is always in thermal equilibrium. In particular, there is chemical equilibrium between its components. This is *not* an adequate approximation: recombination is *not an equilibrium process* [79, 80], it is slightly delayed. We consider inequilibrium recombination in Sec. 6.2, and here we restrict ourselves to the complete thermal equilibrium. In this way we will understand the relevant range of temperatures and cosmological times. We will also obtain the equation of chemical equilibrium (Saha equation) which is used in this and other Chapters. We stress again that one should bear in mind that the picture we consider in this Section is not directly relevant for recombination in the real Universe.

One could naively expect that the temperature of equilibrium recombination is similar to the binding energy of electron in hydrogen atom. This is not the case, however: the recombination occurs at much lower temperature. The physical reason is as follows. At low electron and proton densities, the recombination of a given electron with some proton occurs in time[1] τ_+, inversely proportional to the number density of protons $\tau_+ \propto n_B^{-1}$. The newly formed hydrogen atom is ionized by a photon whose energy exceeds the binding energy of hydrogen atom, Δ_H. There exist thermal photons of these energies even at $T \ll \Delta_H$, though their number is exponentially small. Hence, the lifetime of a hydrogen atom in medium is finite, albeit exponentially long, $\tau_- \propto e^{\Delta_H/T}$. Recombination becomes effective when

$$\tau_+ \sim \tau_-, \tag{6.1}$$

which indeed gives $T \ll \Delta_H$ for small n_B (and small electron velocities which enter τ_+). One could find the equilibrium recombination temperature by elaborating on this kinetic approach (see Problem 6.4). It is simpler, however, to make use of thermodynamical approach which we now present.

We will see that the temperatures of interest are about $0.3\,\text{eV}$. At these temperatures and in chemical equilibrium, electrons and protons are non-relativistic, while most hydrogen atoms are in the ground state (see Problem 6.6). The equilibrium number densities of electrons, protons and hydrogen atoms in the ground state are

$$n_e = g_e \left(\frac{m_e T}{2\pi} \right)^{3/2} e^{(\mu_e - m_e)/T}, \tag{6.2}$$

$$n_p = g_p \left(\frac{m_p T}{2\pi} \right)^{3/2} e^{(\mu_p - m_p)/T}, \tag{6.3}$$

$$n_H = g_H \left(\frac{m_H T}{2\pi} \right)^{3/2} e^{(\mu_H - m_H)/T}. \tag{6.4}$$

These formulas involve yet unknown chemical potentials. The number of spin states of proton and electron equals $g_e = g_p = 2$, while for hydrogen atom $g_H = 4$.

Problem 6.1. *Show that $g_H = 4$.*

The equilibrium recombination temperature T_r^{eq} is determined from the relation

$$n_p(T_r^{eq}) \simeq n_H(T_r^{eq}). \tag{6.5}$$

To find T_r^{eq}, we need three more equations that complete the system of equations on number densities and chemical potentials at given temperature. One is the baryon

---

[1]Superscripts + and − refer to direct and inverse processes — recombination and ionization, respectively.

number conservation,

$$n_p + n_H = n_B. \tag{6.6}$$

Here and in the next Section (and only in these Sections) n_B denotes the baryon number density *without helium* (for total baryon number density we use the notation n_B^{tot}),

$$n_B(T) = \hat{\eta}_B n_\gamma(T), \quad \hat{\eta}_B = 0.75\eta_B. \tag{6.7}$$

The baryon-to-photon ratio is given by (1.21), so that

$$\hat{\eta}_B = 4.5 \cdot 10^{-10}, \tag{6.8}$$

while the number density of photons $n_\gamma(T)$ is a known function of temperature; see (5.16a). Another equation,

$$\mu_p + \mu_e = \mu_H, \tag{6.9}$$

follows from the assumption of thermal equilibrium for the reaction

$$p + e \leftrightarrow H + \gamma. \tag{6.10}$$

The last equation expresses electric neutrality of the Universe,

$$n_p = n_e. \tag{6.11}$$

Thus, we have six equations, (6.2), (6.3), (6.4), (6.6), (6.9) and (6.11), determining three number densities n_e, n_p, n_H and three chemical potentials μ_e, μ_p, μ_H at given temperature T. The quickest way to simplify this system of equations is to multiply Eqs. (6.2) and (6.3):

$$n_p n_e = g_p g_e \left(\frac{m_p T}{2\pi}\right)^{3/2} \left(\frac{m_e T}{2\pi}\right)^{3/2} e^{(\mu_p + \mu_e - m_p - m_e)/T}.$$

Now, using (6.4), (6.9) and (6.11) we get rid of chemical potentials and write the latter equation in the form

$$n_p^2 = \left(\frac{m_e T}{2\pi}\right)^{3/2} n_H e^{-\Delta_H/T}, \tag{6.12}$$

where

$$\Delta_H \equiv m_p + m_e - m_H = 13.6 \text{ eV}$$

is the binding energy of hydrogen atom, and we neglected the difference between m_p and m_H in the pre-exponential factor in (6.12). Together with Eq. (6.6), equation (6.12) determines the equilibrium proton and hydrogen densities n_p and n_H.

It is convenient to introduce dimensionless ratios

$$X_p \equiv \frac{n_p}{n_B}, \quad X_H \equiv \frac{n_H}{n_B}.$$

Then the baryon number conservation equation is

$$X_p + X_H = 1. \tag{6.13}$$

Using Eq. (6.12) we express X_H through X_p, and then Eq. (6.13) becomes the equation for the only unknown X_p,

$$X_p + n_B X_p^2 \left(\frac{2\pi}{m_e T} \right)^{3/2} e^{\frac{\Delta_H}{T}} = 1. \tag{6.14}$$

Equation (6.14) (as well as Eq. (6.12)) is known as the Saha equation. It is convenient to express n_B through the baryon-to-photon ratio $\hat{\eta}_B$ by making use of (6.7) and then use the formula (5.16a) for photon number density $n_\gamma(T)$. This leads to the equation containing dimensionless quantities only,

$$X_p + \frac{2\zeta(3)}{\pi^2} \hat{\eta}_B \left(\frac{2\pi T}{m_e} \right)^{3/2} X_p^2 e^{\frac{\Delta_H}{T}} = 1. \tag{6.15}$$

The second term here is the equilibrium number density of hydrogen atoms expressed in terms of X_p,

$$X_H = \frac{2\zeta(3)}{\pi^2} \hat{\eta}_B \left(\frac{2\pi T}{m_e} \right)^{3/2} X_p^2 e^{\frac{\Delta_H}{T}}. \tag{6.16}$$

We now see explicitly that recombination occurs at temperatures well below the binding energy Δ_H. Indeed, unless the exponential factor in (6.16) is very large, the equilibrium proton abundance X_p is close to 1, while the hydrogen abundance X_H is small. The suppression is due to both smallness of the baryon-to-photon ratio $\hat{\eta}_B$ and the fact that electrons are non-relativistic at temperatures of interest, $T/m_e \ll 1$. In chemical equilibrium, recombination begins when $X_p \sim 1$ and $X_H \sim 1$ at the same time. At that time the suppression by pre-exponential factors in (6.16) is compensated for by the exponential factor $e^{\Delta_H/T}$. This occurs when Δ_H/T is much greater than 1,

$$\frac{\Delta_H}{T_r^{eq}} \simeq -\log \left[\frac{2\zeta(3)}{\pi^2} \hat{\eta}_B \left(\frac{2\pi T_r^{eq}}{m_e} \right)^{3/2} \right]. \tag{6.17}$$

The latter equation is obtained by substituting $X_H(T_r^{eq}) \sim X_p(T_r^{eq}) \sim 1$ into Eq. (6.16). Equation (6.17) determines the equilibrium recombination temperature.

It is useful at this point to discuss solutions of equations of the type (6.17). This equation belongs to the class of equations of the following form,

$$x = \log \left(A x^\alpha \right). \tag{6.18}$$

In our case the unknown x is

$$x = \frac{\Delta_H}{T_r^{eq}},$$

while parameters A and α are given by

$$\alpha = \frac{3}{2}, \quad A = \frac{\sqrt{\pi}}{4\sqrt{2}\zeta(3)} \left(\frac{m_e}{\Delta_H} \right)^{3/2} \hat{\eta}_B^{-1}.$$

It is important that

$$\log A \gg 1. \tag{6.19}$$

In the leading logarithmic approximation, i.e., treating $\log A$ as a large parameter, we can set $x = 1$ in the right-hand side of Eq. (6.18) and immediately obtain the solution

$$x \approx \log A. \tag{6.20}$$

Problem 6.2. *The solution x can be written as*

$$x = (1 + \epsilon) \log A.$$

Find the correction ϵ to the first non-trivial order in $(\log A)^{-1}$ and show that it is indeed small provided that $\alpha \sim 1$.

Coming back to recombination, we obtain in the leading logarithmic approximation

$$T_r^{eq} \approx \frac{\Delta_H}{\log \left[\frac{\sqrt{\pi}}{4\sqrt{2}\zeta(3)} \left(\frac{m_e}{\Delta_H} \right)^{3/2} \hat{\eta}_B^{-1} \right]} \approx 0.38 \text{ eV}. \tag{6.21}$$

The numerical solution of (6.17) with the effective value $\hat{\eta}_B$ presented in (6.8) gives $T_r^{eq} = 0.33$ eV.

By comparing the equilibrium recombination temperature (6.21) to the temperature at equality (4.21b) we see that recombination occurs at matter-dominated stage. This is true also for realistic, out-of-equilibrium recombination, since the absence of thermal equilibrium can only *delay* the process as compared to the equilibrium situation.

Let us find the age of the Universe at $T_r^{eq} = 0.33$ eV. To this end, we use the relation (3.22) valid at matter domination and write

$$t_r^{eq} = \frac{2}{3} H^{-1}(t_r^{eq}) = \frac{2}{3\sqrt{\Omega_M} H_0 (1 + z_r^{eq})^{3/2}}. \tag{6.22}$$

Numerically $t_r^{eq} \simeq 330$ thousand years, while the Hubble time is

$$H^{-1}(T_r^{eq}) \simeq 490 \text{ thousand years} \tag{6.23}$$

for $T_r^{eq} = 0.33$ eV, $\Omega_M = 0.315$. This estimate can be refined by making use of the result (4.27), but we do not need that here.

Later on the proton abundance is small, and we can neglect the first term in the left-hand side of Eq. (6.15). It is then clear that the equilibrium proton abundance exponentially decreases as the Universe cools down, $X_p \propto e^{-\Delta_H/2T}$. More precisely

$$n_p^{eq} = n_e^{eq} = n_B^{1/2} \left(\frac{m_e T}{2\pi} \right)^{3/4} e^{-\Delta_H/2T}. \tag{6.24}$$

This formula will be useful in what follows.

Problem 6.3. *Find chemical potentials μ_e, μ_p, μ_H in equilibrium at temperature $T_r^{eq} = 0.33\, eV$. Compare them with masses of electron, proton and hydrogen atom.*

Problem 6.4. *Give alternative derivation of the equilibrium recombination temperature based on kinetic approach and the relation (6.1).*

Problem 6.5. *Find equilibrium recombination temperature for helium. Hint: Do not forget that helium nucleus has to catch two electrons to make a neutral helium atom.*

Problem 6.6. *Find equilibrium abundances of hydrogen atoms at 2s- and 2p-levels at temperature $T_r^{eq} = 0.33\, eV$, relative to the abundance of atoms in the ground state 1s. Disregard higher levels in this calculation.*

6.2. Photon Last Scattering in Real Universe

In the previous Section we defined the "moment" of recombination as the time at which the equilibrium abundances of free protons and hydrogen atoms are equal. In fact, this is not quite the "moment" of primary interest for cosmology. Really interesting is the epoch at which photons scatter for the last time, and after which relic photons propagate freely through the Universe. We are talking here about Compton scattering of photons off free electrons,

$$\gamma e \rightarrow \gamma e. \tag{6.25}$$

The cross-section of this process is determined by the diagrams of quantum electrodynamics (QED) shown in Fig. 6.1. We are interested in temperatures, and hence

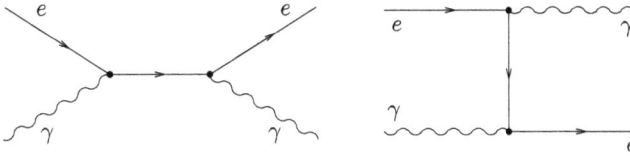

Fig. 6.1.   Feynman diagrams for Compton scattering.

photon frequencies, much lower than the electron mass. In this case the Compton cross-section reduces to the Thomson cross-section known from classical electrodynamics (the study of Compton scattering in QED can be found, e.g., in the book [81]; the derivation of the Thomson cross-section in classical theory is given in the book [83]). The Thomson cross-section is

$$\sigma_T = \frac{8\pi}{3}\frac{\alpha^2}{m_e^2} \approx 0.67 \cdot 10^{-24}\,\text{cm}^2. \qquad (6.26)$$

The mean free time of photon with respect to the Compton scattering is

$$\tau_\gamma = \frac{1}{\sigma_T \cdot n_e(T)}, \qquad (6.27)$$

where n_e is the number density of free electrons. At the time when the abundances of free protons (and hence free electrons) and hydrogen atoms are equal, $n_p = n_e \simeq n_H$, the electron number density is

$$n_e \simeq \frac{n_B}{2} = \frac{\zeta(3)}{\pi^2}\hat\eta_B T^3.$$

For $T = 0.33\,\text{eV}$ this is equal to

$$n_e \simeq 250\,\text{cm}^{-3}. \qquad (6.28)$$

The photon mean free time (6.27) in this situation equals $\tau_\gamma \simeq 2 \cdot 10^{11}\,\text{s} \simeq 6000\,\text{yrs}$. This time is much smaller than the Hubble time (6.23). Hence, *photon last scattering happens somewhat later*, at smaller density of free electrons. We will see that the temperature at last scattering is $T_r = 0.26\,\text{eV}$, and the purpose of this Section is precisely to calculate this temperature.

At temperatures of interest to us, there is *kinetic* equilibrium between photons, electrons, and also protons. We will show that in Sec. 6.3. On the contrary, *chemical* equilibrium does not hold. The latter property can be understood by noticing that there are very few thermal photons capable of ionizing hydrogen from the ground state, i.e., thermal photons with energies $\omega > \Delta_H$. Indeed, the number density of photons of energies above a given value ω is (hereafter the superscript *eq* refers to thermal, equilibrium quantities)

$$n_\gamma^{eq}(\omega) = \int_\omega^\infty F_\gamma^{eq}(\omega')d\omega', \tag{6.29}$$

where

$$F_\gamma^{eq}(\omega) = 8\pi\omega^2 f_{Pl}(\omega)$$

and $f_{Pl}(\omega)$ is the Planck distribution (that is the Bose–Einstein distribution for zero mass; we recall that photon has two polarization states). At $\omega \gg T$ we have

$$F_\gamma^{eq}(\omega) = \frac{\omega^2}{\pi^2}e^{-\omega/T}, \tag{6.30}$$

and

$$n_\gamma^{eq}(\omega) = \frac{\omega^2 T}{\pi^2}e^{-\omega/T}. \tag{6.31}$$

For $\omega = \Delta_H$ and $T = T_r = 0.26\,\mathrm{eV}$ this number density is $n_\gamma^{eq}(\Delta_H) \simeq 10^{-9}\,\mathrm{cm}^{-3}$. So small number of energetic photons is insufficient for maintaining chemical equilibrium. We will see in the end of this Section that there exist more numerous *non-thermal* photons with $\omega > \Delta_H$; this fact will unequivocally prove that the reaction of recombination to the ground state is out of thermal equilibrium.

Reactions of transition to the ground state $1s$ from excited states are also out of equilibrium. We will see that later on. On the other hand, chemical equilibrium *does hold* for free electrons and protons and hydrogen atoms at *excited levels*. Indeed, the number density of photons capable of ionizing, say, $2s$- and $2p$-levels is $n_\gamma^{eq}(\Delta_{2s}) \simeq 10^8\,\mathrm{cm}^{-3}$, where $\Delta_{2s} = \Delta_H/4$ is the binding energy of $2s$- and $2p$-levels. This is much greater than the baryon number density, which serves as a hint towards thermal equilibrium: even if all electrons and protons recombined into $2s$- and $2p$-levels, the photon spectrum in the region $\omega \sim \Delta_{2s}$ would not change, and would remain thermal. Later on we will further discuss the conditions for this partial chemical equilibrium; see Problem 6.8.

So, the system is in thermal equilibrium with respect to all reactions except for the reactions of production and destruction of hydrogen $1s$-atoms. It is useful to note at this point that if $1s$-atoms were not produced at all, the abundances of excited atoms would be small, and practically all electrons would be free at $T = T_r = 0.26\,\mathrm{eV}$.

Problem 6.7. *Prove the last statement. Hint: Make use of the formulas of Sec. 6.1.*

The last remark shows that electrons are mostly either free or bound in $1s$-atoms; as we discussed above, photon last scattering occurs when the number of free electrons in small, so most of the electrons occupy $1s$ states. Therefore, the key processes are production and destruction of $1s$-atoms. In what follows, we will take into account $1s$-, $2s$- and $2p$-levels only, and ignore the higher levels. This is actually a reasonable approximation, since the equilibrium ratio of the occupancy of 3s-levels to 2s is

$$\frac{n_{3s}}{n_{2s}} \simeq e^{-(\Delta_{2s}-\Delta_{3s})/T} \sim 10^{-3}. \tag{6.32}$$

Hereafter n_{2s}, n_{3s}, etc., denote number densities of atoms at the corresponding levels; the atoms themselves will be denoted simply as $1s$, $2s$, etc. When writing (6.32) we used $\Delta_{3s} = \Delta_H/9$.

Within our approximation, $1s$-atoms can be produced and disappear in the following reactions,

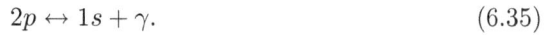

$$e + p \leftrightarrow 1s + \gamma, \tag{6.33}$$

$$2s \leftrightarrow 1s + 2\gamma, \tag{6.34}$$

$$2p \leftrightarrow 1s + \gamma. \tag{6.35}$$

Let us stress that $2s \to 1s$ transition is a two-photon process, while transitions from continuum and from $2p$-state occur by emitting one photon. The $2p \to 1s$ transition produces a photon in the first line of Lyman series, Ly-α.

Let us begin our analysis from the transitions from continuum (6.33). These turn out to be basically irrelevant (!). In the situation of interest, $n_{1s} \gtrsim n_e$, a photon emitted in the process $e + p \to 1s + \gamma$ ionizes a "neighboring" $1s$-atom before losing its energy in collisions with electrons. Indeed, the ionization cross-section near threshold for $1s$-atom is [81]

$$\sigma_{1s} = \frac{2^9 \pi^2}{3e^4} \frac{1}{\alpha m_e^2} = 6.3 \cdot 10^{-18} \text{ cm}^2,$$

and it is much larger than the Thomson cross-section (6.26). Also, we will see in the end of this Section that photon redshift caused by the cosmological expansion is also inefficient for the reactions (6.33). Hence, the processes (6.33) effectively do not change the numbers of free electrons and $1s$-atoms.

Let us now consider the processes (6.34). Unlike in recombination of free electron and proton, there are two photons produced in the direct process, so their appearance in the medium does not trigger the inverse reaction. The width of the direct process $2s \to 1s + 2\gamma$ is [84]

$$\Gamma_{2s} \approx 8.2 \text{ s}^{-1}.$$

This width is small, so the direct process is slow in the sense that $2s$-atom more likely gets ionized by a thermal photon than makes this transition; see Problem 6.8. This means that $2s \to 1s$ transitions are unimportant from the viewpoint of the abundance of $2s$-atoms relative to electrons, and there is indeed partial thermal equilibrium in the medium, as we discussed above.

Problem 6.8. *Show that at $T > 2500\,K$ the probability that $2s$-atom gets ionized in the reaction $2s + \gamma^{th} \to e + p$ is higher than the probability that it becomes*

1s-state due to the process $2s \to 1s + 2\gamma$. Here γ^{th} denotes thermal photon. Hint:
Use ionization cross section near threshold

$$\sigma_{2s} = \frac{2^{16}\pi^2}{3e^8} \frac{1}{\alpha m_e^2} = \frac{2^7}{e^4} \sigma_{1s} \approx 2.34\,\sigma_{1s}.$$

We note that the temperature $2500\,K$ is lower than $T_r = 0.26\,eV = 3000\,K$, so the approximation of partial equilibrium is indeed valid at photon last scattering epoch. We note also that the typical time scales $t \lesssim \Gamma_{2s}^{-1}$ are much smaller than the time scale of the cosmological expansion, so the expansion does not spoil the equilibrium for processes $e + p \leftrightarrow 2s + \gamma$.

Since $2s$-atoms are in equilibrium with free electrons and protons, they obey the Saha relation (cf. Sec. 6.1)

$$n_e n_p = n_e^2 = n_{2s} \cdot \left(\frac{m_e T}{2\pi}\right)^{3/2} \cdot e^{-\frac{\Delta_{2s}}{T}}. \tag{6.36}$$

The transition rate to $1s$-level via $2s \to 1s$ channel is determined by the number of $2s$-atoms and the width of $2s$-$1s$ transition, so the rate at which free electrons leak into $1s$-states via this channel is

$$\left(\frac{dn_e}{dt}\right)_{2s \to 1s} = -\Gamma_{2s} n_{2s}. \tag{6.37}$$

Here we neglect the inverse process which converts $1s$-atom into $2s$-state via absorption of two photons. We justify this approximation in the end of this Section.

Finally, let us discuss the processes (6.35). Similarly to free electron recombination, Ly-α photon emitted in the direct process has high probability to get absorbed by "neighboring" $1s$-atom in the inverse process, leaving no change in the medium. However, redshift of photons due to the cosmological expansion is important in the case (6.35). The point is that there are many Ly-α photons in the medium. Some of them get redshifted away from the Ly-α line and no longer participate in the inverse process $1s + \gamma \to 2p$. This results in the overall increase of the number of $1s$-atoms and decrease of the number of free electrons and protons. We will consider this effect at quantitative level later on, and here we simply quote that the corresponding effective rate is numerically similar to the estimate (6.37). Hence, the number density of free electrons decays according to

$$\frac{dn_e}{dt} = -\Gamma_{eff}(T) n_{2s}, \tag{6.38}$$

where $\Gamma_{eff}(T)$ mildly depends on temperature; see Eq. (6.50), and $\Gamma_{eff}(T_r) \simeq 2\Gamma_{2s}$.

Making use of (6.36) we now write Eq. (6.38) as the equation for the number density of free electrons,

$$\frac{dn_e}{dt} = -\Gamma_{eff} \cdot \left(\frac{2\pi}{T m_e}\right)^{3/2} \cdot e^{\frac{\Delta_{2s}}{T}} \cdot n_e^2.$$

Neglecting the time-dependence of the pre-exponential factor and changing variable from t to $1/T$ according to (3.34) we find the solution

$$\frac{1}{n_e} = \frac{\Gamma_{eff}T}{\Delta_{2s}H} \cdot \left(\frac{2\pi}{Tm_e}\right)^{3/2} \cdot e^{\frac{\Delta_{2s}}{T}} + \text{const.}$$

The first term here grows very fast, so the integration constant is irrelevant at $t \sim t_r$. Recalling that $\Delta_{2s} = \Delta_H/4$, we obtain finally

$$n_e(T) = \frac{\Delta_H H}{4\Gamma_{eff}T} \cdot \left(\frac{Tm_e}{2\pi}\right)^{3/2} \cdot e^{-\frac{\Delta_H}{4T}}. \tag{6.39}$$

Clearly, the result of non-equilibrium analysis is quite different from the equilibrium formula (6.24), as both exponential and pre-exponential factors are different. To avoid confusion we recall that the formula (6.39) is valid at $T > 2500\,\text{K} = 0.22\,\text{eV}$.

Problem 6.9. *Show that at recombination, when $n_e/n_B \ll 1$, the inequilibrium density (6.39) always exceeds the equilibrium one (6.24); consider only temperatures $T > 2500\,K$, for which the above analysis is valid. This means that the inequilibrium recombination is delayed as compared to the equilibrium one, as expected.*

Let us come back to the photon last scattering. The average number of collisions of a photon with free electrons from time t to the present time (the latter may be treated as infinity for this purpose) is

$$N(t) = \int_t^\infty \sigma_T n_e dt'.$$

Let us define the time of last scattering as the time since which photon scatters once, $N(t_r) = 1$. Changing the integration variable to temperature, we have for the temperature of last scattering T_r,

$$1 = \int_0^{T_r} \sigma_T n_e \frac{dT'}{HT'} = \frac{4T_r}{H(T_r)\Delta_H}\sigma_T n_e(T_r); \tag{6.40}$$

when evaluating the integral we made use of the strong dependence of the exponent in integrand on temperature; see (6.39). This gives the equation for T_r,

$$\left(\frac{T_r m_e}{2\pi}\right)^{3/2} \cdot e^{-\frac{\Delta_H}{4T_r}} = \frac{\Gamma_{eff}}{\sigma_T}.$$

Interestingly, T_r depends on cosmological parameters very weakly; such a dependence exists in $\Gamma_{eff}(T)$ only. The result of numerical solution is

$$T_r = 0.26\,\text{eV}.$$

At the last scattering epoch, the Hubble parameter is

$$H(T_r) \sim \sqrt{\Omega_M}H_0(1+z_r)^{3/2} \simeq (720\text{ thousand years})^{-1} = (0.22\text{ Mpc})^{-1} \tag{6.41}$$

and the age of the Universe is $t_r = (2/3)H^{-1} = 480$ thousand years. The precise value is obtained from (4.27), which gives

$$t_r = 370 \text{ thousand years.} \tag{6.42}$$

Problem 6.10. *Obtain the estimates (6.41) and (6.42).*

Problem 6.11. *Show that the Hubble rate at temperature $T_r = 0.26\,eV$, taking into account the contribution of radiation into the energy density, is*

$$H(T_r) = (650 \text{ thousand years})^{-1} = (0.2 \text{ Mpc})^{-1}.$$

The number density of free electrons at last scattering is obtained from (6.40),

$$n_e(t_r) \simeq 30 \text{ cm}^{-3}. \tag{6.43}$$

This is considerably smaller than the electron number density at the beginning of recombination; see (6.28).

Let us make a remark here. Of course, recombination does not happen instantaneously: the electron abundance (6.39) smoothly decreases in time. Yet the main change in the number density of free electrons occurs in a small interval of temperatures,

$$\Delta T \ll T_r.$$

This property is due to fast evolution of the exponential factor in (6.39). The interval ΔT can be defined by the requirement that the exponential factor in (6.39) is e times larger and smaller than the central value at the ends of this interval, i.e.,

$$\left| \frac{\Delta_H}{4(T_r \pm \Delta T)} - \frac{\Delta_H}{4T_r} \right| = 1.$$

This gives

$$\frac{\Delta T}{T_r} \approx \frac{4T_r}{\Delta_H} \sim 0.1.$$

Thus, the number density of free electrons falls considerably when the temperature decreases by several percent. In other words, the process of recombination occurs during a fraction of Hubble time, $\Delta t \ll H^{-1}(t_r)$.

Photons that scattered for the last time at $t \approx t_r$ come to us from a certain distance; by observing CMB we obtain a photographic image of a certain sphere which surrounds us. This sphere is called surface of last scattering; according to our discussion, it has small but finite width.

Problem 6.12. *Show that the mean free time of a photon with respect to scattering off neutral atoms exceeds the Hubble time at recombination and afterwards.*

This means that the presence of neutral hydrogen and helium does not affect CMB photons.

Let us consider the contribution of processes (6.35) into the production rate of 1s-atoms, or, equivalently, the rate of disappearance of free electrons. This contribution is due to the photon redshift caused by the cosmological expansion. The decrease of the number of Ly-α photons created in the direct reaction and participating in the inverse reaction is given by the number of photons leaving Ly-α line, i.e., crossing the low-frequency boundary of this line in unit time,

$$\frac{dn_\gamma(\omega_*)}{dt} = F_\gamma(\omega_*)\frac{d\omega}{dt}(\omega_*) = -\omega_* H F_\gamma(\omega_*), \tag{6.44}$$

where $n_\gamma(\omega)$ is the number density of photons with energy greater than ω, $F_\gamma(\omega) = dn(\omega)/d\omega$ is the photon spectral density (note that the equilibrium values of these quantities enter (6.29)), while ω_* is the lower boundary of the Ly-α line; we will see that the precise value of ω_* is irrelevant for the calculation. Photons of energy below ω_* cannot excite 1s-state into 2p, so the effective increase of the number of 1s-atoms and decrease of the number of free electrons are given by

$$\frac{dn_{1s}}{dt} = -\frac{dn_e}{dt} = -\frac{dn_\gamma(\omega_*)}{dt}. \tag{6.45}$$

Since the Universe expands slowly, very few photons exit the Ly-α line, so the direct and inverse processes are to very good approximation in thermal equilibrium. Thus, the spectral density $F_\gamma(\omega)$ can be calculated in the approximation of equal rates of direct and inverse processes,

$$\Gamma_{2p}\varphi(\omega)n_{2p}d\omega = C(\omega)\varphi(\omega)F_\gamma(\omega)n_{1s}d\omega. \tag{6.46}$$

The left-hand side of this formula gives the number of photons emitted in the process $2p \to 1s + \gamma$ per unit time per unit volume in the frequency interval $(\omega, \omega + d\omega)$, Γ_{2p} is the width of 2p-level, $\varphi(\omega)$ is the shape of Ly-α line. The right-hand side is the absorption rate of photons by 1s-atoms. We used the fact that this rate is proportional to the photon spectral density and to the number density of 1s-atoms, so that the coefficient C depends on ω only. It is given by

$$C(\omega) = \frac{\pi^2\Gamma_{2p}}{\omega^2}. \tag{6.47}$$

Note that $A(\omega) = \Gamma_{2p}\varphi(\omega)$ and $B(\omega) = C(\omega)\varphi(\omega)/\omega$ are nothing but the Einstein coefficients known from the theory of radiation, and the equality (6.47) is the Einstein relation, see the book [81].

Problem 6.13. *Derive the relation (6.47). Hint: Make use of the fact that equality (6.46) is valid in thermal equilibrium and consider the low temperature regime $T \ll \omega$.*

The relations (6.46), (6.47) give

$$F_\gamma(\omega) = \frac{\omega^2}{\pi^2}\frac{n_{2p}}{n_{1s}}. \tag{6.48}$$

This is a smooth function of ω, so one can set $\omega_* = \Delta_H - \Delta_{2p} = 3\Delta_H/4$ in (6.44). Also, most protons are bound in 1s-atoms at last scattering epoch, so we can take $n_{1s} \approx n_B$.

Making use of Eqs. (6.44) and (6.45) we find that the decrease of the number of free electrons due to reactions (6.35) is

$$\left(\frac{dn_e}{dt}\right)_{2p\to1s} = -\frac{H}{n_B\pi^2}\left(\frac{3\Delta_H}{4}\right)^3 n_{2p}. \tag{6.49}$$

Finally, at partial thermal equilibrium we have $n_{2p} = 3\,n_{2s}$ according to the degeneracy of the levels. Adding (6.49) to (6.37) we obtain the formula (6.38) with

$$\Gamma_{\mathit{eff}} = \Gamma_{2s} + 3\frac{H}{n_B\pi^2}\left(\frac{3\Delta_H}{4}\right)^3. \tag{6.50}$$

At $T = T_r = 0.26\,\mathrm{eV}$ and $n_B = 0.75\cdot6.05\cdot10^{-10}n_\gamma$ the second term here equals $7.1\,\mathrm{s}^{-1}$, so that $\Gamma_{\mathit{eff}}(T_r) = 14.3\,\mathrm{s}^{-1} = 1.85\,\Gamma_{1s}$. By the end of recombination, about half of electrons get to $1s$-level via $2s \to 1s$ transitions, while another half of electrons get there via Ly-α channel. This is consistent with the complete analysis of Ref. [82].

An analogous phenomenon exists also in the case of the reactions (6.33). In that case instead of (6.46) we have

$$A(\omega)n_e^2 = \sigma_I(\omega)F_\gamma(\omega)n_{1s}. \tag{6.51}$$

The rate of the direct reaction of recombination of electron and proton is proportional to $n_p n_e = n_e^2$ which is reflected in the left-hand side of (6.51). The rate of inverse reaction is proportional to the number of $1s$-atoms, ionization cross-section σ_I and photon spectral density at energy ω, where $\omega \geq \Delta_H$. Let us recall that the direct and inverse rates are equal in thermal equilibrium, and write the equation for the coefficient $A(\omega)$ (this is nothing but detailed balance equation),

$$A(\omega)\left(n_e^{eq}\right)^2 = \sigma_I(\omega)F_\gamma^{eq}(\omega)n_{1s}^{eq}.$$

We express $A(\omega)$ from this equation and substitute it into (6.51). This gives for the spectral density at $\omega \geq \Delta_H$

$$F_\gamma(\omega) = \frac{n_e^2 n_{1s}^{eq}}{(n_e^{eq})^2 n_{1s}}F_\gamma^{eq}(\omega). \tag{6.52}$$

Redshift of photons away from the region $\omega > \Delta_H$ is given by (6.44) with $\omega_* = \Delta_H$; the same formula gives the decrease of the number of free electrons due to the processes (6.33). The ratio of the latter contribution to that of Ly-α processes is given by the ratio of (6.52) and (6.48) (modulo the ratio of photon energies, $\Delta_H/(\Delta_H - \Delta_{2p}) = 4/3$),

$$\frac{(dn_e/dt)_{ep\to1s}}{(dn_e/dt)_{2p\to1s}} \simeq \frac{F_\gamma^{eq}(\Delta_H)n_{1s}^{eq}}{(n_e^{eq})^2}\frac{\pi^2}{\Delta_H^2}\left(\frac{Tm_e}{2\pi}\right)^{3/2}\cdot e^{-\frac{\Delta_H}{4T}},$$

where we inserted $n_{2p} = 3\,n_{2s}$, made use of (6.36) and omitted factor of order 1. Let us now make use of (6.30), (6.24) and set $n_{1s} \approx n_B$. We obtain finally

$$\frac{(dn_e/dt)_{ep\to1s}}{(dn_e/dt)_{2p\to1s}} \simeq e^{-\frac{\Delta_H}{4T}}.$$

This ratio is small, which means that the processes (6.33) are irrelevant indeed.

We note in passing that in our case of delayed recombination the number of free electrons n_e is considerably larger than its equilibrium value n_e^{eq}, while the number of $1s$-atoms is smaller than in equilibrium. Therefore, the actual spectral density (6.52) exceeds considerably the equilibrium spectral density at $\omega \geq \Delta_H$, as we have claimed in the text.

Finally, let us justify the approximation which neglects the process inverse to $2s \to 1s + 2\gamma$. The inverse process occurs via absorption of two thermal photons. Its rate is proportional to the number of $1s$-atoms, and in thermal equilibrium it must be equal to the rate of direct process (6.37). This immediately gives

$$\left(\frac{dn_e}{dt}\right)_{1s \to 2s} = \Gamma_{2s} n_{2s}^{eq} \frac{n_{1s}}{n_{1s}^{eq}} = \Gamma_{2s} n_{1s} e^{-(\Delta_H - \Delta_{2s})/T}.$$

This process is negligible as compared to the direct one when its rate is smaller than the rate (6.37), i.e., for

$$n_{1s} e^{-(\Delta_H - \Delta_{2s})/T} \ll n_{2s}. \tag{6.53}$$

n_{2s} can be found from (6.36), (6.39), while the estimate for n_{1s} is $n_{1s} \approx n_B = \hat{\eta}_B n_\gamma$. Inserting the numerical values, one obtains that (6.53) indeed holds.

Problem 6.14. *Make the numerical estimate sketched here.*

To conclude this Section, we point out that Lyman-α photons which are emitted, get redshifted and survive, do not lose their energy in scattering off electrons. As we discussed, the number density of these photons is of order $n_{Ly-\alpha} \sim n_e/2$ (their emission leads to recombination of about half of electrons). Temperatures at which they are emitted range, roughly speaking, from $T_r^{eq} = 0,33$ eV to $T_r = 0.26$ eV. Let us compare the number of Lyman-α photons with the number of thermal photons in the same spectral region. We make use of (6.31) and write

$$\frac{n_{Ly-\alpha}}{n_\gamma^{eq}(\omega_{2p})} \sim \frac{n_e}{2} \frac{\pi^2}{\omega_{2p}^2 T} e^{\omega_{2p}/T} \sim \hat{\eta}_B \zeta(3) \frac{T^2}{\omega_{2p}^2} e^{\omega_{2p}/T},$$

where $\omega_{2p} = \Delta_H - \Delta_{2p} = 3\Delta_H/4 = 10.2$ eV is energy of Lyman-α line. Even at temperature $T_r^{eq} = 0,33$ eV this ratio is of order 10, and it is even larger at lower temperatures. Thus, the Lyman-α region is dominated by *non-thermal* photons. Since their emission occurs in a relatively wide range of temperatures (and redshifts), the Lyman-α "line" in the CMB energy spectrum is fairly wide. Its relative width is estimated as follows:

$$\frac{\Delta\omega_{2p}}{\omega_{2p}} \sim \frac{T_r^{eq} - T_r}{T_r} \simeq 0.25.$$

Nevertheless, this "line" is in principle an interesting feature in the energy spectrum. The fact that it has not been detected yet is due to small number of photons in this "line". To quantfy this statement, let us estimate the present brightness of the "line" (energy flux per unit frequency $\nu = \omega/2\pi$ per unit solid angle; it is this quantity that is shown in Fig. 1.4, as well as in Fig. 6.2):

$$\Delta I_\nu = \frac{n_{Ly-\alpha,0} \omega_{2p,0}}{4\pi \Delta\nu_{2p,0}} \sim \frac{0,5 \cdot \hat{\eta}_B n_{\gamma,0}}{4\pi} \frac{2\pi\omega_{2p}}{\Delta\omega_{2p}}$$

$$\simeq 6 \cdot 10^{-24} \; \text{erg} \cdot \text{cm}^{-2} \cdot \text{s}^{-1} \cdot \text{Hz}^{-1} \cdot \text{ster}^{-1}$$

$$= 6 \cdot 10^{-27} \; \text{J} \cdot \text{m}^{-2} \cdot \text{s}^{-1} \cdot \text{Hz}^{-1} \cdot \text{ster}^{-1}.$$

This is much smaller than brightness that can be measured so far, see Fig. 1.4. Nevertheless, it is expected that future experiments will discover not only Lyman-α "line" but also other features in the CMB energy spectrum. The result of the calculation of the CMB energy spectrum [87] is shown in Fig. 6.2.

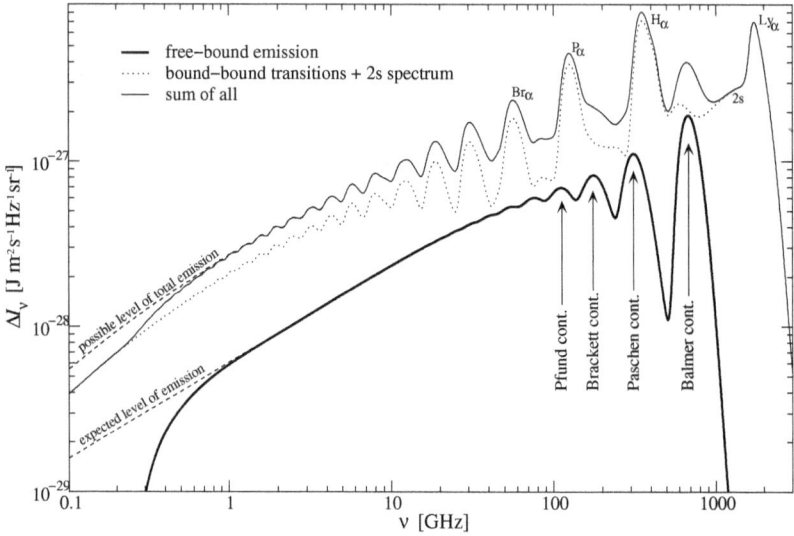

Fig. 6.2. Energy spectrum of CMB photons: excess of the total spectrum over the thermal spectrum. We used standard atomic physics notations for hydrogen lines. "Free-bound emission" denotes emission in transitions from continuum; "bound-bound transitions + 2s spectrum" is emission in transitions between bound states and transition $2s \to 1s$.

6.3. *Kinetic Equilibrium

It is of importance to understand to what extent the electron–proton–photon plasma is in kinetic equilibrium before and during recombination epoch. Indeed, our previous calculations were made under the assumption that the distribution functions have equilibrium forms (6.2)–(6.4) with one and the same temperature equal to the photon temperature.

We will see that electron–electron and electron–proton Coulomb interations are most rapid processes which keep *electrons and protons* in kinetic equilibrium (Sec. 6.3.2). So, electrons and protons do have thermal distribution functions with one and the same temperature.

To make sure that the temperature of electron–proton component coincides with the photon temperature, we have to study the following effect. We have seen in Sec. 2.5 that in the absence of interactions between photons and non-relativistic particles, the distributions of all these particles would have thermal form. However, the effective temperature of photons would decrease in time slower than that of electrons and protons. Thus, we have to check that energy transfer from photons to electrons and protons is sufficiently fast. We will see that this is indeed the case (Sec. 6.3.1).

There is a subtlety, however, which concerns photons. The energy transfer from photons to the electron–proton component distorts the photon distribution function, while energy exchange between *photons* themselves is slow at recombination and somewhat earlier. Thus, generally speaking, photons *do not* have Planckian spectrum. This effect is very small, however, because the number densities of electrons and protons are small. So, from the viewpoint of recombination, the approximation of complete kinetic equilibrium is very good.

Yet the result that the photon spectrum is not exactly Planckian is of interest, since the deviations from the thermal spectrum are potentially detectable in precision CMB studies. Besides the energy transfer from photons to electrons and protons, there are other mechanisms that distort the energy spectrum, see, e.g., the end of Sec. 6.2. We will see in Sec. 6.3.3 that photons get out of *kinetic* equilibrium well before recombination, at $T_{\gamma,kin} \sim 3 \cdot 10^5$ K. Even before that, at $T_{\gamma,chem} \simeq 6 \cdot 10^6$ K, photons get out of *chemical* equilibrium: reactions which change the number of photons become slower than the expansion of the Universe. Thus, in analogy to the sphere of last scattering, one can talk about the "black body photosphere of the Universe" [88] at temperature $6 \cdot 10^6$ K and redshift $2 \cdot 10^6$. Any processes at higher redshifts do not distort the CMB spectrum, while processes at lower redshifts, generally speaking, do lead to the spectrum distortion.

6.3.1. *Energy transfer from photons to electrons*

Let us begin with electrons. They get energy from photons via Compton scattering process (6.25) that occurs with Thomson cross-section (6.26). The time between two subsequent collisions of a given electron with photons is $\tau = 1/(\sigma_T n_\gamma)$. We are interested, however, in the energy transfer. So, we have to find the time τ_E in which an electron obtains kinetic energy of order T due to the Compton scattering. To estimate this time, we note that the typical energy transfer in a collision of a slow electron with a low energy photon is actually suppressed. To see this, let us write down the 4-momentum conservation equation in the following form,

$$p_e + p_\gamma - p'_\gamma = p'_e,$$

where p_e, p_γ are 4-momenta of incident electron and proton, while primed quantities refer to final particles. The square of this equality gives

$$m_e(\omega - \omega') - |\mathbf{p}_e|(\omega \cos\theta_1 - \omega' \cos\theta_2) - \omega\omega'(1 - \cos\theta) = 0, \tag{6.54}$$

where $\mathbf{p}_e$ is 3-momentum of the initial electron, ω, ω' are initial and final photon frequencies, θ_1, θ_2 are angles between the initial electron momentum and momenta of initial and final photon, $\mathbf{p}_\gamma$ and $\mathbf{p}'_\gamma$, respectively, while θ is the photon scattering angle; see Fig. 6.3. If the electron is initially at rest, the second term in Eq. (6.54) vanishes, so the typical energy transfer is

$$\Delta E = \omega' - \omega \sim \frac{\omega^2}{m_e}. \tag{6.55}$$

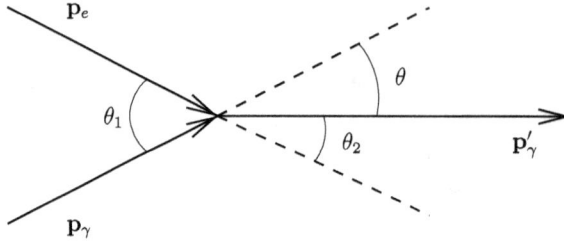

Fig. 6.3.   Kinematics of Compton scattering.

From this we would conclude that the number of collisions after which the electron has energy of order T is

$$N \sim \frac{T}{\Delta E} \sim \frac{m_e}{T}. \qquad (6.56)$$

However, we are interested in the situation where electron kinetic energy E_e before the collision is comparable to the photon frequency. In this situation the electron 3-momentum is

$$|\mathbf{p}_e| = \sqrt{2E_e m_e} \sim \sqrt{\omega m}.$$

Hence, the third term in Eq. (6.54) is negligible, and we obtain the estimate of the energy transfer in a single collision,

$$\Delta E \sim \omega\sqrt{\frac{\omega}{m}}. \qquad (6.57)$$

We note, however, that unlike the third term in (6.54), the second term can have any sign: the angles θ_1 and θ_2 take random values. Therefore, we use the analogy with random walk and estimate the number of scattering events needed to heat up a moving electron,

$$N \sim \left(\frac{T}{\Delta E}\right)^2 \sim \frac{m_e}{T}.$$

This coincides with the estimate (6.56) obtained for electron originally at rest.[2] In this way we obtain the time of electron heating,

$$\tau_E(T) \sim N\tau_e(T) \sim \frac{m_e}{T\sigma_T n_\gamma(T)}. \qquad (6.58)$$

Using numerical values of parameters in this formula we find that at recombination

$$\tau_E(T_r) \simeq 1.8 \cdot 10^8 \text{ s} = 5.7 \text{ years},$$

_____

[2]Note that for fast electron, the second term in (6.54) gives rise to systematic decrease of electron energy, since the number of collisions with photons moving towards electron is larger than that with photons "chasing" electron. This is the mechanism of thermalization of fast electron in photon medium.

which is much smaller than the Hubble time (6.42). Thus, energy transfer from photons to electrons is efficient, so electrons (and protons) have the photon temperature. This is true for earlier Universe too, since the time τ_E rapidly decreases as temperature grows, $\tau_E \propto T^{-4}$.

Energy transfer from photons to electrons leads to slight decrease in energy density of the photon gas. We estimate this effect by writing for the density of kinetic energy of electrons

$$\rho_{e,kin} = \frac{3}{2} T n_e = \hat{\eta}_B \frac{3\zeta(3)}{\pi^2} T^4 .$$

Protons have the same kinetic energy. So, the rough estimate of the energy loss by photons is $|\Delta\rho_\gamma| \sim (\rho_{e,kin} + \rho_{p,kin})$, and the relative loss of photon energy is

$$\frac{|\Delta\rho_\gamma|}{\rho_\gamma} \sim \eta_B \sim 10^{-9} . \tag{6.59}$$

This small loss does not affect the results of the previuos Section, but does lead to potentially observable CMB spectrum distortion; see Sec. 6.3.3 and review [88].

Let us refine the estimate (6.59). The Compton scattering keeps the kinetic energy density of electrons equal to $\rho_e^{kin} = 3n_e T/2$. The number density of electrons decreases due to the expansion of the Universe, $n_e \propto 1/a^3$. In the absence of the Compton scattering, the electron momentum would redshift, $p_e \propto 1/a$, and the expansion would lead to the rapid decrease of the kinetic energy, $\langle p_e^2/(2m) \rangle \propto 1/a^2$. So the effective temperature T_{eff} would fall as $T_{eff} \propto \langle p_e^2/(2m) \rangle \propto 1/a^2$; see Eq. (2.47). It is the Compton scattering that restores the slower dependence $T \propto 1/a$.

Let us denote the energy density of photons that is tansfered to electrons in time δt by $\delta\rho_\gamma < 0$. The above observation leads to the energy balance equation

$$-\delta\rho_\gamma = n_e \frac{3}{2} \delta T - n_e \delta \left\langle \frac{p_e^2}{2m} \right\rangle , \tag{6.60}$$

where the right-hand side involves the changes caused solely by expansion. Since $p, T \propto 1/a$, one has

$$\delta T = -\frac{\dot{a}}{a} T \delta t = -HT\delta t, \quad \delta\frac{p_e^2}{2m} = -2\frac{p_e^2}{2m} H\delta t.$$

We insert these expressions into (6.60) and obtain

$$-\delta\rho_\gamma = -\left(n_e \frac{3}{2} T - 2n_e \delta \left\langle \frac{p_e^2}{2m} \right\rangle \right) H\delta t = n_e \frac{3}{2} TH\delta t,$$

where we use the fact that $\langle p_e^2/(2m) \rangle = 3T/2$ in kinetic equilibrium. Baryons acquire almost the same kinetic energy (charged baryons are in kinetic equilibrium with electrons; see Sec. 6.3.2), so the total energy loss by photons equals $-\delta\rho_\gamma^{tot} = (3/2)n_{nr}HT\delta t$, where n_{nr} is the number density of all charged non-relativistic particles. For simplicity, we include into n_{nr} electrons and protons only,[3] $n_{nr} = 2\hat{\eta}_B n_\gamma$. Thus, the photon energy density obeys

$$\dot{\rho}_\gamma + 4H\rho_\gamma = \frac{\delta\rho_\gamma^{tot}}{\delta t} = -\frac{3}{2}n_{nr}HT = -\frac{3n_{nr}T}{2\rho_\gamma}\rho_\gamma H,$$

where, as usual, the second term in the left-hand side is due to the cosmological expansion. Note that

$$\frac{3n_{nr}T}{2\rho_\gamma} = \frac{90\zeta(3)}{\pi^4}\hat{\eta}_B = \mathrm{const.}$$

This gives for the total change in the photon energy density from time t_i to time t

$$\rho_\gamma(t) = \rho_{\gamma,i}\left(\frac{a_i}{a(t)}\right)^{4+3n_{nr}T/2\rho_\gamma} = \rho_{\gamma,i}\left(\frac{a_i}{a(t)}\right)^4\left[1 - \frac{90\zeta(3)}{\pi^4}\hat{\eta}_B \ln\left(\frac{a}{a_i}\right)\right].$$

So, the relative change in the photon energy density due to energy transfer to electrons and protons is

$$\frac{\Delta\rho_\gamma}{\rho_\gamma} = -\frac{90\zeta(3)}{\pi^4}\hat{\eta}_B \log\left(\frac{T_i}{T(t)}\right) = -1.1 \cdot \hat{\eta}_B \log\left(\frac{T_i}{T(t)}\right). \tag{6.61}$$

The logarithmic enhancement here is related to the fact that electrons and baryons permanently lose their energy due to the cosmological expansion, and photons permanently heat them up. We will see in Sec. 6.3.3 that energy loss by photons distorts the CMB energy spectrum. We will also discuss in Sec. 6.3.3 the relevant interval of temperatures, which enters (6.61).

6.3.2.   *Coulomb scattering: thermal equilibrium of electron-proton component*

Let us now discuss heating of protons. The direct interaction of proton with photons is irrelevant, since the corresponding time for protons is given by Eq. (6.58) but with m_p substituted for m_e. Since the Thomson cross-section is proportional to m_e^{-2}, the time for protons is larger by a factor $(m_p/m_e)^3$; this time is larger than the Hubble time (6.42).

Energy transfer to protons occurs due to elastic scattering of electrons off protons. The non-relativistic cross-section of the latter process in the proton rest frame is given by the Rutherford formula,

$$d\sigma_R = \frac{\pi\alpha^2}{m_e^2 v^4}\frac{\sin\theta}{\sin^4\theta/2}d\theta, \tag{6.62}$$

where v is the electron velocity and θ is the scattering angle. The differential cross-section (6.62) has power-law singularity $d\theta/\theta^3$ at small scattering angle, so the total cross-section diverges, and the mean free time is formally zero. However,

---

[3] This is not true before recombination of helium. Baryons bound in helium nuclei make up 25% of the total number of baryons; half of the baryons in helium are neutrons. Thus, before recombination of helium one has $n_e = n_B - (0.25/2)n_B \simeq 0.88n_B$, $n_p = 0.75n_B$, $n_{He} = 0.25n_B/4 \simeq 0.06n_B$, $n_{nr} = 1.69\eta_B n_\gamma$. This leads to the correction factor in (6.61) equal to $1.69\eta_B/\hat{\eta}_B = 1.13$. This correction is unimportant for our estimates.

we are again interested in energy transfer time τ_E rather than the mean free time. To obtain its estimate we note that the energy transfer to proton initially at rest is

$$\Delta E = 2\frac{m_e}{m_p}E_e(1 - \cos\theta), \tag{6.63}$$

where E_e is the electron energy.

Problem 6.15. *Obtain the formula (6.63).*

Like in the case of Compton scattering, energy transfer to a moving proton has a sign-indefinite contribution which is suppressed by $\sqrt{m_e/m_p}$ rather than by m_e/m_p as in (6.63). However, upon averaging over angles this term again has the same effect as the term (6.63), so we stick to (6.63).

Making use of (6.62) and (6.63) we obtain the estimate

$$\tau_E \sim \left(n_e v \int d\theta \frac{d\sigma_R}{d\theta}\frac{\Delta E(\theta)}{E}\right)^{-1} = \left(\frac{8n_e\pi\alpha^2}{m_p m_e v^3}\int d\theta\, \mathrm{ctg}\frac{\theta}{2}\right)^{-1}. \tag{6.64}$$

The integral in the right-hand side of (6.64) is still divergent at small angles, but now the divergence is logarithmic. This softening of the divergence is natural: the energy transfer is small at small collision angle. To obtain finite value for the integral in (6.64) we recall that the Coulomb interaction in plasma experiences the Debye screening at distance [78]

$$r_D = \left(\frac{T}{4\pi n_e\alpha}\right)^{1/2}.$$

Hence, the integral in (6.64) must be cut off at the scattering angle corresponding to impact parameter equal to r_D,

$$\theta_D \sim \frac{\alpha}{m_e v^2 r_D}. \tag{6.65}$$

Problem 6.16. *Making use of the Rutherford formula (6.62), obtain the estimate (6.65) for θ_D.*

The leading contribution to the integral in (6.64) comes from small scattering angles $\theta \sim \theta_D$, so we obtain the energy transfer time

$$\tau_E(T) \sim \frac{m_p m_e v^3}{16\pi n_e(T)\alpha^2 \log \theta_D^{-1}} \sim \frac{m_p m_e}{16\pi n_e(T)\alpha^2 \log(6Tr_D/\alpha)}\left(\frac{3T}{m_e}\right)^{3/2}. \tag{6.66}$$

Numerically,

$$\tau_E(T_r) \sim 3 \cdot 10^4 \text{ s},$$

which is a very short time compared to the Hubble time at recombination (6.42). Hence, kinetic equilibrium between electrons and protons is an excellent approximation.

The estimate for protons is valid for electrons as well, with m_e substituted for m_p and modulo a numerical fator of order 1. Combined with the results of Sec. 6.3.1 this means that electrons and protons have equilibrium distrbution functions with temperature equal to photon temperature.

6.3.3.   *Thermal (in)equilibrium of photons*

Similarly to electrons, *kinetic* equilibrium of the photon componet is due to the Compton scattering. The characterisctic time of energy exchange is obtained in the same way as in Sec. 6.3.1, but now the mean free time between the collisions of photons with electrons equals $(\sigma_T n_e)^{-1}$ (rather than $(\sigma_T n_\gamma)^{-1}$ for electrons). As a result, we have for photons

$$\tau_E \sim \frac{m_e}{\sigma_T n_e T}. \tag{6.67}$$

This time is much greater than that of electrons (6.58) because of small value of[4] $n_e \simeq \eta_B n_\gamma$. This time increases in the course of expansion as T^{-4}, which is faster than the increase of the Hubble time. Therefore, photons are in kinetic equilibrium at high temperatures, but eventually the kinetic equilibrium breaks down. This happens, as we show immediately, at radiation domination, when $H = T^2/M_{Pl}^*$ where the effective number of degrees of freedom is $\hat{g}_* = 3,36$; see Sec. 4.4. By equating the time (6.67) to the Hubble time H^{-1} we obtain an estimate for the temperature at which photons get out of kinetic equilibrium:

$$T_{\gamma,kin} \simeq \left(\frac{\pi^2}{2\zeta(3)} \frac{m_e}{M_{Pl}^* \sigma_T \eta_B} \right)^{1/2} \sim 3 \cdot 10^5 \text{ K} .$$

This is indeed much larger than the temperature of the transition to matter domination $T_{eq} = 0.8$ eV $\simeq 10^4$ K.

It is of interest also to estimate the time at which an energetic photon of frequency $T \ll \omega \ll m_e$ loses a substantial part of its energy. This happens when the number of its collisions with electrons is of order $\omega/\Delta E$, where energy loss in each collision is given by (6.55). Therefore, the time scale in queston is

$$\tau_\omega \sim \frac{m_e}{\sigma_T n_e \omega} .$$

Let us apply this to a Lyman-α photon, $\omega = 10, 2$ eV, which is emitted at recombination. We obtain

$$\tau_{Ly-\alpha} \simeq 3 \cdot 10^8 \text{ years} ,$$

---

[4]More precisely, $n_e \simeq 0.88\eta_B n_\gamma$ before recombination of helium; see footnote 3. This subtlety is unimportant for our estimates.

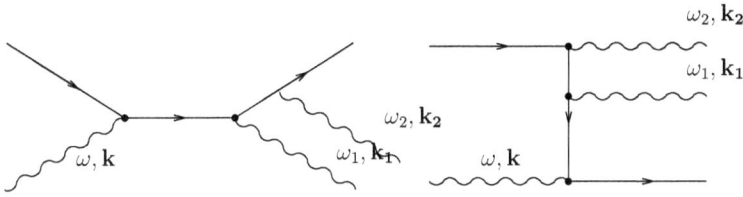

Fig. 6.4. Examples of diagrams of double Compton scattering.

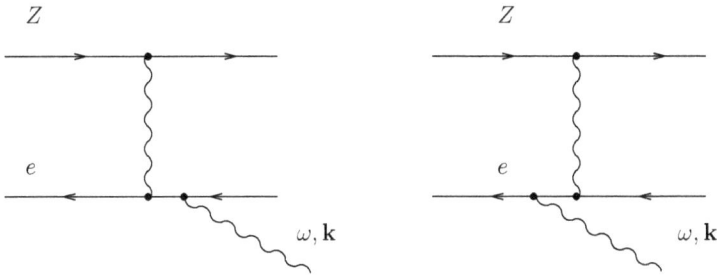

Fig. 6.5. Bremsstrahlung diagram for electron e which scatters off nucleus Z.

which is much greater than the Hubble time at recombination. It is for this reason that we claimed in Sec. 6.2 that Lyman-α photons (and also photons of other lines, $2s-1s$-transitions, etc.) come to us without losing their energies.

Compton scattering changes photon energies, but does not change the total number of photons. This is insufficient for complete thermal equilibrium: it needs processes with emission and absorption of extra photons. These are double Compton process, Fig. 6.4, and bremsstrahlung on protons and helium nuclei, Fig. 6.5.

Let us make a crude estimate of the temperature at which double Compton processes get out of equilibrium. We assume for this purpose that incoming and outgoing photons have energies of order temperature. Then the diagrams of Fig. 6.4 immediately give an estimate for the double Compton cross-section

$$\sigma_{DC} \sim \frac{\alpha^3}{m_e^4}T^2.$$

Hence, the change of the photon number density due to the double Compton processes is roughly estimated as follows:

$$\left(\frac{dn_\gamma}{dt}\right)_{DC} \sim \sigma_{DC} n_e n_\gamma \sim \frac{\alpha^3}{m_e^4}\eta_B T^5 n_\gamma.$$

This gives for the characteristic time

$$\tau_{DC} \sim \left(\frac{\alpha^3}{m_e^4}\eta_B T^5\right)^{-1}.$$

By equating it to the Hubble time $H^{-1} = M_{Pl}^*/T^2$ we obtain a crude estimate for the temperature at which double Compton processes get out of equilibrium:

$$T_{DC} \sim \left(\frac{m_e^4}{M_{Pl}^* \alpha^3 \eta_B} \right)^{1/3} \sim 5 \cdot 10^7 \text{ K} .$$

This estimate, however, is unrealistically high. The point is that the double Compton cross-section is enhanced when the energy of one outgoing photon is low [89] (see also Ref. [90]), $\omega_2 \ll \omega_1 \ll m_e$:

$$d\sigma_{DC} = \frac{4\alpha}{3\pi} \frac{\omega^2}{m_e^2} (1 - \cos\theta_1) d\sigma_T \frac{d\omega_2}{\omega_2} , \qquad (6.68)$$

where θ_1 is the scattering angle of the energetic photon and

$$d\sigma_T = \frac{\alpha^2}{2m_e^2} (1 + \cos^2\theta_1) \sin\theta_1 d\theta_1 d\phi_1$$

is the Thomson differential cross-section. Besides a large numerical factor (which is partially due to integration of the cross-section (6.68) over ω with Planckian distribution), this leads to an effect of qualitative character. Since the cross-section is large at small ω_2, thermal equilibrium is maintained considerably longer at low photon energies (Rayleigh–Jeans region) as compared to the rest of the spectrum. Relatively fast single Compton scattering transfers photons from the Rayleigh–Jeans region to the main part of the spectrum where photon energy is of order T. Thus, the overall chemical equilibrium is maintained longer. An accurate analysis of this phenomenon gives [88]

$$T_{DC} = T_{\gamma,chem} \simeq 6 \cdot 10^6 \text{ K} .$$

The reason for equating the estimate valid for double Compton scattering to overall $T_{\gamma,chem}$ is that the bremsstrahlung processes are somewhat less efficient.

To see that bremsstrahlung is subdominant, we write its differential cross-section for an electron moving with the initial velocity v, which scatters off nucleus of charge Z and emits a photon of frequency ω [81]:

$$d\sigma_{Br} = \frac{16\pi}{3\sqrt{3}} \frac{Z^2\alpha^3}{m_e^2} \frac{g(v,v')}{v^2} \frac{d\omega}{\omega} ,$$

where v' is the final electron velocity, $v'^2 = v^2 - 2\omega/m_e$, and the Gaunt factor $g(v,v')$ has simple forms in limiting cases:

$$g = \begin{cases} \frac{\sqrt{3}}{\pi} \ln \frac{v+v'}{v-v'} , & v,v' \gg \alpha Z \\ 1 , & v' \to 0 \end{cases} .$$

Here the dependence on the photon frequency ω is essentially the same as the dependence of the cross-section (6.68) on ω_2, so the above discussion of soft photons applies. The ratio of times characteristic of double Compton and bremsstrahlung processes is of order[5]

$$\frac{\tau_{DC}}{\tau_{Br}} \simeq \frac{(\langle \sigma_{DC} \rangle n_e n_\gamma)^{-1}}{(\langle \sigma_{Br} v \rangle n_e n_B)^{-1}} \sim \eta_B \left(\frac{m_e}{T} \right)^{5/2},$$

This ratio is small at $T \gtrsim 6 \cdot 10^6$ K, so the double Compton scattering dominates indeed.

To summarize, the processes that change the number of photons get out of thermal equilibrium at the highest temperature $T_{\gamma,chem} \simeq T_{DC}$. This is the temperature of the black body photosphere of the Universe. Any effects that heat up or cool down the photon gas at $T_r < T \lesssim T_{\gamma,chem}$ give rise to the distortion of the observable CMB spectrum. These effects include the heating of electron–proton component which we discussed already, dissipation of inhomogeneities in energy density (Silk effect), and some others. These distortions will hopefully be detected in future precision measurements of the CMB energy spectrum. An exotic possibility is decays of hypothetical long-lived particles (see below in this Section).

Let us now discuss the following situaton. Consider some time before recombination and assume that some photons are injected at that time in a certain spectral region (for definiteness, in the Wien region of high energies); we leave the physics behind this injection unspecified. Since electrons are quickly heated due to the Compton scattering (Sec. 6.3.1) and electron–baryon component rapidly thermalizes by Coulomb interactions (Sec. 6.3.2), its temperature becomes higher than the temperature of photons away from the injection spectral region. The interaction of warmer electrons with photons leads to the distortion of the photon energy spectrum. We are interested in the shape and amplitude of this distortion. Note that instead of injection we could consider absorption of photons (still in the Wien region) or injection of energy directly into electron–baryon component. In all these cases the CMB spectral distortion away from the injection/absorption region is described in the same way. For definiteness we talk about the injection of photons into the Wien region.

The results we are about to discuss are almost literally carried over from the theory developed by Y. Zeldovich, R. Sunyaev and collaborators for describing the heating of CMB photons that travel through hot ionized gas in clusters of galaxies. This is the Zeldovich–Sunyaev (SZ) effect [19]. In cosmological context, the results depend on the epoch of injection.

Suppose first that the injection occured at temperature $T_i > T_{\gamma,chem} \sim 6 \cdot 10^6$ K. Then the Coulomb and double Coulomb scattering lead to complete thermalization of the cosmic plasma, and the CMB spectrum remains Planckian with slightly higher temperature. The only observable consequence in that case is the difference between the baryon-to-photon ratio η_B at Big Bang Nucleosynthesis and at recombination. The only relevant parameter is the total energy of injected photons relative to the energy of photons already present. This completes the analysis of the high injection temperature, $T_i > T_{\gamma,chem}$.

The second possibility is that the injection temperature T_i is in the range $T_{\gamma,kin} < T_i < T_{\gamma,chem}$, i.e., $3 \cdot 10^5$ K $< T_i < 6 \cdot 10^6$ K. In this case, double Coulomb scattering and Bremsstrahlung are switched off, while single Compton scattering establishes kinetic

---

[5]Helium has not yet recombined by the time we discuss, so the bremsstrahlung rate is proportional to $n_e \cdot (n_p + 4n_{He}) = n_e n_B$.

equilibrium between electrons and photons without changing the number of photons. In this situation, the photon energy spectrum has the Bose–Einstein form with non-vanishing chemical potential:

$$f(\omega) = \frac{1}{e^{(\omega-\mu)/T} - 1} \,, \tag{6.69}$$

where the photon temperature is determined by the injected energy, and we introduced the distribution function related to the function $f(\mathbf{p})$ of Sec. 5.1 as follows:

$$f(\omega(\mathbf{p})) = (2\pi)^3 f(\mathbf{p}) \,.$$

This spectral distortion is called μ-type. The chemical potential μ is negative, if energy is injected[6] (there are less photons in the gas as compared to complete thermal equilibrium at temperature T), and, vice versa, μ is positive if energy is absorbed. Note that after the kinetic equilibrium is established, the chemical potential decreases like temperature, $\mu \propto a^{-1}$. This follows from the conservation of the number of photons in comoving volume. The property $\mu \propto a^{-1}$ holds after recombination as well, cf. Sec. 2.5.

Problem 6.17. *Prove the last two statements.*

Problem 6.18. *Show that at small μ photon number density and energy density are*

$$n = \frac{2\zeta(3)}{\pi^2} T^3 + \frac{1}{3}\mu T^2,$$

$$\rho = \frac{\pi^2}{15} T^4 + \frac{6\zeta(3)}{\pi^2} \mu T^3.$$

Note that the first of these relations is the same as (5.21); the factor 1/2, that has to do with the fact that we consider the number of photons rather than the difference between the numbers of particles and antiparticles, is canceled out by the factor 2 coming from two photon polarizations.

Problem 6.19. *Let $\Delta\rho$ be small energy density injected at temperature T_i (where $T_{\gamma,kin} < T_i < T_{\gamma,chem}$),*

1) Making use of the results of Problem 6.18, show that after kinetic equilibrium is established, the photon temperature is $T_i + \delta T$ where δT is related to the chemical potential as follows:

$$\delta T = -\frac{\pi^2}{18\zeta(3)}\mu = -\frac{\mu}{2.19} \,.$$

2) Find the relationship between μ and $\Delta\rho$ and show that, numerically,

$$\frac{\mu}{T} = -1.4\frac{\Delta\rho}{\rho} \,.$$

---

[6]Some authors define the chemical potential with the opposite sign, see, e.g., Ref. [88]. Some other authors use the therm "chemical potential" for a quantity which in our notations is $-\mu/T$; see, e.g., Ref. [91].

The results of the latter Problem together with the results of Sec. 6.3.1 show that the energy transfer from photons to electron–baryon component in the expanding Universe gives rise to the spectral distortion of μ-type at the level

$$\frac{\mu}{T} = +1.4 \cdot 1.1 \cdot 1.13 \cdot 0.75 \eta_B \log\left(\frac{6 \cdot 10^6}{3 \cdot 10^5}\right) \simeq +2.4 \cdot 10^{-9}.$$

(We make use of the correction factor described in footnote 3.) The dissipation of inhomogeneities (Silk effect) produces spectral distortion of μ-type at similar level with negative μ.

Finally, let us consider the case when the injection occurs at $T_i \ll T_{\gamma,kin} \sim 3 \cdot 10^5$. In that case one makes use of an approximate kinetic equation for $\mathfrak{f}(\omega)$, which is obtained from the Boltzmann equation in the limit of small energy transfer in each electron–photon collision. This is the Kompaneets equation, and it reads in the expanding Universe

$$\frac{\partial \mathfrak{f}}{\partial t} - H\omega \frac{\partial \mathfrak{f}}{\partial \omega} = \frac{\sigma_T n_e}{m_e} \frac{1}{\omega^2} \frac{\partial}{\partial \omega}\left[\omega^4\left(T_e \frac{\partial \mathfrak{f}}{\partial \omega} + \mathfrak{f} + \mathfrak{f}^2\right)\right].$$

The derivation of the Kompaneets equation is given in the book [85]; instead of reproducing it here, we make a few comments. First, the second term in the left-hand side describes photon redshift: if the right-hand side vanishes, then the distribution function does not change its form, $\mathfrak{f}(\omega, t) = \mathfrak{f}[\omega a(t)/a_i, t_i]$, cf. Sec. 2.5. Second, one makes use of the fact that the electron component is thermalized at temperature T_e. Third, the evolution of the distribution function at frequency ω depends only of the properties of this function in a neighborhood of ω, i.e., Kompaneets equation is local in ω. This reflects the fact that the energy transfer in each collision is small. Fourth, it follows from this equation that the number of photons in comoving volume is conserved. Indeed this number is

$$na^3 = a^3 \cdot 2 \cdot \int_0^\infty 4\pi\omega^2 d\omega \frac{\mathfrak{f}(\omega, t)}{(2\pi)^3},$$

and we have for its time derivative

$$\frac{1}{a^3}\frac{\partial(a^3 n)}{\partial t} = \frac{\partial n}{\partial t} + 3Hn$$

$$= \int_0^\infty \frac{\omega^2 d\omega}{\pi^2}\left\{3H\mathfrak{f} + H\omega\frac{\partial \mathfrak{f}}{\partial \omega} + \frac{\sigma_T n_e}{m_e}\frac{1}{\omega^2}\frac{\partial}{\partial \omega}\left[\omega^4\left(T_e\frac{\partial \mathfrak{f}}{\partial \omega} + \mathfrak{f} + \mathfrak{f}^2\right)\right]\right\}.$$

The right-hand side is actually equal to zero upon integration by parts. Fifth, the factor m_e^{-1} originates from the energy transfer in each electron–photon collision, which is proportional to ω^2/m_e. Sixth, the right-hand side of the Kompaneets equation vanishes, if the photon distribution function has the Bose–Einstein form (6.69) with $T = T_e$, which has to be the case. Finally, the three terms in the right-hand side are interpreted as follows. The term $T_e\partial\mathfrak{f}/\partial\omega$ is related to the Doppler effect in collisions of photons with moving electrons (it is proportional to T_e rather than $v_e \propto \sqrt{T_e}$ because of angular averaging, see Sec. 6.3.1). The second and third terms describe the retardation effects (electron gains and photon loses energy); the third term is proportional to $\mathfrak{f}^2$ and has to do with induced photon scattering.

Since each electron collides with photons very often, the electron temperature is such that the electron component as a whole does not gain or lose energy (modulo small heating effect in the expanding Universe). So, this temperature obeys

$$\int_0^\infty \frac{\omega^2 d\omega}{\pi^2} \omega \left(\frac{\partial \mathfrak{f}}{\partial t} - H\omega \frac{\partial \mathfrak{f}}{\partial \omega} \right) = 0.$$

This equation means that the total energy density of photons evolves because of the cosmological expansion only. It can be written as follows:

$$T_e = \frac{1}{4} \frac{\int (\mathfrak{f} + \mathfrak{f}^2) \omega^4 d\omega}{\int \mathfrak{f} \omega^3 d\omega}.$$

Since the evolution of the photon spectrum is governed by the small parameter $\sigma_T n_e T/(m_e H)$ (see below), to the zeroth order in this parameter this evolution is solely due to redshift, and hence $T_e \propto a^{-1}$. Having this in mind and introducing the dimensionless frequency

$$x = \frac{\omega}{T_e},$$

we write the Kompaneets equation in terms of it:

$$\frac{\partial \mathfrak{f}}{\partial t} = \frac{\sigma_T n_e T_e}{m_e} \frac{1}{x^2} \frac{\partial}{\partial x} \left[x^4 \left(\frac{\partial \mathfrak{f}}{\partial x} + \mathfrak{f} + \mathfrak{f}^2 \right) \right]. \qquad (6.70)$$

Let us come back to the photon spectral distortion caused by injection of photons at temperature T_i obeying $T_r < T_i \ll T_{\gamma,kin}$. In this temperature range the overall factor in (6.70) is small compared to the Hubble parameter, so the spectral distortion away from the injection region is also small. Photons before injection have Planckian distribution at temperature T,

$$\mathfrak{f}_{Pl}(x_T) = \frac{1}{e^{x_T} - 1},$$

where

$$x_T = \frac{\omega}{T} = x \cdot \frac{T_e}{T};$$

note that $T_e > T$ in the case of injection because of fast electon heating. Therefore,

$$\frac{\partial \mathfrak{f}_{Pl}(x_T)}{\partial x} = \frac{T_e}{T} \frac{\partial \mathfrak{f}_{Pl}(x_T)}{\partial x_T},$$

and the following identity holds:

$$\mathfrak{f}_{Pl}(x_T) + \mathfrak{f}_{Pl}^2(x_T) = -\frac{\partial \mathfrak{f}_{Pl}(x_T)}{\partial x_T}.$$

So, to the leading order in $(T_e - T)$ the right-hand side of Eq. (6.70) is

$$\frac{\sigma_T n_e T_e}{m_e} \left(\frac{T_e}{T} - 1 \right) F(x_T),$$

where

$$F(x) = \frac{1}{x^2} \frac{\partial}{\partial x} \left(x^4 \frac{\partial \mathfrak{f}_{Pl}(x)}{\partial x} \right) = \frac{x e^x}{(e^x - 1)^2} \left(\frac{e^x + 1}{e^x - 1} - 4 \right).$$

(The difference between x and x_T is irrelevant here.) We now write

$$\mathfrak{f} = \mathfrak{f}_{Pl} + \delta \mathfrak{f},$$

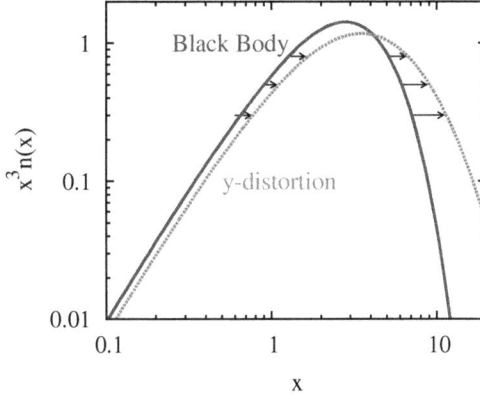

Fig. 6.6.   CMB spectral distortion of y-type [88].

and finally obtain the CMB spectral distortion

$$\delta f = y F\left(\omega/T\right) \, ,$$

where

$$y = \int_{t_i}^{t_r} dt \, \frac{\sigma_T n_e T}{m_e}\left(\frac{T_e}{T} - 1\right) = \int_{T_r}^{T_i} dT \, \frac{\sigma_T n_e}{m_e H}\left(\frac{T_e}{T} - 1\right) \, . \tag{6.71}$$

The spectral distortion has a universal shape (away from the injection region), and the only relevant parameter is its amplitude y. Such a distortion is called y-type; see Fig. 6.6. Since $n_e \propto T^3$, and $H \propto T^2$ ($H \propto T^{3/2}$) at radiation domination (matter domination), the integral in (6.71) is saturated near the injection temperature T_i. The higher T_i, the stronger the distortion at given T_e/T, i.e., at given $\Delta\rho/\rho$. Distortions of y-type caused by the heating of electron–proton component and Silk effect are expected at the level $y = 10^{-8} - 10^{-9}$.

Genuine CMB spectral distortions have not been detected yet (the qualification has to do with Zeldovich–Sunyaev effect). The bound on the distortion of y-type is set by the analysis of the data from FIRAS instrument [92]:

$$|y| \lesssim 1.5 \cdot 10^{-5}$$

at 95% C.L. The bound on the distortion of μ-type is

$$\frac{|\mu_\gamma|}{T_\gamma} < 0.9 \cdot 10^{-4}$$

at 95% C.L.

The distortion of the CMB energy spectrum is possible in extensions of the Standard Model of particle physics in which new particles decay into charged Standard Model particles and/or photons at relatively late time, $z \lesssim 10^7$. Candidates to these new particles are neutralino, charged sleptons, gravitino in supersymmetric extensions of the Standard Model, and also sterile neutrinos and others. Among even more exotic candidates are particles with very small electric charge. The negative results of search for CMB spectral distortion places limits on the parameters of the extensions of the Standard Model. As

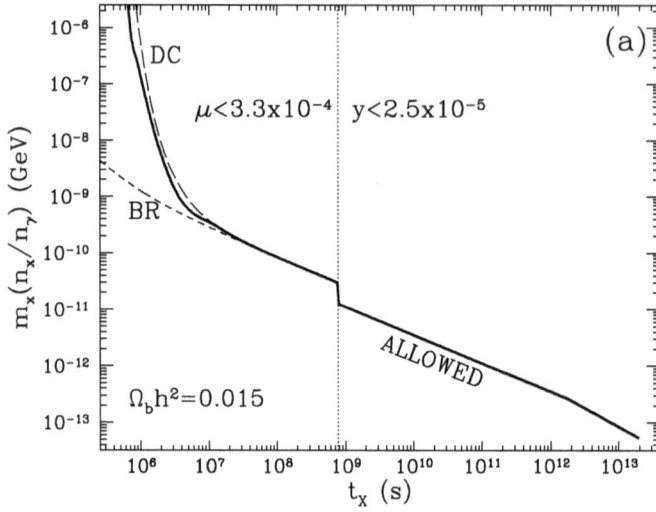

Fig. 6.7. Allowed region in the parameter plane $(m_X n_X / n_\gamma, t_X)$ for a model with new non-relativistic particles of mass m_X, initial number density n_X and lifetime t_X, which decay into photons [93]. The region above the solid line is ruled out by search for CMB spectral distortion; other lines are obtained within various analytical approximations. We note that current bounds on parameters y and μ are somewhat stronger than those used here; they are given in the text.

an example, we show in Fig. 6.7 the limit valid for non-relativistic particles of mass m_X, initial number density n_X and lifetime t_X which decay into photons [93].

Problem 6.20. *Consider a model with new light stable particles $X, \bar{X}$ which have very small electric charge ϵe, $\epsilon \ll 1$, where e is the electron charge. Assuming that they are pair created in electromagnetic interactions in cosmic plasma and that the CMB spectrum is Planckian at the level $\sim 10^{-5}$, find the allowed region in the parameter plane (ϵ, m_X).*

6.4. Horizon at Recombination and its Present Angular Size. Spatial Flatness of the Universe

CMB photons, which last scattered at recombination, give us a photographic picture of the Universe at temperature $T_r = 0.26 \, \text{eV}$ and redshift $z_r = 1100$. The characteristic length scale at recombination epoch is the horizon size at that time, $l_{H,r}$. This scale is of interest for two reasons. First, in the Hot Big Bang theory, regions separated at recombination epoch by distance exceeding $l_{H,r}$ were causally disconnected at that epoch. Second, this scale is imprinted in the properties of CMB photons. We show in the accompanying book that CMB anisotropy and polarization have features in their spectra whose angular sizes are determined, roughly speaking, by the angle at which the horizon at recombination is seen today. In this Section we give the preliminary estimate of this angle and discuss its dependence on cosmological parameters. Our purpose here is to give a general idea of how cosmological parameters are determined from CMB measurements and other observations; this is discussed in more detail in the accompanying book.

In fact, more relevant for CMB is the *sound horizon* at recombination (4.52). This follows from the discussion in Secs. 1.7.1 and 4.7; we will heavily use the notion of the sound horizon in accompanying book. The sound velocity depends on the baryon density, see (5.38), so the sound horizon size depends on η_B. This parameter, however, is well known from the analysis of Big Bang Nucleosynthesis (see Chapter 8) and CMB measurements, so there is basically one-to-one correspondence between the sound horizon and particle horizon. In this Section we will consider particle horizon $l_{H,r}$, having in mind this qualification.

As we have noticed already, recombination occurs at matter dominated epoch. So, we estimate the horizon size at recombination by making use of Eq. (3.26): We write

$$l_{H,r} = \frac{2}{H_r} = \frac{2}{H_0\sqrt{\Omega_M}}\frac{1}{(1+z_r)^{3/2}},\tag{6.72}$$

where $H_r \equiv H(t_r)$ is the Hubble parameter at recombination. Precise value of $l_{H,r}$ is obtained by making use of the result (4.29), which can be written as

$$l_{H,r} = a(t_r)\eta(t_r) = \frac{2}{H_0\sqrt{\Omega_M}}\frac{f}{(1+z_r)^{3/2}},\tag{6.73}$$

where

$$f = \frac{1}{2}\left(\frac{1+z_r}{1+z_{eq}}\right)^{1/2}\mathcal{F}_\eta(T_r/T_{eq}) = \left(1+\frac{T_r}{T_{eq}}\right)^{1/2} - \left(\frac{T_r}{T_{eq}}\right)^{1/2}.$$

Numerically $f = 0.56$ for $\Omega_M = 0.315$.

Problem 6.21. *Neglecting the contribution of radiation, show that the Hubble parameter at recombination can be written as follows:*

$$H(T_r) = \left(\frac{16\zeta(3)}{3\pi}\frac{\Omega_M}{\Omega_B}\eta_B m_p \frac{T_r^3}{M_{Pl}^2}\right)^{1/2},$$

where m_p is proton mass. Thus, the apparent dependence of the expression (6.72) on the parameters of the present Universe is actually not there.

The length interval (6.73) has stretched $(1+z_r)$ times since recombination, so its present size is

$$l_{H,r}(t_0) = \frac{2}{H_0\sqrt{\Omega_M}}\frac{f}{\sqrt{1+z_r}}.$$

Clearly, this size is about $\sqrt{1+z_r} \simeq 30$ times smaller than the present horizon size (4.35); a more precise estimate gives 50 instead of 30. In other words, the Universe visible at present contains about 10^5 regions that were causally disconnected by recombination. (Of course, this is true in the framework of the Hot Big Bang theory only.) Nevertheless, these regions were exactly the same, we know that from both CMB observations and galaxy surveys. How did it happen that despite the absence of causal contact the different regions have exactly the same properties? This question cannot be answered within the Hot Big Bang theory; this is a problem for this theory called horizon problem. The horizon problem finds its elegant solution in inflationary theory.

Let us calculate the present angular size of a region whose spatial size at recombination was equal to $l_{H,r}$. Like in Sec. 4.6, we do not assume that the Universe is spatially flat: as we discuss in this Section at the qualitative level (and quantitatively in the accompanying book), it is the calculation of this sort and its comparison with CMB data that lead to the result that the spatial curvature is close to zero. We assume for simplicity that dark energy density ρ_Λ is constant in time (cosmological constant); the generalization to time-dependent dark energy is straightforward.

Like in Sec. 4.6, let us choose the open cosmological model ($\varkappa = -1$, $\Omega_{curv} > 0$), whose metric is given by (2.10). The last scattering surface is at the coordinate distance

$$\chi_r = \int_{t_r}^{t_o} \frac{dt}{a(t)};$$

(6.74)

CMB photons traveled precisely this distance since recombination. This coordinate distance is still given by Eq. (4.40), where we have to set $z = z_r$. Since $z_r \gg 1$, the integration in (4.40) can be extended to $z_r = \infty$, which corresponds to the limit $t_r \to 0$ in the integral (6.74). In physical terms this means that we neglect the difference between the distance that photon traveled since recombination and the present horizon size. Hence,

$$\chi_r \simeq \chi_{H,0} = \int_0^\infty \frac{dz}{a_0 H_0} \frac{1}{\sqrt{\Omega_M(1+z)^3 + \Omega_\Lambda + \Omega_{curv}(1+z)^2}},$$

(6.75)

which is the coordinate size of the present horizon. Making use of the results of Sec. 4.7, we write the angular size,

$$\Delta\theta_r = \frac{l_{H,r}}{r_a(z_r)},$$

where $r_a(z_r) = (1 + z_r)^{-1} \cdot a_0 \cdot \sinh\chi_r$ is the angular diameter distance to recombination. We finally obtain

$$\Delta\theta_r = \frac{2}{\sqrt{\Omega_M} a_0 H_0 \sinh\chi_r} \frac{f}{\sqrt{z_r + 1}}.$$

(6.76)

When discussing this formula, one should bear in mind the relations (4.45) and (4.46).

Let us begin with the hypothetical case of spatially flat Universe without dark energy, $\Omega_{curv} = 0$, $\Omega_\Lambda = 0$. Spatially flat Universe corresponds to the limit $a_0 \to \infty$, which gives

$$\Delta\theta_r = \frac{f}{\sqrt{z_r + 1}}, \quad \Omega_{curv} = \Omega_\Lambda = 0. \tag{6.77}$$

Of course, this result would be easier to obtain by using directly the formulas for the Universe filled with matter, given in Sec. 3.2.1.

Problem 6.22. *Obtain the formula (6.77) directly in the model of spatially flat Universe filled with non-relativistic matter.*

For $z_r = 1100$ the formula (6.77) gives $\Delta\theta_r = 0.017$, or $\Delta\theta_r = 1°$. A manifestation of the horizon problem is that photons coming from directions separated by more than $1°$ were emitted in causally disconnected regions, and yet they have the same temperature within 0.01%.

Continuing the discussion of formula (6.76) we recall that the angle[7] $\Delta\theta_r$ determines the angular scale of the features in the CMB angular spectrum. Hence, the angle $\Delta\theta_r$ is a measurable quantity. It is therefore of interest to consider the dependence of $\Delta\theta_r$ on cosmological parameters. Since z_r is known, there are only two relevant parameters:[8] due to relations (4.45) and (4.46) this pair may be chosen as $(\Omega_M, \Omega_{curv})$. To see which parameter is more relevant (in the sense that the dependence on it is stronger), let us first consider the case of spatially flat Universe, $\Omega_{curv} = 0$. Unlike in the case (6.77), we now have $\Omega_\Lambda \neq 0$ and therefore $\Omega_M \neq 1$. We take the limit $a_0 \to \infty$, and the expression (6.76) becomes

$$\Delta\theta_r = \frac{f}{\sqrt{z_r + 1}} \frac{1}{I(\Omega_M)}, \quad \Omega_{curv} = 0, \tag{6.78}$$

where

$$I = \frac{\sqrt{\Omega_M}}{2} \int_0^\infty \frac{dz}{\sqrt{\Omega_M(z+1)^3 + \Omega_\Lambda}}, \tag{6.79}$$

and $\Omega_\Lambda = 1 - \Omega_M$. Changing the variable to $(1 + z) = y^{-2}$ we write this integral in the following form,

$$I = \int_0^1 \frac{dy}{\sqrt{1 + \frac{\Omega_\Lambda}{\Omega_M} y^6}}. \tag{6.80}$$

It is clear that if Ω_M is not very small, this integral depends on Ω_Λ/Ω_M rather weakly. At $\Omega_M = 1$, $\Omega_\Lambda = 0$ it is equal to 1, while for $\Omega_M = 0.315$, $\Omega_\Lambda = 0.685$ its

---

[7]More precisely, similar angle determined by the sound horizon; we talk about the angle $\Delta\theta_r$ for definiteness, while the discussion applies to the observable angle as well.
[8]In the realistic case, there is some dependence on other parameters, but one of these parameters, baryon-to-photon ratio, is known from other observations while dependence on others (e.g., neutrino masses) is weak.

value is 0.91. We conclude that the dependence of the angle $\Delta\theta_r$ on the distribution of energy between Ω_M and Ω_Λ is rather weak.

On the other hand, the angle $\Delta\theta_r$ depends on Ω_{curv} quite strongly. To see this, let us consider the hypothetical case $\Omega_\Lambda = 0$, when $\Omega_M + \Omega_{curv} = 1$. In that case the change of variables $(1+z) = y^{-2}$ gives the analytic expression for the integral (6.75), namely

$$\chi_r = 2\,\mathrm{Arsinh}\sqrt{\frac{\Omega_{curv}}{\Omega_M}}, \quad \Omega_\Lambda = 0,$$

where we made use of (4.46). The angle $\Delta\theta_r$ is then given by

$$\Delta\theta_r = \frac{f}{\sqrt{z_r+1}}\frac{1}{\sqrt{1+\Omega_{curv}/\Omega_M}}.$$

Clearly, at $\Omega_{curv} \sim \Omega_M$ this result is quite different from the results (6.77) and (6.78) which are valid for the flat Universe. It becomes clear that the measurement of $\Delta\theta_r$ (more precisely, angular scales related to it) gives strong bound on the spatial curvature of the Universe.

Finally, let us consider the case when Ω_{curv} is small compared to both Ω_M and Ω_Λ, while Ω_M is roughly comparable to Ω_Λ. Then one can neglect Ω_{curv} in the square root in (6.75) (this corresponds to small contribution of curvature into the Friedmann equation, see the discussion in Sec. 4.2), and using (4.46) we obtain

$$\chi_r = 2\sqrt{\frac{\Omega_{curv}}{\Omega_M}}I(\Omega_M,\Omega_\Lambda),$$

where I is the same integral (6.79) or (6.80), but Ω_Λ is no longer equal to $(1-\Omega_M)$. Using (6.76) again, we find

$$\Delta\theta_r = \frac{f}{\sqrt{z_r+1}}\frac{2\sqrt{\Omega_{curv}/\Omega_M}}{\sinh\left(2\sqrt{\Omega_{curv}/\Omega_M}I\right)}. \tag{6.81}$$

The spatial curvature shows up here through the fact that the denominator contains hyperbolic sine rather than linear function; the angle at which a given interval is seen in hyperboloid is smaller than the angle in Euclidean space. The dependence of the right-hand side of (6.81) on Ω_{curv}/Ω_M is rather strong, unlike the dependence on Ω_Λ/Ω_M. This reiterates our conclusion that $\Delta\theta_r$ is particularly sensitive to spatial curvature. Let us note that we again encounter the degeneracy in parameters, as we did in Sec. 4.6, but now the less important parameter is Ω_Λ/Ω_M.

The first measurents of the CMB angular anisotropy at relatively small angular scales (about an angular degree) already gave rise to the conclusion that the spatial curvature of our Universe is small. Together with the SNe Ia observations they lead to the ΛCDM model as the working hypothesis (see Fig. 6.8). Later data are not (yet?) in contradiction with the ΛCDM model, they rather pin down its parameters further. This is illustrated in Fig. 6.9.

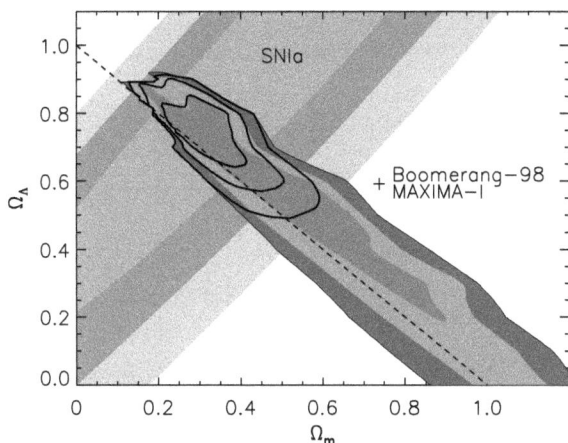

Fig. 6.8. Early bounds on cosmological parameters from the analysis [86] of the CMB anisotropy data together with SNe Ia observations. Regarding contours, see comment to Fig. 4.9 given before Problem 4.9.

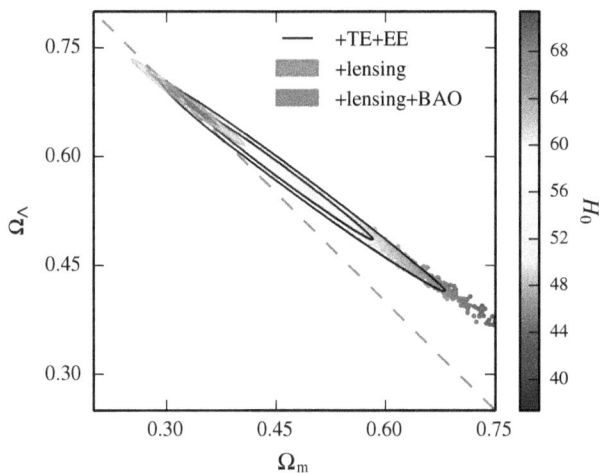

Fig. 6.9. Allowed region in the space of parameters $(\Omega_M, \Omega_\Lambda)$ from CMB anisotropy data (+TE+EE, lensing) and data on baryon acoustic oscilations (BAO), the preferable values of the present Hubble parameter H_0 are indicated in color [94]. Shown here are parameter regions allowed at 68 %–95 % C.L. See Fig. 13.14 for color version.

Problem 6.23. *Calculate the integral* (6.75) *numerically and draw the lines of constant* $\Delta\theta_r$ *on the plane* $(\Omega_\Lambda, \Omega_M)$. *Compare with Fig. 6.8. Hint: Use the relations* (4.45), (4.46).

Problem 6.24. *Find the size of the sound horizon at recombination. Hint: recall that the sound speed in the baryon–photon medium depends on time through the ratio* ρ_B/ρ_γ, *Eq.* (5.38). *At what angle is this sound horizon seen today?*

Chapter 7

Relic Neutrinos

The earlier the period in the history of the Universe, the higher the temperature and density of matter. Hence, those processes which are too weak to be relevant in the present Universe, are important at earlier stages and may leave imprints in the Universe we see today.

We have discussed the phenomenon of this sort in Chapter 6 in the context of relic photons. Here we move further back in time and consider other light particles, neutrinos.

7.1. Neutrino Freeze-Out Temperature

Let us estimate the temperature at which neutrino interactions between themselves and with cosmic plasma switch off ("freeze-out temperature"). We will see that this temperature is of the order of a few MeV. At this epoch, electrons and positrons are still relativistic, and their number density is given by (5.16b). On the other hand, baryons are non-relativistic, so their abundance is suppressed by a factor of order η_B relative to the abundance of e^+e^--pairs. Hence, as far as neutrino freeze-out is concerned, relevant processes are neutrino scattering off electrons, positrons and themselves and neutrino–antineutrino annihilation into e^+e^--pair and neutrino–antineutrino pair of different type, as well as inverse processes. In all these processes, particles are relativistic at temperatures of interest.

We will not need the precise value of the neutrino freeze-out temperature in what follows. So, a dimensional estimate for the cross-sections of the above processes is sufficient for our purposes. Neutrinos participate in weak interactions only (see Appendix C). At energies of interest, the cross-sections are proportional to the Fermi constant squared G_F^2, where

$$G_F = 1.17 \cdot 10^{-5} \, \text{GeV}^{-2}.$$

On dimensional grounds, we immediately obtain for the cross-section of any of the above processes at $E > m_e$,

$$\sigma_\nu \sim G_F^2 E^2,$$

where E is a typical collision energy, $E \sim T$.

The mean free time of neutrino is given, as usual, by

$$\tau_\nu = \frac{1}{\langle \sigma_\nu n v \rangle}, \tag{7.1}$$

where v is the relative velocity of neutrino and a particle it collides with, and n is the number density of the latter particles. In our case all particles are relativistic, the number density is given by (5.16), i.e., $n \sim T^3$ and $v \simeq 1$. In this way we come to the estimate of the mean free time,

$$\tau_\nu \sim \frac{1}{G_F^2 T^5}. \tag{7.2}$$

We now compare τ_ν with the Hubble time (see (3.31))

$$H^{-1} = \frac{M_{Pl}^*}{T^2}. \tag{7.3}$$

We see that as the Universe cools down, τ_ν increases faster than H^{-1}. Thus, in accordance with our expectation, at early stages the mean free time is shorter than the Hubble time, and neutrinos are in thermal equilibrium with matter. Indeed, the number of neutrino collisions since time t is estimated as

$$N(t) \sim \int_t^\infty \frac{dt'}{\tau_\nu(t')} = \int_t^\infty \frac{dt'}{t'} \frac{t'}{\tau_\nu(t')} \sim \frac{t}{\tau_\nu(t)} \sim \frac{1}{H(t)\tau_\nu(t)},$$

where we made use of the fact that

$$\frac{t}{\tau_\nu(t)} \sim \frac{1}{H(t)\tau_\nu(t)}$$

rapidly decreases in time. If $N(t) \gg 1$ then neutrinos are in thermal equilibrium, while for $N(t) \ll 1$ they are non-interacting. Hence, neutrino interactions switch off at

$$\tau_\nu(T) \sim H^{-1}(T).$$

It follows from (7.2) and (7.3) that this happens at freeze-out temperature[1]

$$T_{\nu,f} \sim \left(\frac{1}{G_F^2 M_{Pl}^*} \right)^{1/3} \sim 2 - 3 \, \text{MeV}.$$

Problem 7.1. *Estimate the age of the Universe at neutrino freeze-out.*

---

[1] This temperature weakly depends on the neutrino flavor: electron neutrinos interact with electrons and positrons via exchange of both $W^\pm$- and Z-bosons, while muon and tau neutrinos participate only in reactions with Z-boson exchange. Therefore, the annihilation and scattering cross-sections are somewhat smaller for ν_μ and ν_τ as compared to ν_e, and ν_μ and ν_τ freeze out somewhat earlier. Also, neutrino oscillations play a role. All these subtleties are unimportant here.

Thus, at freeze-out temperature $T_{\nu,f}$ neutrinos interacted for the last time. Since then they freely propagate through the Universe and their number in comoving volume does not change. The very assumption that there existed temperatures of the order of a few MeV leads to the conclusion that the present Universe contains the gas of relic neutrinos analogous to the gas of relic photons, CMB. We will see that the latter assumption is strongly supported by Big Bang Nucleosynthesis theory and observations, and also other data, so there is no doubt that relic neutrinos indeed exist.

7.2. Effective Neutrino Temperature. Cosmological Bound on Neutrino Mass

It follows from the results of Sec. 2.5 that neutrinos after freeze-out are still described by relativistic distribution function with present effective temperature

$$T_{\nu,0} = T_{\nu,f} \frac{a(t_\nu)}{a(t_0)} = \frac{T_{\nu,f}}{1 + z_\nu}, \tag{7.4}$$

where z_ν is the redshift at neutrino freeze-out. At the time of freeze-out, neutrino temperature equals to that of photons. After freeze-out, the photon temperature also decreases due to the cosmological expansion. However, at neutrino freeze-out, the cosmic plasma contained also a lot of relativistic electrons and positrons. As the temperature drops below the electron mass, electrons and positrons annihilate away, injecting energy into the photon component. Due to this process, the photon temperature becomes higher than the effective neutrino temperature. This effect is quantitatively described by making use of entropy conservation of the electron–photon component in comoving volume,

$$g_*^{e\,\gamma}(T) a^3 T^3 = \text{const}, \tag{7.5}$$

where $g_*^{e\,\gamma}(T)$ is the effective number of relativistic degrees of freedom in the electron–photon plasma, see (5.28) and (5.30). Right after neutrino freeze-out, this plasma consists of relativistic electrons, positrons and photons, which gives

$$g_*^{e\,\gamma}(T_{\nu,f}) = 2 + \frac{7}{8}(2 + 2) = \frac{11}{2}.$$

After $e^+ e^-$-annihilation, the entropy is due to photons only, $g_*^{e\,\gamma} = 2$, and we arrive at the following result for temperatures well below electron mass (including present epoch),

$$\frac{T_{\gamma,0}}{T_{\nu,0}} = \left(\frac{g_*^{e\,\gamma}(T_{\nu,f})}{g_*^{e\,\gamma}(T_0)} \right)^{1/3} = \left(\frac{11}{4} \right)^{1/3} \simeq 1.4. \tag{7.6}$$

We conclude that the present neutrino temperature is[2]

$$T_\nu(t_0) \simeq 1.95 \, \text{K}. \tag{7.7}$$

Making use of (5.16) we find the present number density of each species of neutrinos together with its antineutrinos,

$$n_{\nu,0} = \frac{3}{4} \cdot 2 \cdot \frac{\zeta(3)}{\pi^2} T_\nu^3(t_0) \simeq 112 \, \text{cm}^{-3}. \tag{7.8}$$

Problem 7.2. *Let us make a wrong assumption that there is no Z^0-boson in Nature, while $W^\pm$ bosons exist. Furthermore, let us neglect loop processes and neutrino oscillations (which is also wrong). What are the relic abundances of neutrinos of different types?*

Direct detection of relic neutrinos is unsolved and is a very hard problem, in view of extremely small cross-section of low energy neutrino interactions and tiny energy release.

Problem 7.3. *Assuming that neutrinos are massless, estimate the mass of a detector required for having one relic neutrino interaction per year.*

Neutrino oscillation data are consistent with the possibility that one neutrino species is massless. Contribution of massless neutrino to the present energy density would be small. Making use of (5.12) we would find for one type of neutrino plus its antineutrino,

$$\Omega_\nu = 2\frac{7}{8}\frac{\pi^2}{30}\frac{T_{\nu,0}^4}{\rho_c} \approx 10^{-5},$$

so massless neutrinos would make very little effect on the expansion of the present Universe. It is worth noting that this was not always the case: neutrino contribution was sizeable at radiation domination, and, indeed, Big Bang Nucleosynthesis imposes strong bound on the number of neutrino species, as we discuss in Chapter 8.

The situation is quite different for neutrinos of mass $m_\nu > T_{\nu,0}$. Their present energy density is

$$\rho_{\nu,0} = m_\nu n_{\nu,0},$$

and the relative contribution to the total energy density is

$$\Omega_\nu = \frac{\rho_{\nu,0}}{\rho_c} \approx \left(\frac{m_\nu}{1 \, \text{eV}}\right) \cdot 0.01 \, h^{-2}. \tag{7.9}$$

Let us require that the neutrino energy density does not exceed the total energy density of non-relativistic matter. Recalling that there are three types of neutrinos,

---

[2]Let us recall that *massive* neutrinos today have *relativistic* distribution in momenta with effective temperature (7.6). Their abundance is independent of their mass and is given by (5.16).

we obtain the cosmological bound on the sum of their masses [98],

$$\sum_i m_{\nu_i} < 100 \cdot h^2 \Omega_M \text{ eV}. \tag{7.10}$$

With conservative bound of the early 1990s, $\Omega_M < 0.4$ and $h = 0.7$, we obtain

$$\sum_i m_{\nu_i} < 20 \text{ eV}.$$

For a long time a similar bound was the strongest bound on the masses of μ- and τ-neutrino. Presently one can combine the direct limit on electron neutrino mass [1], $m_{\nu_e} < 2$ eV, with the results of neutrino oscillation experiments. The latter show that the differences of masses squared Δm^2 between ν_e, ν_μ and ν_τ are small, $\Delta m^2 \lesssim 5 \cdot 10^{-3}$ eV^2. So, all types of neutrinos have low masses,

$$m_\nu < 2 \text{ eV}, \tag{7.11}$$

see Appendix C. The bound (7.11) and the relation (7.9) show that the contribution of all types of neutrinos into energy density is not very large,

$$\sum_i \Omega_{\nu_i} < 0.12. \tag{7.12}$$

Nevertheless, by comparing this to the relative energy density of all non-relativistic matter, $\Omega_M \approx 0.3$, we see that the bound (7.12) *per se* allows neutrinos to be substantial part of dark matter. However, the data on cosmic structures and CMB exclude the relative contribution of neutrinos at the level

$$\sum_i \Omega_{\nu_i} h^2 < 0.002\text{--}0.005, \tag{7.13}$$

depending on details of the analysis, see Fig. 7.1. This corresponds to the bound on the sum of neutrino masses [94, 99]

$$\sum_i m_{\nu_i} < 0.2\text{--}0.5 \text{ eV} \tag{7.14}$$

and rules out neutrino as a dark matter candidate. We discuss the origin of the bound (7.13) in the accompanying book.

It is worth noting that neutrino oscillation data (Appendix C) tell that the differences in neutrino masses squared are small: the largest of them equals $m_{atm}^2 \approx (0.05 \text{ eV})^2$. Therefore, the limit (7.14) means that the mass of *each type* of neutrino is bounded from above:

$$m_{\nu_i} \lesssim 0.2 \text{ eV}. \tag{7.15}$$

This is substantially stronger than the direct experimental bound (7.11).

To end this Section we note that the above results have been obtained under the assumption that there is essentially no asymmetry between neutrinos and antineutrinos in the Universe. In other words, we assumed that the neutrino chemical

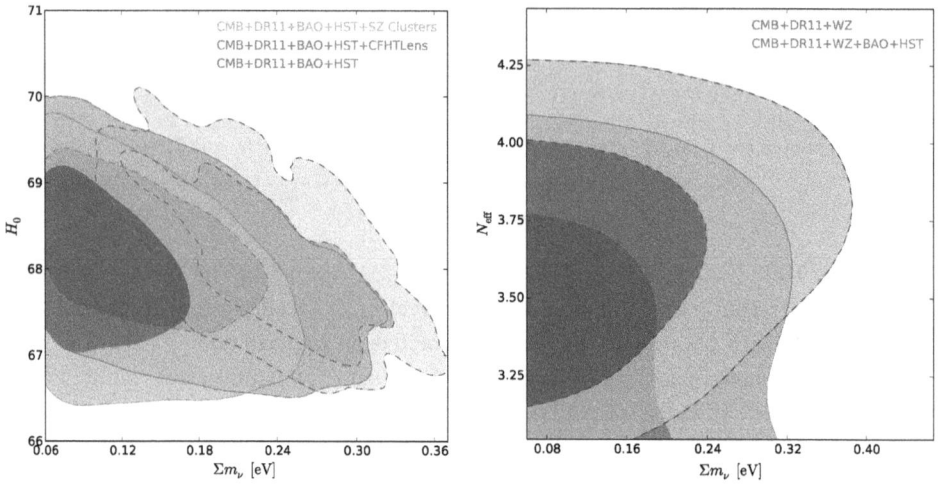

Fig. 7.1.   Limits on the sum of neutrino masses obtained under various assumptions on other cosmological parameters and using different cosmological datasets [104]; H_0 (left) is the present value of the Hubble parameter, N_{eff} (right) is the effective number of neutrino species. See Fig. 13.15 for color version.

potential is close to zero. This assumption is quite plausible, especially in view of the observation that electroweak processes at $T \gtrsim 100$ GeV equalize, modulo factor of order 1, lepton and baryon asymmetries (see Chapter 11), and that the baryon asymmetry is very small, $\eta_B \sim 10^{-9}$. Nevertheless, one cannot rule out completely the possibility of substantial neutrino–antineutrino asymmetry. In that case, neutrino oscillation data (in particular, lower bound on the mass of the heaviest neutrino, $m_\nu > m_{atm} \approx 0.05$ eV), together with the bound (7.13) can be used for placing the bound on the lepton asymmetry in the present and early Universe. This is not a particularly strong bound, and we leave it for the problem.

Problem 7.4. *Making use of the neutrino oscillation results (Appendix C) and the bound (7.13), obtain the bound on neutrino asymmetry in the present Universe defined as*

$$\Delta_{L,0} = \frac{\sum_i (n_{\nu_i} - n_{\bar{\nu}_i})}{s},$$

where s is entropy density.

Stronger bound on the lepton asymmetry is obtained from Big Bang Nucleosynthesis. In Sec. 8.1 we obtain the bound on the chemical potential of electron neutrino at temperature of order 1 MeV at the level $|\mu_{\nu_e}/T| < 0.025$. Precise bound (at 68% confidence level) is

$$-0.023 < \frac{\mu_{\nu_e}}{T} < 0.014, \quad T \sim 1\,\mathrm{MeV}. \tag{7.16}$$

In fact, this bound applies to all types of neutrinos, since neutrino oscillations equalize neutrino abundances before neutrino freeze-out. Indeeed, the typical time for neutrino oscillations at energy E is (see Appendix C)

$$t_{osc} \simeq \pi \frac{4E}{\Delta m^2}.$$

At $E \simeq 3T$, $T \simeq 3$ MeV and even for the smallest value $\Delta m_{sol}^2 \simeq 8 \cdot 10^{-5}$ eV^2 this time is

$$t_{osc} \simeq 5 \cdot 10^{-4} \, \text{s},$$

which is much shorter than the Hubble time at freeze-out, $H^{-1}(T \simeq 3 \, \text{MeV}) \sim 0.1 \, \text{s}$. Note that we neglected matter effects in our estimate; this is indeed a reasonable approximation at temperature of about 3 MeV (see [95, 96] for details). Hence, at neutrino freeze-out

$$\mu_{\nu_e} = \mu_{\nu_\mu} = \mu_{\nu_\tau}, \quad T \sim 3 \, \text{MeV}.$$

This means that the bound (7.16) indeed applies to all types of neutrinos.

Neutrinos excess over antineutrinos is given by (5.22), and the neutrino number density itself by (5.16b). So, for each type of neutrinos we have

$$\frac{n_\nu - n_{\bar{\nu}}}{n_\nu + n_{\bar{\nu}}} = \frac{\pi^2}{9\zeta(3)} \frac{\mu_\nu}{T}.$$

This asymmetry persists until the present time, so we find from (7.16) that for each type of neutrinos

$$\frac{|n_\nu - n_{\bar{\nu}}|}{n_\nu + n_{\bar{\nu}}} < 0.02.$$

We conclude that the excess of neutrinos over antineutrinos (or vice versa) is very small in our Universe.

7.3. Sterile Neutrinos as Dark Matter Candidates

The existence of neutrino oscillations points towards incompleteness of the Standard Model of particle physics (see Appendix C for details). Some extensions of the Standard Model include new particles, sterile neutrinos. These may be described as Majorana fermions which mix with conventional neutrinos. They are sterile in the sense that the new fields do not interact with gauge fields of the Standard Model; in particular, they do not directly participate in weak interactions. Conventional neutrinos, interacting with W- and Z-bosons, are called *active* in this context. The total number of neutrino states is then equal to $3 + N_s$ where N_s is the number of sterile neutrino species. We will restrict ourselves to the case $N_s = 1$ for definiteness. We note here that sterile neutrinos have not been discovered (yet?), although there are anomalies (at not so high confidence levels) in various neutrino oscillation experiments which may be interpreted in terms of sterile neutrinos. In any case, such an extension of the Standard Models is quite reasonable and worth discussing.

We are going to consider the relatively light sterile neutrino, $m_s \lesssim 100$ MeV. In the simplest models [100–102] the creation of sterile neutrino states $|\nu_s\rangle$ in the early Universe occurs due to their mixing with active neutrinos $|\nu_\alpha\rangle$, $\alpha = e, \mu, \tau$. In the approximation of mixing between two states only, we have

$$|\nu_\alpha\rangle = \cos\theta_\alpha |\nu_1\rangle + \sin\theta_\alpha |\nu_2\rangle, \quad |\nu_s\rangle = -\sin\theta_\alpha |\nu_1\rangle + \cos\theta_\alpha |\nu_2\rangle,$$

where $|\nu_\alpha\rangle$ and $|\nu_s\rangle$ are active and sterile neutrino states, $|\nu_1\rangle$ and $|\nu_2\rangle$ are mass eigenstates of masses $m_1 < m_2$, and θ_α is the vacuum mixing angle between sterile and active neutrino. Let us assume that mixing is weak, $\theta_\alpha \ll 1$, so that the heavy state is mostly sterile neutrino $|\nu_2\rangle \approx |\nu_s\rangle$. In this situation the mass of the heavy state is naturally called the sterile neutrino mass, $m_2 \equiv m_s$. Let us assume that the sterile neutrino mass is large compared to the mass of active neutrino, $m_s \gg m_1$. All these assumptions are very natural from particle physics viewpoint.

The calculation of probability of oscillation $\nu_\alpha \leftrightarrow \nu_s$ is done in the same way as in the case of oscillation between different types of active neutrinos (see Sec. C.1). For relativistic neutrino, $E_\nu \gg m_s$, the probability of the transition $\nu_\alpha \to \nu_s$ in time t in vacuo is

$$P(\nu_\alpha \to \nu_s) = \sin^2 2\theta_\alpha \cdot \sin^2\left(\frac{t}{2t_\alpha^{vac}}\right),$$

$$t_\alpha^{vac} = \frac{2E_\nu}{\Delta m^2}, \quad \Delta m^2 = m_s^2 - m_1^2 \simeq m_s^2. \tag{7.17}$$

Cosmic plasma effects, however, are quite important, especially at high temperatures. The plasma affects the propagation of active neutrino $|\nu_\alpha\rangle$, so the Hamiltonian in the active-sterile basis $(|\nu_\alpha\rangle, |\nu_s\rangle)$ is

$$H = U \cdot \text{diag}\left(\frac{m_1^2}{2E_\nu}, \frac{m_2^2}{2E_\nu}\right) \cdot U^\dagger + V_{int}, \tag{7.18}$$

where the mixing matrix U and matrix V_{int} describing matter effects are

$$U = \begin{pmatrix} \cos\theta_\alpha & \sin\theta_\alpha \\ -\sin\theta_\alpha & \cos\theta_\alpha \end{pmatrix}, \quad V_{int} = \begin{pmatrix} V_{\alpha\alpha} & 0 \\ 0 & 0 \end{pmatrix}.$$

The quantity $V_{\alpha\alpha}$ can be calculated for $\nu_\alpha = \nu_e, \nu_\mu, \nu_\tau$ by making use of the methods of quantum field theory at finite temperatures. These methods are presented in Appendix D. Assuming the absence of lepton asymmetry, one finds that the contributions to $V_{\alpha\alpha}$ cancel out to the first order in G. Most important are the contributions of the second order in G_F. At interesting temperatures, $T \lesssim 100$ MeV, there are no τ-leptons and few muons in the medium, but there are numerous relativistic electrons and positrons. For this reason, $V_{\alpha\alpha}$ are quite different for different types of neutrinos. One has the following expression [103] for ν_τ,

$$V_{\tau\tau} = -\frac{14\pi}{45\alpha} \sin^2\theta_W \cos^2\theta_W \cdot G_F^2 T^4 \cdot E_\nu \approx -25 \cdot G_F^2 T^4 \cdot E_\nu.$$

The numerical coefficient is 3.5 times larger for electron neutrino, while the corrsponding factor for muon neutrino is between 1 and 3.5 depending on the ratio of temperature to muon mass.

By diagonalizing the Hamiltonian (7.18) one obtains effective masses and mixing angles of neutrinos in medium, which are different from those in vacuo. As a result, the oscillation probability is given by the formula similar to (7.17) but with different mixing angle $\theta_\alpha^{\text{mat}}$ and period of oscillations t_α^{mat},

$$P(\nu_\alpha \to \nu_s) = \sin^2 2\theta_\alpha^{\text{mat}} \cdot \sin^2\left(\frac{t}{2t_\alpha^{\text{mat}}}\right), \tag{7.19a}$$

$$t_\alpha^{\text{mat}} = \frac{t_\alpha^{vac}}{\sqrt{\sin^2 2\theta_\alpha + (\cos 2\theta_\alpha - V_{\alpha\alpha} \cdot t_\alpha^{vac})^2}}, \tag{7.19b}$$

$$\sin 2\theta_\alpha^{\text{mat}} = \frac{t_\alpha^{\text{mat}}}{t_\alpha^{vac}} \cdot \sin 2\theta_\alpha. \tag{7.19c}$$

Note that $V_{\alpha\alpha}$ has negative sign, so there is no resonance: $\sin 2\theta_\alpha^{\text{mat}}$ monotonously increases in time and tends to the vacuum value $\sin 2\theta_\alpha$: mixing in plasma is suppressed with respect to that in vacuum. At interesting temperatures, when the sterile neutrino mass is in the range $m_1 \ll m_s \lesssim T$, we have the following estimate for the time of oscillations in the medium,

$$t_\alpha^{\text{mat}} = \min(t_\alpha^{vac}, |V_{\alpha\alpha}|^{-1}) = \min(2Tm_s^{-2}, (0.04 - 0.01) \cdot T^{-5} \cdot G_F^{-2}). \tag{7.20}$$

The typical oscillation time is not only smaller than the Hubble time $H^{-1}(T)$, but also smaller than the typical time of weak interactions τ_ν given by (7.2).

Scattering of active neutrino leads to the collapse of its wave function. Hence, during time τ_ν active and sterile neutrinos oscillate, and at the moment of scattering the coherence is destroyed. Every active neutrino ν_α oscillates into sterile neutrino ν_s many times before it collides with a particle in plasma. Therefore, the probability that after time τ_ν this neutrino becomes the sterile neutrino is

$$\langle P(\nu_\alpha \to \nu_s) \rangle = \frac{1}{2} \cdot \sin^2 2\theta_\alpha^{\text{mat}}, \tag{7.21}$$

where we performed averaging of (7.19a) over several oscillation periods. At temperatures $T \gtrsim 3$ MeV, when active neutrinos are in thermal equilibrium with plasma, this process does not reduce the number of active neutrinos but it leads to production of sterile neutrinos. The production rate per unit volume $\Gamma_{\alpha\to s}$ equals the product of averaged oscillation probability (7.21) and half of the rate[3] of active

---

[3]This factor 1/2 is a result of accurate calculation [97]. Yet it admits heuristic interpretation. If both active and sterile neutrinos interacted with plasma at the same rate, then this rate would be

neutrino interactions in plasma $\Gamma_\alpha = \tau_{\nu_\alpha}^{-1}$:

$$\Gamma_{\alpha \to s} = \frac{1}{2}\Gamma_\alpha \cdot \langle P(\nu_\alpha \to \nu_s)\rangle. \tag{7.22}$$

Although we are interested here in neutrinos with energy $E \sim T$, let us write for future reference the expression for Γ_α in the case of τ neutrino with energy E [103]

$$\Gamma_\tau = \frac{7\,\pi}{24} G_F^2 T^4 E. \tag{7.23}$$

Thus, the production rate of sterile neutrinos per unit volume is approximately given by

$$\Gamma_{\alpha \to s} n_{\nu_\alpha} \sim \tau_{\nu_\alpha}^{-1} \cdot \langle P(\nu_\alpha \to \nu_s)\rangle \cdot n_{\nu_\alpha}.$$

This gives for the density of sterile neutrinos n_{ν_s} (assuming that they do not decay and neglecting the inverse process of their oscillation into active neutrinos)

$$\frac{dn_{\nu_s}}{dt} + 3Hn_{\nu_s} = \tau_\nu^{-1} \cdot \langle P(\nu_\alpha \to \nu_s)\rangle \cdot n_{\nu_\alpha}, \tag{7.24}$$

where the second term in the left-hand side is due to the cosmological expansion. Equation (7.24) is conveniently written as the equation for the sterile-to-active ratio,

$$\frac{d(n_{\nu_s}/n_{\nu_\alpha})}{d\log T} = -\frac{\langle P(\nu_\alpha \to \nu_s)\rangle}{H(T)\tau_\nu}, \tag{7.25}$$

where we use the argument T instead of t and disregard the temperature dependence of the effective number of degrees of freedom g_*. The temperature dependence of the mixing angle in matter $\theta_\alpha^{\mathrm{mat}}$ is strong, so the right-hand side of Eq. (7.25) behaves as T^3 at low temperatures and as T^{-9} at high temperatures. This implies that the production of sterile neutrinos takes place mostly in a narrow temperature interval around some critical temperature T_*. The latter is determined from the requirement that the right-hand side of (7.25) is at maximum, i.e., the two expressions in parenthesis in (7.20) are of the same order of magnitude. This gives

$$T_* \sim \left(\frac{m_s}{5G_F}\right)^{1/3} \simeq 200\,\mathrm{MeV} \cdot \left(\frac{m_s}{1\,\mathrm{keV}}\right)^{1/3}. \tag{7.26}$$

As the sterile neutrino production rate has strong power-law dependence on temperature, their number density is estimated as

$$\frac{n_{\nu_s}(T_*)}{n_{\nu_\alpha}(T_*)} \sim \frac{\sin^2 2\theta_\alpha}{H(T_*) \cdot \tau_\nu(T_*)} \sim T_*^3 M_{Pl}^* G_F^2 \cdot \sin^2 2\theta_\alpha$$

$$\sim 10^{-2} \cdot \left(\frac{m_s}{1\,\mathrm{keV}}\right) \cdot \left(\frac{\sin 2\theta_\alpha}{10^{-4}}\right)^2. \tag{7.27}$$

---

equal to the decoherence rate. In reality, the neutrino state is half of the time active and half of the time sterile. Sterile neutrino does not scatter in plasma, so the decoherence rate is twice lower.

We note that our estimate for sterile-to-active ratio (7.27) coincides with the ratio of the rate of sterile neutrino production to the expansion rate entering (7.25) at temperature $T = T_*$ when this ratio is at maximum. This implies that sterile neutrinos are never in thermal equilibrium with plasma, provided that the ratio (7.27) is much smaller than 1.

The number of sterile neutrinos in comoving volume remains nearly constant afterwards, and so does the ratio[4] n_{ν_s}/n_{ν_α}. Making use of the expression (7.8) for the number density of active neutrinos, we find from (7.27) the estimate for the present contribution of sterile neutrinos into energy density,

$$\Omega_{\nu_s} \simeq 0.2 \cdot \left(\frac{\sin 2\theta_\alpha}{10^{-4}} \right)^2 \cdot \left(\frac{m_s}{1 \text{ keV}} \right)^2 . \tag{7.28}$$

Accurate calculations show that this estimate is reasonably precise. Thus, sterile neutrino of mass $m_\nu \gtrsim 1$ keV and small mixing angle $\theta_\alpha \lesssim 10^{-4}$ could serve as dark matter candidate. At $m_s \sim 1$–10 keV this would be *warm* dark matter, since the sterile neutrino momenta at production coincide with momenta of active neutrinos, and energy of sterile neutrinos is of order T until the temperature drops below m_s. Sterile neutrinos of masses $m_s \lesssim 1$ keV are too hot to be dark matter particles (see discussion in Sec. 9.1).

To quantify whether dark matter is cold, warm or hot, one calculates the distribution function of dark matter particles, in our case $f_s(\mathbf{p}, t)$. Neglecting rare processes of sterile neutrino conversion into active neutrino, we write the Boltzmann equation [100]

$$\frac{df_s}{dt} = f_\alpha \cdot \Gamma_{\alpha \to s},$$

where $f_\alpha = f_\alpha(\mathbf{p}, t)$ is the distribution function of active neutrinos ν_α, i.e., the Fermi–Dirac function (5.4) with zero chemical potential, and $\Gamma_{\alpha \to s}$ is given by (7.22). We note that the total time derivative accounts, in particular, for the the time-dependence of momentum, $\mathbf{p} \propto 1/a$, so the Boltzmann equation is written as follows:

$$\frac{\partial}{\partial t} f_s - H\mathbf{p} \frac{\partial}{\partial \mathbf{p}} f_s = \frac{1}{4} \Gamma_\alpha \sin^2 2\theta_\alpha^{\text{mat}} f_\alpha(t, \mathbf{p}) . \tag{7.29}$$

Since particles are relativistic and plasma is isotropic, one makes use of energy $E = |\mathbf{p}|$ as an argument of f instead of momentum $\mathbf{p}$. It is again convenient to use the temperature of cosmic medium T istead of time t. Making use of (3.34) we write

$$-\left(T \frac{\partial f_s}{\partial T} + E \frac{\partial f_s}{\partial E} \right) H = \frac{1}{4} \Gamma_\alpha \sin^2 2\theta_\alpha^{\text{mat}} f_\alpha(T, E). \tag{7.30}$$

We now introduce the dimensionless quantity $y \equiv E/T$ instead of E and note that the equilibrium distribution f_α depends precisely on y, while the expression in parenthesis in Eq. (7.30) is the partial derivative of f_s over temperature at fixed y. This enables us to

---

[4]For the same reason as in the case of photons in Sec. 7.2, the number of active neutrinos in comoving volume increases by a factor of order 1 as the effective number of relativistic degrees of freedom decreases until the epoch of neutrino decoupling. This does not significantly modify our order-of-magnitude estimate.

integrate the Boltzmann equation and find explicitly the distribution function of sterile neutrinos. As an example, for τ neutrino transitions we get, making use of (7.23),

$$f_s(y,T) = f_\tau(y)\frac{7\pi}{96}G_F^2\sin^2 2\theta_\tau \int_T^\infty \frac{y\,T'^4\,dT'}{H(T')\cdot\left(1 + \frac{7\pi}{45\alpha}\sin^2 2\theta_W\,G_F^2 T'^6 y^2\right)^2}. \qquad (7.31)$$

At finite T this gives a non-trivial distribution function. Note, however, that the final distribution function has the same shape as $f_\tau(E/T)$. Indeed, since $H \propto T^2$, we introduce a new integration variable $z = y\,T^3$ in (7.31) and set the lower limit of integration equal to zero. Then the integral becomes a dimensionful constant, and hence the final function is

$$f_s(y,T) = f_s(|\mathbf{p}|/T) = \text{const} \cdot f_\tau(|\mathbf{p}|/T). \qquad (7.32)$$

Problem 7.5. *Find the value of constant in (7.32). Using this value, refine the estimate (7.28).*

This is an example of a situation in which an out-of-equilibrium component has equilibrium shape of the energy spectrum. The only consquence of inequilibrium is that the constant in (7.32) is much smaller than in equilibrium. The shape of the distribution function immediately gives the average momentum of sterile neutrinos, which coincides with (5.17):

$$\langle p_s \rangle = 3.15\,T. \qquad (7.33)$$

We see that all bounds on dark matter whose particles decouple being relativistic are valid for sterile neutrinos which have never been in thermal equilibrium. The strongest bound comes from the analysis of the evolution of phase space density. This is discussed in the accompanying book. For sterile neutrinos with the distribution function (7.32) the bound is [135]

$$m_s > 5.7 \text{ keV}. \qquad (7.34)$$

The sterile neutrino is, generally speaking, unstable due to mixing with active neutrinos. The main channel is the decay into three active neutrinos. The requirement that the lifetime is larger than the present age of the Universe gives the limit on the parameters of the model [101]:

$$\theta_\alpha^2 < 1.6 \cdot 10^{-7}\left(\frac{50\text{ keV}}{m_s}\right)^5. \qquad (7.35)$$

Problem 7.6. *Derive the limit (7.35).*

Sterile neutrino decays also into active neutrino and photon; this occurs at one loop. This process leads to much stronger bound on the parameters: the model should not contradict measurements of the flux of cosmic X-rays and higher energy photons. This bound is about $6 \cdot 10^6$ times stronger [106] than (7.35) in the sterile neutrino mass range 1–10 keV. It is shown in Fig. 7.2.

This bound is actually so strong that, taken together with the limit (7.34), it disfavors the non-resonant production of sterile neutrino dark matter discussed so

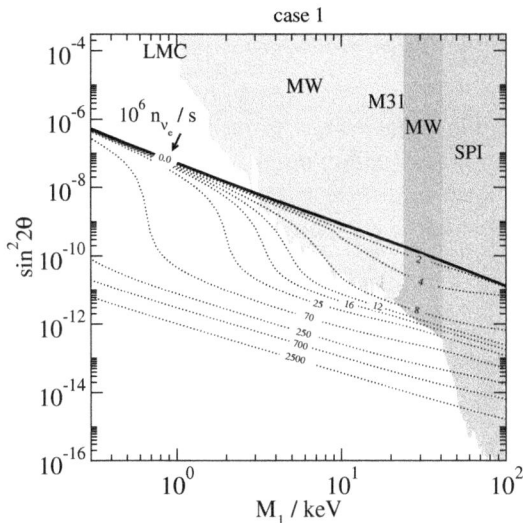

Fig. 7.2.   Bounds on the parameters of sterile neutrinos (mass M_1 and mixing angle θ) from the data of X-ray telescopes, together with predictions for sterile neutrinos as dark matter particles, $\Omega_s = \Omega_{DM}$. The thick line corresponds to the production in non-resonant oscillations (cf. (7.28)); in this case the bound (7.34) applies. The dashed lines correspond to the resonance mechanism, numbers in units 10^{-6} are the values of the present electron neutrino asymmetry $(n_{\nu_e} - n_{\bar{\nu}_e})/s$ [105].

far [135]. Also, sterile neutrinos obeying this bound make very small contribution to active neutrino masses via see-saw mechanism (see Sec. C.4 for details), $\theta_\alpha^2 m_s \ll m_{sol}$, i.e., if light sterile neutrinos are dark matter particles then *they are irrelevant for the generation of active neutrino masses*.

We emphasize that the results discussed here are valid for long lived sterile neutrinos only. Note also that even though we studied sterile neutrino mixing with one active neutrino species, adding two other active neutrinos does not grossly modify the results.

Let us point out that there are more complicated mechanisms of sterile neutrino production in the early Universe. In this sense our estimate (7.28) is, generally speaking, a lower bound. As an example of a realistic, albeit somewhat exotic mechanism, let us consider sterile neutrino production in oscillations in lepton-asymmetric medium [107]. It follows from Big Bang Nucleosynthesis theory and observations that the lepton asymmetry at $T \sim 1$ MeV must be below one per cent. On the other hand, under assumption that the maximum temperature in the Universe was higher than 100 GeV, sphaleron processes equilibrate the baryon and lepton asymmetries at that time (see Sec. 11.2.1), so that the lepton asymmetry at $T \sim 100$ GeV is of order $\eta_B \sim 10^{-9}$. This is insufficient for production of sterile neutrino dark matter, as we will see soon. Thus, the mechansim we are about to discuss is indeed exotic, as it requires the production of lepton asymmtery at $T < 100$ GeV.

In analogy to the baryon-to-entropy ratio, let us introduce the asymmetries

$$\Delta_{\nu_\alpha} = \frac{n_{\nu_\alpha} - n_{\bar\nu_\alpha}}{s}, \quad \Delta_\alpha = \frac{n_\alpha - n_{\bar\alpha}}{s}, \quad \Delta_{q_i} = \frac{n_{q_i} - n_{\bar{q}_i}}{s},$$

where $\alpha = e, \mu, \tau$ and q_i denotes the ith quark flavor. Due to neutrino oscillations, one has $\Delta_{\nu_e} = \Delta_{\nu_\mu} = \Delta_{\nu_\tau}$. Individual lepton asymmetries $\Delta_{L_\alpha} = \Delta_{\nu_\alpha} + \Delta_\alpha$ are not conserved, again due to neutrino oscillations. What is conserved at $T < 100$ GeV is the total lepton asymmetry $\Delta_L = \sum_\alpha \Delta_{L_\alpha}$. Other conserved numbers are baryon number, i.e., $\Delta_B = \sum_i \Delta_{q_i}/3$, and electric charge. Neglecting tiny baryon asymmetry $\Delta_B \sim 10^{-9}$ and recalling that cosmic medium is elecrically neutral, one sets the latter two numbers equal to zero. This does not forbid non-vanishing individual asymmetris in charged lepton and quark sectors, Δ_α and Δ_{q_i}, provided that charged leptons α and quarks q_i are relativistic at temperature considered, $m_\alpha, m_{q_i} \lesssim T$ (otherwise they annihilate out). In fact, asymmetries in the charged lepton sector are natural at non-zero Δ_L, since charged leptons carry lepton number. Asymmetries in quark sector are then required for the overall electric neutrality. Asymmetries Δ_{ν_α}, Δ_α and Δ_{q_i} are proportional to each other (and to the total lepton asymmtery Δ_L), with proportionality coefficients depending on temperature: the lower the temperature, the more species disappear, as their masses exceed T. Needless to say, quark (or rather hadron) asymmetries, and, by elecric neutrality, charged lepton asymmetries, vanish well below the QCD transition temperature $T_{QCD} \sim 200$ MeV, when quarks and gluons are confined in hadrons which annihilate out because their masses are larger than temperature.

As an example, let us find the relations between the lepton and quark asymmetries slightly before the QCD transition, $T > T_{QCD} \sim 200$ MeV, but $T < m_c, m_\tau$. In this case, relativistic are three neutrino species, electron, muon and u-, d-, s-quarks. We introduce the chemical potentials for lepton number μ_L, baryon number μ_B and electric charge μ_Q and write the chemical potentials (5.3) of relativistic particles in plasma:

$$\mu_{\nu_\alpha} = \mu_L, \quad \mu_u = \frac{1}{3}\mu_B + \frac{2}{3}\mu_Q,$$

$$\mu_e = \mu_\mu = \mu_L - \mu_Q, \quad \mu_d = \mu_s = \frac{1}{3}\mu_B - \frac{1}{3}\mu_Q.$$

We then make use of Eq. (5.22) to obtain the asymmetries

$$\Delta_{\nu_\alpha} = \frac{T^2}{6s} \cdot \mu_L, \quad \Delta_u = \frac{T^2}{6s} \cdot 2 \cdot 3 \left(\frac{1}{3}\mu_B + \frac{2}{3}\mu_Q\right), \tag{7.36a}$$

$$\Delta_e = \Delta_\mu = \frac{T^2}{6s} \cdot 2 (\mu_L - \mu_Q), \quad \Delta_d = \Delta_s = \frac{T^2}{6s} \cdot 2 \cdot 3 \left(\frac{1}{3}\mu_B - \frac{1}{3}\mu_Q\right), \tag{7.36b}$$

where we recall that charged leptons and quarks have two spin states and quarks exist in three colors. Electric charge and baryon number densities are then proportional to

$$Q \propto -\Delta_e - \Delta_\mu + \frac{2}{3}\Delta_u - \frac{1}{3}\Delta_d - \frac{1}{3}\Delta_s, \quad B \propto \frac{1}{3}(\Delta_u + \Delta_d + \Delta_s).$$

These should vanish, so we have two relations between the three chemical potentials μ_L, μ_Q, μ_B. Explicitly, the solution to $Q = B = 0$ is $\mu_Q = \mu_L/2$, $\mu_B = 0$, and the relations

(7.36) give finally

$$\Delta_e = \Delta_\mu = \Delta_{\nu_\alpha}, \qquad \Delta_d = \Delta_s = -\Delta_{\nu_\alpha} \qquad \Delta_u = 2\Delta_{\nu_\alpha}. \qquad (7.37)$$

Interactions of active neutrinos with matter give rise to additional contribution to the interaction matrix (7.18), which at the relevant temperatures $T \lesssim 1\,\mathrm{GeV}$ reads [103]

$$
\begin{aligned}
V_{\alpha\alpha} = \sqrt{2}\,G_F \bigg[& 2\left(n_{\nu_\alpha}(T) - n_{\bar\nu_\alpha}(T)\right) + \sum_{\beta\neq\alpha}\left(n_{\nu_\beta}(T) - n_{\bar\nu_\beta}(T)\right) \\
& + \left(\frac{1}{2} + 2\sin^2\theta_W\right)(n_\alpha(T) - n_{\bar\alpha}(T)) + \left(-\frac{1}{2} + 2\sin^2\theta_W\right) \\
& \times \sum_{\beta\neq\alpha}(n_\beta(T) - n_{\bar\beta}(T)) + \left(\frac{1}{2} - \frac{4}{3}\sin^2\theta_W\right) \\
& \times \sum_{q=u,c}(n_q(T) - n_{\bar q}(T)) + \left(-\frac{1}{2} + \frac{2}{3}\sin^2\theta_W\right)\sum_{q=d,s}(n_q(T) - n_{\bar q}(T)) \bigg].
\end{aligned}
$$

These are obtained in complete analogy to the calculation of the potential due to electron asymmetry in the Sun, Eq. (C.28). One makes use of the Fermi theory (see Appendix B.5), and finds that the potential term equals

$$\langle V \rangle = -\langle \mathcal{L}_C \rangle - \langle \mathcal{L}_N \rangle,$$

where the thermal averaging over all fermionic components is done with the help of relations

$$\langle \bar{f}\gamma^0 f \rangle = n_f, \qquad \langle \bar{f}\gamma^i f \rangle = 0.$$

The latter relation implies the absence of macroscopic currents in the plasma.

Problem 7.7. *Obtain the explicit form of the neutrino effective potential $V_{\alpha\alpha}$ at temperatures above the QCD phase transition.*

Taking the electron–neutrino asymmetry Δ_{ν_e} as a reference parameter, we define

$$V_{\alpha\alpha} = \frac{2\sqrt{2}\pi^2}{45}\,g_* \, G_F\, T^3 \cdot \left(2\Delta_{\nu_\alpha} + \sum_{\beta\neq\alpha}\Delta_{\nu_\beta} + \dots\right) \equiv \frac{2\sqrt{2}\pi^2}{45}\,g_* \, G_F\, T^3 \cdot A_{\alpha,e} \cdot \Delta_{\nu_e},$$

where the numerical factor $A_{\alpha\,e}$ for each neutrino flavor α depends on temperature for the reason explained above. In particular, $A_{e\,e} = 7 - 2\sin^2\theta_W \approx 6.5$ when c-quark is still relativistic, and $A_{e\,e} = 6$ at lower temperatures but above the QCD transition.

Now, the oscillation probability formula (7.19) shows that there are resonant transitions when

$$V_{\alpha\alpha}t_{\alpha}^{vac} = 1. \tag{7.38}$$

At that time mixing becomes large, $\sin^2 2\theta_{\alpha}^{mat} = 1$ (we assume small vacuum mixing angle). The transitions into sterile neutrinos occur most rapidly at the resonant temperature, which is found from (7.38):

$$T_{res} \simeq 50 \cdot \left(\frac{m_s}{1\,\text{keV}}\right)^{1/2} \left(\frac{10^{-4}}{\Delta_{\nu_e}}\right)^{1/4} \left(\frac{160}{A_{\alpha,e}g_*}\right)^{1/4} y^{-1/4}\,\text{MeV}, \tag{7.39}$$

where

$$y = \frac{E}{T}.$$

The simplest possibility which is realized at fairly large vacuum mixing angles is that active neutrinos of resonance energies get rapidly converted into sterile neutrinos. On the other hand, antineutrinos do not experience the resonant conversion because of the opposite sign of $V_{\alpha\alpha}$. As active neutrinos, but not their antineutrinos, get converted, lepton asymmetry decreases and eventually becomes so small that the resonant transitions terminate. This happens when the total lepton asymmetry Δ_L becomes considerably smaller than the initial one, $\Delta_{L,i}$: almost all initial lepton asymmetry leaks into the sterile neutrino sector. Hence, after the resonant transitions, the number of sterile neutrinos per unit entropy is of the order of the initial lepton asymmetry, $n_s/s \sim \Delta_{L,i}$. This gives

$$\Omega_s = \frac{\Delta_{L,i}s_0 m_s}{\rho_c}.$$

Numerically

$$\Omega_s \simeq 0.2 \cdot \left(\frac{m_s}{3\,\text{keV}}\right)\left(\frac{\Delta_{L,i}}{10^{-4}}\right). \tag{7.40}$$

This corresponds to nearly vertical parts of dashed lines in Fig. 7.2: the present mass density of sterile neutrinos is almost independent of the mixing angle θ_{α}. Note that the initial lepton asymmetry $\Delta_{L,i}$ is by a factor of 3 to 5 greater than the initial electron neutrino asymmetry; see (7.37). Note also that the numbers near the dashed lines are the values of the *present* electron–neutrino asymmetry, which, depending on the mixing angle, may be substantially (by a factor of several) smaller than the initial asymmetry because of the wash-out process resulting in the sterile neutrino production.

It is worth noting that in this scenario, the average momentum of sterile neutrinos produced by the resonant mechanism is lower than that in the case of non-resonant oscillations (7.33). The resonance conditions are reached earlier (at higher temperature) in the low energy part of the spectrum, see (7.39); so, the production of sterile neutrinos begins at low $y = E/T$ and ends when the lepton asymmetry gets washed out. Thus, the maximum energy of sterile neutrinos is estimated as follows:

$$\int_0^{y_{max}} f_\alpha(y) \cdot 4\pi y^2 dy \simeq \frac{2\pi^2}{45} g_* \Delta_L . \tag{7.41}$$

For realistic values of the initial asymmetry $\Delta_L \lesssim 10^{-3}$ this gives $y_{max} = E_{max}/T < 0.7$.

For small vacuum mixing angles the above simple picture is incorrect. Only a small fraction of the initial lepton asymmetry is reprocessed into sterile neutrinos, so the sterile neutrino mass must be larger at smaller θ. Nevertheless, the resonant mechanism is realistic in a large range of vacuum mixing angles and masses. At small vacuum mixing angles there are three relevant time scales: oscillation time in vacuo t_α^{vac}, oscillation time at resonance $t_\alpha^{mat}(T_{res}) = t_\alpha^{vac}/\sin 2\theta_\alpha$ and neutrino mean free time τ_ν. They obey

$$t_\alpha^{mat}(T_{res}) \ll \tau_\nu \ll t_\alpha^{vac}. \tag{7.42}$$

The Boltzmann equation written for neutrino of thermal energy $E \sim 3T$ has the form (7.25) with

$$\langle P(\nu_\alpha \to \nu_s) \rangle = \sin^2 \left(\frac{\tau_\nu}{2 t_\alpha^{mat}(T)} \right) \left(\frac{t_\alpha^{mat}(T)}{t_\alpha^{vac}} \right)^2 \sin^2 2\theta_\alpha$$

$$= \left(\frac{\tau_\nu}{2 t_\alpha^{vac}} \right)^2 \sin^2 2\theta_\alpha \qquad \text{at} \quad \tau_\nu \ll t_\alpha^{mat}(T) \tag{7.43}$$

and

$$\langle P(\nu_\alpha \to \nu_s) \rangle = \frac{1}{2} \left(\frac{t_\alpha^{mat}(T)}{t_\alpha^{vac}} \right)^2 \sin^2 2\theta_\alpha \qquad \text{at} \quad \tau_\nu \gg t_\alpha^{mat}(T), \tag{7.44}$$

where we make use of (7.19a) and (7.19c) (see also (7.21)). Note that Eq. (7.43) is valid away from the resonance. As a function of $t_\alpha^{mat}(T)$, this conversion probability increases as long as $t_\alpha^{mat}(T) \ll \tau_\nu$ and stays constant at $t_\alpha^{mat}(T) \gg \tau_\nu$. Hence, the conversion is efficient only at $t_\alpha^{mat}(T) > \tau_\nu$, i.e., near the resonance.[5] The overall number of sterile neutrinos is obtained from (7.25) and (7.43):

$$\frac{n_s}{s} \simeq \frac{n_{\nu_\alpha}}{s} \frac{\tau_\nu}{4H(t^{vac})^2} \sin^2 2\theta_\alpha \cdot \frac{\Delta T}{T}, \tag{7.45}$$

where ΔT is such that

$$t_\alpha^{mat}(T \pm \Delta T) \sim \tau_\nu, \tag{7.46}$$

---

[5]In fact, the process we consider is closer to scattering rather than to oscillations, as the weak scattering rate exceeds the effective oscillation rate.

and all other quantities are evaluated at $T = T_{res}$. To estimate ΔT, we make use of (7.19b) and (7.38) and write

$$t_\alpha^{mat}(T \pm \Delta T) = t_\alpha^{vac} \left(\frac{d \log(V_{\alpha\alpha} t_\alpha^{vac})}{d \log T} \cdot \frac{|\Delta T|}{T_{res}} \right)^{-1}.$$

Both $V_{\alpha\alpha}$ and t_α^{vac} have power-law dependence on T, so the relation (7.46) gives

$$\frac{\Delta T}{T_{res}} \sim \frac{t_\alpha^{vac}}{\tau_\nu}.$$

In this way we obtain

$$\frac{n_s}{s} \sim \frac{n_{\nu_\alpha}}{s} \frac{1}{H t_\alpha^{vac}} \frac{\sin^2 2\theta}{4}, \qquad (7.47)$$

where the quantities in the right-hand side are calculated at the resonance temperature (7.39). The numerical estimate of Ω_s is fairly complicated, since the parameters g_* and $A_{\alpha,e}$ in (7.39) and in $H = T_{res}^2/M_{Pl}^*$ depend on temperature; its simplified form is

$$\Omega_s = 0.2 \cdot \left(\frac{\sin^2 2\theta}{10^{-12}} \right) \left(\frac{m_s}{1 \text{ keV}} \right)^{3/2} \left(\frac{\Delta_{\nu_\alpha}}{2.5 \cdot 10^{-3}} \right)^{3/4}. \qquad (7.48)$$

Numerical predictions made in the case of electron neutrino mixing with sterile neutrino are shown in Fig. 7.2. As we already pointed out, nearly vertical parts of dashed lines there correspond to the case (7.40). (Ω_s is nearly independent of $\sin 2\theta$.) To the left of these parts the resonant mechanism is inefficient because of too small lepton asymmetry and too small sterile neutrino mass, while to the right of these parts one needs heavier sterile neutrinos for lower $\sin 2\theta$. In this region our simple analytical approximation works reasonably well. In the latter case the sterile neutrino spectrum strongly depends on Δ_L; in general it is shifted towards lower energies [105], in some similarity with the simplest case of Eq. (7.41). This modifies the cosmological bounds on sterile neutrino mass: sterile neutrinos are not as warm as in the non-resonant case. So, in the case of resonant production, the sterile neutrino mass may be considerably smaller than in the non-resonant production model.

In the case of fairly light sterile neutrino, $m_s \lesssim 1$ keV, there exists another relevant time scale over and beyond the scales listed in (7.42). This is the time in which the resonance is traversed due to the expansion of the Universe. To this end, let us discard the active neutrino scattering for the time being and consider the effective Hamiltonian

$$H = \begin{pmatrix} \omega_1 & \omega_{12} \\ -\omega_{12} & \omega_2 \end{pmatrix}, \qquad (7.49)$$

where the entries are read off from (7.18):

$$\omega_1 = V_{\alpha\alpha}, \quad \omega_2 = \frac{m_s^2}{2E}, \quad \omega_{12} = \frac{m_s^2}{4E}\sin 2\theta_\alpha. \tag{7.50}$$

(We neglect terms of order $\sin^2 2\theta_\alpha$; these can be straighforwardly taken care of.) In our case the off-diagonal matrix element ω_{12} is small. All entries in (7.49) depend on time, so the system spends finite time in the resonance region. This sets the new time scale. To estimate it, let us consider the off-diagonal part of the Hamiltonian as perturbation and write $H = H_0 + H_1$, where

$$H_0 = \begin{pmatrix} \omega_1 & 0 \\ 0 & \omega_2 \end{pmatrix}, \quad H_1 = \begin{pmatrix} 0 & \omega_{12} \\ -\omega_{12} & 0 \end{pmatrix}.$$

Importantly, $|\omega_1 - \omega_2|$ is large at almost all times except for a short time near the resonance, while at the resonance one has $\omega_1 = \omega_2$. We are interested in the probability of transition $1 \to 2$. To the first order in perturbation theory, the relevant solution to the propagation equation

$$i\partial_t \psi = H\psi \tag{7.51}$$

is

$$\psi = e^{-i\int_0^t \omega_1(t')dt'}\begin{pmatrix} 1 \\ 0 \end{pmatrix} + C(t)e^{-i\int_0^t \omega_2(t')dt'}\begin{pmatrix} 0 \\ 1 \end{pmatrix}, \tag{7.52}$$

where $C(-\infty) = 0$ and $C(+\infty)$ is the transition amplitude. Equation for $C(t)$ is obtained by plugging (7.52) into (7.51):

$$\dot{C} = -i\omega_{12}\, e^{-i\int_0^t [\omega_1(t')-\omega_2(t')]dt'},$$

which gives

$$C(+\infty) = -i\int_{-\infty}^{+\infty} dt\, \omega_{12}\, e^{-i\int_0^t [\omega_1(t')-\omega_2(t')]dt'}.$$

The integrand here rapidly oscillates at almost all times except for the resonance region, hence the integral is saturated near the resonance. In the resonance region we approximate $\omega_{12} = \omega_{12}(t_{res})$, $\omega_1(t) - \omega_2(t) = (\dot{\omega}_1 - \dot{\omega}_2)(t_{res}) \cdot (t - t_{res})$, set $t_{res} = 0$ (Otherwise we would have to keep track of an irrelevant phase factor in C.) we obtain finally.[6]

$$C(+\infty) = -i\omega_{12}\int_{-\infty}^{+\infty} dt\, e^{-i\frac{1}{2}[\dot{\omega}_1-\dot{\omega}_2]t^2} = -i^{1/2}\,\omega_{12}\sqrt{\frac{2\pi}{|\dot{\omega}_1 - \dot{\omega}_2|}} \tag{7.53}$$

where all quantities are evaluated at the resonance and we recall that $(\dot{\omega}_1 - \dot{\omega}_2) < 0$ in our case. The probability of transition $1 \to 2$ (i.e., $\nu_\alpha \to \nu_s$) is

$$P(\nu_\alpha \to \nu_s) \equiv P(1 \to 2) = |C(+\infty)|^2 = \frac{2\pi\omega_{12}^2}{|\dot{\omega}_1 - \dot{\omega}_2|}. \tag{7.54}$$

This result is valid provided that $\omega_{12}^2 \ll |\dot{\omega}_1 - \dot{\omega}_2|$. In fact, it is a particular case of the Landau–Zener formula; see Refs. [108–110] for its application to neutrinos. The formula

---

[6]The integral is implicitly regularized by considering ω_{12} decaying at large $|t|$, e.g., like $\omega_{12} \propto e^{-\epsilon t^2}$, where the regularization parameter ϵ is set equal to zero after the calculation.

gives the transition probability for small ω_{12} but arbitrary $\omega_{12}^2/|\dot\omega_1 - \dot\omega_2|$:

$$P(1 \to 2) = 1 - e^{-2\pi\gamma},$$

where

$$\gamma = \frac{\omega_{12}^2}{|\dot\omega_1 - \dot\omega_2|}(t_{res}).$$

The point of the above calculation is the estimate of the effective time scale of the traversal through the resonance. This time scale is read off from (7.53): the integral is saturated at

$$t_{trans} \sim \frac{1}{\sqrt{|\dot\omega_1 - \dot\omega_2|}}.$$

In the cosmological setting we have

$$\frac{d(\omega_1 - \omega_2)}{dt} = -\frac{d(\omega_1 - \omega_2)}{d\log T}H = -4\omega_2 H \qquad (7.55)$$

(since $\omega_1 \propto T^3$ while $\omega_2 \propto T^{-1}$). For neutrinos of thermal energy $E \sim T$ this gives

$$t_{trans} \sim \frac{1}{\sqrt{\omega_2 H}} \sim \sqrt{\frac{M_{Pl}^*}{m_s^2 T_{res}}}.$$

For $m_s \lesssim 1$ keV this scale is shorter than the active neutrino free time τ_ν, so the resonant transition occurs between the collisions of active neutrinos, i.e., in free flight. In this situation the formulas (7.43) and (7.44) do not work, and the transition probability is given by (7.54).

Curiously, in the latter case the parametric dependence of the sterile neutrino abundance is the same as in (7.47). Indeed, we recall (7.50), make use of (7.54), (7.55) and obtain (modulo factor of order 1)

$$\frac{n_s}{s} \sim \frac{n_{\nu_\alpha}}{s}\frac{2\pi\omega_{12}^2}{|\dot\omega_1 - \dot\omega_2|} \sim \frac{n_{\nu_\alpha}}{s}\frac{\omega_2}{H}\sin^2 2\theta_\alpha,$$

which is parametrically the same as (7.47) given that $\omega_2 = (t_\alpha^{vac})^{-1}$. So, the effect we have studied does not grossly modify the estimate (7.48).

Chapter 8

Big Bang Nucleosynthesis

The earliest epoch in the hot Universe which has been tested observationally is the Big Bang Nucleosynthesis epoch (BBN). As we will see, it begins at temperature of about 1 MeV and lasts until the temperature drops to a few dozen keV. At this time, neutrons in cosmic medium combine with protons into light nuclei, mostly helium-4 (^{4}He = α-particle) with small but measurable admixture of deuterium (D$\equiv^2$H), helium-3 (^{3}He) and lithium-7 (^{7}Li). The main thermonuclear reactions of BBN are listed in the beginning of Sec. 8.3. We will see that at relevant temperatures neutrons are less abundant than protons; "extra" protons remain free and in the end form hydrogen atoms.

Measurements of the primordial chemical composition not only confirm the theory of the hot Universe, but also provide the determination of an important cosmological parameter, baryon-to-photon ratio η_B. Furthermore, BBN constraints models pretending to extend the Standard Model of particle physics.

Precise calculation of light element abundances produced at the BBN epoch is a complicated task. An appropriate tool here is the numerical analysis of kinetic equations with account of numerous thermonuclear reactions. See, e.g., Ref. [115]. In this Chapter, like in many other Chapters of this book, we limit ourselves by order-of-magnitude estimates having in mind our main purpose: to discuss physics of processes in the early Universe and explain, at qualitative level, the dependence of the results on cosmological parameters.

8.1. Neutron Freeze-Out. Neutron-Proton Ratio

The first stage of BBN is neutron freeze-out. We will see in a moment that it occurs at a temperature of about 1 MeV, which is still too high for light nuclei production.

In the early Universe, neutrons are produced and destroyed in weak interaction processes

$$p + e \leftrightarrow n + \nu_e, \tag{8.1}$$

and crossing processes. The energies relevant here are the mass difference between neutron and proton,

$$\Delta m \equiv m_n - m_p = 1.3\,\text{MeV}$$

and electron mass $m_e = 0.5\,\text{MeV}$. Let us assume for simplicity that the temperature is high enough,

$$T \gtrsim \Delta m, m_e. \tag{8.2}$$

Then, like in Chapter 7, the mean free time of a neutron *with respect to the process* (8.1) can be estimated on dimensional grounds,

$$\tau_n = \Gamma_n^{-1}, \quad \Gamma_n = C_n G_F^2 T^5, \tag{8.3}$$

where C_n is a constant of order 1. The processes (8.1) and crossing processes terminate when the time τ_n becomes of the order of the Hubble time, i.e.,

$$\Gamma_n(T) \sim H(T) = \frac{T^2}{M_{Pl}^*}. \tag{8.4}$$

As before,

$$M_{Pl}^* = \frac{M_{Pl}}{1.66\sqrt{g_*}}, \tag{8.5}$$

and the effective number of relativistic degrees of freedom is

$$g_* = 2 + \frac{7}{8} \cdot 4 + \frac{7}{8} \cdot 2 \cdot N_\nu. \tag{8.6}$$

The first and the second terms here are due to photons and electrons/positrons, while the third term comes from neutrinos. We temporarily denote the number of neutrino species by N_ν, the actual value is $N_\nu = 3$. Let us recall here (see Chapter 7) that at temperatures $T > m_e$, i.e., before electron–positron annihilation, neutrinos have the same temperature as photons.

Equations (8.3) and (8.4) determine the neutron freeze-out temperature,

$$T_n = \frac{1}{(C_n M_{Pl}^* G_F^2)^{1/3}}. \tag{8.7}$$

The constant C_n entering (8.3) is known: the process (8.1) originates from the four-fermion vertex shown in Fig. 8.1(a); the same vertex describes neutron decay, Fig. 8.1(b). Hence, the constant C_n is determined from neutron lifetime; numerically, $C_n = 1.2$. Thus, freeze-out temperature (8.7) does not contain unknown parameters. It depends, however, on the number of light neutrino species; see (8.5) and (8.6).

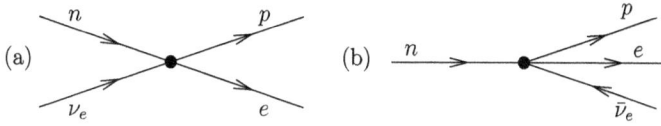

Fig. 8.1. Feynman diagrams for processes (a) $n + \nu_e \leftrightarrow p + e$ and (b) $n \to pe\bar{\nu}_e$.

Recalling that the Fermi constant is $G_F = 1.17 \cdot 10^{-5}$ GeV (see Sec. B.5) and $g_* = 43/4$ for $N_\nu = 3$, we find numerically

$$T_n \approx 1.4 \text{ MeV.} \tag{8.8}$$

It is worth noting that our assumption (8.2) is not well justified, so a more accurate calculation is needed. We quote the result below.

It is amazing that neutron freeze-out temperature and the mass difference Δm almost coincide. This property would not hold if the masses of u- and d-quarks, or the Fermi constant, or Newton's gravity constant were different from what they are. This is one of the coincidences which makes the chemical composition of primordial plasma interesting at all. Due to this coincidence, the neutron abundance at freeze-out is rather large, so the abundances of light nuclei are considerable in the end: if it turned out that $\Delta m \gg T_n$, then the neutron abundance at freeze-out would be exponentially small, $n_n \propto \exp(-\Delta m/T_n)$; see (8.13). On the other hand, for $\Delta m \ll T_n$ neutrons and protons would be equally abundant at freeze-out, so all of them would end up in ^{4}He nuclei and the primordial plasma would lack hydrogen — such a hydrogen-free Universe would hardly be habitable.

Let us estimate the neutron abundance after freeze-out. With reasonable accuracy, it is equal to the neutron abundance just before freeze-out. It is useful for what follows to write the general formula for the number density of particles of type A (protons, neutrons, light nuclei) in chemical equilibrium at temperature $T \ll m_A$ (see Chapter 5):

$$n_A = g_A \left(\frac{m_A T}{2\pi} \right)^{3/2} e^{(\mu_A - m_A)/T}, \tag{8.9}$$

where μ_A is the chemical potential for particle A. We now apply this formula for protons and neutrons and make use of the fact that reaction (8.1) is in equilibrium just before freeze-out, hence $\mu_p + \mu_e = \mu_n + \mu_\nu$, i.e.,

$$\mu_n = \mu_p + \mu_e - \mu_\nu. \tag{8.10}$$

Electrons and neutrinos are relativistic, and we have from (5.22)

$$n_{e^-} - n_{e^+} \sim \mu_e T^2, \tag{8.11}$$

so that

$$\frac{\mu_e}{T} \sim \frac{n_{e^-} - n_{e^+}}{T^3}.$$

The difference between number densities of electrons and positrons is equal to the number density of protons by electric neutrality,

$$n_{e^-} - n_{e^+} = n_p,$$

while n_p/T^3 is of the order of the baryon-to-photon ratio, $n_p/T^3 \sim \eta_B \sim 10^{-9}$. We conclude that the chemical potential for electrons is negligibly small,

$$\frac{\mu_e}{T} \sim 10^{-9}.$$

Let us assume here that lepton asymmetry of the Universe is small (we will discuss the opposite situation in the end of this Section), i.e.,

$$n_\nu - n_{\bar\nu} \ll n_\nu + n_{\bar\nu} \sim T^3.$$

Then μ_ν/T is also negligible, and we obtain from (8.10) that

$$\mu_n = \mu_p. \tag{8.12}$$

It then follows from (8.9) that neutron-proton ratio at freeze-out is

$$\frac{n_n}{n_p} = e^{-(m_n - m_p)/T_n} \equiv e^{-\Delta m/T_n}. \tag{8.13}$$

(We used the fact that both proton and neutron have two spin states, and set $m_n = m_p$ in the pre-exponential factors.)

Neutron–proton ratio (8.13) is roughly of order 1, so one needs to calculate the freeze-out neutron abundance with good precision. The result is [112, 113]

$$\frac{n_n}{n_p} = 0.18, \tag{8.14}$$

which corresponds to the effective freeze-out temperature $T_n \simeq 0.75\,\mathrm{MeV}$, cf. (8.8). Note that the neutron–proton ratio depends on the number of light neutrino species N_ν. More generally, the dependence is on the effective number of relativistic degrees of freedom in the plasma at $T \sim 1\,\mathrm{MeV}$. On the other hand, there is practically no dependence on other cosmological parameters.

Let us calculate the age of the Universe at neutron freeze-out. According to (3.29) we have

$$t = \frac{1}{2H(T_n)} = \frac{M_{Pl}^*}{2T_n^2}.$$

For $T_n = 0.75\,\mathrm{MeV}$ and $N_\nu = 3$ we get numerically

$$t = 1.1\ \mathrm{s}.$$

Thus, BBN theory deals with the cosmological epoch beginning at one second after the Big Bang.

To conclude this Section, we recall that the above calculation has been done under the assumption of negligible lepton asymmetry. If this assumption is not valid, then instead of (8.12) we have $\mu_n = \mu_p - \mu_{\nu_e}$, and instead of (8.13) we obtain

$$\frac{n_n}{n_p} = \exp\left(-\frac{\Delta m}{T_n} - \frac{\mu_{\nu_e}}{T_n}\right). \tag{8.15}$$

The neutron–proton ratio in the end determines the ^{4}He abundance (see Sec. 8.2),

$$n_{^4\text{He}} \propto \frac{n_n}{n_p}.$$

By comparing BBN theory predictions with observations, one finds that the deviation of the ratio n_n/n_p from the standard value should not be large,

$$\frac{|\Delta(n_n/n_p)|}{n_n/n_p} \lesssim 0.025.$$

This leads to the bound on chemical potential for the electron neutrino,

$$\left|\frac{\mu_{\nu_e}}{T}\right| \lesssim 0.025.$$

A more accurate bound is given in (7.16).

Big Bang Nucleosynthesis is also sensitive to the effective number of relativistic degrees of freedom in the cosmic plasma at $T \sim 1$ MeV. This number is traditionally parametrized in terms of the number of neutrino species N_{eff}. In the Standard Model $N_{eff} = N_\nu = 3$, but N_{eff} may take a different value in extensions of the Standard Model. One of the reasons[1] for the dependence of helium abundance on N_{eff} is that the neutron freeze out temperature (8.7) depends on g_* (see Eq. (8.5)):

$$T_n \propto (1/M_{Pl}^*)^{1/3} \propto g_*^{1/6}.$$

Hence, the deviation of n_n/n_p from the Standard Model value (8.13) is, to the leading order in Δg_*,

$$\frac{\Delta(n_n/n_p)}{n_n/n_p} = \frac{\Delta m}{T_n}\frac{\Delta T_n}{T_n} = \frac{\Delta m}{T_n}\frac{\Delta g_*}{6 g_*}.$$

Now, from Eq. (8.6) we find

$$\frac{\Delta g_*}{g_*} = \frac{(7/8)\cdot 2\Delta N_{eff}}{g_*^{SM}} = \frac{7}{43}\Delta N_{eff},$$

and obtain finally

$$\frac{\Delta(n_n/n_p)}{n_n/n_p} = \frac{7}{258}\frac{\Delta m}{T_n}\Delta N_{eff}. \tag{8.16}$$

Requiring this to be less than 0.025, we get a limit

$$|\Delta N_{eff}| \lesssim 0.5, \tag{8.17}$$

which estimates the sensitivity of BBN to N_{eff}.

---

[1] Another reason, which we ignore here, is that the time of the creation of light nuclei t_{NS} also depends on g_*; see Problem 8.2.

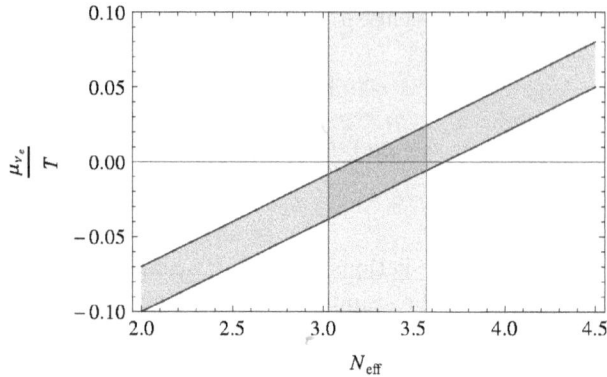

Fig. 8.2. BBN data allow the range in the parameter plane $(N_{eff}, \mu_{\nu_e}/T)$ inside the inclined strip [118]) (68% C.L.). This illustrates the degeneracy in these parameters. The degeneracy is lifted by CMB data (vertical strip, 68% C.L.), under an assumption that neither lepton asymmetry nor abundance of new relativistic particles change during the evolution from the BBN epoch to recombination.

There is, however, a degeneracy between N_{eff} and the chemical potential of the electron neutrino, as they have the same effect on n_n/n_p. According to Eqs. (8.15) and (8.16), the degeneracy is along the lines

$$-\frac{\mu_{\nu_e}}{T} + \frac{7}{258}\frac{\Delta m}{T_n}\Delta N_{eff} \approx -\frac{\mu_{\nu_e}}{T} + 0.05\Delta N_{eff} = \text{const.}$$

So, if there are both hypothetical light particles and lepton asymmetry in cosmic plasma, their effects on the helium abundance can cancel out. This is illustrated in Fig. 8.2. The degeneracy is lifted by invoking other cosmological data, notably, on CMB.[2]

8.2. Beginning of Nucleosynthesis. Direction of Nuclear Reactions. Primordial ^{4}He

The chain of thermonuclear reactions in the early Universe begins with deuterium production in the reaction

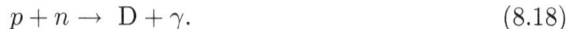

$$p + n \rightarrow D + \gamma. \tag{8.18}$$

This reaction is denoted in nuclear physics as

$$p(n, \gamma)D. \tag{8.19}$$

---

[2]To be accurate, the degeneracy remains there to some extent. The nonzero chemical potential contributes also to the sum of neutrino and antineutrino number densities and thus mimics $\Delta N_{eff} = (90/7\pi^2)(\mu_\nu/T)^2$, as follows from integrating the Fermi–Dirac distribution (5.4). So, the presence at the recombination epoch of a tiny number of new relativistic particles is indistinguishable from the presence of a small lepton asymmetry.

Other nuclear reactions are denoted in a similar way. To calculate the temperature at which this reaction begins, we use the following approach. *We assume* that the reaction (8.18) is fast, so there is chemical equilibrium between deuterium, protons and neutrons. We also switch off other thermonuclear reactions. Under these assumptions we calculate equilibrium abundance of deuterium at temperature T. This abundance is small at high temperatures, and we conclude that deuterium is not produced (cf. Sec. 6.1). The physics behind this phenomenon is that the deuterium nuclei, that are created in neutron–proton collisions, dissociate very fast back into neutrons and protons by absorbing hard photons from the tail of thermal distribution. The deuterium production begins when its equilibrium abundance calculated under above assumptions becomes comparable to abundances of neutrons and protons (the latter are of the same order of magnitude, see (8.14)).

This "equilibrium" approach enables us to determine the *direction* of thermonuclear reactions. Whether or not these reactions are actually fast depends on their cross sections and the cosmological expansion rate. In fact, the latter is rather high, and chemical equilibrium is not reached. Because of the latter property, the relic abundances of D, ^{3}He, ^{7}Li are not negligibly small: in chemical equilibrium, all neutrons would end up in ^{4}He, the most tightly-bound light nucleus.

Continuing with the equilibrium approach, let us write (8.9) as the Saha equation. Because of thermonuclear reactions, chemical potentials of neutrons and protons are different. The formula (8.9) gives for neutrons and protons

$$n_n = 2 \left(\frac{m_p T}{2\pi} \right)^{3/2} e^{(\mu_n - m_n)/T}, \tag{8.20}$$

$$n_p = 2 \left(\frac{m_p T}{2\pi} \right)^{3/2} e^{(\mu_p - m_p)/T}, \tag{8.21}$$

where we neglected mass difference of neutron and proton in pre-exponential factors. If a nucleus of atomic weight A and charge Z is in chemical equilibrium,[3] then its chemical potential is

$$\mu_A = \mu_p \cdot Z + \mu_n \cdot (A - Z).$$

Indeed, chemical equilibrium means that there is a chain of fast reactions leading to production of a nucleus (A, Z) from Z protons and $(A - Z)$ neutrons.

Proceeding as in Sec. 6.1, we obtain from (8.9), (8.20), (8.21) that

$$n_A = n_p^Z n_n^{A-Z} 2^{-A} g_A A^{3/2} \left(\frac{2\pi}{m_p T} \right)^{\frac{3}{2}(A-1)} e^{\Delta_A/T}, \tag{8.22}$$

---

[3] Hereafter subscript A labels nucleus (A, Z); to simplify formulas, we do not use more accurate notations like $\mu_{A,Z}$.

where we set $m_A = Am_p$ in the pre-exponential factor and introduced the binding energy of nucleus (A, Z),

$$\Delta_A = Zm_p + (A - Z) m_n - m_A.$$

Let us also introduce dimensionless ratio of the number of baryons in nuclei (A, Z) to the total number of baryons,

$$X_A = \frac{An_A}{n_B}.$$

Then Eq. (8.22) is written in the form of the Saha equation,

$$X_A = X_p^Z X_n^{A-Z} n_B^{A-1} 2^{-A} g_A A^{5/2} \left(\frac{2\pi}{m_p T}\right)^{\frac{3}{2}(A-1)} e^{\Delta_A/T}.$$

The number of baryons is

$$n_B = \eta_B \cdot n_\gamma = \eta_B \cdot \frac{2\zeta(3)}{\pi^2} T^3 = 0.24\eta_B T^3.$$

So, we have finally

$$X_A = X_p^Z X_n^{A-Z} 2^{-A} g_A A^{5/2} \eta_B^{A-1} \left(\frac{2.5T}{m_p}\right)^{\frac{3}{2}(A-1)} e^{\Delta_A/T}. \qquad (8.23)$$

Like in Sec. 6.1, the right-hand side of this equation contains a small factor $\eta_B^{A-1} (T/m_p)^{\frac{3}{2}(A-1)}$, so the equilibrium abundance of nuclei becomes sizeable at $T \ll \Delta_A$ only, i.e., when the temperature becomes much smaller than nuclear binding energy.

Nucleosynthesis begins when production of deuterium becomes thermodynamically favored. We find from Eq. (8.23) that modulo a numerical factor of order one

$$\frac{X_D}{X_n} \simeq \eta_B \left(\frac{2.5T}{m_p}\right)^{\frac{3}{2}} e^{\frac{\Delta_D}{T}}, \qquad (8.24)$$

where $\Delta_D = 2.23$ MeV, and $A = 2$, $Z = 1$ for deuterium. Let us introduce temperature T_D at which equilibrium abundances of neutrons and deuterium are equal:

$$\frac{X_D}{X_n}(T_D) = 1. \qquad (8.25)$$

We find from (8.24) and using $\eta_B = 6.05 \cdot 10^{-10}$ that

$$T_D = 65 \text{ keV}.$$

In fact, nucleosynthesis begins somewhat earlier, since the cross sections of relevant thermonuclear reactions are fairly large, and deuterium burns already at the time when its abundance is small, $X_D/X_n \ll 1$. We discuss this point in Sec. 8.3.2, and here we just state that the characteristic temperature is (see also [112, 114])

$$T_{NS} = 75 \text{ keV}.$$

We note that this temperature depends on η_B weakly (logarithmically).

Even though the temperatures T_D and T_{NS} are numerically close to each other, the physical situations at these temperatures are quite different. At $T = T_{NS}$, there are many more neutrons than deuterium nuclei in cosmic medium; duterium, once created, rapidly burns into heavier nuclei. On the other hand, the medium at $T = T_D$ contains roughly equal number of free neutrons and deuterium nuclei, but their abundances are already small, as we will see below.

In fact, at these temperatures the production of ^{4}He is thermodynamically favored. Let us see this, making use of the equilibrium approach. If practically all neutrons at $T \approx T_{NS}$ are indeed in ^{4}He, then Eq. (8.23) must give $X_{^4\text{He}} \sim 1$, while abundances of all other light nuclei, including free neutrons, must be small. Let us write Eq. (8.23) for ^{4}He ($A = 4$, $Z = 2$, $g_A = 4$),

$$X_{^4\text{He}} = X_p^2 X_n^2 \cdot 8\eta_B^3 \left(\frac{2.5T}{m_p}\right)^{9/2} e^{\Delta_{^4\text{He}}/T}.$$

There are more protons than neutrons in plasma (see (8.14)), so extra protons give $X_p \sim 1$. Again omitting factors of order 1, we express the neutron abundance in terms of $X_{^4\text{He}}$,

$$X_n = X_{^4\text{He}}^{1/2} \eta_B^{-3/2} \left(\frac{2.5T}{m_p}\right)^{-9/4} e^{-\Delta_{^4\text{He}}/2T}. \tag{8.26}$$

By substituting (8.26) into (8.23), and taking $X_{^4\text{He}} \sim 1$ we find for other nuclei, modulo factors of order 1,

$$X_A = \left[\eta_B \cdot \left(\frac{2.5T}{m_p}\right)^{3/2}\right]^{\frac{3}{2}Z - \frac{1}{2}A - 1} e^{\frac{\Delta_A - \Delta_{^4\text{He}}(A-Z)/2}{T}}$$

$$\simeq 10^{7.3(A+2-3Z)} e^{\frac{\Delta_A - \Delta_{^4\text{He}}(A-Z)/2}{T}}, \tag{8.27}$$

where numerical value of the pre-exponential factor is given for $\eta_B = 6.05 \cdot 10^{-10}$, $T = 75$ keV. Note that the sign in the exponent depends on binding energy per neutron, $\Delta_A/(A - Z)$. Among light nuclei, this quantity is the largest for ^{4}He, that is why ^{4}He is predominantly produced in the early Universe.

Binding energies of relevant stable or sufficiently long lived light nuclei are given in Table 8.1. Making use of these data, we find from (8.27) the following estimates at $T = 75$ keV: $X_n \sim 10^{-60}$, $X_D \sim 10^{-62}$, $X_{^3\text{H}} \sim 10^{-100}$, $X_{^3\text{He}} \sim 10^{-45}$, $X_{^6\text{Li}} \sim 10^{-68}$, $X_{^7\text{Li}} \sim 10^{-101}$, $X_{^7\text{Be}} \sim 10^{-50}$, $X_{^8\text{B}} \sim 10^{-64}$. Thus, at $T = T_{NS}$ equilibrium abundances of light nuclei are much smaller than that of ^{4}He.

Table 8.1.   Binding energies of some stable or long lived nuclei (MeV).

Z	Nucleus	Δ_A	Δ_A/A	$\Delta_A/(A-Z)$
1	$^2\mathrm{H} \equiv \mathrm{D}$	2.23	1.11	2.23
	$^3\mathrm{H} \equiv \mathrm{T}$	8.48	2.83	4.24
2	$^3\mathrm{He}$	7.72	2.57	7.72
	$^4\mathrm{He} \equiv \alpha$	28.30	7.75	14.15
3	$^6\mathrm{Li}$	31.99	5.33	10.66
	$^7\mathrm{Li}$	39.24	5.61	9.81
4	$^7\mathrm{Be}$	37.60	5.37	12.53
5	$^8\mathrm{B}$	37.73	4.71	12.58
6	$^{12}\mathrm{C}$	92.2	7.68	15.37

Let us make two points here. First, applying (8.27) to ^{12}C we would get $X_{^{12}\mathrm{C}} \gg 1$. This means that if carbon could be produced, our assumption about domination of ^{4}He would be wrong: almost all neutrons would end up in ^{12}C (and heavier nuclei). However, one can see from Table 8.1 that ^{12}C cannot be produced in two-body reactions involving lighter stable nuclei[4]: fusion of two ^{6}Li isotopes is very seldom because of tiny abundance of this isotope. (This isotope is produced only in the fusion of helium-3 and tritium, and this process is strongly suppressed as compared to production of helium-4 in collision of the same nuclei.) Elements with $A = 5$ and $A = 8$ cannot be produced in two-body reactions of abundant nuclei. This is why nucleosynthesis chain does not reach carbon in the early Universe. Thermonuclear reactions proceed towards production of ^{4}He.

Second, if matter in the Universe were in chemical equilibrium with respect to production of ^{4}He at $T > 75\,\mathrm{keV}$, the nucleosynthesis would occur at higher temperatures. This is because binding energy of ^{4}He is greater than that of deuterium. However, ^{4}He is produced by deuterium burning, rather than directly from protons and neutrons. Hence, it is the production of deuterium that determines the nucleosynthesis temperature ("deuterium bottleneck"). In this sense the *nucleosynthesis is delayed in the early Universe.*

Problem 8.1. *Find nucleosynthesis temperature in a hypothetical case of fast production of ^{4}He directly from protons and neutrons.*

Let us now find the age of the Universe at the epoch of thermonuclear reactions, i.e., at $T_{NS} = 75\,\mathrm{keV}$. According to (3.29) we have

$$t_{NS} = \frac{1}{2H(T_{NS})} = \frac{M_{Pl}^*}{2T_{NS}^2}. \tag{8.28}$$

---

[4]Element ^{12}C is produced in stars in the processes of creation of short-lived ^{8}B in collision of two ^{4}He nuclei and subsequent fast absorption of yet another ^{4}He. This mechanism does not work in the early Universe because of low density of ^{4}He nuclei.

The expansion rate is determined by photons and neutrinos, and the latter are already frozen out. So, we have

$$M_{Pl}^* = \frac{M_{Pl}}{1.66\sqrt{\hat{g}_*}}, \quad \text{where } \hat{g}_* = 2 + \frac{7}{8} \cdot 2 \cdot N_\nu \cdot \left(\frac{4}{11}\right)^{4/3}. \tag{8.29}$$

The age of the Universe at $T = 75$ keV is, therefore (for $N_\nu = 3$),

$$t_{NS} \approx 230\,\text{s} \approx 4\text{ min}.$$

Using this estimate we now calculate the primordial abundance of helium-4. We will see in Sec. 8.3 that thermonuclear reactions proceed rapidly at $T \simeq T_{NS}$. After BBN, most neutrons are collected in helium-4, so the abundance of helium-4 is half of that for neutrons,

$$n_{^4\text{He}}(T_{NS}) = \frac{1}{2} n_n(T_{NS}).$$

The latter, in turn, is related to proton abundance as follows,

$$\frac{n_n(T_{NS})}{n_p(T_{NS})} = \frac{n_n(T_n) e^{-t_{NS}/\tau_n}}{n_p(T_n) + n_n(T_n)\left(1 - e^{-t_{NS}/\tau_n}\right)} \approx 0.14, \tag{8.30}$$

where we modified the relation (8.14) by taking into account neutron lifetime, $\tau_n \approx 886\,\text{s}$. As a result, we obtain the mass fraction of ^{4}He,

$$X_{^4\text{He}} = \frac{m_{^4\text{He}} \cdot n_{^4\text{He}}(T_{NS})}{m_p[n_p(T_{NS}) + n_n(T_{NS})]} = \frac{2}{\frac{n_p(T_{NS})}{n_n(T_{NS})} + 1} \approx 25\%. \tag{8.31}$$

As we discussed in the end of Sec. 8.1, this mass fraction depends on the number of relativistic degrees of freedom through the neutron freeze-out temperature T_n. We see that there is an extra source of this dependence: the nucleosynthesis time t_{NS} which enters Eq. (8.30), also depends on g_*. This modifies somewhat our estimate (8.16), although our crude numerical bound (8.17) remains almost intact.

Overall, barring large lepton asymmetry, BBN is sensitive to the number of relativistic degrees of freedom and to the baryon-to-photon ratio. (The latter is particularly relevant for residual deuterim fraction, see Sec. 8.3.2.) The BBN theory is consistent with the data in the allowed range of these parameters shown in Fig. 8.3. These results constrain extensions of the Standard Model with exotic light particles.

Problem 8.2. *Refine the estimate (8.16) by taking into account that $t_{NS} \propto H^{-1}(T_{NS} = 75\ keV)$ depends on g_*.*

Problem 8.3. *Find the lowest possible freeze-out temperature of hypothetical massless particles, assuming 20% uncertainty in g_*. Do the same for uncertainty of 5%.*

Some uncertainty in the theoretical prediction of primordial helium abundance is due to uncertainty in measured neutron lifetime. The latter is relevant for both the temperature T_n (see (8.7), (8.13), (8.14)) and the number of neutrons remaining in the plasma at time t_{NS} (see (8.30)).

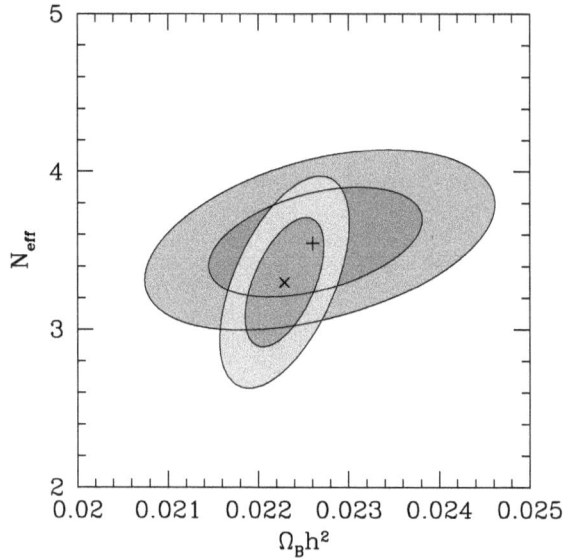

Fig. 8.3. Allowed ranges of the baryon density and number of relativistic species (68% and 95% C.L.) from BBN (larger ellipses) and CMB [119]. Crosses show the best fit values.

8.3. Kinetics of Nucleosynthesis

We have seen in the previous Section that nuclear reactions proceed towards production of helium-4. In this Section we discuss the rates of the most relevant reactions and estimate the residual abundances of other light elements.

The direction of nuclear reactions that we found in the previous Section suggests for us to divide these reactions into several categories:

(1) $p(n, \gamma)$D, production of deuterium, initial stage.
(2) D$(p, \gamma)^3$He, D$(D, n)^3$He, D(D, p)T, ^{3}He(n, p)T, preliminary reactions preparing material for ^{4}He production.
(3) T$(D, n)^4$He, ^{3}He$(D, p)^4$He, production of ^{4}He.
(4) T$(\alpha, \gamma)^7$Li, ^{3}He$(\alpha, \gamma)^7$Be, ^{7}Be$(n, p)^7$Li, production of the heaviest elements.
(5) ^{7}Li$(p, \alpha)^4$He, burning of ^{7}Li.

We note that the reaction rates are proportional to abundances of colliding nuclei, so among all possible reactions the most relevant are those involving at least one of the abundant nuclei, i.e., p, n, D, ^{4}He.

Let us consider these reactions in turn, with the purpose to estimate their rates in the early Universe. By comparing these rates with the expansion rate at nucleosynthesis,

$$H(T_{NS} = 75 \, \text{keV}) = 2 \cdot 10^{-3} \, \text{s}^{-1},$$

we will find the residual *inequilibrium* abundances. These are of course much higher than abundances that would be present in thermal equilibrium.

8.3.1. *Neutron burning,* $p + n \rightarrow D + \gamma$

As we have seen, deuterium production becomes thermodynamically favored at temperature $T = T_{NS} \approx 75\,\text{keV}$. However, the Universe expands rather fast, so some neutrons could in principle be not burned out. Let us show that, in fact, the fraction of neutrons which are not burned is negligibly small. To this end, we compare the rate of neutron burning with the expansion rate at $t = t_{NS}$.

The cross section of deuterium production can be roughly estimated as the geometric cross section,

$$(\sigma v)_{p(n,\gamma)D} \sim \frac{\alpha}{m_\pi^2} \simeq \frac{1}{137}\frac{1}{(200\,\text{MeV})^2} = 2 \cdot 10^{-18}\,\frac{\text{cm}^3}{\text{s}},$$

where m_π is the pion mass determining the typical spatial range of nuclear interactions, $r \sim m_\pi^{-1}$, while the fine structure constant α accounts for suppression related to photon emission. Note that this estimate does not depend on the velocity of colliding particles, i.e., on temperature. In fact, the temperature dependence exists, and the corresponding corrections change the cross section by a factor of 1.5–2 at $T \sim T_{NS}$. Furthermore, since deuterium is a loosely bound nucleus, there is an additional factor ω_γ/p_D, where $\omega_\gamma \sim \Delta_D$ is the photon energy and p_D is the typical center-of-mass momentum of neutron and proton in deuterium. The latter can be found from the virial theorem,

$$\frac{p_D^2}{M_D} \simeq \Delta_D, \tag{8.32}$$

where we assumed for the estimate that the interaction potential between proton and neutron is inversely proportional to the distance. The final estimate is

$$(\sigma v)_{p(n,\gamma)D} \approx 6 \cdot 10^{-20}\,\frac{\text{cm}^3}{\text{s}}. \tag{8.33}$$

Neutron burning occurs in collisions of neutrons with protons leading to deuterium production. Its rate per neutron is given by

$$\Gamma_{p(n,\gamma)D} = n_p \cdot (\sigma v)_{p(n,\gamma)D} = \eta_B \cdot 2\frac{\zeta(3)}{\pi^2}T^3 \cdot (\sigma v)_{p(n,\gamma)D}$$

$$= 0.5\,\text{s}^{-1}, \quad \text{for } \eta_B = 6.05 \cdot 10^{-10}, \quad T = T_{NS} = 75\,\text{keV}, \tag{8.34}$$

where we expressed the proton number density through baryon-to-photon ratio η_B and photon number density at $T = T_{NS}$. Since this rate is much higher than the cosmological expansion rate, $\Gamma_{p(n,\gamma)D} \gg H(T_{NS})$, neutrons indeed burn out, and practically all of them combine into deuterium.[5]

---

[5]That would not be the case for $\eta_B < 10^{-11}$.

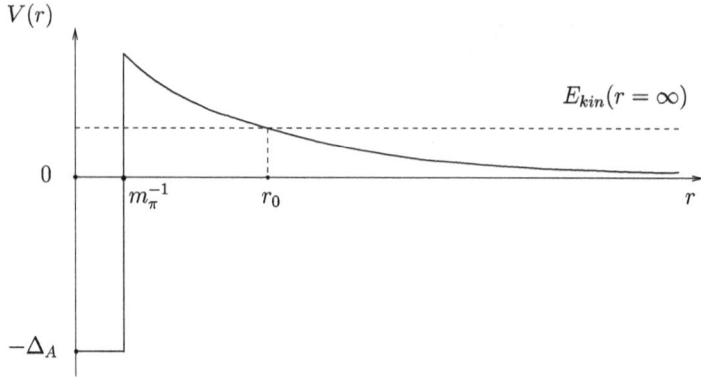

Fig. 8.4. Sketch of the potential between colliding nuclei.

Problem 8.4. *Neglecting neutron decays, estimate the concentration of free neu-*
trons towards the end of BBN, i.e. at $T \ll T_{NS}$, where $T_{NS} \simeq 75$ keV.
Hint: At $T < T_{NS}$ deuterium rapidly burns into heavier nuclei (^4He in the end),
which are not destroyed at these temperatures.

8.3.2. Deuterium burning

Deuterium is the material from which tritium and helium-3 are produced. The cross
sections of the reactions

$$D + D \to {}^3He + n \quad \text{and} \quad D + D \to T + p \tag{8.35}$$

could be estimated as geometric cross sections, but we have to take into account the
Coulomb barrier: both colliding nuclei carry positive electric charge, so they repel
each other. This repulsion dominates at long distances, $r \gg 1/m_\pi$, and inhibits the
reactions. The form of the potential is schematically shown in Fig. 8.4.

Because of the Coulomb barrier, nuclear reactions occur due to quantum tunneling. To estimate the corresponding suppression factor for nuclei of charges Z_1
and Z_2, masses M_1 and M_2 and velocities v_1 and v_2, let us work in the center-
of-mass frame. In this frame, the incident kinetic energy is $E_{kin} = Mv^2/2$, where
$M = M_1 M_2/(M_1 + M_2)$ is the reduced mass and $v = v_1 - v_2$ is the relative velocity.
(Even though we are currently interested in deuterium burning, it is useful for what
follows to consider general case.)

The tunneling amplitude is exponentially suppressed. For s-wave scattering we
have

$$A \propto \exp\left[-\int_0^{r_0} \sqrt{2M(V(r) - E_{kin})}\,dr\right],$$

where the turning point r_0 is determined by the relation

$$E_{kin}(r = r_0) = E_{kin}(r = \infty) - V(r_0) = \frac{1}{2}Mv^2 - \frac{\alpha Z_1 Z_2}{r_0} = 0;$$

in writing the exponent we assumed that $r_0 \gg 1/m_\pi$. Thus, the exponent is

$$-\sqrt{2\pi\alpha Z_1 Z_2} \int_0^{r_0} \sqrt{\frac{1}{r} - \frac{1}{r_0}} \, dr = -\frac{\pi\alpha Z_1 Z_2}{v}.$$

As a result, the cross section is suppressed as $\sigma \propto \exp\left(-2\pi\alpha Z_1 Z_2/v\right)$, and including pre-exponential factor we have

$$\sigma v = \sigma_0 \cdot \frac{2\pi\alpha Z_1 Z_2}{v} \cdot e^{-2\pi\alpha Z_1 Z_2/v}, \tag{8.36}$$

where σ_0 is the geometric cross section in the absence of the Coulomb suppression.

The expression (8.36) is to be averaged with the Maxwell–Boltzmann distribution,

$$\langle\sigma v\rangle = \sigma_0 \cdot 2\pi\alpha Z_1 Z_2 \cdot \frac{\int_0^\infty \exp\left(-\frac{Mv^2}{2T} - \frac{2\pi\alpha Z_1 Z_2}{v}\right) v \, dv}{\int_0^\infty \exp\left(-\frac{Mv^2}{2T}\right) v^2 \, dv}. \tag{8.37}$$

The normalization integral in the denominator is straightforwardly calculated; it is equal to $\sqrt{\pi/2}(T/M)^{3/2}$. We evaluate the integral in the numerator by the saddle point method and obtain

$$\int_0^\infty \exp\left(-\frac{Mv^2}{2T} - \frac{2\pi\alpha Z_1 Z_2}{v}\right) v \, dv$$

$$\approx v_0 \sqrt{\frac{2\pi}{\frac{M}{T} + \frac{4\pi\alpha Z_1 Z_2}{v_0^3}}} \exp\left(-\frac{Mv_0^2}{2T} - \frac{2\pi\alpha Z_1 Z_2}{v_0}\right),$$

where the saddle point v_0 is determined by

$$\frac{Mv_0}{T} = \frac{2\pi\alpha Z_1 Z_2}{v_0^2}.$$

As a result, (8.37) becomes

$$\langle\sigma v\rangle \approx \sigma_0 \cdot \frac{2}{\sqrt{3}} \cdot (2\pi\alpha Z_1 Z_2)^{4/3} \cdot \left(\frac{M}{T}\right)^{2/3} \cdot \exp\left[-\frac{3}{2} (2\pi\alpha Z_1 Z_2)^{2/3} \left(\frac{M}{T}\right)^{1/3}\right].$$

We now introduce convenient quantities: dimensionless reduced mass of incident nuclei $\bar{A} \equiv M/m_p$ and temperature in units of billion Kelvin, $T_9 \equiv T/(10^9\,\mathrm{K}) = T/(86\,\mathrm{keV})$. In these notations, the final result is

$$\langle\sigma v\rangle = 9.3 \cdot \sigma_0 \cdot (Z_1 Z_2)^{4/3} \bar{A}^{2/3} T_9^{-2/3} \cdot e^{-4.26 \cdot (Z_1 Z_2)^{2/3} \bar{A}^{1/3} T_9^{-1/3}}. \tag{8.38}$$

We note that this estimate for $\langle\sigma v\rangle$ assumes that σ_0 is independent of momenta of colliding nuclei at energies in the interval $E_{kin} \sim 10{-}100$ keV. This is often not the case and such a dependence leads to more cumbersome expressions instead of (8.38). In particular, the pre-exponential factor often has different dependence on temperature as compared to (8.38). Furthermore, sometimes the expression (8.38) is not relevant at all, since the cross section is dominated by intermediate resonance states. Finally, the production of new nuclei may not occur in s-wave scattering; in these cases the non-zero angular momentum gives important contribution to the effective potential determining the tunneling exponent. This yields contributions to the cross section that have different temperature dependence in the exponent as compared to (8.38). We omit these "details" here, and will use the correct expressions in appropriate places. We discuss the calculations of burning rates in more detail in the end of Sec. 8.3.4.

Coming back to deuterium burning reactions $D(D,p)T$ and $D(D,n)^3$He, let us roughly estimate σ_0 by making use of the typical range of nuclear force,

$$\sigma_0 \sim m_\pi^{-2} \sim 10^{-26} \text{ cm}^2 \sim 3 \cdot 10^{-16} \frac{\text{cm}^3}{\text{s}}.$$

As a result we obtain for these reactions ($\bar{A} = Z_1 = Z_2 = 1$ for DD initial state)

$$\langle \sigma v \rangle_{DD} \simeq 3 \cdot 10^{-15} \frac{\text{cm}^3}{\text{s}} \cdot T_9^{-2/3} \cdot e^{-4.26 \cdot T_9^{-1/3}}. \tag{8.39}$$

In fact, this is a reasonably good estimate for the combined rate of reactions (8.35) at temperatures of interest, including $T_9 \sim 1$.

Already at the beginning of deuterium burning this process is fast. As an example, at deuterium abundance of 10^{-3} relative to protons, the burning rate per deuterium nucleus is estimated as

$$\Gamma_D \sim \langle \sigma v \rangle_{DD} \cdot 10^{-5} \eta_B \frac{2\zeta(3)}{\pi^2} T_{NS}^3 \sim 0.2 \text{ s}^{-1},$$

which is much greater than the Hubble parameter. As we will see momentarily, it is for this reason that the charactristic nucleosynthesis temperature $T_{NS} = 75$ keV is somewhat higher than the temperature $T_D = 65$ keV introduced in (8.25): deuterium burns when its abundance is relatively small, and free neutrons get bound, predominantly in ^{4}He, already at higher temperatures. For approximate description of deuterium burning, let us introduce the quantity

$$n_{Dn} = n_n + n_D,$$

which is the density of neutrons either in free state or in deuterium. We write the Boltzmann equation for this quantity, which accounts for deuterium burning and cosmological expansion[6]

$$\frac{dn_{Dn}}{dt} + 3H n_{Dn} = -\langle \sigma v \rangle_{DD} n_D^2. \tag{8.40}$$

The right-hand side here descibes irreversible transition of neutrons into ^{3}H and ^{3}He, and then into ^{4}He. Note that we ignore here deuterium burning in reactions

---

[6] Here we use the fact that the total number of DD reactions per unit time per unit volume is $(1/2) \cdot \langle \sigma v \rangle_{DD} n_D^2$. Indeed, the reaction rate (inverse lifetime) of a deuterium nucleus equals $\langle \sigma v \rangle_{DD} n_D$, but one has to count half of all nuclei. However, each DD reaction kills two deuterium nuclei, and this compensates for the factor $1/2$.

with heavier nuclei, to be discussed in Sec. 8.3.3. This approximation is sufficient for our estimates.

The process $p(n\gamma)D$ and inverse process $D(\gamma n)p$ are fast; see Eq. (8.34). Therefore, neutrons and deuterium are in approximate chemical equilibrium with each other. We substantiate this claim later on; See Eq. (8.48). In this approximation, the abundances obey the Saha equation (8.24), hence

$$n_D = \left(\frac{n_D}{n_D + n_n}\right)^{eq} \cdot n_{Dn} \equiv S(T) \cdot n_{Dn},$$

where

$$S = \frac{X_D/X_n}{1 + X_D/X_n}, \quad \frac{X_D}{X_n} = \eta_B \left(\frac{2.5T}{m_p}\right)^{\frac{3}{2}} e^{\frac{\Delta_D}{T}}. \tag{8.41}$$

It is convenient to study the ratio n_D/T^3, and, more precisely, introduce

$$\nu_{Dn} = \frac{n_{Dn}}{\alpha_p T^3},$$

where

$$\alpha_p = 0.75\frac{n_B}{T^3} = 0.75\eta_B \cdot 2\zeta(3)/\pi^2 = 1.1 \cdot 10^{-10} \quad \text{for } \eta_B = 6.05 \cdot 10^{-10}. \tag{8.42}$$

Towards the end of BBN epoch, the quantity $\alpha_p T^3$ is equal to the number density of free protons (factor 0.75 accounts for the fact that a quarter of all baryons get bound ito helium), and n_{Dn} is the final number density of deuterium nuclei. Thus,

$$\nu_{Dn}(T \to 0) = \frac{n_D}{n_p}(T \to 0).$$

On the other hand, before the BBN epoch we have

$$n_{Dn} = n_n = n_B \frac{n_n/n_p}{1 + n_n/n_p}$$

and

$$\nu_{Dn}(T > T_{NS}) \equiv \nu_{Dn}^{(i)} = \frac{1}{0.75}\frac{n_n/n_p}{1 + n_n/n_p} = 0.16,$$

where we use the estimate (8.30).

Equation (8.40) leads to the Boltzmann equation for ν_{Dn}:

$$\frac{d\nu_{Dn}}{dt} = -\langle\sigma v\rangle_{DD} \, \alpha_p T^3 \, S^2(T) \, \nu_{Dn}^2.$$

As usual, let us use temperature instead of time and recall the relation (3.34) to write

$$\frac{d\nu_{Dn}}{dT} = \langle\sigma v\rangle_{DD} \, \alpha_p M_{Pl}^* \, S^2(T) \, \nu_{Dn}^2. \tag{8.43}$$

This gives

$$\frac{1}{\nu_{Dn}} = \frac{1}{\nu_{Dn}^{(i)}} + I_D(T),\tag{8.44}$$

where

$$I_D(T) = \int_T^\infty \langle\sigma v\rangle_{DD}(T') \cdot \alpha_p M_{Pl}^* \cdot S^2(T')\, dT'.\tag{8.45}$$

As temperature decreases, the integral $I_D(T)$ grows and ν_{Dn} decreases, as expected.

By making use of the formula (8.44), we first estimate the characteristic temperature at which free neutrons and deuterium burn and most of neutrons get bound in ^{3}H, ^{3}He and in the end in ^{4}He. We define this temperature T_{NS} as the temperature at which half of neutrons remain free or bound in deuterium nuclei, while another half are bound in heavier nuclei. This happens when $\nu_{Dn}(T_{NS}) = \nu_{Dn}^{(i)}/2$, i.e.,

$$I_D(T_{NS}) = \frac{1}{\nu_{Dn}^{(i)}}.\tag{8.46}$$

We will see momentarily that $T_{NS} > T_D = 65$ keV, so that deuterium is less abundant than free neutrons at $T = T_{NS}$, i.e., $X_D \ll X_n$, and therefore

$$S(T) = \frac{X_D}{X_n} = \eta_B \left(\frac{2.5T}{m_p}\right)^{\frac{3}{2}} e^{\frac{\Delta_D}{T}}.$$

At temperatures of interest the parameter $\langle\sigma v\rangle_{DD}$ is almost independent of temperature and is given by

$$\langle\sigma v\rangle_{DD}(T > T_D) = 3.4 \cdot 10^{-17}\frac{\text{cm}^3}{\text{s}},\tag{8.47}$$

in accordance with the estimate (8.39). The integral $I_D(T_{NS})$ is saturated at lower limit of integration. Because of strong (exponential) dependence of S^2 on temperature, this integral at $T > T_D$ can be calculated by using the variable T^{-1} and integrating only the exponential factor $\exp(2\Delta_D/T)$. As a result, we obtain approximate expression

$$I_D(T > T_D) \simeq \langle\sigma v\rangle_{DD}(T > T_D) \cdot \alpha_p M_{Pl}^* \cdot \frac{T^2}{2\Delta_D} \cdot \eta_B^2 \left(\frac{2.5T}{m_p}\right)^3 e^{\frac{2\Delta_D}{T}}.$$

With $M_{Pl}^* = 4 \cdot 10^{18}$ GeV, we get from (8.46)

$$T_{NS} = 75 \text{ keV}.$$

We have used this estimate in our previous analysis.

As the temperature decreases, the integral $I_D(T)$ rapidly grows, and the relation (8.44) tells that the abundances of free neutrons and deuterium rapidly decrease; neutrons get bound in heavier nuclei. As a example, at temperature $T_D = 65$ keV, when abundances of free neutrons and deuterium nuclei are equal to each other, the formula (8.44) gives[7] $\nu_D(T_D) = \nu_n(T_D) = [2I_D(T_D)]^{-1} \simeq 4 \cdot 10^{-4}$.

Let us check that free neutrons and deuterium are indeed in chemical equilibrium with each other at $T_{NS} = 75$ keV. Chemical equilibrium holds if the deuterium creation rate in reaction $p(n\gamma)D$ (which in chemical equilibrium is equal to the rate of inverse reaction) is higher than the rate of deuterium burning in reactions (8.35):

$$\left(\frac{dn_D}{dt}\right)_{p(n\gamma)D} = \langle\sigma v\rangle_{p(n\gamma)D} \cdot n_p n_n \gg \left|\frac{dn_D}{dt}\right|_{DD} = \langle\sigma v\rangle_{DD} \cdot n_D^2.$$

The ratio of these rates is given by

$$R \equiv \frac{\langle\sigma v\rangle_{DD} \cdot n_D^2}{\langle\sigma v\rangle_{p(n\gamma)D} \cdot n_p n_n} = \frac{\langle\sigma v\rangle_{DD}}{\langle\sigma v\rangle_{p(n\gamma)D}} \cdot \frac{n_n}{n_p} \left(\frac{n_D}{n_n}\right)^2. \tag{8.48}$$

At $T = T_{NS}$ one has $n_n/n_p \simeq \nu_{Dn}^{(i)}/2 \simeq 0.1$, while n_D/n_n is given by Eq. (8.24) and its numerical value is $n_D/n_n = 1.4 \cdot 10^{-2}$. Making use of the cross sections (8.33) and (8.47), we get $R \simeq 10^{-2} \ll 1$, which is the desired result.

We now use the result (8.44) to estimate the final abundance of deuterium. To this end, the lower limit of integration in (8.45) is set to $T = 0$. To estimate this integral, we replace $S^2(T)$ in the integrand by step function $\theta(T_D - T)$: at $T \approx T_D$, the parameter $\langle\sigma v\rangle_{DD}$ depends on temperature weakly, while $S^2(T)$ rapidly grows from zero to one as the temperature decreases and crosses T_D. Thus, the integral is estimated as follows:

$$I_D(T = 0) = \int_0^{T_D} \langle\sigma v\rangle_{DD}(T) \cdot \alpha_p M_{Pl}^* dT. \tag{8.49}$$

The dependence of $\langle\sigma v\rangle_{DD}$ is fairly important at $T < T_D$. To take this into account, we change the integration variable to $y = T_9^{-1/3}$ and write

$$I_D(T = 0) = 9.7 \cdot 10^6 \cdot \int_{y_D}^{\infty} \frac{3}{y^2} e^{-4.26y}\, dy \quad \text{for} \quad \eta_B = 6.05 \cdot 10^{-10}, \tag{8.50}$$

where $y_D = T_{D\,9}^{-1/3} = 1.1$. This integral is much greater than $(\nu^{(i)})^{-1}$, and according to (8.44) the final abundance of deuterium is practically independent of the initial neutron abundance. We have for the final deuterium abundance

$$\frac{n_D}{n_p} = \nu_{Dn}(T = 0) = I_D^{-1}(T = 0) \tag{8.51a}$$

$$= 2 \cdot 10^{-5} \quad \text{for} \quad \eta_B = 6.05 \cdot 10^{-10}. \tag{8.51b}$$

This ratio remains constant after the BBN epoch.

---

[7]The approximation of chemical equilibrium between free neutrons and deuterium is not very accurate at $T = T_D$, so the results for $\nu_D(T_D)$ and $\nu_n(T_D)$ should be considered as order-of-magnitude estimates.

Since T_D depends on η_B weakly (logarithmically), the main dependence of the deuterium-to-proton ratio (8.51a) on η_B comes from the parameter α_p, see Eq. (8.42). It follows from the results (8.44), (8.49) that

$$\frac{n_D}{n_p} \propto \frac{1}{\alpha_p} \propto \frac{1}{\eta_B},$$

i.e., this fraction is inversely proportional to η_B. The observational determination of the deuterium-to-proton ratio in the present Universe enables one to find the baryon density with good precision.

Note that the integral in (8.50) is saturated in the interval $\Delta y \sim 0.25$, i.e., the relevant temperature range extends from $T_D = 65$ keV to

$$T_9 \sim \left(T_{D\,9}^{-1/3} + \Delta y\right)^{-3} \sim 0.4, \quad T \sim 0.4 \cdot 10^9 \text{ K} \sim 35 \text{ keV}.$$

In this temperature range the free neutron abundance is small, $n_n \ll n_D$, so that $n_{Dn} = n_D$, and Eq. (8.40) describes deuterium burning without its production in the reaction $p(n\gamma)D$. Therefore, our assumption of chemical equilibrium between deuterium and free neutrons is actually irrelevant for the estimate of the final deuterium abundance.

It is worth pointing out that an order-of-magnitude estimate of the final deuterium abundance can be obtained in a very simple way. Namely, let us write the Boltzmann equation (8.43) at $T > T_D$ (with ν_D substituted for ν_{Dn}) as follows:

$$\frac{d\ln\nu_D}{d\ln T} = \langle\sigma v\rangle_{DD}\, \alpha_p M_{Pl}^*\, T\nu_D.$$

Deuterium burning terminates when the right-hand side of this equation decreases and becomes of order 1. This determines the final abundance

$$\frac{n_D}{n_p} \sim (\langle\sigma v\rangle_{DD}\, \alpha_p M_{Pl}^*\, T)^{-1}. \tag{8.52}$$

By inserting here the typical cross section (8.47) and temperature $T_D \sim 65$ keV, we obtain an estimate which has the same order of magnitude as the result (8.51b).

Let us now consider one more reaction involving deuterium, $D + p \longrightarrow \gamma + {}^3\text{He}$. Its cross section is much smaller than the cross section of the DD reactions we discussed above. The reason is that photon emission leads to the electromagnetic suppression. Hence, $\sigma_0 \sim 10^{-21}$ cm^3/s and

$$\langle\sigma v\rangle_{D(p,\gamma)^3\text{He}} = 8 \cdot 10^{-21} \frac{\text{cm}^3}{\text{s}} \cdot T_9^{-2/3} \cdot e^{-3.7 \cdot T_9^{-1/3}},$$

where we made use of the general formula (8.38) with $\bar{A} = 2/3$, $Z_1 = Z_2 = 1$. The rate of deuterium burning via this channel is proportional to proton abundance,

$$\Gamma = n_p \cdot \langle\sigma v\rangle_{D(p,\gamma)^3\text{He}}.$$

For $\eta_B = 6.05 \cdot 10^{-10}$ and $T \lesssim T_{NS}$ this rate is well below the expansion rate. This reaction would be important at large η_B: the larger is the number of baryons[8] the more deuterium is burned out.

8.3.3. *Primordial ^{3}He and ^{3}H*

Helium-3 and tritium produced in collisions of deuterium nuclei, then burn into helium-4. The simple estimate (8.38) does not work for helium-3 burning reaction $^3\text{He} + \text{D} \to p + {}^4\text{He}$. (See discussion in Sec. 8.3.2 and in the end of Sec. 8.3.4). Instead, the reaction rate in the energy range of interest is well described by the formula

$$\langle \sigma v \rangle_{^3\text{He}(D,p)^4\text{He}} \equiv \langle \sigma v \rangle_{^3\text{He}D} = 10^{-15} \frac{\text{cm}^3}{\text{s}} \cdot T_9^{-1/2} e^{-1.8T_9^{-1}}. \tag{8.53}$$

We give here an elementary analysis similar to that leading to the estimate (8.52). Like in previous Section, let us write the Boltzmann equation

$$\frac{d\nu_{^3\text{He}}}{d\ln T} = \alpha_p M_{Pl}^* T \left(\langle \sigma v \rangle_{^3\text{He}D} \, \nu_D \cdot \nu_{^3\text{He}} - \frac{1}{2} \langle \sigma v \rangle_{DD} \, \nu_D^2 \right), \tag{8.54}$$

where the first term in parenthesis describes burning of ^{3}He, while the second term is due to ^{3}He production in DD reaction. We have taken into account that the prduction cross section of ^{3}He is approximately twice smaller than the total deuterium burning cross section. At high temperatures, when $\alpha_p M_{Pl}^* \langle \sigma v \rangle_{^3\text{He}D} T \nu_D \gg 1$, produced ^{3}He rapidly burns out, and its abundance is kept at such a level that the right-hand side of Eq. (8.54) is small, i.e.,

$$\nu_{^3\text{He}} \simeq \frac{1}{2} \frac{\langle \sigma v \rangle_{DD}}{\langle \sigma v \rangle_{^3\text{He}D}} \cdot \nu_D. \tag{8.55}$$

Burning of ^{3}He terminates when

$$\alpha_p M_{Pl}^* \langle \sigma v \rangle_{^3\text{He}D} T \nu_D \sim 1. \tag{8.56}$$

The meaning of this relation is that the burning rate per ^{3}He nucleus becomes of the order of the Hubble parameter, $\langle \sigma v \rangle_{^3\text{He}D} n_D \sim H$ (recall that $\alpha_p \nu_D = n_D/T^3$). We insert $\nu_D \sim 2 \cdot 10^{-5}$ into Eq. (8.56) and obtain an estimate for the temperature

$$T_{^3\text{He}} \simeq 35 \text{ keV}, \qquad T_{^3\text{He}\,9} = 0.4.$$

Note that at $T \simeq T_{^3\text{He}}$ the abundance of deuterium ν_D is of the order of the final deuterium abundance (8.51b), so we can indeed make use of Eq. (8.51b) in (8.56).

---

[8]Our emphasis in this section is on BBN determination of η_B, so we keep this parameter free. We note, however, that irrespectively of BBN, this parameter is determined with good precision from CMB observations; see Fig. 8.2.

Problem 8.5. *Show that at $T \gg T_{^3\mathrm{He}}$ the evolution rate of $\nu_{^3\mathrm{He}}$, calculated according to (8.55), is such that the left-hand side of Eq. (8.54) is small compared to each of the terms in the right-hand side. This justifies the result (8.55).*

At $T = T_{^3\mathrm{He}}$ the abundance of ^{3}He is still estimated as in Eq. (8.55). At $T < T_{^3\mathrm{He}}$ the right-hand side of Eq. (8.54) is dominated by the second term: the reaction $D + D \to {}^3\mathrm{He} + n$ still proceeds, albeit at low rate. The estimate for production of ^{3}He at $T < T_{^3\mathrm{He}}$ gives, roughly speaking, the same result as in (8.55). In this way we obtain an estimate for the final abundance

$$\nu_{^3\mathrm{He}} = \frac{n_{^3\mathrm{He}}}{n_p} \sim \frac{\langle \sigma v \rangle_{DD}}{2 \langle \sigma v \rangle_{^3\mathrm{He}\,D}} \cdot \nu_D \sim \frac{1}{2} \alpha_p M_{Pl}^* T \langle \sigma v \rangle_{DD} \nu_D^2, \tag{8.57}$$

where the right-hand side is calculated at $T = T_{^3\mathrm{He}}$. Numerically,

$$\frac{n_{^3\mathrm{He}}}{n_p} \sim 10^{-5},$$

which is of the same order of magnitude as deuterium abundance. The latter property has to do with the fact that the burning rates $\langle \sigma v \rangle$ are almost equal for ^{3}He and deuterium at temperature $T = 35$ keV, when burning of both deuterium and ^{3}He terminates.

Let us turn to tritium. Tritium nucleus is unstable, so its final abundance cannot be measured directly. However, tritium participates in production of other elements, and its abundance is important for BBN. Tritium burning in reaction

$$T + D \to {}^4\mathrm{He} + n.$$

is treated in a similar way as for ^{3}He. The production rate of tritium is approximately the same as that of ^{3}He, whereas the burning rate is

$$\langle \sigma v \rangle_{T(D,\,n)^4\mathrm{He}} = 10^{-15} \frac{\mathrm{cm}^3}{\mathrm{s}} \cdot T_9^{-2/3} \cdot e^{-0.5 \cdot T_9^{-1}}. \tag{8.58}$$

Note that the temperature dependence of this rate is weak. Tritium burning rate is higher than for ^{3}He, so tritium burns longer than ^{3}He, and its final abundance is lower. We make use of the relation similar to (8.56) and find that tritium burning terminates at

$$T_T \simeq 12 \text{ keV}, \qquad T_{T\,9} = 0.14.$$

The formula analogous to (8.57) gives

$$\frac{n_T}{n_p} \simeq 3 \cdot 10^{-7} \quad \text{for} \quad \eta_B = 6.05 \cdot 10^{-10}. \tag{8.59}$$

Both ratios $n_{^3\mathrm{He}}/n_p$ and n_T/n_p, are, roughly speaking, inversely proportional to η_B. More careful analysis gives $n_{^3\mathrm{He}}/n_p \propto \eta_B^{-0.6}$; see Fig. 8.5.

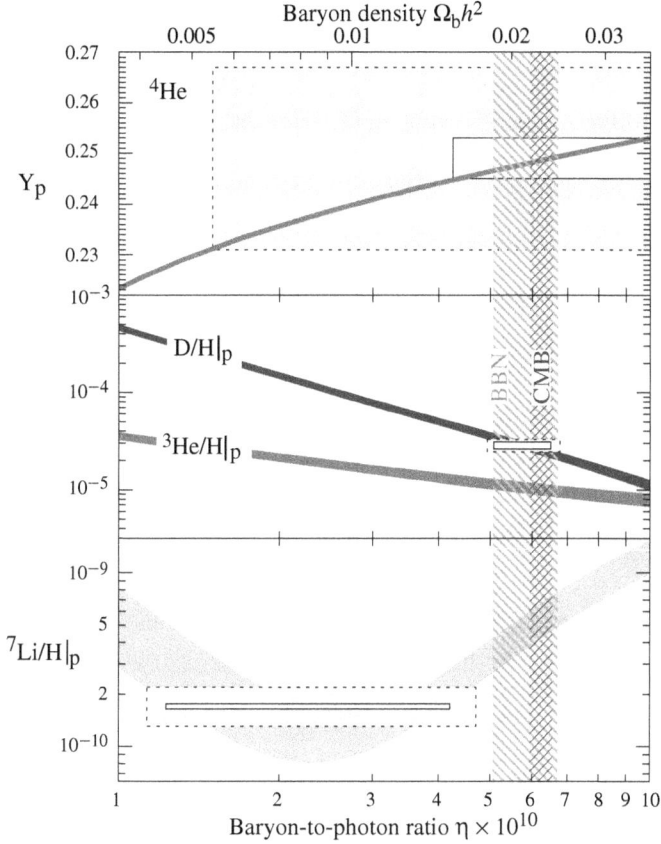

Fig. 8.5. Predictions of BBN theory for primordial abundances of ^{4}He, D, ^{3}He, ^{7}Li and the observational data at 2σ confidence level [1]: statistical errors (solid lines) and statistical and systematic errors together (dashed lines). Theoretical uncertainties are shown by thickness of the lines. Vertical strip "CMB" is the CMB result for η_B. The parameter on horizontal axis is $\eta_{10} = \eta_B \cdot 10^{10}$. On vertical axis are: $Y_p = \frac{n_{^4\text{He}} \cdot m_{^4\text{He}}}{n_H \cdot m_H + n_{^4\text{He}} \cdot m_{^4\text{He}}}$, mass fraction of ^{4}He (the same as $X_{^4\text{He}}$ in the text); n_D/n_H, $n_{^3\text{He}}/n_H$ and $n_{^7\text{Li}}/n_H$, relative abundances of other elements. Subscript p in notations refers to primordial abundances.

Let us make one point here. We have claimed that practically all neutrons end up in ^{4}He, and this is true. There is a non-trivial reason for that, however. Namely, burning of helium-3 and tritium occurs faster than their production from deuterium. Were this not the case, deuterium would burn out first, and reactions ^{3}He + D → ^{4}He + p, T + D → ^{4}He + n would terminate at the stage when neutrons are bound in helium-3 and tritium, rather than in ^{4}He. On the other hand, freeze-out abundances of ^{3}He and T are sizeable, since their burning rates are not vastly higher than deuterium burning rate at $T \simeq T_{NS}$. A certain diversity of light elements in cosmic medium after BBN is the result of rather accidental coincidences between low energy thermonuclear cross sections.

8.3.4.   *Production and burning of the heaviest elements in primordial plasma*

As an example of reactions involving the heaviest elements of primordial plasma, let us consider production and burning of ^{7}Li in reactions $T(\alpha, \gamma)^7$Li and ^{7}Li$(p, \alpha)^4$He, respectively.

Production reaction is reasonably well described by the formula (8.38), in which $\sigma_0 \propto m_\pi^{-2} \cdot \alpha$ (the factor of α is due to photon emission). Numerically,

$$\langle \sigma v \rangle_{T(\alpha,\gamma)^7\text{Li}} \sim 10^{-18} \frac{\text{cm}^3}{\text{s}} \cdot T_9^{-2/3} e^{-8.0 T_9^{-1/3}}. \tag{8.60}$$

The rate of tritium burning via this channel is

$$\langle \sigma v \rangle_{T(\alpha,\gamma)^7\text{Li}} \cdot n_\alpha \simeq 1.5 \cdot 10^{-4}\, \text{s}^{-1}, \quad T_9 = 0.75.$$

This rate is small as compared to the Hubble parameter. The burning reaction is also described by formula (8.38), and we obtain

$$\langle \sigma v \rangle_{^7\text{Li}(p,\alpha)^4\text{He}} \sim 10^{-15} \frac{\text{cm}^3}{\text{s}} \cdot T_9^{-2/3} e^{-8.5 T_9^{-1/3}}.$$

Here the parameter σ_0 is determined by strong interactions only, so the rate is much higher than the production rate (8.60). Numerically, the burning rate of lithium-7 is

$$\langle \sigma v \rangle_{^7\text{Li}(p,\alpha)^4\text{He}} \cdot n_p \simeq 0.7\, \text{s}^{-1}, \quad \text{at} \quad T_9 = 0.75, \quad \eta_B = 6.05 \cdot 10^{-10},$$

which is higher than the Hubble parameter. Hence, burning of ^{7}Li terminates rather late, when the number density of protons gets diluted substantially due to the cosmological expansion.

Temperature at which burning of ^{7}Li terminates, is determined from the relation (cf. (8.56); we recall that $\nu_p = 1$)

$$\alpha_p M_{Pl}^* \langle \sigma v \rangle_{^7\text{Li}(p,\alpha)^4\text{He}} \sim 1.$$

We obtain numerically

$$T_{^7\text{Li}} \simeq 18 \text{ keV}, \quad T_{^7\text{Li}\, 9} \simeq 0.2.$$

The final abundance of ^{7}Li produced in the way we consider is estimated as (cf. (8.57))

$$\nu_{^7\text{Li}} \simeq \frac{\langle \sigma v \rangle_{T(\alpha,\gamma)^7\text{Li}}}{\langle \sigma v \rangle_{^7\text{Li}(p,\alpha)^4\text{He}}} \cdot \nu_T \nu_\alpha \sim 6 \cdot 10^{-11}, \tag{8.61}$$

where $\nu_\alpha = n_\alpha/n_p = (0.25/4)/0.75 \simeq 0.08$ is the abundance of ^{4}He, and we use (8.59) for tritium abundance.[9]

---

[9] Since $T_{^7\text{Li}} > T_T$, using (8.59) for tritium abundance is not quite legitimate. Instead, one should make use of the relation similar to (8.55). However, this does not grossly modify the result, since $T_{^7\text{Li}}$ and T_T are close to each other.

The production of ^{7}Be is also important. This isotope is unstable, so its abundance is not directly measurable. Beryllium-7 transforms into lithium-7 in cosmic plasma either via electron capture ^{7}Be$(e^-, \nu_e)^7$Li, or in the reaction ^{7}Be$(n, p)^7$Li. Thus, lithium-7 is produced either directly, in tritium-α fusion, or through beryllium-7. The existence of the two production mechanisms gives rise to non-monotonic dependence of the primordial lithium-7 abundance on η_B. Formula (8.61) is valid for low η_B where the abundance of ^{7}Li decreases as η_B increases, while realistic value of η_B corresponds to the range where the abundance increases with η_B. The latter behavior is due to processes involving ^{7}Be.

Let us describe in some details the calculation of burning rates $\langle \sigma v \rangle$ (see Ref. [115] for further details). Averaging over energies proceeds with the Maxwell–Boltzmann distribution,

$$\langle \sigma v \rangle = \frac{2}{T} \cdot \sqrt{\frac{2}{\pi M T}} \cdot \int_0^\infty \sigma(E) \cdot E \cdot e^{-E/T} dE, \qquad (8.62)$$

where v is the relative velocity of colliding nuclei, M is reduced mass, E is kinetic energy in the center-of-mass frame, $\sigma(E)$ is the cross section of the reaction of interest; processes $2 \to 2$ are by far dominant.

If one of the particles is a neutron, Coulomb barrier is absent. Assuming the s-wave reaction, i.e., $\sigma \sim v^{-1}$, the neutron reaction cross section far from resonance energies is natural to write as follows,

$$\sigma(E) \equiv \frac{R(E)}{v(E)} = \sqrt{\frac{M}{2E}} R(E).$$

The function $R(E)$ depends on E weakly in the interesting energy range $E \lesssim 1\,\mathrm{MeV}$, so that it can be approximated by a few terms of the Taylor series in velocity, i.e., $\sqrt{E}$,

$$R(E) = \sum_{n=0}^{n=m} \frac{R^{(n)}(0)}{n!} E^{n/2},$$

where $R^{(n)}(0)$ are determined by fitting experimental data at low energies. Then the integral (8.62) can be evaluated analytically, and the neutron burning rate in the channel of interest is given by

$$\langle \sigma v \rangle(T) = \sum_{n=0}^{n=m} \frac{R^{(n)}(0)}{n!} \frac{\Gamma\left(\frac{n+3}{2}\right)}{\Gamma\left(\frac{3}{2}\right)} \cdot T^{n/2}.$$

If the reaction has resonance character (like the reaction ^{7}Be$(n, p)^7$Li which has resonances at $E_R \simeq 0.32\,\mathrm{MeV}$ and $E_R \simeq 2.7\,\mathrm{MeV}$), then in the case of isolated resonance the cross section in the resonance region has the Breit–Wigner form,

$$\sigma(E) = \frac{\pi}{2ME} \frac{(2J+1)(1+\delta_{ij})}{(2J_i+1)(2J_j+1)} \frac{\Gamma_{in}(E)\Gamma_{out}(E)}{(E-E_R)^2 + (\Gamma_R/2)^2},$$

where E_R and Γ_R are energy and width of the resonance, J_i, J_j and J are total angular momenta of the initial nuclei and the resonance, and $\Gamma_{in}(E)$ and $\Gamma_{out}(E)$ are partial decay widths of the resonance state into initial and final states, respectively. The functions $\Gamma_{in}(E)$ and $\Gamma_{out}(E)$ are also determined from experiment. In the narrow resonance case, $\Gamma \ll E_R$, the integral (8.62) is well approximated by

$$\langle \sigma v \rangle (T) \simeq \left(\frac{2\pi}{MT} \right)^{3/2} \frac{(2J+1)\,(1+\delta_{ij})}{(2J_i+1)\,(2J_j+1)} \frac{\Gamma_{in}\Gamma_{out}}{\Gamma_R} \cdot e^{-E_R/T}. \tag{8.63}$$

If both incident nuclei are charged, the important phenomenon is the exponential Coulomb suppression. As we have seen, the exponent is determined by the Sommerfeld parameter

$$\zeta = \alpha Z_i Z_j \sqrt{\frac{M}{2E}} \equiv \frac{1}{2\pi} \sqrt{\frac{E_g}{E}},$$

where E_g is known as the Gamow energy. The cross section is conveniently represented as

$$\sigma(E) = \frac{S(E)}{E} \cdot e^{-2\pi\zeta}.$$

Assuming that $S(E)$ is a polynomial in E, the integral (8.62) can be evaluated by the saddle point method, which gives

$$\langle \sigma v \rangle (T) = \frac{2\sigma_0}{T} \sqrt{\frac{2}{MT\pi}} e^{-3\left(\frac{E_g}{4T} \right)^{1/3}} \cdot S_0(E_0), \tag{8.64}$$

where

$$E_0 = E_g \cdot \left(\frac{T}{2E_g} \right)^{2/3}, \qquad \sigma_0 = \frac{2E_g}{\sqrt{3}} \left(\frac{T}{2E_g} \right)^{5/6},$$

are the saddle point value and width, while $S_0(E_0)$ is a polynomial in $(\sigma_0/E_0)^2 \propto (T/E_g)^{1/3}$. The latter function can be determined from experiment. Note that the saddle point parameter $\sigma_0/E_0 = (T/E_g)^{1/6}$ is fairly large, so obtaining good accuracy requires employing high order polynomials. In practice one often makes use of other semi-analytical approximations to compute the rates $\langle \sigma v \rangle$. As an example, reactions $T(D,n)^4He$ and $^3He(D,p)^4He$ proceed through resonances. Even though these resonances are wide, the resonance contributions can be approximated by (8.63). The latter approximation is quite good numerically, while the resonance contributions turn out to be dominant in the temperature range of interest. We have used this approximation in the text; this is why the temperature dependence of the rates given in (8.53), (8.58) is different from the dependence that would follow from (8.64).

Thus, the calculation of the rates $\langle \sigma v \rangle$ involves, in an important way, experimental data on nuclear reactions. In some cases, including the reactions $p(n,\gamma)D$, $D(p,\gamma)^3He$, the data are scarce, which leads to uncertainties in predictions of element abundances. The first reaction, $p(n,\gamma)D$, is, however, well understood theoretically, and this knowledge is often used in real calculations.

We mention in the end that effects due to excitations of nuclei and Debye screening of nuclei by free electrons are not very relevant for BBN.

8.4. Comparison of Theory with Observations

BBN theory is well-developed. Numerical analysis gives precise predictions for light element abundances in primordial plasma. These predictions are tested by measuring the chemical composition of matter in those places in the present Universe where the composition is thought to be primordial despite evolution.

The evolution effects are very strong in most places in the Universe. Primordial matter is processed in stellar thermonuclear reactions occurring in the recent Universe, at $z \sim 0-10$. Some of the primordial nuclei transform into heavier elements, while others are destroyed by hard γ-quanta emitted in the star formation processes. These γ-quanta destroy also heavier elements already produced, so that more light elements appear. All these processes lead to considerable changes in the light element abundances as compared to the primordial ones.

In some regions of the Universe, however, local abundances of some elements are thought to remain unchanged. These are regions with low star formation rate: very distant (high redshift) regions where star formation has not happened yet and/or low-metallicity regions. The latter can be found by analyzing absorption spectra.

Deuterium is special among light nuclei: it has very small binding energy and hence is not produced in stellar nucleosynthesis; it predominantly gets destroyed. No substantial sources of deuterium are known, so any measurement of local deuterium abundance sets a lower bound on its primordial abundance. Recently, deuterium abundance has been determined by spectroscopy of high-z, low-metallicity clouds which absorb light of distant quasars.

Primordial helium-4 abundance is measured by spectroscopy of low-metallicity clouds of ionized hydrogen in dwarf galaxies. Production of helium-4 in stars is accompanied by production of heavier elements, metals in astrophysics terminology, so absence of the latter in clouds suggests that helium-4 is mostly of BBN origin there.

Lithium-7 abundance is determined by spectroscopy of low-metallicity old stars in globular clusters of our Galaxy.

No regions in the Universe have been found so far, where ^{3}He abundance could be measured and where ^{3}He would be mostly of primordial origin. Primordial abundance of this element is not as sensitive to the parameter η_B as that of deuterium, and measurements of ^{3}He tell more about evolution of stars and Galaxy than about BBN.

Spectroscopic measurements of relative local element abundances are quite precise by themselves. The major uncertainties are systematic and, roughly speaking, have to do with limited confidence on the primordial origin of these abundances. The predictions of BBN theory together with observational data [1] are shown in Fig. 8.5. These results are in overall agreement with each other and with the value $\eta_B = 6.05 \cdot 10^{-10}$ obtained from CMB data, although systematic uncertainties are still pretty high.

Still, the data is precise enough to make an impact on particle physics. First, as we have already discussed, BBN imposes limits on the density of new relativistic particles at $T \sim 1\,\mathrm{MeV}$; see (8.17).

Second, there should be no decays or annihilations of new heavy particles with emission of numerous hard photons at BBN epoch and somewhat later. If the main decay channel of a heavy particle X is

$$X \to \gamma Y,$$

where Y is a particle with $m_Y \ll m_X$, then the energy of produced photon is $E_\gamma \approx m_X/2$. It is the emission of these photons that is dangerous for BBN. There are two effects. The first one is the destruction of light elements by the hard photons. Particularly sensitive to this destruction is deuterium. The second effect has to do with the disintegration of helium-4. Even though this nucleus is tightly bound, hard photons may well break it up into lighter nuclei. Since helium-4 is very abundant, this would lead to overproduction of lighter elements, notably, ^{3}He. The corresponding bounds [116] in the space of parameters τ_X and ζ_X, where τ_X is X-particle lifetime and

$$\zeta_X = m_X \frac{n_X}{n_\gamma} = \frac{\Omega_X}{\Omega_B} m_p \eta_B,$$

are shown in Fig. 8.6.

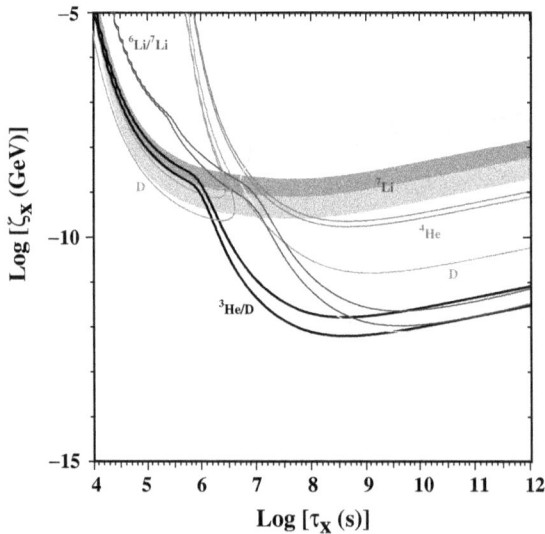

Fig. 8.6.   Model-independent BBN bounds on models with long-lived particles decaying into high energy photons [116]. Models with parameters above solid lines are ruled out by measurements of light element abundances. Upper lines corresponds to the most conservative bounds.

Examples of models with X- and Y-particles include some supersymmetric extensions of the Standard Model, with the decaying X-particle being, e.g., neutralino and stable Y-particle being gravitino (see Sec. 9.7). Similar bounds [117] exist on models with long-lived particles decaying into hadrons.

Yet another source of bounds on extensions of the Standard Model is the observation that there should be (almost) no entropy production at BBN epoch. Otherwise the nucleosynthesis temperature would be different, and the abundance of helium-4 would be inconsistent with observations.

Chapter 9

Dark Matter

As we have discussed already, large contribution to the total energy density in the present Universe (about 25%) comes from dark matter which consists, most likely, of new massive particles absent in the Standard Model of particle physics. These particles must be non-relativistic and should have practically no interactions with photons.[1] The latter requirement comes from the observations showing that dark matter halos are much larger than baryon parts of galaxies, meaning that dark matter experiences very little, if any, photon cooling (see Fig. 1.7 in Chapter 1).

Clearly, dark matter particles must be very long lived. The minimal requirement is that their lifetime exceeds the age of the Universe. The point is that the dark matter density is known both at the present epoch (from mass distributions in clusters of galaxies and other data) and at early epoch of intense structure formation and even earlier, at recombination. By comparing these data one concludes that the variation of dark matter density in comoving volume, if any, must be small. If dark matter particles decay into Standard Model particles, then the limit on lifetime is typically much stronger: null (for now?) results of searches for photons, positrons, antiprotons, neutrinos produced in decays of dark matter particles imply that the decay rates through these channels exceed the age of the Universe by many orders of magnitude. In such a situation a natural assumption is that the dark matter particles are absolutely stable. In elementary particle physics, stability is often guaranteed by the existence of a conserved quantum number (or several numbers). Therefore, an explanation of the dark matter phenomenon requires introducing a new particle and, likely (though not necessarily), a new conserved quantum number.

---

[1]An interesting possibility is that dark matter particles experience fairly strong interactions between themselves. Their long-ranged interactions would lead, among other things, to formation of spherical halos, while observationally most halos of clusters of galaxies are ellipsoidal. This contradiction is avoided provided that the mass and cross section obey [161] $\sigma/M_X \ll 10^{-24}$ cm^2/GeV. At the same time, we mention that considerable elastic cross section of dark matter particles would be helpful in explaining the distribution of dark matter in galactic centers: numerical simulations of non-interacting dark matter predict cuspy profiles, with strong increase of dark matter density towards the center. These cusps apparently are not observed.

210 Dark Matter

In this Chapter we consider several mechanisms of the dark matter generation in the Universe and some extensions of the Standard Model which contain dark matter particle candidates. Let us make an important comment right away: none of these mechanisms explains approximate (valid within a factor of 5) equality

$$\rho_{B,0} \sim \rho_{DM,0}, \qquad (9.1)$$

where $\rho_{B,0}$ and $\rho_{DM,0}$ are energy (mass) densities of baryons and dark matter in the present Universe. This approximate equality was also valid at earlier stages, since the times dark matter and baryon asymmetry were generated. Several suggestions have been made in literature on how mechanisms of baryon asymmetry and dark matter generation may be related and lead to (9.1), but none of them appears compelling. The approximate equality (9.1) may be accidental indeed.

9.1. Cold, Hot and Warm Dark Matter

Let us make, for the sake of concreteness, a rather natural assumption that dark matter particles X were in kinetic equilibrium with conventional matter in the early Universe.[2] At some moment of time these particles get out of equilibrium and since then they propagate freely. If the corresponding decoupling temperature T_d is much smaller than the mass of the dark matter particle M_X, these particles decouple being non-relativistic. In this case dark matter is *cold*. In the opposite case, $T_d \gtrsim M_X$, there are two possibilities, $M_X \lesssim 1\,\text{eV}$ and $M_X \gtrsim 1\,\text{eV}$. The former corresponds to *hot dark matter*: its particles remain relativistic at matter-radiation equality (recall that equality occurs at $T_{eq} \sim 1\,\text{eV}$, see Sec. 4.4); this is the case, e.g., for neutrino, as we have seen in Chapter 7. In the latter case dark matter is called *warm*: it is non-relativistic by equality epoch. We see in the accompanying book that density perturbations grow differently at radiation domination and matter domination, and that this growth strongly depends on whether dark matter is relativistic or not at equality. This is the reason for distinguishing hot and warm dark matter.

One effect specific to hot and warm dark matter is as follows. Let dark matter have primordial density perturbations and its particles be free and relativistic in the temperature interval $T_d \gtrsim T \gtrsim M_X$. At that time dark matter particles escape potential wells and fill in underdense regions of sizes up to current horizon size. Due to this *free streaming*, dark matter density perturbations of these sizes get washed out. Hence, hot and warm dark matter have low amplitudes of density perturbations at relatively short scales.

Free streaming terminates at $T \sim M_X$. The horizon size at that time, stretched by a factor $(1 + z) = T/T_0$, is the present maximum size of suppressed density perturbations. In the warm dark matter case the equality $T \sim M_X$ occurs at radiation

---

[2]This assumption, in fact, might not hold, some examples are given in Sec. 9.4.2–9.8.

domination, so the horizon size at that time is

$$l_H \sim \frac{M^*_{Pl}}{T^2} \sim \frac{M^*_{Pl}}{M^2_x}.$$

The corresponding present size is

$$l_{x,0} \sim l_H \frac{T}{T_0} \sim \frac{M^*_{Pl}}{T_0 M_x}. \tag{9.2}$$

Thus, models with warm dark matter predict the suppression of density perturbations of the present size $l_0 < l_{x,0}$. For $M_x \sim 1\,\text{keV}$ we take $\hat{g}_*(T \sim M_X) = 3.36$ (see (4.22)), so that $M^*_{Pl} = M_{Pl}/(1.66\sqrt{\hat{g}_*}) = 4 \cdot 10^{18}\,\text{GeV}$. Then Eq. (9.2) gives

$$M_x \sim 1\,\text{keV}: \quad l_{x,0} \sim 3 \cdot 10^{23}\,\text{cm} = 0.1\,\text{Mpc},$$

while

$$M_x \sim 1\,\text{eV}: \quad l_{x,0} \sim 100\,\text{Mpc}. \tag{9.3}$$

Hence, models with hot dark matter predict the suppression of density perturbations of present sizes up to 100 Mpc. We refine these estimates in the accompanying book.

In hot dark matter models, largest structures — superclusters of galaxies — get formed first, and then they fragment into smaller structures, clusters of galaxies. Galaxies are the latest objects in these models. This sequence of events is in strong disagreement with observations.

Probably the best option is cold dark matter. Hot dark matter particles (e.g., neutrinos) should make a small contribution into the total dark matter density.

Spatial size of order 0.1 Mpc is typical for perturbations that developed into small structures like dwarf galaxies.[3] The studies of structures of this and somewhat larger sizes gives the lower bound on the mass of dark matter particle,

$$M_X \gtrsim 1\,\text{keV}. \tag{9.4}$$

Similar bound comes from quite different argument based on the phase space density of dark matter particles in dwarf galaxies. We discuss this bound in some details in the accompanying book, and give numerical result in (7.34).

Thus, warm dark matter is still a viable possibility. We emphasize that the above bounds on the mass of dark matter particle apply to dark matter which was in kinetic equilibrium with the usual matter at some early epoch. For non-thermal momentum distribution the estimate (9.4) gets modified by a factor of order $\langle |\mathbf{p}| \rangle / \langle |\mathbf{p}| \rangle^{eq}$, where $\langle |\mathbf{p}| \rangle$ and $\langle |\mathbf{p}| \rangle^{eq}$ are actual average momentum and thermal one, respectively. It is worth noting that the comparison of CMB data and data on structures has the following general outcome: if dark matter particles were in kinetic equilibrium with the baryon–electron–photon plasma, they decoupled at

---

[3]Mass density in galaxies exceeds the average mass density by a factor 10^5–10^6. This means that matter in a galaxy clumped from a region whose size exceeds the size of the galaxy itself by a factor of 50–100. This leads to the estimate given in the text.

temperature $T_d \gtrsim 1$ keV irrespectively of their mass. (We discuss this point in the accompanying book.)

There exist model independent bounds on the masses of dark matter particles, which apply equally well to dark matter that had never been in kinetic equilibrium with the usual matter. Of course, these bounds are very weak. They come from the fact that dark matter particles must be confined in galaxies. For *bosons* the latter requirement implies that their de Broglie wavelength $\lambda = 2\pi/(M_X v_X)$ must be smaller than the dwarf galaxy size, 1 kpc. Making use of the fact that velocities in galaxies are $v_X \sim 10^{-4}$, we find

$$M_X \gtrsim 4 \cdot 10^{-22} \text{ eV}.$$

The bound is much stronger for *fermions*, due to Pauli principle. Assuming Maxwell distribution of dark matter fermions in galactic halo (this in fact is a reasonable assumption), we find for their phase space density

$$f(\mathbf{p}, \mathbf{x}) = \frac{\rho_X(\mathbf{x})}{M_X} \cdot \frac{1}{(\sqrt{2\pi} M_X v_X)^3} \cdot e^{-\frac{\mathbf{p}^2}{2M_X^2 v_X^2}},$$

where $\rho_X(\mathbf{x})/M_X$ and v_X^2 are the number density and velocity dispersion of dark matter particles in a halo. The maximum of the phase space density as function of momentum occurs at $\mathbf{p} = 0$ where $f(\mathbf{p}, \mathbf{x})$ is given by

$$f^{\max}(\mathbf{p}, \mathbf{x}) = \frac{\rho_X(\mathbf{x})}{M_X^4} \cdot \frac{1}{(2\pi)^{3/2} v_X^3}.$$

This maximum value cannot exceed the maximum value allowed by the Pauli principle (see (5.4)),

$$f_f = \frac{g_X}{(2\pi)^3}.$$

Taking $g_X = 2, v_X \sim 10^{-3}$ and $\rho(\mathbf{x}) \sim 0.5 \text{ GeV/cm}^3$ (typical mass density in a halo) we obtain the bound

$$M_X \gtrsim 25 \text{ eV}.$$

Stronger bound is obtained from the existence of dwarf galaxies. There, the mass density reaches $\sim 15 \text{ GeV/cm}^3$, which gives the bound

$$M_X \gtrsim 750 \text{ eV}.$$

We discuss this and similar bounds in the accompanying book.

We note that there is also rather formal *upper* bound on the mass of dark matter particles, of order of a thousand solar masses (see, e.g., [120, 121]),

$$M_X \lesssim 10^3 M_\odot \sim 10^{61} \text{ GeV}.$$

This bound comes from the stability of stellar clusters in the Galaxy, which would be destroyed by gravitational tidal forces induced by dark matter "particles" passing nearby.

9.2. Freeze-Out of Heavy Relic

Let us turn to one of the most attractive scenarios of the dark matter generation. We will discuss concrete examples in the following Sections, and now we calculate the residual abundance of heavy relic particles in general form.

To this end, let us consider the following situation. Let there exist stable heavy particles X and their antiparticles $\bar{X}$ which are in thermal (including chemical) equilibrium with cosmic plasma at sufficiently high temperatures. Let their interactions with the rest of the plasma be strong enough, so that they remain in equilibrium at temperatures somewhat below M_X. This assumption should, of course, be justified by the calculation of freeze-out temperature. Stability of the X-particle suggests that the X-particle can be created together with its antiparticle only. Let us ignore possible complications and assume that this is indeed the case. Finally, let us assume that there is no asymmetry between X-particles and their antiparticles, i.e.,

$$n_X - n_{\bar{X}} = 0. \tag{9.5}$$

This assumption is very important: the results of this Section are not valid for the Universe asymmetric with respect to X-particles. Our task is to calculate the present abundance Ω_X of X- and $\bar{X}$-particles.[4]

Another possibility leading to essentially the same results is that X is a truly neutral particle, i.e., $\bar{X}$ coincides with X, while X-particles are produced and annihilate in pairs. In that case the condition (9.5) is satisfied automatically. We meet this situation in Sec. 9.6.1. In what follows we consider for definiteness theories with X and $\bar{X}$.

The number densities in thermal equilibrium at temperature $T < M_X$ are

$$n_X^{eq} = n_{\bar{X}}^{eq} = g_X \left(\frac{M_X T}{2\pi} \right)^{3/2} e^{-M_X/T}. \tag{9.6}$$

Here we made use of the relation (9.5) to set the chemical potential of X-particles equal to zero. Under the above assumptions, the number of X-particles in comoving volume changes due to annihilation and pair creation only,

$$X\bar{X} \to \text{light particles.}$$

When the rate of annihilation is higher than the cosmological expansion rate, the abundance of X is given by the equilibrium formula (9.6). At some moment of time the pair production terminates: there is not enough light particles of energies of order M_X. The X-$\bar{X}$ annihilation terminates little later, when the the number density of X-particles becomes too small. At that time the abundance of X-particles

---

[4]We note that the calculation is much simpler, if there is asymmetry between X and $\bar{X}$ in primordial plasma. Assuming that X-$\bar{X}$ annihilation is fast, the cosmic medium at low temperatures contains only particles (assuming positive asymmetry). Their abundance is determined by the asymmetry $\eta_X \equiv (n_X - n_{\bar{X}})/s$. The present relative mass density is thus $\Omega_X = M_X s_0 \eta_X / \rho_c$. It does not depend on the annihilation cross-section in a wide range of parameters.

freezes out; after that, the number of particles is constant in comoving volume. We note in passing that *kinetic* equilibrium typically persists long after the freeze-out.

The evolution of the number density of X-particles is governed by the Boltzmann equation (cf. Sec. 5.4),

$$\frac{dn_X}{dt} + 3Hn_X = -\langle \sigma^{ann} \cdot v \rangle \cdot \left(n_X^2 - n_X^{eq\,2}\right). \tag{9.7}$$

Here $\langle \sigma^{ann} \cdot v \rangle$ is the product of the annihilation cross-section and relative velocity v of X-particles, averaged with equilibrium distribution functions (asuming kinetic equilibrium) and summed over all annihilation channels.

A simple way to obtain Eq. (9.7) is as follows. The annihilation probability of a given X-particle per unit time in medium with the antiparticle density $n_{\bar{X}} = n_X$ is

$$\Gamma_{ann} = \langle \sigma^{ann} \cdot v \rangle \cdot n_X. \tag{9.8}$$

This gives for the decrease of X-particle number in comoving volume a^3

$$\left[\frac{d(n_X a^3)}{dt}\right]_{ann} = -\Gamma_{ann} \cdot n_X a^3 = -\langle \sigma^{ann} \cdot v \rangle \cdot n_X^2 a^3. \tag{9.9}$$

In thermal equilibrium, i.e., for $n_X = n_X^{eq}$, this decrease is compensated by pair creation, so the increase of the X-particle number due to pair creation is given by

$$\left[\frac{d(n_X a^3)}{dt}\right]_{creation} = +\langle \sigma^{ann} \cdot v \rangle \cdot n_X^{eq\,2} \cdot a^3. \tag{9.10}$$

The total change in the X-particle number in comoving volume is given by the sum of (9.9) and (9.10), which leads precisely to Eq. (9.7). We emphasize that the contribution (9.10) is independent of the actual abundance of X-particles; the only condition of its validity is that other particles are in thermal equilibrium.

The annihilation of non-relativistic particles often occurs in s-wave. In that case the velocity dependence of the non-relativistic annihilation cross-section is given by (Bethe's law)

$$\sigma_{ann} = \frac{\sigma_0}{v}, \tag{9.11}$$

where σ_0 is a constant which does not depend on velocity and which is determined by interactions responsible for annihilation. If the annihilation occurs in p-wave instead, then the cross-section has the form $\sigma^{ann} = \sigma_1 v$. Contributions of higher angular momenta are additionally suppressed by powers of v^2. s-wave contribution is leading for non-relativistic particles, unless it is very small for some reason.

Let us briefly remind the reader of the way the law (9.11) emerges. (Detailed analysis can be found, e.g., in the book [111].) It applies not only to the annihilation process but also to any inelastic s-wave reaction. The main property which is assumed to hold is that the relevant interaction is not long ranged, i.e., it occurs inside a region of a certain size a. As usual, let us consider the flux of non-relativistic particles X incident on particle $\bar{X}$ at rest. Then the reaction probability per unit time is

$$P = Ca^3 |\psi(a)|^2,$$

where $\psi(a)$ is the wave function of X-particles in the collision region near $\bar{X}$-particle, and the constant C is determined by the details of the interaction. It is important here that particles annihilate in s-wave state, so that there is no centrifugal barrier. To obtain the cross-section, one divides the probability P by the flux of X-particles

$$\mathbf{j} = \frac{i}{2m} (\psi \boldsymbol{\nabla} \psi^* - \psi^* \boldsymbol{\nabla} \psi).$$

Away from the interaction region, the wave function is the plane wave of momentum p describing motion along the third axis,

$$\psi = e^{ipz},$$

then the flux is equal to velocity v. Since the interaction is short ranged, the modulus of the wave function in the reaction region is $|\psi(a)| = \text{const}$ with velocity-independent constant of order 1. The result for the cross-section is

$$\sigma = \frac{P}{|\mathbf{j}|} = \text{const} \cdot \frac{Ca^3}{v},$$

which is precisely (9.11). Note that the annihilation of heavy particles occurs with large energy release, $\Delta E \sim M_X$, so the size of the interaction region is indeed small

$$a \lesssim \frac{1}{M_X}.$$

It is worth noting that in the case of *electrically charged* particles of opposite charges (and in other cases of Coulomb-like attraction), the wave function in the reaction region may be considerably different from the asymptotic one (see the book [111] for details). This occurs at kinetic energies smaller than the binding energy of X–$\bar{X}$ atom, $E < \alpha^2 M_X$. In this case the annihilation cross-section scales with velocity as $\sigma \propto 1/v^2$. This effect will be irrelevant in examples below.

Additional factor v^2 in p-wave annihilation cross section emerges because of the centrifugal barrier. In the case of annihilation in the state of angular momentum l similar factor equals v^{2l}.

It is sufficient for our purposes to consider s- and p-wave contributions to annihilation cross-section, so we write

$$\langle \sigma^{ann} v \rangle = \sigma_0 + \sigma_1 \langle v^2 \rangle = \sigma_0 + 2\sigma_1 \langle v_X^2 \rangle = \sigma_0 + 6 \frac{T}{M_X} \sigma_1, \qquad (9.12)$$

where we use the fact that the average relative velocity squared equals[5] $\langle v^2 \rangle = 2 \langle v_X^2 \rangle$, and that the average kinetic energy of non-relativistic particles is related to

---

[5]Indeed, one has $v^2 = (\mathbf{v_1} - \mathbf{v_2})^2 = \mathbf{v_1}^2 - 2\mathbf{v_1} \cdot \mathbf{v_2} + \mathbf{v_2}^2$, and the cross term averages to zero.

temperature as follows:

$$\langle E_k \rangle = \frac{M_X \langle v^2 \rangle}{2} = \frac{3}{2}T. \tag{9.13}$$

Let us come back to the Boltzmann equation (9.7). It follows from (9.6) that at high temperatures both terms in the right-hand side of Eq. (9.7) are large, chemical equilibrium is maintained and n_X equals to n_X^{eq}. On the other hand, at low temperatures the term with n_X^{eq} is small, and X-particles get out of chemical equilibrium. Chemical equilibrium breaks down at temperature T_f, such that

$$\frac{dn_X^{eq}}{dt} \sim \langle \sigma^{ann} v \rangle n_X^{eq\,2}. \tag{9.14}$$

We accurately justify the relation (9.14) below, and here we point out that at $T \ll M_X$ the most rapidly evolving function is the exponential factor in (9.6), and therefore

$$\frac{dn_X^{eq}}{dt} \approx \frac{M_X \dot{T}}{T^2} n_X^{eq} = -\frac{M_X}{T} H(T) n_X^{eq}, \tag{9.15}$$

where we make use of (3.34). Hence, the temperature at which chemical equilibrium breaks down obeys

$$\frac{M_X}{T_f} H(T_f) \simeq \langle \sigma^{ann} v \rangle n_X^{eq} \tag{9.16}$$

or

$$\frac{M_X}{T_f} = \log \left[g_X (2\pi)^{-3/2} (1,66\sqrt{g_*})^{-1} (M_X T_f)^{1/2} M_{Pl} \langle \sigma^{ann} v \rangle \right], \tag{9.17}$$

where $\langle \sigma^{ann} v \rangle = \sigma_0 + 6\sigma_1 T_f / M_X$. This equation has the form (6.18). The parameter $\langle \sigma^{ann} v \rangle$ can be very roughly estimated as $\langle \sigma^{ann} v \rangle \sim M_X^{-2}$. This estimate is sufficient to understand that the logarithmic approximation works for $M_X \ll M_{Pl}^*$. Then with logarithmic accuracy we have

$$T_f = M_X \cdot \left[\log \left(\frac{g_X M_X M_{Pl} \langle \sigma^{ann} v \rangle}{(2\pi)^{3/2} \sqrt{g_*}} \right) \right]^{-1}. \tag{9.18}$$

We see that the temperature T_f of the exit from chemical equilibrium weakly (logarithmically) depends on the annihilation cross-section. This temperature is smaller than M_X by factor

$$L^{-1} \equiv \left[\log \left(\frac{g_X M_X M_{Pl} \langle \sigma^{ann} v \rangle}{(2\pi)^{3/2} \sqrt{g_*}} \right) \right]^{-1}. \tag{9.19}$$

This justifies our assumption that X-particles freeze out being non-relativistic.

After creation of X-particles terminates (the term with n_X^{eq} in the Boltzmann equation (9.7) becomes small), the annihilation of X-particles continues. To study this effect and develop an approach to the exact solution, we rewrite Eq. (9.7) in terms of X-entropy ratio. To this end, we introduce the variables

$$\Delta_X \equiv \frac{n_X}{s}, \quad \Delta_X^{eq} \equiv \frac{n_X^{eq}}{s},$$

where $s = (2\pi^2/45)g_* T^3$ is the entropy density at temperature T. Using the entropy conservation in comoving volume, Eq. (5.30), we write

$$\frac{dn_X}{dt} = s \cdot \frac{d\Delta_X}{dt} - 3H n_X.$$

This leads to the following form of equation for Δ_X:

$$\frac{d\Delta_X}{dt} = -\langle \sigma^{ann} v \rangle \cdot s \cdot \left(\Delta_X^2 - \Delta_X^{eq\,2} \right). \tag{9.20}$$

It is now convenient to change the variable t to

$$x \equiv \frac{T}{M_X}.$$

This variable shows to what extent X-particles are non-relativistic (see (9.13)). In view of Eq. (3.34), Eq. (9.20) takes the form

$$\frac{d\Delta_X}{dx} = \frac{\langle \sigma^{ann} v \rangle}{Hx} \cdot s \cdot \left(\Delta_X^2 - \Delta_X^{eq\,2} \right)$$

or, using explicit expressions for $s = s(T)$ and $H = H(T)$,

$$\frac{d\Delta_X}{dx} = \langle \sigma^{ann} v \rangle \cdot \frac{\sqrt{\pi g_*}}{3\sqrt{5}} \cdot M_X \cdot M_{Pl} \cdot \left(\Delta_X^2 - \Delta_X^{eq\,2} \right). \tag{9.21}$$

Making use of (9.12), we get finally

$$\frac{d\Delta_X}{dx} = (\sigma_0 + 6\sigma_1 \cdot x) \cdot \frac{\sqrt{\pi g_*}}{3\sqrt{5}} \cdot M_X \cdot M_{Pl} \cdot \left(\Delta_X^2 - \Delta_X^{eq\,2} \right). \tag{9.22}$$

This equation is suitable for numerical integration. We, however, pursue analytic approach and solve this equation at $T < T_f$ approximately, neglecting the term with $\Delta_X^{eq\,2}$ in the right-hand side of Eq. (9.22). This corresponds to switching off the pair creation of X-particles at $T < T_f$. At $T = T_f$ we have $n_X \approx n_X^{eq}$, where $n_X^{eq}(T_f)$ is found from[6] (9.16):

$$n_X^{eq}(T_f) \simeq \frac{M_X T_f}{\langle \sigma^{ann} v \rangle M_{Pl}^*}.$$

---

[6] We emphasize that inserting the temperature (9.18) into the expression (9.6) for number density would lead to large error, since temperature enters (9.6) exponentially, while the formula (9.18) has logarithmic accuracy only.

This gives the initial condition for simplified equation (9.22):

$$\Delta_X(x_f) \approx \frac{n_X^{eq}(T_f)}{s(T_f)} = \frac{3\sqrt{5}}{\sqrt{\pi g_*}} \frac{L^2}{M_X M_{Pl} \langle \sigma^{ann} v \rangle}, \tag{9.23}$$

where L is large logarithm entering (9.19).

Equation (9.22) without the term with Δ_X^{eq} is straightforward to integrate:

$$\Delta_X^{-1}(x) = \left[\sigma_0(x_f - x) + 3\sigma_1(x_f^2 - x^2) \right] \cdot \frac{\sqrt{\pi g_*}}{3\sqrt{5}} \cdot M_X \cdot M_{Pl} + \Delta_X^{-1}(x_f).$$

Because of the presence of *square* of large logarithm in (9.23), the last term in the right-hand side is small, so we obtain at low temperature, $x \to 0$,

$$\Delta_X(0) = \frac{3\sqrt{5}}{\sqrt{\pi g_*}} \frac{1}{M_X \cdot M_{Pl} \cdot x_f \langle \sigma^{ann} v \rangle_{\text{eff}}}, \tag{9.24}$$

where

$$\langle \sigma^{ann} v \rangle_{eff} = \sigma_0 + 3\sigma_1 x_f.$$

This is our final result. Recall that the parameter $x_f = T_f/M_X$ is determined by Eq. (9.17) with logarithmic accuracy,

$$x_f \simeq L^{-1}.$$

A few remarks are in order. First, we observe that $\Delta_X(0)/\Delta_X(x_f) \sim L^{-1}$. This means that the annihilation of X-particles after exiting from thermal equilibrium is important: it reduces their abundance by a factor of L^{-1}. Second, the result (9.24) could be obtained, modulo a factor of order 1, in a very simple way. Indeed, the annihilation terminates when the lifetime of X-particle with respect to annihilation becomes of the order of the Hubble time,[7]

$$\langle \sigma^{ann} v \rangle n_X \sim H. \tag{9.25}$$

If annihilation is effective, the number density of X-particles freezes out at temperature close ot T_f, which gives (9.24). Finally, the result (9.24) is valid with logarithmic accuracy only, since it is obtained by neglecting n_X^{eq} in Eq. (9.22). To obtain exact value of $\Delta_X(T = 0)$, one has to solve Eq. (9.22) exactly. We do not need this solution in what follows.

Problem 9.1. *Show that using the approximation* (9.25) *gives exactly* (9.24) *in the case of s-wave annihilation. This coincidence is absent for p-wave annihilation.*

---

[7]This condition is weaker than the condition for exit from chemical equilibrium: the rate of evolution of $\log n_X^{eq}$ is higher than the Hubble rate at $T \sim T_f$; see (9.15).

After freeze-out, the number density n_X changes only because of the cosmological expansion, and Δ_X does not change at all. So, the present number density of X-particles is

$$n_X(t_0) = s_0 \Delta_X(0),$$

where $s_0 = 2.9 \cdot 10^3$ cm^{-3} is the effective entropy density at present; see (5.34). Then our result (9.24) gives for the present mass density of X and $\bar{X}$

$$\Omega_X = 2 \cdot \frac{M_X n_X(t_0)}{\rho_c} = \frac{7.6 s_0}{\rho_c x_f \langle \sigma^{ann} v \rangle_{eff} M_{Pl} \sqrt{g_*(t_f)}}. \tag{9.26}$$

and, with logarithmic accuracy,

$$\Omega_X h^2 = 1.8 \cdot 10^{-10} \left(\frac{\text{GeV}^{-2}}{\langle \sigma^{ann} v \rangle_{eff}} \right) \frac{1}{\sqrt{g_*(t_f)}} \cdot \log \left(\frac{g_X M_X M_{Pl} \langle \sigma^{ann} v \rangle}{(2\pi)^{3/2} \sqrt{g_*}} \right). \tag{9.27}$$

Clearly, the most relevant parameter is the annihilation cross-section at temperature T_f. The dependence of Ω_X on M_X is logarithmic only, while the effective number of degrees of freedom g_* does not change much during most of the history of the Universe.

Our main emphasis in this Section is freeze-out of X-particles which has to do with breaking of *chemical* equilibrium. It is of interest, however, to figure out at what temperature *kinetic* equilibrium breaks down. Indeed, one of our assumptions was that the distribution functions have equilibrium form. Kinetic equilibrium holds due to scattering of X-particles off light particles, so the time between collisions of a given X-particle is idependent of the number density of X-particles. Therefore, it is shorter than the lifetime with respect to annihilation Γ_{ann}^{-1}; see (9.8). This means that kinetic equilibrium holds much longer than chemical equilibrium, i.e., it breaks down at temperature $T_{kin} \ll T_f$. As an example, if X-particles participate in weak interacions, then the cross-section of their elastic scattering off, say, electrons at $E \ll 100$ GeV is estimated on dimensional grounds as

$$\sigma_{el} \sim G_F^2 E^2. \tag{9.28}$$

The mean free time of X-particles is roughly

$$\tau_{el} \sim (n_e \cdot \sigma_{el} \cdot v)^{-1},$$

where n_e is electron number density and v is the relative velocity of X-particles and electrons, $v \simeq 1$ at $T \gg 1$ MeV. To estimate the temperature at which the kinetic equilibrium breaks down, we equate τ_{el} to the Hubble time $H^{-1}(T)$ and obtain (this estimate is similar to that made in Sec. 7.1) $T_{kin} \sim 1$ MeV. This is of course a crude estimate but it shows that the kinetic equilibrium of X-particles breaks down fairly late.

Problem 9.2. *Refine the estimate of T_{kin} in the case when the cross-section of X-prticles off electrons is given by (9.28). Hint: make use of arguments given in the beginning of Sec. 6.3.*

Let us show that the exit from chemical equilibrium indeed occurs at temperature given by (9.16). To this end, instead of variable $x = T/M_X$ we use the variable $y = M_X/T$ and write the Boltzmann equation (9.20) as follows:

$$\frac{d\Delta_X}{dy} = -\frac{1}{y^2} \cdot \langle \sigma^{ann} v \rangle \cdot \kappa_* M_X M_{Pl}^* \cdot \left(\Delta_X^2 - \Delta_X^{eq\,2} \right),$$

where $\kappa_* = 2\pi^2 g_*/45$. Our purpose is to find temperature at which the relative abundance Δ_X starts to deviate from the equilibrium abundance Δ_X^{eq}. We write

$$\Delta_X = \Delta_X^{eq}(1 + \delta).$$

The exit from equilibrium happens when δ ceases to be small. Before that, the Boltzmann equation can be linearized,

$$\frac{d\delta}{dy} + \left(2 \frac{\langle \sigma^{ann} v \rangle M_X \kappa_* M_{Pl}^*}{y^2} \Delta_X^{eq} - 1 \right) \delta = 1, \qquad (9.29)$$

where we use the fact that due to strong dependence of the the exponential factor on time one has $d\Delta_X^{eq}/dy = -\Delta_X^{eq}$, cf. (9.15). For relatively small y (high temperatures), the first term in parenthesis in (9.29) is large, therefore

$$\delta(y) = \frac{y^2}{2\langle \sigma^{ann} v \rangle M_X \kappa_* M_{Pl}^*} [\Delta_X^{eq}(y)]^{-1}. \qquad (9.30)$$

(Note that $d\delta/dy \sim \delta$, and the first term in the left-hand side of Eq. (9.29) is small.) We see that $\delta(y)$ ceases to be small when $y \simeq y_f$, where

$$y_f^2 = \langle \sigma^{ann} v \rangle M_X \kappa_* M_{Pl}^* \Delta_X^{eq}(y_f).$$

This determines the temperature in question. In terms of physical quantities the latter formula gives

$$\langle \sigma^{ann} v \rangle n_{eq}(T_f) = \frac{M_X}{T_f} H(T_f),$$

which yields (9.16).

Problem 9.3. *By explicitly solving Eq. (9.29) check the relation (9.30). Show that the initial value $\delta(y_i)$ is rapidly washed out, provided that $y_i \gg y_f$: for any initial value of n_X the number density of X-particles rapidly become thermal.*

Let us now give the general discussion of the Boltzmann equation which is widely used in cosmology.

In general, the system of Boltzmann equations describes the balance of interacting particles, and Eq. (9.7) is a particular example, applicable to particles of a certain type which are pair created and annihilate. More accurate treatment of this situation is as follows. Let us consider for diversity the case of intrinsically neutral particles which can be created and annihilate in pairs (this possibility was mentioned in the beginning of this Section), and study these processes in Minkowski space. Let $\mathbf{p_1}, \mathbf{p_2}$ be 3-momenta of incoming particles. The number of particles in the interval of momenta $(\mathbf{p}, \mathbf{p} + d\mathbf{p})$ in the volume element $d^3\mathbf{x}$ is

$$dN = n(t, \mathbf{x})F(t, \mathbf{p})d^3\mathbf{x}d^3\mathbf{p},$$

where $n(t, \mathbf{x})$ is the number density, and the distribution function $F(t, \mathbf{p})$ is normalized at each moment of time as

$$\int F(t, \mathbf{p})d^3\mathbf{p} = 1. \tag{9.31}$$

Let us consider a particle of momentum $\mathbf{p_1}$. In unit time it annihilates with

$$\sigma \cdot v \cdot n(t, \mathbf{x})F(t, \mathbf{p_2})d^3\mathbf{p_2} \tag{9.32}$$

particles whose momenta are in the interval $(\mathbf{p_2}, \mathbf{p_2}+d\mathbf{p_2})$. Here v is the relative velocity of the two annihilating particles and $\sigma = \sigma(\mathbf{p_1}, \mathbf{p_2})$ is the annihilation cross-section. This gives the annihilation rate in volume $d^3\mathbf{x}$ for particles of momenta in the intervals $(\mathbf{p_1}, \mathbf{p_1}+d\mathbf{p_1})$ and $(\mathbf{p_2}, \mathbf{p_2} + d\mathbf{p_2})$,

$$\frac{1}{2}dN(\mathbf{p_1}, \mathbf{x}) \cdot \sigma \cdot v \cdot n(t, \mathbf{x})F(t, \mathbf{p_2})d^3\mathbf{p_2}$$

$$= \frac{1}{2}n(t, \mathbf{x})F(t, \mathbf{p_1})d^3\mathbf{x}d^3\mathbf{p_1} \cdot \sigma \cdot v \cdot n(t, \mathbf{x})F(t, \mathbf{p_2})d^3\mathbf{p_2}, \tag{9.33}$$

where the factor $1/2$ accounts for identical particles. (Otherwise the contribution of one and the same state $(\mathbf{p_1}, \mathbf{p_2})$ and $(\mathbf{p_2}, \mathbf{p_1})$ would be double-counted.)

Since each annihilation reduces the number of particles by 2, the rate of decrease of particles due to annihilation equals twice the expression (9.33),

$$n^2(t, \mathbf{x})F(t, \mathbf{p_1})d^3\mathbf{x}d^3\mathbf{p_1} \cdot \sigma \cdot v \cdot F(t, \mathbf{p_2})d^3\mathbf{p_2}. \tag{9.34}$$

There is also the inverse process of pair creation in collisions of other particles. The latter are in thermal equilibrium, and repeating the argument before Eq. (9.10), we write for pair creation rate

$$n^{eq\,2}(t, \mathbf{x})F^{eq}(t, \mathbf{p_1})d^3\mathbf{x}d^3\mathbf{p_1} \cdot \sigma \cdot v \cdot F^{eq}(t, \mathbf{p_2})d^3\mathbf{p_2},$$

where n^{eq}, F^{eq} are equilibrium quantities. Finally, we integrate over momenta $\mathbf{p_2}$ to obtain the equation for the number density in phase space volume $d^3\mathbf{p}d^3\mathbf{x}$,

$$\left[\frac{\partial(n(t, \mathbf{x})F(t, \mathbf{p_1}))}{\partial t} \right] d^3\mathbf{p_1}d^3\mathbf{x}$$

$$= -\left\{ \int [n^2(t, \mathbf{x})F(t, \mathbf{p_1})F(t, \mathbf{p_2}) \right.$$

$$\left. - n^{eq\,2}(t, \mathbf{x})F^{eq}(t, \mathbf{p_1})F^{eq}(t, \mathbf{p_2})] \cdot v \cdot \sigma\, d^3\mathbf{p_2} \right\} d^3\mathbf{p_1}d^3\mathbf{x}. \tag{9.35}$$

The right-hand side here is the collision integral.

This general equation is simplified in the case of interest, when the number density is homogeneous in space,

$$n(t, \mathbf{x}) = n(t), \quad n^{eq}(t, \mathbf{x}) = n^{eq}(t),$$

while the distribution over momenta corresponds to kinetic equilibrium and is time-independent in Minkowski space,

$$F(t, \mathbf{p}) = F^{eq}(\mathbf{p}).$$

Then one integrates Eq. (9.35) over momentum $\mathbf{p_1}$ and obtains

$$\frac{\partial n(t)}{\partial t} = -\langle \sigma v \rangle (n^2 - n^{eq\,2}). \tag{9.36}$$

Here we introduced the notation

$$\langle \sigma v \rangle = \int d^3 \mathbf{p_1} d^3 \mathbf{p_2} F^{eq}(\mathbf{p_1}) F^{eq}(\mathbf{p_2}) \cdot v \cdot \sigma, \tag{9.37}$$

which is the thermal average of the relative velocity and annihilation cross-section (recall the normalization (9.31)). Note that the right-hand side of (9.36) is such that the system relaxes to thermal equilibrium for any sign of $(n - n^{eq})$.

In the case of annihilation of non-relativistic X-particles into relativistic particles, the normalized equilibrium distribution function is

$$F^{eq}(\mathbf{p}) = \frac{1}{4\pi\, m_X^2 T\, K_2(\frac{m}{T})} \frac{e^{-\frac{E}{T}}}{},$$

where $E \equiv \sqrt{\mathbf{p}^2 + m_X^2}$ is particle energy and $K_2(x)$ is modified Bessel function. The integral in (9.37) is conveniently wtitten in terms of the Mandelstam variable $s \equiv (p_1 + p_2)^2$. To this end, one replaces $v \cdot \sigma$ by Lorentz-invariant quantity

$$W(s) = 4E_1 E_2 v \cdot \sigma = 4\sigma \sqrt{(p_1 p_2)^2 - m_X^4} = 2\sigma(s)\sqrt{s(s - 4m_X^2)}.$$

Then one inserts unity

$$1 = \int ds\, \delta(s - (p_1 + p_2)^2)$$

in the integrand and integrates over angular variables. The non-trivial integral over the angle between 3-momenta is evaluated by making use of the above δ-function. As a result one obtains

$$\langle \sigma v \rangle = \frac{1}{16T^2 m_X^4 K_2^2(\frac{m}{T})} \int ds\, W(s) \int dE_1\, dE_2\, e^{-\frac{E_1 + E_2}{T}}.$$

The integral over energies is evaluated by introducing variables $E_\pm = E_1 \pm E_2$. Then

$$\int dE_1\, dE_2\, e^{-\frac{E_1 + E_2}{T}} = 2\sqrt{1 - \frac{4m_X^2}{s}} \int_{\sqrt{s}}^{\infty} dE_+ \sqrt{E_+^2 - m_X^2}\, e^{-\frac{E_+}{T}}$$

$$= 2T\sqrt{s - 4m_X^2}\, K_1\left(\frac{\sqrt{s}}{T}\right),$$

and finally

$$\langle \sigma v \rangle = \frac{1}{8T\, m_X^4\, K_2^2\left(\frac{m}{T}\right)} \int_{4m_X^2}^{\infty} ds\, W(s)\sqrt{s - 4m_X^2}\, K_1\left(\frac{\sqrt{s}}{T}\right).$$

The Boltzmann equation (9.36) should be modified when applied to matter in expanding Universe. This modification takes into account the increase of the volume and redshift of momenta, and is given in (9.7), see also Sec. 5.4.

9.3. Weakly Interacting Massive Particles, WIMPs

Let us discuss the possibility that new stable heavy particles make dark matter. Then the formula (9.27) is used to find their parameters (annihilation cross-section in the first place). The dark matter density today is $\Omega_{DM} \approx 0.25$. To estimate the annihilation cross-section we set $\sigma_0 \sim 1/M_X^2$ in the argument of logarithm in (9.27), on dimensional grounds. Setting $M_X = 100\,\text{GeV}$ and $g_* = 100$ for the estimate, we obtain the numerical value of the logarithmic factor in (9.27),

$$\log \frac{g_X\, M_{Pl}^*\, M_X\, \sigma_0}{(2\pi)^{3/2}} \sim \log \frac{g_X\, M_{Pl}^*}{(2\pi)^{3/2} M_X} \sim 30.$$

Since logarithm is a slowly varying function, this estimate is valid for wide range of masses M_X and cross-sections σ_0. As an example, one finds the value of $\langle \sigma^{ann} v \rangle$ from Eq. (9.27) and obtains for $M_X = 100$ GeV

$$\log\left(\frac{g_X\, M_{Pl}^*\, M_X\, \langle\sigma^{ann}v\rangle}{(2\pi)^{3/2}}\right) \sim 20. \tag{9.38}$$

The uncertainty in the parameter $\sqrt{g_*(t_f)}$ is also moderate: at $T \gtrsim 100\,\text{GeV}$ and $T \sim 100\,\text{MeV}$ we have, respectively, $\sqrt{g_*(T)} \sim 10$, and $\sqrt{g_*(T)} \sim 3$ (see Appendix B). Thus, formula (9.27) gives the following estimate for the annihilation cross-section of cold dark matter candidates,

$$\langle \sigma^{ann} v \rangle \sim \frac{1.8 \cdot 10^{-10} \cdot 20}{(3 \div 10) \cdot 0.25 \cdot h^2}\,\text{GeV}^{-2} = (0.3 \div 1.0) \cdot 10^{-8}\,\text{GeV}^{-2}$$

$$= (1.1 \div 3.7) \cdot 10^{-36}\,\text{cm}^2, \tag{9.39}$$

where the lower value refers to higher mass M_X, an is, therefore, more realistic. In the case of s-wave annihilation this estimate gives the parameter σ_0 in (9.11). Notably, this value is comparable to weak interaction cross-sections at energies of order $100\,\text{GeV}$, namely $\sigma_W \sim \alpha_W^2/M_W^2 \sim 10^{-7}\,\text{GeV}^{-2}$.

The result (9.39) has several important consequences. One is that it gives a cosmological lower bound on the annihilation cross-section of hypothetical heavy stable particles that may be predicted by extensions of the Standard Model. If the cross-section is below the value (9.39), mass density of these particles is unacceptably high. The main assumption behind this bound is that X-particles were in

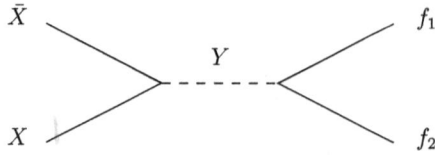

Fig. 9.1.   Annihilation of stable hypothetical X-particles into Standard Model particles f_1 and f_2.

thermal equilibrium at some early epoch. We note here that there is a bound on the annihilation cross-section

$$\sigma_0 \lesssim \frac{4\pi}{M_X^2}.\tag{9.40}$$

In perturbative regime, this bound comes from the fact that the annihilation is described by diagrams like that shown in Fig. 9.1. For slow X-particles, the virtual particle Y has energy $E = 2E_X = 2M_X$ in the center-of-mass frame. Its propagator gives the factor $1/M_X^2$ in the cross-section.[8] Furthermore, there is additional suppression due to small coupling constant.

The bound (9.40) may actually be violated in strongly coupled theories. This occurs when X-particle is a bound state of elementary particles, and the size of this bound state is large compared to the Compton wavelength. An example here is given by the proton. It is difficult, however, to imagine that the bound (9.40) is violated by many orders of magnitude.

With this reservation, one makes use of (9.39) and (9.40) to obtain the cosmological upper bound on the mass of stable particles,

$$M_X \lesssim 100\,\text{TeV}.\tag{9.41}$$

This bound is valid, if their interactions are sufficiently strong and the maximum temperature in the Universe exceeded $M_X/20$; see (9.18) and (9.38).

Problem 9.4. *Let us extend the Standard Model by adding new real scalar field X which interacts with the Englert–Brout–Higgs doublet H only. Let us add the following term to the Standard Model Lagrangian,*

$$\Delta\mathcal{L} = \frac{1}{2}\partial_\mu X \partial^\mu X - \frac{\kappa}{2}H^\dagger H X^2 - \frac{m^2}{2}X^2.$$

The discrete symmetry $X \to -X$ ensures the stability of the scalar particle X, so it is a dark matter candidate. Find the range of parameters (m,κ) in which X-particles constitute all of dark matter.

Let us mention that there is recent activity in discussing the possibility that all dark matter or its substantial part consists of much heavier particles. This scenario is realistic provided that these particles were never in thermal equilibrium and

---

[8]If $M_Y > M_X$, then the propagator of Y-particles is suppressed by M_Y^{-2}, which makes the bound (9.40) even stronger.

were created in small number in the early Universe. We will briefly discuss possible mechanisms of superheavy particle creation in Sec. 9.8.1.

By far more interesting is the possibility that X-particles *are* dark matter particles. These dark matter candidates are called weakly interacting massive particles, WIMPs. The estimate (9.39) suggests the energy scale of X-particle interactions, $\sigma_0^{-1/2} \sim 10\,\text{TeV}$. In fact, this scale is somewhat lower, since the annihilation cross-section is suppressed by coupling constant in realistic models, which we denote as α_X. Assuming that the energy scale of X-particle interactions does not much exceed their mass, we get

$$\sigma_0 \sim \frac{\alpha_X^2}{M_X^2}. \tag{9.42}$$

As an example, for $\alpha_X \sim 1/30$ (W-boson coupling in the Standard Model) we obtain from (9.39) the following estimate:

$$M_X \sim 200\text{--}400\,\text{GeV}. \tag{9.43}$$

This estimate shows that there is a real chance to discover dark matter particles at colliders. The relevant processes at hadron colliders are quark–antiquark annihilation into dark matter particles plus visible Standard Model particles, and creation of dark matter particles in gluon–gluon collisions. The signature of dark matter is missing transverse energy (energy flowing transverse to the incident beam direction), which is carried away by undetected dark matter particles. The diagrams of this sort are similar to that shown in Fig. 9.1, if the latter is looked at from right to left and emission of a Standard Model particle (say, gluon or photon) is added in one of the legs f_1, f_2.

If intermediate paticles Y are much heavier than X, then low energy physics is described by effective contact interaction between dark matter and Standard Model particles, $\bar{X}Xf_1f_2$. The precise structure of this interaction is model dependent. Search for dark matter particles gives limits which can be interpreted as limits on elastic cross-sections of scattering of the dark matter particles off nucleon, if f_1 and f_2 are gluons or first generation quarks. The diagrams for this scattering are again similar to Fig. 9.1, but now viewed from top to bottom (or from bottom to top). Figure 9.2 shows the results of the searches for dark matter interpreted in this way.

It is worth noting that unlike the estimate (9.39) for the annihilation cross-section, our estimate (9.43) for the mass is very crude. Indeed, the dark matter particle may well be lighter or heavier than given in (9.43). Accurate calculations in concrete models show that the mass of the dark matter particle may be considerably smaller than $100\,\text{GeV}$, which makes collider searches even more promising. We also note that our estimate is particularly interesting from the viewpoint of supersymmetric extensions of the Standard Model which often predict the existence of a stable particle of mass in $100\,\text{GeV}$ range; see Sec. 9.4.2.

Thus, there are good reasons to expect that dark matter particles will be found in reasonably near future. Experimental determination of their properties, together

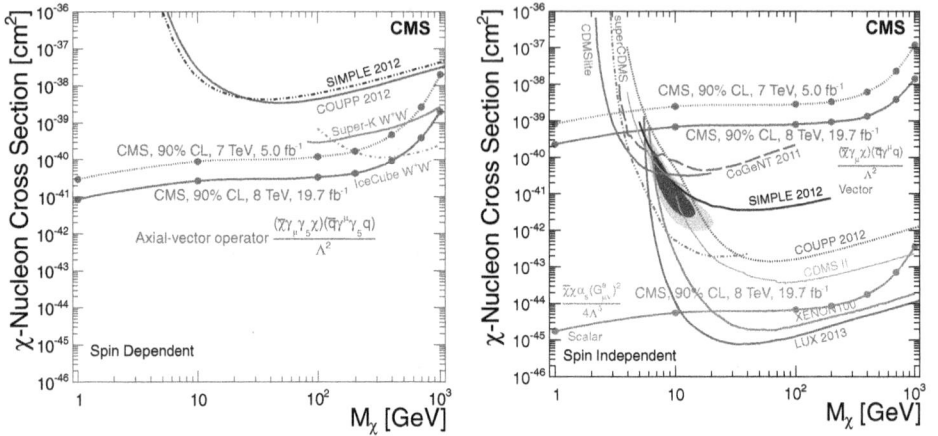

Fig. 9.2. Excluded regions in parameter plane (M_X, σ_{pX}) [162] for interactions independent of nucleon spin (left) and dependent on nucleon spin (right). Regions above the curves are excluded at 90% C.L. CMS: search at Large hadron collider (under assumption of contact interaction $\bar{X}Xf_1f_2$, see the main text); IceCube and Super-K: search for a signal from dark matter annihilation in the Sun; other data: direct search for elastic scattering of cosmic dark matter in low background detectors. Shaded region in the central part of right figure corresponds to possible signal announced by CDMS experiment. See Fig. 13.16 for color version.

with the precise calculation of their abundance and comparison with observational data will then provide a handle on the Universe at temperatures of order of tens GeV and age of 10^{-9} s. This is to be compared with the present knowledge: the Universe has been probed at BBN epoch, $T \simeq 1$ MeV, $t \simeq 1$ s; see Chapter 8.

Search for dark matter particles is underway, but no conclusive evidence has been obtained so far. Direct search for relic WIMPs is carried out in experiments trying to detect energy deposition in a detector caused by elastic scattering of a WIMP off a nucleus. Dark matter, like the usual matter, is more dense in galaxies; it is expected that its mass density near the Earth is similar to that of usual matter,

$$\rho_{DM}^{local} \simeq 0.3 \, \frac{\text{GeV}}{\text{cm}^3}.$$

The velocity of dark matter particles is about $v_X \sim 0.5 \cdot 10^{-3}$ (orbital velocity around the Galactic center at the position of the Sun). Hence, the energy deposition in collision with a nucleus of mass M_A is estimated as

$$\Delta E \sim \frac{1}{2\,M_A} \left(\frac{2v_X M_A M_X}{M_A + M_X} \right)^2 = 2M_X v_X^2 \frac{M_A/M_X}{(1 + M_A/M_X)^2}$$

$$= 2M_A v_X^2 \frac{1}{(1 + M_A/M_X)^2}.$$

(9.44)

The second of these formulas shows that the best choice of the target nucleus for given M_X is $M_A = M_X$, while the third implies that for given detector material, the energy deposition grows as M_X increases. Numerically,

$$\Delta E \simeq 50 \cdot \left(\frac{\min(M_A, M_X)}{100\,\text{GeV}}\right)^2 \frac{100\,\text{GeV}}{M_A}\,\text{keV}.$$

We see that the energy deposition is of order of tens keV, which is very small. Nevertheless, there are methods to detect it. The event rate is

$$\nu \simeq v_X n_X \cdot N_A \cdot \sigma_{AX},$$

where σ_{AX} is the elastic X-nucleus cross-section, v_X is the relative velocity, $n_X = \rho_{DM}^{local}/M_X$ is the local number density of dark matter particles and N_A is the number of nuclei in the detector. As an example, for elastic cross-section $\sigma_{AX} \sim 10^{-38}\,\text{cm}^2$ and X-particle mass $M_X = 100\,\text{GeV}$, the number of events in a detector of mass $10\,\text{kg}$ filled with nuclei of atomic number $A = 100$ is of order

$$\nu \sim 0.5 \cdot 10^{-3} \cdot \left(0.3\frac{\text{GeV}}{\text{cm}^3} \cdot \frac{1}{100\,\text{GeV}}\right) \cdot \left(\frac{6 \cdot 10^{23}\,\text{GeV} \cdot 10^4}{100\,\text{GeV}}\right) \cdot 10^{-38}\,\text{cm}^2 \sim 3 \cdot 10^{-8}\,\text{s}^{-1},$$

$$(9.45)$$

which is about 1 event per year. The absence of a signal gives bounds in the space of model-independent parameters — dark matter mass M_X and dark-matter-nucleon elastic cross-section σ_{NX}. To compare different experiments, one uses, instead of the cross-section σ_{AX} on nucleus A, the cross-section on a single nucleon σ_{pX}. It is important at this point that the cross-sections σ_{AX} *is not equal* to the sum of cross-sections on individual nucleons: for slow dark matter particle there is a coherence effect that increases the cross-section by a factor of order A. More precisely, for interaction independent of the nucleon spin there is a relation $\sigma_{pX} = (\sigma_{AX}/A^2)(m_p^2/m_A^2)(m_X + m_A)^2/(m_X + m_p)^2$ (modulo nuclear form-factor that dumps the cross-section at large A), where the additional mass-dependent factor accounts for the fact that the eleastic cross-section is proprotional to the squared momentum transfer, or (for the nonrelativistic scattering) the energy deposition (9.44). Another important point is that the elastic cross-section may depend on nucleon and nucleus spin. Examples of limits in the parameter plane (M_X, σ_{pX}) are shown in Fig. 9.2.

Problem 9.5. *Find the density of relic massive neutrinos of the hypothetical fourth generation of the Standard Model, assuming that their main annihilation channel is s-wave annihilation through virtual Z-boson, and that their mass obeys $m_\nu > M_Z/2$. Estimate the elastic cross-section of scattering of these neutrinos off nuclei and find the limit on their mass from the data shown in Fig. 9.2.*

Besides the direct search, there are ways to search for dark matter particles indirectly. These include search for products of WIMP annihilation at present time. In this case too, no compelling dark matter signal has been observed so far.

One of the promising indirect ways is to search for monoenergetic photons which are emitted in two-body annihilation processes $X\bar{X} \to \gamma\gamma, X\bar{X} \to Z\gamma$. In this case

Dark Matter

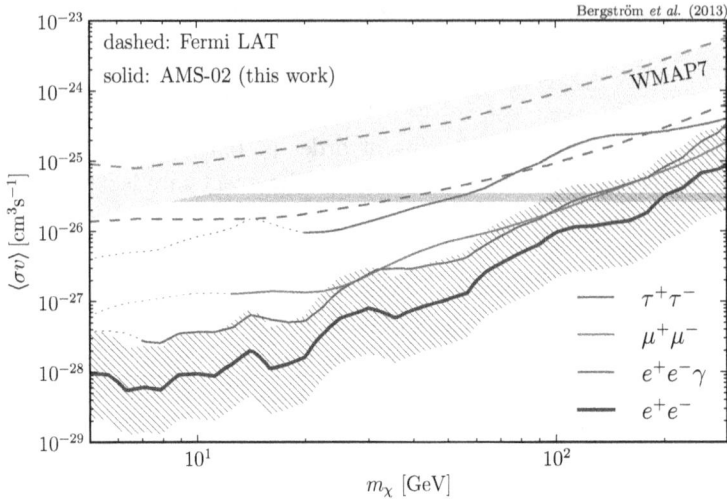

Fig. 9.3.   Exclusion regions in the plane (M_X, σ_{XX}^{ann}) [318] for dark matter particles annihilating through indicated channels. The light gray line is the cross-section at freeze-out in the early Universe. Regions above the lines are ruled out at 90% C.L. AMS-02: search for antiparticles; Fermi LAT: search for photons from dark matter annihilation to $\mu^+\mu^-$ and $\tau^+\tau^-$ (upper and lower lines), WMAP7: limit from possible annihilation into charged lepton pair before recombination (upper range $\tau^+\tau^-$, lower range e^+e^-), which would affect CMB. As an example, the uncertainty for the channel e^+e^- is shown as shadow region. See Fig. 13.17 for color version.

the signal should be enhanced in the direction towards dense regions in and outside the Galaxy, such as Galactic center, Magellanic Clouds, Andromeda galaxy, etc. Also, charged particles produced in dark matter annihilation may emit photons in their own annihilation, via Compton scattering and synchrotron radiation. Another way is to detect *anti*particles created in the annihilation: positrons and antiprotons.

Figure 9.3 shows limits of the dark matter annihilation cross-section for different channels. We see that these channels cannot dominate at dark matter freeze-out in the early Universe for $M_X > 100$ GeV. We note, however, that the bounds actually depend on the assumptions on the dark matter distribution in galaxies, since the annihilation rate is proportional to number density squared. In particular, the signal is enhanced, if dark matter forms small clumps inside galaxies.

Problem 9.6. *Prove the last statement.*

The annihilation cross-section of dark matter particles may be quite different from the cross-section responsible for freeze-out in the early Universe. This happens, in particular, if the cross-section strongly depends on velocity of annihilating particles: velocities of dark matter particles in halos v_h are lower by one-three orders of magnitude as compared to their velocities v_x at freeze-out. If annihilation occurs from a state of anglar momentum l, then the annihilation rate in galaxies is suppressed by a factor $(v_h/v_x)^{2l} \ll 1$. On the contrary, the annihilation in galaxies can be enhanced by a light particle coupled to the dark matter particles. This particle induces the interaction between the two dark matter particles in t-channel, which attracts one dark matter particle to another. Then the annihilation proceeds in two stages: at the first stage two dark matter particles approach each other forming a kind of bound state; at the next stage this state annihilates in s-channel (or in t-channel by exchange of a heavy particle) into a pair of light (Standard Model) particles. The annihilation cross-section of this process gets significant amplification at the first stage, where the mechanism similar to the Coloumb enchancement works as discussed in Section 9.2. For slow dark matter particles, $v_h \ll \alpha'$, where α' is the analog of the fine-structure constant of new interaction, the amplification factor is $\pi\alpha'/v_h \gg 1$ for annihilation from s-channel initial state and behaves like $\propto 1/v_h^{2l+1}$ for the initial state with angular momentum l.

Heavy dark matter particles can also concentrate in the centers of the Earth and the Sun. This is possible if the interactions of dark matter particles with usual matter is strong enough, so that these particles occasionally lose energy when passing through the Earth or Sun, and eventually get caught by them. As a result of the high concentration, the dark matter annihilation may be strongly enhanced. The annihilation often leads to production of high energy neutrinos, an example being $X\bar{X} \to \bar{\nu}\nu$. These neutrinos propagate away without colliding with matter, and can be detected by neutrino detectors. Search for high energy neutrinos from the centers of the Earth and the Sun is being performed at underground, underwater and under-ice detectors, "neutrino telescopes". Limits from these indirect searches are shown in Figure 9.4.

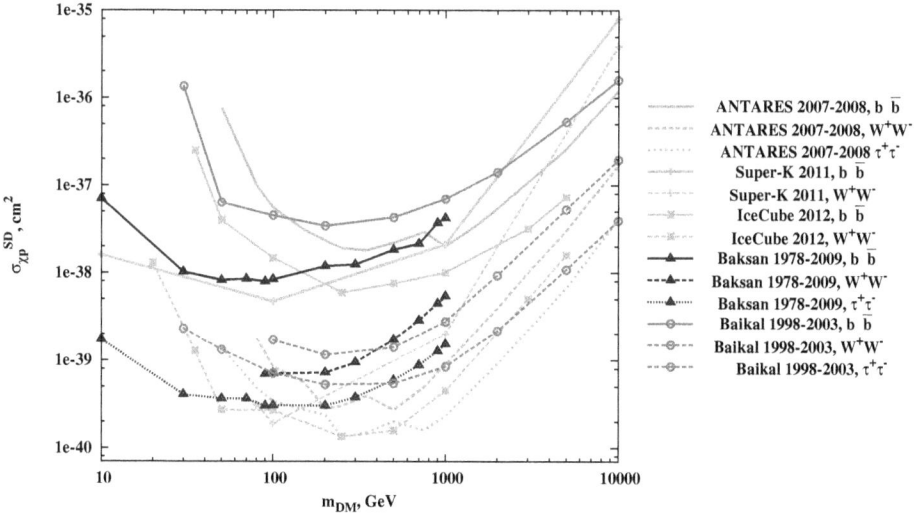

Fig. 9.4. Exclusion regions in the parameter plane (M_X, σ_{pX}) from the data of neutrino telescopes [163] on dark matter annihilation in the center of the Sun. The annihilation channels are shown in the right part. Regions above the curves are ruled out at 90% C.L. These limits are obtained under the assumption of dynamical equilibrium between the capture of dark matter particles by the Sun and their annihilation inside the Sun.

Problem 9.7. *Let us assume that there exist stable, electrically charged parti-cles $X^{\pm}$ much heavier than proton. Let us recall that baryons in the early Uni-verse are predominantly either protons or α-particles (^{4}He nuclei). Mass fraction of α-particles is 25%.*

(1) *Find the binding energy of an "atom" consisting of X^--particle and proton. The same for X-α atom.*
(2) *Assuming that the number density of X^--particles is small compared to that of baryons, find out in which state they predominantly exist today: bound state with proton, bound state with α-particle or free state.*
(3) *Assuming that $X^{\pm}$-particles were in thermal equilibrium in the early Universe, and that the asymmetry between X^- and X^+ equals η_x, find the present mass density of relic X-particles as function of their mass M_X and the asymmetry η_x.*
(4) *Place the bounds on the asymmetry η_x making use of the fact that searches for anomalous heavy isotopes ("wild hydrogen" and "wild helium") set upper bounds on the present mass density of these isotopes at the level [123, 124] $\Omega_x < 10^{-10} \cdot (10 \, TeV/M_x)$ at $M_x < 10 \, TeV$ (see also [1, 124] for other mass intervals).*

9.4.   *Other Applications of the Results of Section 9.2

Before considering concrete particle physics models with dark matter candidates, let us give two examples of the use of the results of Sec. 9.2.

9.4.1.   *Residual baryon density in baryon-symmetric Universe*

Let us find baryon (proton and neutron) number density after baryon–antibaryon annihilation in a toy Universe having no baryon asymmetry. In such a Universe, densities of baryons and antibaryons are equal to each other. We recall the baryon–antibaryon annihilation cross-section at low energies[9]

$$\sigma_0 \approx (100 \, \mathrm{MeV})^{-2}. \tag{9.46}$$

Using this value in (9.18), we find for freeze-out temperature

$$T_f \approx 20 \, \mathrm{MeV}, \tag{9.47}$$

where we calculated M_{Pl}^* with $g_* = 43/4$ (photons, electrons, positrons and three neutrino species). Formula (9.27) gives the mass density of baryons in baryon-symmetric Universe,

$$\Omega_B \approx 5 \cdot 10^{-11}.$$

---

[9] Note that the cross-section (9.46) exceeds by an order of magnitude the bound (9.40). The reason is that baryons are bound states of quarks and gluons.

This is about nine orders of magnitude smaller than the baryon density in real Universe. Substantial amount of baryons in our Universe exists entirely due to the baryon asymmetry.

9.4.2. *Heavy neutrino*

As another example, let us obtain a cosmological bound on the mass of hypothetical heavy neutrino of the fourth generation. In the absence of mixing with other neutrinos, such a neutrino does not experience charged current interactions with usual leptons, but participates in neutral current interactions. Let us assume that the heavy neutrino mass is not very large, $m_\nu \ll M_{Z,W}$. The analysis in Sec. 7 applies to light neutrino only, for which $T_f \gg m_\nu$. We consider here the opposite case in which neutrino annihilates being non-relativistic. In this case the annihilation cross-section is estimated as

$$\sigma_0 \sim G_F^2 m_\nu^2; \tag{9.48}$$

it does not depend on temperature. Using this cross-section in (9.27) we obtain the estimate for the present mass density of heavy neutrinos,

$$\Omega_\nu \approx 0.55 \left(\frac{3\,\text{GeV}}{m_\nu} \right)^2, \tag{9.49}$$

where we set $m_\nu = 3\,\text{GeV}$ in the argument of logarithm and used the following value for the effective number of degrees of freedom,

$$g_* = 2 \cdot (1+8) + \frac{7}{8} \cdot (2 \cdot 4 + 2 \cdot 3 + 3 \cdot 4 \cdot 3) = 61\frac{3}{4},$$

(photons, $e^\pm, \mu^\pm$, three neutrino species, gluons and quarks[10] u, d and s).
     We see that the existence of heavy stable neutrino of mass $m_\nu \lesssim 3\,\text{GeV}$ would be inconsistent with cosmology [125]. For a long time this had been the strongest bound on the mass of heavy neutrino. Present bound comes from the collider data on the Z-boson decay into undetected particles and reads $m_\nu > M_Z/2$.

Problem 9.8. *Check that neutrino of mass $m_\nu = 3\,GeV$ indeed freezes out being non-relativistic. Find the values of the neutrino mass at which this property no longer holds. Show, nevertheless, that cosmologically excluded is the entire mass range of stable neutrino*

$$20\,eV \lesssim m_\nu \lesssim 3\,GeV,$$

where the lower value has been obtained in Chapter 7, formula (7.10).

---

[10]For $m_\nu > 20\,\text{GeV}$, freeze-out temperature exceeds $1\,\text{GeV}$, and g_* receives contributions from τ-lepton and heavy quarks. This modifies the estimate (9.49) only slightly.

9.5. Dark Matter Candidates in Particle Physics

We discuss in the following Sections some popular models of particle physics containing dark matter particle candidates.[11] There are no such candidates in the Standard Model, so the cosmological data on dark matter require that the Standard Model must be extended. This fact is very important for particle physics.

Among the candidates we discuss, the natural one is neutralino. It is automatically stable in many viable supersymmetric extensions of the Standard Model and its present abundance is automatically in the right ballpark. Other candidates are less natural in the sence that their present mass density may *a priori* vary within large margin, and the correct value is obtained by adjusting the parameters of a model.

From the viewpoint of direct experimental searches, all candidates can be classified by using two parameters only, the particle mass M_X and its elastic scattering cross section[12] off nuclei σ_{int}. These parameters are shown in Fig. 9.5 for most popular candidates [164]. It is clear that the candidates have very diverse properties. Search for candidates from different regions in this plot is performed, as a

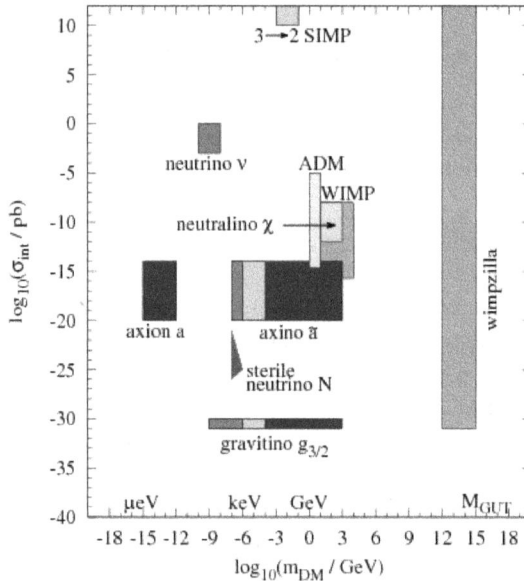

Fig. 9.5. Parameter ranges for various dark metter candidates [164]: their masses M_X and elastic scattering cross-sections off nucleon σ_{int}; $1\,\mathrm{pb} = 10^{-36}\,\mathrm{cm}^2$ (conversion cross-sections for axion and axino). Red, rose, blue and dark blue show hot, warm, cold and ultracold dark matter, respectively. See Fig. 13.18 for color version.

---

[11] One of the candidates, massive sterile neutrino, is considered in Sec. 7.3.

[12] In the case of axion and axino, the relevant parameter is the cross-section of conversion on nucleon into other particles, since this cross-section exceeds considerably the elastic cross-section.

rule, by making use of different detection techniques. In some cases, in particular, for very heavy and/or very weakly interacting particles (wimpzilla and gravitino, respectively), no realistic detection methods are known.

The zoo of proposed dark matter candidates is huge. In what follows we do not intend to discuss even a sizeable fraction of inhabitants of this zoo. We leave aside numerous candidates which we think are rather exotic. These include primordial black holes, strongly interacting dark matter, axino (superpartner of axion), mirror matter and many others.

9.6. *Stable Particles in Supersymmetric Models

Supersymmetry (SUSY) is the symmetry between bosons and fermions. The simplest (so called $N = 1$) SUSY theories in $(3 + 1)$ dimensions contain both particles and their *superpartners*, particles of opposite statistics whose spin differs by $1/2$ from that of particles. Superpartners participate in the same interactions as particles. In other words, particles and their superpartners have the same quantum numbers with respect to gauge group of the theory, while coupling constants of other interactions (e.g., Yukawa) are related in a well defined way. Each vector particle (e.g., gluon) has spin-$1/2$ superpartner (gluino), while each fermion (e.g., quark) has scalar superpartner (squark). More precisely, the number of spin degrees of freedom must be the same, so there are two scalar particles (two states) per each quark (two spin states).

Let us illustrate these properties using the supersymmetric generalization of Quantum Electrodynamics (QED). In fact, this was the first model illustrating the theoretically discovered supersymmetry [127]. QED itself is a theory of massive Dirac fermion ψ (electron) interacting with Abelian gauge field A_μ (photon). The QED Lagrangian is

$$\mathcal{L}_{QED} = -\frac{1}{4}F_{\mu\nu}F^{\mu\nu} + i\bar{\psi}\gamma^\mu(\partial_\mu + ieA_\mu)\psi - m\bar{\psi}\psi.$$

SUSY QED also contains four scalar degrees of freedom (corresponding to four spin states of electron and positron), which are described by two complex scalar fields ϕ_+ and ϕ_-. These carry charges $+1$ and -1 of the gauge group $U(1)$. There is an obvious contribution to the Lagrangian,

$$\mathcal{L}_1 = \mathcal{D}_\mu\phi_+^*\mathcal{D}^\mu\phi_+ - m_+^2\phi_+^*\phi_+ + \mathcal{D}_\mu\phi_-^*\mathcal{D}^\mu\phi_- - m_-^2\phi_-^*\phi_-,$$

$$\mathcal{D}_\mu\phi_\pm = (\partial_\mu \mp ieA_\mu)\phi_\pm,$$

where m_+ and m_- are masses of scalars. In the case of unbroken supersymmetry, these masses are equal to the electron mass, $m_+ = m_- = m$. (This is not the case if supersymmetry is spontaneously broken, see below.) There is also photino, a superpartner of photon. This is electrically neutral Majorana fermion (two spin degrees of freedom), which can be described by left spinor λ_L. Its free Lagrangian is

$$\mathcal{L}_2 = i\bar{\lambda}_L \gamma^\mu \partial_\mu \lambda_L.$$

Besides the gauge interactions between the gauge field A_μ and charged fields, there are "supersymmetry completions", the gauge-invariant Yukawa interaction between photino, electron and scalars,

$$\mathcal{L}_3 = i\sqrt{2}e\bar{\lambda}_L \psi \cdot \phi_- - i\sqrt{2}e\bar{\psi}\lambda_L \cdot \phi_+ + h.c.,$$

and self-interaction in the scalar sector,

$$\mathcal{L}_4 = -\frac{e^2}{2}(\phi_+^* \phi_+ - \phi_-^* \phi_-)^2.$$

The coupling constants of these additional interactions are related to the gauge coupling in a unique way. Unambiguous relations between various couplings is a general property of SUSY Lagrangians.

$N = 1$ supersymmetric extensions of other $(3+1)$-dimensional field theories have similar properties. In particular, supersymmetry completion of the usual Yukawa interaction is interaction between scalars, and vice versa.

If supersymmetry is unbroken, masses of particles are equal to masses of their superpartners. This makes unbroken supersymmetry phenomenologically unacceptable: no superpartners of the Standard Model particles have been discovered so far. Hence, SUSY extensions of the Standard Model assume that supersymmetry is spontaneously broken. In that case masses of superpartners may be arbitrary. The negative results of numerous searches place lower bounds on these masses [1]. These bounds are roughly at the level $M_S \gtrsim 100$ GeV $\div$ 1.5 TeV, depending on superpartner. On the other hand, there are hints from the theory[13] that the superpartner mass range is 30 GeV–3 TeV. In this regard, search for superpartners is one of important directions in high energy physics, with high expectations raised by the proton–proton collider LHC at CERN of the nominal total center-of-mass energy 14 TeV.

SUSY extensions of the Standard Model give rise, generally speaking, to processes with baryon and/or lepton number violation. No processes of this sort have been observed so far, and there are severe bounds on their rates. (The strongest one comes from non-observation of proton decay.) To get rid of these unwanted processes, one usually imposes a symmetry called R-parity which forbids baryon and lepton number violation at low energies.

R-parity is a discrete symmetry, each particle can be either even or odd under it (positive and negative R-parity, respectively); R-parity of a state with several particles is a product of R-parities of all these particles. All Standard Model particles (and also additional Higgs bosons required by SUSY) are R-even, while their superpartners are R-odd. Once all interactions conserve R-parity, superpartners are

---

[13]The most important hint is the partial cancellation of quadratic corrections to the Higgs boson mass, which provides the solution, in a technical sense, of the gauge hierarchy problem (Why $M_W \ll M_{Pl}$?). Another hint is the gauge coupling unification.

produced and disappear *in pairs*; in general, a pair consists of different superpartners. R-parity conservation thus implies that the lightest superpartner — LSP — is stable. This is the new particle that serves as the dark matter candidate. Since electrically charged particles of mass 30 GeV–10 TeV cannot be dark matter (see Problem 9.7), the potential candidates in SUSY theories are neutralino, sneutrino and gravitino.[14] Let us discuss them in turn.

9.6.1. Neutralino

A popular dark matter candidate is the neutralino. In fact, most direct searches for dark matter are oriented towards WIMPs, while the neutralino belongs precisely to this category. There are three reasons for the popularity. The first is that the neutralino is predicted by SUSY extensions of the Standard Model and the lightest neutralino is a stable LSP in a wide range of parameters of the theory. Second, like for any WIMP, the neutralino mass density in the Universe is more or less automatically in the right ballpark. Third, the neutralino participates in weak interactions, so its elastic scattering cross-section, albeit small, may be sufficient for its direct or indirect detection in dark matter searches (see Fig. 9.5).

Neutralino is a term used for electrically neutral Majorana fermion which is a linear combination of superpartners of Z-boson, photon and neutral Higgs bosons. There are four such combinations in the minimal SUSY extensions of the Standard Model. The point is that these theories necessarily contain two Higgs doublets, so there are two neutral higgsinos, superpartners of the Higgs bosons. The other two neutral fermions, photino and zino, are collectively called neutral gaugino.

Like their Standard Model partners, neutralinos participate in gauge interactions. Provided that the lightest neutralino is LSP, its present abundance is crudely estimated along the lines of Sec. 9.2. We set $g_X = 2$, $g_*(t_f) \simeq 100$, and use the estimate (9.42) for the annihilation cross-section. Then the formula (9.27) gives

$$T_f \simeq \frac{M_N}{20}, \tag{9.50}$$

$$\Omega_N = 3 \cdot 10^{-4} \frac{10^{-3}}{\alpha_W^2} \left(\frac{M_N}{100\,\text{GeV}} \right)^2 \log \left(10^{12} \cdot \frac{100\,\text{GeV}}{M_N} \right) \tag{9.51}$$

$$\approx 0.8 \cdot 10^{-2} \cdot \left(\frac{M_N}{100\,\text{GeV}} \right)^2, \tag{9.52}$$

where $\alpha_W \approx 1/30$ is the gauge coupling of $SU(2)_W$ interaction. For the neutralino in the interesting mass range

$$100\,\text{GeV} < M_N < 3\,\text{TeV},$$

---

[14]We restrict ourselves to minimal extensions of the Standard Model and leave aside other dark matter candidates like axino, singlino, etc.

our preliminary estimate is

$$0.01 \lesssim \Omega_N \lesssim 10,$$

which is indeed in the right ballpark.

We are going to refine the estimate (9.52) in this Section. We will make a simplifying assumption that the neutralino is the only new particle present in the cosmic plasma at relevant temperatures. This is the case if the masses of other superpartners are sufficiently larger than the neutralino mass. In this case the only processes of interest to us are neutralino pair creation and annihilation. We note, however, that our assumption is not innocent. If other superpartners have masses close to that of neutralino, they also exist in the plasma at the epoch of neutralino freeze-out, and freeze-out occurs in a more complicated way: neutralino are created and annihilate in pairs with other superpartners, with possible resonant enhancement of some of these processes; also, neutralinos are created in decays of superpartners, etc. We will see that in some versions of the theory the latter phenomena are important and, in fact, they help to obtain the right neutralino abundance.

Still, let us proceed under our simplifying assumption. Since the neutralino is an intrinsically neutral particle, i.e., it is its own antiparticle, the factor 2 in the first of Eq. (9.26) is absent. Therefore, the contribution of relic neutralinos into the present energy density is

$$\Omega_N h^2 \approx 0.9 \cdot 10^{-10} \frac{1}{x_f \sqrt{g_*}} \frac{\text{GeV}^{-2}}{\sigma_0 + 3\sigma_1 x_f}. \tag{9.53}$$

We recall that the parameter $x_f = T_f/M_N$ is the solution to Eq. (9.17), i.e.,

$$x_f^{-1} = \ln\left[\frac{3 \cdot \sqrt{10}}{8\pi^3} \frac{g_N}{\sqrt{g_* x_f}} \cdot M_N \cdot M_{Pl} \cdot (\sigma_0 + 6\sigma_1 x_f)\right]. \tag{9.54}$$

To obtain quantitative estimates, we have to find the parameters σ_0 and σ_1 in terms of the parameters entering the action of the SUSY model. This is done in different ways for different neutralino masses.

Our first example is the light neutralino, $M_N \lesssim M_Z$, and we begin with very low masses, $M_N \ll M_Z$. The region in parameter space where this option is viable is very restricted. Indeed, the kinematically allowed decay $Z \to NN$ should not give too large a contribution into the well measured invisible Z-decay width. Hence, the lightest neutralino in this case must be predominantly bino (superpartner of the gauge boson of the weak hypercharge group $U(1)_Y$), with small admixture ξ of higgsino and the other neutral gaugino.

One possibility is that the dominant annihilation channel for the very light neutralino is the s-channel annihilation into two b-quarks through exchange of virtual scalar or pseudoscalar particle A; see Fig. 9.6. In the minimal SUSY extensions of the Standard Model, A is parity-odd Higgs boson A. The annihilation cross-section

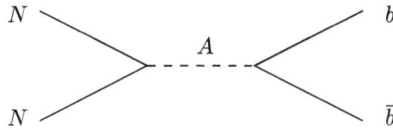

Fig. 9.6.   Light neutralino annihilation.

is estimated as follows,

$$\langle \sigma v \rangle \sim \xi^2 y_b^2 \frac{\alpha M_N^2}{m_A^4},$$

where y_b is the Yukawa coupling of b-quarks with the boson A. Sufficiently large cross-section is obtained for large $y_b \sim 1$. In that case

$$\langle \sigma v \rangle = \sigma_0 \sim 10^{-9} \cdot \left(\frac{\xi^2}{0.1} \right) \cdot \left(\frac{M_N}{10\,\mathrm{GeV}} \right)^2 \cdot \left(\frac{100\,\mathrm{GeV}}{m_A} \right)^4 \mathrm{GeV}^{-2}.$$

We estimate the freeze-out temperature from (9.54) and obtain $x_f \sim 1/20$. The neutralino mass density today is then found from (9.53). Numerically

$$\Omega_N \sim 0.2 \cdot \left(\frac{0.1}{\xi^2} \right) \cdot \left(\frac{10\,\mathrm{GeV}}{M_N} \right)^2 \cdot \left(\frac{m_A}{100\,\mathrm{GeV}} \right)^4. \qquad (9.55)$$

We see that a very light neutralino of mass $M_N \simeq 10\,\mathrm{GeV}$ can indeed be a dark matter particle, provided that $m_A \sim 100$ GeV and $\xi^2 \sim 0.1$.

This mechansim is problematic in the minimal SUSY extensions of the Standard Model. The lower limits on the mass of the axial Higgs boson A are in the range of several hundred GeV, depending on other parameters. So, the mechanism we discuss is typically insufficient for annihilating enough neutralinos in the early Universe. This mechanism does work, however, in models with extended Higgs sector (e.g., with extra scalar field which is a singlet under the Standard Model gauge group): these models contain more (pseudo)-scalar particles, some of which may be light enough.

Let us continue with the case $M_N < M_Z/2$. Another neutralino annihilation mechanism exists in models with light superpartners of the Standard Model fermions, $M_S \lesssim M_Z$: this is the annihilation through t-chanel exchange of these superpartners. With $M_S \sim M_Z$, the estimate of neutralino abundance is similar to (9.55) where one replaces $(\xi y_b)^2 \to \alpha$ and $m_A \to M_S \sim M_Z$. Hence, neutralino of mass $M_N \simeq 30$ GeV is indeed a viable dark matter candidate. The relevant superpartner S in realistic models is stau (superpartner of tau-lepton), and the dark matter neutralino is a mixture of bino and higgsino.

Data from the Large Hadron Collider (LHC) suggest, however, that realistic SUSY models have mass hierarchy in the Englert–Brout–Higgs sector: one Hoggs boson has mass 125 GeV and properties close to those of the Standard Model Higgs boson, while others, including A, have much higher masses. Also, masses of

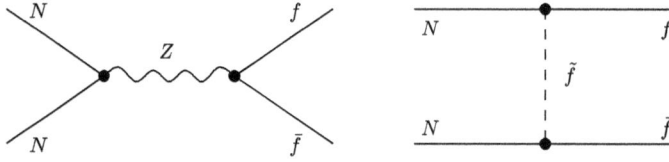

Fig. 9.7. Neutralino annihilation into the Standard Model fermions.

superpartners of the Standard Model fermions are much greater than M_Z. In that case very light neutralinos would be overproduced, as we see from Eq. (9.55). So, let us consider the case of heavy Higgs bosons and neutralino of intermediate mass $M_Z/2 < M_N \lesssim M_Z$. The annihilation cross-section in this case is often suppressed as compared to the estimate leading to (9.51). Indeed, the neutralino in our case predominantly annihilates into the Standard Model fermions,

$$NN \to \bar{f}f. \tag{9.56}$$

These processes occur via the s-channel Z-boson exchange, and also possibly via t-channel exchange of superpartners of the fermions; see Fig. 9.7. Since the neutralino is non-relativistic and its mass is smaller than that of the Z-boson, momenta of virtual particles in these diagrams are small compared to their masses, so they can be neglected. This means that the annihilation process (9.56) is described by effective four-fermion interaction

$$\mathcal{L} = \sum_f \bar{N}\gamma^\mu \gamma_5 N \cdot \bar{f}\gamma_\mu(a_f + b_f \gamma^5)f, \tag{9.57}$$

where the parameters a_f and b_f have dimension GeV^{-2}. These parameters are model-dependent. (In the case of Z-boson exchange they depend on the mixing matrix in the neutralino sector.) An order of magnitude estimate for the Z-boson exchange contribution is

$$a_f, \, b_f \sim G_F = 1 \cdot 10^{-5} \, \text{GeV}^{-2}. \tag{9.58}$$

Superpartners of fermions are typically heavier than Z-boson, so the t-channel exchange diagrams give smaller contributions.

The interaction (9.57) gives for the non-relativistic annihilation cross-section (recall that neutralino is Majorana fermion) to the leading orders in v^2 and m_f^2

$$\sigma v \approx \frac{1}{2\pi} \sum_f \left\{ b_f^2 \cdot m_f^2 + \frac{1}{3} \left[(a_f^2 + b_f^2) M_N^2 \right] \cdot v^2 \right\}, \tag{9.59}$$

where v is the relative velocity of the two incident neutralinos. This means that the parameters σ_0 and σ_1 defined in (9.12) are equal to

$$\sigma_0 = \frac{1}{2\pi} \sum_f b_f^2 \cdot m_f^2, \quad \sigma_1 = \frac{1}{6\pi} \sum_f (a_f^2 + b_f^2) M_N^2.$$

The s-wave annihilation is suppressed by the fermion mass: we see that $\sigma_0 \to 0$ as $m_f \to 0$. This is due to the following property. In the limit of massless fermions, only left-handed Standard Model fermions and right-handed antifermions interact with Z-boson. The projection of the angular momentum of such a pair onto their flight direction z is either $J_z = 1$ or $J_z = -1$. Non-relativistic spin-1/2 Majorana neutralinos should be in the state of angular momentum 1 (p-wave) to produce a pair with $J_z = \pm 1$: the state of angular momentum 0 (s-wave) and total spin $J_z = \pm 1$ (spins directed in the same way) is forbidden by the Pauli principle.

Making use of the above estimates for the parameters a_0 and a_1 we find from (9.53) that

$$\Omega_N \sim 0.2 \quad \text{for} \quad M_N \simeq 60\,\text{GeV}.$$

So, relatively light neutralinos are reasonable dark matter candidates.

The problem with these scenarios, however, is that in concrete models, the neutralino mass is related to masses of other superpartners. The latter are strongly constrained by experiment, so in many models neutralino cannot be light.

Thus, let us turn to the heavier neutralino. In that case, momenta in internal lines of the annihilations diagrams (e.g., diagrams shown in Fig. 9.7) are not small, $Q^2 = 4M_N^2$, so the cross-sections are generically suppressed by a factor of order $M_Z^4 / 4 M_N^4$ as compared to the case of light neutralino. With an account of p-wave suppression of annihilation into fermions this means that the annihilation cross-section is generically too small, i.e., heavy neutralinos are typically overproduced in the Universe.

Heavy neutralinos, $M_N \gtrsim M_Z$, still remain dark matter candidates in a certain (typically narrow) ranges of parameters of concrete models. The annihilation may be enhanced, e.g., by the presence of other superpartners in the cosmic plasma at neutralino freeze-out. This occurs when these superpartners are approximately degenerate in mass with neutralino. Whether this is a natural solution to the neutralino overproduction problem, is another issue.

Problem 9.9. *Consider minimal SUSY extension of the Standard Model in which all superpartners, except for sleptons and photino, are very heavy. Let the masses of sleptons and photino be of order 100 GeV, and photino is LSP. Neglecting slepton mixing, find the relic photino abundance. Assuming that one, two or three sleptons are light and have one and the same mass, find the range of slepton and photino masses in which photino is dark matter. Hint: The Lagrangian of photino interaction with leptons and sleptons is given in the beginning of Sec. 9.6.*

Even the Minimal Supersymmetric Standard Model (MSSM), which is a minimal SUSY extension, contains many free parameters, in addition to those of the Standard Model. Fairly large parameter regions are already ruled out by experiments at the LHC and elsewhere. Yet there are parameter regions where neutralino serves as dark matter particle, WIMP. In most of these regions the elastic cross

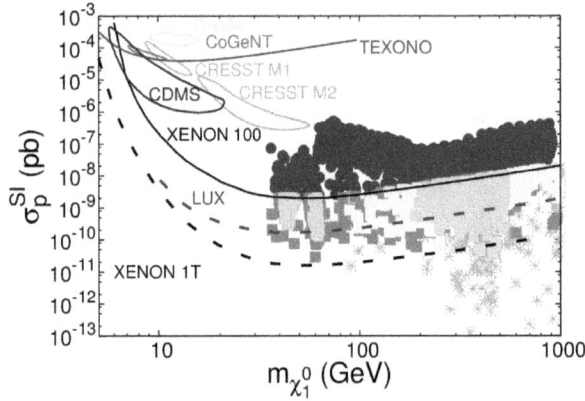

Fig. 9.8. Predictions of MSSM for spin-independent cross-section of elastic scattering of neutralino off nucleons [165]. Color regions correspond to parameters consistent with accelerator data and yielding $\Omega_N \approx 0.25$. The blue region is ruled out by direct search experiment XENON 100. The neutralino annihilation cross-section is enhanced in green regions by resonant annihilation through Z-boson or one of the Higgs bosons of MSSM, when $M_N \approx M_Z/2$ or $M_N \approx m_H/2$; see discussion below. The suppression of neutralino abundance in yellow and magenta regions is due to approximate degeneracy of neutralino and some other superpartner leading to efficient co-annihillation of the neutralino with this superpartner. Other mechanisms that suppress neutralino abundance operate in gray regions; they require fine tuning of parameters too. Shown here are also the regions of possible signals in DAMA, CoGENT, CRESST and CDMS experiments and planned sensitivity of LUX and XENON 1T. See Fig. 13.19 for color version.

sections of neutralino scattering off nuclei is high enough, so that cosmic neutralino can be detected in direct experimental searches, which we talked about in Sec. 9.3, see Fig. 9.2. This is further illustrated in Fig. 9.8.

Let us describe a consistent and phenomenologically viable example with rather few new parameters, called mSUGRA (minimal Supergravity). The model assumes that supersymmetry is broken in some "hidden" sector, and this breaking is transferred to the Standard Model fields and their superpartners by *gravitational interactions* (for a review see, e.g., Ref. [129]). Then at certain energy scale M *all* scalars of the visible sector, irrespective of their quantum numbers, obtain one and the same SUSY breaking mass m_0, while all gaugino (superpartners of gauge bosons) obtain one and the same SUSY breaking mass $M_{1/2}$. Also, trilinear interactions between the Higgs doublets and other scalars are generated; these are assumed to be proportional to the Yukawa couplings with one and the same proportionality coefficient A. These mass terms and trilinear couplings break supersymmetry. At the scale M this breaking is therefore *universal*. From the viewpoint of phenomenology, the advantage of this universality is that dangerous flavor changing neutral current interactions are suppressed.

As we mentioned above, there are two Higgs doublets; one of them, H_U, interacts with up quarks while the other, H_D interacts with down quarks and charged leptons. They both obtain vacuum expectation values, so that there is one more parameter,

$$\tan\beta = \frac{\langle H_U\rangle}{\langle H_D\rangle}.$$

Finally, there is yet another, discrete parameter, which is called sign of μ for technical reasons.[15]

In the simplest, and yet realistic models, the mass scale M is set equal to the Grand Unification scale $M_{GUT} \sim 10^{16}$ GeV [131]. At this scale all three gauge couplings of the Standard Model are equal (modulo normalization factor for weak hypercharge). We discuss Grand Unified Theories in Sec. 11.2.2, and here we only notice that this scale is very high.

The class of models just described is often called constrained Minimal Supersymmetric Standard Model (CMSSM), while the term mSUGRA is reserved for even more constrained model (where, in particular, $\tan \beta$ is not a free parameter, and gravitino mass equals m_0).

Quantum corrections modify all mass parameters $\{m_{0i}, A^U_{ij}, A^D_{ij}, M_i\}$ and coupling constants $\{\alpha_i, Y^U_{ij}, Y^D_{ij}\}$ as the energy scale Q changes from M to M_Z. This is described by renormalization group. At the scale M_Z supersymmetry breaking is no longer universal, but all new (with respect to the Standard Model) parameters are expressed through M, $m_0, M_{1/2}, A, \tan \beta$ and sign μ.

The renormalization group evolution in mSUGRA model [130] is shown in Fig. 9.9. Due to large color $SU(3)$ gauge coupling, masses of squarks and gluinos run faster than other masses. The lightest among gauginos turns out to be bino. The situation is more complicated in the scalar sector, where the gauge and Yukawa interactions compete. As a result, the lightest slepton is a superpartner of τ.

A nice feature of mSUGRA model is that the electroweak symmetry breaking occurs automatically, due to quantum corrections. As Q decreases, one of the Higgs masses squared becomes negative, which means spontaneous symmetry breaking. This is shown

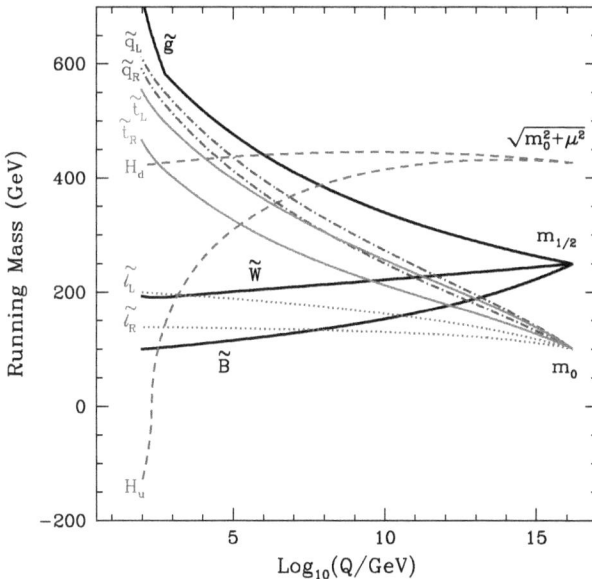

Fig. 9.9. An example of the renormalization group running of the mass parameters in mSUGRA model (plot by T. Falk from [130]). The universal parameters at the scale $M_{GUT} \approx 10^{16}$ GeV are: $M_{1/2} = 250$ GeV, $m_0 = 100$ GeV, $A = 0$, $\tan \beta = 3$ and sign $\mu = -1$.

---

[15] In fact, the parameter μ, the supersymmetric mass of the Higgs doublets, may be complex. This property serves as a potential source of CP-violation in the Higgs sector. This is quite interesting from the viewpoint of the baryon asymmetry generation; see Chapter 11.

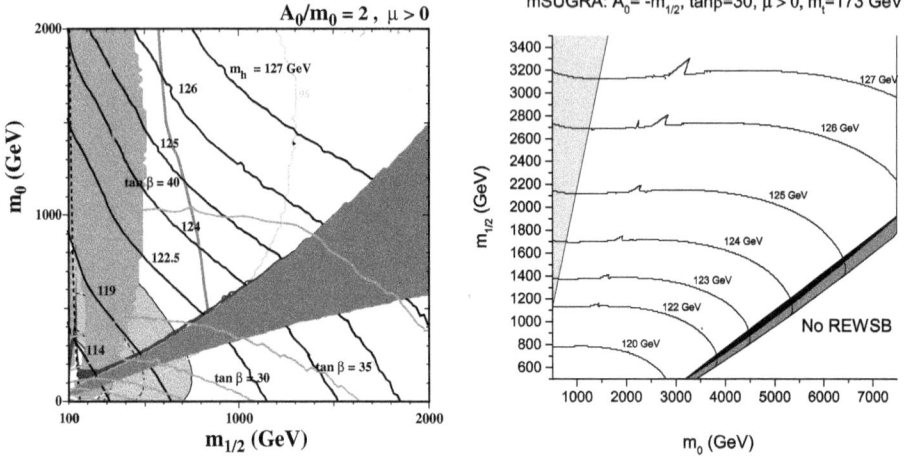

Fig. 9.10. Experimentally forbidden and cosmologically favored regions in the plane $(M_{1/2}, m_0)$ in mSUGRA model [166] with $A_0 = m_0$ and $\mu > 0$ (left) and CMSSM [167] with $A_0 = -m_{1/2}$, $\tan\beta = 30$ and $\mu > 0$ (right). Shown here are lines of constant mass of the lightest Higgs boson (realistic interval is 125–126 GeV. In the left panel, the elongated blue region is consistent with neutralino dark matter and has $m_{1/2} \approx 1300$ GeV, $m_0 \approx 850$ GeV; the region above the blue strip and brown region below are cosmologically ruled out (neuralinos are overproduced and LSP is a charged superpartner of tau-lepton, respectively); below the brown region LSP is gravitino (see disussion in Sec. 9.6.3), the region to the left of solid green line is excluded by the LHC data; gray lines show the values of $\tan\beta$. In the right panel, the cosmologically allowed region is shown in red, in the black region where $m_{1/2} \approx 1800$ GeV, $m_0 \approx 7000$ GeV neutralino abundance coincides with that of dark matter; in the green region LSP is the electrically charged superpartner of t-quark. See Fig. 13.20 for color version.

in Fig. 9.9, where the mass of the Higgs field is given by the combination $\sqrt{m_0^2 + \mu^2}$. Since the renormalization group running is logarithmic in Q, electroweak symmetry breaking occurs many orders of magnitude below M_{GUT} for small μ and SUSY breaking terms, which explains the hierarchy $M_W \ll M_{GUT}$.

Given the tight experimental constraints on the parameters of mSUGRA, there are only three narrow regions in the parameter space where relic neutralino has right abundance to serve as dark matter; see Figs. 9.10 and 9.11. The reason is that the region in parameter space where neutralino is light is experimentally excluded. So, the light neutralino scenarios we discussed above are impossible. On the other hand, heavy neutralinos tend to be overproduced in the early Universe. The cosmologically favored regions in Fig. 9.10 are special in the sense that the neutralino annihilation cross section is considerably enhanced in one or another way. This way of obtaining the correct dark matter abundance does not appear particularly plausible, although the high-energy values of the parameters are not particularly contrived.

One mechanism of enhancement of the neutralino annihilation cross-section exists for large $\tan\beta$ only. It occurs when the masses of scalar or pseudoscalar Higgs bosons (which are automatically almost equal in the relevant parameter region) are close to the mass of a pair of neutralinos $2M_N \approx m_H$ or $2m_N \approx m_A$. Then the decay channel

$$NN \to A^*, H^* \to \text{Standard Model particles},$$

has resonance enhancement. Of couse this occurs due to tuning of parameters.

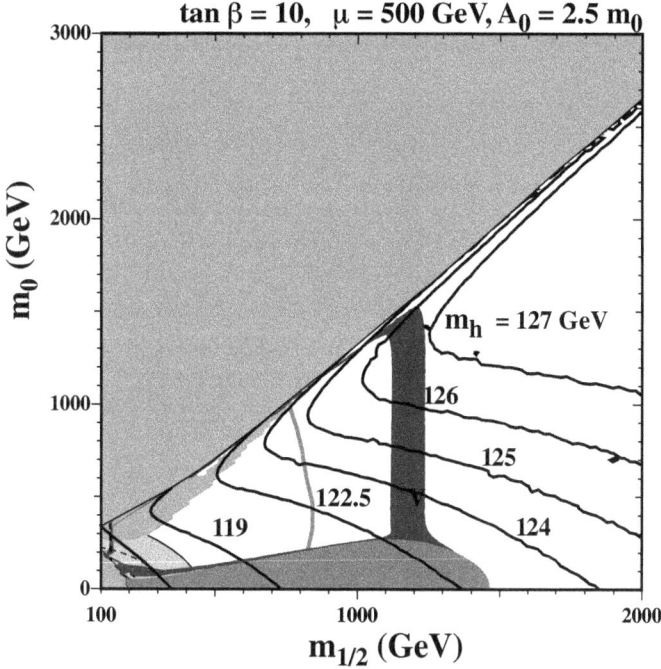

Fig. 9.11. Same as in Fig. 9.10, left, but for model [166] with an additional parameter in the Higgs sector as compared to CMSSM. The magenta region in the left upper corner is ruled out by the requirement of spontaneous breaking of the electroweak symmetry. The neutralino is dark matter particle in the blue region. See Fig. 13.21 for color version.

Two other mechanisms exist in models with co-LSP, where the next-to-lightest super-partner, NLSP, is almost degenerate in mass with neutralino LSP. In one case (strip in Fig. 9.10, left panel, along the cosmologically disfavored region with charged LSP) NLSP is the charged slepton (mostly right stau), while in the other case (narrow strip in Fig. 9.11 along the region without electroweak symmetry breaking) it is the lightest superpartner in the quark sector, the superpartner of t-quark.

Due to approximate degeneracy in masses, the freeze-out epoch is one and the same for reactions

$$LSP + LSP \rightarrow \text{Standard Model particles,}$$

$$NLSP + LSP \rightarrow \text{Standard Model particles,}$$

$$NLSP + NLSP \rightarrow \text{Standard Model particles.}$$

The larger number of channels, and also resonant enhancement of t-channel annihilation reduces the freeze-out abundance of neutralinos. The extra channels are indeed important in a narrow region of the parameter space where

$$\frac{m_{NLSP} - m_{LSP}}{m_{LSP}} \equiv \frac{\Delta m}{m_{LSP}} \lesssim 0.1. \tag{9.60}$$

The latter estimate follows from the fact that NLSP density is suppressed with respect to LSP by the Boltzmann factor

$$e^{-\frac{\Delta m}{T_f}} = e^{-20 \cdot \frac{\Delta m}{m_{LSP}} \cdot \left(\frac{m_{LSP}/20}{T_f}\right)},$$

This factor is less than 15%, unless the inequality (9.60) holds.

To conclude this Section we mention that in some models neutralino is long lived but not absolutely stable. The stable particles — dark matter candidates — may then be produced in neutralino decays. This helps to solve the overproduction problem, as the number of dark matter particles is the same as the freeze-out number of neutralinos, while the mass of the dark matter particles is smaller. We discuss this possibility in the context of gravitino dark matter, Sec. 9.6.3, since gravitino is the most interesting and realistic partner of neutralino in the pair LSP–NLSP.

9.6.2.   *Sneutrino*

The sneutrino is a superpartner of the neutrino. SUSY models contain three generations of sneutrinos described by complex, electrically neutral scalar fields. If the lightest sneutrino is LSP, one might wonder whether it can be a dark matter candidate. Indeed, weak interactions of sneutrinos are the same as those of neutrinos, so the order of magnitude estimate of the present sneutrino mass density would coincide with (9.27). (Unlike in the case of the neutralino, there is no p-wave suppression of the annihilation cross-section.) However, the sneutrino as a dark matter particle is excluded by direct searches for dark matter. Indeed, due to weak interactions, the cross-section of elastic scattering of sneutrino off nuclei is two to three orders of magnitude larger than experimental bounds shown in Fig. 9.2. We note in this regard, that the sneutrino is LSP in a much narrower class of models as compared to the neutralino.

A lot less problematic are right sneutrinos which exist in SUSY theories with right neutrino masses in the 100 GeV range. The right sneutrinos interact with the Standard Model particles very weakly, so their elastic cross-sections are well below the bounds shown in Fig. 9.2.

9.6.3.   *Gravitino*

Consistent supersymmetric theories that include gravity are theories with *local* supersymmetry — supergravities. Supersymmetry is local in the usual gauge theory sense: the action is invariant under supersymmetry transformations depending on space–time point. In these theories the graviton has its superpartner, the gravitino $\tilde{G}_\mu$. If supersymmetry were unbroken, the gravitino would be a massless spin-3/2 particle with 2 polarization states, the same number as for the graviton.

Supersymmetry, however, must be broken. In theories with spontaneously broken global supersymmetry (no supergravity), the Goldstone theorem implies the existence of massless particles whose properties are dictated by the properties of broken symmetry generators. In the case of supersymmetry, the generators are fermionic (Grassmannian) and carry spin 1/2 (they generate the transformation boson ↔ fermion and change spin by 1/2), so the massless particle is a Majorana fermion ψ, *goldstino*. Like any Nambu–Goldstone field, goldstino interacts with other particles through its coupling to the relevant current, which is supercurrent in the case at hand,

$$\mathcal{L} = \frac{1}{F}\partial^\mu \psi \cdot J^{SUSY}_\mu. \tag{9.61}$$

This is a direct analog of the Goldberger–Treiman formula which describes interactions of pions, the (pseudo-)Nambu–Goldstone bosons of broken chiral symmetry. The parameter F in (9.61) has dimension of mass squared and is determined by the vacuum expectation value that breaks supersymmetry. Hence, $\sqrt{F}$ is of the order of the supersymmetry breaking scale (in complete analogy to chiral model where the pion decay constant f_π is of order of the chiral symmetry breaking scale).

Once supersymmetry is *local*, the super-Higgs mechanism is at work. The goldstino becomes the longitudinal component of gravitino,

$$\tilde{G}_\mu \to \tilde{G}_\mu + i\sqrt{4\pi}\frac{M_{Pl}}{F}\partial_\mu \psi, \tag{9.62}$$

and gravitino acquires mass $m_{3/2}$ proportional to F. The relation is

$$m_{3/2} = \sqrt{\frac{8\pi}{3}\frac{F}{M_{Pl}}}. \tag{9.63}$$

Phenomenologically, the value of $\sqrt{F}$ may be anywhere in the interval

$$1\ \text{TeV} \lesssim \sqrt{F} \lesssim M_{Pl},$$

where the lower bound corresponds to the superpartner mass scale. Hence the wide range of possible gravitino masses,

$$2\cdot 10^{-4}\ \text{eV} \lesssim m_{3/2} \lesssim M_{Pl}.$$

The formulas (9.61) and (9.62) where ψ is the longitudinal component of gravitino, imply that gravitino interactions with matter have in principle a rather simple form. Indeed, the interactions of "original" gravitino components are suppressed by the Planck scale. On the other hand, the vast majority of SUSY theories have relatively low SUSY breaking scale, $\sqrt{F} \ll M_{Pl}$, so the massive gravitino interacts with matter mostly through its longitudinal component, the would-be goldstino. This interaction has the form (9.61). To estimate the interaction strength, let us integrate the action by parts and obtain

$$\mathcal{L} = -\frac{1}{F}\psi \cdot \partial^\mu J^{SUSY}_\mu.$$

At low energies, the divergence of the supercurrent is due to explicit breaking of supersymmetry in the effective action. This breaking manifests itself in the mass

differences between particles and superpartners. Hence, the effective gravitino interactions with other particles are[16]

$$\frac{m_S^2}{F} \bar{\tilde{f}} \psi \tilde{f},$$ (9.64a)

$$\frac{M_\lambda}{F} \bar{\lambda} \gamma^\mu \gamma^\nu \psi F_{\mu\nu},$$ (9.64b)

where $\tilde{f}$ and λ are the sfermion and gaugino, respectively, and m_S and M_λ are their
masses. The interesting and phenomenologically viable range of parameters is

$$100 \, \text{GeV} \lesssim m_S, M_\lambda \lesssim 10 \, \text{TeV}.$$

Hence, the effective couplings (9.64) are small, and often very small. This is the
reason for peculiar phenomenology and cosmology of the gravitino. Yet there is one
property of the gravitino in common with other superpartners. It is clear from (9.61)
and (9.62) that in theories with conserved R-parity, the gravitino is odd under
R-parity. Hence, it is created and annihilate in a pair with itself or with another
superpartner. Furthermore, the gravitino is LSP provided that its mass is roughly
in the range

$$2 \cdot 10^{-4} \, \text{eV} \lesssim m_{3/2} \lesssim 100 \, \text{GeV},$$

This range corresponds to fairly low SUSY breaking scale, see (9.63), namely

$$\sqrt{F} \lesssim 10^{10} \, \text{GeV}.$$

In theories of this sort gravitino is stable and may be a dark matter candidate. We
note, though, that direct detection of gravitino dark matter is very unlikely in the
foreseeable future because of very weak gravitino interactions with matter.

A class of realistic SUSY models with gravitino LSP makes use of the gauge mechanism of
mediation of SUSY breaking to the SUSY Standard Model (for reviews see, e.g., Refs. [133,
134]). SUSY is assumed to be broken in the secluded sector, while the superpartners of the
Standard Model particles obtain masses due to the usual gauge interactions of the Standard
Model. The interaction with the secluded sector is due to messengers, heavy fields that
are charged under the Standard Model gauge group and at the same time couple to the
secluded sector. As a result of the latter interactions, the masses of scalar messengers q
get split

$$M_q^2 = M^2 \left(1 \pm \frac{\Lambda}{M} \right), \quad \Lambda < M,$$

where M is the mass of fermionic messengers, and the parameter Λ^2 is proportional to
SUSY breaking vacuum expectation value F.

---

[16]One might question (9.64) as it does not appear to have smooth limit as $F \to 0$ in which
supersymmetry is restored. This is not the case: superpartner mass parameters m_S^2 are themselves
proportional to F. The ratio (9.64) is in fact determined by the couplings of those interactions in
the full theory which provide masses to superpartners after SUSY breaking.

Messengers induce one loop corrections to the gaugino masses,

$$M_\lambda(M) \sim \frac{\alpha}{4\pi}\Lambda, \tag{9.65}$$

where α is the relevant gauge coupling of the Standard Model. The fields from the scalar sector of SUSY Standard Model obtain contributions to their masses at two loops,

$$m_{\tilde{f}}^2(M) \sim \left(\frac{\alpha}{4\pi}\right)^2 \Lambda^2. \tag{9.66}$$

The proportionality coefficients in (9.65) and (9.66) are in general different for different fields.

Hence, the non-supersymmetric masses of scalar and fermion superpartners at the scale M depend on their quantum numbers only, and are of the same order,

$$M_\lambda \sim m_{\tilde{f}} \sim \frac{\alpha}{4\pi}\Lambda.$$

Flavor-independence of the gauge interactions ensures the absence of new flavor-changing parameters at the scale M. Gravitational interactions between the secluded sector and SUSY Standard Model fields lead, generally speaking, to the appearance of these parameters. In particular, the gravitational effects generate flavor-nondiagonal masses for squarks and sleptons, whose estimate is

$$\Delta m_{\tilde{f}_{ab}} \sim F/M_{Pl} \sim m_{3/2}.$$

These masses are phenomenologically acceptable (they do not lead to contradictions with data on rare decays of leptons and mesons), provided that

$$\frac{\Delta m_{\tilde{f}_{ab}}}{m_{\tilde{f}}} \lesssim 10^{-2} - 10^{-3}, \quad m_{\tilde{f}} \sim 500\,\text{GeV}.$$

This gives the upper bound on the gravitino mass in gauge mediation models,

$$m_{3/2} \lesssim (10^{-2} - 10^{-3}) \cdot m_{\tilde{f}} \sim 1\,\text{GeV},$$

as well as on the SUSY breaking scale,

$$\sqrt{F} \lesssim 10^9\,\text{GeV}.$$

On the other hand, the lower bounds on $m_{3/2}$ and $\sqrt{F}$ come from the experimental bounds on the superpartner masses; roughly speaking,

$$M_\lambda, m_{\tilde{f}} \gtrsim 300\,\text{GeV}.$$

Making use of this estimate, we find from (9.65) and (9.66) that

$$\sqrt{F} \gtrsim \Lambda \gtrsim 30\,\text{TeV},$$

$$m_{3/2} \gtrsim 1\,\text{eV}.$$

While the gravitino is LSP in the gauge mediation models, NLSP is either a neutralino or right τ-slepton. The latter case occurs more often.

We note that gravitino production mechanisms in this class of models include processes with messengers and with particles from the secluded sector. These interactions may change the estimates given in the text.

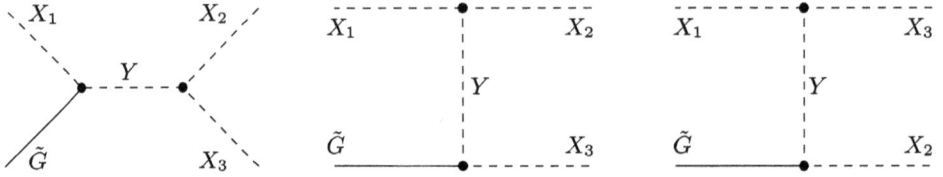

Fig. 9.12. Processes $2 \to 2$ with gravitino annihilation or production; Y, X_i, $i = 1, 2, 3$ are the Standard Model particles and their superpartners. In models with conserved R-parity at least one of the particles X_i must be a superpartner.

Let us discuss cosmology with stable gravitino. Because of very weak gravitino interactions, processes involving two gravitinos (including creation and annihilation of two gravitinos) are negligible. Let us begin with the possibility that gravitino was in thermal equilibrium at high temperatures.

Gravitino $\tilde{G}$ is created and destroyed in reactions

$$X_1 + \tilde{G} \leftrightarrow X_2 + X_3 \tag{9.67}$$

where X_i, $i = 1, 2, 3$ are other particles in the cosmic plasma. These reactions proceed via s-, t- and u-channel exchange of virtual particles Y; see Fig. 9.12. Each diagram involves the coupling between the particles X_1, X_2, Y (e.g., gauge coupling) and the gravitino coupling (9.64). At high tempeartures, when $T \gg m_S, M_\lambda$, processes due to gravitino–gaugino interactions (9.64b) dominate, and the cross-sections of the processes (9.67) are estimated as follows:

$$\sigma_{\tilde{G}} \simeq \alpha \frac{M_\lambda^2}{F^2}, \qquad \alpha = \frac{g^2}{4\pi}.$$

(The cross-sections involving gravitino-sfermion interactions (9.64a) are additionally suppressed by a factor m_S^2/T^2.) Gravitino freeze-out temperature is determined, as usual, from the relation

$$\sum_{X_1} \sigma_{\tilde{G}} \cdot n_{X_1} \cdot v_{\tilde{G}} \simeq H(T_f) = \frac{T_f^2}{M_{Pl}^*},$$

where n_X is the equilibrium number density of particles in the initial state of the reaction (9.67) and $v_{\tilde{G}}$ is the gravitino velocity. Let us assume that freeze-out occurs when gravitinos are still relativistic. Then durect and inverse processes (9.67) freeze out at the same time. We take $n_X \sim T^3$, $v_{\tilde{G}} = 1$ and estimate the number of channels as $g_*(T_f)$. This gives the following estimate:

$$T_f \sim m_{3/2} \cdot \frac{1}{\sqrt{g_*(T_f)}\alpha} \cdot \frac{F}{M_\lambda^2}, \tag{9.68}$$

where one of the factors F has been expressed through the gravitino mass according to (9.63).

Since $F \gg M_\lambda^2$, gravitino freeze-out indeed occurs when gravitino is relativistic.[17] Hence, the estimate of the present gravitino abundance is made in the same way as the estimate for neutrino, see Sec. 7.2. The expression for gravitino is

$$n_{3/2,0} = \frac{3}{4} \cdot \left(\frac{g_{3/2}}{2} \right) \cdot \frac{43}{11} \cdot \frac{1}{g_*(T_f)} \cdot n_{\gamma,0}, \qquad (9.69)$$

where we used the expression (5.33) for entropy at $T \ll 1\,\mathrm{MeV}$. Here $g_{3/2}$ is the number of active degrees of freedom of gravitino; since the interaction of the transverse degrees of freedom is suppressed by negative power of M_{Pl}, we should set $g_{3/2} = 2$.

Problem 9.10. *Derive the formula (9.69).*

Problem 9.11. *Find freeze-out temperature of transverse degrees of freedom of gravitino.*

The result (9.69) shows that the present gravitino mass density is

$$\Omega_{3/2} = \frac{m_{3/2} \cdot n_{3/2}}{\rho_c} = 0.2 \left(\frac{m_{3/2}}{200\,\mathrm{eV}} \right) \left(\frac{g_{3/2}}{2} \right) \cdot \left(\frac{210}{g_*(T_f)} \right) \cdot \frac{1}{2h^2}. \qquad (9.70)$$

This means that stable gravitinos of masses $m_{3/2} \gtrsim 200\,\mathrm{eV}$ are cosmologically forbidden, if these gravitinos were in thermal equilibrium at high temperatures. This translates into the bound on the SUSY breaking scale,

$$\sqrt{F} \lesssim 10^3\,\mathrm{TeV}.$$

Gravitino of mass $m_{3/2} \simeq 200\,\mathrm{eV}$ would make the whole of dark matter. However, this mass is below the lower bound (9.4) dictated by the analysis of structures. Thus, if gravitinos were in thermal equilibrium in the early Universe, they cannot serve as dark matter candidates.

Problem 9.12. *Show that stable gravitinos of mass $m_{3/2} < 200\,eV$ that were in thermal equilibrium at high temperatures are safe from the viewpoint of primordial nucleosynthesis.*

Gravitino as dark matter is still an option if the freeze-out temperature (9.68) was never reached in the Universe. One of the possible scenarios is as follows. Let us assume that the thermal history of the Universe began at some temperature $T_{\max}$. Let us also assume that gravitinos were not produced before the hot stage, so that there are no gravitinos at $T = T_{\max}$, while the Standard Model particles and their superpartners are at thermal equilibrium. The gravitino production then occurs in the conventional hot Big Bang regime.

---

[17] For $m_{3/2} \lesssim 10\,\mathrm{keV}$, the estimate (9.68) gives the freeze-out temperature below the scale of superpartner masses, $T_f \lesssim M_\lambda$. In this case the processes (9.67) terminate, roughly speaking, at $T_f \sim M_\lambda$, since at least one of the particles X_1, X_2, X_3 must be a superpartner. Freeze-out still occurs in the regime of relativistic gravitino, so that the argument that follows remains valid.

In fact, this situation is quite realistic in inflationary theory. According to this theory, energy density before the hot stage was due to the inflaton field, and the hot plasma was produced as a result of the decay of this field (see details in the accompanying book). Naturally, this picture implies that the maximum temperature is well below the Planck mass. This temperature is actually model dependent and varies a lot in concrete inflationary models. Gravitino interactions with inflaton may be very weak, so gravitinos were not produced prior to the hot stage in this picture.

At the hot stage, gravitinos are predominantly produced in two-body decays of superpartners,

$$\tilde{X}_i \to \tilde{G} + X_i, \quad i = 1, \ldots \tag{9.71}$$

and in two-particle scattering processes,

$$X_i + X_j \to X_k + \tilde{G}, \quad i, j, k = 1, \ldots \tag{9.72}$$

With these two production channels, the Boltzmann equation for the gravitino number density is

$$\frac{dn_{3/2}}{dt} + 3H n_{3/2} = \sum_i \Gamma_{\tilde{X}_i} \cdot \gamma_i^{-1} \cdot n_{\tilde{X}_i} + \sum_{ij} \langle \sigma v \rangle_{ij} \cdot n_{X_i} n_{X_j}, \tag{9.73}$$

where $\Gamma_{\tilde{X}_i}$ is the width of the decay (9.71), γ_i is the Lorentz-factor of $\tilde{X}_i$-particles that gives rise to the decrease of their decay rate, n_{X_i} is the number density of particles X_i and $\langle \sigma v \rangle_{ij}$ is the production rate in the reaction (9.72) (thermal average of the product of cross-section and relative velocity.) We neglected here the inverse processes $X_k + \tilde{G} \to X_i + X_j$ and $X_i + \tilde{G} \to \tilde{X}_i$; this is the valid approximation for gravitino abundance well below its equilibrium value. In terms of gravitino-entropy ratio $\Delta_{3/2} = n_{3/2}/s$ and temperature, the Boltzmann equation reads (we follow the logic as in Sec. 9.6.1)

$$\frac{d\Delta_{3/2}}{d\log T} = \sum_i \frac{\Gamma_{\tilde{X}_i} \gamma_i^{-1}}{H} \cdot \frac{n_{\tilde{X}_i}}{s} + \sum_{ij} \langle \sigma v \rangle_{ij} \cdot \frac{n_{X_i} n_{X_j}}{sH}. \tag{9.74}$$

We now make use of the formulas (9.64) for the gravitino coupling to estimate the width and cross-section,

$$\Gamma_{\tilde{X}_i} \simeq \frac{M_{\tilde{X}_i}^5}{16\pi F^2} = \frac{M_{\tilde{X}_i}^5}{6m_{3/2}^2 M_{Pl}^2}, \tag{9.75}$$

$$\langle \sigma v \rangle_{ij} = \text{const} \cdot \alpha \frac{M_\lambda^2}{F^2} = \text{const} \cdot \alpha \frac{8\pi}{3} \cdot \frac{M_\lambda^2}{m_{3/2}^2 M_{Pl}^2}. \tag{9.76}$$

As we already poined out, at $T \gg M_{\tilde{X}}$ the gravitino production in scattering is dominated by its interaction with the gauge particles, Eq. (9.64b). Note also that neither the width (9.75) nor the cross-section (9.76) depends on temperature. The constant in (9.76) is determined by the number of possible final states.

The two terms on the right-hand side of Eq. (9.74) behave in a very different way as functions of temperature. At high temperatures one has $n_{\tilde{X}_i} \sim n_\gamma \propto T^3$ and $\gamma_i^{-1} \propto T^{-1}$. Hence, the first and the second terms are proportional to T^{-3} and T, respectively. We see that the gravitino production in decays is not particularly sensitive to T_{max}, while the number of gravitinos produced in scattering increases linearly with T_{max}.

Let us first discuss gravitino production in decays of superpartners. There are two types of processes relevant here. One is the decays of all superpartners at high temperatures when they are relativistic or almost relativistic. The other is possible delayed decay of next-to-lightest superpartner, NLSP. Let us consider them in turn.

The contribution of the superpartner $\hat{X}$ into the first term in (9.74) behaves as T^{-3} at $T \gg M_{\tilde{X}}$ and $\exp(-M_{\tilde{X}}/T)$ at $T \ll M_{\tilde{X}}$ (assuming that $\tilde{X}$ is in thermal equilibrium; the opposite situation may be relevant for NLSP and will be discussed later.) Hence, the gravitino production occurs predominantly at $T \sim M_{\tilde{X}}$, and we obtain for this contribution

$$\Delta_{3/2} \sim \frac{\Gamma_{\tilde{X}}}{g_* H(T = M_{\tilde{X}})} \sim \frac{M_{\tilde{X}}^3}{g_*^{3/2} m_{3/2}^2 M_{Pl}}. \tag{9.77}$$

We see that the largest contribution comes from the heaviest superpartners (provided that $T_{max} \gtrsim M_{\tilde{X}}$) and that the abundance is larger for *lighter* gravitino. The latter property is due to the fact that lighter gravitinos correspond to lower SUSY breaking scale, see (9.63), and hence stronger gravitino interaction with matter. We find from (9.77) the contribution to the present gravitino mass density,

$$\Omega_{3/2} \sim \frac{s_0}{\rho_c} \frac{M_{\tilde{X}}^3}{g_*^{3/2} m_{3/2} M_{Pl}}.$$

Plugging in numbers, $M_{\tilde{X}} \lesssim 10\,\text{TeV}$, $g_* \simeq 200$, we find that the first decay mechanism is capable of producing gravitino dark matter for $m_{3/2} \lesssim 1\,\text{MeV}$. Interestingly, the right amount of gravitino *warm* dark matter, $m_{3/2} \sim 1\text{--}10\,\text{keV}$ (see Sec. 9.1), is obtained for fairly light superpartners, $M_{\tilde{X}} \lesssim 300\,\text{GeV}$; see Ref. [135] for details.

Let us turn to the second decay mechanism, delayed decays of NLSP. Neglecting for the moment its interactions with gravitino, we apply the analysis of Sec. 9.6.1. The NLSP freeze-out temperature is

$$T_{f,NLSP} \sim \frac{M_{NLSP}}{20}.$$

NLPS do not decay by that time provided that

$$\Gamma_{NLSP} < H(T_{f,NLSP}) = \frac{T_{f,NLSP}^2}{M_{Pl}^*}.$$

Applying the estimate (9.75) to NLSP we see that this condition holds for gravitino of mass $m_{3/2} \gtrsim 10\,\text{keV}$. Let us concentrate on this case. After NLSP decays, the number of gravitinos in comoving volume is the same as freeze-out number of NLSP. This immediately gives

$$\rho_{0,3/2} = \frac{m_{3/2}}{M_{NLSP}}\rho_{0,NLSP},$$

where $\rho_{0,NLSP}$ is a would-be mass density of stable NLSP. The latter has been estimated in Sec. 9.6.1, where we have seen that the mass density is often higher than the critical density,

$$\rho_{0,NLSP} \sim (10-1000)\rho_c.$$

So, the second decay mechanism of gravitino production leads to the right gravitino dark matter density for relatively heavy gravitino,

$$m_{3/2} \sim (0.1-10)\,\text{GeV} \cdot \left(\frac{100\,\text{GeV}}{M_{NLSP}}\right).$$

This scenario is somewhat problematic, however, since according to (9.75), the NLSP lifetime is large,

$$\tau_{NLSP} \equiv \Gamma_{NLSP}^{-1} \sim 5 \cdot 10^4\,\text{s} \cdot \left(\frac{m_{3/2}}{1\,\text{GeV}}\right)^2 \cdot \left(\frac{100\,\text{GeV}}{M_{NLSP}}\right)^5.$$

For $\tau_{NLSP} \gtrsim 100\,\text{s}$, NLSP decays occur at or after BBN epoch. The decay products in the end contain hard photons, electrons and other particles which may destroy light nuclei in primordial plasma, hence spoiling the predictions of BBN theory, see the end of Chapter 8. This problem does not occur if gravitino is sufficiently light ($m_{3/2} \lesssim 100\,\text{MeV}$) and/or NLSP is sufficiently heavy ($M_{NLSP} \gtrsim 300\,\text{GeV}$).

We note that in the latter scenario gravitinos are born relativistic. Hence, they effectively serve as *warm* dark matter. Their distribution in momenta is far from thermal; this, generally speaking, is relevant for structure formation.

Problem 9.13. *Find the distribution over momenta of gravitinos produced in NLSP decays after NLSP freeze-out.*

Problem 9.14. *Let gravitino of mass 100 MeV be dark matter particle, and the mass and lifetime of NLSP be 200 GeV and 10 s. Estimate the present spatial size of density perturbations suppressed as compared to the CDM case (see Sec. 9.1).*

As an example of BBN bounds, let us consider mSUGRA model of Sec. 9.6.1 and add a light gravitino to it. Let gravitino be LSP. Let us treat the gravitino mass as a free parameter (which, strictly speaking, is not the case in simple models of gravity mediation). In this model, NLSP is either neutralino or superpartner of right tau-lepton. The bounds on the parameters of this model [136] are shown in Fig. 9.13 and Fig. 13.9 on color pages.

Let us now turn to gravitino production in scattering, the second term in (9.74). It is important for $T_{\max} \gtrsim M_S$. At $T \sim T_{\max}$ matter particles are thermalized, and

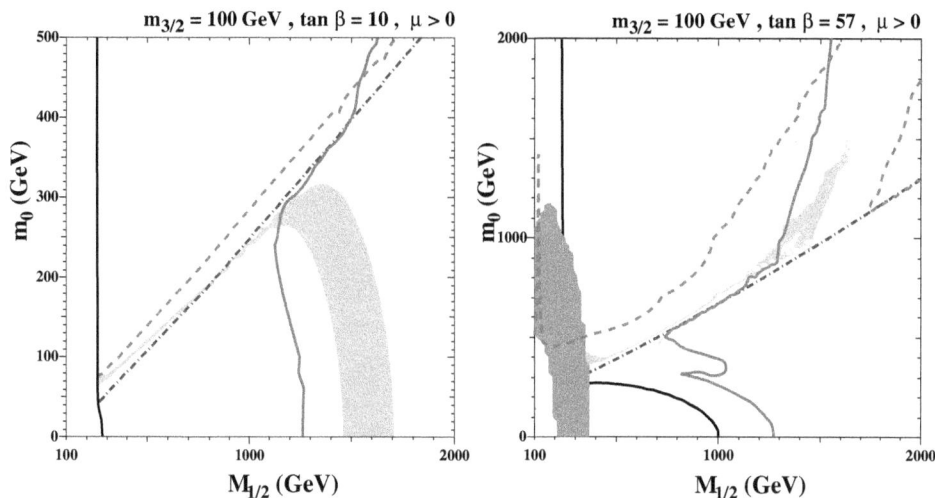

Fig. 9.13. Bounds [136] on the plane $(M_{1/2}, m_0)$ in mSUGRA model with $m_{3/2} = 100$ GeV and two values of $\tan \beta$. See Fig. 13.9 for color version.

their number densities are all of order of the photon number density. We parameterize the sum in the right-hand side of (9.74) as

$$\sum_{ij} \langle \sigma v \rangle_{ij} \cdot n_{X_i} n_{X_j} = \langle \sigma v \rangle_{tot} n_\gamma^2,$$

and make use of the estimate (9.76). The largest contribution to $\langle \sigma_{tot} \rangle$ comes from colored particles; numerically const $\sim$100 in (9.76). Using this estimate for $\langle \sigma_{tot} \rangle$, we obtain the scattering contribution to gravitino-to-entropy ratio as the second term in the right-hand side of Eq. (9.74) at $T = T_{\max}$. This gives

$$\Omega_{3/2} = \frac{m_{3/2} T_{\max}}{\rho_c} \cdot n_{\gamma,0} \cdot \frac{g_*(T_0)}{g_*(T_{\max})} \cdot \langle \sigma_{tot} \rangle \cdot \frac{n_\gamma(T_{\max})}{T_{\max} H(T_{\max})}.$$

Numerically,

$$\Omega_{3/2} \sim \left(\frac{200 \, \text{keV}}{m_{3/2}} \right) \cdot \left(\frac{T_{\max}}{10 \, \text{TeV}} \right) \cdot \left(\frac{M_S}{1 \, \text{TeV}} \right)^2 \cdot \left(\frac{15}{\sqrt{g_*(T_{\max})}} \right) \cdot \frac{1}{2h^2}. \qquad (9.78)$$

Since there is linear dependence on $T_{\max}$, the maximum temperature in the Universe must be bounded from above.

The estimate (9.78) can be refined by integrating the Boltzmann equation (9.74) numerically. The result [137] for a SUSY model with bino NLSP and the mass hierarchy $M_{NLSP} = 50 \, \text{GeV} \ll M_S = 1 \, \text{TeV}$ is shown in Fig. 9.14. This type of cosmological bounds is very restrictive, since in simple models of reheating the maximum temperature exceeds $T \sim 10^8 \, \text{GeV}$. We note that in this model, relic gravitinos make all of dark matter for parameters just below the line shown in Fig. 9.14. It is clear both from this Figure and from Eq. (9.78) that obtaining the

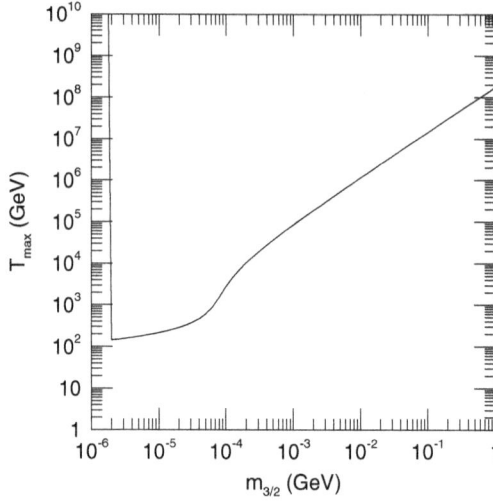

Fig. 9.14.   Bounds [137] on the parameters $(T_{max}, m_{3/2})$. The region above the line is excluded because of gravitino overproduction. The solid line corresponds to $\Omega_{3/2}h^2 \simeq 1$.

right dark matter density with gravitinos requires tuning parameters of different nature: the parameter $m_{3/2}$ has to do with particle physics while T_{max} characterizes the early cosmology.

To conclude this Section, let us consider one possibility that exists in SUSY models with unstable but not very heavy gravitino. This case is typical for simple mSUGRA models. The gravitino lifetime is large, so these models may lead to the cosmological scenario with intermediate domination of non-relativistic gravitino. This is the first of the two scenarios considered in Sec. 5.3. For heavy gravitino, the main contribution to the decay rate comes from "original" transverse components (not goldstino), and the estimate for the width is

$$\Gamma_{3/2} \simeq \frac{m_{3/2}^3}{M_{Pl}^2}.$$

Let the maximum temperature in the Universe be so high that gravitinos were in thermal equilibrium with the rest of plasma (this assumption may be relaxed, see Sec. 5.3). Since gravitino freeze-out occurs in relativistic regime, gravitino mass density at $T \lesssim m_{3/2}$ is given by

$$\rho_{3/2}(T) \sim m_{3/2}T^3.$$

So, the intermediate matter-dominated epoch is indeed possible for long-lived gravitino. The requirement from BBN, that (see (5.49))

$$\Gamma_{3/2} \gtrsim \frac{T_{BBN}^2}{M_{Pl}^*(T_{BBN})}$$

gives rather strong lower bound on the gravitino mass,

$$m_{3/2} > \left[\frac{T_{NS}^2 M_{Pl}^2}{M_{Pl}^*(T_{NS})} \right]^{1/3}, \quad \text{i.e.,} \quad m_{3/2} > 45 \,\text{TeV}.$$

Moreover, the self-consistency of the cosmological model requires either quite light LSP (if stable), $M_{LSP} \sim 10 \,\text{MeV}$, or even much heavier gravitino.

This bound is fairly restrictive in models with gravity mediated SUSY breaking. This bound, however, does not apply to cosmological scenarios in which gravitino mass density was never high.

9.7. Axions and Other Long-lived Particles

Many extensions of the Standard Model contain light scalar or pseudoscalar particles. In some models these new particles are so light and so weakly interacting that their lifetime exceeds the present age of the Universe. Hence, they may serve as dark matter candidates. Depending on the context, they are called axions (see below), dilatons, familons, sgoldstino, etc.

Let us consider general properties of models with light scalars or pseudoscalars. These particles should interact with the usual matter very weakly, so they must be neutral with respect to the Standard Model gauge interactions. This implies that interactions of scalars S and pseudoscalars P with gauge fields are of the form

$$\mathcal{L}_{SFF} = \frac{C_{SFF}}{4\Lambda} \cdot S F_{\mu\nu} F^{\mu\nu}, \quad \mathcal{L}_{PFF} = \frac{C_{PFF}}{8\Lambda} \cdot P F_{\mu\nu} F_{\lambda\rho} \epsilon^{\mu\nu\lambda\rho}, \quad (9.79)$$

where $F_{\mu\nu}$ is the field strength of $SU(3)_c$, or $SU(2)_W$, or $U(1)_Y$ gauge group. The parameter Λ has dimension of mass and can be interpreted as the scale of new physics related to S- and/or P-particle. This parameter has to be large, then the interactions of S and P with gauge bosons are indeed weak at low energies. Because of that, the Lagrangians (9.79) contain gauge-invariant operators of the lowest possible dimension; in principle, one could add to (9.79) terms like $\Lambda^{-5} S (F_{\mu\nu})^4$, but their effects would be negligible at energies well below Λ. Dimensionless constants C_{SFF}, C_{PFF} are determined by the fundamental theory valid at energies above Λ. These parameters run with the energy scale Q^2 characterizing the processes considered. However, the latter property will not be important for our estimates, and we will assume that these parameters are numbers of order 1. In the standard basis of the Standard Model gauge fields, the terms (9.79) describe interactions of (pseudo)scalars with pairs of photons, gluons, as well as with $Z\gamma$-, ZZ- and W^+W^--pairs.

Interactions with fermions can also be written on symmetry grounds. Since S and P are singlets under $SU(3)_c \times SU(2)_W \times U(1)_Y$, no combinations like $S\bar{f}f$ or $P\bar{f}\gamma^5 f$ are gauge invariant, so they cannot appear in the Lagrangian (hereafter f denotes the Standard Model fermions). Gauge invariant operators of the lowest dimension have the form $H\bar{f}f$, where H is the Higgs field. Hence, the interactions with fermions are

$$\mathcal{L}_{SH\!f\!f} = \frac{Y_{SH\!f\!f}}{\Lambda} \cdot SH\bar{f}f, \quad \mathcal{L}_{PH\!f\!f} = \frac{Y_{PH\!f\!f}}{\Lambda} \cdot PH\bar{f}\gamma^5 f.$$

It often happens that the couplings $Y_{SH\!f\!f}$ and $Y_{SP\!f\!f}$ are of the order of the Standard Model Yukawa couplings, so upon electroweak symmetry breaking the low energy Lagrangians have the following structure,

$$\mathcal{L}_{S\!f\!f} = \frac{C_{S\!f\!f}m_f}{\Lambda} \cdot S\bar{f}f, \quad \mathcal{L}_{P\!f\!f} = \frac{C_{P\!f\!f}m_f}{\Lambda} \cdot P\bar{f}\gamma^5 f, \qquad (9.80)$$

where we assume that dimensionless couplings $C_{S\!f\!f}$ and $C_{P\!f\!f}$ are also[18] of order 1.

Making use of (9.79) and (9.80) we estimate the partial widths of decays of P and S into the Standard Model particles (of course, we are talking about kinematically allowed decays)

$$\Gamma_{P(S)\to AA} \sim \frac{m_{P(S)}^3}{64\pi\Lambda^2}, \quad \Gamma_{P(S)\to f\!f} \sim \frac{m_f^2 m_{P(S)}}{8\pi\Lambda^2}, \qquad (9.81)$$

where A denotes vector bosons, and we omitted threshold factors. By requiring that the lifetime of new particles exceeds the present age of the Universe, $\tau_{S(P)} = \Gamma_{S(P)}^{-1} > H_0^{-1}$ we find a bound on the mass of the dark matter candidates,

$$m_{P(S)} < (16\pi\Lambda^2 H_0)^{1/3}. \qquad (9.82)$$

Assuming that the new physics scale is below the Planck scale, $\Lambda < M_{Pl}$, we obtain an (almost) model-independent bound,

$$m_{P(S)} < 100\,\text{MeV}. \qquad (9.83)$$

Hence, the kinematically allowed decays are $P(S) \to \gamma\gamma$, $P(S) \to \nu\bar{\nu}$ and $P(S) \to e^+e^-$. It follows from (9.81) that the two-photon decay mode dominates, unless the mass of the new particles is close to that of electron. The decay $P(S) \to \gamma\gamma$ enables one to search for the dark matter particles $P(S)$: they would produce a photon line of energy equal to half of the mass of $P(S)$. The relative line width must be of order 10^{-3}, which is roughly the velocity disperion in a galaxy. The results of this search by cosmic telescopes are shown in Fig. 9.15. It is clear from this Figure that the lifetime of a dark matter particle must exceed considerably the age of the Universe: the typical value 10^4 on vertical axis corresponds to the lifetime $\tau_{P(S)} \sim 10^{26}$ s $\sim 3 \cdot 10^{18}$ years.

Problem 9.15. *Making use of Fig. 9.15 and assuming that the dominant decay channel is $P(S) \to \gamma\gamma$, refine the estimate (9.83).*

---

[18]In some cases the sets of constants $\{C_{s\!f\!f}\}$, $\{C_{P\!f\!f}\}$ have hierarchical structure. To simplify the discussion, we do not consider these cases in what follows.

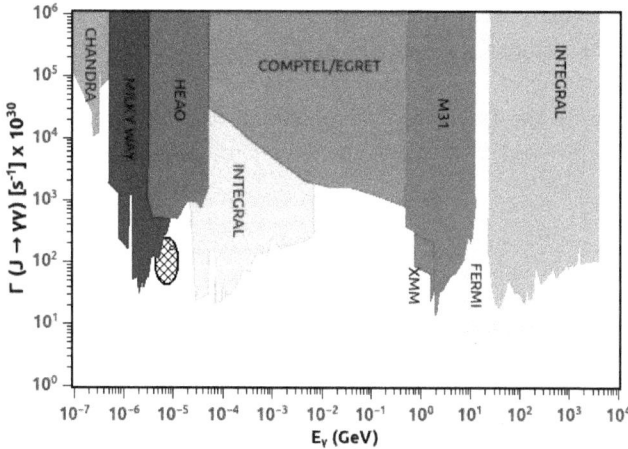

Fig. 9.15. Observational bounds on partial decay width of dark matter particle to two photons [168].

Let us now consider generation of relic (pseudo)scalars in the early Universe. There are several generation mechanisms; two of them are fairly generic for the class of models we discuss. These are generation in decays of condensates and thermal generation. (We will consider yet another mechanism later, in the model with axion.) Let us briefly discuss the two mechanisms in turn. We will denote the light particle by S for definiteness.

Let some scalar field ϕ be in condensate in the early Universe; we recall here that the condensate is the homogeneous scalar field that oscillates at relatively late times, when $m_\phi > H$. In other words, the condensate is a collection of ϕ-particles at rest; see Sec. 4.8.1. Let both particles, ϕ and S, interact with matter so weakly that they never get into thermal equilibrium, and let the interaction between ϕ and S have the form $\mu \phi S^2 / 2$, where μ is the coupling constant. Then the width of the decay $\phi \to SS$ is estimated as[19]

$$\Gamma_{\phi \to SS} \sim \frac{\mu^2}{16 \pi m_\phi}. \tag{9.84}$$

If the widths of other decay channels do not exceed the value (9.84), the decay of ϕ-condensate occurs at temperature T_ϕ determined by[20]

$$\Gamma_{\phi \to SS} \sim H(T_\phi) = \frac{T_\phi^2}{M_{Pl}^*}.$$

---

[19] As we discuss in the accompanying book, the decay of condensate may be faster due to the Bose enhancement. We leave this possibility aside for simplicity.
[20] We assume for definiteness that the energy density of ϕ-condensate is small compared to that of hot matter.

Let the energy density of ϕ-condensate at that time be equal to ρ_ϕ, so that the number density of decaying ϕ-particles is $n_\phi \sim \rho_\phi/m_\phi$; see Sec. 4.8.1. Immediately after the epoch of ϕ-particles decays, the number density of S-particles is of order $\epsilon\rho_\phi/m_\phi$, where ϵ is the fraction of the condensate that decayed into S-particles. After S-particles become non-relativistic, their mass density is of order

$$\rho_S \sim \epsilon\rho_\phi \cdot \frac{m_S T^3}{m_\phi T_\phi^3},$$

where we omitted the dependence on g_* for simplicity. In this way we estimate the mass fraction of S-particles today,

$$\Omega_S = \frac{\rho_S}{\rho_c} \sim \frac{m_S T_0^3}{\rho_c} \cdot \frac{\epsilon\rho_\phi}{m_\phi T_\phi^3} \sim 0.2 \cdot \left(\frac{m_S}{1\,\mathrm{eV}}\right) \cdot \frac{\epsilon\rho_\phi}{m_\phi T_\phi^3}. \tag{9.85}$$

With appropriate choice of parameters, the right value $\Omega_S \simeq 0.25$ can be obtained indeed. We note that the last factor on the right-hand side of (9.85) must be small.

Let us now consider production of light (pseudo)scalars in collisions of the Standard Model particles in primordial plasma. Since the interactions (9.79) and (9.80) are very weak, and since they involve three particles, the dominant processes of S-particle creation and annihilation at temperatures $T \ll \Lambda$ are[21]

$$S + X_1 \leftrightarrow X_2 + X_3. \tag{9.86}$$

These are analogous to the processes (9.72) which we discussed in the gravitino case; the relevant diagrams are basically the same as in Fig. 9.12 (with S substituted for $\tilde{G}$). Making use of (9.79) and (9.80) we obtain the estimate for the cross-section in the relevant case $m_S < T < \Lambda$,

$$\sigma_S \sim \frac{\alpha}{\Lambda^2},$$

where $\alpha = g^2/(4\pi)$, and g is the Standard Model coupling entering the vertex $Y X_2 X_3$ (see Fig. 9.12). We now equate the Hubble time to the mean free time of the Standard Model particles with respect to the emission of S-particle,

$$\tau_S \sim \frac{1}{\sigma_{S(P)} n_\gamma(T)} \sim \frac{1}{H(T)} \sim \frac{M_{Pl}^*}{T^2},$$

and obtain the freeze-out temperature of S-particles,

$$T_S \sim \alpha^{-1}\Lambda \cdot \frac{\Lambda}{M_{Pl}^*}. \tag{9.87}$$

If the temperature in the Universe had ever exceeded $T_{S(P)}$, the $S(P)$-particles were in thermal equilibrium. In that case their properties are the same as the properties of

---

[21] At $T > m_h$ there are also processes $S + h \leftrightarrow \bar{f}f$ and crossing processes. The order-of-magnitude estimates below apply to these processes as well.

thermally produced gravitino. The estimate (9.70) applies, and one concludes that the mass of the dark matter S-particle must be unacceptably small, $m_S \sim 200\,\mathrm{eV}$.

Problem 9.16. *Neglecting nucleosynthesis constraints, estimate the fractional mass density Ω_S in models where S-particles freeze out being non-relativistic, i.e., may pretend to make cold dark matter.*

If the maximum temperature in the Universe was lower than T_S, the number density of thermally produced S-particles is smaller than in the above case. The mass m_S may then be in the acceptable range of 1 keV and larger. The corresponding analysis is similar to that for gravitino, and we do not make it here.

Let us now turn to a concrete class of models with Peccei–Quinn symmetry and axions. This symmetry provides a solution to the *strong CP-problem*, and the existence of axions is an inevitable consequence of the construction.

Let us briefly discuss the strong CP-problem [138–140] and its solution in models with axion. The starting point is that one can extend the Standard Model Lagrangian (see Appendix B) by adding the following term,

$$\Delta \mathcal{L}_0 = \frac{\alpha_s}{8\pi} \cdot \theta_0 \cdot G^a_{\mu\nu} \tilde{G}^{\mu\nu\,a}, \tag{9.88}$$

where $\alpha_s = g_s^2/(4\pi)$ is the $SU(3)_c$ gauge coupling, $G^a_{\mu\nu}$ is the gluon field strength, $\tilde{G}^{\mu\nu\,a} = \frac{1}{2}\epsilon^{\mu\nu\lambda\rho}G^a_{\lambda\rho}$ is the dual tensor and θ_0 is an arbitrary dimensionless parameter (the factor $\alpha_s/(8\pi)$ is introduced for later convenience). The interaction term (9.88) is invariant under gauge symmetries of the Standard Model, but it violates P and CP. The term (9.88) can be written as the divergence of the vector (we consider the theory in Minkowski space for definiteness)

$$\Delta \mathcal{L}_0 = \frac{\alpha_s}{4\pi} \cdot \theta_0 \cdot \partial_\mu K^\mu,$$

where

$$K^\mu = \epsilon^{\mu\nu\lambda\rho} \cdot \left(G^a_\nu \partial_\lambda G^a_\rho + \frac{1}{3} f^{abc} G^a_\nu G^b_\lambda G^c_\rho \right),$$

and G^a_μ is the gluon vector potential. Hence, the term (9.88) does not contribute to the classical field equations, and its contribution to the action is reduced to the surface integral. For any perturbative gauge field configurations (small perturbations about $G^a_\mu = 0$) this contribution is equal to zero. However, this is not the case for configurations of instanton type. This means that CP is violated in QCD at non-perturbative level.

Furthermore, quantum effects due to quarks give rise to the anomalous term in the Lagrangian[22] which has the same form as (9.88) with proportionality coefficient determined

---

[22] We have in mind the anomaly in the axial current corresponding to $U(1)_A$ transformations, $\partial_\mu J^\mu_A \propto \frac{\alpha_s}{2\pi} G^a_{\mu\nu} \tilde{G}^{\mu\nu\,a}$.

by the phase of the quark mass matrix $\hat{M}_q$. The latter enters the Lagrangian as

$$L_m = \bar{q}_L \hat{M}_q q_R + \text{h.c.}$$

By chiral rotation of quark fields one makes quark masses real (i.e., physical), but that rotation induces a new term in the Lagrangian,

$$\Delta \mathcal{L}_m = \frac{\alpha_s}{8\pi} \cdot \text{Arg}(\text{Det}\hat{M}_q) \cdot G^a_{\mu\nu} \tilde{G}^{\mu\nu\,a}. \tag{9.89}$$

There is no reason to think that $\text{Arg}(\text{Det}\hat{M}_q) = 0$. Neither there is a reason to think that the "tree-level" term (9.88) and anomalous contribution (9.89) cancel each other. Indeed, the former term is there even in the absence of quarks, while the latter comes from the Yukawa sector, as the quark masses are due to their Yukawa interactions with the Englert–Brout–Higgs field.

Thus, the Standard Model Lagrangian should contain the term

$$\Delta \mathcal{L}_\theta = \Delta \mathcal{L}_0 + \Delta \mathcal{L}_m = \frac{\alpha_s}{8\pi}(\theta_0 + \text{Arg}(\text{Det}\hat{M}_q))G^a_{\mu\nu} \tilde{G}^{\mu\nu\,a} \equiv \frac{\alpha_s}{8\pi} \cdot \theta \cdot G^a_{\mu\nu} \tilde{G}^{\mu\nu\,a}. \tag{9.90}$$

This term violates CP, and off-hand the parameter θ is of order 1.

The term (9.90) has non-trivial phenomenological consequences. One is that it generates electric dipole moment (EDM) of neutron[23] d_n, which is estimated as [141, 142]

$$d_n \sim \theta \cdot 10^{-16} \cdot e \cdot \text{cm}. \tag{9.91}$$

Neutron EDM has not been found experimentally, and the searches place strong bound [1]

$$d_n \lesssim 3 \cdot 10^{-26} \cdot e \cdot \text{cm}. \tag{9.92}$$

This leads to the bound on the parameter θ,

$$|\theta| < 0.3 \cdot 10^{-9}.$$

The problem to explain such small value of θ is precisely the strong CP-problem.

A solution to this problem apparently does not exist within the Standard Model.[24] The solution is offered by models with axion. These models make use of the following observation. If at the classical level the quark Lagrangian is invariant under phase rotations of quark fields

$$q_{Ln} \to e^{ie_n^{(PQ)}\beta/2} q_{Ln}, \quad q_{Rn} \to e^{-ie_n^{(PQ)}\beta/2} q_{Rn}, \tag{9.93}$$

where $e_n^{(PQ)}$ is the axion charge of nth quark flavor, and $\sum_n e_n^{(PQ)} \neq 0$, then the θ-term would be rotated away by applying this transformation. Indeed, under this transformation

---

[23] Neutron EDM corresponds to the interaction of neutron spin $\mathbf{S}$ with electric field $\mathbf{E}$ described by the Hamiltonian $H = d_n \mathbf{E} \mathbf{S}/|\mathbf{S}|$.

[24] If at least one of the light quarks (e.g., u-quark) were massless, there would be no effects due to the θ-term. At the classical level, there would exist a global $U(1)_A$-symmetry of γ^5 phase rotations of u-quark, $u_L \to e^{i\beta} u_L$, $u_R \to e^{-i\beta} u_R$. This classical symmetry could be employed to set θ equal to zero. However, experimental data imply that all quarks are actually massive, so this solution of the strong CP-problem is most likely ruled out.

one has $\mathrm{Arg}(\mathrm{Det}\hat{M}_q) \to \mathrm{Arg}(\mathrm{Det}\hat{M}_q) + \beta$, and hence $\theta \to \theta + \beta$, where the axion charges are normalized in such a way that

$$\sum_n e_n^{(PQ)} = 1.$$

This global symmetry is called Peccei–Quinn symmetry [143] $U(1)_{PQ}$. Peccei–Quinn (PQ) symmetry is explicitly broken in the Standard Model by Yukawa couplings of quarks to the Englert–Brout–Higgs field (see Sec. B.1; we omit group and generation indices),

$$L_Y = Y^d \bar{Q}_L H D_R + Y^u \bar{Q}_L i\tau^2 H^* U_R. \tag{9.94}$$

Indeed, the first term in (9.94) would be invariant under transformations (9.93) with $e_n^{(PQ)} = 1/6$ supplemented with the phase rotation $H \to e^{i\beta/6} H$, while the second term would be invariant if the phase of the scalar field was rotated in the opposite way, $H \to e^{-i\beta/6} H$.

One can, however, extend the Standard Model in such a way that the PQ symmetry is exactly at the classical level. Quark masses are not invariant under the PQ transformations (9.93), so PQ symmetry is spontaneously broken. At the classical level, this leads to the existence of massless Nambu–Goldstone field $a(x)$, axion. As for any Nambu–Goldstone field, its properties are determined by its transformation law under the PQ-symmetry

$$a(x) \to a(x) + \beta \cdot f_{PQ}, \tag{9.95}$$

where β is the same parameter as in (9.93) and f_{PQ} is a constant of dimension of mass, the energy scale of $U(1)_{PQ}$ symmetry breaking. The mass terms in the low energy quark Lagrangian must be symmetric under the transformations (9.93), (9.95), so the quark and axion fields enter the Lagrangian in the combination

$$\mathcal{L}_m = \bar{q}_{R_n} m_{q_n} e^{-ie_n^{(PQ)} \frac{a}{f_{PQ}}} q_{L_n} + h.c. \tag{9.96}$$

Making use of (9.89) we find that at the quantum level the low energy Lagrangian contains the term

$$\mathcal{L}_a = \frac{\alpha_s}{8\pi} \cdot \frac{a}{f_{PQ}} G^a_{\mu\nu} \tilde{G}^{\mu\nu\,a}. \tag{9.97}$$

Clearly, PQ symmetry (9.93), (9.95) is *explicitly* broken by quantum effects of QCD, and axion is a *pseudo*-Nambu–Goldstone boson.

Hence, θ-parameter multiplying the operator $G^a_{\mu\nu} \tilde{G}^{\mu\nu\,a}$ obtains a shift depending on space–time point and proportional to the axion field,

$$\theta \to \bar{\theta}(x) = \theta + \frac{a(x)}{f_{PQ}}. \tag{9.98}$$

Strong interactions would conserve CP provided the axion vacuum expectation value is such that $\langle\bar{\theta}\rangle = 0$. The QCD effects indeed do the job. They generate non-vanishing quark condensate $\langle \bar{q}q \rangle \sim \Lambda^3_{QCD}$ at the QCD energy scale $\Lambda_{QCD} \sim 200\,\mathrm{MeV}$. This condensate breaks chiral symmetry and in turn generates the axion effective potential[25]

$$V_a \sim -\frac{1}{2}\bar{\theta}^2 \frac{m_u m_d}{m_u + m_d}\langle \bar{q}q \rangle + \mathcal{O}(\bar{\theta}^4) = \frac{1}{2}\frac{m_u m_d}{(m_u + m_d)^2}\bar{\theta}^2 \cdot m_\pi^2 f_\pi^2 + \mathcal{O}(\bar{\theta}^4), \qquad (9.99)$$

where $m_\pi = 135\,\mathrm{MeV}$ and $f_\pi = 93\,\mathrm{MeV}$ are pion mass and decay constant, and $m_u/m_d \approx 0.5$. The potential has the minimum at $\langle \bar{\theta} \rangle = 0$, so the strong CP-problem finds an elegant solution. It follows from (9.98) and (9.99) that axion has a mass

$$m_a \approx 0.5\frac{m_\pi f_\pi}{f_{PQ}}, \qquad (9.100)$$

i.e., it is indeed *pseudo*-Nambu–Goldstone boson.

The simplest way to implement PQ mechanism is to add the second Englert–Brout–Higgs doublet to the Standard Model and choose the Yukawa interaction as

$$Y^d \bar{Q}_L H_1 D_R + Y^u \bar{Q}_L i\tau^2 H_2^* U_R. \qquad (9.101)$$

Then the axion charge of every quark equals $e_n^{(PQ)} = 1/6$, while the two scalar fields transform under $U(1)_{PQ}$-transformation (9.93) as follows,

$$H_1 \to e^{i\beta/6}H_1, \quad H_2 \to e^{-i\beta/6}H_2,$$

This ensures $U(1)_{PQ}$-invariance of the Lagrangian (9.101), and hence the absence of the θ-term.

Unless other new fields are added, spontaneous breaking of PQ symmetry in this simplest case occurs due to the Englert–Brout–Higgs expectation values. Then the classically massless field (would-be Nambu–Goldstone boson) is the relative phase of the two fields H_1 and H_2. To obtain the low energy Lagrangian let us write

$$H_1 = e^{i\beta(x)/6}\begin{pmatrix} 0 \\ \frac{v_1}{\sqrt{2}} \end{pmatrix}, \quad H_2 = e^{-i\beta(x)/6}\begin{pmatrix} 0 \\ \frac{v_2}{\sqrt{2}} \end{pmatrix}, \qquad (9.102)$$

where v_1 and v_2 are the Englert–Brout–Higgs expectation values. They both contribute to the W- and Z-boson masses, hence

$$\sqrt{v_1^2 + v_2^2} \equiv v = 246\,\mathrm{GeV}.$$

The kinetic term for the field $\beta(x)$ comes from the kinetic terms of the Higgs doublets,

$$\mathcal{L}_{kin,H} = \partial_\mu H_1^\dagger \partial^\mu H_1 + \partial_\mu H_2^\dagger \partial^\mu H_2.$$

Inserting (9.102) into this expression, we find

$$\mathcal{L}_{kin,\beta} = \frac{f_{PQ}^2}{2}\partial_\mu \beta \partial^\mu \beta,$$

where

$$f_{PQ} = \frac{\sqrt{v_1^2 + v_2^2}}{6} = \frac{v}{6}. \qquad (9.103)$$

---

[25] Since (9.93) is the symmetry under phase rotations, the axion potential must be periodic in $\bar{\theta}$ with period 2π. The simplest periodic generalization of the expression (9.99) is $V_a = m_a^2 f_{PQ}^2 \cdot (1 - \cos\bar{\theta})$, while the computation within chiral perturbation theory gives a more cumbersome formula for V_a [160]. This point will not be important for us in what follows.

The axion field is related to $\beta(x)$ by

$$a(x) = f_{PQ} \cdot \beta(x);$$

with this definition the field $a(x)$ has the standard ("canonical") kinetic term. The axion is rather heavy in this model: we find from (9.100) that

$$m_a \sim 150 \, \text{keV}.$$

The interaction of this axion with quarks, gluons and photons (see below) is rather strong, and this particle, called Weinberg–Wilczek axion [144, 145], is experimentally ruled out.

This problem is solved in models with "invisible axion", in particular, in Dine–Fischler–Srednicki–Zhitnitsky [146, 147] (DFSZ) model and Kim–Shifman–Vainshtein–Zakharov [148, 149] (KSVZ) model. The idea is to make the scale of PQ symmetry breaking independent of the electroweak symmetry breaking scale. In DFSZ model this occurs in the following way. One adds into the model with the Lagrangian (9.101) complex scalar field S which is singlet under the Standard Model gauge group. One also adds interactions involving PQ invariants

$$S^\dagger S, \quad H_1^\dagger H_2 \cdot S^2.$$

The field S transforms under $U(1)_{PQ}$ as

$$S \to e^{i\beta/6} S. \tag{9.104}$$

The axion field is now a linear combination of the phases of fields H_1, H_2 and S. By repeating the calculation leading to (9.103) we find

$$f_{PQ} = \frac{\sqrt{v_1^2 + v_2^2 + v_s^2}}{6}, \tag{9.105}$$

where v_s is the vacuum expectation value of the field S. The latter can be large, so it is clear from (9.105) that the mass of axion is small and, most importantly, its couplings to the Standard Model fields are weak: these couplings are inversely proportional to $f_{PQ} \sim v_s$. Note that DFSZ axion interacts with both quarks and leptons.

The KSVZ mechanism makes use of additional quark fields Ψ_R and Ψ_L which are triplets under $SU(3)_c$ and singlets under $SU(2)_W \times U(1)_Y$. Only these quarks transform non-trivially under $U(1)_{PQ}$, $e_\Psi^{(PQ)} = 1$, while the usual quarks have zero PQ charge. One also introduces a complex scalar field S which is a singlet under the Standard Model gauge group. One writes PQ-invariant Yukawa interaction of the new fields,

$$\mathcal{L} = y_\Psi S \bar{\Psi}_L \Psi_R + \text{h.c.},$$

so that S transforms under $U(1)_{PQ}$ as $S \to e^{i\beta} S$. PQ symmetry is spontaneously broken by the vacuum expectation value $\langle S \rangle = v_s/\sqrt{2}$. The axion here is the phase of the field S, therefore

$$f_{PQ} = v_s. \tag{9.106}$$

The KSVZ model does not contain explicit interaction of axion with usual quarks and leptons.

The Peccei–Quinn solution to the strong CP problem has its own drawbacks. First, one expects that any global symmetry is broken by gravitational interactions. This may give rise to terms in the low energy Lagrangian which break $U(1)_{PQ}$ explicitly and have the form $\Psi_{PQ}^{(N)}/M_{Pl}^{N-4}$, where $\Psi_{PQ}^{(N)}$ is a (pseudo)scalar combination of the fields with non-zero PQ charge and mass dimension N. These terms would generically shift the vacuum expectation value $\langle \bar{\theta} \rangle$ from zero to a value of order 1. In other words, axion solution to the strong CP problem is unstable against the addition of operators which break $U(1)_{PQ}$, even though they are suppressed by the Planck scale [169]. Second, models with spontaneously broken global $U(1)$ symmetry generically have topological defects similar to those studied in Chapter 12. If this symmetry is broken down completely, these defects are cosmic strings. If the symmetry is broken down to a discrete subgroup (e.g., $\mathbb{Z}_2$), which is often the case in theories with axions, the situation is even worse, since there appear domain walls. The latter are dangerous from the cosmological viewpoint, as we will discuss in Sec. 12.4.

Thus, the axion is a light particle whose interactions with the Standard Model fields are very weak. Its mass is related to PQ symmetry breaking scale f_{PQ} by (9.100). The property that its interactions are weak relates to the fact that it is pseudo-Nambu–Goldstone boson of a global symmetry spontaneously broken at high energy scale $f_{PQ} \gg M_W$. Like for any Nambu–Goldstone field, the interactions of axion are described by the generalized Goldberger–Treiman formula

$$\mathcal{L}_a = \frac{1}{f_{PQ}} \cdot \partial_\mu a \cdot J^\mu_{PQ}. \tag{9.107}$$

Here

$$J^\mu_{PQ} = \sum_f \frac{e_f^{(PQ)}}{2} \cdot \bar{f} \gamma^\mu \gamma^5 f. \tag{9.108}$$

The contributions of fermions to the current J^μ_{PQ} are proportional to their PQ charges $e_f^{(PQ)}$; these charges are model-dependent. Besides the interaction (9.107), there are also interactions of axions with gluons, see (9.97), and photons,

$$\mathcal{L}_{ag} = \frac{\alpha_s}{8\pi} \cdot \frac{a}{f_{PQ}} \cdot G^a_{\mu\nu} \tilde{G}^{\mu\nu\, a}, \quad \mathcal{L}_{a\gamma} = C_\gamma \frac{\alpha}{8\pi} \cdot \frac{a}{f_{PQ}} \cdot F_{\mu\nu} \tilde{F}^{\mu\nu}, \tag{9.109}$$

where the dimensionless constant C_γ is also model-dependent and, generally, is of order 1 (although may be fairly small numerically). In accordance with (9.96), the action (9.107) can be integrated by parts and we obtain instead

$$\mathcal{L}_a = -\frac{1}{f_{PQ}} \cdot a \cdot \partial_\mu J^\mu_{PQ} = -\frac{a}{f_{PQ}} \cdot \sum_f e_f^{(PQ)} m_f \cdot \bar{f} \gamma^5 f, \tag{9.110}$$

plus anomalous interactions (9.109). The interaction terms (9.109) and (9.110) indeed have the form (9.79), (9.80) (with $P(x) = a(x)$), i.e., models with axions belong to the class of models with light, weakly interacting pseudoscalars. The axion mass, however, is not a free parameter: we find from (9.100) that

$$m_a \approx 0.6\,\text{eV} \cdot \left(\frac{10^7\,\text{GeV}}{f_{PQ}} \right). \tag{9.111}$$

The main decay channel of the light axion is decay into two photons. The lifetime τ_a is found from (9.81) by setting $\Lambda = 2\pi f_{PQ}/\alpha$ and using (9.111),

$$\tau_a = \frac{1}{\Gamma_{a \to \gamma\gamma}} = \frac{64\pi^3 m_\pi^2 f_\pi^2}{\alpha^2 m_a^5} \simeq 4 \cdot 10^{24}\,\text{s} \cdot \left(\frac{\text{eV}}{m_a} \right)^5.$$

By requiring that this lifetime exceeds the age of the Universe, $\tau_a > t_0 \approx 14$ billion years, we find the bound on the mass of axion as dark matter candidate,

$$m_a < 25\,\text{eV}. \tag{9.112}$$

There are astrophysical bounds on the strength of axion interactions f_{PQ}^{-1} and hence on the axion mass. Axions in theories with $f_{PQ} \lesssim 10^9\,\text{GeV}$, which are heavier than $10^{-2}\,\text{eV}$ would be intensely produced in stars and supernovae explosions. This would lead to contradictions with observations. So, we are left with very light axions, $m_a \lesssim 10^{-2}\,\text{eV}$.

As far as dark matter is concerned, thermal production of axions is irrelevant. Indeed, the estimate of the mass density of thermally produced axions basically coincides with that of gravitino (9.70), which gives way too low Ω_a.

It may seem that axion cannot serve as dark matter candidate. This is not the case. There are at least two mechanisms of axion production in the early Universe that can provide not only right axion abundance but also small initial velocities of axions. The latter property makes axion a *cold* dark matter candidate, despite its very small mass. One mechanism has to do with decays of global strings [150] — topological defects that exist in theories with spontaneously broken global $U(1)$ symmetry ($U(1)_{PQ}$ in our case; for a discussion of this mechanism see. e.g., Ref. [151]). Another mechanism employs axion condensate [152–154], homogeneous axion field that oscillates in time after the QCD epoch. Let us consider the second mechanism in some details.

As we have seen in (9.99), the axion potential is proportional to the quark condensate $\langle \bar{q}q \rangle$. This condensate breaks chiral symmetry. The chiral symmetry is in fact restored at high temperatures.[26] Hence, one expects that the axion potential is negligibly small at $T \gg \Lambda_{QCD}$. This is indeed the case: the effective potential for the field $\bar{\theta} = \theta + a/f_{PQ}$ vanishes at high temperatures, and this field can take any value,

$$\bar{\theta}_i \in [0 , 2\pi),$$

where we recall that the field $\bar{\theta}$ is a phase. There is no reason to think that the initial value $\bar{\theta}_i$ is zero. As the temperature decreases, the axion mass starts to

---

[26] We have here an analogy to phase transitions considered in Chapter 10.

get generated, and the field $\bar{\theta}$, remaining homogeneous, starts to roll down from $\bar{\theta}_i$ towards its value $\bar{\theta} = 0$ at the minimum of the potential. This homogeneous evolution is described by the Lagrangian

$$\mathcal{L} = \frac{f_{PQ}^2}{2} \cdot \left(\frac{d\bar{\theta}}{dt}\right)^2 - \frac{m_a^2(T)}{2} f_{PQ}^2 \bar{\theta}^2,$$

where $m_a(T)$ is a function of temperature, so that

$$m_a(T) \simeq 0 \quad \text{at} \quad T \gg \Lambda_{QCD},$$
$$m_a(T) \simeq m_a \quad \text{at} \quad T \ll \Lambda_{QCD}.$$

Hereafter m_a denotes the zero-temperature axion mass.

The evolution of a scalar field in expanding Universe is studied in Sec. 4.8.1. It follows from that study that axion field practically does not evolve when $m_a(T) \ll H(T)$ and at the time when $m_a(T) \sim H(T)$ it starts to oscillate. Let us estimate the present energy density of axion field in this picture, without using the concrete form of the function $m(T)$ for the time being.

The oscillations start at the time t_{osc} when

$$m_a(t_{osc}) \sim H(t_{osc}). \tag{9.113}$$

At this time, the energy density of the axion field is estimated as

$$\rho_a(t_{osc}) \sim m_a^2(t_{osc}) f_{PQ}^2 \bar{\theta}_i^2.$$

According to the discussion in the end of Sec. 4.8.1, the oscillating axion field is the same thing as a collection of axions at rest. Their number density at the beginning of oscillations is estimated as

$$n_a(t_{osc}) \sim \frac{\rho_a(t_{osc})}{m_a(t_{osc})} \sim m_a(t_{osc}) f_{PQ}^2 \bar{\theta}_i^2 \sim H(t_{osc}) f_{PQ}^2 \bar{\theta}_i^2.$$

This number density, as any number density of non-relativistic particles, then decreases as a^{-3}. (We will explicitly see this in the end of this Section.)

Axion-to-entropy ratio at time t_{osc} is

$$\frac{n_a}{s} \sim \frac{H(t_{osc}) f_{PQ}^2}{\frac{2\pi^2}{45} g_* T_{osc}^3} \cdot \bar{\theta}_i^2 \simeq \frac{f_{PQ}^2}{\sqrt{g_*} T_{osc} M_{Pl}} \cdot \bar{\theta}_i^2.$$

The axion-to-entropy ratio remains constant after the beginning of oscillations, so the present mass density of axions is

$$\rho_{a,0} = \frac{n_a}{s} m_a s_0 \simeq \frac{m_a f_{PQ}^2}{\sqrt{g_*} T_{osc} M_{Pl}} s_0 \cdot \bar{\theta}_i^2. \tag{9.114}$$

In fact, it is a decreasing function of m_a. Indeed f_{PQ} is inversely proportional to m_a, see (9.100); at the same time, axion obtains its mass near the epoch of QCD transition, i.e., at $T \sim \Lambda_{QCD}$, so T_{osc} depends on m_a rather weakly.

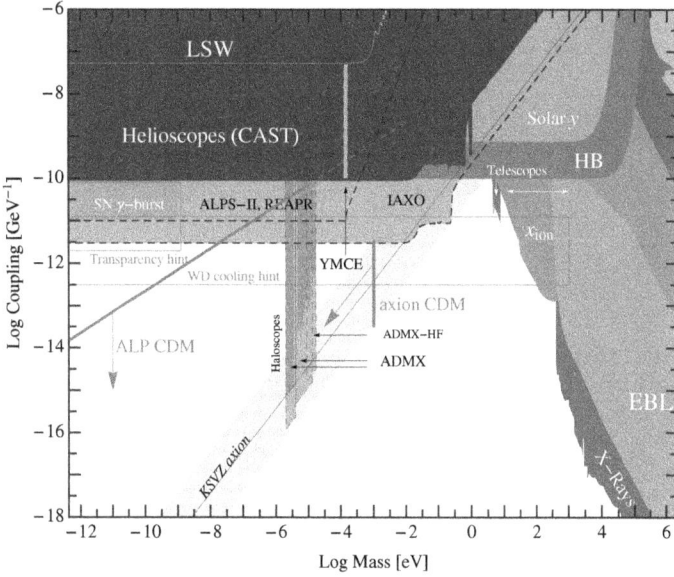

Fig. 9.16. Experimental constraints on the parameters of models with dark matter axions [170]. The inclined line "KSVZ axion" is the prediction of the KSVZ model; the yellow band around it shown predictions of the majority of axion models. See Fig. 13.22 for color version.

To obtain preliminary estimate, let us set $T_{osc} \sim \Lambda_{QCD} \simeq 200\,\text{MeV}$ and make use of (9.100) with $C_g \sim 1$. We find

$$\Omega_a \equiv \frac{\rho_{a,0}}{\rho_c} \simeq \left(\frac{10^{-6}\,\text{eV}}{m_a}\right)\bar{\theta}_i^2. \qquad (9.115)$$

The natural assumption about the initial phase is $\bar{\theta}_i^2 \sim 0.1 - 10$. Hence, axion of mass 10^{-5}–$10^{-7}\,\text{eV}$ is a good dark matter candidate.[27] This is *cold* dark matter: we have seen in Sec. 4.8.1 that effective pressure of the oscillating field is zero. This is of course consistent with the fact that oscillating field corresponds to axions at rest.

Let us note here that search for relic axions with masses $m_a \sim 10^{-5} - 10^{-7}\,\text{eV}$ is a difficult but not hopeless experimental problem [155]. Limits on cosmic axions are shown in Fig. 9.16.

To refine the above estimate, let us make use of the explicit formula for the axion mass at $T > \Lambda_{QCD}$. It reads [156]

$$m_a(T) \simeq 0.1 \cdot m_a(0) \cdot \left(\frac{\Lambda_{QCD}}{T}\right)^{3.7}, \quad T > \Lambda_{QCD}. \qquad (9.116)$$

---

[27]We note that axion of lower mass $m_a < 10^{-7}\,\text{eV}$ may also serve as dark matter particle, if for some reason the initial phase $\bar{\theta}_i$ is much smaller than $\pi/2$. We note also that the mechanism of axion production in decays of topological defects is capable of producing right axion abundance for $m_a > 10^{-5}\,\text{eV}$ as well.

We then find from (9.113) that the temperature at which axion field oscillations start is

$$T_{osc} \sim 200\,\text{MeV} \cdot \left(\frac{m_a}{10^{-9}\,\text{eV}}\right)^{0.2} \cdot \left(\frac{\Lambda_{QCD}}{200\,\text{MeV}}\right)^{0.7}. \tag{9.117}$$

Note that the assumption $T_{osc} > \Lambda_{QCD}$ is justified for $m_a > 10^{-9}\,\text{eV}$. Inserting the estimate (9.117) into (9.114) we get

$$\Omega_a \simeq 0.2 \cdot \bar{\theta}_i^2 \cdot \left(\frac{4 \cdot 10^{-6}\,\text{eV}}{m_a}\right)^{1.2} \cdot \frac{1}{2h^2}. \tag{9.118}$$

We see that our earlier estimate (9.115) is quite reasonable, and the dependence on the axion mass is close to inverse proportionality. This is related to strong dependence on temperature in[28] (9.116). Interestingly, similar dependence of Ω_a on m_a is obtained in models employing string mechansim of the axion production.

   Let us discuss in more detail what can be the value of the initial phase $\bar{\theta}_i$ entering the estimate (9.118). In inflationary models, this phase becomes spatially homogeneous either after the inflationary epoch or at that epoch. The former case occurs if $U(1)_{PQ}$ breaks down after inflation. Then the phase $\bar{\theta}_i$ is homogeneous over spatial regions of size comparable to the Hubble size or smaller, while on larger scales the phase is randomly distributed. Naive averaging over large scales gives

$$\langle \bar{\theta}_i^2 \rangle = \frac{1}{2\pi} \int_{-\pi}^{\pi} \bar{\theta}^2 \, d\bar{\theta} = \frac{\pi^2}{3}.$$

However, non-linear terms in the axion potential modify this estimate and yield a larger value [160] $\langle \bar{\theta}_i^2 \rangle \approx 4.5$. Making use of this value in (9.118), one obtains the prediction for the mass of dark matter axions:

$$m_a \simeq 1 \cdot 10^{-5}\,\text{eV}. \tag{9.119}$$

Under the above assumption of late breaking of $U(1)_{PQ}$, lighter axions are ruled out by cosmology while heavier axions can make only part of dark matter.

   The second case occurs if $U(1)_{PQ}$ is broken already at inflationary stage. As we point out in Sec. 1.7 and further discuss in accompanying book, an initially small, causally connected region is inflated at that stage to a region larger than the present horizon. Therefore, the initial phase $\bar{\theta}_i$ is the same in the visible Universe, but its value cannot be predicted. Hence, the axion mass cannot be predicted either.

   In fact, in the latter case the initial phase $\bar{\theta}_i$ is not quite homogeneous in space. At inflationary stage, vacuum fluctuations of all massless or light scalar fields get

---

[28]The estimate (9.116) for the temperature dependent axion mass is obtained within dilute instanton gas approximation. Recent lattice calculations, Ref. [159] and references therein, reveal larger numerical factor and much milder dependence on temperature, $m_a(T) \propto T^{-1.5}$, although Ω_a is still close to inverse proportionality to the axion mass. However, the window of m_a is shifted with respect to estimate (9.118) by a factor of 5 towards lighter axions. Likewise, the estimate (9.119) becomes $m_a \simeq 2 \cdot 10^{-6}$ eV.

enhanced. As a result, scalar fields become inhomogeneous on scales exceeding the inflationary Hubble scale H_{infl}^{-1}. The amplitudes of these inhomogeneities (for canonically normalized fields) are equal to $H_{\mathrm{infl}}/(2\pi)$; this property is studied in the detail in the accompanying book. Applying this to axion, we observe that the initial configuration is

$$\bar{\theta}_i(\mathbf{x}) = \bar{\theta}_i^{(0)} + \delta\bar{\theta}_i(\mathbf{x}),$$

where $\bar{\theta}_i^{(0)}$ is homogeneous in space, while $\delta\bar{\theta}_i(\mathbf{x})$ is the Gaussian random field which exists at all cosmological scales and has amplitude squared

$$\langle \delta\bar{\theta}_i^2 \rangle \sim \frac{H_{\mathrm{infl}}^2}{(2\pi f_a)^2}.$$

Phase perturbations give rise to perturbations of axion dark matter energy density, which are uncorrelated with perturbations of conventional matter. These uncorrelated dark matter perturbations are called isocurvature (or entropy) modes. Cosmological observations show that their contribution cannot exceed a few per cent of the dominant adiabatic mode; see the accompanying book for details.

Both of these cases are illustrated in Fig. 9.17, which shows the parameter space $(m_a, \bar{\theta}_i, H_{\mathrm{infl}})$. The boundary between the two cases is the line $f_a = T_{\mathrm{GH}} \equiv H_{\mathrm{infl}}/(2\pi)$.

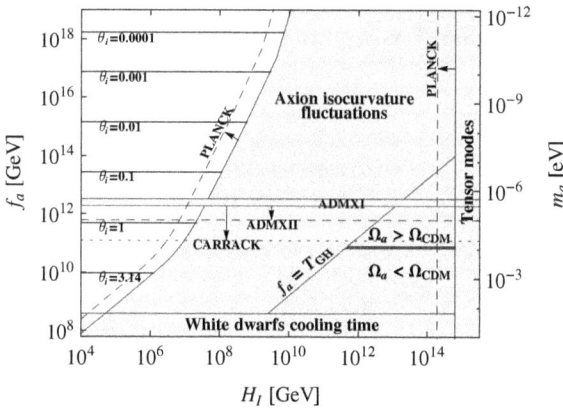

Fig. 9.17. Cosmological bounds on parameters of inflationary models with axion dark matter [171]. Axion can be dark matter particle in white regions. In the case of post-inflationary PQ symmetry breaking, this is the narrow strip between the regions $\Omega_a < \Omega_{CDM}$ and $\Omega_a > \Omega_{CDM}$ in the right part. In the case of PQ symmetry breaking at inflation, allowed is the white region in the left part. Besides the cosmological bounds, shown are astrophysical bound (axion effect on white dwarf cooling), bound on the Hubble parameter at inflation from non-observation of tensor modes and experimental limits (data from ADMX-I experiment and projected sensitivities of ADMX II and CARRACK).

To end this Section, let us check explicitly that the homogeneous oscillating field of variable mass has the property that

$$n_a(t) = \frac{\rho_a(t)}{m_a(t)}$$

decays as a^{-3}. Let us still use the notation $\bar{\theta}$ for this field. Let us write the field equation in expanding Universe,

$$\frac{d^2\bar{\theta}}{dt^2} + 3H(T)\frac{d\bar{\theta}}{dt} + m_a^2(T)\bar{\theta} = 0. \tag{9.120}$$

Let us multiply this equation by $d\bar{\theta}/dt$ and find

$$\frac{1}{2}\frac{d}{dt}\left(\frac{d\bar{\theta}}{dt}\right)^2 + 3H\cdot\left(\frac{d\bar{\theta}}{dt}\right)^2 + \frac{m_a^2(t)}{2}\frac{d}{dt}\bar{\theta}^2 = 0. \tag{9.121}$$

We solve this equation approximately, making use of the fact that for $m_a(T) \gg H(T)$, the following equality for averages over the oscillation period holds,

$$\left\langle\left(\frac{d\bar{\theta}}{dt}\right)^2\right\rangle = m_a^2(t)\langle\bar{\theta}^2\rangle. \tag{9.122}$$

So, we obtain the equation

$$\frac{d\langle\bar{\theta}^2\rangle}{dt} + \left(3H + \frac{1}{m_a(t)}\frac{dm_a(t)}{dt}\right)\langle\bar{\theta}^2\rangle = 0.$$

It gives

$$m(t)\langle\bar{\theta}^2\rangle(t) = \frac{\text{const}}{a^3}.$$

The left-hand side here coincides with n_a, in view of the fact that

$$\rho_a = \text{const}\cdot\left\langle\left(\frac{d\bar{\theta}}{dt}\right)^2\right\rangle = \text{const}\cdot m_a^2(t)\langle\bar{\theta}^2\rangle.$$

(The constant here equals f_{PQ}^2 for the field $\bar{\theta}$, and equals to 1 for canonically normalized field.)

9.8. Other Candidates

As other dark matter candidates, we consider sterile neutrinos in Sec. 7.3 and Q-balls in Sec. 12.7. Let us briefly discuss one more proposal.

9.8.1.   *Superheavy relic particles*

Less natural dark matter candidates are particles of very large mass (we will call them X-particles):

$$M_X \gg 100 \, \text{TeV}.$$

We recall that the assumption that these particles were in thermal equilibrium at $T \gtrsim M_X$ would lead to overproduction of these particles; see (9.41). Hence, one has to assume that thermal (more precisely, chemical) equilibrium had been never reached. The superheavy non-thermal relic particles are sometimes called "wimpzillas", see, e.g., Ref. [157].

Let us briefly discuss several production mechanisms of superheavy dark matter. We begin with production in collisions of light particles in hot plasma. This mechanism works if the maximum temperature in the Universe $T_{\max}$ is smaller, but not much smaller than M_X. Generalizing the analysis of Sec. 9.2, we find that the right abundance, $\Omega_X \approx 0.2$ is obtained when there is a very definite relation between $T_{\max}$ and M_X,

$$\frac{M_X}{T_{\max}} = 25 + \frac{1}{2} \cdot \log(M_X^2 \langle \sigma \rangle), \tag{9.123}$$

where σ is the production cross-section of X-particles in the plasma. The suppression of the relic mass density, as compared to the equilibrium case, is due to the Boltzmann factor here. Note that the right mass density of X-particles can be obtained at the expense of fine-tuning between the particle physics parameter M_X and cosmology-related parameter $T_{\max}$.

Problem 9.17. *Derive the relation (9.123).*

Heavy particles can be created before the hot stage, at the very process of formation of hot plasma. Such a process, reheating, occurs, in particular, in successful models with inflationary stage. We discuss reheating in the accompanying book. Here we only mention that production of particles at reheating may be efficient up to masses $M_X \sim 10^{16} \, \text{GeV}$, even though the maximum temperature of nearly equilibrium plasma after reheating is several orders of magnitude smaller.

Of even more exotic possibility is that inflation ends by vacuum phase transition. It occurs via spontaneous creation of bubbles of new vacuum, expansion of these bubbles and collisions of bubble walls. The collision of two walls may be viewed locally as a collision of particles of mass m and Lorentz-factor γ. The scale of m is the energy scale of the phase transition; the same scale determines the final temperature, $T_{\max} \lesssim m$. On the other hand, the Lorentz-factor γ may be large, and one expects that particles of mass up to $M_X \sim \gamma m$ can be produced in wall collisions. So, this mechanism is also capable of producing a heavy relic of very large mass.

Let us also note that time-dependent gravitational field produces particles as well. This mechanism can work both at inflationary stage and afterwards. The most efficient production occurs at the time when $M_X \sim H$. That can happen at the radiation-dominated epoch as well. In the latter case, the number density of X-particles produced when $M_X \sim H$ is given at later times by [158]

$$\rho_X \simeq 5 \cdot 10^{-4} \cdot M_X \cdot \left(\frac{M_X}{t} \right)^{3/2} .$$

The present mass fraction of the heavy relic produced in this way is then

$$\Omega_X \sim \left(\frac{M_X}{10^9 \, \text{GeV}} \right)^{5/2} .$$

We see that this mechanism is successful for $M_X \sim 10^9$ GeV and that it overproduces heavy particles of larger masses. Detection of superheavy dark matter with $M_X \gg 10^9$ GeV would probably mean that the Hubble parameter at the hot stage was never as large as M_X.

9.8.2. *Exotica*

To end this chapter, we notice that we have not considered many other dark matter candidates that have been proposed. These include relic black holes, new strongly self-interacting particles, axino (axion superpartner), mirror matter, etc. These exotic possibilities often require fine tuning of parameters and/or rather contrived cosmological scenarios. In any case, the predictions for the dark matter density are strongly model depenedent in the majority of the models.

Chapter 10

Phase Transitions in
the Early Universe

As we mentioned in Chapter 1, there are no direct experimental indications yet that temperatures above a few MeV existed in the Universe. Nevertheless, it is natural to assume that the Universe in the past had much higher temperatures.[1] Properties of cosmic medium were very different at different temperatures. In particular, there could occur phase transitions associated with the rearrangement of the ground state.

At temperatures above 200 MeV, quarks and gluons do not form bound states — hadrons — and the medium is in the phase of quark–gluon matter. At these temperatures, there is no quark condensate, i.e., the phase of unbroken approximate chiral symmetry is realized. If temperature in the Universe ever exceeded 200 MeV, then at some moment of time there was the transition from quark–gluon plasma to hadronic matter comprised of colorless particles: pions, kaons, nucleons and other hadrons. Moreover, at the same or almost the same time there must have occurred the chiral transition, responsible for the formation of the quark condensate.

It is quite likely that there was an era of even higher temperatures, $T \gtrsim M_{EW} \sim 100$ GeV. Oversimplifying the situation, we can say that at these temperatures electroweak symmetry was unbroken, and the Englert–Brout–Higgs expectation value was zero. When the temperature dropped, the electroweak transition [30–33] may have occurred, which resulted in the non-zero Englert–Brout–Higgs expectation value and spontaneous breaking of electroweak symmetry $SU(2)_W \times U(1)_Y$ down to the electromagnetic $U(1)_{em}$.

Depending on the maximum temperature at the hot stage of the cosmological expansion, and on physics at ultra-short distances and at ultra-high energies, phase transitions could occur at even higher temperatures. Namely, if the Universe had temperature of about 10^{16} GeV (which is very strong and hardly realistic assumption), and physics at these energies is described by Grand Unified Theory,

---

[1] For instance, in Chapter 9 we noted that a simple and efficient (and therefore plausible) mechanism of non-baryonic dark matter generation works at temperatures of tens of GeV or above. Many mechanisms generating the baryon asymmetry of the Universe (though not all) require even higher temperatures, from 100 GeV up to 10^{15} GeV (see Chapter 11), depending on specific mechanism.

then there was the Grand Unified phase transition at temperature comparable to $M_{GUT} \sim 10^{16}$ GeV. It is not excluded that phase transitions occurred also at intermediate temperatures, $M_{EW} \ll T \ll M_{GUT}$.

It is important that properties of cosmic medium can in principle change in either a smooth or abrupt way. In the latter case one speaks about phase transition proper, while the former is known as smooth crossover. The notions of different phases and phase transions is particularly well defined when there exists an exact global symmetry,[2] and the phases differ by their properties with respect to this symmetry. In most of such cases there is an order parameter which is nonzero in one phase and which vanishes in the other. A condensed matter example is the phase transition to ferromagnetic state in metals: the symmetry in this case is invariance under rotations in space, and the order parameter is spontaneous magnetization. If there is no symmetry and hence no order parameter, then the system may exhibit either phase transition or smooth crossover, depending on internal or external parameters. An example is vapor–liquid transition, which is the first order phase transition at low pressure and is not a phase transition at all at high pressure. Slightly above the critical pressure, the properties of the medium change rapidly as the temperature decreases, so there is a smooth but fast crossover.

Both possibilities — smooth crossover and phase transition — are of importance for cosmology. Particularly interesting is first order phase transition, which, as we will see in this Chapter, is a strongly out-of-thermal-equilibrium process. We emphasize, however, that lattice data show that the transition from quark–gluon plasma to hadronic matter in QCD is actually a smooth crossover. Likewise, electroweak transition we mentioned above is a crossover too [173]. The point is that neither QCD no Standard Model have gauge invariant order parameter [180, 181] that distinguishes different "phases". The absence of the phase transition in QCD is due to the fact that quarks have nonzero masses, and chiral symmetry is not exact. For the electroweak transition, it is important that the Higgs boson is quite heavy, $m_h = 125$ GeV.

The study of phase transitions in the Universe is not only of academic interest; it also sheds light on some of the mysteries of cosmology. Among them are the problems of the baryon asymmetry and dark matter. Also, phase transitions are responsible for possible formation of topological defects in the early Universe. first order phase transitions in some models lead to the generation of gravitational waves, which may be detected by gravity wave detectors.

In this Chapter we recall the general classification of phase transitions and introduce methods for describing phase transitions in the early Universe. We discuss predominantly theories with a scalar field and are interested in transitions leading to the generation of nonzero expectation value of this field. In other words, we

---

[2]Gauge invariance ("local symmetry") is not actually a symmetry; rather, it is a redundancy in the description which enables one to formulate the theory of spin-1 particles in a Lorentz-invariant way.

consider models that have the same structure as the Standard Model. As usual in field theory, the applicability of analytical methods is limited to theories with small couplings, but we will see that this is not enough: detailed analytical description of phase transitions is possible only when the vacuum value of the mass of the Higgs boson is sufficiently small.

In theories where couplings are not small, analytical study of phase transitions from "first principles" is usually impossible; the most reliable source of information here are numerical methods based on lattice field theory. An important example is the confinement–deconfinement transition and the transition with chiral symmetry breaking in QCD. They occur at temperatures $T \sim 200\,\mathrm{MeV}$, when the QCD gauge coupling $\alpha_s(T)$ is large. We will not present any detailed study of QCD transitions in this book, although there is no doubt that they actually occurred in the early Universe (assuming that temperatures $T \gtrsim 200\,\mathrm{MeV}$ were indeed realized). The reason is that these transitions apparently left no traces in the present Universe (with the exception of rather exotic proposals such as the formation of quark nuggets with large number of strange quarks [172]). We also mention in this context axions as candidates for dark matter: one of the mechanisms of their generation is based on the very fact that the chiral phase transition occurred in the early Universe, while the results are practically insensitive to the dynamics of this transition; see Sec. 9.7.

10.1. Order of Phase Transition

Phase transitions occur because of mismatch between the properties of the ground states of the theory at zero and nonzero temperatures. As we show below, in theories with Englert–Brout–Higgs scalar field(s) this is caused by non-trivial temperature-dependent terms in the effective potential. At finite temperature, the equilibrium state of the medium corresponds to the minimum of the Grand thermodynamic potential (Landau potential). As we discussed in Chapter 5, chemical potentials are negligibly small in the early Universe at interesting temperatures $T \gtrsim 1\,\mathrm{GeV}$, and the Grand potential reduces to the (Helmholtz) free energy F. Hence, we consider below the free energy of primordial plasma. To find the expectation value of the scalar field $\langle \phi \rangle_T$ at temperature T, one considers a system in which the average value of the field is fixed and equal to ϕ everywhere in space, but otherwise there is thermal equilibrium. The free energy of such a system depends, of course, on the chosen value ϕ, as well as on temperature. Because of spatial homogeneity, the free energy is proportional to the spatial volume Ω,

$$F = \Omega V_{eff}(T, \phi). \tag{10.1}$$

The free energy density of medium at temperature T with the average scalar field uniform and equal to ϕ is called the effective potential $V_{eff}(T, \phi)$. In thermal equilibrium, the free energy is at minimum with respect to all macroscopic parameters,

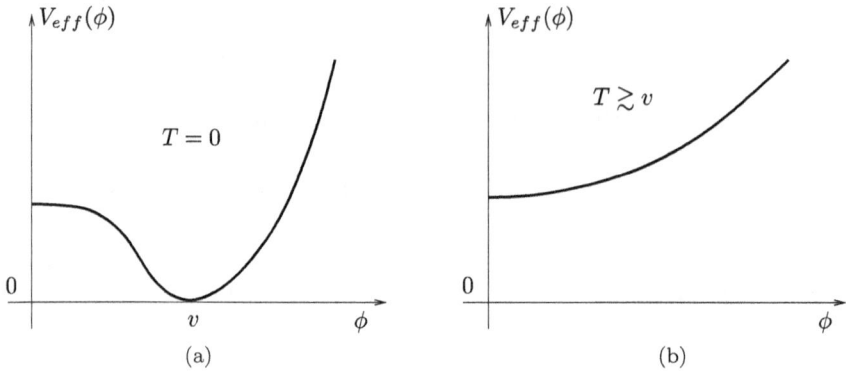

Fig. 10.1.   The effective potential at (a) zero and (b) high temperatures.

including the average scalar field. Therefore, $\langle\phi\rangle_T$ is the absolute minimum of the effective potential $V_{\text{eff}}(T,\phi)$ at given temperature (we will often omit the argument T in V_{eff}).

At zero temperature, the free energy reduces to the energy of the system, and the effective potential coincides with the scalar potential $V(\phi)$ entering the field theory action,[3] see Fig. 10.1(a); in theories we are considering, the field ϕ has nonzero expectation value v.

At finite temperature, $V_{\text{eff}}(T,\phi)$ does not coincide with $V(\phi)$. As a result, it often happens that the expectation value of the field ϕ vanishes; in this sense the symmetry is restored; see Fig. 10.1(b).

Let us emphasize one point here, which applies to *gauge* theories with the Englert–Brout–Higgs mechanism; let us talk about the Standard Model for definiteness. The Englert–Brout–Higgs field ϕ is not gauge invariant in the Standard Model, so in strict sense, its expectation value is not a legitimate object. Gauge invariant operators like $\phi^\dagger\phi$ (more precisely, $H^\dagger H$, see Appendix B) are invariant under all symmetries of the Lagrangian, and they cannot serve as order parameters. Thus, "phases" with "unbroken" and "broken" symmetry are in fact indistinguishable. A related observation is that the notion of the effective potential $V(T,\phi)$ makes sense, with reservations, only in perturbation theory. We discuss perturbation theory at finite temperature in Sec. 10.2, and consider its applicability in Sec. 10.3. It suffices to say here that perturbation theory *is not applicable* at temperatures and field values which are relevant for the electroweak transition. It is precisely for this reason that the transition is actually a smooth crossover, as we already mentioned. An adequate method of its study is lattice field theory.

---

[3]In fact, even at zero temperature the effective potential does not coincide with the scalar potential entering the classical action. This is due to quantum corrections. In weakly coupled theories, quantum corrections to the effective potential are often small.

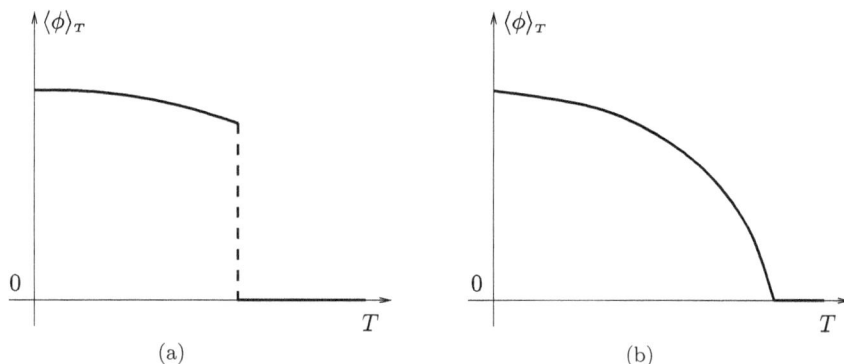

Fig. 10.2. The expectation value $\langle\phi\rangle_T$ as a function of temperature for systems with (a) 1st order and (b) 2nd order phase transition.

We consider in this Chapter the opposite situation, when the perturbation theory is applicable. In that case, there is a phase transition from $\langle\phi\rangle_T = 0$ to $\langle\phi\rangle_T \neq 0$ which occurs at critical temperature T_c.

Two types of phase transitions are most common; these are phase transitions of the first and second order. From the standpoint of the general formalism, first order phase transition is accompanied by jumps in quantities which directly characterize the system, like density. In field theory this corresponds to a jump in the expectation value $\langle\phi\rangle_T$ as a function of temperature, see Fig. 10.2(a). On the contrary, 2nd order phase transition is characterized by continuous behavior of the expectation value $\langle\phi\rangle_T$, see Fig. 10.2(b). This difference is illustrated in Fig. 10.3 where the families of effective potentials $V_{eff}(\phi, T)$ as functions of ϕ at various temperatures T are shown. The left panel of Fig. 10.3 shows the first order phase transition, culminating in an abrupt change of $\langle\phi\rangle_T$. The right part of Fig. 10.3 corresponds to the second order phase transition: the expectation value $\langle\phi\rangle_T$ is a smooth function of temperature.

The famous example of the first order phase transition is boiling of liquid. Examples of the second order phase transition are transitions in ferromagnets, order-disorder transitions in alloys of metals, transitions into superconducting and superfluid states.

The way the transition proceeds is quite different for the first and second order phase transitions. We are interested in the case where the rate at which temperature changes in time is low compared to the typical rate of particle interactions in the medium; this is the case for the early Universe. For the second order phase transition, the medium properties (for example, the expectation value $\langle\phi\rangle_T$) change slowly and homogeneously over entire space. At every moment of time the medium is in a state close to thermal equilibrium. The same applies to smooth crossover. The situation is different in the case of the first order phase transition. Before the phase transition, the expectation value $\langle\phi\rangle_T$ equals zero, but as soon as the minimum of the effective potential at $\phi = \langle\phi\rangle_T \neq 0$ becomes deeper than the minimum at

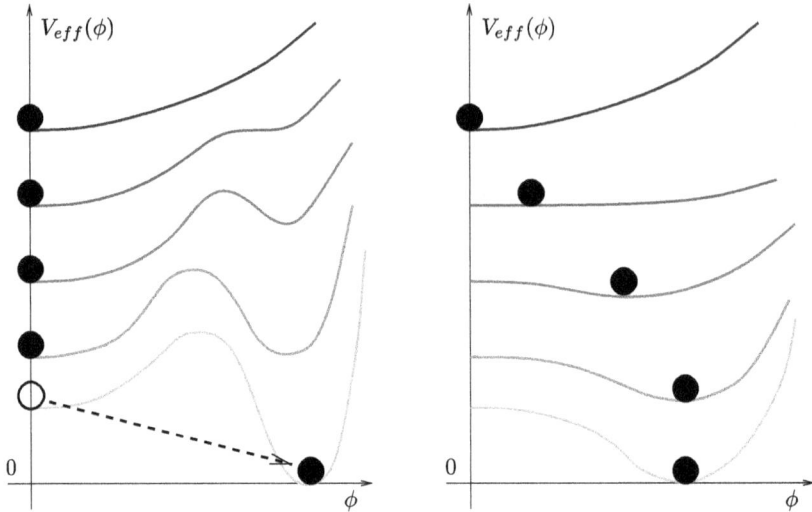

Fig. 10.3.   Shapes of the effective potential $V_{\text{eff}}(\phi)$ at various temperatures: upper darker curves correspond to higher temperatures. Left and right panels describe systems with 1st and 2nd order phase transition, respectively. Black circles show the expectation value $\langle\phi\rangle_T$.

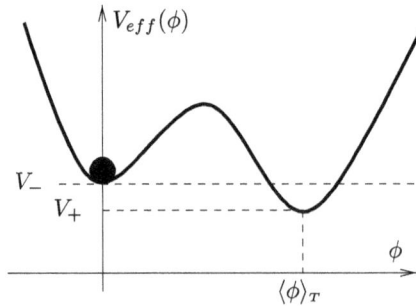

Fig. 10.4.   Shape of the effective potential of the system undergoing the first order phase transition.

$\phi = 0$, the ground state with $\langle\phi\rangle_T \neq 0$ becomes thermodynamically favorable, see Fig. 10.4. The transition from the state $\phi = 0$ to the state $\phi = \langle\phi\rangle_T$ cannot occur homogeneously over entire space: the field value ϕ in such a process would evolve homogeneously from $\phi = 0$ to $\phi = \langle\phi\rangle_T$, and free energy (10.1) in infinite volume would be infinitely large in the intermediate states as compared to its initial value at $\phi = 0$. The transition proceeds via spontaneous nucleation of bubbles of the new phase, their subsequent expansion and mergers, see Fig. 10.5. Nucleation of a bubble with $\phi = \langle\phi\rangle_T \neq 0$ in the medium with $\phi = 0$ is local in space and may occur due to thermal fluctuations.[4] Bubbles expand, their walls collide, the new phase percolates, and after this "boiling" the system eventually returns to spatially

---

[4]Sometimes the dominant process is quantum tunneling.

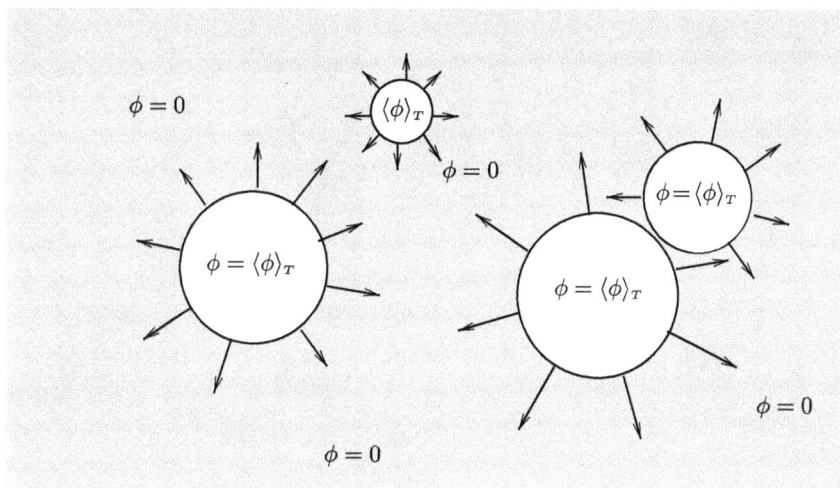

Fig. 10.5. Spontaneous nucleation and subsequent growth of bubbles of the new phase at the first order phase transition.

homogeneous state of thermal equilibrium, but with $\phi = \langle \phi \rangle_T \neq 0$; the released free energy converts into heat.

This boiling is a highly inequilibrium process. We have already noted that the most important stages of the evolution of the hot Universe are those when the cosmic plasma is out of thermal equilibrium. Therefore, the first order phase transitions are of particular interest for cosmology.

Let us estimate the nucleation probability of a bubble of the new phase at temperature T. Let $V_- = V_{\text{eff}}(T, \phi = 0)$ and $V_+ = V_{\text{eff}}(T, \phi = \langle \phi \rangle_T)$ be free energy densities of the old and new phase, respectively, $V_+ < V_-$, see Fig. 10.4. The free energy of a bubble of size R, relative to the free energy of medium with $\phi = 0$ without a bubble, contains the volume and surface terms. The former is due to the fact that the free energy density inside the bubble is smaller than the free energy density of the surrounding medium; it is negative and equal to

$$\frac{4}{3}\pi R^3 \left(V_+ - V_-\right).$$

The surface term exists because the field ϕ near the surface is inhomogeneous and differs from both zero and $\langle \phi \rangle_T$; the contributors here are effective potential $V_{\text{eff}}(\phi)$ and the gradient term in the free energy, the latter being now a functional $F[\phi(\mathbf{x})]$. The surface term is proportional to the bubble area and equal to $4\pi R^2 \cdot \mu$, where μ is the free energy per unit area (surface tension). Thus, the free energy of the bubble, relative to the free energy of the old phase, is (see Fig. 10.6)

$$F(R) = 4\pi R^2 \mu - \frac{4\pi}{3} R^3 \cdot \Delta V, \tag{10.2}$$

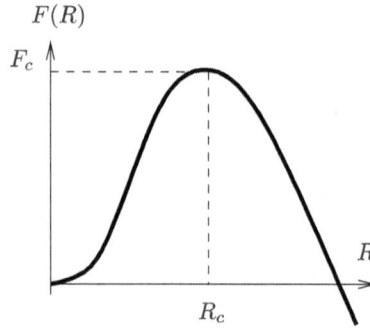

Fig. 10.6.   Free energy of a bubble of the new phase as a function of its radius.

where

$$\Delta V = V_- - V_+ > 0$$

is the difference between the free energy densities of the old and new phases (latent heat of the phase transition). From (10.2) we see that at sufficiently small sizes the free energy of the bubble decreases with *decrease* of R; this means that spontaneously nucleated small bubble collapses due to surface tension force, and the system returns to the initial homogeneous state with $\phi = 0$. On the contrary, for sufficiently large R the free energy decreases with *increase* of R, i.e., the bubble expands and the system converts to the new phase. The minimum size at which the bubble begins to expand is determined by the equation

$$\frac{\partial F}{\partial R} = 0.$$

Hence, it is equal to

$$R_c = \frac{2\mu}{\Delta V}. \tag{10.3}$$

The bubble of this size is called the critical bubble; its free energy is positive and equal to

$$F_c = 4\pi R_c^2 \mu - \frac{4\pi}{3} R_c^3 \cdot \Delta V = \frac{16\pi}{3} \frac{\mu^3}{(\Delta V)^2}. \tag{10.4}$$

Importantly, both the size of the critical bubble and its free energy are larger for smaller ΔV.

Spontaneous nucleation of bubbles of the new phase in hot medium occurs via thermal fluctuations, i.e., thermal jumps on top of the barrier shown in Fig. 10.6. The probability of such a jump per unit time per unit spatial volume is mainly determined by the Boltzmann factor $e^{-F_c/T}$:

$$\Gamma = AT^4 e^{-\frac{F_c}{T}} \tag{10.5}$$

(the Arrhenius formula), where the factor T^4 is introduced on dimensional grounds, and pre-exponential factor A does not depend very strongly on temperature and other parameters. The formula (10.5) is valid for $F_c \gg T$, i.e., when the probability of the bubble nucleation is small. This formula together with Eq. (10.4) imply immediately that for finite cooling rate the medium remains in supercooled state with $\phi = 0$ for some time, even though the new phase is already thermodynamically favorable. This occurs when $\Delta V = V_- - V_+$ is still so small that the bubble nucleation rate is smaller than the cooling rate. In the cosmological context, the nucleation of bubbles is efficient when the probability of the bubble nucleation per Hubble volume per Hubble time is roughly of order 1, i.e.,

$$AT^4 e^{-\frac{F_c}{T}} \sim H^4(T) = \left(\frac{T^2}{M_{Pl}^*}\right)^4 . \tag{10.6}$$

In specific models, this equation determines the extent to which the cosmic plasma is supercooled before the phase transition, and the latent heat ΔV released as a result of the phase transition. These properties are model-dependent. However, one can make a general statement concerning the picture of the first order phase transition in the Universe: the phase transition begins when *handful* of bubbles have nucleated in the entire Hubble volume. Their size at the time of nucleation is determined by microscopic physics, and it is much smaller than the Hubble scale $H^{-1}(T)$, while the distance between their centers is comparable to the Hubble size. Bubbles have time to expand by many orders of magnitude before they begin to percolate, and only a small number of new bubbles are produced during that time. We consider this picture at quantitative level in the end of this Section.

For instance, at $T \sim 100\,\mathrm{GeV}$ (the electroweak scale) the Hubble size is of order

$$H^{-1} = \frac{M_{Pl}^*}{T^2} \sim 1\,\mathrm{cm}.$$

The bubble size at the time of nucleation is, roughly speaking, of order T^{-1}, i.e., $R_c \sim 10^{-16}\,\mathrm{cm}$ (in fact, it is one to two orders of magnitude larger). Thus, the phase transition in the Universe occurs through the formation of several bubbles of subnuclear size in a cubic centimeter of cosmic plasma, their expansion up to macroscopic size and merger as a result of collisions of their walls.

Note that theoretically there is a possibility that the phase transition does not complete at all in the expanding Universe, despite the fact that the effective potential has the form shown in Fig. 10.4. The zero-temperature scalar potential may have a local minimum at $\phi = 0$, the false vacuum. It has positive energy density V_-, so that even in the absence of particles, the Universe filled with false vacuum expands at the Hubble rate

$$H_- = \sqrt{\frac{8\pi}{3} G V_-}.$$

If the rate of bubble nucleation per Hubble time per Hubble volume is small, $\Gamma \ll H_-^4$, the bubble walls do not collide, because the centers of neighboring bubbles move from each other with velocities exceeding the speed of light. (In other words, nucleated bubbles are outside the event horizons of each other; for definition of event horizon see Sec. 3.2.3.) Regions of false vacuum grow faster than regions of the new phase, and the phase transition does not complete.

We also note another theoretical possibility. Namely, because of the gravitational inter-actions, the false vacuum decay may not occur since bubbles of the true vacuum do not nucleate at all [174]. This takes place when the energy density of the true vacuum is neg-ative, and the gravitational effects are strong. The study of this effect is beyond the scope of this book.

Let us discuss in general terms how to calculate the surface tension of the bubble wall. Let us neglect the curvature of the wall (i.e., consider the bubble of large size R) and take the difference between the free energies of the old and the new phase ΔV small. In this case, the configuration of the field $\phi_W(r)$ inside the wall is the minimum of the free energy $F[\phi(r)]$ considered as a functional of the inhomogeneous field $\phi(r)$. At the one side of the wall, $r \ll R$ (i.e., inside the bubble), the field tends to $\phi = \phi_+$, while at the other side, $r \gg R$, it approaches $\phi = 0$. If R is sufficiently large, then the coordinate $(r - R)$ inside the bubble can be formally extended to $-\infty$, and we write the boundary conditions as

$$\phi_W(x) \to \langle\phi\rangle_T \quad \text{as} \quad (r - R) \to -\infty, \tag{10.7}$$

$$\phi_W(x) \to 0 \quad \text{as} \quad (r - R) \to +\infty. \tag{10.8}$$

We assume that the temperature corrections to the gradient term in the energy functional are small. (This assumption is indeed valid in weakly coupled theories.) Then the free energy (relative to the free energy of the old phase) can be written as a functional of the field $\phi(r)$,

$$F[\phi] = \int_0^\infty 4\pi r^2 dr \left[\frac{1}{2}\left(\frac{d\phi}{dr}\right)^2 + V_{\text{eff}}(\phi) - V_- \right]. \tag{10.9}$$

The wall thickness is small compared to R for large bubble, and a slowly varying factor $4\pi r^2$ can be treated as a constant inside the wall. Thus,

$$F[\phi] = 4\pi R^2 \int_{-\infty}^{+\infty} d\tilde{r} \left[\frac{1}{2}\left(\frac{d\phi}{d\tilde{r}}\right)^2 + V_{\text{eff}}(\phi) - V_- \right], \tag{10.10}$$

where

$$\tilde{r} = r - R,$$

and we formally extended the integration over this variable to $-\infty$ (cf. (10.7)). The field configuration obeys the Euler–Lagrange equation for the extremum of the functional (10.10),

$$\frac{d^2\phi}{d\tilde{r}^2} = \frac{\partial V_{\text{eff}}(\phi)}{\partial \phi}. \tag{10.11}$$

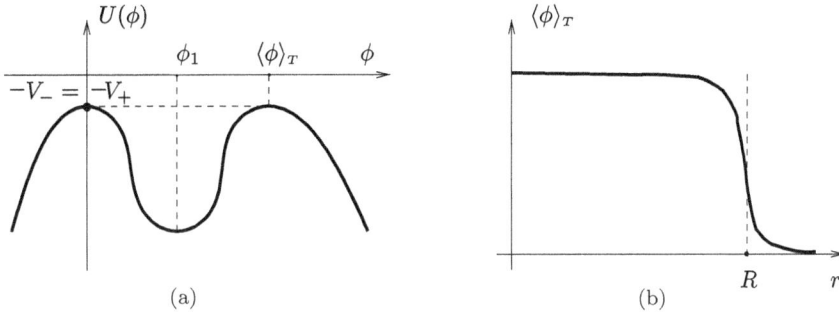

Fig. 10.7. (a) The potential of the analogous classical mechanics problem; (b) The field configuration for the bubble of the new phase.

This equation is formally identical to the equation of one-dimensional classical mechanics of a particle in the potential

$$U(\phi) = -V_{eff}(\phi),$$

where $\tilde{r}$ plays the role of time. We can now neglect ΔV and set $U(\langle\phi\rangle_T) = U(0)$. Then the potential $U(\phi)$ has two equally high maxima; see Fig. 10.7(a). The solution $\phi_W(r)$ to Eq. (10.11) describes the roll down of a particle from right hump, in accordance with (10.7), and subsequent roll up to left hump. The whole process occurs in infinite time; see (10.8). Using the analogy with the classical particle, it is straightforward to find the solution in quadratures to Eq. (10.11) with boundary conditions (10.7), (10.8),

$$\int_{\phi_1}^{\phi_W} \frac{d\phi}{\sqrt{2\left(V_{eff} - V_-\right)}} = -(R - r), \qquad (10.12)$$

where the limit of integration is chosen in such a way that at $r = R$ the field $\phi(r)$ takes an intermediate value ϕ_1 between $\phi = 0$ and $\phi = \langle\phi\rangle_T$. The configuration $\phi_W(r)$ is shown in Fig. 10.7(b). Note that in the one-dimensional scalar field theory with degenerate minima of the scalar potential, this solution is called "kink". In view of (10.12), the free energy (10.10) of the wall reads

$$F_W = 4\pi R^2 \mu,$$

where

$$\mu = \int_0^{\langle\phi\rangle_T} \sqrt{2\left[V_{eff}(\phi) - V_-\right]} d\phi. \qquad (10.13)$$

Note that the surface tension μ is finite in the limit $\Delta V \to 0$.

The expression (10.2) for the free energy of the bubble, as well as the analysis of the field behavior near the wall are valid when the wall thickness is small compared to the bubble size R, i.e., thin-wall approximation is applicable. According to (10.3), it really works if the difference of free energies ΔV is a small parameter.

Otherwise, the configuration of the critical bubble should be obtained by finding
the extremum (saddle point configuration) of the free energy functional (10.9) with
the only boundary condition $\phi(r \to \infty) = 0$. Details can be found in [175].

Problem 10.1. *Check the formulas* (10.12), (10.13).

Problem 10.2. *Let the effective potential be*

$$V_{e\!f\!f}(\phi) = \frac{\lambda}{4}\phi^2 (\phi - v)^2 - \epsilon\phi^2,$$

*where λ, v and ϵ are positive parameters. Find the conditions at which the thin-wall
approximation is valid. Find the surface tension and wall thickness in the thin-wall
approximation, as well as the size of the critical bubble R_c; estimate the probability of
nucleation of a bubble of the new phase inside the phase with $\phi = 0$ in the thin-wall
approximation at temperature T.*

Let us make a brief comment on the false vacuum decay at *zero* temperature. We have in
mind scalar field models in which the scalar potential (at zero temperature) has a local
minimum (e.g., at $\phi = 0$), i.e., it has the form shown in Fig. 10.8. The state in which the
field expectation value is spatially homogeneous and equal to zero is metastable; it is the
false vacuum. False vacuum decay also occurs via spontaneous formation of bubbles of
the new phase, but in contrast to the medium at finite temperature, the bubble emerges not
due to thermal fluctuations, but as a result of tunneling process [176–178]. The description
of tunneling in the semiclassical approximation is given, for example, in [175]. In weakly
coupled theories, the probability of the bubble nucleation is exponentially small,

$$\Gamma \propto e^{-\frac{\mathrm{const}}{\alpha}},$$

where α is the small coupling constant. Finally, in a certain range of temperatures the
bubble formation is dominated by a combination of thermal fluctuation and tunneling.

Let us discuss the percolation process in some detail. Let us assume that the bubble nucleation occurs in the thin wall approximation regime; this is indeed the case
in the weak coupling limit. We will soon see that in weakly coupled theories, the

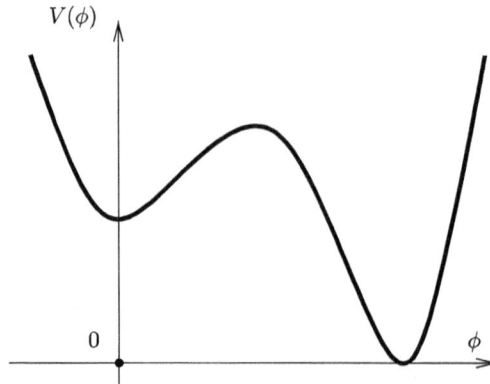

Fig. 10.8.　Shape of scalar potential with two non-degenerate minima.

phase transition occurs when the temperature is only slightly below the temperature T_c at which the depths of the two minima are equal. (The latter temperature is denoted by T_{c1} in Sec. 10.2.) Then the difference between the free energy densities of the old and new phases is

$$\Delta V = \left[-\frac{\partial \Delta V}{\partial T}(T_c) \right] \cdot (T_c - T),$$

and the nucleation probability per unit time per unit volume is parametrized as follows [185]

$$\Gamma = A T_c^4 e^{-\kappa \frac{T_c^2}{(T_c - T)^2}}, \tag{10.14}$$

where

$$\kappa = \frac{16\pi}{3} \frac{\mu^3}{T_c^3 \left[\frac{\partial \Delta V}{\partial T}(T_c) \right]^2}.$$

Our analysis is valid if κ is not large, $\kappa \lesssim 1$. This is the case for one-loop effective potential in weakly coupled theories; see Sec. 10.2.

We will see that the duration of the phase transition is shorter than the Hubble time. Therefore, we neglect the expansion of space, and treat space–time as Minkowskian. On the other hand, it is important that *temperature* decreases in time.

Let a bubble be nucleated in the time interval $(t', t' + dt')$. By the time t its size is $R(t, t') = u(t - t')$, where u is the wall velocity (soon after nucleation, the bubble expands in stationary regime $u = \text{const}$). Since the bubble expands in hot and dense matter, this velocity is somewhat lower than the speed of light; typically, $u \sim 0.1$. Let $x(t)$ be the fraction of volume occupied by the old phase at time t. We neglect the wall collisions and obtain the following equation:

$$x(t) = 1 - \int_{t_c}^{t} dt' \Gamma(t') \frac{4\pi}{3} R^3(t, t') x(t'), \tag{10.15}$$

where time t_c corresponds to temperature T_c. Percolation occurs at time $t = t_p$ when the right-hand side of Eq. (10.15) becomes of order 1 (but somewhat smaller than 1). To estimate this integral, we set $x(t') = 1$ in the integrand and write the pecolation condition:

$$\int_{t_c}^{t_p} dt \Gamma(t) \cdot \frac{4\pi}{3} u^3 (t_p - t)^3 \sim 1. \tag{10.16}$$

We now recall that $t = (2H)^{-1} = M_{Pl}^*/(2T^2)$. The integral in (10.16) is saturated, as we will see soon, at T close to T_p, and both of these temperatures are close to T_c. Therefore

$$t_p - t = \frac{M_{Pl}^*}{T_c^2} \frac{T - T_p}{T_c}.$$

We rewrite the integral (10.16) as an integral over variable

$$\tau = \frac{T_c - T}{T_c},$$

and obtain, by making use of (10.14),

$$\frac{4\pi}{3} u^3 A \left(\frac{M_{Pl}^*}{T_c}\right)^4 \int_0^{\tau_p} d\tau \, (\tau_p - \tau)^3 e^{-\frac{\kappa}{\tau^2}} \sim 1.$$

We see that percolation occurs in the regime $e^{-\frac{\kappa}{\tau^2}} \ll 1$, i.e., when the nucleation rate is still small. This precisely means that $(T_c - T_p) \ll T_c$ (recall that we assume that κ is not large). The integral is saturated at $(\tau_p - \tau) \ll \tau_p$; we write $\tau = \tau_p(1 - \xi)$, where $\xi \ll 1$, and obtain

$$\frac{4\pi}{3} u^3 A \left(\frac{M_{Pl}^*}{T_c}\right)^4 \tau_p^4 e^{-\frac{\kappa}{\tau_p^2}} \int_0^\infty d\xi \, \xi^3 e^{-2\frac{\kappa}{\tau_p^2}\xi}$$

$$= \frac{4\pi}{3} u^3 A \left(\frac{M_{Pl}^*}{T_c}\right)^4 \tau_p^4 e^{-\frac{\kappa}{\tau_p^2}} \cdot 6 \left(\frac{\tau_p^2}{2\kappa}\right)^4 \sim 1. \tag{10.17}$$

We see that, indeed, the growth of bubbles and percolation occur in a narrow temperature interval, which is determined by the relation

$$\Delta\xi = \frac{\Delta\tau}{\tau_p} \sim \frac{\tau_p^2}{\kappa}.$$

The formula (10.17) gives for the percolation temperature

$$\tau_p \equiv \frac{T_c - T_p}{T_c} = \sqrt{\frac{\kappa}{L}},$$

where the logarithmically large parameter L obeys

$$L^6 e^L = u^3 A \kappa^2 \left(\frac{M_{Pl}^*}{T_c}\right)^4. \tag{10.18}$$

We solve this equation in the logarithmic approximation.[5] We obtain finally (leading logarithm and first correction)

$$L = \log\left[u^3 A \kappa^2 \left(\frac{M_{Pl}^*}{T_c}\right)^4\right] - 6\log 6 - 6\log\left\{\frac{1}{6}\log\left[u^3 A \kappa^2 \left(\frac{M_{Pl}^*}{T_c}\right)^4\right] - \log 6\right\}.$$

The relevant temperature interval is determined by the relation $\Delta\tau/\tau_p \sim L^{-1}$.

---

[5] It is convenient to set $L = 6x$, rewrite Eq. (10.18) as

$$6x e^x = \left[u^3 A \kappa^2 \left(\frac{M_{Pl}^*}{T_c}\right)^4\right]^{1/6}$$

and make use of the formulas (6.18)–(6.20) and the result of Problem 6.2 in Chapter 6.

Finally, let us estimate the bubble size at percolation. The volume of each bubble at percolation is estimated as $n^{-1}(t_p)$ where $n(t)$ is the number density of bubbles. (All bubbles occupy, roughly speaking, half of space.) Proceeding as before, we write

$$n(t_p) \sim \int_{t_c}^{t_p} dt \ \Gamma(t).$$

This integral is calculated in the same way as the integral in (10.16), and we obtain

$$n(t_p) \sim AT_c^2 M_{Pl}^* e^{-\frac{\kappa}{\tau_p^2}} \frac{T_p^3}{2\kappa}.$$

We make use of (10.18) to express $\exp\left(-\frac{\kappa}{\tau_p^2}\right) \equiv \exp(-L)$, and get

$$n(t_p) \sim \frac{T_c^6}{M_{Pl}^{*3}} \frac{L^{9/2}}{u^3 \kappa^{3/2}}, \tag{10.19}$$

so that the typical bubble size at percolation is

$$R(t_p) \sim n^{-1/3}(t_p) \sim \frac{M_{Pl}^*}{T_c^2} \frac{u\kappa^{1/2}}{L^{3/2}} = H^{-1}(T_c) \frac{u\kappa^{1/2}}{L^{3/2}}.$$

This is smaller than the Hubble size by a few orders of magnitude, since the velocity u is smaller than 1, and the logarithmic factor is large (as an example, $L^{3/2} \sim 10^3$ at $T_c \sim 100$ GeV).

10.2. Effective Potential in One-Loop Approximation

In accordance with (10.1), the effective potential is the free energy density of the plasma in a state where the average Higgs field takes one and the same value ϕ everywhere in space. The free energy F of the system is related to its energy E and entropy S by the thermodynamical relation $F = E - TS$, so that for the free energy density we have

$$f = \rho - Ts,$$

where, as usual, ρ and s are energy density and entropy density, respectively. We know from Sec. 5.2 that entropy density is expressed in terms of energy density and pressure,

$$s = \frac{\rho + p}{T},$$

hence the free energy density is[6]

$$f = -p.$$

---

[6]The fact that the medium chooses the phase with the lowest free energy has a simple physical interpretation: in this phase pressure is at maximum, and sufficiently large region of this phase, spontaneously created inside the phase of lower pressure, will expand, "pushing away" the phase of lower p.

So, to calculate the effective potential one needs to find the pressure of the system under the constraint that the average Higgs field is equal to ϕ in the whole space.

In theories with scalar fields, particle masses depend, generally speaking, on the expectation values of these scalar fields. An example is the Standard Model where all particles (except for the Higgs boson) acquire masses due to the Englert–Brout–Higgs mechanism. If the field value equals ϕ, then quarks and charged leptons, denoted collectively by f, and $W^{\pm}$- and Z-bosons have masses

$$m_f(\phi) = y_f\phi, \quad M_W(\phi) = \frac{g}{\sqrt{2}}\phi, \quad M_z(\phi) = \frac{\sqrt{g^2 + g'^2}}{\sqrt{2}}\phi, \tag{10.20}$$

where y_f are Yukawa couplings, and g and g' are gauge couplings (see Appendix B for notations and details). The vacuum value is $\phi = v/\sqrt{2}$, and we return to the zero-temperature formulas for the particle masses (see Appendix B), namely,

$$m_f = \frac{y_f}{\sqrt{2}}v, \quad M_W = \frac{g}{2}v, \quad M_z = \frac{\sqrt{g^2 + g'^2}}{2}v. \tag{10.21}$$

In what follows we consider models of precisely this type, and for the time being we just assume that particle masses depend in some way on the average value of a field ϕ, which we assume to be the only relevant scalar field at hand. Our nomenclature is that $\phi \neq 0$ and $\phi = 0$ correspond to broken and unbroken symmetry; what is this symmetry depends on a concrete model.

Let us study theories with small couplings. In this Section we neglect the interactions between particles in the cosmic plasma, i.e., we consider the free energy of an ideal gas of elementary particles. The free energy of the gas depends non-trivially on the average value ϕ, since the latter determines particle masses, and, consequently, particle contributions to pressure. For reasons given in Appendix D, the ideal gas approximation is called one-loop approximation in this context.

In this approximation, the pressure is a sum of the contributions of the homogeneous field ϕ itself and of each type of particles and antiparticles,

$$f = V_{\text{eff}}(T, \phi) = V(\phi) + \sum_i f_i, \tag{10.22}$$

where $V(\phi)$ is scalar potential entering the scalar field action[7]

$$S_\phi = \int [\partial_\mu\phi\partial^\mu\phi - V(\phi)]\, d^4x. \tag{10.23}$$

The first term in (10.22) arises from the fact that the energy–momentum tensor for time-independent and spatially homogeneous scalar field is

$$T_{\mu\nu}(\phi) = g_{\mu\nu} \cdot V(\phi),$$

---

[7] In the Standard Model of particle physics, the Englert–Brout–Higgs field is a complex doublet with kinetic term in Lagrangian given in (B.8). This explains the choice of the coefficient in front of the kinetic term in (10.23). The relation between the field ϕ considered here and the Standard Model Higgs boson is $\phi(x) = \frac{v+h(x)}{\sqrt{2}}$.

i.e., the homogeneous scalar field gives a contribution to the pressure equal to $p(\phi) = T_{11} = T_{22} = T_{33} = -V(\phi)$. Of course, this is a reformulation of the fact that the free energy in vacuo coincides with the energy and its density equals $V(\phi)$ for homogeneous scalar field. In what follows, we assume that the scalar potential is given by the standard expression

$$V(\phi) = \lambda \left(\phi^2 - \frac{v^2}{2} \right)^2,$$

where v is the vacuum expectation value at zero temperature, λ is the Higgs self-coupling, and $\lambda \ll 1$.

The second term in (10.22) is the medium contribution in the ideal gas approximation,

$$f_i = -p_i[T, m_i(\phi)],$$

where $p_i[T, m_i(\phi)]$ is the contribution to pressure coming from particles and antiparticles of ith type, whose mass equals m_i and depends on ϕ. According to Sec. 5.1, we have

$$f_i = -p_i = -\frac{g_i}{6\pi^2} \int_0^\infty \frac{k^4 dk}{\sqrt{k^2 + m_i^2}} \frac{1}{e^{\frac{\sqrt{k^2+m_i^2}}{T}} \mp 1}, \qquad (10.24)$$

where g_i is the number of spin states, upper (lower) sign corresponds to bosons (fermions). Contributions of heavy particles with $m_i \gg T$ are exponentially small, therefore the interesting case is $m_i \lesssim T$.

Integral (10.24) cannot be evaluated analytically. We analyze it for the particular case of high temperature, $T \gg m$, and make use of the expansion in m/T (subscript i will be omitted wherever possible). This approach is called high-temperature expansion. In dimensionless variables

$$x = k/T \quad \text{and} \quad z_i = m_i/T,$$

the expression (10.24) takes the form

$$f_i = -\frac{g_i}{6\pi^2} T^4 \cdot I(z_i)_\mp, \qquad I(z)_\mp = \int_0^\infty \frac{x^4 dx}{\sqrt{x^2 + z^2}} \frac{1}{e^{\sqrt{x^2+z^2}} \mp 1}. \qquad (10.25)$$

We are interested in the behavior of these integrals at small z.

To the zeroth order in z, the contributions f_i correspond to pressures of free gases of massless particles (see Sec. 5.1); they do not depend on ϕ, and will be omitted. The integrand in Eq. (10.25) is a function of z^2, so one might expect that $I(z)$ is a series in z^2. The first term in this series is

$$I(z) = z^2 \left(\frac{dI}{dz^2} \right)_{z^2=0} = -\frac{z^2}{2} \left(\int_0^\infty \frac{x dx}{e^x \mp 1} + \int_0^\infty \frac{x^2 e^x dx}{(e^x \mp 1)^2} \right). \qquad (10.26)$$

The integrals here are finite, so the first non-trivial term in the high-temperature expansion is indeed quadratic in z. Performing the integration with the use of formulas given in the end of Sec. 5.1 (in the process, it is convenient to integrate the second term in (10.26) by parts), we obtain in this order

$$V_{\textit{eff}}(\phi) = \lambda \left(\phi^2 - \frac{v^2}{2} \right)^2 + \frac{T^2}{24} \left[\sum_{\text{bosons}} g_i m_i^2(\phi) + \frac{1}{2} \sum_{\text{fermions}} g_i m_i^2(\phi) \right]. \qquad (10.27)$$

This expression has a particularly simple form in models where particles acquire masses due to the scalar field condensate,

$$m_i(\phi) = h_i \phi, \qquad (10.28)$$

where h_i are coupling constants. This is precisely what happens in the Standard Model. The only exception is the Higgs boson itself. We will soon see that its contribution into the effective potential in our approximation also has the form (10.27), (10.28) with

$$g_h h_h^2 = 6\lambda. \qquad (10.29)$$

Let us first pretend that the effective potential contains the terms (10.27) only, and the masses are given by (10.28). Then the effective potential is

$$V_{\textit{eff}}(\phi) = \left(-\lambda v^2 + \frac{\alpha}{24} T^2 \right) \phi^2 + \lambda \phi^4 \qquad (10.30)$$

(ϕ-independent terms are omitted), where

$$\alpha = \sum_{\text{bosons}} g_i h_i^2 + \frac{1}{2} \sum_{\text{fermions}} g_i h_i^2 \qquad (10.31)$$

is a positive quantity. The formula (10.30) shows that when the average scalar field equals ϕ, the effective Higgs boson mass squared is[8] (in the unitary gauge)

$$m_h^{\textit{eff}\,2}(\phi) = 6\lambda \phi^2 - \lambda v^2 + \frac{\alpha}{24} T^2. \qquad (10.32)$$

Together with Eq. (10.27) this leads to the relation (10.29).

At low temperatures, the expression (10.30) has a minimum at $\phi \neq 0$ (symmetry is broken), while at high temperatures the only minimum is $\phi = 0$, corresponding to the restored symmetry. Minimum at $\phi = 0$ disappears and turns into maximum

---

[8]We do not discuss here an important and difficult problem of gauge-dependence of the effective potential. We note only, that even though the shape of the effective potential, including the positions of its extrema, is gauge-dependent, the values of the effective potential at extrema do not depend on gauge choice [317]. Similar property holds for the effective action, which generalizes the effective potential to coordinate-dependent background field $\phi(x)$; the latter property ensures, in particular, gauge-independence of the rate of false vacuum decay. The issue of gauge-dependence of the effective potential as applied to concrete calculations is considered, e.g., in Ref. [186].

when the first term in (10.30) flips sign, which happens at temperature (the notation will be clarified later)

$$T_{c2} = 2v \left(\frac{6\lambda}{\alpha} \right)^{1/2}. \tag{10.33}$$

In what follows, when discussing weakly coupled theories in general terms, we assume that the relevant couplings h_i are small, $h \ll 1$, and the following order-of-magnitude relation holds,

$$\lambda \sim h^2. \tag{10.34}$$

The latter ensures that the Higgs boson mass $m_h \sim \sqrt{\lambda} v$ is of the same order as masses of all other particles contributing noticeably to the effective potential. Formally, the relation (10.34) means that the ratio λ/h^2 is finite in the limit $h \to 0$. This does not prevent us to have $\lambda/h^2 \ll 1$, which is necessary for the applicability of perturbation theory for studying phase transition; see Sec. 10.3.

With this prescription, the estimate for the critical temperature reads

$$T_{c2} \sim v.$$

This follows from (10.31) in the case of not too large number of particle species.

In the Standard Model, the main contributions to α come from the heaviest particles, $W^{\pm}$- and Z-bosons, t-quark and Higgs boson. Equations (10.28), (10.20) and (10.21) yield, in terms of zero-temperature masses,

$$\alpha = \frac{1}{v^2} \left(12M_W^2 + 6M_Z^2 + 12m_t^2 + 3m_h^2 \right). \tag{10.35}$$

Here we used the fact that W^+- and W^--bosons together have six polarizations, Z-boson has three polarizations, and t-quark with its antiparticle have four; also, t-quark has three color states. Recalling that the Higgs boson mass is

$$m_h = \sqrt{2\lambda} v,$$

we arrive at the following one-loop expression for the critical temperature T_{c2} in the Standard Model,

$$T_{c2} = \left(\frac{4m_h^2}{4M_W^2 + 2M_Z^2 + 4m_t^2 + m_h^2} \right)^{1/2} \cdot v = 146 \text{ GeV}. \tag{10.36}$$

(Be reminded that $M_W = 80.4 \text{ GeV}$, $M_Z = 91.2 \text{ GeV}$, $m_t \approx 173 \text{ GeV}$, $m_h = 125 \text{ GeV}$, $v = 246 \text{ GeV}$; see Appendix B). As we already mentioned, perturbation theory is not adequate for describing the electroweak transition, so the result (10.36) is to be considered as an approximate estimate of the relevant temperature. Numerical calculations give the estimate $T_c \approx 160 \text{ GeV}$ for the transition temperature [199].

If the high-temperature expansion of the integrals (10.24) were really a series in $z^2 \equiv m^2(\phi)/T^2$, one would conclude that we are dealing with the second order phase transition: corrections of the fourth and higher orders in ϕ are small compared to terms written in (10.30) (see below), and the position of the minimum of the expression (10.30) smoothly moves from $\phi = 0$ towards large ϕ as temperature decreases from T_{c2} to zero. In other words, the behavior of the expression (10.30) corresponds to the right plot in Fig. 10.3. However, integrals (10.24) are *not analytic* in z^2, and the one-loop effective potential actually corresponds to the first order phase transition. The lack of analyticity can be seen from the behavior of contributions to the integrals (10.25) coming from the low-momentum region $k \ll T$, i.e., $x \ll 1$ (infrared region). For small z and x, the expansion of the exponential terms in the integrand gives

$$I_-^{(IR)} = \int_0^\Lambda \frac{x^4 dx}{x^2 + z^2}, \qquad \text{bosons}$$

$$I_+^{(IR)} = \frac{1}{2} \int_0^\Lambda \frac{x^4 dx}{\sqrt{x^2 + z^2}}, \qquad \text{fermions,}$$

(10.37)

where $\Lambda \ll 1$ is a fictitious parameter separating the infrared region. Formal expansion of the integrands in these formulas in z^2 would result in the order z^4 contributions of the form

$$z^4 \int_0^\Lambda \frac{dx}{x^2}, \qquad \text{bosons}$$

$$z^4 \int_0^\Lambda \frac{dx}{x}, \qquad \text{fermions.}$$

The first of them linearly diverges at the lower limit of integration, while the second one diverges logarithmically. We can therefore expect that besides the above terms of order z^2, bosons and fermions give contributions of order z^3, and $z^4 \log z$, respectively. Contributions of the latter type give the terms in the effective potential of the form

$$m_i^4(\phi) \log \frac{\phi}{T} = h_i^4 \phi^4 \log \frac{\phi}{T}.$$

(10.38)

These are of little interest from the viewpoint of the phase transition, because at $\lambda \gg h_i^4$ (which holds for not too small Higgs boson mass), they are small compared to $\lambda \phi^4$ coming from the scalar potential $V(\phi)$. (Only contributions due to t-quark are important in determining certain parameters of the phase transition.) On the contrary, terms of order z^3 are very important: it is due to these terms that the transition (in the one-loop approximation) is of the first order.

To calculate the term of order z^3 in the boson integral I_-, we divide this integral into two parts using the fictitious parameter Λ,

$$I_- = \int_\Lambda^\infty \frac{x^4 dx}{\sqrt{x^2 + z^2}} \frac{1}{e^{\sqrt{x^2+z^2}} - 1} + I_-^{(IR)}.$$

The first term is analytic in z^2, while for the second term the approximation (10.37) is sufficient, i.e.,

$$I_-^{(IR)} = \int_0^\Lambda \left(x^2 - z^2 \right) dx + z^4 \int_0^\Lambda \frac{dx}{x^2 + z^2}.$$

The first term here is again analytic in z^2, while the second term gives the contribution of order z^3 we are after,

$$I_-^{(IR)} \longrightarrow \frac{\pi}{2} z^3 + \mathcal{O}\left(\frac{z^4}{\Lambda} \right).$$

As a result, the effective potential in the one-loop approximation reads

$$V_{\text{eff}}(\phi) = \lambda \left(\phi^2 - \frac{v^2}{2} \right)^2 + \frac{T^2}{24} \left(\sum_{\text{bosons}} g_i m_i^2(\phi) + \frac{1}{2} \sum_{\text{fermions}} g_i m_i^2(\phi) \right)$$

$$- \frac{T}{12\pi} \sum_{\text{bosons}} g_i m_i^3(\phi) + \mathcal{O}\left(m_i^4(\phi) \log \frac{m_i(\phi)}{T} \right).$$

In models where particle masses are related to the expectation value of the scalar field by (10.28), this expression is rewritten as

$$V_{\text{eff}}(\phi) = \frac{\alpha}{24} \left(T^2 - T_{c2}^2 \right) \phi^2 - \gamma T \phi^3 + \lambda \phi^4, \tag{10.39}$$

where the parameter γ is positive and equal to[9]

$$\gamma = \frac{1}{12\pi} \sum_{\text{bosons}} g_i |h_i|^3 = \frac{\sqrt{2}}{6\pi} \sum_{\text{bosons}} g_i \left(\frac{m_i}{v} \right)^3. \tag{10.40}$$

Here we use the notations introduced in (10.31) and (10.33).

Problem 10.3. *Calculate the terms of order $\phi^4 \log \frac{\phi}{T}$ in the high-temperature expansion of the effective potential, using the relation (10.28). Show that at $\lambda \sim h_i^2$ and $h_i \ll 1$ these terms are small compared to those written in (10.39) in the entire interesting range of ϕ, namely, $0 < \phi \lesssim v$.*

The behavior of effective potential (10.39) corresponds to the left plot in Fig. 10.3, i.e., to the first order phase transition. Extrema of the effective potential

---

[9]Effective mass squared of the Higgs boson $m_h^{\text{eff}\,2} = \alpha \cdot (T^2 - T_{c2}^2)/24$ is small near the phase transition, so the Higgs boson itself gives small contribution to the cubic part of the effective potential, i.e., to the parameter γ. Another qualification is that the time components of gauge fields acquire the Debye mass $m_D \sim gT \gg g\Phi_c$, see (10.46), so their contributions to γ are also absent. In other words, one should set $g_i = 2$ in Eq. (10.40) for each of the vector bosons (rather than $g_i = 3$ as in Eq. (10.31)).

are determined by the equation

$$\frac{\partial V_{eff}}{\partial \phi} = \frac{\alpha}{12} \left(T^2 - T_{c2}^2 \right) \phi - 3\gamma T \phi^2 + 4\lambda \phi^3 = 0. \tag{10.41}$$

At temperature T_{c0} such that

$$9\gamma^2 T_{c0}^2 = \frac{4\alpha\lambda}{3} \left(T_{c0}^2 - T_{c2}^2 \right),$$

the effective potential acquires two extrema at $\phi \neq 0$: minimum and maximum. This temperature exceeds T_{c2} only slightly: for $\lambda \sim h^2$ we have $\gamma \sim h^3$, $\alpha\lambda \sim h^4$, and thus

$$\frac{T_{c0}^2 - T_{c2}^2}{T_{c2}^2} = \frac{27\gamma^2}{4\alpha\lambda} \sim h^2. \tag{10.42}$$

Due to the small difference of T and T_{c2} in the interesting temperature range, one can replace T by T_{c2} in the second term in (10.41). It is important that the second minimum of the effective potential (if $\phi = 0$ is treated as the first minimum) appears at nonzero $\phi = \Phi_c(T_{c0})$,

$$\Phi_{c0} = \Phi_c(T_{c0}) = \frac{3\gamma}{8\lambda} T_{c0}. \tag{10.43}$$

As temperature decreases, the second minimum becomes deeper, and the values of the effective potential at this minimum and at the minimum $\phi = 0$ become equal at temperature T_{c1}, such that both Eq. (10.41) and equation[10] $V_{eff} = 0$ are satisfied. The solution to the latter system of equations gives the first critical temperature T_{c1},

$$\frac{T_{c1}^2 - T_{c2}^2}{T_{c2}^2} = \frac{6\gamma^2}{\alpha\lambda},$$

and the position of the second minimum at this temperature,

$$\Phi_{c1} = \Phi_c(T_{c1}) = \frac{\gamma}{2\lambda} T_{c1}. \tag{10.44}$$

In view of (10.40), this value is much smaller than the critical temperature, $\Phi_c(T_{c1}) \sim hT_{c1}$ in the weak coupling limit $h \ll 1$. Once temperature decreases down to the second critical temperature T_{c2}, the minimum of the effective potential at $\phi = 0$ disappears and turns into maximum. At that instant the second minimum is at

$$\Phi_{c2} = \Phi_c(T_{c2}) = \frac{3\gamma}{4\lambda} T_{c2}. \tag{10.45}$$

Thus, the evolution of the one-loop effective potential shown in the left plot in Fig. 10.3 takes place in weakly coupled theories in a narrow temperature interval near the critical temperature (10.33), $T_{c2} \leq T \leq T_{c0}$, where T_{c0} is defined by (10.42).

---

[10]Recall that we dropped ϕ-independent terms in the effective potential, i.e., we set $V_{eff}(\phi = 0) = 0$.

Immediately after the phase transition, the expectation value of the Higgs field is much smaller than its vacuum value,

$$\Phi_c \sim \frac{\gamma}{\lambda} T_{c2} \sim h T_{c2} \sim h v \ll v. \tag{10.46}$$

Note that the high-temperature expansion is justified in weakly coupled theories, since

$$m_i(\Phi) = h_i \Phi \sim h^2 T_{c2} \ll T_{c2}.$$

Another point concerns the latent heat of the phase transition. At $T = T_{c2}$, the value of the effective potential at the minimum (10.45), relative to its value at $\phi = 0$, is

$$V_{e\!f\!f}(T_{c2}, \Phi_c(T_{c2})) = -\frac{27}{256} \frac{\gamma^4}{\lambda^3} T_{c2}^4.$$

This value (with opposite sign) gives the maximum latent heat of the transition, which is roughly equal to

$$-V_{e\!f\!f} \sim h^6 T_{c2}^4 \ll T_{c2}^4. \tag{10.47}$$

Thus, the energy released during the phase transition is small compared to the energy of particles in the plasma, whose density is of order T_c^4. The cosmic plasma is heated by the phase transition only slightly.

Let us estimate the surface tension of a bubble of a new phase. The effective potential at $T = T_{c1}$ has the form

$$V_{e\!f\!f} = \lambda \phi^2 \left(\phi - \frac{\gamma}{2\lambda} T_{c1} \right)^2.$$

The expression (10.13) has to be modified, since the kinetic term is non-canonical in (10.23), so we get

$$\mu = 2 \int_0^{\Phi_{c1}} \sqrt{V_{e\!f\!f}} \, d\phi = \frac{\sqrt{\lambda}}{24} \frac{\gamma^3}{\lambda^3} T_{c1}^3.$$

The difference between free energy densities of old and new phases obeys

$$\frac{d\Delta V}{dT}(T_{c1}) = \frac{\partial V(\Phi_c)}{\partial \Phi_c} \frac{\partial \Phi_c}{\partial T} + \frac{\partial V(\Phi_c)}{\partial T} = \frac{\partial V(\Phi_c)}{\partial T} = \frac{\alpha \gamma^2}{48 \lambda^2} \left(1 - \frac{6\gamma^2}{\alpha \lambda} \right) T_{c1}^3.$$

This gives the following expression for the parameter κ in (10.14):

$$\kappa = \frac{8\pi}{9} \frac{\gamma^5}{\lambda^{7/2} \alpha^2 \left(1 - \frac{6\gamma^2}{\alpha \lambda} \right)^2}.$$

In the case we consider, we have $\alpha \sim h^2$, $\lambda \sim h^2$, $\gamma \sim h^3$, hence $\kappa \sim h^4$, i.e., the parameter κ is small in the weak coupling limit. We have used this fact in Section 10.1.

Problem 10.4. *Estimate critical temperatures, field values at minima of the effective potential, latent heat and bubble surface tension in weakly coupled theory whose gauge and Yukawa couplings are of order h, while the scalar self-coupling is of the third order, $\lambda \sim h^3$.*

High-temperature expansion of the integrals (10.24) obviously does work at small values of ϕ. Therefore, the result that within the one-loop approximation the phase transition is of the first order, is justified from this point of view. Of course, the use of the high-temperature expansion is not necessary for one-loop calculations. Integrals (10.24) are easily computed numerically, and thus the exact one-loop effective potential is straightforwardly obtained. The corresponding graphs for the theory coinciding with the Standard Model but with wrong Higgs boson masses are shown in Figs. 10.9 and 10.10. It is seen that at $\phi \lesssim T$, results for the effective potential obtained by using the high-temperature expansion are in reasonable agreement with the exact one-loop calculation. At the same time, the results for a number of characteristics of the phase transition agree only qualitatively, within a factor of 3 or so (see Fig. 10.11). For analytical estimates of these characteristics, the omitted higher order terms in m/T are important. Indeed, the ratios $\Phi_c(T_{c1})/T_{c1}$ and $\Phi_c(T_{c2})/T_{c2}$ are inversely proportional to the coefficient in front of ϕ^4 in the effective potential (see formulas (10.44), (10.45)), and this coefficient gains a significant contribution when the omitted higher-order terms in (10.38) are taken into account.

Problem 10.5. *Making use of the high-temperature expansion, calculate the values of $\Phi_c(T_{c1})/T_{c1}$ and $\Phi_c(T_{c2})/T_{c2}$ taking into account the contributions of the form (10.38). Check that the results are in better agreement with exact numerical results.*

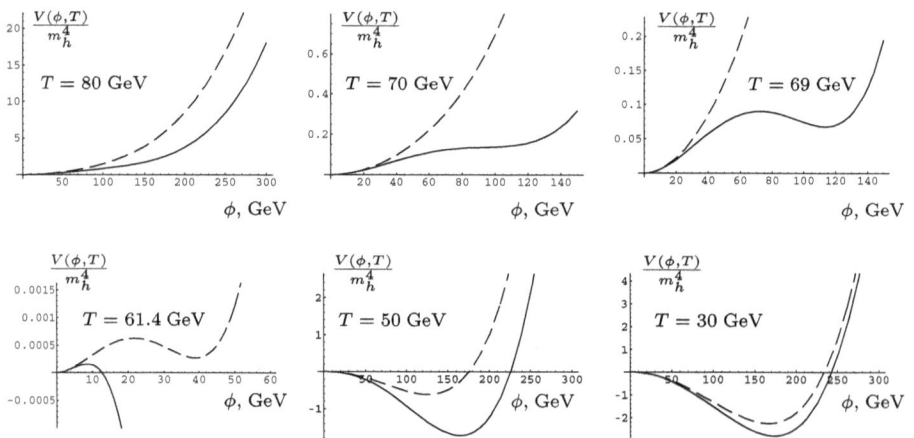

Fig. 10.9. One-loop effective potentials at different temperatures, obtained numerically (solid lines) and analytically within the high-temperature expansion (dashed lines) for the Standard Model with the Higgs boson mass $m_h = 50\,\text{GeV}$. Note the difference in scales of the axes for different temperatures.

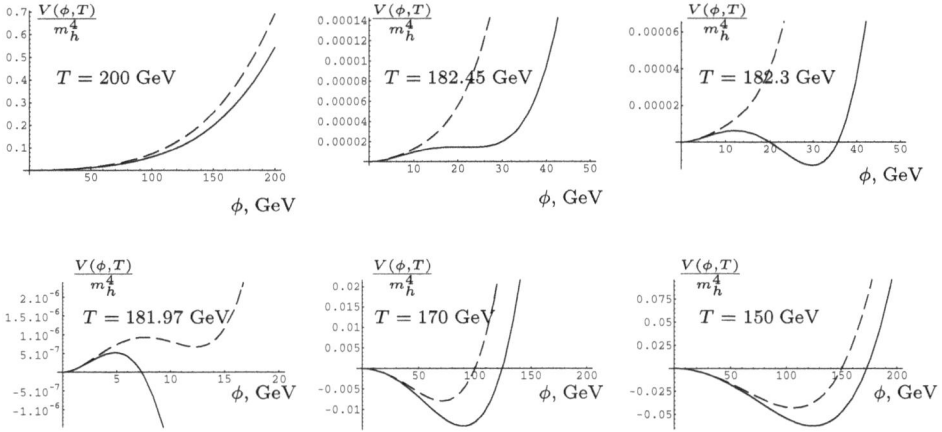

Fig. 10.10.   The same as in Fig. 10.9, but for the Higgs boson mass $m_h = 150\,\text{GeV}$.

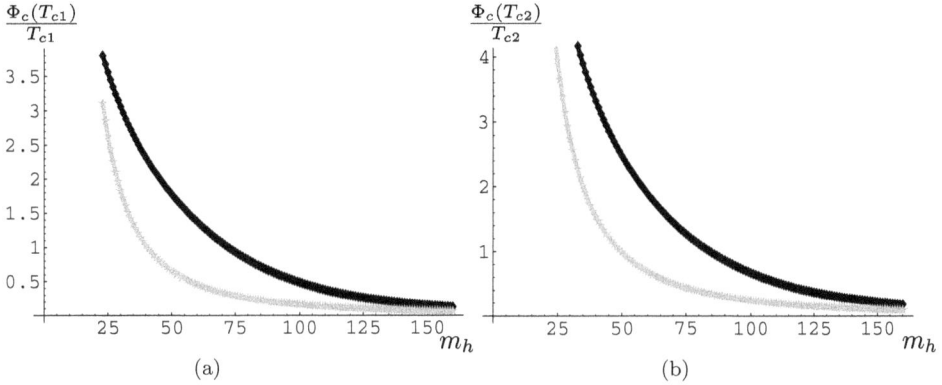

(a)   (b)

Fig. 10.11.   Comparison of numerical results (black lines) with the analytical results obtained using the high-temperature expansion (gray lines) for the one-loop values of (a) $\Phi_c(T_{c1})/T_{c1}$ and (b) $\Phi_c(T_{c2})/T_{c2}$ at various Higgs boson masses m_h.

Much more serious issue is the applicability of the one-loop approximation itself, and, more generally, the applicability of the perturbation theory at finite temperatures. These issues are discussed in Sec. 10.3. An important conclusion is that in the Standard Model, the one-loop approximation *does not* describe the transition correctly, given the experimental value $m_h = 125\,\text{GeV}$.

Let us describe a simple extension of the Standard Model where the Englert–Brout–Higgs state emerges as a result of the first order phase transition [187]. The idea is that the elctroweak transition may occur from a state with non-vanishing condensate of a new scalar field, rather than from a state with vanishing condensates. The phase with the condensate of the new field may be separated from the electroweak phase by a potential barrier, hence the first order phase transition.

We add a new scalar field s to the Standard Model, and take s to be a singlet under the Standard Model gauge group. We require for simplicity that the action is invariant under discrete symmetry $s \to -s$, then the only additional renormalizable terms in the Lagrangian are

$$L_s = \frac{1}{2} (\partial_\mu s)^2 - \frac{\lambda_s}{4} (s^2 - v_s^2)^2 - \varkappa H^\dagger H s^2,$$

where H is the Englert–Brout–Higgs doublet, λ_s and $\varkappa$ are dimensionless couplings and v_s is a new parameter of dimension of mass. Hence, the scalar potential at zero temperature is

$$V(H,s) = \lambda \left(H^\dagger H - \frac{v^2}{2} \right)^2 + \frac{\lambda_s}{4} (s^2 - v_s^2)^2 + \varkappa H^\dagger H s^2, \qquad (10.48)$$

where the first term is the scalar potential of the Standard Model. The potential $V(H,s)$ has extrema at

$$H = \begin{pmatrix} 0 \\ \frac{v}{\sqrt{2}} \end{pmatrix}, \qquad s = 0 \qquad (10.49)$$

and

$$H = 0, \qquad s = v_s. \qquad (10.50)$$

Let us impose the following conditions on the parameters:

$$\lambda v^2 > \varkappa v_s^2, \qquad (10.51\text{a})$$
$$\varkappa^2 > \lambda \lambda_s. \qquad (10.51\text{b})$$

Then there are no other nonzero extrema of the potential (10.48), the extremum (10.50) is unstable (saddle point of the potential), and the standard electroweak extremum (10.49) is the only minimum.

Problem 10.6. *Prove above statements. Show that the conditions* (10.51) *ensure the stability of the vacuum* (10.49).

Let us now consider the one-loop potential at finite temperatures. It is sufficient for our purposes to include only terms which are quadratic in temperature, and consider the effective potential along the axes $(H \neq 0, s = 0)$ and $(H = 0, s \neq 0)$. The terms quadratic in temperature are in general given by the second term in (10.27). At $(H \neq 0, s = 0)$, we have to include contributions due to $W^\pm$-, Z-bosons, t-quark and also the Higgs boson and s-boson. It is convenient to introduce canonically normalized field

$$\chi = \sqrt{2}\phi = \sqrt{2}|H|.$$

Then the mass squared of the Higgs boson at $\chi \neq 0, s = 0$ is

$$m_h^2 = \frac{\partial^2 V}{\partial \chi^2} = 3\lambda \chi^2 + \cdots, \qquad (10.52)$$

where we do not include terms independent of χ. We have for s-boson

$$m_s^2 = \varkappa \chi^2, \qquad (10.53)$$

where we again omit χ-independent terms. Then the quadratic in T part of the effective potential along the axis $(\chi \neq 0, s = 0)$ is

$$\Delta V_{\text{eff}} = \frac{1}{2}a_\chi T^2 \chi^2, \tag{10.54}$$

where

$$a_\chi = \frac{1}{48}\left(9g^2 + 3g'^2 + 12y_t^2 + 12\lambda + 4\varkappa\right).$$

The four first terms here agree with (10.30), (10.35), and the last term comes from (10.53). It is worth noting that the terms omitted in (10.52), (10.53) do not yield contributions into ΔV_{eff} which depend on χ. Thus, the effective potential along the axis $(\chi \neq 0, s = 0)$ reads

$$V_{\text{eff}}(\chi, s = 0) = \frac{1}{2}a_\chi(T^2 - T_\chi^2)\chi^2 + \frac{\lambda}{4}\chi^4, \tag{10.55}$$

where

$$T_\chi = \left(\frac{\lambda v^2}{a_\chi}\right)^{1/2}$$

is the critical temperature for this axis, and we normalized the potential in such a way that $V_{\text{eff}}(0,0) = 0$.

We now calculate the effective potential along the axis $(\chi = 0, s \neq 0)$. At $s \neq 0$ there are four bosons making the complex doublet H, whose masses are $m_H^2 = \varkappa s^2$, while for s-boson we have $m_s^2 = 3\lambda_s s^2 + \dots$. Other particles do not acquire masses. Therefore,

$$V_{\text{eff}}(\chi = 0, s) = \frac{1}{2}a_s(T^2 - T_s^2)s^2 + \frac{\lambda_s}{4}s^4,$$

where

$$a_s = \frac{1}{12}(3\lambda_s + 4\varkappa),$$

$$T_s = \left(\frac{\lambda_s v_s^2}{a_s}\right)^{1/2}.$$

Let us choose the parameters in such a way that

$$T_s > T_\chi. \tag{10.56}$$

Due to large t-quark mass, i.e., large y_t, this condition is not particularly restrictive. As an example, one can choose $\lambda_s \approx \lambda$, $\varkappa \approx \lambda$, $v_s \approx v$ (and also satisfy the inequalities (10.51)). Then

$$\frac{T_s^2}{T_\chi^2} \approx \frac{9g^2 + 3g'^2 + 12y_t^2 + 16\lambda}{28\lambda} \approx 5.$$

This shows that the inequalities (10.51) and (10.56) can be satisfied without fine tuning of parameters.

Once the condition (10.56) holds, then the cosmic medium first (at $T \approx T_s$) evolves into the state with $\chi = 0, s \neq 0$. As the temperature drops down to $T \approx T_\chi$, the second minimum of the effective potential emerges at $\chi \neq 0, s = 0$. The field values at co-existing minima and their depths are

$$s = \left[\frac{a_s}{\lambda_s}(T_s^2 - T^2)\right]^{1/2}, \quad V_{\text{eff}}(\chi = 0, s) = -\frac{a_s^2}{4\lambda_s}(T_s^2 - T^2)^2$$

and

$$\chi = \left[\frac{a_\chi}{\lambda}(T_\chi^2 - T^2)\right]^{1/2}, \quad V_{\text{eff}}(\chi, s = 0) = -\frac{a_\chi^2}{4\lambda}(T_\chi^2 - T^2)^2.$$

The minimum $\chi \neq 0, s = 0$ has the same depth as the minimum $\chi = 0, s \neq 0$ at

$$T_c^2 = \frac{T_\chi^2 a_\chi/\sqrt{\lambda} - T_s^2 a_s/\sqrt{\lambda_s}}{a_\chi/\sqrt{\lambda} - a_s/\sqrt{\lambda_s}} = \frac{v^2\sqrt{\lambda} - v_s^2\sqrt{\lambda_s}}{a_\chi/\sqrt{\lambda} - a_s/\sqrt{\lambda_s}}.$$

At temperature below T_c, the minimum $\chi \neq 0, s = 0$ is deeper, and the first order phase transition occurs. Importantly, $\chi(T_c)$ is not small compared to T_c in a large region of the parameter space, so our perturbative treatment is legitimate; see Sec. 10.3 in this regard.

Problem 10.7. *Let us set* $\varkappa = \lambda_s$ *for simplicity. Find the range of parameters* λ_s *and* v_s/v *where the relations* (10.51) (10.56) *hold, and, furthermore,* $\chi(T_c) > T_c$. *We note that models where the latter inequality is satisfied are of particular interest from the viewpoint of electroweak baryogenesis; see Sec. 11.5.*

Problem 10.8. *Find the range of parameters where the relations* (10.51) (10.56) *and inequality* $\chi(T_c) > T_c$ *hold, and the scalar* s *is a sizeable component of dark matter; see Problem 9.4 in Chapter 9. Can scalars* s *make all of dark matter?*

10.3. Infrared Problem

In this Section we show that the finite-temperature perturbation theory is not always applicable, even if couplings are small [156, 182]. Let us discuss this issue at the qualitative level first. The physical reason for failure of the perturbation theory is that the distribution function of *bosons* $f_B = [\exp(\omega/T)-1]^{-1}$ is large at low particle energies ω. Owing to that, the interaction between the bosons at low energies is enhanced in the medium. Indeed, for small momenta and masses of particles, $p \ll T$, $m \ll T$, the bosonic distribution function has the form

$$f_B(p) = \frac{T}{\omega_p}.$$

In quantum field theory, this means that number densities of low-momentum particles,

$$\langle\langle a_{\mathbf{p}}^\dagger a_{\mathbf{p}'}\rangle\rangle = f_B(p)\delta(\mathbf{p} - \mathbf{p}'),$$

are large in comparison with the commutator

$$\langle\langle [a_{\mathbf{p}}, a_{\mathbf{p}'}^\dagger]\rangle\rangle = \delta(\mathbf{p} - \mathbf{p}'),$$

provided that the particles are light, $m(T) \ll T$. Therefore, the infrared part of a light boson field Φ is a *classical field*. The linearized expression for it is

$$\Phi(x) = \frac{1}{(2\pi)^{3/2}} \int \frac{d^3p}{\sqrt{2\omega_p}} \left(e^{-ipx} a_{\mathbf{p}} + e^{-ipx} a_{\mathbf{p}}^\dagger \right),$$

where $a_{\mathbf{p}}$, $a_{\mathbf{p}}^\dagger$ can be treated as c-numbers at low momenta. Hence, omitting numerical factors, we obtain for the field fluctuation

$$\langle\langle \Phi^2(x) \rangle\rangle = \int \frac{d^3p}{\omega_p} f_B(p).$$

We are interested in the contribution from the infrared (IR) region,

$$\langle\langle \Phi^2(x) \rangle\rangle_{IR} = \int \frac{p^2 dp}{\omega_p} \frac{T}{\omega_p} = T \int \frac{dp\, p^3}{p\, \omega_p^2}.$$

Thus, the amplitude of the field fluctuations with momenta of order p is estimated as

$$\langle\langle \Phi^2(x) \rangle\rangle_p = T \frac{p^3}{\omega_p^2}.$$

The field is in the linear regime, if the quadratic (free) contribution to the free energy is larger than the contribution due to interactions. At $p^2 \gtrsim m^2$, the free contribution can be estimated as $(\nabla \Phi)^2 = p^2 \Phi^2$. Following the convention adopted in this Chapter, we choose the self-interaction of the field Φ as $h^2 \Phi^4$, where h is a small coupling. Comparison of the interaction and free contributions to the free energy gives

$$\frac{h^2 \Phi^4}{(\nabla \Phi)^2} \sim \frac{h^2 T^2 p^6}{\omega_p^4} \left(p^2 T \frac{p^3}{\omega_p^2} \right)^{-1} \sim \frac{h^2 T p}{\omega_p^2}. \tag{10.57}$$

This implies for the massless field

$$\frac{h^2 \Phi^4}{(\nabla \Phi)^2} \sim \frac{h^2 T}{p}.$$

The field is in the linear regime if this ratio is small, which is true only at $p \gg h^2 T$. In the infrared region, $p \lesssim h^2 T$, on the contrary, the regime is strongly non-linear, and the perturbation theory is inapplicable. For massive field, the ratio (10.57) is small for all momenta, only if $m_\Phi \gg h^2 T$. Otherwise, the infrared region is at strong coupling. Note that this is the strong coupling regime in the classical field theory at finite temperature.

In gauge theories with Englert–Brout–Higgs nechanism, this argument applies to *spatial components of non-Abelian gauge field*, whose interaction contains, in particular, the commutator term of the type $g^2 A^4$. Thus, instead of h^2 we have g^2, and the condition of applicability of the perturbation theory has the form

$$M_W(T) \gg g^2 T,$$

where $M_W(T)$ is the mass of the gauge field. For the component A_0 the argument is inapplicable because of the Debye mass $m_D \sim gT \gg g^2 T$.

Thus, the perturbative calculation of the effective potential is justified when $M_W(T) \gg g^2 T$, i.e.,

$$\phi \gg gT. \tag{10.58}$$

We see that the effective potential cannot be perturbatively calculated near $\phi = 0$. Moreover, the description of the first order phase transition presented in the previous Section is valid only when the position of the second minimum of the effective potential given by Eqs. (10.43), (10.44) or (10.45), satisfies $\Phi_c \gg gT_c$. In terms of couplings, this restriction has the form

$$\lambda \ll \frac{\gamma}{g}.$$

Given that $\gamma \sim g^3$ (see (10.40)), and omitting numerical factors (they actually work towards stronger restriction), we get from this $\lambda \ll g^2$, i.e., in terms of the zero-temperature mass

$$m_h^2 \ll M_W^2. \tag{10.59}$$

It is difficult to refine this estimate. In other words, it is impossible to find analytically up to what values of m_h the phase transition is of the first order. Still, the restriction (10.59) suggests that the one-loop results of Sec. 10.2 most likely have nothing to do with reality in the Standard Model, taking into account the actual value $m_h = 125\,\text{GeV}$.

Since the difficulty is due to interactions of low-momentum particles, it is called the infrared problem. The main role here is played by the self-interactions of gauge bosons inherent in any non-Abelian gauge theory.

Let us see at the formal level that the applicability of the perturbation theory for calculating the effective potential is indeed limited to the values of the scalar field obeying (10.58). As shown in Appendix D, the effective potential is given by the functional integral

$$e^{-\beta V(\phi)} = \int \mathcal{D} A_\mu e^{-S^{(\beta)}[A]}, \tag{10.60}$$

where we neglected all but gauge fields. Here $\beta = T^{-1}$, the functional $S^{(\beta)}[A]$ is the Euclidean action in the Euclidean time interval $0 \leq \tau \leq \beta$,

$$S^{(\beta)}[A] = \int_0^\beta d\tau \int d^3 x \left[\frac{1}{4} F_{\mu\nu}^b F_{\mu\nu}^b + \frac{M^2(\phi)}{2} A_\mu^b A_\mu^b \right], \tag{10.61}$$

summation over indices is performed with the Euclidean metric. The integration in (10.60) is over fields $A_\mu^b(\mathbf{x}, \tau)$ periodic in τ with period β.

Because of the periodicity, the field $A_\mu^b(\mathbf{x}, \tau)$ can be represented as a discrete sum

$$A_\mu(\mathbf{x}, \tau) = \frac{1}{\sqrt{\beta}} a_\mu(\mathbf{x}) + \sum_{n=\pm 1,\dots} \frac{1}{\sqrt{\beta}} a_\mu^{(n)}(\mathbf{x}) e^{i\omega_n \tau}, \tag{10.62}$$

where

$$\omega_n = \frac{2\pi n}{\beta} \equiv 2\pi n T$$

are the Matsubara frequencies, and we omitted the group index and extracted explicitly the terms in (10.62) with zero Matsubara frequency. Upon substituting expansions (10.62) into (10.61), we arrive at the action of 3-dimensional Euclidean theory with an infinite set of fields $a_\mu(\mathbf{x})$, $a_\mu^{(n)}(\mathbf{x})$. We are interested in the infrared region, more precisely, in the region of spatial momenta

$$|\mathbf{p}| \ll gT.$$

At these momenta, only light 3-dimensional fields are relevant, whose masses are much smaller than gT. Note first, that a_0, $a_0^{(n)}$ are not light fields: they acquire the Debye mass $m_D \sim gT$ (see Sec. D.5). Therefore, the fields a_0, $a_0^{(n)}$ can be ignored. Now, the fields $a_i^{(n)}$ with $n \neq 0$ are also heavy: the term $F_{0i}^a F_{0i}^a$ in the original Lagrangian leads to the term in the action

$$\sum_{n=\pm 1, \pm 2,\dots} \int d^3 x\, \omega_n^2 a_i^{(n)} a_i^{(-n)},$$

i.e., the mass term in the 3-dimensional theory with large masses $|\omega_n|$. As a result, the only light fields are $a_i(\mathbf{x})$, i.e., homogeneous in Euclidean time components of spatial vector potentials A_i. Substituting $A_i(\mathbf{x}) = \beta^{-1/2} a_i(\mathbf{x})$ into the action (10.61), we obtain the effective 3-dimensional action describing the infrared properties of the theory at finite temperature,

$$S^{\text{eff}} = \int d^3 x \left(\frac{1}{4} f_{ij}^b f_{ij}^b + \frac{1}{2} M^2(\phi) a_i^b a_i^b \right), \tag{10.63}$$

where

$$f_{ij}^b = \partial_i a_j^b - \partial_j a_i^b + g\sqrt{T} f^{bcd} a_i^c a_j^d \tag{10.64}$$

and f^{bcd} are structure constants of the non-Abelian gauge group. (In the case of gauge group $SU(2)$ these are $f^{bcd} = \epsilon^{bcd}$.) The factor $T^{1/2} = \beta^{-1/2}$ emerges in (10.64) due to the normalization in (10.62), chosen in such a way that quadratic part of the 3-dimensional action (10.63) has canonical form.

As a digression, we make an observation regarding fermions. Within the Euclidean approach *all* 3-dimensional fermions are heavy: they are antiperiodic in β, and their Matsubara frequencies $\omega_{n'} = 2\pi T n'$, $n' = \pm\frac{1}{2}, \pm\frac{3}{2},\dots$ are all of order T. Therefore, fermionic fields are insignificant for the infrared properties of the theory. This also follows from the consideration given in the beginning of this Section: the distribution function of fermions $f_F = [\exp(\omega/T) + 1]^{-1}$ does not diverge as $\omega \to 0$, in accordance with the Pauli principle.

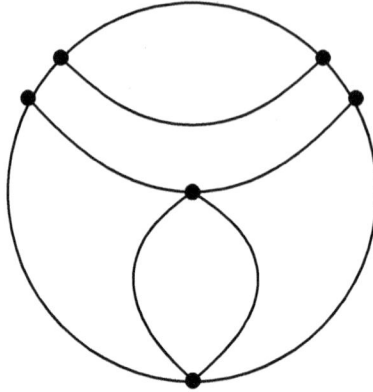

Fig. 10.12.   An example of a diagram contributing to the effective potential.

Coming back to the action (10.63), we note that it is the action of 3-dimensional vector fields with mass $M(\phi)$ and dimensionful coupling

$$g^{(3)} = g\sqrt{T}. \tag{10.65}$$

The ratio of dimensionful quantities $[g^{(3)}]^2/M(\phi)$ is the effective coupling constant. Indeed, in the framework of perturbation theory, the effective potential (more precisely, $-\beta V(\phi)$) is given by the sum of one-particle irreducible diagrams without external lines (see Appendix D) of the type shown in Fig. 10.12. For our purpose, these are diagrams in 3-dimensional theory with the action (10.63). Diagrams with n loops give contributions to βV_{eff} proportional to $[g^{(3)}]^{2(n-1)}$. On dimensional grounds, these contributions are of order

$$\beta V_{\text{eff}}^{(n)} \sim \frac{[g^{(3)}]^{2(n-1)}}{[M(\phi)]^{n-4}} = [M(\phi)]^3 \left(\frac{[g^{(3)}]^2}{M(\phi)}\right)^n.$$

Thus, the expansion parameter in the perturbation theory is $[g^{(3)}]^2/M(\phi)$, i.e., the perturbation theory is applicable at $M(\phi) \gg g^2T$. This confirms the result of the qualitative analysis given in the beginning of this Section.

To conclude this Section, we make a few remarks. First, we note that at momenta and energies small compared to g^2T, the theory is effectively reduced to a 3-dimensional gauge theory (in general, with scalar fields) with gauge coupling (10.65). Such a theory exhibits quite non-trivial nonperturbative properties. These have been explored in a number of lattice studies.

Second, for the fields $A_i(\mathbf{x})$ independent of Euclidean time τ, the partition function (10.60) is

$$e^{-\beta F} = \int \mathcal{D}A_i(\mathbf{x}) e^{-\beta \mathcal{H}[A_i]},$$

where

$$\mathcal{H}[A_i] = \int d^3x \left(\frac{1}{4} F_{ij}^b F_{ij}^b + \cdots\right),$$

and dots denote terms involving possible scalar fields. (Here we have preserved the four-dimensional normalization of the vector field.) The expression on the right side is the partition function of *classical four-dimensional* gauge field theory (in general, with scalar fields) with the Hamiltonian $\mathcal{H}[A_i]$. Thus, our analysis leads to understanding [183, 184] that the behavior of *quantum* theory at high temperatures at length and time scales comparable to or in excess of $(g^2 T)^{-1}$, basically coincides with the behavior of *classical field theory*.[11] This fact is useful for studying static, and especially dynamical (showing up in time evolution) properties of the theory at high temperatures.

10.4. *Electroweak Vaccum at Zero Temperature

This Section is a digression from the theory of the early Universe: we are going to discuss an intriguing property of the Standard Model *at zero temperature*.

Classical (tree level) scalar potential is not, generally speaking, the right object to study when considering vacuum of the theory. The correct one is the zero temperature effective potential, which includes radiative corrections [189]. If one assumes that the effective potential for the Englert–Brout–Higgs field can be calculated within the Standard Model,[12] then it is quite likely that it has a minimum at high field value $\phi \gg 100$ GeV, and that this minimum is deeper than "our" minimum $\phi = 246$ GeV. This would mean that the Universe resides in false, metastable vacuum.

The effective potential that incudes one-loop corrections has the same form as in (10.22), where the contributions of various particles are calculated in Section D.4 (see Eqs. (D.38), (D.43)),

$$f_i(T = 0, \phi) = \pm \frac{g_i}{2} \int \frac{d^3 \mathbf{p}}{(2\pi)^3} \sqrt{\mathbf{p}^2 + m_i^2(\phi)}, \qquad (10.66)$$

where the positive (negative) sign is characteristic of bosons (fermions). Here g_i denotes the number of internal degrees of freedom (each Dirac fermion has $g_i = 4$). Integral in (10.66) is ultraviolet (UV) divergent. This is rather general situation in quantum field theory; the UV divergence is removed by renormalization of "bare" parameters in the Lagrangian; see, e.g., Refs. [268, 269, 271].

In the case at hand, we regularize the integral in (10.66) by cutting it off at $|\mathbf{p}| = \Lambda_{UV}$, where Λ_{UV} is cutoff parameter which will eventually be sent to infinity. This procedure is known as momentum cutoff regularization (we use its

---

[11] There is a subtlety associated with the need for ultraviolet cutoff in the classical field theory: for the Rayleigh–Jeans distribution, the energy density diverges in the ultraviolet region.
[12] This is a very strong assumption, implying, in particular, that new physics, responsible for neutrino oscillations, dark matter and baryon asymmetry of the Universe, does not induce sizeable corrections to the parameters of the Standard Model even at high energies.

3-dimensional version). We find the regulatized contribution

$$f_i = \pm \frac{g_i}{4\pi^2} \int_0^{\Lambda_{UV}} |\mathbf{p}|^2 d|\mathbf{p}| \sqrt{|\mathbf{p}|^2 + m_i^2(\phi)} \tag{10.67a}$$

$$= \pm \frac{g_i}{16\pi^2} \left[\Lambda_{UV}^4 + \Lambda_{UV}^2 m_i^2(\phi) + \frac{m_i^4(\phi)}{4} \left(\ln \frac{m_i^2(\phi)}{4\Lambda_{UV}^2} + \frac{1}{2} \right) + \mathcal{O}(1/\Lambda_{UV}^2) \right]. \tag{10.67b}$$

In renormalizable theories, the "bare" (tree level) potential is a fourth order polynomial

$$V^{(0)}(\phi) = \rho_\Lambda^{(0)} + c_1^{(0)}\phi + m^{2\,(0)}\phi^2 + c_3^{(0)}\phi^3 + \lambda^{(0)}\phi^4.$$

Parameters $\rho_\Lambda^{(0)}$, $m^{2\,(0)}$, $\lambda^{(0)}$ (as well as $c_{1,3}^{(0)}$) are independent of ϕ and are bare cosmological constant, bare mass parameter and bare self-coupling of the field ϕ. There are no linear and cubic terms in the Standard Model, $c_1^{(0)} = c_3^{(0)} = 0$. In renormalizable theories, mass squared $m_i^2(\phi)$ in (10.67b) is a polynomial of at most second order in ϕ, so the effective potential with onee-loop corrections has the following structure:

$$V = V^{(0)} + \sum_i f_i = V^{(0)} + V^{div} + V^{(1)},$$

where

$$V^{div} = \sum_i \pm \frac{g_i}{16\pi^2} \left[\Lambda_{UV}^4 + \Lambda_{UV}^2 m_i^2(\phi) + \frac{m_i^4(\phi)}{4} \left(\ln \frac{\mu^2}{4\Lambda_{UV}^2} + \frac{1}{2} \right) \right] \tag{10.68a}$$

$$= \delta\rho_\Lambda + \delta c_1 \cdot \phi + \delta m^2 \cdot \phi^2 + \delta c_3 \cdot \phi^3 + \delta\lambda \cdot \phi^4 \tag{10.68b}$$

is UV divergent part with parameters $\delta\rho_\Lambda, \ldots, \delta\lambda$, depending on Λ_{UV} (the second and fourth terms in (10.68b) are again absent in the Standard Model) and the term

$$V^{(1)} = \sum_i \pm \frac{g_i}{64\pi^2} m_i^4(\phi) \ln \frac{m_i^2(\phi)}{\mu^2}$$

is finite[13] as $\Lambda \to \infty$. Note that we introduced an arbitrary parameter μ of dimension of mass. This is the renormalization scale, which enters also $\delta\rho_\Lambda$, δm^2, $\delta\lambda$ and $\delta c_{1,3}$. Physical quantities must be independent of μ.

---

[13]Note that *finite* (in the limit $\Lambda_{UV} \to \infty$) polynomial terms which have the structure (10.68b), can be reshuffled between V^{div} and $V^{(1)}$. This arbitrariness corresponds to arbitrary choice of the renormalization scheme.

We define the renormalized parameters

$$m^2(\mu) = m^{2\,(0)} + \delta m^2$$

$$\lambda(\mu) = \lambda^{(0)} + \delta\lambda$$

(and similarly for ρ_Λ and $c_{1,3}$) and treat them as finite in the limit $\Lambda_{UV} \to \infty$ (the bare parameters are divergent in this limit). This is precisely the renormalization of the one-loop effective potential. The renormalized effective potential

$$V = \rho_\Lambda(\mu) + c_1(\mu)\phi + m^2(\mu)\phi^2 + c_3(\mu)\phi^3 + \lambda(\mu)\phi^4 + V^{(1)}$$

is finite as the regularization is removed. The parameters entering $V^{(1)}$ are to be treated as renormalized: the difference between bare and renormalized parameters in $V^{(1)}$ is the second loop effect which is neglected in one-loop approximation.

We are interested in the behavior of the effective potential at large ϕ. We have

$$V(\phi) = \lambda(\mu)\phi^4 + \sum_i \pm \frac{g_i\, m_i^4(\phi)}{64\pi^2} \ln \frac{m_i^2(\phi)}{\mu^2}. \tag{10.69}$$

The explicit dependence on μ in this formula is compensated for by the dependence of self-coupling $\lambda(\mu)$. This is the simplest manifestation of the renormalization group.

Since the particle masses are proportional to couplings to the scalar field, the second term in (10.69) leads to corrections to the effective potential which are proportional to powers of these couplings: the heavier the particle, the larger its couping to the scalar field and the larger its contribution to $V(\phi)$. In the context of the Standard Model, the main contributions come from t-quark, Z- and $W^\pm$-bosons and the Higgs boson itself. To the leading order in ϕ, we find in the Landau gauge[14]

$$V(\phi) = \lambda\phi^4 + \frac{3M_Z^4(\phi)}{64\pi^2} \ln \frac{M_Z^2(\phi)}{\mu^2} + \frac{6M_W^4(\phi)}{64\pi^2} \ln \frac{M_W^2(\phi)}{\mu^2} - \frac{12m_t^4(\phi)}{64\pi^2} \ln \frac{m_t^2(\phi)}{\mu^2}$$

$$+ \frac{36\lambda^2\phi^4}{64\pi^2} \ln \frac{6\lambda\phi^2}{\mu^2} + \frac{12\lambda^2\phi^4}{64\pi^2} \ln \frac{2\lambda\phi^2}{\mu^2}, \tag{10.70}$$

where particle masses depend on the scalar field as in (10.20).

---

[14]Here we deviate from our analysis in Sec. 10.2 where we use the unitary gauge. This subtlety is not very important, however.

The last two terms in (10.70) come from the scalar field itself. Let us explain their origin. We decompose the Englert–Brout–Higgs field into background part ϕ and quantum part:

$$H = \begin{pmatrix} 0 \\ \phi \end{pmatrix} + \frac{1}{\sqrt{2}} \begin{pmatrix} \xi_1 + i\xi_2 \\ h + i\xi_3 \end{pmatrix} \equiv \Phi + \varphi.$$

The quadratic part of the kinetic term in the Englert–Brout–Higgs Lagrangian is

$$(D_\mu H)^\dagger D_\mu H = \frac{1}{2}(\partial_\mu h)^2 + \sum_{a=1}^{3}\frac{1}{2}(\partial_\mu \xi_a)^2 - \Phi^\dagger \hat{A}_\mu \partial_\mu \varphi + \partial_\mu \varphi^\dagger \hat{A}_\mu \Phi + \Phi^\dagger \hat{A}_\mu \hat{A}_\mu \Phi, \quad (10.71)$$

where we write in somewhat sloppy way $D_\mu = \partial_\mu + \hat{A}_\mu$ with anti-Hermitean $\hat{A}_\mu$. The last term in (10.71) is the mass term for $W^\pm$- and Z-bosons. The Landau gauge $\partial_\mu \hat{A}_\mu = 0$ is particularly convenient here, since the cross terms (the third and fourth in (10.71)) vanish upon integrating by parts, and the fields h and ξ_a have canonical kinetic terms. At large background field ϕ, the tree level scalar potential, up to quadratic order in h and ξ_a, reads[15]

$$\lambda(H^\dagger H)^2 = \lambda\phi^4 + \frac{m_h^2(\phi)}{2}h^2 + \sum_{a=1}^{3}\frac{m_{\xi_a}^2(\phi)}{2}\xi_a^2,$$

where

$$m_h^2(\phi) = 6\lambda\phi^2, \quad m_{\xi_a}^2(\phi) = 2\lambda\phi^2.$$

We substitute these masses into the general formula (10.69), and obtain the second line in (10.70).

The one loop effective potential (10.70) can be written as follows:

$$V(\phi) = \lambda(\mu)\phi^4 + \frac{3M_Z^4 + 6M_W^4 - 12m_t^4 + 3m_h^4}{16\pi^2 v^4} \cdot \phi^4 \ln\frac{\phi^2}{\mu^2}, \quad (10.72)$$

where the correction is written in the leading logarithmic approximation, and the couplings are expressed through "our" particle masses. Because of the large t-quark mass, the second term here is negative, so the effective potential becomes negative at large ϕ. Thus, within the one loop approximation, "our" vacuum $\phi = v/\sqrt{2}$ is metastable, rather than absolutely stable.

One loop approximation is, however, insufficient for deciding whether "our" vacuum is stable or metastable. Let us set $\mu^2 = v^2/2$ in (10.72) and find that modulo small corrections $\lambda|_{\mu=v/\sqrt{2}} = m_h^2/2v^2 = 0.13$. Then the one loop effective potential (10.72) becomes negative at $\phi \gtrsim 10^5$ GeV. For so large field values, the higher loop effects become important. These can be incorporated in the following

---

[15] The term linear in h does not contribute to the action, since the homogeneous part of the scalar field is set equal to Φ, and therefore $\int d^4x\, h(x) = 0$.

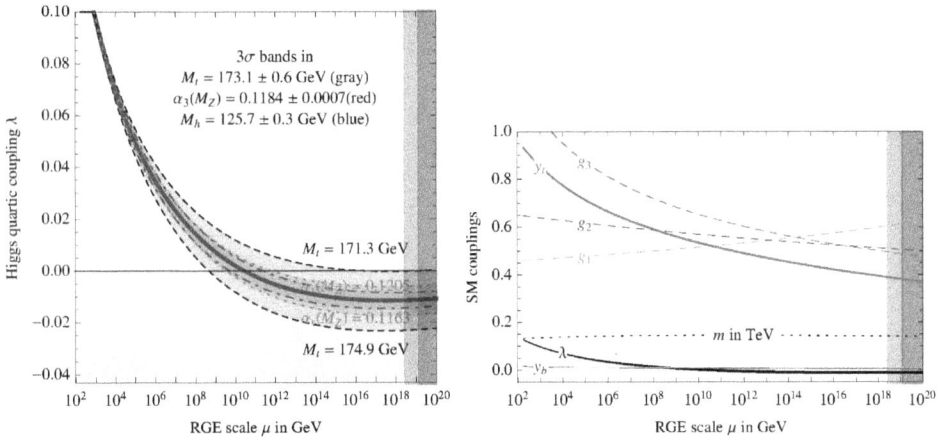

Fig. 10.13. *Left*: Evolution of running self-coupling of the Englert–Brout–Higgs field at three loop level in the Standard Model. *Right*: Evolution of the most relevant couplings in the Standard Model [190].

way. Let us choose $\mu = \phi$ in (10.72). Then the effective potential becomes

$$V(\phi) = \lambda(\phi)\phi^4.$$

Written in this form, the effective potential is determined by the running self-coupling[16] $\lambda(\phi)$. In particular, the potential becomes negative if the self-coupling changes sign.

The self-coupling $\lambda(\mu)$ together with other couplings obey the renormalization group equations. Higher order corrections in this formalism show up as higher order perturbative contributions to the renormalization group functions (beta-functions). The behavior of the running self-coupling $\lambda(\mu)$ at three loop level is shown in Fig. 10.13 (left). For central values of the parameters of the Standard Model, the scalar self-coupling becomes negative at $\phi > 10^{11}$ GeV. This suggests that "our" vacuum is metastable. The situation is not completely clear yet, however: Fig. 10.13 shows that uncertainties in the determination of couplings at the electroweak scale allow for the absolute stability of "our" vacuum at 3σ C.L. If at some $\phi_* \ll M_{Pl}$ the potential is indeed negative, then the cosmological scenario should be such that the field never reaches ϕ_*, otherwise the field would stay in the "non-standard" vacuum. In particular, in inflationary scenario every scalar field takes values of the order of the inflationary Hubble parameter H_{inf}. Therefore, the inflationary dynamics must be such that $H_{\text{inf}} < \phi_*$.

Note that neglecting gravity, the self-coupling flattens out at large $\phi \sim 10^9$–10^{22} GeV, while other couplings do not; see Fig. 10.13 (right). In fact, the low energy values of the Standard Model couplings, in particular, t-quark Yukawa

---

[16]There is a subtlety here which has to do with the anomalous dimension of the field ϕ. This is not important for our purposes.

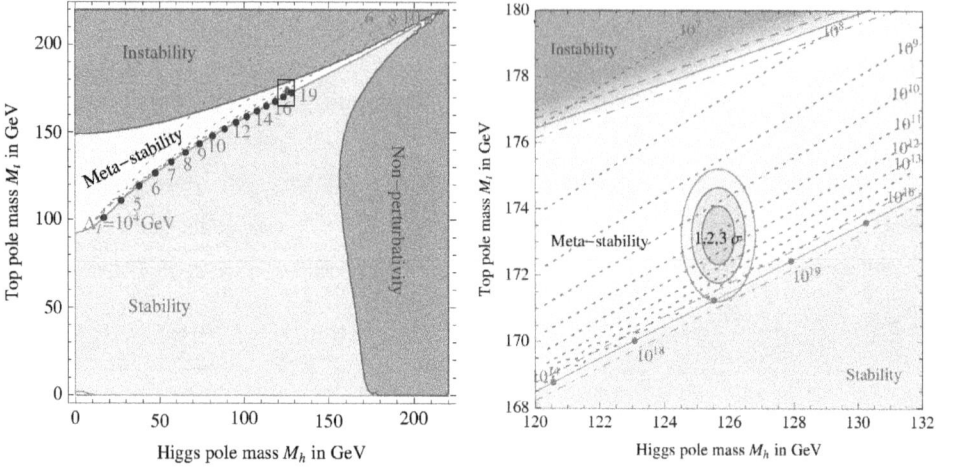

Fig. 10.14. *Left:* Regions of strong coupling, stability, instability and metastability of the Standard Model vacuum. Black dots with numbers show the field values at which the scalar self-coupling, and hence the effective potential, becomes negative. *Right:* Rectangle shown in the left part. Ellipses surround the region favored by the current data on the Standard Model parameters at 1σ, 2σ, 3σ C.L. [190].

coupling and Higgs self-coupling, are rather special. Namely, the electroweak vacuum is likely false but long lived, while couplings remain small up to the Planck energy; see Fig. 10.14.

Since it is likely that the elctroweak vacuum $\phi = v/\sqrt{2}$ is metastable, it is of interest to estimate its lifetime. The false vacuum decay occurs through the spontaneous nucleation of bubbles, which at zero temperature is a tunneling process. The nucleation probability per unit time per unti volume has the following form:

$$\Gamma = \frac{dP}{dt\,dV} \simeq \frac{e^{-S[R_c]}}{R_c^4}, \tag{10.73}$$

where $S[R_c]$ is the Euclidean action on classical Euclidean configuration in four dimensions, and R_c is the size of this configuration; see details in [175]. The relevant configuration is an extremum of the Euclidean action

$$S_\phi = \int d^4x \, \left[(\partial_\mu \phi)^2 + V(\phi) \right],$$

where $V(\phi)$ is the effective potential, summation over μ is done with Euclidean metric. This extremum obeys the Euclidean field equation

$$-2\partial_\mu\partial_\mu\phi + \frac{\partial V}{\partial \phi} = 0 \tag{10.74}$$

with the boundary condition

$$\phi(|x| \to \infty) = \phi_-,$$

where ϕ_- is the field value in the false vacuum.

In the Standard Model we deal with very large fields, so we neglect the mass term of the field ϕ. Also, we neglect weak (logarithmic) dependence of the self-coupling on ϕ, and set

$$V(\phi) = -|\lambda|\phi^4 \tag{10.75}$$

with $\lambda = \lambda(\mu) = \text{const}$. Then Eq. (10.74) takes simple form

$$\partial_\mu\partial_\mu\phi + 2|\lambda|\phi^3 = 0, \tag{10.76}$$

and the boundary condition is $\phi(|x| \to \infty) = 0$. It follows from Eq. (10.76) that the solution depends on λ as $\phi \propto |\lambda|^{-1/2}$, so the action on the solution is $S = \text{const}/|\lambda|$. To find the constant here, we solve Eq. (10.76). The solution is [188]

$$\phi = \frac{2}{\sqrt{|\lambda|}} \frac{R_c}{x^2 + R_c^2}, \tag{10.77}$$

where $x^2 = x^\mu x^\mu$, R_c is an arbitrary parameter, and the self-coupling λ is to be computed at the scale $\mu \sim R_c^{-1}$. The Euclidean action on this solution is

$$S[R_c] = \frac{8\pi^2}{3|\lambda|}. \tag{10.78}$$

Had we neglected the dependence of λ on energy scale, the theory with the potential (10.75) would be scale-invariant, and this is precisely the reason for the appearance of the arbitrary parameter R_c. To find the value of this parameter, we search for the maximum of the tunneling probability (10.73), and since the dependence of the tunneling exponent on self-coupling is strong, this means that we have to find the maximum of $|\lambda|$, which is determined by the equation $d\lambda/d\mu(\mu = R_c^{-1}) = 0$.

Figure 10.13 shows that $|\lambda|$ is small for negative λ, so the tunneling probability is strongly suppressed. To estimate the probability of the bubble nucleation in our part of the Universe, we recall that the size and lifetime of this part are of order H_0^{-1}, so the relevant 4-dimensional volume is of order $\Omega_4(t_0) \sim H_0^{-4}$. The bubble size R_c is about thirty times smaller than the Planck length; see Fig. 10.13. In this way we get

$$P(t_0) = \Gamma \cdot \Omega_4(t_0) \sim \left(\frac{3 \cdot 10^{17}\,\text{GeV}}{H_0}\right)^4 e^{-\frac{8\pi^2}{3|\lambda|}}, \tag{10.79}$$

and for $|\lambda| = 2 \cdot 10^{-2}$ (see Fig. 10.13) we have $P \sim 10^{-340}$. The electroweak vacuum did not have time to decay.

Problem 10.9. *Refine the estimate of the four-volume $\Omega_4(t_0)$ in (10.79).*

Let us look into the future. For $t \gg t_0$ the estimate for the tunneling time t_{tunn} depends on the properties of dark energy. This time is greater if dark energy is the cosmological constant as compared to the Universe with decaying dark energy, which eventually becomes dominated by dark matter; see Fig. 10.15. Indeed, the relevant spatial scale in the Universe dominated by the cosmological constant is the

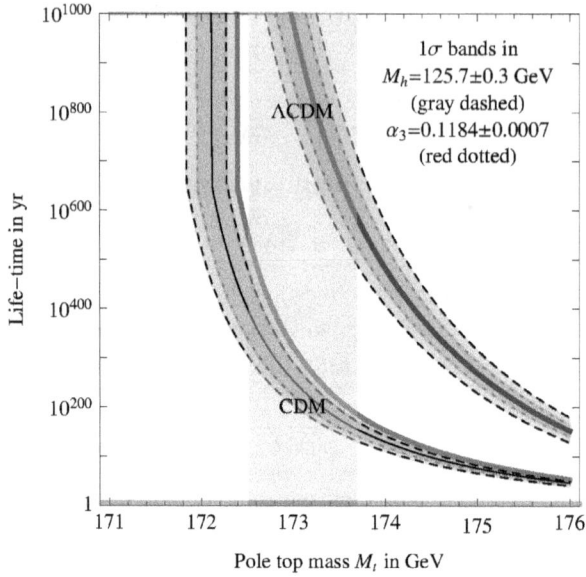

Fig. 10.15.   Lifetime of the false electroweak vacuum, as a function of t-quark mass, in the Universe with the cosmological constant (right curve) and the Universe which is matter dominated in far future [190]. Vertical strip shows the experimental uncertainty in m_t.

event horizon size $H_\Lambda^{-1} = H_0^{-1}\Omega_\Lambda^{-1/2} \sim H_0^{-1}$. Therefore, the tunneling time for this Universe is estimated from

$$\frac{t_{\text{tunn}}^{(\Lambda)}}{R_c} \frac{1}{(H_0 R_c)^3} \, e^{-\frac{8\pi^2}{3|\lambda|}} \sim 1. \tag{10.80}$$

If the Universe is matter dominated in far future, then the relevant spatial scale is the particle horizon size $H^{-1}(t) \sim t$, and the tunneling time obeys

$$\left(\frac{t_{\text{tunn}}^{(M)}}{R_c}\right)^4 e^{-\frac{8\pi^2}{3|\lambda|}} \sim 1. \tag{10.81}$$

By comparing (10.80) with (10.81) we see that $t_{\text{tunn}}^{(\Lambda)} \gg t_{\text{tunn}}^{(M)}$ indeed.

Chapter 11

Generation of Baryon Asymmetry

As we discussed in Secs. 1.3 and 5.2, there are baryons and no antibaryons in the present Universe.[1] The present baryon abundance is characterized by the baryon-to-photon ratio $\eta_B = n_{B,0}/n_{\gamma,0} \simeq 6.05 \cdot 10^{-10}$. Another way to quantify the baryon asymmetry is to use the baryon-to-entropy ratio (see Sec. 5.2)

$$\Delta_B = \frac{n_B - n_{\bar{B}}}{s} \simeq 0.86 \cdot 10^{-10}. \tag{11.1}$$

This ratio stays constant at the hot stage of the cosmological evolution provided there are no processes that violate baryon number or produce large entropy. One of the problems for cosmology is to explain the baryon asymmetry [34, 35]. According to our discussion in Secs. 1.5.5 and 5.2, for the initial state it is natural to assume baryon-symmetry.

Let us stress right away that there is no unambiguous answer to the question of the origin of the baryon asymmetry. It could be produced either at hot stage or at earlier post-inflationary reheating epoch. We discuss reheating in the accompanying book, and here we consider several mechanisms that could work at hot stage. Before discussing concrete mechanisms, let us study general conditions necessary for baryogenesis.

Besides the baryon asymmetry, there can be, and most likely there is, lepton asymmetry. If it is not large, its value does not seem to be possible to measure. Lepton asymmetry resides today in the excess of neutrinos over antineutrinos, or vice versa (electron number density equals that of protons by electric neutrality), while measuring relic neutrino abundance will be impossible in the foreseeable future.

---

[1]Modulo exotica like possible existence of compact objects made of antimatter. The latter are predicted in some models [192] and do not contradict observational data [193].

11.1. Necessary Conditions for Baryogenesis

Baryon asymmetry generation at a certain cosmological epoch is possible only if three conditions are met. These are dubbed Sakharov conditions. Namely, three properties must hold simultaneously:

(1) Baryon number non-conservation (with qualification, see below)
(2) C- and *CP*-violation.
(3) Thermal inequilibrium.

The fact that the first condition is necessary to produce asymmetry is self-evident. Now, if C- or *CP*-invariance is exact, processes with quarks and leptons occur in the same way as processes with their antiparticles, and no asymmetry is generated. Hence the second condition.

The latter result follows, at formal level, from the evolution law of the density matrix,

$$\rho(t) = e^{-i\hat{H}(t-t_i)}\rho(t_i)e^{i\hat{H}(t-t_i)}, \tag{11.2}$$

where $\hat{H}$ is the Hamiltonian of the system, t_i is initial moment of time. C- or *CP*-invariance means that the corresponding unitary operator U_C or U_{CP} commutes with the Hamiltonian. Say, for *CP*-invariance

$$U_{CP}\hat{H}U_{CP}^{-1} = \hat{H}.$$

This gives

$$U_{CP}\rho(t)U_{CP}^{-1} = \rho(t),$$

provided that the initial state is symmetric, i.e., $U_{CP}\rho(t_i)U_{CP}^{-1} = \rho(t_i)$. Together with the fact that the baryon number operator is *CP*-odd, $U_{CP}\hat{B}U_{CP}^{-1} = -\hat{B}$, the latter property gives

$$\langle \hat{B}(t) \rangle = \text{Tr}[\hat{B}\rho(t)] = 0.$$

The medium remains baryon symmetric.

Finally, the third condition is also fairly obvious. In thermal equilibrium with respect to baryon number violating reactions, the system is in a state with zero baryon chemical potential, i.e., zero baryon number density. Baryon asymmetry tends to get washed out, rather than generated, as the system approaches thermal equilibrium. (There are, however, special cases in which this rule is violated; we discuss an example later in this Section.)

The latter conclusion needs qualification. It would be literally valid if the baryon number were the only relevant quantum number. We will see that lepton numbers are also important in the Standard Model of particle physics: the baryon number itself is not conserved at temperatures above $T \sim 100$ GeV, while its linear combinations with lepton numbers are. In particular, conserved quantum number is

$(B-L)$ where $L = L_e + L_\mu + L_\tau$ is the total lepton number. In thermal equilibrium at nonzero $(B-L)$ both baryon and lepton numbers do not vanish,

$$B = C \cdot (B-L), \quad L = (C-1) \cdot (B-L), \tag{11.3}$$

where the constant C is of order 1 (but somewhat less than 1). Thus, the baryon number does not vanish. This property is used in leptogenesis mechanisms: they produce lepton asymmetry (and hence $(B-L)$) at very high temperature by processes beyond the Standard Model, and then this asymmetry is partially reprocessed into baryon asymmetry by electroweak physics.

Each of the three Sakharov conditions, generally speaking, is associated with one or another small parameter. Since all three should work at the same time, the net baryon asymmetry is naturally small in most models. We will see how these conditions are fulfilled in concrete mechanisms.

The third Sakharov condition should not necessarily hold at the time the asymmetry is produced, although it must be satisfied at some cosmological epoch amyway. Namely, dynamics may be such that the *equilibrium* value of a global charge (B, L, $(B-L)$, etc.) is nonzero for some time. This charge is generated if processes that violate this charge, as well as processes violating C and CP are in thermal equilibrium.

To illustrate this idea, let us suppose that there is interaction of the global (say, baryonic) current J_B^μ with *non-stationary* background field Φ,

$$\mathcal{L}_{\rm int} = j_\mu^B \, \partial^\mu \Phi. \tag{11.4}$$

This interaction breaks CP, and in the case of homogeneous but time-dependent Φ it breaks also invariance under time reversal (T-invariance), so CPT is also broken. This induces disbalance between particles and antiparticles: the interaction term (11.4) make the following contribution to the Hamiltonian density

$$\mathcal{H}_{\rm int} = -\dot{\Phi} \, n_B \,. \tag{11.5}$$

In thermal equilibrium with respect to baryon number and CP-violating processes there is asymmetry

$$\frac{n_B - n_{\bar{B}}}{n_B + n_{\bar{B}}} \sim \frac{\dot{\Phi}}{T}. \tag{11.6}$$

Indeed, having the term (11.5) is equivalent to adding baryon chemical potential $\mu_B = \dot{\Phi}$.

If $\dot{\Phi}$ itself depends on time, the equilibrium asymmetry also varies in time. It freezes out at temperature T_B, when baryon number violating processes switch off. Equations (5.22) and (11.6) give for the resulting asymmetry

$$\Delta_B \equiv \frac{n_B - n_{\bar{B}}}{s} \sim \frac{\dot{\Phi}(T_B)}{T_B}.$$

This mechanism of the baryon asymmetry generation is called spontaneous (or gravitational) baryogenesis [215, 216].

To end this Section, let us see that irrespective of the baryogenesis, the very fact that baryon asymmetry exists implies that there is no relic antimatter in the Universe.

To estimate the relic abundance of antibaryons, let us write the Boltzmann equation for antibaryon number density $n_{\bar{B}}$:

$$\frac{d(n_{\bar{B}}a^3)}{dt} = -\langle\sigma_{ann}v\rangle \cdot \left(n_B n_{\bar{B}} a^3 - n_B^{eq} n_{\bar{B}}^{eq} a^3\right), \tag{11.7}$$

where σ_{ann} is the annihilation cross-section. (We neglect the difference between the annihilation cross sections of protons and neutrons.) In the non-relativistic situation we have $\sigma_{ann} = \sigma_0/v$ (see Sec. 9.3), where

$$\sigma_0 \sim 1 \ (\text{Fermi})^2 = 10^{-26} \ \text{cm}^2 \sim 25 \ \text{GeV}^{-2}. \tag{11.8}$$

Equation (11.7) is obtained in the same way as Eq. (9.7) for dark matter particles.

Problem 11.1. *Derive Eq. (11.7) by making use of eq. (5.59). Hint:* $\langle\sigma_{ann}v\rangle \cdot n_B = \tau_{ann}^{-1}$ *is the annihilation probability per one antibaryon per unit time (i.e., τ_{ann} is the lifetime of an antibaryon in the cosmic medium). Make use of Eq. (5.51) and the fact that the baryon density greatly exceeds the antibaryon density.*

When the annihilation rate exdceeds the cosmological expansion rate, the antibaryon abundance is nearly equilibrium. This situation occurs at high enough temperature, as we will soon see. Chemical equilibrium breaks down when the annihilation and creation rates become insufficient to keep chemical equilibrium. The change of regimes occurs when (cf. Eq. (9.14))

$$\left|\frac{d(n_{\bar{B}}^{eq}a^3)}{dt}\right| \sim \langle\sigma_{ann}v\rangle \cdot \cdot n_B^{eq} n_{\bar{B}}^{eq} a^3. \tag{11.9}$$

At about this time the antibaryon abundance gets frozen out.

It remains to find the freeze-out temperature and the equilibrium antibaryon density at this temperature. Since chemical potentials of particles and antiparticles differ only by sign in thermal equilibrium, we write at $T \ll m_p$ (we neglect the difference between proton and neutron masses and do not write spin factors),

$$n_B^{eq} = \left(\frac{m_p T}{2\pi}\right)^{3/2} e^{-\frac{m_p - \mu_B}{T}}, \quad n_{\bar{B}}^{eq} = \left(\frac{m_p T}{2\pi}\right)^{3/2} e^{-\frac{m_p + \mu_B}{T}}. \tag{11.10}$$

Note that at $T \ll m_p$ the baryon density n_B^{eq} is not exponentially small only for $\mu_B = m_p + O(T)$. Using $n_B = \eta_B \cdot n_\gamma$, we find from (11.10)

$$n_{\bar{B}}^{eq} = \frac{1}{n_B}\left(\frac{m_p T}{2\pi}\right)^3 e^{-\frac{2m_p}{T}} \sim \frac{m_p^3}{\eta_B} \cdot e^{-\frac{2m_p}{T}}. \tag{11.11}$$

The right-hand side rapidly changes with temperature, so

$$\left|\frac{d(n_{\bar{B}}^{eq}a^3)}{dt}\right| = \left|a^3\frac{dn_{\bar{B}}^{eq}}{dt}\right|.$$

We now use $|\dot{T}/T| = H(T) = T^2/M^*_{Pl}$ and obtain from (11.11) that

$$\left| a^3 \frac{dn_{\bar{B}}^{\text{eq}}}{dt} \right| \sim a^3 \frac{m_p}{T} H \cdot n_{\bar{B}}^{\text{eq}}.$$

The freeze out relation (11.9) takes the form

$$\frac{m_p}{T} H \equiv \frac{m_p}{T} \frac{T^2}{M^*_{Pl}} \sim \langle \sigma_{ann} v \rangle n_B \sim \sigma_0 \eta_B T^3.$$

In this way we obtain freeze out temperature,

$$T \sim \left(\frac{m_p}{M^*_{Pl} \cdot \eta_B \cdot \sigma_0} \right)^{1/2} \sim 10 \text{ keV}. \tag{11.12}$$

The antibaryon abundance is fantastically small at this temperature: we find from (11.11)

$$n_{\bar{B}} \sim 10^{-10^5}, \tag{11.13}$$

no matter in which units. The visible Universe contains no single relic antibaryon.

Clearly, one naturally questions the validity of statistical physics methods, which we used in our calulation, for the system with so small number of antibaryons. Still, there is no doubt that all antibaryons annihilate away in baryon-asymmetric Universe.

Problem 11.2. *Solve the Boltzmann equation (11.7) in quadratures. Compute the integral by saddle point method and find the temperature that gives the largest contribution into the present value of $n_{\bar{B}}$, thus refining the estimate (11.12). Find the present value of $n_{\bar{B}}$. Is it consistent with (11.13)?*

Problem 11.3. *Find the relic positron abundance at the present epoch.*

11.2. Baryon and Lepton Number Violation in Particle Interactions

In this Section we discuss two mechanisms of baryon number violation. One indeed works in Nature (provided that the temperature in the Universe much exceeded 100 GeV), since it exists already in the Standard Model [194]. Another is inherent in Grand Unified Theories, GUTs (see, e.g., Ref. [195]). Even though there is no direct experimental evidence for Grand Unification, this hypothesis is very plausible. Finally, we will see in what follows that lepton number violation is also relevant for baryogenesis; we discuss one of the mechanisms of this violation at the end of this Section.

11.2.1. *Electroweak mechanism*

Baryon and lepton number violation are non-perturbative in the Standard Model; they are not visible in Feynman diagrams. We will only give here sketch of what is going on; interested reader may find details in Ref. [175] or in appropriate reviews.

The phenomenon we are going to discuss is called 't Hooft effect [138]; it is due to gauge interactions corresponding to the subgroup $SU(2)_W$ of the Standard Model group $SU(3)_c \times SU(2)_W \times U_Y$ (see Appendix B). These interactions involve *left* quarks and leptons. The classical Lagrangian is invariant under common phase rotations of all quark fields; this symmetry would give rise to baryon number conservation. It is also invariant under phase rotations of leptons of each generation separately (we neglect neutrino masses and mixing at this point); these would correspond to the conservation of three lepton numbers L_n, $n = e, \mu, \tau$, see Appendix B. However, the corresponding currents j_μ^B, $j_\mu^{L_n}$ are anomalous at the quantum level,

$$\partial^\mu j_\mu^B = 3 \frac{g^2}{32\pi^2} V^{\mu\nu\,a} \tilde{V}_{\mu\nu}^a, \tag{11.14}$$

$$\partial^\mu j_\mu^{L_n} = \frac{g^2}{32\pi^2} V^{\mu\nu\,a} \tilde{V}_{\mu\nu}^a, \quad n = 1, 2, 3, \tag{11.15}$$

where $V_{\mu\nu}^a = \partial_\mu V_\nu^a - \partial_\nu V_\mu^a + g\epsilon^{abc} V_\mu^b V_\nu^c$ is the field strength of the $SU(2)_W$ gauge field, $\tilde{V}_{\mu\nu}^a = \frac{1}{2}\epsilon_{\mu\nu\lambda\rho} V^{\lambda\rho a}$ is the dual tensor, g is $SU(2)_W$ gauge coupling. The reason for this anomaly is that left and right fermions interact with the field V_μ^a in different ways[2] (in fact, right fermions do not interact with V_μ^a at all).

Equations (11.14) and (11.15) show that baryon and lepton numbers are not conserved,[3] if there emerge non-zero gauge fields in vacuo or in medium,

$$\Delta B = B(t_f) - B(t_i) = \int_{t_i}^{t_f} dt \int d^3\mathbf{x} \partial^\mu j_\mu^B = 3 \int_{t_i}^{t_f} d^4 x \frac{g^2}{32\pi^2} V^{\mu\nu\,a} \tilde{V}_{\mu\nu}^a, \tag{11.16}$$

where t_i and t_f denote initial and final time. Similar relation is valid for each of the lepton numbers. Baryon number violation requires strong fields, $V_{\mu\nu}^a \propto \frac{1}{g}$, making the integral (11.16) different from zero (this integral takes integer values only). Energy of very strong fields is proportional to $\frac{1}{g^2}$. Thus, we conclude that baryon and lepton number violation occur when the system overcomes energy barrier; see Fig. 11.1. The estimate for its height is

$$E_{sph} \sim \frac{M_W}{g^2},$$

---

[2]In QCD, left and right quarks interact with gluon field in the same way, and baryon number is conserved in strong interactions.
[3]We do not discuss the actual physics leading to non-conservation; see in this regard [175] and references therein.

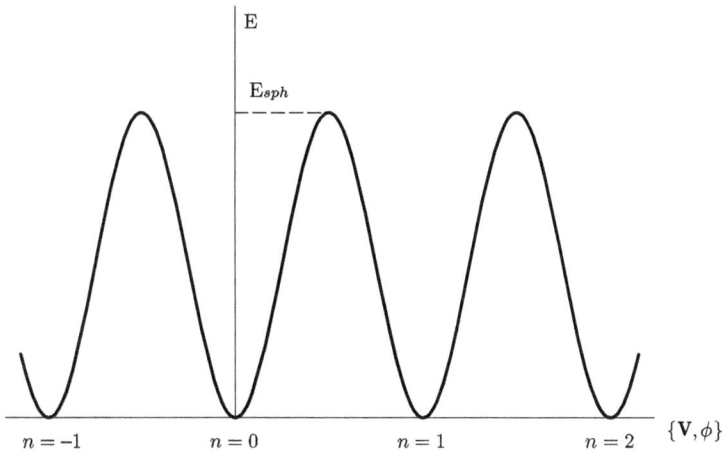

Fig. 11.1.   Schematic plot of static energy as functional of classical gauge and Higgs fields. The horizontal line corresponds to (infinite-dimensional) space of all field configurations $\{\mathbf{V}, \phi\}$. Absolute minima labeled by an integer $n = 0, \pm 1, \pm 2, \ldots$ are pure gauge configurations of zero energy, in which the fields $\mathbf{V}(\mathbf{x})$, $\phi(\mathbf{x})$ have different topological properties (topologically distinct vacua). The integral in (11.16) is equal to 1 for the fields that evolve from the vacuum with topological number n to the vacuum $(n+1)$ as time runs from t_i to t_f. This integral is zero for fields evolving near one vacuum. The "maximum" with energy E_{sph} is in fact a saddle point: energy decreases along one direction in the configuration space and increases along all other directions.

where the factor M_W is inserted on dimensional grounds.[4] The calculation of the barrier height (sphaleron energy) gives [196]

$$E_{sph} = \frac{2 M_W}{\alpha_W} B\left(\frac{m_h}{M_W}\right),$$

where

$$\alpha_W \equiv \frac{g^2}{4\pi},$$

the function $B(m_h/M_W)$ takes values from $B = 1.56$ at $m_h/M_W \ll 1$ to $B = 2.72$ at $m_h/M_W \gg 1$, and $B = 2.4$ for $m_h = 125$ GeV. Thus, the barrier height in the Standard Model is about 11 TeV.

At zero temperature and zero fermion density the transmission through the energy barrier is only possible via quantum tunneling. This tunneling is described by instanton [197]. The tunneling probability is exponentially small,

$$\Gamma \propto e^{-\frac{4\pi}{\alpha_W}}.$$

Since $\alpha_W \sim 1/30$, the suppression factor is extremely small, $\Gamma \propto 10^{-165}$. Baryon number violating processes practically do not occur in usual circumstances.

---

[4]The reason for notation E_{sph} is as follows. The height of energy barrier is equal to energy of saddle point configuration, which extremizes the static energy. This configuration is called sphaleron, from Greek $\sigma\phi\alpha\lambda\epsilon\rho o\nu$, unreliable, ready to fall.

Another situation occurs at finite temperatures [194]. In that case, the barrier may be overcome via thermal jumps on its top (better to say, on the saddle point). At relatively low temperatures a naive estimate for the suppression factor in the rate is given by the Boltzmann formula for configuration of energy E_{sph},

$$\Gamma_{sph} \propto e^{-\frac{E_{sph}}{T}}.$$

This estimate, however, is incorrect at interesting temperatures when the suppression factor is not extremely small. The probability of the realization of a given configuration is determined by its *free* energy, rather than energy. In our case the main effect is that the Englert–Brout–Higgs expectation value, and hence M_W, depends on temperature; see Chapter 10. So, the better estimate is

$$\Gamma_{sph} = C \cdot T^4 e^{-\frac{F_{sph}(T)}{T}}, \qquad (11.17)$$

where

$$F_{sph}(T) = \frac{2M_W(T)}{\alpha_W} B\left(\frac{m_h}{M_W}\right), \qquad (11.18)$$

and one can use zero-temperature masses in the argument of the function B. (In fact, the latter statement is not quite correct, but we will not need further refinement, in view of weak dependence of B on its argument.) Γ_{sph} is the over-barrier transition rate per unit time per unit volume, so we have written the factor T^4 in (11.17) on dimensional grounds. The pre-exponential factor C is dimensionless; it depends on temperature, the Englert–Brout–Higgs expectation value and couplings. The most important in (11.17) is the exponential factor, so we will set

$$C \sim 1$$

in our estimates.

Thus, at low enough temperatures the rate of electroweak baryon number violation is given by (11.17). This formula does not work, however, at high temperatures, when $F_{sph} \lesssim T$ and the exponential suppression is absent. The latter situation occurs in unbroken phase,[5] when $\langle \phi \rangle = 0$ and hence $M_W(T) = 0$. In that situation one makes use of the estimate (modulo logarithm)

$$\Gamma_{sph} = \varkappa' \alpha_W^5 T^4, \qquad (11.19)$$

where $\varkappa'$ is a numerical coefficient. This estimate follows, to a certain extent, from the results of Sec. 10.3. We have seen there that at high temperatures and $M_W(T) = 0$ the theory possesses the non-perturbative parameter $g_3^2 = g^2 T$. The sphaleron transition rate is mostly determined by this parameter; using it, we would obtain dimensional estimate $\Gamma_{sph} \sim g_3^4 \sim (\alpha_W T)^4$. The additional factor α_W is due

---

[5] Our terminology here is standard, though not quite appropriate, see discussion in Sec. 10.

to specific plasma effects [198]. The coefficient $\varkappa'$ has been found by lattice simulations and it turned out to be rather large [199],

$$\varkappa' \approx 18.$$

Note that if one substitutes the non-perturbative scale g_3^2 into (11.18) instead of $M_W(T)$, the sphaleron free energy becomes of the order of temperature, so the Boltzmann suppression is indeed absent.

Let us make use of the estimates (11.17) and (11.19) to find the range of temperatures in which electroweak baryon number violating processes are in thermal equilibrium in the early Universe. This is the case when a particle participates in at least one of these processes per Hubble time, i.e., $\Gamma_{sph} \gtrsim nH$, where $n \sim T^3$ is the number density of particles of a given type. Thus, the condition for equilibrium with respect to the sphaleron processes is

$$\frac{\Gamma_{sph}}{T^3} \gtrsim H(T) = \frac{T^2}{M_{Pl}^*}. \tag{11.20}$$

At high temperatures, we use (11.19) and find

$$T \lesssim T_{sph} \sim 10^{12} \text{ GeV}. \tag{11.21}$$

At relatively low temperatures the condition (11.20) gives

$$\frac{M_W(T)}{T} \lesssim \frac{\alpha_W}{2B(m_h/M_W)} \log \frac{M_{Pl}^*}{T}.$$

Making use of the crude estimate $T \sim 100$ GeV in the argument of logarithm, we find numerically

$$\frac{M_W(T)}{T} \lesssim \frac{0.66}{B(m_h/M_W)} \approx 0.28. \tag{11.22}$$

Thus, the sphaleron transitions switch off after electroweak transition only, i.e., at $T \lesssim 100$ GeV. We conclude that electroweak baryon number violation is in thermal equilibrium in wide temperature interval extending from 10^{12} GeV to^6 about 100 GeV.

---

[6] Numerical calculation shows that the expansion rate starts to dominate the sphaleron rate in the Standard Model at $T \approx 132$ GeV [199].

Let us now discuss selection rules for electroweak baryon number violating processes. They follow from the relations (11.14) and (11.15) and have the form

$$\Delta B = 3\Delta L_e = 3\Delta L_\mu = 3\Delta L_\tau.$$

In other words, there are three conserved combinations of baryon and lepton numbers, which can be chosen as follows,

$$(B - L), \quad (L_e - L_\mu), \quad (L_e - L_\tau). \tag{11.23}$$

At temperatures $100 \text{ GeV} \lesssim T \lesssim 10^{12} \text{ GeV}$ these numbers may be non-vanishing, while baryon and lepton numbers themselves are adjusted in such a way that the Grand canonical potential is at its minimum.

Let us find baryon and lepton number densities at temperatures above the electroweak transition but below (11.21) at given $(B-L)$ number density, assuming that the three lepton number densities are equal to each other. Let us make the calculation for ν_f fermionic generations and ν_s Englert–Brout–Higgs doublets. Quantum numbers of all particles are given in Appendix B. Above the electroweak transition, it is convenient to work in terms of the scalar doublets with components h^+ and h^0, their antiparticles and gauge bosons of two polarizations. To calculate particle number densities, we introduce chemical potentials to all conserved quantum numbers. Relevant quantum numbers are $(B - L)$ and weak hypercharge Y (see Problem 5.4). Then particles of type I have chemical potential

$$\mu_I = \mu(B_I - L_I) + \mu_Y \frac{Y_I}{2},$$

while for antiparticles $\mu_{\bar{I}} = -\mu_I$. Here B_I, L_I and Y_I are baryon number, lepton number and weak hypercharge of particle I, μ and μ_Y are chemical potentials to $(B - L)$ and $Y/2$. As an example, for left electron and neutrino we write

$$\mu_\nu = \mu_{e_L} = -\mu - \frac{1}{2}\mu_Y,$$

while for charged and neutral scalars (components of one of the Englert–Brout–Higgs doublets) we have

$$\mu_{h^+} = \mu_{h^0} = \frac{1}{2}\mu_Y,$$

etc. We make use of the result of Problem 5.4 and write for asymmetry of fermions of type F of all generations

$$n_F - n_{\bar{F}} = \Delta n_F = \frac{1}{2}\nu_f \cdot \mu_F \frac{T^2}{3},$$

while the asymmetry of scalars of type H is

$$n_H - n_{\bar{H}} = \Delta n_H = \nu_s \cdot \mu_H \cdot \frac{T^2}{3}.$$

(We assume here that asymmetries are small, so that $\mu_I \ll T$.) This gives

$$\Delta n_{h^+} + \Delta n_{h^0} = \nu_s \cdot \mu_Y \cdot \frac{T^2}{3},$$

$$\Delta n_\nu + \Delta n_{l_L} = \nu_f \left(-\frac{1}{2}\mu_Y - \mu\right) \cdot \frac{T^2}{3},$$

$$\Delta n_{l_R} = \frac{1}{2}\nu_f(-\mu_Y - \mu) \cdot \frac{T^2}{3},$$

$$\Delta n_{u_L} + \Delta n_{d_L} = \nu_f \left(\frac{1}{6}\mu_Y + \frac{1}{3}\mu\right) \cdot 3 \cdot \frac{T^2}{3},$$

$$\Delta n_{u_R} = \frac{1}{2}\nu_f \left(\frac{2}{3}\mu_Y + \frac{1}{3}\mu\right) \cdot 3 \cdot \frac{T^2}{3},$$

$$\Delta n_{d_R} = \frac{1}{2}\nu_f \left(-\frac{1}{3}\mu_Y + \frac{1}{3}\mu\right) \cdot 3 \cdot \frac{T^2}{3},$$

$$(11.24)$$

where the factor 3 in the last three formulas accounts for quark colors. Gauge bosons carry zero baryon number, lepton number and weak hypercharge, so there is no asymmetry in them.

The system is neutral with respect to all gauge charges, including weak hypercharge (this is the analog of electric neutrality of the usual plasma), therefore

$$\sum_I Y_I \cdot \Delta n_I = 0.$$

Making use of (11.24) we find

$$\nu_f \left[\frac{5}{3}\mu_Y + \frac{4}{3}\mu\right] + \frac{1}{2}\nu_s \cdot \mu_Y = 0. \qquad (11.25)$$

This explains why we introduced the chemical potential μ_Y: had we set $\mu_Y = 0$ in the beginning, the medium would not be neutral with respect to weak hypercharge at $\mu \neq 0$.

We now eliminate one of the chemical potentials, say, μ, by making use of Eq. (11.25) and express all asymmetries (11.24) in terms of the only remaining chemical potential. In this way we obtain for baryon number density, which we denote simply by B,

$$B \equiv \frac{1}{3}(\Delta n_{u_L} + \Delta n_{d_L} + \Delta n_{u_R} + \Delta n_{d_R}) = -\frac{T^2}{3}\left(\frac{1}{2}\nu_f + \frac{1}{4}\nu_s\right)\mu_Y, \qquad (11.26)$$

while the lepton asymmetry is

$$L = \frac{T^2}{3}\left(\frac{7}{8}\nu_f + \frac{9}{16}\nu_s\right)\mu_Y.$$

Thus, the value of $(B - L)$ is related to μ_Y by

$$B - L = -\frac{T^2}{3}\left(\frac{11}{8}\nu_f + \frac{13}{16}\nu_s\right)\mu_Y.$$

Let us again use (11.26) and obtain finally

$$B = \frac{8\nu_f + 4\nu_s}{22\nu_f + 13\nu_s} \cdot (B - L). \qquad (11.27)$$

This gives the constant C entering (11.3). There are three generations of fermions which are effectively massless at electroweak temperature, so in the theory with one scalar doublet we find

$$C = \frac{8\nu_f + 4\nu_s}{22\nu_f + 13\nu_s}(\nu_f = 3,\ \nu_s = 1) = \frac{28}{79}. \tag{11.28}$$

We stress that this value, as well as the whole analysis, is valid above the electroweak transition only; at the tarnsition epoch and later, the parameter C depends on temperatue and is somewhat (but not dramatically) different from (11.28); see, e.g., [200].

Problem 11.4. *Considering temperatures above electroweak transition, introduce, besides μ and μ_Y, the chemical potential μ_3 to the diagonal (third) component T^3 of weak isospin. Find asymmetries in all particles including vector bosons and show that the requirement of neutrality with respect to the weak isospin is equivalent to $\mu_3 = 0$ at any μ and μ_Y. Note: This result is not valid below electroweak transition temperature.*

Problem 11.5. *For temperatures above electroweak transition, consider general situation in which the densities of all conserved quantum numbers (11.23) are nonzero. Show that the baryon asymmetry is still given by (11.27).*

11.2.2.   *Baryon number violation in Grand Unified Theories*

Another mechanism of baryon and lepton number violation exists in Grand Unified Theories (GUTs). These contain new superheavy particles, vectors and scalars (and also fermions in supersymmetric theories) whose interactions with the Standard Model particles violate baryon and lepton numbers already at the level of perturbation theory. As an example, there is a vertex shown in Fig. 11.2(a). This vertex describes the interaction of vector boson V with two quarks (unlike the gauge vertices of the Standard Model which contain both quark and antiquark). The existence of this vertex does not yet mean that baryon number is violated: were this vertex the only one, we could assign baryon number $2/3$ to the boson V, and baryon number would still be conserved. However, there is also a vertex shown

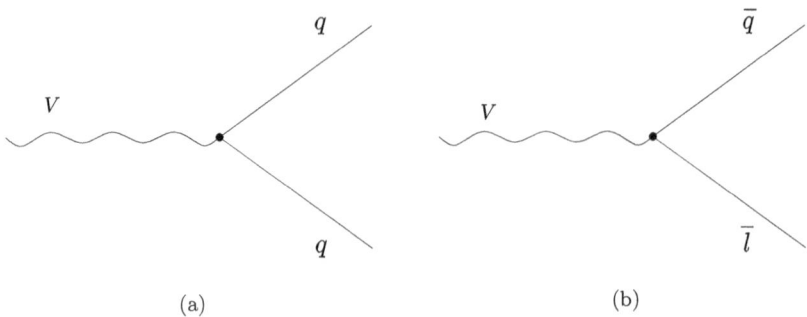

Fig. 11.2.   Vector boson interaction with quarks q, antiquarks $\bar{q}$ and antileptons $\bar{l}$.

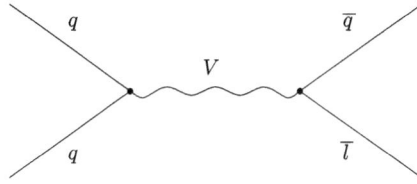

Fig. 11.3. V-boson exchange leading to baryon number violation.

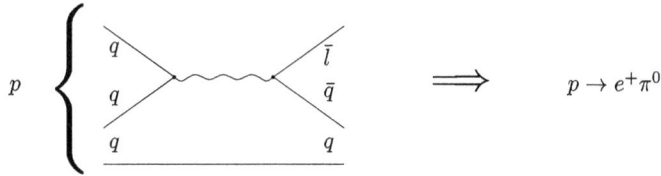

Fig. 11.4. Proton decay diagram.

in Fig. 11.2(b) (we will explain shortly why it involves antilepton and not lepton). Therefore, baryon number is indeed violated: V-boson exchange leads to the process shown in Fig. 11.3.

Interactions of Fig. 11.3 give rise to proton decay, see Fig. 11.4. There are very strong experimental bounds on the proton lifetime: depending on the decay mode

$$\tau_p > 10^{32} - 10^{33} \text{ yrs.} \tag{11.29}$$

This leads to strong bounds on baryon number violating interactions. The diagram 11.4 implies the following estimate for the proton width,

$$\Gamma_p \equiv \frac{1}{\tau_p} \sim \frac{\alpha_V^2}{M_V^4} m_p^5,$$

where $\alpha_V = \frac{g_V^2}{4\pi}$, g_V is the coupling constant in each vertex; the factor M_V^{-2} in amplitude, and hence M_V^{-4} in the width, comes from the V-boson propagator, and the factor m_p^5 is introduced on dimensional grounds. Together with (11.29) this gives the bound

$$M_V \gtrsim 10^{16} \text{ GeV.} \tag{11.30}$$

Thus, we are dealing with interactions that could be of interest for cosmology only if the temperatures in the Universe were as high as the GUT energy scale

$$M_{GUT} \sim 10^{16} \text{ GeV.} \tag{11.31}$$

If the theory does not involve new fermions beyond the Standard Model ones, then the baryon number violating interactions are indeed given only by diagrams of the types shown in Fig. 11.2, with the diagram 11.2(b) containing precisely antilepton. This is true both for new vector boson V and possible new scalar boson S.

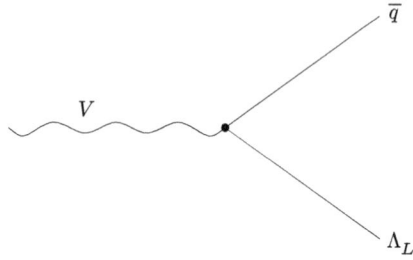

Fig. 11.5. $(B - L)$-violating vertex.

This follows from the invariance under the Standard Model gauge group; we will see that in the end of this Section. This result is important, since all these interactions conserve $(B - L)$ if we assign $(B - L) = \frac{2}{3}$ to vectors V and scalars S. Hence, baryon asymmetry cannot be due to these interactions only: they would not generate non-vanishing $(B - L)$, and electroweak processes studied in the previous Section would wash the baryon asymmetry out. We note that similar situation occurs in supersymmetric models as well.

If there are new fermions in the theory, $(B - L)$ may be violated simultaneously with B. A simple example is given by a theory with extra fermion Λ_L which is neutral with respect to the Standard Model gauge interactions and whose baryon and lepton numbers are zero. Then the theory may contain both vertices shown in Fig. 11.2 and vertex of Fig. 11.5. Final states in Fig. 11.2 have $(B - L) = \frac{2}{3}$, while the final state in Fig. 11.5 has $(B - L) = -\frac{1}{3}$, so $(B - L)$ is not conserved indeed. Another possibility is that Λ_L is lepton, i.e., its lepton number is $L = 1$.

The reason for the existence of interactions shown in Fig. 11.2 in GUTs is as follows. Grand Unification of strong and electroweak interactions assumes that at ultra-high energies, all these gauge interactions are one and the same force. In other words, the Standard Model gauge group $SU(3)_c \times SU(2)_W \times U_Y$ is a subgroup of simple gauge group G, and known fermions combine (possibly together with new fermions) into representations of G. This immediately implies that a multiplet containing the lepton doublets (as well as possibly different multiplet containing right charged lepton) must also contain fermions that transform non-trivially under $SU(3)_c$, i.e., carry color. If one does not introduce many new fermions, the color partners of leptons are identified with quarks (or antiquarks). Now, since quarks and leptons are in one gauge multiplet, there must be a gauge boson interacting as shown in Fig. 11.2(b). A multiplet with colored $SU(2)_W$-doublets may also contain colored $SU(2)_W$-singlets. In other words, a multiplet under group G with left quark doublets may contain also $SU(2)_W$-singlet colored fermions, which should also be left. These are left *anti*-quarks. So, there may exist multiplets of the full gauge group G that contain both quarks and antiquarks. This leads to interaction shown in Fig. 11.2(a).

The simplest, but probably unrealistic example is given by the theory with gauge group $G = SU(5)$ [201], whose algebra includes the Standard Model gauge algebras in the following way,

$$SU(3)_c \; : \; \begin{pmatrix} SU(3) & 0_{3\times2} \\ 0_{2\times3} & 0_{2\times2} \end{pmatrix},$$

$$SU(2)_w \; : \; \begin{pmatrix} 0_{3\times3} & 0_{3\times2} \\ 0_{2\times3} & SU(2) \end{pmatrix},$$

$$(11.32)$$

$$U(1)_Y : Y = \sqrt{\frac{5}{3}} T^{24}, \quad T^{24} = \frac{1}{2\sqrt{15}} \cdot \operatorname{diag}(2,2,2,-3,-3). \qquad (11.33)$$

(Subscripts in (11.32) denote dimensions of matrices; $0_{m\times n}$ is matrix with zero entries.) New fermions are not needed to complete $SU(5)$ multiplets, while all Standard Model fermions of one generation are comfortably placed in representations $\bar{5}$ (antifundamental representation of $SU(5)$) and 10 (antisymmetric representation of $SU(5)$):

$$\bar{5} = (d^c_{L1}, d^c_{L2}, d^c_{L3}, e^-_L, \nu_e),$$

$$10 = \begin{pmatrix} 0 & u^{c\,3}_L & -u^{c\,2}_L & u^1_L & d^1_L \\ -u^{c\,3}_L & 0 & u^{c\,1}_L & u^2_L & d^2_L \\ u^{c\,2}_L & -u^{c\,1}_L & 0 & u^3_L & d^3_L \\ -u^1_L & -u^2_L & -u^3_L & 0 & e^+_L \\ -d^1_L & -d^2_L & -d^3_L & -e^+_L & 0 \end{pmatrix},$$

where superscript refers to color and u^c_L and d^c_L denote left antiquark fields (antitriplets under $SU(3)_c$ and singlets under $SU(2)_W$).

Problem 11.6. *Show that all fermions have correct quantum numbers with respect to the Standard Model gauge group $SU(3)_c \times SU(2)_W \times U_Y$ embedded in $SU(5)$ according to (11.32), (11.33).*

Gauge bosons, as always, belong to adjoint representation of $SU(5)$ (24-plet). Besides the Standard Model gauge bosons corresponding to the algebras (11.32), (11.33), there are 12 extra gauge bosons whose fields are embedded in $SU(5)$ algebra as follows,

$$\begin{pmatrix} 0 & 0 & 0 & V^1_\mu & U^1_\mu \\ 0 & 0 & 0 & V^2_\mu & U^2_\mu \\ 0 & 0 & 0 & V^3_\mu & U^3_\mu \\ V^{1*}_\mu & V^{2*}_\mu & V^{3*}_\mu & 0 & 0 \\ U^{1*}_\mu & U^{2*}_\mu & U^{3*}_\mu & 0 & 0 \end{pmatrix}.$$

Here every field U^a_μ, V^a_μ is complex and describes two vector bosons. Their interactions with quarks and leptons have precisely the form shown in Fig. 11.2.

Problem 11.7. *Write all terms in the Lagrangian describing the interactions of V_μ- and U_μ-bosons with quarks and leptons. Hint: To simplify notations, notice that (V_μ, U_μ) together form $SU(2)_W$-doublet and that they are $SU(3)_c$-triplets.*

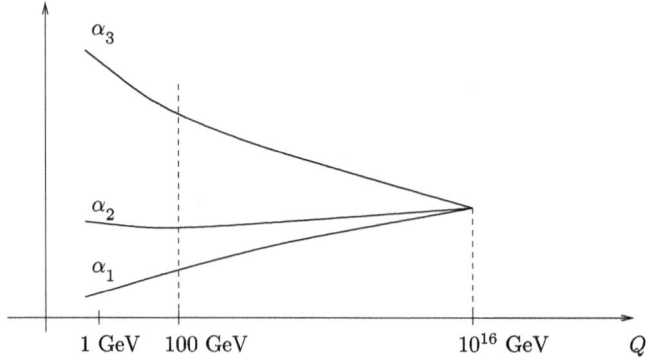

Fig. 11.6. Sketch of the evolution of the Standard Model gauge couplings with momentum transfer in the supersymmetric extension of the Standard Model. Here $\alpha_3 \equiv \alpha_s \equiv \frac{g_s^2}{4\pi}$, $\alpha_2 \equiv \alpha_w \equiv \frac{g^2}{4\pi}$, where g_s and g are gauge couplings of $SU(3)_c$ and $SU(2)_w$; $\alpha_1 = \frac{5}{3}\frac{g'^2}{4\pi}$, where g' is gauge coupling of $U(1)_Y$; the factor $\frac{5}{3}$ is due to the fact that the normalization of the generator $U(1)_Y$ is different from the standard normalization of the generators of GUT gauge group. (As an example, trace of the generator T^{24} squared in $SU(5)$ equals the standard value $1/2$, while trace of Y^2 equals $5/6$, see (11.33); hence, gauge couplings of $SU(5)$ and $U(1)_Y$ are related by $g_{SU(5)} = \sqrt{\frac{5}{3}}g'$.)

The energy scale (11.31) emerges in GUTs in a natural way [202]. As we pointed out in Sec. 9.4.2, couplings logarithmically depend on energy (more precisely, on momentum transfer Q). The gauge couplings of the groups $SU(3)_c$, $SU(2)_w$ and $U(1)_Y$ are very different at low energies, but as the energy increases, they get closer, as schematically shown in Fig. 11.6. In *supersymmetric extension* of the Standard Model (but *not* in the Standard Model itself), all three couplings become equal to $\alpha_{GUT} \approx 1/25$ at $Q = M_{GUT} \approx 10^{16}$ GeV [131]. This is precisely what should happen in a theory with a single gauge interaction above energy M_{GUT} with single coupling. Gauge coupling unification is a very strong argument both for Grand Unification and for the supersymmetric extension of the Standard Model at relatively low energies.

Let us continue our general discussion and find the constraints that the Standard Model gauge symmetries impose on the structure of baryon number violating interactions. Let us assume for the time being that there are no new fermions. It is convenient in this Section to treat all fermions as left, i.e., instead of right quark and lepton fields U_R, D_R and E_R consider left fields denoted by U_L^c, D_L^c and E_L^c and describing left antiquarks and antileptons. These are singlets under $SU(2)_w$, antitriplets (U_L^c, D_L^c) and singlets (E_L^c) under color $SU(3)_c$; their weak hypercharges are opposite in sign to hypercharges of the respective particles, i.e,

$$Y_{U_L^c} = -\frac{4}{3}, \quad Y_{D_L^c} = \frac{2}{3}, \quad Y_{E_L^c} = 2.$$

Let us also recall the weak hypercharges of left doublets,

$$Y_{Q_L} = \frac{1}{3}, \quad Y_{L_L} = -1.$$

In terms of left fields, the only Lorentz-invariant renormalizable interaction with vector field has the Lorentz structure

$$\bar{\psi}_{Li}\gamma^\mu V_\mu \psi_{Lj} \propto V\bar{\psi}_i\psi_j, \tag{11.34}$$

while Yukawa interaction with scalar field S can only have the form (compare with Majorana mass term for neutrino, Appendix C)

$$\bar{\psi}_L^c S \psi_L \propto S \psi_i \psi_j \quad \text{or} \quad \bar{\psi}_L S \psi_L^c \propto S \bar{\psi}_i \bar{\psi}_j, \tag{11.35}$$

where ψ_i, ψ_j are various Standard Model fermion fields. We stress that (11.34) contains both ψ and $\bar{\psi}$, while (11.35) contains only ψ or only $\bar{\psi}$.

Interactions (11.34), (11.35) must be invariant under all gauge symmetries of the Standard Model. Let us begin with color $SU(3)_c$. If there exists vertex of Fig. 11.2(a), then V is either color antitriplet or sextet (since $SU(3)_c$ representations obey $\mathbf{3} \times \mathbf{3} = \mathbf{6} + \bar{\mathbf{3}}$). $SU(3)_c$ forbids other vertices in the sextet case, so this case is not interesting. For antitriplet V, there are two more vertices allowed by $SU(3)_c$: this is the vertex of Fig. 11.2(b) and similar vertex with lepton instead of antilepton. The same holds true for a scalar S. We see that interactions with (anti)triplet bosons exhaust all possible baryon number violating interactions, unless new fermions are introduced.

Further constraints come from the requirement of invariance under $SU(2)_W$ and weak hypercharge. Let us begin with vector particles V. Since U_L^c and D_L^c are *anti*quark fields, there are two combinations of the fields that can enter the interactions of the type (11.34) and give rise to the vertex of Fig. 11.2(a):

$$\bar{U}_L^c Q \quad \left(\mathbf{2}, \frac{5}{3}\right), \tag{11.36}$$

$$\bar{D}_L^c Q \quad \left(\mathbf{2}, -\frac{1}{3}\right). \tag{11.37}$$

(We show in parenthesis the representations of $SU(2)_W$ — doublet $\mathbf{2}$ in our case — and weak hypercharge of each operator.) The combination of antiquark and lepton or antilepton interacting with V must have the same quantum numbers. The doublet under $SU(2)_W$ combinations are

$$\bar{Q}_L E_L^c \quad \left(\mathbf{2}, \frac{5}{3}\right), \quad \bar{L} U_L^c \quad \left(\mathbf{2}, -\frac{1}{3}\right), \quad \bar{L} D_L^c \quad \left(\mathbf{2}, \frac{5}{3}\right). \tag{11.38}$$

(Recall that doublet representation of $SU(2)$ coincides with its conjugate.) We see that the vector boson V must be a doublet under $SU(2)_W$. There are two possibilities: either it has weak hypercharge $-\frac{1}{3}$ and has interactions

$$V^\dagger \bar{D}_L^{(c)} Q_L + V^\dagger \bar{L} U_L^{(c)} + h.c.,$$

or its weak hypercharge is $\frac{5}{3}$ and the interactions are

$$V^\dagger \bar{U}_L^c Q_L + V^\dagger \bar{Q}_L E_L^c + V^\dagger \bar{L} D_L^c + h.c.,$$

where we do not write Lorentz structure and couplings. The corresponding vertices for the second case are shown in terms of usual quarks and leptons in Fig. 11.7.

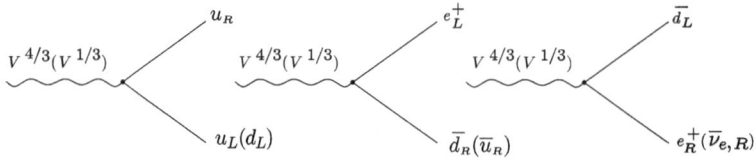

Fig. 11.7.   Vector boson interactions that violate baryon number. $V^{4/3}$ and $V^{1/3}$ denote upper and lower components of the weak doublet, 4/3 and 1/3 are electric charges. Quarks and leptons are assumed to belong to the first generation; in fact, each vertex may contain arbitrary linear combination of quarks and leptons of all generations with quantum numbers shown here.

Table 11.1.   Baryon number violating interactions in terms of the usual quark and lepton fields.

	Quantum numbers with respect to $SU(3)_c \times SU(2)_W \times U(1)_Y$	Interacts with
vector V_1	$\bar{\mathbf{3}}, \mathbf{2}, -\frac{1}{3}$	$D_R Q_L,\ \bar{L} \bar{U}_R$
vector V_2	$\bar{\mathbf{3}}, \mathbf{2}, \frac{5}{3}$	$U_R Q_L,\ \bar{Q}_L \bar{E}_R,\ \bar{L} \bar{D}_R$
scalar S_1	$\bar{\mathbf{3}}, \mathbf{3}, \frac{2}{3}$	$Q_L Q_L,\ \bar{Q}_L \bar{L}$
scalar S_2	$\bar{\mathbf{3}}, \mathbf{1}, \frac{2}{3}$	$Q_L Q_L,\ U_R D_R,\ \bar{Q}_L \bar{L},\ \bar{U}_R \bar{E}_R$
scalar S_3	$\bar{\mathbf{3}}, \mathbf{1}, \frac{8}{3}$	$U_R U_R,\ \bar{D}_R \bar{E}_R$

Let us now turn to scalars. Now we have to consider interactions with Lorentz structure (11.35). The combinations leading to the vertex of type of Fig. 11.2(a) are

$$Q_L Q_L \quad \left(\mathbf{3}, \frac{2}{3} \right) \tag{11.39}$$

$$Q_L Q_L \ \left(\mathbf{1}, \frac{2}{3} \right), \quad \bar{U}_L^c \bar{U}_L^c \ \left(\mathbf{1}, \frac{8}{3} \right), \quad \bar{U}_L^c \bar{D}_L^c \ \left(\mathbf{1}, \frac{2}{3} \right), \quad \bar{D}_L^c \bar{D}_L^c \ \left(\mathbf{1}, -\frac{4}{3} \right). \tag{11.40}$$

There is one triplet combination of the type (11.35) involving antiquark and giving rise to the vertex of the type of Fig. 11.2(b),

$$\bar{Q}_L \bar{L} \quad \left(\mathbf{3}, \frac{2}{3} \right), \tag{11.41}$$

and three singlet combinations,

$$\bar{Q}_L \bar{L} \ \left(\mathbf{1}, \frac{2}{3} \right), \quad U_L^c E_L^c \ \left(\mathbf{1}, \frac{2}{3} \right), \quad D_L^c E_L^c \ \left(\mathbf{1}, \frac{8}{3} \right). \tag{11.42}$$

Thus, triplet under $SU(2)_W$ scalar of weak hypercharge $\frac{2}{3}$ can interact with combinations (11.39), (11.41), while siglet scalars can interact with appropriate combinations from (11.40), (11.42). All these possibilities are listed in Table 11.1; structure of all interaction vertices is the one shown in Fig. 11.8. We see that $(B - L)$ is conserved in the absence of new fermions.

The structure of vertices for all these interactions is shown in Fig. 11.8. Thus, we again see that symmetries of the Standard Model ensure $(B - L)$ conservation unless there are new fermions.

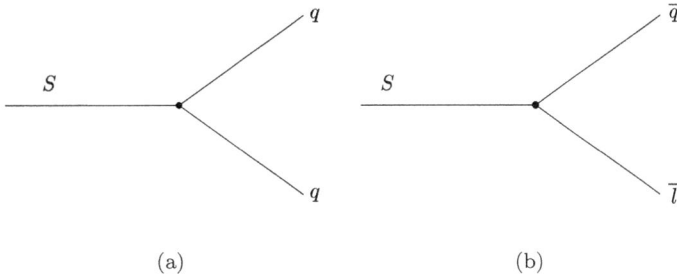

Fig. 11.8. Interactions with scalars also conserve $(B - L)$.

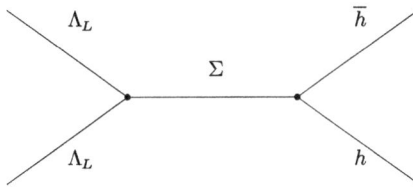

Fig. 11.9. Annihilation of two Λ_L.

Let us now include into the theory a new fermion Λ_L. Let it be a singlet under the Standard Model gauge interactions. Color anti-triplet combinations that could interact with vectors and lead to vertices shown in Fig 11.5 are

$$\bar{Q}_L \Lambda_L \ \left(2, -\frac{1}{3}\right), \quad \bar{\Lambda}_L U_L^c \ \left(1, -\frac{4}{3}\right), \quad \bar{\Lambda}_L D_L^c \ \left(1, \frac{2}{3}\right).$$

The first of them has quantum numbers (11.37), so that color antitriplet vector boson with these quantum numbers can have vertices shown both in Fig. 11.2(a) and in Fig. 11.5. The scalar combinations of antiquarks and new fermions are

$$\bar{Q}_L \bar{\Lambda}_L \ \left(2, -\frac{1}{3}\right), \quad U_L^c \Lambda_L \ \left(1, -\frac{4}{3}\right), \quad D_L^c \Lambda_L \ \left(1, \frac{2}{3}\right).$$

The second and third of them match one of the combinations in (11.40), so there are two types of scalars that have vertices shown both in Fig. 11.8(a) and in Fig. 11.5.

The neutral fermion Λ_L may have Majorana mass; see Appendix C. If it is stable, it is cosmologically allowed and for appropriate values of parameters it can be dark matter candidate, provided it pair annihilates via, say, exchange of a new neutral scalar Σ, see Fig. 11.9, and the annihilation cross-section is sufficiently large.[7] On the other hand, if baryon asymmetry is generated in the processes shown in Figs. 11.2 and 11.5, then there should be practically no interactions with leptons and Englert–Brout–Higgs doublets like

$$h_\Lambda \tilde{H}^\dagger \Lambda_L L + h.c. \tag{11.43}$$

These would give rise to the vertex of Fig. 11.10 that violates lepton number and $(B - L)$, and hence in combination with electroweak processes would wash out baryon asymmetry.

---

[7]The existence of these processes means that the lepton number of Λ_L is naturally set equal to zero.

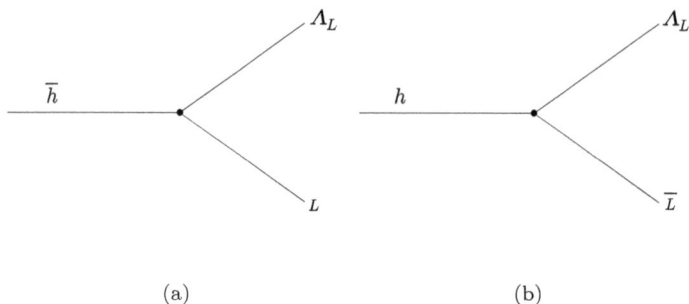

Fig. 11.10.   Lepton number violation in interactions with the Higgs boson.

Problem 11.8. *Assuming that masses of Λ_L- and Σ-particles are small compared to masses of V- and S-bosons participating in baryon number violating processes, and that at energies below these masses Λ_L-particles participate only in interactions shown in Fig. 11.9, find cosmological bounds on masses m_Λ, m_Σ and Yukawa couplings. Consider the cases $m_\Lambda \gg m_\Sigma$ and $m_\Lambda \ll m_\Sigma$ separately. At what values of these parameters is the fermion Λ_L a dark matter candidate?*

Problem 11.9. *Let the fermion Λ_L have Majorana mass $m_\Lambda \gg m_h$, where m_h is the mass of the Standard Model Higgs boson.*

 (1) *At what values of the coupling h_Λ in (11.43) is the interaction of Fig. 11.10 irrelevant for cosmology? Assume that the annihilation processes of Fig. 11.9 lead to negligible number of Λ-particles at low temperatures.*
 (2) *The interactions (11.43) contribute to masses of the conventional neutrinos via seesaw mechanism (see Appendix C). Using the previous result, find bounds on these contributions.*

Problem 11.10. *Let the interactions of Fig. 11.9 be absent, while interactions of Fig. 11.10 be non-negligible. Let the baryon asymmetry be generated at $T \gg m_\Lambda$ (say, in decays of super-heavy scalar bosons $S \to qq$ and $S \to \bar{q}\Lambda_L$; see Figs. 11.5 and 11.8). Assuming that fermions Λ_L have large Majorana mass $m_\Lambda \gg m_h$, find bounds on m_Λ and h_Λ from the following requirements: (a) baryon asymmetry must be preserved until the present epoch; (b) fermions Λ_L must have decayed long before BBN epoch. Estimate in this case the contribution of interactions (11.43) to Majorana masses of conventional neutrinos.*

Let us now turn to another case mentioned above, namely, that the neutral fermion Λ_L has lepton number $+1$, while lepton number is conserved below M_{GUT} modulo electroweak effects, and new lepton has short enough lifetime. The minimal possibility is that Λ_L is the left component of a Dirac fermion Λ. Then the Standard Model Lagrangian can be extended by adding terms (we do not write kinetic term for Λ),

$$M_\Lambda \bar{\Lambda}\Lambda + h_\Lambda \bar{\Lambda}\tilde{H}^\dagger L + h.c.$$

If the mass of Λ is large compared to the mass of the Higgs boson, $M_\Lambda > m_h$, then the main decay channels of the lepton Λ are $\Lambda \to h^0 \nu$, $\Lambda \to h^\pm l^\mp$, while for $M_\Lambda < m_h$ its decay proceeds via the Higgs boson exchange. Heavy leptons of this type are cosmologically acceptable.

Problem 11.11. *Are there massless neutrinos in the latter extension of the Standard Model?*

11.2.3.   *Violation of lepton numbers and Majorana masses of neutrino*

As we discuss in Appendix C, one of the plausible ways to explain small neutrino masses is the see-saw mechanism [203–206]. One adds new fields N_L^α, heavy sterile neutrinos, which we treat as left fermions. These fields are neutral with respect to the gauge interactions of the Standard Model, hence the name "sterile". The superscript α labels the species of these new fields; in what follows we assume, although this is not completely necessary, that $\alpha = 1, 2, 3$, i.e., there are three new fields, according to the number of Standard Model generations. The fields N_L^α have Majorana masses and interact with the Standard Model fields, so additional terms in the Lagrangian are

$$\mathcal{L} = -\frac{M_\alpha}{2} \bar{N}_{L\alpha}^c N_{L\alpha} - y_{\alpha\beta}^* \bar{N}_{L\alpha}^c \tilde{H}^\dagger L_\beta + h.c., \tag{11.44}$$

where L_α are left leptonic doublets of the Standard Model, $\tilde{H}$ is related to the Standard Model field H by (see Appendix B) $\tilde{H}_i = \epsilon_{ij} H_j^*$ (i, j are the indices of the doublet representation of $SU(2)_W$), summation over α, β is assumed. The Yukawa couplings $y_{\alpha\beta}$ are in general complex while the masses M_α are real. The second term in (11.44) is the only interaction of $N_{L\alpha}$ consistent with the Standard Model gauge symmetries, unless other new fields are introduced.

Neglecting the second term in (11.44) (this is an excellent approximation in the present context) one observes that the fields N_L^α describe three fermions of masses M_α. We will assume that $M_\alpha \gg v$; this range of masses is indeed preferred from the viewpoint of baryogenesis. The Yukawa interaction in (11.44) leads to the instability of N-particles. We will be interested in high temperature situation, when the Englert–Brout–Higgs expectation value is zero (see Chapter 10). Then the doublet H describes one electrically neutral and one electrically charged scalar particle plus their antiparticles. Assuming that $M_\alpha \gg m_h$, the Yukawa interaction explicitly written in (11.44) gives rise to the decay (Fig. 11.11(a)),

$$N_\alpha \to h l_\beta, \tag{11.45}$$

where l_β is a charged lepton or neutrino of generation β, while h is one of the Higgs particles. The Hermitean conjugate interaction term gives rise to the decay into antilepton in the final state[8] (Fig. 11.11(b)),

$$N_\alpha \to h \bar{l}_\beta. \tag{11.46}$$

---

[8]The scalar particles in (11.45) and (11.46) are actually different: say, in the case of charged lepton $h = h^+$ and $h = h^-$ in (11.45) and (11.46), respectively. This will be unimportant for what follows, and we will use somewhat vague notation h.

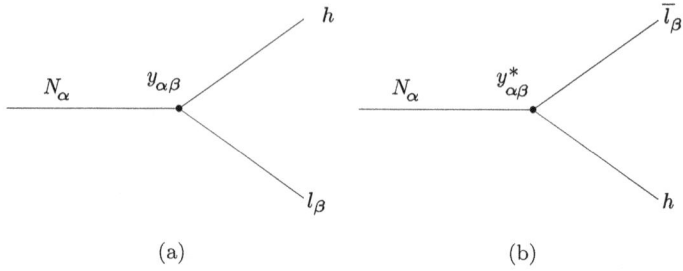

Fig. 11.11.   Decays of heavy neutrino.

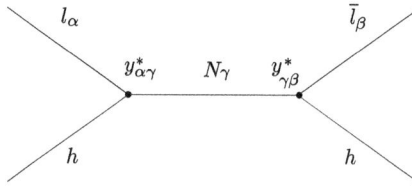

Fig. 11.12.   Lepton–Higgs scattering with lepton number violation.

Clearly, the existence of both of these processes means that lepton numbers are not conserved, whatever lepton numbers are assigned to N_α. The same conclusion follows from the existence of a process (Fig. 11.12)

$$hl_\alpha \to h\bar{l}_\beta,$$

as well as neutrinoless double-β decay. It is useful for what follows to estimate the decay width,

$$\Gamma_{N_\alpha} \sim \frac{y^2}{8\pi} M_\alpha. \tag{11.47}$$

More precisely, the total decay width is given by (assuming $M_\alpha \gg v$)

$$\Gamma_{N_\alpha} = \sum_\beta \frac{|y_{\alpha\beta}|^2}{8\pi} M_\alpha.$$

Let us stress that *in the tree level approximation,* partial widths of the decays (11.45) and (11.46) are equal to each other,

$$\Gamma(N_\alpha \to hl_\beta) = \Gamma(N_\alpha \to h\bar{l}_\beta). \tag{11.48}$$

This is *not* the case once loop corrections are included. This is very important from the viewpoint of the baryon asymmetry generation; see discussion in Sec. 11.4.

Problem 11.12. *Consider a theory of one fermion field N_L (the subscript L will be omitted in this and next problem) whose Lagrangian is*

$$\mathcal{L} = i\bar{N}\gamma^\mu\partial_\mu N + \frac{M}{2}\bar{N}^c N.$$

Obtain the field equation and its positive-energy solutions for $|\mathbf{p}| \ll M$ and $|\mathbf{p}| \gg M$, where $\mathbf{p}$ is spatial momentum. Wave functions of which particles do they describe?

Problem 11.13. *Quantize the model of the previous problem. Upon adding the Yukawa term $(y\bar{N}^c\psi\varphi + h.c.)$, where ψ and φ are massless left fermion and scalar, respectively, find the decay widths of $N \to \psi\varphi$ and $N \to \bar{\psi}\varphi^*$ at the tree level, thus confirming* (11.47) *and* (11.48)*.*

11.3. Asymmetry Generation in Particle Decays

Theories extending the Standard Model sometimes provide a simple mechanism of the baryon asymmetry generation due to particle decays. As we discussed in Secs. 11.2.2 and 11.2.3, these theories may contain new particles whose decays violate baryon and/or lepton numbers. In fact, what is important is $(B-L)$-violation: these decays occur before the electroweak transition, so the electroweak processes discussed in Sec. 11.2.1 make the baryon asymmetry equal[9] to the generated $(B-L)$ asymmetry up to a numerical factor of order 1.

As we discussed in Sec. 11.2.2, scalar bosons of GUTs, S (or vector bosons V), may have decay channels

$$\begin{aligned}
(1): & \quad S \to qq \\
(2): & \quad S \to \bar{q}\Lambda
\end{aligned} \tag{11.49}$$

where q denotes conventional quarks and Λ is a new fermion neutral with respect to $SU(3)_c \times SU(2)_W \times U(1)_Y$. To simplify formulas below we assume (though this is not necessary) that the decay $S \to \bar{q}\bar{l}$ has negligible partial width. Lepton number of Λ is either 0 or +1; we will take $L_\Lambda = 0$ for definiteness. The values of $(B-L)$ of the final states of the first and second type are, respectively,

$$(B-L)_{(1)} = \frac{2}{3} \quad \text{and} \quad (B-L)_{(2)} = -\frac{1}{3}. \tag{11.50}$$

The boson S is color antitriplet, so there exists its antiparticle, triplet $\bar{S}$. Its decay channels are

$$\begin{aligned}
(\bar{1}): & \quad \bar{S} \to \bar{q}\bar{q}, \\
(\bar{2}): & \quad \bar{S} \to q\bar{\Lambda},
\end{aligned} \tag{11.51}$$

---

[9]The possibility that the baryon asymmetry is generated in the electroweak processes themselves is studied in Sec. 11.5.

and $(B - L)$ of final states are given by

$$(B - L)_{(\bar{1})} = -\frac{2}{3} \quad \text{and} \quad (B - L)_{(\bar{2})} = \frac{1}{3}. \tag{11.52}$$

Suppose that at temperatures exceeding the S-boson mass m_s, bosons S and $\bar{S}$ are in thermal equilibrium. Then their number densities, modulo color and spin factors, are the same as those of other particles. As the Universe cools down, thermal equilibrium breaks down: S- and $\bar{S}$-bosons decay, while the inverse processes of their production do not occur. These decays produce $(B - L)$-asymmetry, provided that the probabilities of decays (1) and (2) are not the same as probabilities of $(\bar{1})$ and $(\bar{2})$, respectively. Let us denote partial widths of decays (1), (2), $(\bar{1})$ and $(\bar{2})$ by $\Gamma_{(1)}$, $\Gamma_{(2)}$, $\Gamma_{(\bar{1})}$ and $\Gamma_{(\bar{2})}$. $(B - L)$-asymmetry produced in decays of one S-boson and one $\bar{S}$-boson is

$$\delta = \frac{1}{\Gamma_{tot}} \left[\left(\frac{2}{3}\Gamma_{(1)} - \frac{1}{3}\Gamma_{(2)} \right) - \left(\frac{2}{3}\Gamma_{(\bar{1})} - \frac{1}{3}\Gamma_{(\bar{2})} \right) \right],$$

where Γ_{tot} is the total S-boson width. The latter equals the total $\bar{S}$ boson width due to CPT-theorem,

$$\Gamma_{tot} = \Gamma_{(1)} + \Gamma_{(2)} = \Gamma_{(\bar{1})} + \Gamma_{(\bar{2})}.$$

(We assume for simplicity that there are no other channels.) Making use of the latter equality, we obtain for "microscopic asymmetry"

$$\delta = \frac{\Gamma_{(1)} - \Gamma_{(\bar{1})}}{\Gamma_{tot}}. \tag{11.53}$$

The difference between partial widths of the decays (1) and $(\bar{1})$ is possible only if C and CP are violated; this is the way the second condition of Sec. 11.1 shows up.

   Partial widths of a particle and antiparticle coincide at the tree level, e.g. (see Fig. 11.8(a)),

$$\Gamma^{tree}(S \rightarrow qq) = \Gamma^{tree}(\bar{S} \rightarrow \bar{q}\bar{q}). \tag{11.54}$$

Indeed, modulo one and the same kinematic factor, the tree level probabilities are equal to the modulus squared of the coupling $g_{(1)}$ in the interaction vertex, which is the same for particle and antiparticle. However, the equality (11.54) does not hold, generally speaking, at one loop level (cf. Sec. B.4). To obtain nonzero microscopic asymmetry (11.53), it is sufficient to have other bosons S' with the same quantum numbers as S, with the couplings to all these bosons being complex. Note that complex couplings mean CP-violation in models we consider here. At the one loop level, the amplitude of decay $S \rightarrow qq$ is given then by the sum of the diagrams shown in Fig. 11.13. (There are other diagrams, but considering these is sufficient for the sake of argument.)

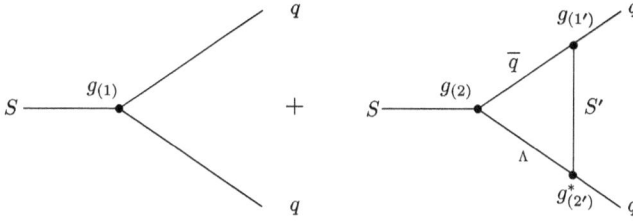

Fig. 11.13.   Diagrams for $S \to qq$ decay at the one loop level. Bosons S' have the same quantum numbers as S. Shown here are couplings in interaction vertices.

Modulo common kinematical factor, partial width $\Gamma_{(1)}$ is, at the one loop level,

$$\Gamma_{(1)} = \text{const} \cdot |g_{(1)} + Dg_{(2)}g^*_{(2')}g_{(1')}|,$$

where D is the one loop Feynman integral for the diagram of Fig. 11.13, and summation over all S'-bosons is assumed. The analogous amplitude for antiparticle contains complex conjugate couplings, therefore

$$\Gamma_{(\bar{1})} = \text{const} \cdot |g^*_{(1)} + Dg^*_{(2)}g_{(2')}g^*_{(1')}|.$$

As a result, the microscopic asymmetry is[10]

$$\delta = -2\text{Im}(D) \cdot \frac{\text{Im}(g_{(1)}g^*_{(2)}g_{(2')}g^*_{(1')})}{|g_{(1)}|^2 + |g_{(2)}|^2}. \tag{11.55}$$

Crudely speaking, this is proportional to coupling constant squared, while $\text{Im}(D)$ contains small loop factor. Assuming that the phases of couplings $g_{(i)}$ and $g_{(i')}$ are neither small nor correlated, we obtain crude estimate,

$$\delta \sim \frac{g^2}{4\pi} f\left(\frac{m_S}{m_{S'}}\right), \tag{11.56}$$

where the function f of mass ratio $m_S/m_{S'}$ is of order 1 at $m_S \gtrsim m_{S'}$ and decays as $m_S/m_{S'}$ decreases (fermion masses have been neglected).

Problem 11.14. *Find the asymmetry δ in the above case assuming that S and S' are scalars. Find its dependence on the ratio of masses of S- and S'- bosons assuming $m_S \lesssim m_{S'}$ (but not necessarily $m_S \ll m_{S'}$).*

Let us now turn to cosmological generation of $(B - L)$-asymmetry. The simplest case is when S- and $\bar{S}$-particles are in thermal equilibrium at $T \gg m_S$, while at $T \lesssim m_S$ their decays are the dominant processes. Both of these properties are non-trivial. The first implies that the temperature in the Universe was indeed as high as $T \gg m_S$, and, furthermore, that production and annihilation of $S\bar{S}$-pairs is fast at these temperatures. The second is valid only if pair creation and annihilation switch

---

[10]Note that in a theory with one S-boson one would get $\delta = 0$ at the one loop level. Indeed, in that case one should set $g_{(1')} = g_{(1)}$ and $g_{(2')} = g_{(2)}$ in (11.55).

off at $T \lesssim m_S$, and if production of S and $\bar{S}$ in processes inverse to (11.49) and (11.51) (inverse decays) is negligible at those temperatures. The latter requirement is indeed satisfied when the decay width of S-particles is small compared to the expansion rate at $T \sim m_S$,

$$\Gamma_{tot} \lesssim H(T \sim m_S) = \frac{m_S^2}{M_{Pl}^*}. \tag{11.57}$$

If all these conditions are fulfilled, then at $T \gtrsim m_S$ the number density of S-particles and their antiparticles is the same as of all other species, $n_S \sim T^3$, i.e.,

$$\frac{n_S}{s} \sim \frac{1}{g_*},$$

where s is the entropy density. The decays of S and $\bar{S}$ at later times produce the asymmetry

$$\Delta_{B-L} \sim \delta \frac{n_S}{s} \sim \frac{\delta}{g_*}. \tag{11.58}$$

Clearly, the asymmetry is quite large in this simplest case: the required value $\Delta_{B-L} \sim 10^{-10}$ is obtained for $\delta \sim 10^{-8}$ (assuming that the number of degrees of freedom g_* at $T \sim m_S$ is similar to its value $g_*^{SM} \sim 100$ at $T \sim m_S$).

The requirement (11.57) implies, in fact, that the S-particle mass is very large. Indeed, we have an estimate

$$\Gamma_{tot} \sim \frac{g^2}{4\pi} m_S. \tag{11.59}$$

Then the requirement (11.57) gives

$$m_S \gtrsim \frac{g^2}{4\pi} M_{Pl}^*, \tag{11.60}$$

As an example, for $g^2/4\pi \sim 10^{-2}$ one obtains $m_S \gtrsim 10^{16}$ GeV. The possibility that the Universe had such a large temperature is rather problematic (see accompanying book). For smaller couplings the requirement (11.60) is easier to fulfill, but the mass of S-particles must be large in this case too. Indeed, without fine-tuning of parameters the estimate for the microscopic asymmetry is given by (11.56), i.e., $\delta \lesssim g^2/4\pi$. Making use of (11.58) one obtains the value $\Delta_{B-L} \sim 10^{-10}$ for $g^2/4\pi \gtrsim 10^{-8}$, i.e.,

$$m_S \gtrsim 10^{10} \text{ GeV}.$$

The general statement is that the asymmetry generation in particle decays can occur only for large masses and hence at high temperatures.

We note that the departure from thermal equilibrium (the third necessary condition of Sec. 11.1) is achieved rather trivially in the scenario just described: at $T \lesssim m_S$ decay processes occur while inverse decays do not.

Let us now consider the case in which the inequality (11.57) *is not* valid. Let us introduce the parameter

$$K = \frac{\Gamma_{tot}}{H(T = m_s)} = \frac{\Gamma_{tot} M_{Pl}^*}{m_s^2} \qquad (11.61)$$

and let us take it large,

$$K \gg 1.$$

The dependence on this parameter is not particularly strong, so in this case the generation of the required $(B - L)$-asymmetry is also possible. Again, the mass m_s must be quite large. Unlike in the previous case, we now need not assume that S and $\bar{S}$ were in thermal equilibrium at $T \gg m_s$; furthermore, the maximum temperature in the Universe may be even somewhat smaller than m_s.

We will see in the end of this Section that at $K \gg 1$ the estimate for $(B - L)$-asymmetry is

$$\Delta_{B-L} = \text{const} \cdot \frac{\delta}{g_* K \log K}, \qquad (11.62)$$

where constant is of order 1. Similar to the small width case (11.57) this estimate implies that S-particle mass is large. If all relevant couplings are of the same order, then the microscopic asymmetry is again estimated by (11.56), i.e., $\delta \lesssim \frac{g^2}{4\pi}$, while the estimate for the width is given by (11.57). Making use of the definition (11.61) we find, modulo logarithm,

$$m_s \gtrsim \Delta_{B-L} \cdot g_* M_{Pl}^*,$$

i.e., we again obtain

$$m_s \gtrsim 10^{10} \text{ GeV}.$$

This bound is weaker in models where couplings differ by orders of magnitude (e.g., couplings $g_{(1')}$ and $g_{(2')}$ in Fig. 11.13 are considerably greater than $g_{(1)}$ and $g_{(2)}$), but in any case the mass of S-particles must be large.

To summarize, heavy particle decays provide efficient mechanism of generation of $(B - L)$ asymmetry and hence baryon asymmetry. This mechanism may work in $(B - L)$-violating GUTs. It may work also in supersymmetric GUTs, and in that case S-particles may be not only bosons but also fermions. Some (but not fatal) problem with GUTs is that their mass scale is extremely high (see (11.29)), so the maximum temperature in the Universe could be lower than this scale (this is indeed the case in most inflationary models).

Let us obtain the estimate (11.62). Let us assume for the time being that at $T \lesssim m_s$ the dominant processes are decays and inverse decays of S and $\bar{S}$. One might think that inverse decays of the type

$$qq \to S \qquad (11.63)$$

340   *Generation of Baryon Asymmetry*

should be very rare, since the center-of-mass energy of the two incoming particles must be adjusted to the mass of S-particle with precision of order of its width Γ_{tot}. Still, the probability of inverse decays is not small in thermal equilibrium, as the rates of direct and inverse processes are equal. Hence, for quarks and leptons in thermal equilibrium the number of inverse decays per unit time in comoving volume is

$$\frac{d(na^3)^{\text{inv.}}}{dt} = \Gamma_{tot} \cdot n_S^{eq} \cdot a^3, \tag{11.64}$$

where n_S^{eq} is the equilibrium number density of S-particles.

Problem 11.15. *Let us neglect cosmological expansion. Show by explicit calculation that the rate of the inverse processes (11.63) is*

$$\left(\frac{dn}{dt}\right)_{qq \to S} = \Gamma(S \to qq) \cdot n_S^{eq},$$

where $\Gamma(S \to qq)$ is the partial width of the decay $S \to qq$. Simplifications: Consider low temperatures $T \ll m_S$; do not account for color of quarks and S-particles.

S-particle decays give the following contribution to the evolution of their number in comoving volume,

$$\frac{d(na^3)^{\text{dec}}}{dt} = -\Gamma_{tot} \cdot n_S \cdot a^3. \tag{11.65}$$

By adding (11.64) and (11.65) we obtain the Boltzmann equation for the number density,

$$\frac{d(n_S a^3)}{dt} = -\Gamma_{tot} \cdot (n_S \cdot a^3 - n_S^{eq} \cdot a^3). \tag{11.66}$$

Let us now consider $(B-L)$ density in comoving volume $n_{B-L}a^3$. Without loss of generality we assign $(B-L) = 0$ to S- and $\tilde{S}$-particles. The Boltzmann equation for $(B-L)$ density in comoving volume reads

$$\frac{d(n_{B-L} \cdot a^3)}{dt} = \delta \cdot \Gamma_{tot} \cdot (n_S \cdot a^3 - n_S^{eq} \cdot a^3) - c\Gamma_{tot} \cdot n_S^{eq} \cdot a^3 \cdot \frac{n_{B-L}}{n_q^{eq}}. \tag{11.67}$$

The first term in the right-hand side of Eq. (11.67) describes the asymmetry production in decays of S- and $\tilde{S}$-particles, while the second and third terms wash out the asymmetry. The second term is induced by CP-violation in the inverse decay processes. It is non-trivial even if $(B-L)$ is zero in the medium, and its form is given by the requirement to cancel the first term in termal equilibrium.[11] The third term is non-trivial even in the absence of the microscopic CP-violation. It has to do with the following dynamics: if there is $(B-L)$ in the medium, it gets washed out due to production of S-particles. Indeed, if the medium contains more quarks than antiquarks, then there are more inverse decays $qq \to S$ than

---

[11]We simplify the situation slightly: the second term also includes the contribution of resonant scattering of fermions, e.g., $qq \to S \to \bar{q}\Lambda$. Still, the term is correct, since it follows from the requirement that $(B-L)$ is not generated in thermal equilibrium when $n_S = n_S^{eq}$.

the conjugate processes $\bar{q}\bar{q} \to \bar{S}$, which reduces the asymmetry. Evolution of $(B-L)$ due to the latter mechanism is determined by the equation

$$\frac{d(n_{B-L} \cdot a^3)}{dt} = -\langle \sigma_{qq \to S} \rangle \cdot n_q^2 \cdot a^3 + \langle \sigma_{\bar{q}\bar{q} \to \bar{S}} \rangle \cdot n_{\bar{q}}^2 \cdot a^3 = -\langle \sigma_{qq \to S} \rangle \cdot a^3 \cdot (n_q^2 - n_{\bar{q}}^2),$$

where we neglected the CP-asymmetry (it would give the second term in right-handside of Eq. (11.67)). To the leading order in asymmetry, $n_{B-L} \ll n_q$, one obtains,

$$n_q^2 - n_{\bar{q}}^2 = \frac{4}{3} \cdot n_{B-L} \cdot n_q^{eq},$$

where we recall $n_q - n_{\bar{q}} = (2/3)n_{B-L}$. Since in the termal equilibrium the rates of decays and the inverse decays coincide, then

$$\Gamma_{tot} \cdot n_S^{eq} = \langle \sigma_{qq \to S} \rangle \cdot n_q^{eq\,2},$$

hence

$$\langle \sigma_{qq \to S} \rangle \cdot n_q^{eq} = \Gamma_{tot} \cdot \frac{n_S^{eq}}{n_q^{eq}},$$

and we arrive at the third term in Eq. (11.67) with parameter $c = 4/3 \sim 1$ accounting for the number of decay channels.

It is convenient to make use of Eq. (11.66) and write

$$\frac{d(n_{B-L}a^3)}{dt} = -\delta \cdot \frac{d(n_S a^3)}{dt} - c\Gamma_{tot} \cdot n_S^{eq} \cdot a^3 \cdot \frac{n_{B-L}}{n_q}. \tag{11.68}$$

The system of equations (11.66) and (11.68) determines number density of S-particles and $(B-L)$ asymmetry at all times if they are known at initial time t_i.

It is convenient to introduce the variables

$$N_S = \frac{n_S}{T^3} \quad \text{and} \quad N_{B-L} = \frac{n_{B-L}}{T^3}, \tag{11.69}$$

so that $n_S a^3 = \text{const} \cdot N_S$, $n_{B-L}a^3 = \text{const} \cdot N_{B-L}$ with one and the same constant. Also, let us use, instead of time, the variable

$$z = \frac{m_S}{T}$$

and make use of the relation (3.34)

$$-\frac{\dot{T}}{T} = H(T) = H(T = m_S) \cdot \frac{T^2}{m_S^2}.$$

Recalling that $n_q \propto T^3$ we obtain the following form of Eqs. (11.66) and (11.68)

$$\frac{dN_S}{dz} = -Kz(N_S - N_S^{eq}), \tag{11.70}$$

$$\frac{dN_{B-L}}{dz} = -\delta \cdot \frac{dN_S}{dz} - \tilde{K}z \cdot N_S^{eq} \cdot N_{B-L}, \tag{11.71}$$

where the parameter K is defined in (11.61) and

$$\tilde{K} = c\frac{T^3}{n_q} \cdot K \sim K.$$

We are interested in the case $K \gg 1$, when the relevant temperatures are low, $T \ll m_S$, i.e., $z \gg 1$. The equilibrium number density of S-particles in this regime is (spin and color factors are omitted)

$$n_S^{eq} = \left(\frac{m_S T}{2\pi}\right)^{3/2} e^{-\frac{m_S}{T}},$$

i.e., modulo irrelevant constant $\hat{c}$, we have

$$N_S^{eq} = \hat{c} z^{3/2} e^{-z}. \tag{11.72}$$

Hence, we have explicit expressions for all quantities entering (11.70), (11.71).
Let us begin with Eq. (11.70). Its solution is

$$N_S(z) = \int_{z_i}^{z} e^{-\frac{K}{2}(z^2 - z'^2)} K z' N_S^{eq}(z') dz' + e^{-\frac{K}{2}(z^2 - z_i^2)} N_S(z_i),$$

where $z_i \equiv \frac{m_S}{T}$ is the initial value of the variable z, at which the initial value of the
S-particle relative number density is $N_S(z_i)$. This solution shows that for $K \gg 1$ the
initial value of the number density gets irrelevant soon (the second term rapidly tends to
zero). Now, at $K \gg 1$ the integral is saturated at z' near z (if $N_S^{eq}(z)$ is not very small),
therefore

$$N_S(z) = N_S^{eq} + \mathcal{O}\left(\frac{1}{K}\right). \tag{11.73}$$

Both properties are quite obvious: large K means high rates of decays and inverse decays,
so S-particles are close to thermal equilibrium.
Let us turn to Eq. (11.71). Neglecting corrections of order K^{-1} and making use of
(11.73), we write it in the following form,

$$\frac{dN_{B-L}}{dz} = -\delta \cdot \frac{dN_S^{eq}}{dz} - \tilde{K} z \cdot N_S^{eq} \cdot N_{B-L}. \tag{11.74}$$

Departure from thermal equilibrium that leads to the asymmetry generation is now rather
subtle: even though the density of S-particles is close to equilibrium at each moment of
time, it changes in time (and hence in z), so the equilibrium is incomplete.
The solution to Eq. (11.74) is

$$N_{B-L}(z) = -\delta \cdot \int_{z_i}^{z} e^{-I(z',z)} \frac{dN_S^{eq}}{dz'} dz' + e^{-I(z_i,z)} N_{B-L}(z_i), \tag{11.75}$$

where

$$I(z, z') = \int_{z}^{z'} N_S^{eq}(z'') \tilde{K} z'' dz''.$$

We see that initial anisotropy is irrelevant at late times, just like the initial S-particle
density, provided that the initial temperature is of order of m_S or only slightly less than
m_S (so that $N_S^{eq}(z_i)$ is not very small).

Problem 11.16. *Let the maximum temperature in the Universe be T_i, and $K \gg 1$. Estimate the maximum value of T_i such that most of the initial $(B-L)$-asymmetry is preserved
up to the present epoch.*

Assuming that the initial temperature is sufficiently high, we neglect the second term in (11.75) and write for the resulting asymmetry

$$N_{B-L} = -\delta \cdot \int_0^\infty e^{-\hat{I}(z)} \frac{dN_S^{eq}}{dz} dz, \tag{11.76}$$

where

$$\hat{I}(z) \equiv I(z, \infty) = \int_z^\infty N_S^{eq}(z') \tilde{K} z' dz'.$$

The integral in (11.76) is saturated at rather large z (i.e., at temperatures well below m_S): the asymmetry produced at $z \sim 1$ (i.e., $T \sim m_S$) due to the first term in (11.74) is washed out in the course of evolution. We find from (11.72) that at large z

$$\frac{dN_S^{eq}}{dz} = -\hat{c} z^{3/2} e^{-z} = -N_S^{eq}(z). \tag{11.77}$$

Hence, the integrand in (11.76) is the product of two exponential factors: the decaying factor e^{-z} from (11.77) (the decreasing density of S-particles makes the generation of $(B - L)$ slower) and increasing factor $e^{-\hat{I}(z)}$ (wash-out becomes less efficient). Such an integral is determined by the behavior of the exponent

$$f(z) = \hat{I}(z) + z.$$

This function has a minimum at $z = z_*$ such that

$$N_S^{eq}(z_*) \tilde{K} z_* = 1. \tag{11.78}$$

We recall (11.77) and obtain[12] (see Sec. 6.1 regarding equations of the latter type)

$$z_* = \log \tilde{K} + \mathcal{O}(\log \log \tilde{K}) = \log K + \mathcal{O}(\log \log K). \tag{11.79}$$

Both the integral $\hat{I}(z_*)$ and its first and second derivatives are of order one at the saddle point, so

$$f(z_*) = z_* + \mathcal{O}(1),$$

and

$$\frac{d^2 f}{dz^2} = \mathcal{O}(1).$$

The latter property means that the integral (11.76) is saturated in the region around z_* whose size is

$$\Delta z \sim 1 \ll z_*.$$

We obtain in the saddle point approximation

$$N_{B-L} = \delta \cdot \text{const} \cdot N_S^{eq}(z_*),$$

and finally, making use of (11.78), we find

$$N_{B-L} = \text{const} \cdot \frac{\delta}{\tilde{K} z_*} = \text{const} \cdot \frac{\delta}{K \log K}, \tag{11.80}$$

with constant of order 1. This is precisely the quoted result (11.62).

---

[12]The approximation $z_* = \log K$ works poorly at moderate K but this is unimportant for our purposes.

Let us now discuss the result obtained and calculation performed. First, the temperature of the asymmetry generation, T_*, is determined by (11.79), i.e.,

$$T_* = \frac{m_S}{\log K}. \tag{11.81}$$

Wash-out effect is not very strong at this temperature ($\hat{I}(z_*) \sim 1$), while the density of S-particles and the number of their decays are still sizeable,

$$N_S^{eq} \simeq \frac{dN_S^{eq}}{dz} \sim \frac{1}{K \log K}.$$

It is these properties that make the resulting asymmetry suppressed rather mildly at large K. Second, the result (11.80) is valid provided that the maximum temperature in the Universe exceeded T_* which is somewhat lower than m_S. Third, the asymmetry is generated in the temperature interval which is small compared to temperature itself,

$$\frac{\Delta T}{T_*} = \frac{\Delta z}{z_*} = \frac{1}{\log K}.$$

Finally, let us point out that the equation we actually used, Eq. (11.74), does not contain information on the processes that keep the number density of S-particles in thermal equilibrium. Instead of decays and inverse decays (or together with them) these may be processes of pair creation and annihilation of S and $\bar{S}$. It is this situation that happens for colored S and $\bar{S}$ which we considered in Sec. 11.2.2. Still, the result (11.62) remains valid for $K \gg 1$. What we have *not* included into our analysis is non-resonant scattering of the type $qq \to \bar{q}\Lambda$ occurring via S-boson exchange. The latter process tends to wash out $(B - L)$ and its effect is sometimes important; see Sec. 11.4.

To end this Section, let us briefly discuss the possibility that baryon asymmetry and dark matter are generated by one and the same mechanism. An example is decays of hypothetical particles in the early Universe into quarks and dark matter particles. This mechanism is known as *hylogenesis* [217]; various models are considered in the review [218]. The idea is that dark matter particles carry negative baryon number, and matter in the Universe is neutral with respect to the baryon number. Decays of new heavy particles produce the asymmetry in both visible and dark sectors. The model has to be constructed in such a way that the lightest particles in both sectors are stable.

The appealing feature of this scenario is that the relic abundances of baryons and dark matter particles are of the same order, $n_B \sim n_{DM}$, so the relation $\Omega_B \sim \Omega_{DM}$ is naturally obtained provided the masses of the stable particles are also of the same order, $m_B \sim m_{DM}$. In the simplest versions one has $n_B = q_{DM} n_{DM}$, where $q_{DM} \sim 1$ is the baryonic charge of dark matter particle. Then the dark metter particle mass is $m_{DM} = m_p q_{DM} \Omega_{DM}/\Omega_B \approx 5\, q_{DM}$ GeV. These particles may be produced at the LHC, and not necessarily in pairs (as opposed to WIMPs). They can annihilate with protons in galaxies and give signals in cosmic rays. Standard indirect search for these particles is not particularly promising, but their annihilation in an underground detector would yield similar signal as proton decay. Search for these processes is underway in, e.g., Super-K experiment.

11.4. Baryon Asymmetry and Neutrino Masses: Leptogenesis

As we discussed in Sec. 11.2.3, violation of lepton number, and hence $(B - L)$ may be related to nonzero neutrino masses. Hence, it is appealing to explain the baryon asymmetry within the same approach that deals with neutrino masses. The lepton asymmetry may indeed be generated in decays of N-particles considered in Sec. 11.2.3. This asymmetry is then partially reprocessed into baryon asymmetry by electroweak processes of Sec. 11.2.1. This scenario [207] is known as leptogenesis.

The mechanism of the lepton asymmetry generation is basically the same as in Sec. 11.3. The only difference is that S-particles are now replaced by Majorana fermions N_α. The simplest possibility is that the lepton asymmetry is produced in decays of the *lightest* of these particles, otherwise one should worry about wash-out of the asymmetry generated in decays of heavier particles by processes involving the lightest ones.[13] Let N_1 be the lightest of N-particles. Then the relevant decays are

$$N_1 \to lh, \tag{11.82}$$

and

$$N_1 \to \bar{l}h. \tag{11.83}$$

As we discussed in Sec. 11.3, the microscopic asymmetry is generated if their partial widths are different. This occurs at one loop, if couplings in the Lagrangian (11.44) are complex, and hence CP is violated. The relevant diagrams are shown in Fig. 11.14. Thus, one loop partial width of decay $N_1 \to lh$ is given by

$$\Gamma(N_1 \to lh) = \text{const} \cdot \sum_\alpha \left| y_{1\alpha} + \sum_{\beta,\gamma} D\left(\frac{M_1}{M_\gamma}\right) \cdot y^*_{1\beta} y_{\gamma\alpha} y_{\gamma\beta} \right|^2, \tag{11.84}$$

where M_γ is mass of particle N_γ (we neglect masses of the Standard Model particles: we will see that leptogenesis requires $M_\alpha \gg 100$ GeV), $D(M_1/M_\gamma)$ is the sum of loop integrals shown in Fig. 11.14.

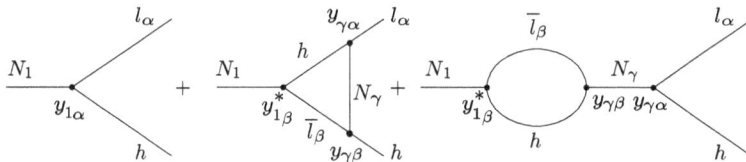

Fig. 11.14. Amplitude of the decay $N_1 \to l_\alpha h$ is the sum of tree-level and one loop diagrams. Summation over generation indices β and γ is assumed in the latter. Yukawa couplings entering the vertices are shown explicitly.

---

[13] Generation of the asymmetry in decays of the next-to-lightest N-particle has been considered, e.g., in Ref. [214].

The partial width of $N_1 \to \bar{l}h$ decay is obtained from (11.84) by replacing the Yukawa couplings by their complex conjugates, $y_{\alpha\beta} \to y^*_{\alpha\beta}$. We denote

$$\mathrm{Im}\, D\left(\frac{M_1}{M_\gamma}\right) = \frac{1}{8\pi} f\left(\frac{M_1}{M_\gamma}\right),$$

and find the microscopic lepton asymmetry,

$$\delta \equiv \frac{\Gamma(N_1 \to lh) - \Gamma(N_1 \to \bar{l}h)}{\Gamma_{tot}}$$

$$= -\frac{1}{8\pi} \sum_{\gamma=2,3} f\left(\frac{M_1}{M_\gamma}\right) \cdot \frac{\mathrm{Im}(\sum_\alpha y_{1\alpha} y^*_{\gamma\alpha})^2}{\sum_\alpha |y_{1\alpha}|^2}. \tag{11.85}$$

We used here the fact that $\mathrm{Im}(y_{1\alpha} y^*_{\gamma\alpha}) = 0$ for $\gamma = 1$, so the diagrams with N_1 in the loop do not contribute. We will concentrate on the mass hierarchy $M_1 \ll M_{2,3}$, when

$$f\left(\frac{M_1}{M_\gamma}\right) = -\frac{3}{2}\frac{M_1}{M_\gamma}. \tag{11.86}$$

So, the microscopic asymmetry becomes

$$\delta = \frac{3M_1}{16\pi} \frac{1}{\sum_\alpha |y_{1\alpha}|^2} \sum_{\alpha\beta\gamma} \mathrm{Im}\left[y_{1\alpha} y_{1\beta} \left(y^*_{\gamma\alpha} \frac{1}{M_\gamma} y^*_{\gamma\beta} \right) \right]. \tag{11.87}$$

Problem 11.17. *Show by calculating Feynman diagrams that* (11.86) *is indeed valid for* $M_\gamma \gg M_1$.

Let us make the following comment regarding the expressions (11.85) and (11.87). They contain combinations of the Yukawa couplings which are *different* from the combinations entering neutrino mass matrix (see (C.69)),

$$m_{\alpha\beta} = -\frac{v^2}{2} \sum_\gamma y_{\gamma\alpha} \frac{1}{M_\gamma} y_{\gamma\beta}. \tag{11.88}$$

As an example, the unitary transformation of the form

$$y \to yU \tag{11.89}$$

changes the neutrino mass matrix $m_{\alpha\beta}$ but leaves intact the expressions (11.85) and (11.86). This is not surprising: the transformation (11.89) corresponds to the change of basis for lepton fields l_α, while the asymmetry is insensitive to this basis. (We neglect masses of the Standard Model leptons, so all choices of the basis are equivalent.) This shows that, generally speaking, the asymmetry δ is not directly related to the parameters of the Pontecorvo–Maki–Nakagawa–Sakata (PMNS) matrix which describes mixing between conventional neutrinos and responsible for neutrino oscillations. Therefore, measurements of neutrino oscillation parameters does not tell whether there is asymmetry in N-particle decays. Still, the very fact that oscillations exist suggests that the matrix of Yukawa couplings $y_{\alpha\beta}$ has non-trivial structure. Additional hint towards asymmetry in N-particle decays would be given by the observation of CP-violation in neutrino oscillations: it would show that elements of neutrino mass matrix, and hence Yukawa couplings, are complex at least in the gauge basis.

We will get back to the expression (11.87) for the microscopic asymmetry, and now we turn to the cosmological lepton asymmetry generation in decays of particles N_1. The analysis repeats the treatment in Sec. 11.3 word by word, so we simply use the results of that Section. The generation is most efficient for

$$\Gamma_{tot}(M_1) \ll H(T = M_1), \tag{11.90}$$

but one has to assume in that case that particles N_1 are produced in cosmic plasma at $T \gg M_1$ by interactions other than Yukawa interactions (11.44). We recall that

$$\Gamma_{tot} = \frac{M_1}{8\pi} \sum_\alpha |y_{1\alpha}|^2,$$

and that $H = T^2/M^*_{Pl}$, and find that the relation (11.90) can be written as

$$\tilde{m}_1 \ll \frac{4\pi}{M^*_{Pl}} \cdot v^2 \sim 10^{-3} \text{ eV}, \tag{11.91}$$

where

$$\tilde{m}_1 = \sum_\alpha \frac{|y_{1\alpha}|^2}{2M_1} \cdot v^2 \tag{11.92}$$

is the sum of absolute values of the N_1-particle contribution into neutrino mass matrix (see Section (C.4)). We see that the case under study requires strong hierarchy of Yukawa couplings:[14] if all $y_{\alpha\beta}$ were of the same order, then the contributions of the *lightest* particle N_1 into mass matrix (11.88) would be the largest, and all neutrinos would have masses below 10^{-3} eV, in contradiction to experiment (see the discussion in Appendix C before Eq. (C.57)). Still, let us continue with this case and obtain the estimate for the mass M_1, assuming for definiteness the direct neutrino mass hierarchy (C.59) with small mass of the lightest mass eigenstate. Once the inequality (11.90) holds, the estimate for the lepton (and hence baryon) asymmetry is (see (11.58))

$$\Delta_L \sim \frac{\delta}{g_*}.$$

---

[14]This possibility is not particularly natural, but there is no reason to discard it; recall that there *is* strong hierarchy of the Yukawa couplings of quarks and leptons in the Standard Model.

Now, the formula (11.87) can be written as

$$\delta = -\frac{3M_1}{8\pi v^2}\frac{1}{\sum_\alpha |y_{1\alpha}|^2}\sum_{\alpha\beta}\mathrm{Im}\left(y_{1\alpha}y_{1\beta}\sum_{\gamma=2,3}m_{\alpha\beta}^{(\gamma)*}\right),\qquad(11.93)$$

where

$$m_{\alpha\beta}^{(\gamma)} = -y_{\alpha\gamma}\frac{v^2}{2M_\gamma}y_{\gamma\beta}\qquad(11.94)$$

is the contribution of N_γ-particle into neutrino mass matrix. Making use of (C.59) we find

$$\delta \lesssim \frac{3M_1}{8\pi v^2}m_{atm}.$$

The asymmetry $\Delta_L \sim 10^{-10}$ is obtained for $\delta \sim 10^{-8}$ (assuming that $g_* \sim 100$ like in the Standard Model), so

$$M_1 \gtrsim 10^9 \text{ GeV}.$$

Without fine tuning of parameters this gives the minimum mass scale of N-particles in the leptogenesis scenario; the maximum temperature in the Universe must exceed 10^9 GeV.

Probably the most natural possibility is that there is direct neutrino mass hierarchy (C.59) related to inverse mass hierarchy of N_α, so that $m_3 \propto M_1^{-1}$, $m_2 \propto M_2^{-1}$, $m_1 \propto M_3^{-1}$ with $M_1 \ll M_2 \ll M_3$. In that case the right-hand side of (11.92) is estimated as

$$\tilde{m}_1 \sim m_{atm} \simeq 0.05 \text{ eV},\qquad(11.95)$$

so that the inequality (11.91) and hence (11.90) do not hold. Then the generation of lepton asymmetry is suppressed. Let us use the estimate (11.62) and write

$$\Delta_L \simeq \text{const} \cdot \frac{\delta}{g_* K \log K},$$

where the constant is of order 1, and

$$K = \frac{\Gamma_{tot}}{H(T \sim M_1)} = \frac{\tilde{m}_1 M_{Pl}^*}{4\pi v^2}.\qquad(11.96)$$

Using the estimate (11.95) we see that the suppression factor is of order

$$K \sim 100,$$

so the correct value $\Delta_L \sim 10^{-10}$ is obtained for larger microscopic asymmetry

$$\delta \gtrsim 10^{-5}.\qquad(11.97)$$

On the other hand, for $\gamma = 2, 3$ the right-hand side of (11.94) is estimated as $m_{\alpha\beta}^{(\gamma)} \sim m_{sol}$, so (11.93) and (11.97) give

$$M_1 \gtrsim 10^{12} \text{ GeV}.$$

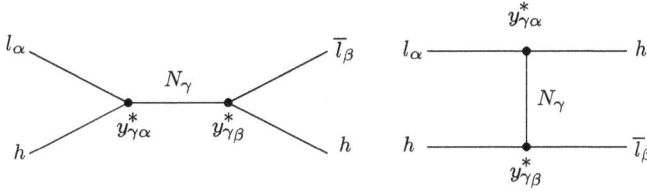

Fig. 11.15.   Lepton scattering off scalars that washes out the lepton asymmetry.

We see that even in the least favorable case the required mass scale is not unacceptably large. We note that in the case we study here, one does not need to invoke extra mechanisms of N-particle production: they are efficiently produced in inverse decays. What is required is that the maximum temperature in the Universe exceeds (see (11.81))

$$T \gtrsim \frac{M_1}{\log K}.$$

Thus, leptogenesis scenario indeed works in a wide range of parameters. Interestingly, the range of neutrino masses suggested by the oscillation experiments is not very far from $\frac{4\pi v^2}{M_{Pl}^*} \sim 10^{-3}$ eV, so that the suppression factor (11.96) is not very large in any case. In other words, neutrino masses $m \lesssim 1$ eV are such that there is crude (within two or three orders of magnitude) equality between the cosmological expansion rate at leptogenesis and the width of the lightest N-particle, $\Gamma_{tot} \sim (1-1000) \cdot H(T \sim M_1)$. This coincidence is a rather strong argument in favor of leptogenesis.

If neutrino masses are much larger than $m_{atm} \simeq 0.05$ eV, they must be degenerate. In that case one obtains interesting bounds from the non-resonant scattering shown in Fig. 11.15,

$$lh \to \bar{l}\bar{h}, \tag{11.98}$$

and crossing processes. These tend to wash out the lepton asymmetry. The new contribution to the Boltzmann equation for lepton number density, as compared to Sec. 11.3, is

$$\frac{d(n_L \cdot a^3)}{dt} \propto \Gamma_{lh} \cdot (n_L \cdot a^3),$$

where Γ_{lh} is the rate of processes of the type (11.98). In terms of the variable $z = M_1/T$ this contribution reads

$$\frac{dN_L}{dz} \propto -\frac{M_{Pl}^*}{M_1 T}\Gamma_{lh}(z) \cdot N_L,$$

where $N_L = \frac{n_L}{T^3}$. Wash-out of the lepton number is important after the time at which the asymmetry is generated, i.e., at $z \gtrsim z_* = \log K$ (see Sec. 11.3). Since z_*

is typically large, the N-particles are non-relativistic at $z > z_*$ and the cross-sections of the processes (11.98) are estimated as

$$\sigma_{lh} = \text{const} \cdot \sum_{\alpha\beta\gamma} \left| \frac{y_{\gamma\alpha} y_{\gamma\beta}}{M_\gamma} \right|^2 .$$

(Recall that N-particles are fermions; their propagator at low momenta is $\frac{1}{M_\gamma}$.) Making use of the expression (C.69) for neutrino mass matrix, the cross-section is cast into the form

$$\sigma_{lh} = \text{const} \cdot \frac{\text{Tr}(mm^\dagger)}{v^4} = \text{const} \cdot \frac{1}{v^4} \sum m_\nu^2 . \tag{11.99}$$

It follows already from dimensional argument that

$$\Gamma_{lh} = \text{const} \cdot \sigma_{lh} \cdot T^3 .$$

As a result, we find that at $z > z_*$, when decays and inverse decays are switched off, lepton asymmetry obeys

$$\frac{dN_L}{dz} = -\text{const} \cdot \frac{M_{Pl}^* T^2}{M_1} \frac{\sum m_\nu^2}{v^4} \cdot N_L = -\text{const} \cdot \frac{M_{Pl}^* M_1}{z^2} \frac{\sum m_\nu^2}{v^4} \cdot N_L , \tag{11.100}$$

where the constant is of order 1. We see that the scattering processes suppress the asymmetry by a factor

$$\exp\left(-\int_{z_*}^{\infty} dz \frac{\text{const}}{z^2} \cdot M_{Pl}^* M_1 \cdot \frac{\sum m_\nu^2}{v^4} \right)$$

$$= \exp\left(-\frac{\text{const}}{z_*} \cdot M_{Pl}^* M_1 \cdot \frac{\sum m_\nu^2}{v^4} \right) . \tag{11.101}$$

Let us require that this factor is not very small. This gives

$$\sum m_\nu^2 \lesssim \frac{v^4 z_*}{M_{Pl}^* M_1} .$$

Even for relatively small $M_1 \sim 10^9$ GeV when $z_* \sim 1$ (the case (11.91)) the resulting bound is quite strong (we take into account neutrino degeneracy)

$$m_\nu \lesssim 1 \text{ eV}. \tag{11.102}$$

In the case $M_1 \gtrsim 10^{12}$ GeV (and $z_* \sim 10$) this bound is even stronger,

$$m_\nu \lesssim 0.1 \text{ eV}.$$

Accurate calculations of the wash-out effect in scattering give [208]

$$m_\nu < 0.12 \text{ eV} \tag{11.103}$$

for most values of the parameters of the model. This bound does not contradict existing experimental and cosmological bounds and again suggests that the neutrino masses are just right for leptogenesis. On the other hand, if double-beta decay experiments would measure neutrino mass above the bound (11.103), the leptogenesis scenario[15] will be much less attractive.

Problem 11.18. *Discarding experimental bounds on neutrino masses and neutrino oscillation data, and assuming that all Yukawa couplings are of the same order, show that successful leptogenesis is possible only for*

$$m_\nu \lesssim 1 \ eV.$$

There is an alternative to the leptogenesis in decays. Lepton asymmetry may instead be generated in oscillations of active and sterile neutrinos in the Universe. This mechanism does not require very heavy sterile neutrinos and, in fact, works if their masses are well below the electroweak scale. The idea is to redistribute the lepton number, by employing CP-violation, between active and sterile neutrinos [219]. (The role of lepton number is played in the case of sterile neutrinos by their helicity, which is conserved at high energies.) The oscillations do not produce the net lepton number, but part of lepton number of active sector is reprocessed into baryon number by sphalerons.

This mechanism is at work if the sterile neutrinos are not in thermal equilibrium by the epoch of the electroweak transition, otherwise the asymmetry would be washed out. The production rate of sterile neutrinos N_γ in particle collisions is estimated on dimensional grounds as $\Gamma_N \sim T|y_{\gamma\alpha}|^2/(8\pi)$, cf. (11.47). As usual, the sterile neutrinos enter the thermal equilibrium when this rate becomes comparable with the Hubble rate $H = T^2/M_{Pl}^*$, i.e., at temperature

$$T_N^{\rm eq} \sim M_{Pl}^* |y_{\gamma\alpha}|^2/(8\pi).$$

The condition $T_N^{eq} < 100$ GeV is satisfied for small Yukawa couplings

$$|y_{\gamma\alpha}|^2 \ll 10^{-14}. \tag{11.104}$$

If sterile neutrinos give masses to active ones via the see-saw mechanism, then, without fine tuning of parameters, one has the estimate (C.68), which gives $|y_{\gamma\alpha}|^2 \sim M_N m_\nu/v^2$. The latter shows that the condition $T_N^{eq} < 100$ GeV holds for sterile neutrino masses below the electroweak scale, $M_N \lesssim 10$ GeV.

The generation of asymmetry requires fast CP-violating processes due to complex Yukawa matrix $y_{\gamma\alpha}$. Like in the case of decays, these processes occur at one loop level; see Eq. (11.85). So, they are suppressed by the same parameter $|y_{\gamma\alpha}|^2$ which, by the requirement of inequilibrium (11.104), must be very small. Thus, one needs the enhancement of the CP-violation in oscillations. A mechanism for this enhancement is the resonance in the coherent system of two sterile neutrinos N_1, N_2, which should be degenerate in mass, $\Delta M \ll M_{1,2} \approx M$. The resonance occurs when the oscillation frquency

$$\omega \sim |M_1^2 - M_2^2|/T \sim \Delta M \cdot M/T$$

is of order of the expansion rate, $\omega(T) \sim H(T)$. This gives for the temperature of the resonance

$$T_{\rm res} \sim M \cdot (M_{Pl}^* \Delta M/M^2)^{1/3}.$$

At earlier times, when $\omega \ll H$, the resonance does not have time to develop, while later on, when $\omega \gg H$, the oscillations are too fast, and the small factor $|y_{\gamma\alpha}|^2$ is not compensated

---

[15]We talk here about leptogenesis at the hot Big Bang stage ("thermal leptogenesis"). The analysis in the text does not apply to leptogenesis at reheating which we do not consider here.

for. It has been shown that the mechanism is efficient for T_{res} in the hundred GeV range; in that case the asymmetry is large, $\Delta_L \sim 1$ [220], but this requires strong degeneracy in sterile neutrino masses, $\Delta M/M \ll 1$.

We note that the generation of *lepton* asymmetry may happen even after the electroweak transition. This asymmetry is not reprocessed into baryon asymmetry, but it may help to produce sterile neutrinos serving as dark matter candidatesl; see Sec. 7.3.

11.5.   Electroweak Baryogenesis

It is natural to ask whether electroweak baryon number violation discussed in Sec. 11.2.1 can by itself generate baryon asymmetry at temperature of order 100 GeV. Such a possibility — electroweak baryogenesis — is of particular interest because the relevant energy range of 100 GeV – 1 TeV is accessible to collider experiments. Examining this energy range will either rule out or give strong arguments in favor of electroweak baryogenesis in our Universe.

Both the requirement of strong enough CP-violation and the thermal inequilibrium condition are quite non-trivial in this scenario. CP-violation in the CKM matrix of the Standard Model is insufficient for baryogenesis; strong deviation from thermal equilibrium is also not inherent in the Standard Model. However, some extensions of the Standard Model have both extra sources of CP-violation and possibility of enhancement of thermal inequilibrium at the electroweak transition, so the electroweak baryogenesis still remains an option.

11.5.1.   *Departure from thermal equilibrium*

The Universe expansion at temperatures of order 100 GeV is pretty slow. The Hubble time is

$$t_U \sim H^{-1} = \frac{M_{Pl}^*}{T^2} \sim 10^{14}\ \mathrm{GeV}^{-1} \sim 10^{-10}\ \mathrm{s}.$$

This is much longer than the time scale of electroweak interactions between particles in the medium,

$$t_{int} \sim \frac{1}{\alpha_w T} \sim 1\ \mathrm{GeV}^{-1} \sim 10^{-24}\ \mathrm{s}.$$

Hence, one of the necessary conditions of Sec. 11.1, strong deviation from thermal equilibrium, is very non-trivial. It appears that the only way this condition can be satisfied is that the electroweak phase transition is *first order*. As we discussed in Sec. 10.1, first order phase transition occurs via spontaneous creation of bubbles of the new phase, their subsequent growth and percolation. This boiling of matter in the Universe is a strongly inequilibrium phenomenon. We will see that baryon asymmetry may indeed be generated in this process.

Even the requirement that the electroweak phase transition is first order is insufficient. Medium *after* the phase transition quickly equilibrates, and the asymmetry generated in the course of transition may be washed out. This does not happen only if the electroweak baryon number violation rate is smaller than the expansion rate after the transition. As we will now see, the latter requirement is not fulfilled in the Standard Model (in fact, there is no phase transition in the Standard Model at all), but can be satisfied in some of its extensions.

Electroweak baryon number violation is switched off in the broken phase after the phase transition, if the inequality inverse to (11.22) holds,

$$\frac{M_W(T)}{T} \gtrsim \frac{0.66}{B(m_h/M_W)}.$$

Recalling that $M_W(T) = g\phi(T)/\sqrt{2}$, $\alpha_W = g^2/4\pi \approx 1/30$ and $B \approx 2.4$, we find that the latter inequality gives[16]

$$\frac{\sqrt{2}\Phi_c}{T_c} \gtrsim 1, \tag{11.105}$$

where T_c and Φ_c are the phase transition temperature and the Englert–Brout–Higgs expectation value just after the transition. The requirement (11.105) strongly restricts the physics at the electroweak scale. To see this, we make use of the one-loop results of Sec. 10.2. In the one loop approximation we have

$$\frac{\Phi_c}{T_c} = c \cdot \frac{\gamma}{\lambda},$$

where λ is the Higgs self-coupling, the parameter γ is defined by (10.40), and the constant c ranges from $1/2$ to $3/4$ depending on how delayed the phase transition is (see Eqs. (10.44), (10.45)). Making use of the relation between the Higgs boson mass and self-coupling, $m_h = \sqrt{2\lambda}v$, we obtain from (11.105) a bound on the Higgs boson mass,

$$m_h^2 < c \cdot \frac{2}{3\pi} \cdot \frac{1}{v} \cdot \sum_{\text{bosons}} g_i m_i^3, \tag{11.106}$$

where the realistic value is $c = 1/2$. Within the Standard Model with one Englert–Brout–Higgs doublet, the right-hand side obtains contributions from W- and Z-bosons of masses 80.4 GeV and 91.2 GeV and numbers of spin states $g = 6$ and $g = 3$, respectively. (The contribution of the Higgs boson itself may be neglected for estimates.) Recalling that $v = 246$ GeV, we obtain from (11.106) that the bound in the Standard Model is [209]

$$m_h < 50 \text{ GeV.}$$

This is not the case in Nature: the mass of the Higgs boson is $m_h = 125$ GeV. Thus, electroweak baryogenesis is impossible within the Standard Model.

---

[16]We recall that the normalization of the field ϕ adopted in Sec. 10 is such that $\langle\phi\rangle = v/\sqrt{2}$.

In fact, a less restrictive requirement of the first order phase transition is not satisfied in the Standard Model as well. We pointed out in Sec. 10.3 that the electroweak transition is a smooth crossover. The cosmic medium always remains close to thermal equilibrium, and baryon asymmetry is not generated at all.

This problem can be overcome in some extensions of the Standard Model. One example is given in Sec. 10.2. Another possibility exists (yet [221]) in supersymmetric extension of the Standard Model with light superpartners $\tilde{t}$ of t-quark [210]. Their explicit ("soft") masses are small, and $\tilde{t}$ acquire masses due to interactions with the Englert–Brout–Higgs field. Their couplings with this field are equal to the t-quark Yukawa coupling, and hence

$$m_{\tilde{t}}(\phi) = y_t \phi.$$

Superpartners $\tilde{t}$ form two complex, color-triplet scalars. They contribute to the right-hand side of Eq. (11.106) with $g_{\tilde{t}} = 12$. Due to this contribution, the inequality like (11.106) can hold for $m_h = 125$ GeV. Other extensions of the Standard Model with scalar singlets and/or additional scalar doublets may have similar properties.

11.5.2. * Thick wall baryogenesis

As we discussed in Sec. 10.1, bubbles of the new phase spontaneously created at the first order phase transition expand to macroscopic sizes before their walls collide. Hence, most of the matter in the Universe interacts with the walls, while the regions where the walls collide have small volume. (The ratio of the corresponding volumes is d/R, where d is wall thickness and R is the bubble size at percolation.) Thus, when calculating the baryon asymmetry one neglects processes occurring in wall collisions; the baryon asymmetry is generated in interactions of matter with bubble walls. Importantly, the walls rapidly move through the medium; their velocity is determined by "friction" produced by the medium and in most models is of order 0.1 of the speed of light.

In this and next Sections we describe mechanisms of baryon asymmetry generation in interactions of cosmic plasma particles with moving bubble wall. Our study will be rather sketchy, since we will not take into account a number of effects that are of various degree of importance for this quite complicated dynamical process. For details see, e.g., Ref. [211].

A fairly simple, though not completely realistic mechanism of electroweak baryogenesis works in so called adiabatic regime. Let us assume that the thickness of a bubble wall is much larger than the mean free path of particles in cosmic plasma. Let us also assume that the wall moves through the plasma sufficiently slowly. Then at any given time and everywhere in space (including the region inside the wall), the medium is in local thermal equilibrium with respect to fast processes of elastic scattering, pair creation and annihilation, etc. On the other hand, the electroweak baryon number violating processes are not so fast even in the unbroken phase, as can be seen from (11.19). So, it is natural that the latter processes may be out of thermal equilibrium. This may lead to the baryon asymmetry generation.

We are going to illustrate this mechanism making use of a simplified model with the gauge group $SU(2)_L$, two Englert–Brout–Higgs doublets H_1 and H_2 and one pair of fermions[17] Q_L (doublet) and q_R (singlet), whose Yukawa interactions are similar to those of the Standard Model,

$$\mathcal{L}_{int} = h_1 \bar{Q}_L H_1 q_R + h.c. \tag{11.107}$$

In what follows we take the coupling h_1 real (by redefining the scalar doublets one can always set the Yukawa coupling with H_2 equal to zero and h_1 real). The theory with two or more scalar doublets contains extra, with respect to the Standard Model, source of CP-violation in the Englert–Brout–Higgs sector. Both in broken phase and in the wall region, the fields H_1 and H_2 have expectation values

$$\langle H_{1,2} \rangle = \begin{pmatrix} 0 \\ \phi_{1,2} \end{pmatrix}, \tag{11.108}$$

where $\phi_{1,2}$ are complex functions of the coordinate normal to the wall. One of the phases of these complex functions is unphysical, and another is a source of CP-violation in the wall region. To see this, we recall that under gauge transformations with gauge function $e^{-i\alpha \frac{\tau^3}{2}}$ the fields transform as follows:

$$\phi_1 \to e^{i\frac{\alpha}{2}} \phi_1, \quad \phi_2 \to e^{i\frac{\alpha}{2}} \phi_2.$$

Therefore, one of the phases is would-be Nambu–Goldstone field which is "eaten up" due to the Englert–Brout–Higgs mechanism and becomes the longitudinal component of massive vector field (Z-boson in the Standard Model). Another phase is a physical field. To see how the phases θ_1 and θ_2 are expressed through this physical field, we consider the terms in the Lagrangian which lead to the mass term of vector boson,

$$L_m = \mathcal{D}_\mu H_1^\dagger \mathcal{D}_\mu H_1 + \mathcal{D}_\mu H_2^\dagger \mathcal{D}_\mu H_2,$$

where $\mathcal{D}_\mu H_{1,2} = \partial_\mu H_{1,2} - ig\frac{\tau^3}{2} A_\mu^3 H_{1,2}$. (We write the term with an analog of the field Z_μ only.) Near the vacuum $\phi_1 = v_1/\sqrt{2}$, $\phi_2 = v_2/\sqrt{2}$ the fields have the

---

[17]In realistic extensions of the Standard Model the major effect comes from the Yukawa interactions of t-quark, and our simplified model is thus adequate.

following form:

$$H_{1,2} = \begin{pmatrix} 0 \\ \frac{v_{1,2}}{\sqrt{2}} e^{i\theta_{1,2}} \end{pmatrix},$$

therefore

$$L_m = \frac{g^2}{8}(A_\mu^3)^2 - \frac{g}{2}A_\mu^3(v_1^2\partial_\mu\theta_1 + v_2^2\partial_\mu\theta_2) + \frac{1}{2}\left[v_1^2(\partial_\mu\theta_1)^2 + v_2^2(\partial_\mu\theta_2)^2\right].$$

Let us choose the unitary gauge in which the field A_μ^3 describes physical vector boson. To this end, we require that there is no mixing between A_μ^3 and $\theta_{1,2}$. This is achieved by imposing the gauge condition $v_1^2\theta_1 + v_2^2\theta_2 = 0$. Then the phases are [223]

$$\theta_1 = \frac{v_2^2}{v_1^2 + v_2^2}\theta,$$

$$\theta_2 = -\frac{v_1^2}{v_1^2 + v_2^2}\theta,$$

where $\theta = \theta_1 - \theta_2$ is the relative phase of the fields ϕ_1 and ϕ_2, which is the physical field.

In the presence of the background fields (11.108) the quadratic fermion Lagrangian, besides the kinetic term, contains the term

$$\mathcal{L}_f = h_1\bar{q}_L\rho_1 e^{i\theta_1}q_R + h.c.,$$

where q_L is the lower component of the doublet Q_L.

If there is CP-violation in the scalar sector, then both ρ_1 and θ_1 vary across the wall. Since the wall moves, both ρ_1 and θ_1 depend on time at a given point of space. Therefore, the fermion Lagrangian depends on time,

$$\mathcal{L}_f = h_1\bar{q}_L\rho_1(t)e^{i\theta_1(t)}q_R + h.c. \tag{11.109}$$

Time-dependent phase $\theta_1(t)$ leads to CP-violation in the fermion sector and in the end to the baryon asymmetry, once Q_L and q_R are identified with quarks.

The simplest way to analyse this mechanism is to perform time-dependent phase rotation of the field[18] q_R

$$q_R \to e^{-i\theta_1(t)}q_R. \tag{11.110}$$

This phase rotation makes the Yukawa coupling θ_1-independent, but induces extra term in the fermion Lagrangian, as the kinetic term gives

$$i\bar{q}_R\gamma^\mu\partial_\mu q_R \to i\bar{q}_R\gamma^\mu\partial_\mu q_R + \bar{q}_R\gamma^0 q_R\dot{\theta}_1.$$

---

[18]The phase rotation of Q_L is an anomalous symmetry, so if we performed the rotation of left quarks rather than right quarks, there would appear θ_1-dependent term in the effective Lagrangian of gauge fields. This would be a different, though equivalent approach. We are not going to use it.

The latter term modifies the Hamiltonian as follows,

$$H \to H - \dot{\theta}_1 N_R, \tag{11.111}$$

where

$$N_R = \int \bar{q}_R \gamma^0 q_R d^3 x$$

is the right quark number operator.

We are assuming that transitions from right to left quark (e.g., $q_R \to Q_L + H$) are fast processes as compared to the rate of time-variation of $\dot{\theta}_1$, while the anomalous baryon number processes are not. If we neglect the latter, the baryon number density vanishes. In the case of the Hamiltonian (11.111) this means that there is chemical potential μ_B to baryon number, which is the only conserved quantum number in our model. Hence, the calculation of the free energy proceeds by replacing the Hamiltonian (11.111) with

$$H - \dot{\theta}_1 N_R - \mu_B (N_R + N_L), \tag{11.112}$$

where $(N_R + N_L) = B$ is the baryon number. Hence, the effective chemical potential for right quarks equals $(\mu_B + \dot{\theta}_1)$, while it is equal to μ_B for left quarks. Using the result of Problem 5.4, we find the expression for the baryon number,

$$\Delta_B = \Delta_R + \Delta_L = \frac{T^2}{6} \left[\left(\mu_B + \dot{\theta}_1 \right) + 2\mu_B \right],$$

where we accounted for two left quark species, hence the factor 2 in the last term. We now require $\Delta_B = 0$ and obtain

$$\mu_B = -\frac{1}{3} \dot{\theta}_1.$$

Let us now turn on the electroweak processes that violate the baryon number. Since there is non-vanishing chemical potential μ_B, they generate the baryon number by partially washing out μ_B. Let us make use of Eqs. (5.53) and (5.54) and write

$$\frac{dn_B}{dt} = -\frac{\Delta F \cdot \Delta B}{T} \Gamma_{sph},$$

where ΔF and ΔB are the changes of free energy and baryon number at one sphaleron process, and Γ_{sph} is the rate of the sphaleron processes. For one left quark doublet (color triplet) $\Delta B = 1$ and $\Delta F = \mu_B \cdot \Delta B = \mu_B$. Thus,

$$\frac{dn_B}{dt} = -\frac{\mu_B}{T} \Gamma_{sph} = \frac{1}{3} \frac{\dot{\theta}_1}{T} \Gamma_{sph}. \tag{11.113}$$

Here we use the assumption that the sphaleron processes are slow, and hence we neglect their effect on the evolution of μ_B. In this way we obtain baryon number density generated by the wall passing through a given region of space,

$$n_B = \frac{1}{3T} \int \dot{\theta}_1 \Gamma_{sph}(t) dt. \tag{11.114}$$

Here Γ_{sph} depends on time, since the background values of the scalar fields at a given point in space depend on time. As a reasonable approximation we take Γ_{sph} given by (11.19),

$$\Gamma_{sph} = \varkappa' \alpha_W^5 T^4, \quad \varkappa' \approx 18$$

so long as the combination $\sqrt{|\phi_1|^2 + |\phi_2|^2}$, which determines the vector boson masses, is relatively small, $\sqrt{|\phi_1|^2 + |\phi_2|^2} < T$, and $\Gamma_{sph} = 0$ after that (see Eq. (11.105); the W-boson mass in the two-Higgs doublet model is determined by $v = \sqrt{|\phi_1|^2 + |\phi_2|^2}$). We recall the formula for entropy density $s = (2\pi^2/45)g_* T^3$ and obtain the estimate

$$\frac{n_B}{s} \simeq \varkappa' \frac{\alpha_W^5}{g_*} \Delta\theta_1 = \varkappa' \frac{\alpha_W^5}{g_*} \frac{v_2^2}{v_1^2 + v_2^2} \Delta\theta, \tag{11.115}$$

where $\Delta\theta$ is the change in the relative phase of the scalar fields from the unbroken phase to the point where sphaleron transitions switch off. For $g_* \sim 100$, $\alpha_W \simeq 1/30$ and $v_1 \sim v_2$ we find

$$\frac{n_B}{s} \simeq 10^{-8} \cdot \Delta\theta, \tag{11.116}$$

which is quite acceptable for generating the required asymmetry (11.1). This estimate is indeed valid in realistic extensions of the Standard Model modulo the numerical factor in (11.114). The dominant effect there often comes from t-quark.

The necessary conditions of asymmetry generation are fulfilled here in the following way:

(1) Baryon number is violated because Γ_{sph} is nonzero.
(2) The source of CP-violation is the time-dependent phase θ. This is an additional source of CP-violation as compared to the Standard Model where CP is violated only in the quark mixing matrix. This is a general situation: electroweak baryogenesis requires extra source of CP-violation.
(3) Departure from thermal equilibrium occurs due to the time-dependence of the phase θ and low rate of electroweak baryon number violating processes.

Problem 11.19. *Consider the interaction of the $SU(2)_W$ gauge field with a new scalar field φ described by the Lagrangian*

$$\mathcal{L} = f(\varphi) \cdot \frac{g^2}{32\pi^2} V^{\mu\nu\,a} \tilde{V}_{\mu\nu}^a.$$

Let the background field f rapidly evolve at the electroweak transition epoch. Estimate the baryon asymmetry produced in that epoch.

To end this Section, let us see that the time-dependence of the relative phase of the Higgs fields is indeed possible. Let us consider the scalar potential of the form

$$V(H_1, H_2) = V_1(H_1^\dagger H_1) + V_2(H_2^\dagger H_2)$$
$$+ \lambda_+[\mathrm{Re}(H_2^\dagger H_1) - v_1 v_2 \cos 2\xi]^2 + \lambda_-[\mathrm{Im}(H_2^\dagger H_1) - v_1 v_2 \sin 2\xi]^2,$$

where the minima of the functions V_1 and V_2 are at $|\phi_1| = v_1$ and $|\phi_2| = v_2$, respectively, and $\lambda_\pm, \xi$ are dimensionless parameters. For $\lambda_\pm > 0$ the scalar potential has its minimum at

$$\phi_1 = e^{i\xi} v_1, \quad \phi_2 = e^{-i\xi} v_2,$$

which corresponds to the vacuum phase

$$\theta_{vac} = \xi.$$

Let the phase transition occur in such a way that both Higgs fields develop the expectation values. ϕ_1 and ϕ_2 are small at the beginning of the transition, so the relevant part of the potential is quadratic in the fields,

$$V_{eff} = V_{1,eff}(|\phi_1|^2) + V_{2,eff}(|\phi_2|^2) - v_1 v_2 \mathrm{Re}(\phi_2^* \phi_1 \cdot \lambda e^{2i\zeta}), \qquad (11.117)$$

where

$$\lambda e^{2i\zeta} = \lambda_+ \cos 2\xi + i\lambda_- \sin 2\xi,$$

i.e.,

$$\tan 2\zeta = \frac{\lambda_-}{\lambda_+} \tan 2\xi.$$

The minimum of the potential (11.117) with respect to the relative phase is at

$$\theta_i = \zeta.$$

This phase gives the direction in the space of the Higgs fields along which the fields roll in the beginning of the phase transition, i.e., in the wall region near the unbroken phase. Hence, the phase θ changes across the wall from $\theta_i = \zeta$ (front, near the unbroken phase) to $\theta_{vac} = \xi$ (back, near the broken phase), as required.

In fact, the above analysis is not quite realistic. The medium in and around the domain wall region is strongly inhomogeneous; this is seen from (11.112), where the chemical potentials μ_B and $(\mu_B + \dot\theta_1)$ depend on both time and spatial location. An important phenomena in inhomogeneous medium are diffusion processes which we totally ignored. Diffusion changes the overall picture of the baryon asymmetry generation [222, 224]. Nevertheless, an order-of-magnitude estimate (11.115) remains valid, although the dependence of the baryon asymmetry on parameters (e.g., particle masses) is more complicated. The state-of-art approaches to the analysis of the baryon asymmetry generation are described, e.g., in the review [225].

11.5.3.   * Thin wall case

The case when the wall thickness is small compared to the mean free path of particles is more realistic, but at the same time more complicated. We will discuss here the most important physical processes leading to the thin wall baryogenesis, leaving aside the detailed analysis.

We are going to use the same model as in previous Section, but with different values of parameters. As before, we assume that *CP*-violation occurs due to the phase of the scalar field that varies in space and time. We will work in the wall rest frame for the time being, so instead of (11.109) we have now the following Lagrangian for fermions interacting with the wall,

$$\mathcal{L}_f = h_1 \bar{q}_L \rho(z) e^{i\theta(z)} q_R + h.c., \qquad (11.118)$$

where the coordinate z is normal to the wall. The function $\rho(z)$ changes from zero (as $z \to -\infty$, unbroken phase, we consider the wall moving from right to left) to Φ_c (as $z \to +\infty$, broken phase); according to (11.105) we have

$$\Phi_c \gtrsim T. \qquad (11.119)$$

Let us consider fermion with small Yukawa coupling,[19] $h_1 \ll 1$. Then its effective mass in the broken phase is also small, $m_f = h_1 \Phi_c \ll T$. We will make our order-of-magnitude estimates for the wall thickness of the order of inverse temperature,

$$l_w \sim T^{-1}. \qquad (11.120)$$

We will also assume (and this is indeed the case) that the wall velocity is small, $v_w \ll 1$. Finally, without loss of generality we set the phase $\theta(z)$ equal to zero in the broken phase, $\theta(z) = 0$ as $z \to \infty$.

Some fraction of fermions moving towards the wall from the unbroken phase gets reflected by the wall. Since the phase $\theta(z)$ depends on z, the reflection coefficients R_L and $\bar{R}_L$ for the left fermion and its antiparticle are different. Particles whose momenta p_z well exceed the inverse wall thickness do not experience the reflection (they are well described in the WKB approximation), while particles with lower momenta do. Still, the reflection coefficient for the latter is small, $R \ll 1$: the height of the energy barrier is equal to m_f, which is small compared to p_z for typical momenta $p_z \sim l_w^{-1}$. On the other hand, particles with $p_z < m_f$ do not penetrate through the wall at all. Hence, the important range of momenta is

$$m_f < p_z < l_w^{-1}. \qquad (11.121)$$

In this range, we can use perturbation theory in h_1, so that the reflection amplitude is of order h_1, and reflection probability (amplitude squared) is of order h_1^2. The

---

[19]This is not the case for *t*-quark, which requires special treatment.

asymmetry of reflection is thus roughly estimated as

$$R_L - \bar{R}_L \sim h_1^2 \theta_{CP}, \tag{11.122}$$

where θ_{CP} is determined by the variation of the phase θ in the wall region.

It is useful to note at this stage that left fermion after having been reflected becomes right fermion; see below.

If the wall is at rest, the system is in thermal equilibrium, and CP-asymmetric flux of left and right particles vanishes. This is because the asymmetries of fluxes of reflected and transmitted particles cancel each other. This is no longer the case for moving wall: the flux of right particles from the wall to the symmetric phase is different from the flux of their antiparticles (or vice versa). The asymmetry is proportional to the wall velocity. To estimate this asymmetry, we note that momenta p_x, p_y of the reflected particles are arbitrary, while p_z is in the range (11.121). Hence, we have the estimate for the asymmetry in the flux of right particles and their antiparticles

$$J_R \sim v_w T^2 l_w^{-1} \left[R_L - \bar{R}_L \right]_{p_z \sim l_w^{-1}}, \tag{11.123}$$

where we take into account that right particles are left before reflection. Here $T^2 l_w^{-1}$ is the number density of particles of relevant momenta, and the factor v_w accounts for the fact that the asymmetry vanishes in the static limit $v_w \to 0$. The flux of left reflected particles is

$$J_L = -J_R. \tag{11.124}$$

Hence, there is an excess of right particles and deficit of left ones (if $J_R > 0$) in unbroken phase in front of the wall. The opposite situation takes place behind the wall in broken phase. The moving wall separates particles of different types.

Once the inequality (11.105) holds, the asymmetry behind the wall does not lead to baryon and lepton number violation. In front of the wall it does: even though the overall fluxes of baryon and lepton numbers are zero according to (11.124), there is deficit of *left* fermions there. These are precisely the particles that participate in electroweak sphaleron processes. Hence, the latter processes tend to wash out the deficit of left fermions, thus partially equilibrating the system and generating the net baryon asymmetry.

The asymmetric fermion reflection changes the medium locally, at distance l from the wall. (We will soon estimate the value of l.) At a given point of space, the time interval from the appearance of the asymmetry to the moment when the wall passes through that point equals $t = l/v_w$. The baryon number violating processes occur precisely during this time interval. To estimate l we notice that in time t, the reflected particle experiences t/t_f collisions and, according to the Brownian law, moves away to distance

$$l = l_f \sqrt{\frac{t}{t_f}}, \tag{11.125}$$

where t_f and l_f are mean free time and mean free path of the particle; $l_f = t_f$ in our relativistic case. Together with $t = l/v_w$, Eq. (11.125) gives

$$l = \frac{l_f}{v_w}, \quad t = \frac{t_f}{v_w^2}. \tag{11.126}$$

The excess of left particles generated in the region of size l by collisions with the wall in time t (per unit wall area) is of order

$$N_L = J_L \cdot t,$$

so the excess number density is

$$n_L^{ind} = J_L \cdot \frac{t}{l} = \frac{J_L}{v_w}. \tag{11.127}$$

The sphaleron processes are typically so slow that they wash out a small fraction of the excess (11.127). Then (see (11.114))

$$\frac{dn_B}{dt} \sim -\frac{\mu_L}{T} \Gamma_{sph},$$

where for relativistic particles $\mu_L \sim n_L/T^2$. This gives

$$n_B \sim \frac{n_L^{ind}}{T^3} \Gamma_{sph} \cdot t,$$

where the time interval during which the sphaleron processes work is given by (11.126). Combining formulas (11.123), (11.124), (11.126), (11.127) and recalling that the entropy density is $s \sim g_* T^3$, we obtain the baryon asymmetry

$$\Delta_B \equiv \frac{n_B}{s} \sim \frac{1}{v_w^2 g_*} \frac{t_f}{l_w} \frac{\Gamma_{sph}}{T^4} \left[R_L - \bar{R}_L \right]_{p_z \sim l_w^{-1}}. \tag{11.128}$$

Now, the mean free time of left leptons is of order

$$t_f \sim (\alpha_W^2 T)^{-1}.$$

The mean free time of quarks is shorter due to collisions caused by strong interactions, so the most important[20] is τ-lepton: indeed, according to (11.122), the

---

[20] Our analysis does not apply to t-quark, in particular because left t-quark quickly becomes right due to the Yukawa interaction. Hence, t-quark does not play major role in the mechanism under study.

asymmetry is larger for larger $h_1 = \frac{m}{v}$. We use $\Gamma_{sph} = \varkappa' \alpha_W^5 T^4$, $l_w \sim T^{-1}$ as well
as the estimate (11.122) in (11.128) and find for τ-lepton

$$\Delta_B \sim \frac{\varkappa' \alpha_W^3}{v_w^2 g_*} h_\tau^2 \theta_{CP},$$

where $h_\tau^2 = \frac{m_\tau^2}{v^2} \sim 10^{-4}$. For $v_w \sim 3 \cdot 10^{-2}$ (which is quite realistic) and $\varkappa' \simeq 18$ we
obtain numerically

$$\Delta_B \sim 10^{-7} \theta_{CP}. \tag{11.129}$$

This is sufficient for the generation of the realistic asymmetry.

In our analysis, we have not taken into account a number of important effects
such as dynamical masses of particles, conservation of $(B-L)$ and weak hypercharge,
Debye screening of gauge charges in the plasma, etc. Like in Sec. 11.5.2, diffusion is
also important. Nevertheless, our order of magnitude estimate (11.129) is roughly
correct, so the mechanism we described is indeed efficient.

Problem 11.20. *At what wall velocities is the approximation of slow sphaleron
processes, made in the text, valid?*

Problem 11.21. *At what wall velocities are the transitions of left τ-lepton into
right one, $\tau_L + Z \to \tau_R + h$, irrelevant?*

Let us refine the estimate (11.122). To do that, we need to solve the Dirac equation for the
fermion that interacts with the wall according to (11.118). Proceeding to work in the wall
rest frame, we make the Lorentz transformation along the wall after which the fermion
motion is normal to the wall. We use the chiral (Weyl) representation for γ-matrices. In
our reference frame $\partial_1 \psi = \partial_2 \psi = 0$, so the term (11.118) leads to the following form of
the Dirac equation,

$$i\partial_0 q_R + i\sigma^3 \partial_z q_R + m^*(z) q_L = 0, \tag{11.130}$$
$$i\partial_0 q_L - i\sigma^3 \partial_z q_L + m(z) q_R = 0, \tag{11.131}$$

where $m(z) = h_1 \rho e^{i\theta}$. The wave function of *left* fermion incident on the wall from the left
(i.e., from unbroken phase where $m(z) = 0$), is

$$q_L^{(in)} = e^{-i\omega t + ipz} \cdot \begin{pmatrix} 0 \\ 1 \end{pmatrix}, \quad z \to -\infty, \tag{11.132}$$

where $p = \omega > 0$. Indeed, this is precisely the solution to Eq. (11.131) with $m = 0$. To calculate the reflection coefficient of left fermion, we have to find the solution to Eqs. (11.130),
(11.131), which contains the incoming wave (11.132), outgoing wave at $z \to -\infty$ that moves
to the left, and only outgoing wave at $z \to +\infty$. We see from Eqs. (11.130), (11.131) that

the solution has the structure

$$q_L = e^{-i\omega t}\psi_L(z) \cdot \begin{pmatrix} 0 \\ 1 \end{pmatrix}, \quad q_R = e^{-i\omega t}\psi_R(z) \cdot \begin{pmatrix} 0 \\ 1 \end{pmatrix}.$$

The complex functions ψ_L and ψ_R obey

$$-i\partial_z\psi_R + \omega\psi_R + m^*(z)\psi_L = 0, \tag{11.133}$$

$$i\partial_z\psi_L + \omega\psi_L + m(z)\psi_R = 0. \tag{11.134}$$

The reflected wave behaves at $z \to -\infty$ as $\psi \propto e^{-ipz}$. We see that the reflected wave is right, since Eq. (11.134) with $m = 0$ does not admit solutions of this form. This general result is actually a consequence of angular momentum conservation. Hence, the reflected wave is

$$\psi_R = Ae^{-ipz}, \quad z \to -\infty, \tag{11.135}$$

where A is the reflection amplitude, and $p = \omega$. To find the amplitude A we make use of perturbation theory in $m(z)$. At zeroth order we have $\psi_L = e^{-ipz}$, $\psi_R = 0$. At the first order, ψ_R is determined from Eq. (11.133), which gives

$$-i\partial_z\psi_R + \omega\psi_R = -m^*(z)e^{ipz}.$$

The general solution of the latter equation is

$$\psi_R = e^{-ipz}\left[-i\int_{z_0}^z m^*(z')e^{2ipz'}dz' + c\right], \tag{11.136}$$

where z_0 and c are arbitrary constants. We choose the constant c in such a way that there is no left-moving wave beyond the wall, i.e., at $z \to +\infty$. We choose z_0 to be in the region behind the wall. There we have $m^* = m_f$ (without loss of generality, h_1 is real and $\theta(z \to +\infty) = 0$), and the solution at $z \to +\infty$ is

$$\psi_R = e^{-ipz}\left[-\frac{m_f}{2p}(e^{2ipz} - e^{2ipz_0}) + c\right].$$

Hence, the requirement of the absence of incoming wave at $z \to +\infty$ gives

$$c = -\frac{m_f}{2p}e^{2ipz_0}.$$

The solution (11.136) indeed has the form (11.135), and we find

$$A = -i\int_{z_0}^{-\infty} m^*(z)e^{2ipz}dz - \frac{m_f}{2p}e^{2ipz_0}, \tag{11.137}$$

where we have to take the limit $z_0 \to +\infty$. The reflection amplitude $\bar{A}$ for antiparticle has the same form with $m^*(z)$ substituted for $m(z)$.

Problem 11.22. *Prove the last statement.*

The reflection coefficient for left fermion R_L is equal to the modulus of the amplitude (11.137) squared. We are interested in the asymmetry

$$R_L - \bar{R}_L = |A|^2 - |\bar{A}|^2. \tag{11.138}$$

To proceed further, we write the amplitude (11.137) in the following form,

$$A = i \int_{-\infty}^{+\infty} [m^*(z) - m_f \Theta(z)] e^{2ipz} dz - \frac{m_f}{2p},$$

where $\Theta(z)$ is the usual step function, and we have taken the limit $z_0 \to \infty$. This is the final analytical result for the amplitude. To estimate the asymmetry (11.138), let us assume that the function $\mathrm{Re}[m(z)] - m_f \Theta(z)$ is symmetric in z. Then

$$A = -\frac{M}{2p} + \int_{-\infty}^{+\infty} \mathrm{Im}[m(z)] e^{2ipz} dz, \tag{11.139}$$

where

$$M = m_f + 2 \int_{-\infty}^{+\infty} \{\mathrm{Re}[m(z)] - m_f \Theta(z)\} \sin(2pz) p \, dz.$$

We note that there is an order of magnitude relation

$$M \sim m_f. \tag{11.140}$$

From (11.139) we obtain the asymmetry

$$R_L - \bar{R}_L = -\frac{2M}{p} \int_{-\infty}^{+\infty} \mathrm{Im}[m(z)] \cos 2pz \, dz.$$

Finally, we introduce

$$\theta_{CP}(p) = \frac{1}{l_w m_f} \int_{-\infty}^{+\infty} \mathrm{Im}[m(z)] \cos 2pz \, dz, \tag{11.141}$$

and obtain

$$R_L - \bar{R}_L = -\frac{2M m_f l_w}{p} \theta_{CP}(p).$$

Because of the oscillatory factor $\cos 2pz$, the integral in (11.141) rapidly tends to zero at large momenta, $p \gg l_w$, while at $p \lesssim l_w^{-1}$ it is determined by the phase $\theta(z)$ of the function $m(z)$. Making use of (11.140) we obtain the estimate

$$R_L - \bar{R}_L \sim \frac{2m_f^2 l_w}{p_z} \theta_{CP}, \quad p_z \lesssim l_w^{-1}, \tag{11.142}$$

where we recalled that p is the momentum normal to the wall. This estimate is independent of p_x, p_y, as the Lorentz boost along the wall does not change p_z. We stress that (11.141) contains fermion mass in the broken phase, $m_f = h_1 \Phi_c$. The estimate (11.122) is obtained from (11.142) at $p_z \sim l_w^{-1}$ by assuming that the wall thickness is of order Φ_c^{-1} (i.e., by assuming the validity of (11.119) and (11.120)).

The asymmetry in reflection of right particles is

$$R_R - \bar{R}_R = -(R_L - \bar{R}_L). \tag{11.143}$$

Problem 11.23. *Prove the last statement in the general case by making use of the properties of Eqs. (11.130) and (11.131).*

11.5.4. *Electroweak baryogenesis, CP-violation and neutron EDM*

To end this Section we notice that both electroweak mechanisms we have discussed employ extra sources of CP-violation as compared to the Standard Model. This extra CP-violation is in fact interesting from the viewpoint of precision particle physics experiments.

One class of experiments sensitive to new CP-violating phases is the search for electric dipole moments (EDMs) of electron and neutron, d_e and d_n. EDM d is the parameter of the Hamiltonian describing the interaction of spin $\mathbf{S}$ with electric field $\mathbf{E}$,

$$H = -d \cdot \mathbf{E} \cdot \frac{\mathbf{S}}{|\mathbf{S}|}.$$

The corresponding relativistic Lagrangian for electromagnetic interaction of fermion ψ is

$$\mathcal{L} = -d\frac{i}{2}\bar{\psi}\gamma^\mu\gamma^\nu\gamma^5\psi F_{\mu\nu}. \tag{11.144}$$

EDM violates P and T (see Sec. B.3) and therefore CP. EDM of composite particle, neutron, is of the order of EDMs of quarks in it, $d_n \sim d_u, d_d$. The present experimental limits are [1]

$$d_e < 1.05 \cdot 10^{-27} \cdot e \cdot \text{cm} = 0.5 \cdot 10^{-12} \cdot e \cdot \text{GeV}^{-1}, \tag{11.145}$$

$$d_n < 2.9 \cdot 10^{-26} \cdot e \cdot \text{cm} = 0.32 \cdot 10^{-11} \cdot e \cdot \text{GeV}^{-1}, \tag{11.146}$$

where e is the electron charge. In the two-doublet model (11.107), the additional CP-violation in the Englert–Brout–Higgs sector induces quark EDM via one-loop diagram of Fig. 11.16, where H denotes all Higgs fields, and summation over these fields is assumed. The estimate for this diagram is

$$d_u \sim \theta_{CP}\frac{e}{(4\pi)^2}\frac{m_{q_i}U_{uq_i}U_{q_iu}^* Y_u Y_{q_i}}{m_H^2}, \tag{11.147}$$

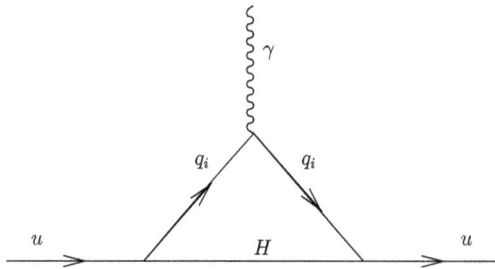

Fig. 11.16.   EDM of u-quark at one loop level.

where m_H is the mass scale of the Higgs bosons, θ_{CP} is the CP-violating phase, U is the quark mixing matrix (so that the Yukawa matrix is $Y = Y_{diag} \cdot U$), $(4\pi)^{-2}$ is the loop factor, and the quark mass m_{q_i} comes about due to the fact that the Lagrangian (11.144) violates helicity.

The dominant contribution into (11.147) comes from virtual t-quark, so for $U_{ut} \sim 1$ we obtain

$$d_n \sim d_u \sim \theta_{CP} \cdot 1.6 \cdot 10^{-11} \cdot \left(\frac{1 \text{ TeV}}{m_H}\right)^2 e \cdot \text{GeV}^{-1}. \qquad (11.148)$$

By comparing (11.148) to (11.146) we see that the phase θ_{CP} must be fairly small, $\theta_{CP} \lesssim 10^{-3}$ for $m_H \sim 100$ GeV, provided that mixing in Higgs-quark interactions is of order 1. Similar bounds on CP-violating phases exist in other models. These bounds are completely independent of the baryon asymmetry. On the other hand, electroweak baryogenesis requires sizeable CP-violating phases. Thus, most models with successful electroweak baryogenesis predict EDMs comparable to the limits (11.145), (11.146). As an example, the result (11.116) for the asymmetry in the model of the previous Section suggests that $\theta_{CP} \sim 10^{-2}$. Making use of the estimate (11.148) we obtain the prediction

$$d_n \sim \left(10^{-13} - 10^{-14}\right) \cdot \left(\frac{1 \text{ TeV}}{m_H}\right)^2 \cdot e \cdot \text{GeV}^{-1}.$$

This is close to the experimental limit for realistic mass scale of the Higgs bosons, say $m_H \sim 200$ GeV. The general conclusion is that the new mechanisms of CP-violation required for electroweak baryogenesis will most probably be indirectly discovered in the EDM experiments of the next generation.

11.6. *Affleck–Dine Mechanism

11.6.1. *Scalar fields carrying baryon number*

Some extensions of the Standard Model contain not only quarks, but also scalar particles that carry baryon number. Scalars carrying lepton numbers are also possible. These particles have not been found experimentally, so they must be quite heavy;[21] roughly speaking, their masses must exceed a few hundred GeV. Yet the dynamics of the corresponding scalar fields in expanding Universe may lead to baryon asymmetry generation; the class of baryogenesis scenarios of this sort is generically called the Affleck–Dine mechanism [212]. Certainly, the necessary conditions of Sec. 11.1 must be satisfied in this case as well. In particular, baryon number should not be exact symmetry, now of the Lagrangian of these scalar fields; CP must be violated as well.

---

[21] Assuming that their interactions with Standard Model particles are not particularly weak.

To construct a prototype model, let us extend the Standard Model by adding a complex scalar field ϕ carrying baryon number, $B_\phi \neq 0$, and a fermion ψ with zero baryon number. The kinetic and mass terms in the action for ψ have the standard form, while the action for the scalar field is chosen as

$$S_\phi = \int d^4 x \sqrt{-g}\, [g^{\mu\nu} \partial_\mu \phi^* \partial_\nu \phi - V(\phi)], \qquad (11.149)$$

where

$$V(\phi) = m^2 \phi^* \phi + \frac{\lambda}{2}(\phi^* \phi)^2 + \frac{\lambda'}{4}(\phi^4 + \phi^{*4}). \qquad (11.150)$$

The parameters λ and λ' are real and positive,[22] and we assume that $\lambda' \ll \lambda$. Finally, there is a Yukawa term involving ϕ, ψ and a spinor combination q of quark fields,

$$\mathcal{L}_{int} = h\bar{q}\psi\phi + h.c., \qquad (11.151)$$

where h is the Yukawa coupling. In fact, q may be (and often is) a composite operator, a product of the Standard Model fields; it may or may not carry color. The only important property is that the operator q carries baryon number B_q such that

$$B_\phi = B_q,$$

and the Yukawa term (11.151) conserves baryon number. We do not discuss here the transformation properties of the fields ϕ and ψ under the Standard Model gauge group $SU(3)_c \times SU(2)_w \times U(1)_Y$; the relevant representations can be straightforwardly found.

Problem 11.24. *Find the representations of ϕ and ψ under $SU(3)_c \times SU(2)_w \times U(1)_Y$ and the composite operator q such that the full action is gauge invariant.*

Were the last term in (11.150) absent, the model would be invariant under global phase rotations[23]

$$\phi \to e^{i\alpha B_\phi}\phi, \quad q \to e^{i\alpha B_\phi}q, \quad \psi \to \psi.$$

The corresponding conserved quantum number is nothing but baryon number, and the baryon number density is

$$n_B = iB_\phi(\partial_t \phi^* \cdot \phi - \phi^* \cdot \partial_t \phi) + n_{B,q}, \qquad (11.152)$$

where $n_{B,q} = \frac{1}{3}(n_q - n_{\bar{q}})$ is the baryon number density of quarks. For small but finite constant λ' in (11.150) baryon number is conserved only approximately.

Problem 11.25. *Obtain the expression (11.152) from the Noether theorem.*

---

[22] The third term in (11.150) is, generically, $\frac{1}{4}(\lambda'\phi^4 + h.c.)$ with complex λ'. However, by redefining the field ϕ it can be cast into the form (11.150).
[23] Modulo electroweak anomaly which is irrelevant here.

The situation similar to what has been just described occurs in supersymmetric extensions of the Standard Model; see Sec. 9.6. The composite field ϕ is a combination of squarks, sleptons and Higgs fields, while the field ψ is a combination of gauginos. There is indeed the interaction of the form (11.152), and up to numerical factor, the coupling h coincides with the gauge coupling g_s of the color group $SU(3)_c$. The interaction term $\lambda'(\phi^4 + h.c.)$ is forbidden by gauge invariance under $SU(3)_c$, but higher order interactions violating baryon number and $(B-L)$ are allowed (e.g., $\lambda'\phi^6$). $(B-L)$-violation is important, since the generation of $(B-L)$ asymmetry is necessary and sufficient for the generation of the present baryon asymmetry, once one employs mechanisms operating at temperatures above 100 GeV.

A generic feature of SUSY extensions of the Standard Model is the existence of *flat directions* (moduli) in the field space, along which the scalar potential is small up to large values of the scalar fields. In terms of the potential (11.150) in which ϕ is the field parameterizing flat direction, this means that the mass m is low while the constants λ and λ' are extremely small.

11.6.2. *Asymmetry generation*

Let us continue with the model described above. Let the field ϕ at initial time (right after the end of inflation in inflationary theory) be spatially homogeneous and take some complex value $\phi_i = r_i e^{i\theta_i}/\sqrt{2}$. The analysis of further evolution is simple in the case

$$m^2|\phi_i|^2 \gg \lambda|\phi_i|^4. \tag{11.153}$$

Let us discuss this case in some details; the opposite case is the subject of Problem 11.27. We know from Sec. 4.8.1 that the field stays practically constant for some time during which the slow roll conditions are satisfied. Once these conditions get violated, the field, remaining homogeneous, evolves towards the minimum of the potential, $\phi = 0$, and in a few Hubble times comes to the vicinity of this minimum. Near the minimum, real and imaginary parts of the field evolve independently (the potential is quadratic), and each of them evolves according to (4.63), i.e.,

$$\operatorname{Re}\phi \equiv \phi_R = \frac{C_R}{a^{3/2}(t)}\cos(mt + \beta_R),$$
$$\operatorname{Im}\phi \equiv \phi_I = \frac{C_I}{a^{3/2}(t)}\cos(mt + \beta_I). \tag{11.154}$$

Note that for time-independent scale factor a the trajectory (11.154) in the field space is an ellipse in complex plane (for $\beta_I \neq \beta_R$). In reality, the scale factor $a(t)$ grows, so the ellipse turns into elliptical spiral, see Fig. 11.17. The presence of the baryon number violating term in (11.150) (the term with λ') is important: were it not for this term, the phases β_R and β_I would be equal, and the ellipse would degenerate into interval.

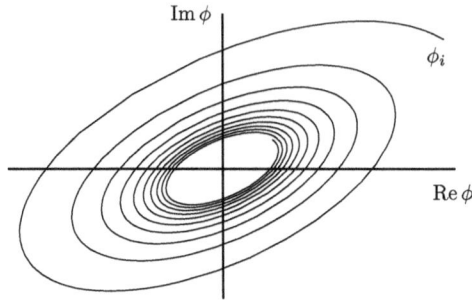

Fig. 11.17.   Trajectory $\phi(t)$ in complex plane.

The field (11.154) carries baryon number density

$$n_B = iB_\phi(\partial_t\phi^* \cdot \phi - \phi^*\partial_t\phi) = 2B_\phi(\phi_R\partial_t\phi_I - \phi_I\partial_t\phi_R) \tag{11.155}$$

which is equal to

$$n_B = 2B_\phi \frac{mC_IC_R}{a^3(t)} \sin(\beta_R - \beta_I). \tag{11.156}$$

In the absence of interactions with quarks, baryon number density decreases as $n_B \propto a^{-3}$, so baryon number is conserved in comoving volume (baryon number violation due to the last term in (11.150) is negligible at small ϕ). Interaction with quarks (11.151) transmits this baryon number into quarks, and the Universe becomes baryon asymmetric in the end.

As we noticed in Sec. 4.8.1, the field (11.154), in quantum language, describes a coherent state of ϕ-bosons and their antiparticles at rest. The ϕ-boson and its antiparticle carry baryon number B_ϕ and $(-B_\phi)$, respectively. Non-vanishing baryon number (11.156) means that the numbers of ϕ-bosons and their antiparticles are different. The decay of the oscillating field ϕ into quarks and fermions ψ may, with reservations, be viewed as decays of ϕ-particles and their antiparticles. In this language, the generation of asymmetry between quarks and antiquarks in the decay of the field (11.154) is completely obvious.

Let us now turn to the estimate of the generated asymmetry. It is convenient to rewrite the action (11.149) in terms of variables r and θ, defined by

$$\phi = \frac{r}{\sqrt{2}}e^{i\theta}.$$

The action for the homogeneous field in expanding Universe is

$$S_{\rho,\theta} = \int dt a^3(t) \left(\frac{1}{2}\dot{r}^2 + \frac{1}{2}r^2\dot{\theta}^2 - V(r,\theta)\right),$$

where

$$V(r,\theta) = \frac{m^2}{2}r^2 + \frac{\lambda}{8}r^4 + \frac{\lambda'}{8}r^4\cos 4\theta. \tag{11.157}$$

This gives the equation for θ,

$$\frac{1}{a^3}\frac{\partial}{\partial t}(a^3 r^2 \dot\theta) = -\frac{\partial V}{\partial \theta}. \tag{11.158}$$

The expression for the baryon number density (11.155) has the following form in terms of the variables r and θ,

$$n_B = B_\phi r^2 \dot\theta.$$

Hence, Eq. (11.158) reads

$$\frac{1}{a^3}\frac{\partial}{\partial t}(a^3 n_B) = -B_\phi \frac{\partial V}{\partial \theta}. \tag{11.159}$$

If the potential V were independent of θ, this would be the baryon number conservation law for comoving volume. For the θ-dependent potential Eq. (11.159) is the equation describing the generation of baryon number.

Let us assume that the initial phase θ_i is nonzero. This means that the initial state is CP-asymmetric; it is in this place where one of the necessary conditions for asymmetry generation shows up. This assumption is important, as the evolution for $\theta_i = 0$ proceeds along $\text{Im}\,\phi = 0$ (recall that λ' is real), and baryon number density (11.155) is always zero. For non-vanishing θ_i, the baryon number density at later time t is

$$a^3(t)n_B(t) = -B_\phi \int_{t_i}^t \frac{\partial V}{\partial \theta} a^3(t')dt' = \frac{B_\phi \lambda'}{2}\int_{t_i}^t r^4(t')a^3(t')\sin 4\theta(t')\,dt'. \tag{11.160}$$

Since $r^4 a^3 \propto a^{-3} \propto t^{-3/2}$, the integral in (11.160) is saturated at lower limit of integration, i.e., in a few Hubble times[24] after the field starts rolling down towards $\phi = 0$: before that, $\partial V/\partial\theta$ is constant but $a^3(t)$ is small; after that $\partial V/\partial\theta \propto r^4$ is small (proportional to a^{-6}, see (11.154)). Note that we assume that the evolution of the field ϕ occurs at the hot epoch. We proceed to work under this assumption, which implies, in particular, that the energy of the field ϕ never dominates in the Universe. We discuss the condition for that later on.

Equation (11.160) immediately gives the estimate

$$n_B(t) \sim \frac{a^3(t_r)}{a^3(t)}\cdot\frac{\partial V}{\partial\theta}(t_r)\cdot\frac{1}{H(t_r)},$$

where t_r is the time at which the field starts to roll down. We recall that $a(t)\propto T^{-1}$ and obtain the estimate for asymmetry,

$$\Delta_B \equiv \frac{n_B}{s} \sim \frac{1}{g_* T_r^3 H(T_r)}\cdot\frac{\partial V}{\partial\theta}(t_r) \sim \frac{\lambda'}{g_*}\frac{r_i^4}{H_r(M_{Pl}^* H_r)^{3/2}}\sin 4\theta_i. \tag{11.161}$$

---

[24]It is at this point that the assumption (11.153) simplifies the analysis.

The relation (11.153) shows that the slow roll conditions get violated when

$$m^2 \sim H_r^2. \tag{11.162}$$

Making use of this relation, we write the estimate (11.161) as

$$\Delta_B \sim \lambda' g_*^{-1/4} \left(\frac{r_i}{m}\right)^{5/2} \left(\frac{r_i}{M_{Pl}}\right)^{3/2} \sin 4\theta_i. \tag{11.163}$$

Clearly, the resulting asymmetry strongly depends both on parameters of the model (mass m and coupling λ') and on the initial amplitude r_i and phase θ_i of the scalar field ϕ. Hence, the value of Δ_B appears to be the result of a choice of initial data.

To find out under what conditions the expression for the asymmetry (11.163) gives the required value $\Delta_B \sim 10^{-10}$, we have to specify the initial data r_i and θ_i. One possible (but not at all unique) choice is to assume that the initial phase is not small, $\sin 4\theta_i \sim 1$, while the initial amplitude is of the order of the Planck scale, $r_i \sim M_{Pl}$. Then one finds from (11.163) and $m \ll M_{Pl}$ that the required asymmetry is obtained only for very weak baryon number violation, $\lambda' \ll 10^{-10}$. Otherwise the asymmetry is too large. This result illustrates rather general property of the Affleck–Dine mechanism: it often (though not always) generates too large asymmetry.

At $r_i \sim M_{Pl}$, the assumption that the slow roll regime terminates when the Universe is already at the hot stage of its evolution is valid for quite small m and very small λ only. Let T_{max} be the maximum temperature in the Universe. Then the relation (11.162) indeed holds at $T_i < T_{max}$ only if

$$m < \frac{T_{max}^2}{M_{Pl}^*}.$$

For $T_{max} \sim 10^{15}$ GeV (which is the order of magnitude of the maximum possible reheating temperature in realistic inflationary models; see the accompanying book) and $r_i \sim M_{Pl}$ we have

$$m < 10^{12} \text{ GeV}. \tag{11.164}$$

Furthermore, our assumption that $m^2|\phi|^2 \gg \lambda|\phi|^4$ holds only for $\lambda \ll 10^{-14}$, while the required baryon asymmetry is obtained from (11.163) at $\lambda' \lesssim 10^{-27}$. These estimates illustrate another property of the Affleck–Dine mechanism: it needs almost flat scalar potential $V(\phi)$ for the field ϕ. As we pointed out, almost flat potentials are rather natural in supersymmetric theories.

Let us discuss one more property of the Affleck–Dine mechanism, still assuming that r_i is roughly of order M_{Pl}. It is clear from (11.162) that at the beginning of oscillations, the energy of the field ϕ is roughly comparable to the energy of hot matter,

$$\frac{\rho_\phi(t_r)}{\rho_{rad}(t_r)} \sim \frac{m^2 r_i^2}{H^2 M_{Pl}^2} \sim \frac{r_i^2}{M_{Pl}^2}.$$

The energy density of the oscillating field (11.154) decreases in the same way as that of non-relativistic matter, $\rho_\phi \propto a^{-3}(t)$, see (4.69), and the decay of oscillations occurs at $H \sim \Gamma$ (see Sec. 5.3), where Γ is the decay width of the ϕ-boson. For $\Gamma \ll m$ (which is consistent with the fact that ϕ has tiny self-interaction) this implies that the energy of the oscillating field starts to dominate in the Universe soon after the oscillations begin. The existence of the intermediate matter-dominated stage is another characteristic property of the Affleck–Dine mechanism.

Problem 11.26. *If there is indeed the intermediate matter-dominated stage, the estimate (11.163) is not valid. Estimate Δ_B in that case, still assuming that slow roll conditions get violated at radiation domination.*

Problem 11.27. *Consider the case*

$$\lambda |\phi_i|^4 \gg m^2 |\phi_i^2|.$$

(a) *Show that the amplitude of oscillations in the case of quartic potential, $V = \lambda \phi^4$, decreases as a^{-1} (as opposed to $a^{-3/2}$ for quadratic potential). This means that $\partial V / \partial \theta$ in the integrand in (11.160) decreases as a^{-4} until the time when $\lambda r^4 \sim m^2 r^2$, and only afterwards it decreases as a^{-6}.*

(b) *Making use of this observation and assuming that the energy of the field ϕ never dominates in the Universe, show that the resulting asymmetry is of order*

$$\Delta_B \sim \lambda' g_*^{-1/4} \frac{r_i}{m\lambda^{3/4}} \left(\frac{r_i}{M_{Pl}} \right)^{3/2} \sin 4\theta_i.$$

(c) *Show that for r_i roughly of order M_{Pl} the required properties of the theory (small m, very small λ', etc.) are qualitatively the same as in the case considered in the text.*

Thus, the Affleck–Dine mechanism can successfully work at the hot stage, albeit in a rather restricted range of parameters. The scalar potential must be quite flat, and the maximum temperature in the Universe quite high. An interesting possibility here is that the Universe is temporarily dominated by the oscillating scalar field, and hence experiences intermediate matter dominated stage.

If the maximum temperature in the Universe is not very high, and the mass of the field ϕ is relatively large, then the field ϕ rolls down to the minimum of its potential before the hot stage. This is a fairly realistic situation. In that case the Affleck–Dine mechanism may work at post-inflationary reheating epoch. The corresponding estimates are different from those given above, but the general conclusion on the possibility of the baryon asymmetry generation remains valid.

Let us now discuss a version of the Affleck–Dine mechanism [213], in which the initial value r_i is determined dynamically, while the flatness requirements for the scalar potential are less severe. Let us assume that besides the terms written in

(11.149), there is one more term in the action,

$$S_{R\phi} = -\int d^4x \sqrt{-g} cR \, |\phi|^2,$$ (11.165)

where c is a positive dimensionless constant which we choose somewhat larger (but not much larger) than 1. For the FLRW Universe one has

$$R = -8\pi G T^{\mu}_{\mu} = -8\pi G(\rho - 3p).$$

The initial value of ϕ is determined by its dynamics at inflationary epoch, when $p \approx -\rho$ and $R = -12H^2$, so the extra term effectively changes the mass term in the potential,

$$m^2 \, |\phi|^2 \rightarrow (m^2 - 12cH^2) \, |\phi|^2.$$ (11.166)

At inflation one has $H \gg m$, the potential has *maximum* at $\phi = 0$, and in the limit $\lambda' \rightarrow 0$ there is a valley of minima on a circle in complex plane,

$$|\phi|^2 \equiv \frac{r^2}{2} = \frac{12cH^2}{\lambda}.$$ (11.167)

Slow roll conditions are *not* satisfied in this case for radial motion (this is the reason for the choice $c > 1$), so $r(t)$ is approximately given by (11.167). At small but finite λ' the valley (11.167) is slightly inclined, the potential depends on the phase θ, but this phase is almost flat direction. The evolution along this direction occurs in the slow roll regime.

Problem 11.28. *Show that for $H \gtrsim m$ and $\lambda' \ll \lambda$ the evolution of the phase θ occurs in the slow roll regime.*

At radiation domination one has $p \approx \frac{1}{3}\rho$, so that $R \ll H^2$, the contribution (11.165) is irrelevant and the dynamics is described by the scalar potential (11.150). If the reheating after inflation is fast and happens in a few Hubble times, then right after inflation the field ϕ rolls down towards $\phi = 0$, as described in Problem 11.27. The initial value of ϕ is determined by the Hubble parameter at the end of inflation, which in the instant reheating approximation coincides with the Hubble parameter at the beginning of the hot epoch,

$$r_i^2 \sim \frac{24cT_{max}^4}{\lambda M_{Pl}^{*2}}.$$ (11.168)

As long as the fourth order term dominates in the scalar potential, $r(t)$ decays as $a^{-1} \propto T$; see Problem 11.27. At time t_m, when $\lambda r^2 \sim m^2$, this regime terminates, the field starts to decay faster, $r \propto a^{-3/2}$, and the asymmetry generation terminates. At that time, the temperature is estimated as follows:

$$T_m \sim \frac{mT_{max}}{\lambda^{1/2}r_i}. \tag{11.169}$$

As a result, we obtain from (11.160) that the asymmetry is

$$\Delta_B \simeq \frac{B_\phi \lambda'}{2g_*} r_i^4 \sin 4\theta_i \cdot \int_{t_{reh}}^{t_m} \left(\frac{T}{T_{max}}\right)^4 \frac{dt}{T^3} \sim \frac{B_\phi \lambda'}{g_*} \frac{r_i^4}{T_{max}^4} \frac{M_{Pl}^*}{T_m} \sin 4\theta_i,$$

where t_{reh} is the time the hot epoch begins and temperature equals T_{max}. To evaluate the integral, we have used the relation (3.34). We recall Eq. (11.169) and then Eq. (11.168) and obtain finally

$$\Delta_B \sim (24c)^{5/2} B_\phi \frac{\lambda'}{\lambda^2} \frac{T_{max}^5}{mM_{Pl}^4}.$$

We see that if T_{max} is not very low, we do not need very flat potential; as an example, even for $T_{max} \sim 10^{12}$ GeV, $\lambda' = 0.1\lambda$, $m = 1$ TeV, the required baryon asymmetry is obtained at $\lambda \sim 10^{-4}$.

Problem 11.29. *Show that in the latter scenario, the initial value r_i is typically relatively low, $r_i \ll M_{Pl}$. This means, in particular, that the term (11.165) is small compared to the Einstein–Hilbert term, and gravity does not get modified.*

Problem 11.30. *Study the latter scenario at $c \gg 1$, where c is the parameter in the action (11.165). Can this mechanism work at the hot epoch and yield the required baryon asymmetry for $T_{max} \lesssim 10^9$ GeV?*

11.7.   Concluding Remarks

The baryon asymmetry generation mechanisms considered in this Chapter by no means exhaust the possibilities proposed in literature. Among other things, the asymmetry generation may occur at post-inflationary reheating epoch, rather than at the hot stage. Unfortunately, many mechanisms (like those studied in Secs. 11.3 and 11.4) make use of physics beyond the energy scale accessible at future colliders. Hence, direct experimental evidence in favor of one of these mechanisms will be hard if not impossible to obtain. The exception is the electroweak mechanism that will be supported or falsified by collider experiments in near future. Evidence for leptogenesis through neutrino oscillations could come from the discovery of sterile neutrinos with masses below the electroweak scale; this would be possible at experiments of new generation. As to the Affleck–Dine mechanism, it may be supported by the discovery of baryon isocurvature perturbations in the spectrum of density perturbations in the Universe.[25] Search for isocurvature perturbations is one of the goals of the studies of the CMB anisotropy and polarization, galaxy surveys, etc. We consider these issues in the accompanying book.

---

[25] Generation of baryon isocurvature perturbations is a possible, but not necessary consequence of the Affleck–Dine mechanism, so this mechanism cannot be ruled out by the cosmological data.

Chapter 12

Topological Defects and Solitons in the Universe

In this Chapter we consider cosmological aspects of field theory models which admit soliton or soliton-like solutions. These solutions are specific (sometimes macroscopic) field configurations whose stability is ensured either by non-trivial topology of the space of vacua (topological defects) or by the existence of conserved global quantum numbers (non-topological solitons, Q-balls). Of interest are both particle-like solitons (monopoles, Q-balls) and extended objects (cosmic strings, domain walls); we will jointly call them defects or solitons in what follows. These objects are absent in the Standard Model (modulo exotic proposals like quark nuggets), but they naturally exist in some extensions of the Standard Model. Generally speaking, there may be also unstable solutions whose lifetime is comparable to the Hubble time.

We point out right away that no undisputable evidence for defects of any sort in our Universe has been found so far. Nevertheless, their study is of considerable interest. We will see that there are numerous effects occurring in models with defects, including peculiarities of the expansion history, new mechanisms of structure formation, lensing of distant sources, features in CMB angular anisotropy, new processes generating the baryon asymmetry, etc. The energy scale of the defect formation is considerably higher than the electroweak scale. Experimental discovery of defects would thus give evidence that the temperature in the Universe reached that energy scale.[1] Furthermore, this discovery would show the existence of physics beyond the Standard Model and give a hint on what this new physics is. This would open up a window to energy scales well exceeding the reach of particle colliders.

A general reason for the existence of *topological* defects in field theoretic models is the non-trivial topology of the space of vacua in these models. This means that the ground state — vacuum — is not unique, and, furthermore, some homotopy group π_N of the manifold of vacua $\mathcal{M}$ is non-trivial,

$$\pi_N(\mathcal{M}) \neq 0. \tag{12.1}$$

---

[1]In fact, there exist mechanisms of defect production that do not require so high temperature. But in any case, the energy density at the time of defect production must be very high.

In other words, there exist non-trivial mappings from N-dimensional sphere S^N to the vacuum manifold $\mathcal{M}$. In most cases the defect configuration corresponds to a non-trivial mapping from spatial asymptotics (sphere S^N in the general case) to vacuum manifold (since the fields tend to vacuum values at spatial infinity, otherwise energy would be infinite). In space–time of dimension $d + 1$ the property (12.1) suggests the existence of topological defects of space–time dimension $d - N$. Stability of these defects is due to the fact that destruction of a defect would require rearrangement of the fields at spatial infinity; that would cost infinite energy. Non-trivial configurations thus have topological charges which are conserved. In the case of 4-dimensional space–time, three types of defects are possible, whose dimensions are $2 + 1$ (walls), $1 + 1$ (strings) and $0 + 1$ (particle-like defects, e.g., monopoles). There may exist hybrid defects, including combinations of defects of different dimensionality.

We note that the relevant symmetry of a theory may be local (gauge) or global. Accordingly, the defects are called local or global. In the former case the energy is localized. On the contrary, in the global defect case, the gradient energy falls off with distance so slowly that the energy integral diverges. In the physically relevant situation with numerous defects this means that one cannot neglect the interaction between them. These interactions render the energy density finite.

12.1.   Production of Topological Defects in the Early Universe

A rather general property of field theory models with topological defects is that the defects exist only in the phase with spontaneously broken symmetry, whereas they are absent in the unbroken phase. We will encounter this situation in Secs. 12.2–12.5. As we discussed in Chapter 10, symmetry is usually restored at high temperatures, and the broken phase appears in the Universe at lower temperatures as a result of phase transition. Thus, the existence of topological defects is possible only after this phase transition, i.e., at $T < T_c$, where T_c is the phase transition temperature.[2]

Defects may be produced after the phase transition thermally, in collisions of particles. This mechanism is often inefficient. If defects equilibrate, their density is small. Indeed, considering for definiteness particle-like defects like monopoles (see Sec. 12.2), we quote the result that their mass M_{TD} is typically much larger than the critical temperature, so the equilibrium abundance is suppressed by the Boltzmann factor

$$n_{TD}^{(eq)} \propto e^{-\frac{M_{TD}}{T_c}}. \tag{12.2}$$

Strings and domain walls are suppressed even stronger.

There exists, however, another way of producing topological defects, which is called Kibble mechanism [226]. We will consider this mechanism in detail in Sec. 12.2 where we study magnetic monopoles, but we will see that it is quite general. This mechanism works at the phase transition epoch for *topological* defects. The phase

---

[2]In fact, the production of topological defects is possible at both thermal phase transitions occurring at the hot Big Bang stage and non-thermal phase transitions that may take place, e.g., at post-inflationary reheating.

transition proceeds independently in regions of the Hubble size. Inside a Hubble region the field configurations may be correlated, so the system may end up in one and the same vacuum. On the other hand, vacua in different Hubble regions, generally speaking, are different points of the vacuum manifold. Hence, the topology of the field configuration at the spatial scale exceeding the Hubble scale may coincide with the defect topology. The relaxation processes after the phase transition do not change the topology, so the result is the formation of a defect. This is precisely the Kibble mechanism. It is clear from the above reasoning that this mechanism is quite universal; it works for defects of different types and depends rather weakly on M_{TD}/T_c.

More precisely, the number density of defects produced by the Kibble mechanism is determined, right after the phase transition, by the correlation length l_{cor}, which definitely does not exceed the Hubble size,[3] $H^{-1}(t_c)$. So, the number of defects is at least of order 1 per the Hubble volume,

$$n_{TD}(t_c) \sim l_{cor}^{-3} \gtrsim H^3(t_c) = \frac{T_c^6}{M_{Pl}^{*\,3}}. \tag{12.3}$$

This estimate is general for point-like topological defects, modulo a numerical factor. (The latter may be quite different from unity, though.) For extended defects (cosmic strings, domain walls), the above argument applies to the distance $l_D(t_c)$ between them, i.e., one has $l_D(t_c) \lesssim H^{-1}(t_c)$.

Further evolution is specific for defects of different types. As a rule, this evolution is out of thermal equilibrium: the equilibrium number density of defects falls off exponentially, while the actual density exhibits power-law behavior. The evolution of various defects will be considered in the following sections, and here we mention one property. Right after the phase transition, defects do not affect the expansion rate, since their energy density is smaller than that of hot matter by a factor T_c^α/M_{Pl}^α where the positive exponent α depends on the type of defects. (It follows from (12.3) that $\alpha = 3$ for point-like defects.) However, at later times the energy density of defects may become significant and may even dominate the expansion.

12.2. 't Hooft–Polyakov Monopoles

12.2.1. *Magnetic monopoles in gauge theories*

The simplest model admitting monopole (and antimonopole) solution [227, 228] is the Georgi–Glashow model. This is the gauge model with gauge group $SU(2)$ and triplet of scalar fields $\phi^a, a = 1, 2, 3$ which transform under adjoint representation.

---

[3]In the case of the first order phase transition, l_{cor} is the typical size of bubbles at percolation. For the second order phase transition, l_{cor} is the size of a region in which the energy needed to unwind the defect is of the order of temperature.

The Lagrangian is

$$\mathcal{L} = -\frac{1}{4}F^a_{\mu\nu}F^{a\mu\nu} + \frac{1}{2}\mathcal{D}_\mu\phi^a\mathcal{D}^\mu\phi^a - \frac{\lambda}{4}(\phi^a\phi^a - v^2)^2, \qquad (12.4)$$

where

$$F^a_{\mu\nu} = \partial_\mu A^a_\nu - \partial_\nu A^a_\mu + g\epsilon^{abc}A^b_\mu A^c_\nu,$$

$$\mathcal{D}_\mu\phi^a = \partial_\mu\phi^a + g\epsilon^{abc}A^b_\mu\phi^c.$$

The last term in (12.4) is self-interaction of the scalar fields leading to spontaneous symmetry breaking. The vacuum energy density is determined by minimizing the scalar potential, so that

$$\langle\phi^a\phi^a\rangle = v^2. \qquad (12.5)$$

This equation determines the vacuum manifold; in our case it is a 2-sphere S^2_{vac}. Each point in this manifold is invariant only under the subgroup $U(1)$ of the gauge group $SU(2)$. Hence, the symmetry breaking pattern is $SU(2) \to U(1)$. We identify the unbroken $U(1)$ as the gauge group of electromagnetism in this toy model.

By choosing the vacuum as

$$\langle\phi^a\rangle = \delta^a_3 v, \qquad (12.6)$$

we find that the field A^3_μ remains massless (gauge field of unbroken $U(1)$, photon), while the fields

$$W^\pm_\mu = \frac{1}{\sqrt{2}}(A^1_\mu \pm iA^2_\mu) \quad \text{and} \quad h = \phi^3 - v$$

obtain masses

$$m_V = gv \quad \text{and} \quad m_h = \sqrt{2\lambda}v,$$

respectively.

Problem 12.1. *By writing the quadratic action for small perturbations about the vacuum (12.6), find the spectrum of the theory, thus confirming the above statements.*

Let us study static field configurations of the form

$$A^a_0 = 0, \quad A^a_i = A^a_i(\mathbf{x}), \quad \phi^a = \phi^a(\mathbf{x}),$$

and require that the energy is finite. The energy functional is

$$E = \int d^3\mathbf{x} \left[\frac{1}{4}F^a_{ij}F^a_{ij} + \frac{1}{2}\mathcal{D}_i\phi^a\mathcal{D}_i\phi^a + \frac{\lambda}{4}(\phi^a\phi^a - v^2)^2\right]. \qquad (12.7)$$

Hence, at spatial infinity ($r \to \infty$, where $r^2 \equiv \mathbf{x}^2$) the fields tend to vacuum values, $\phi^a\phi^a = v^2$, $A^a_\mu = 0$, up to gauge transformation. Furthermore, the convergence of the energy integral requires sufficiently fast decay at infinity of the gauge-covariant quantities F^a_{ij}, $\mathcal{D}_i\phi^a$ and $(\phi^a\phi^a - v^2)$.

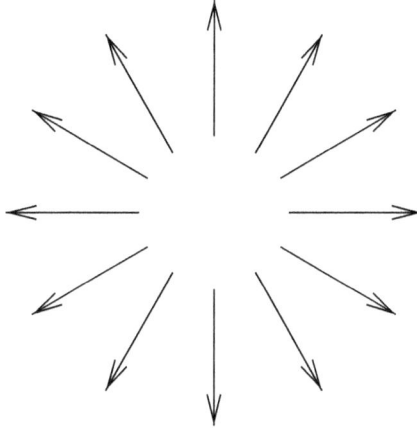

Fig. 12.1. Asymptotic behavior of the Englert–Brout–Higgs field for the monopole configuration. The direction of the scalar field in internal space (shown by arrows) is the same as the direction of the radius vector in the physical space.

The length of the vector (ϕ^1, ϕ^2, ϕ^3) is equal to v at spatial infinity. Hence, any finite energy configuration defines the mapping from a sphere S^2_∞ at spatial infinity to vacuum manifold S^2_{vac}. Since

$$\pi_2(S^2) = \mathbb{Z} \neq 0,$$

this mapping may be topologically non-trivial, and the configuration topologically stable.

The simplest non-trivial mapping is given by the hedgehog configuration of the scalar field, see Fig. 12.1,

$$\phi^a(r \to \infty) = v n^a, \quad n^a \equiv \frac{x^a}{r}. \tag{12.8}$$

The energy (12.7) is finite provided that the asymptotics of the vector field is

$$A_i^a(\mathbf{x}) = \frac{1}{gr} \epsilon^{aij} n_j. \tag{12.9}$$

Thus, we write the following Ansatz,

$$\phi^a = v n^a \cdot (1 - f(r)),$$
$$A_i^a = \frac{1}{gr} \epsilon^{aij} n^j (1 - a(r)). \tag{12.10}$$

This Ansatz is invariant under spatial rotations supplemented by rotations in the internal space, i.e., by global $SU(2)$-transformations. The functions $f(r)$ and $a(r)$ must obey the requirements of the absence of singularity at the origin and finiteness of energy. This gives

$$f(r \to \infty) = a(r \to \infty) = 0,$$

$$[1 - f(r \to 0)] \propto r, \quad [1 - a(r \to 0)] \propto r^2.$$

The actual configuration of the soliton is obtained by solving the field equations in terms of $f(r)$ and $a(r)$.

Problem 12.2. *Show that the Ansatz (12.10) is consistent with the field equations. Show that at $g \sim \sqrt{\lambda}$, the functions $f(r)$ and $a(r)$ decay exponentially at spatial infinity.*

The soliton just described is a magnetic monopole. Indeed, according to (12.9), the electric components of the field strength vanish, while the magnetic components are

$$B_i^a = -\frac{1}{2}\epsilon_{ijk}F_{jk}^a = \frac{1}{gr^2}n_i n_a. \tag{12.11}$$

Their direction in the internal space coincides with the direction of the scalar field, so this field is actually the magnetic field of the unbroken electromagnetic $U(1)$ gauge group. The gauge-invariant field strength

$$B_i = B_i^a \frac{\phi^a}{v} = \frac{1}{g}\frac{n_i}{r^2} \tag{12.12}$$

equals the magnetic field of magnetic monopole of magnetic charge $g_m = 1/g$. The massive vector and scalar fields decay exponentially away from the monopole center, and far from the center the field coincides with that of the Dirac monopole. The soliton we discuss is known as the 't Hooft–Polyakov monopole.

It is useful to note that there is also an antimonopole solution. It has the different behavior of the scalar field,

$$\phi^a = -vn^a(1 - f(r)),$$

whereas the field A_i^a is the same as for monopole. The magnetic field $(B_i^a \phi^a / v)$ has the sign opposite to that in (12.12).

For $m_V \sim m_h$, the monopole mass can be estimated on dimensional grounds. One rewrites the integral (12.7) by changing the variables

$$x = (gv)^{-1}\xi, \quad A_\mu^a = v\mathcal{A}_\mu^a, \quad \phi^a = v\varphi^a.$$

Then the energy (12.7) becomes

$$E = \frac{m_V}{g^2}\int d^3\xi \left[\frac{1}{4}\mathcal{F}_{ij}^a\mathcal{F}_{ij}^a + \frac{1}{2}\mathcal{D}_i\varphi^a\mathcal{D}_i\varphi^a + \frac{m_h^2}{8m_V^2}(\varphi^a\varphi^a - 1)^2\right],$$

where $\mathcal{D}_i\varphi^a = \partial_i\varphi^a + \epsilon^{abc}\mathcal{A}_i^b\varphi^c$, etc. The monopole configuration minimizes the energy functional, and for $m_V \sim m_h$ the integrand does not contain large or small parameters. Hence, the monopole energy (mass) is estimated as

$$m_M \simeq \frac{4\pi m_V}{g^2} = \frac{4\pi v}{g},$$

where the factor 4π corresponds to angular integration. Thus, the monopole mass exceeds the energy scale v of symmetry breaking $SU(2) \to U(1)$.

The existence of magnetic monopoles is a general property of Grand Unified Theories, GUTs. In GUT context, monopoles have masses of order $m_M \sim 10^{17}\,\mathrm{GeV}$. They are produced at the GUT phase transition which occurs at $T_c \sim 10^{16}\,\mathrm{GeV}$. Of course, this is true under the assumption that so high temperatures indeed existed in the Universe.

Let us describe the necessary (and very often sufficient) condition for the existence of magnetic monopoles in a general gauge theory with the Englert–Brout–Higgs mechanism. Let the Lagrangian be invariant under gauge group G and the ground state be invariant under subgroup H of the group G. In other words, the symmetry breaking pattern is $G \to H$. The manifold $\mathcal{M}$ of vacua is then the coset space G/H (here we assume that G acts on $\mathcal{M}$ transitively, i.e., all vacua are related to each other by symmetry transformations; this assumption is not, in fact, completely necessary). Stable monopole configurations are possible if the vacuum manifold $\mathcal{M}$ contains non-contractible 2-spheres, i.e., if its second homotopy group is non-trivial,

$$\pi_2(G/H) \neq 0. \qquad (12.13)$$

If the gauge group G is simple or semi-simple (gauge groups are always compact), and H includes one factor $U(1)$, then

$$\pi_2(G/H) = \pi_1(H) = \mathbb{Z},$$

so monopoles exist. In the case of the Standard Model, the gauge group $G = SU(3) \times SU(2) \times U(1)$ is not semi-simple, the unbroken subgroup is $H = SU(3) \times U(1)$ and $\pi_2(G/H) = 0$; thus, there are no monopoles. The situation is entirely different in GUTs. There, the gauge group is, as a rule, simple (sometimes semi-simple), and the symmetry gets broken at high energies down to $SU(3)_c \times SU(2)_w \times U(1)_Y$, and at lower energies to $H = SU(3)_c \times U(1)_{em}$ (an example is the $SU(5)$-theory, discussed in Sec. 11.2.2). Hence, $\pi_2(G/H) = \mathbb{Z}$, and monopoles always exist.

12.2.2. Kibble mechanism

Let us make use of the Georgi–Glashow model and the 't Hooft–Polyakov monopole to discuss the Kibble mechanism. The Englert–Brout–Higgs expectation value vanishes at high temperatures, the system is in symmetric phase and monopoles do not exist. The Englert–Brout–Higgs expectation value becomes nonzero as a result of the phase transition. Just after the phase transition, the directions of the scalar field are uncorrelated at distances exceeding the correlation length l_{cor}. As a result, there are regions in the Universe with the scalar field shown in Fig. 12.2(a), as well as regions where the scalar field configurations show the patterns of Figs. 12.2(b) and 12.2(c). The configuration of Fig. 12.2(a) is topologically trivial, and in the course of evolution it relaxes to a state without a monopole. On the other hand,

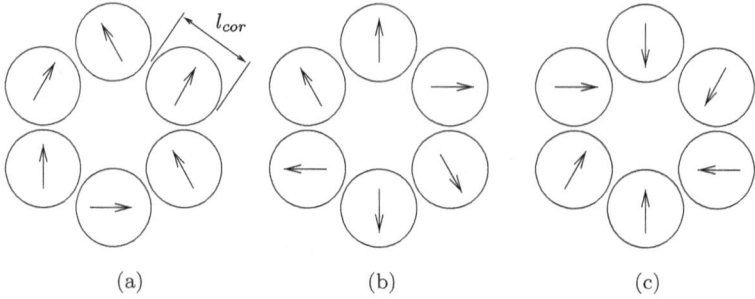

Fig. 12.2. Possible Englert–Brout–Higgs field configurations just after the phase transition. Circle radii are of order l_{cor}.

the configuration of Fig. 12.2(b) has hedgehog topology, cf. Fig. 12.1; it relaxes to a monopole. The configuration of Fig. 12.2(c) becomes an antimonopole in the end. The probabilities of all three types of configurations are roughly the same (and precisely the same for configurations of Figs. 12.2(b) and 12.2(c)), so the number densities of monopoles and antimonopoles just after the phase transition are estimated as $n_M = n_{\bar{M}} \sim l_{cor}^{-3}$. The Englert–Brout–Higgs condensate is definitely formed in an independent way at separation exceeding the cosmological horizon, so that $l_{cor} \lesssim H^{-1}(T_c)$, as we noticed in Sec. 12.1. Hence, the abundance of monopoles produced by the Kibble mechanism is indeed estimated as in (12.3).

12.2.3. *Residual abundance: the monopole problem*

Monopoles are non-relativistic at temperatures below the phase transition in which they are produced. If the monopole–antimonopole annihilation were negligible, the ratio of their number density to entropy density would remain constant. Hence, the mass density of monopoles today would be equal to

$$\rho_{M,0} = m_M n_{M,0} = m_M \frac{n_{M,T_c}}{s(T_c)} \cdot s_0 \sim \sqrt{g_*} \cdot m_M \frac{T_c^6}{M_{Pl}^3} \frac{g_{*,0} T_0^3}{T_c^3}$$

$$\sim 10^{12} \frac{m_M}{10^{16}\,\mathrm{GeV}} \left(\frac{T_c}{10^{16}\,\mathrm{GeV}}\right)^3 \sqrt{\frac{g_*}{10^2}}\,\mathrm{GeV\,cm^{-3}}, \qquad (12.14)$$

where we made use of the expression (5.28) for entropy density and estimated the monopole number density according to (12.3). The scale $10^{16}\,\mathrm{GeV}$ is typical for GUTs, so we see that the mass density of monopoles would exceed the critical density $\rho_c \sim 10^{-5}\,\mathrm{GeV\,cm^{-3}}$ by 17 orders of magnitude. As we will see momentarily, monopole–antimonopole annihilation does not change the result qualitatively: there remain too many monopoles. This is the essence of the monopole problem [229, 230], as we should either suppose that the temperature in the Universe was never as high as $T_c \sim 10^{16}\,\mathrm{GeV}$, or give up the idea of Grand Unification.

Let us now take into account monopole–antimonopole annihilation. First, let us check that monopoles are in kinetic equilibrium with plasma. A non-relativistic monopole in plasma experiences effective friction force due to the electromagnetic interactions with plasma particles. This force is estimated as $f \sim n(T) \cdot \sigma \cdot \Delta p$, where $n(T)$ is the number density of charged particles in plasma, $\sigma \sim \alpha g_m^2 / T^2$ is the interaction cross-section and $\Delta p \sim T v_M$ is the momentum that monopole of velocity v_M looses in each interaction. So, we have an estimate

$$f = -\kappa T^2 v_M,$$

where $\kappa \sim \alpha g_m^2 g_* \sim g_*$. We now write the second Newton law neglecting the cosmological expansion,

$$m_M \frac{dv_M}{dt} = -\kappa T^2 v_M.$$

We see that the monopole velocity changes significantly in the time interval $t_M \sim m_M / (\kappa T^2)$. This time is always smaller than the Hubble time M_{Pl}^* / T^2, so the monopoles are indeed in kinetic equilibrium with plasma. They have thermal velocity of order $v_M \sim v_T = \sqrt{T / m_M}$, and mean free path is

$$l_M \sim v_T \cdot t_M = \frac{1}{\kappa T} \sqrt{\frac{m_M}{T}}.$$

The mean free path grows as temperature decreases.

The efficient monopole–antimonopole annihilation occurs at high temperatures when the mean free path is short. The monopole and antimonopole attract each other, while interactions with plasma damp their velocities. As a result, they form *monopolonium*, the monopole–antimonopole bound state. Monopolonium then annihilates into conventional particles. Unlike direct annihilation, which has very small cross-section, this two-stage process reduces the monopole abundance significantly.

The bound state can be formed when the electromagnetic interaction energy of monopole and antimonopole exceeds temperature, i.e., at $r \lesssim r_0 = g_m^2 / T$. This determines the cross-section of monopolonium production, which subsequently leads to annihilation,

$$\sigma_{\text{ann}} \sim r_0^2 \equiv g_m^4 / T^2.$$

This estimate is valid at high temperatures, when l_M is small compared to r_0, otherwise monopolonium is not formed, and annihilation practically does not occur.

The monopole abundance at high temperatures is such that the monopole mean free time with respect to annihilation is of order of the Hubble time,

$$\sigma_{\text{ann}} n_M v_M \sim \frac{T^2}{M_{Pl}^*}.$$

We recall that monopoles have thermal velocities and obtain

$$\frac{n_M}{s} \cdot g_* g_m^4 \frac{M_{Pl}^*}{\sqrt{m_M T}} \sim 1, \tag{12.15}$$

where we used $s \sim g_* T^3$. We see that annihilation dilutes n_M/s as the Universe expands. This regime, however, terminates at $l_M \sim r_0$, and monopole abundance freezes out. The latter relation gives freeze-out temperature,

$$T_f \sim \frac{m_M}{g_m^4 \kappa^2}.$$

The monopole abundance at freeze-out is then found from (12.15),

$$\frac{n_M}{s} \sim \frac{1}{\kappa g_m^6} \frac{m_M}{M_{Pl}^* g_*}. \tag{12.16}$$

This ratio stays constant until today; using $g_m^2 \sim \alpha^{-1} \sim 100$, $\kappa \sim g_* \sim 100$ we get the present mass density

$$\rho_{M,0} = m_M \cdot \frac{n_M}{s} \cdot s_0 \sim 10^7 \cdot \left(\frac{m_M}{10^{16}\,\text{GeV}}\right)^2 \text{GeV cm}^{-3}.$$

This is much smaller than the estimate (12.14), but still way too large for GUT monopoles with $m_M \sim 10^{17}\,\text{GeV}$.

In retrospect, monopole problem was a strong reason for not extrapolating the hot Big Bang theory up to temperature $10^{16}\,\text{GeV}$. This was (and still is) one of the arguments in favor of inflation.

Problem 12.3. *Show that at $T \ll T_f$ the monopole–antimonopole annihilation practically does not change the ratio (12.16). Hint: Assume that the effect of plasma on annihilation process is negligible. Take into account Coulomb enhancement of the annihilation cross-section.*

Problem 12.4. *Find the temperature in an unrealistic Universe where monopoles of mass $m_M \sim 10^{17}\,GeV$ were produced at GUT phase transition at $T \sim 10^{16}\,GeV$, at the time when the Hubble parameter takes the same value as in our present Universe.*

12.3. Cosmic Strings

12.3.1. *String solutions*

The minimal model admitting cosmic strings — dimension-1 topological defects — is the Abelian Higgs model.[4] Its Lagrangian is

$$\mathcal{L} = D^\mu \phi^* D_\mu \phi - \frac{1}{4} F_{\mu\nu} F^{\mu\nu} - \lambda \left(\phi^* \phi - \frac{v^2}{2}\right)^2, \tag{12.17}$$

$$D_\mu \phi = \partial_\mu \phi - ie A_\mu \phi, \quad F_{\mu\nu} = \partial_\mu A_\nu - \partial_\nu A_\mu,$$

---

[4]Models with *global* Abelian symmetry group may admit global strings.

where ϕ is complex scalar field, A_μ is the gauge field of $U(1)$ gauge group. The Lagrangian (12.17) is invariant under $U(1)$ gauge transformations

$$\phi \to \phi e^{i\alpha(x)}, \quad \phi^* \to \phi^* e^{-i\alpha(x)}, \quad A_\mu \to A_\mu + \frac{1}{e}\partial_\mu \alpha(x).$$

The $U(1)$ symmetry is spontaneously broken in the theory (12.17): the vacua are obtained by minimizing the scalar potential and obey

$$\langle \phi^* \phi \rangle = \frac{v^2}{2}. \tag{12.18}$$

Neither of this vacua is invariant under $U(1)$; the symmetry $G = U(1)$ is broken completely ($G \to H$, $H = I$). According to (12.18) the vacuum manifold is a circle S^1 (This is in accordance with the relation $G/H = U(1)$, as $U(1)$ is itself a circle.),

$$\langle \phi \rangle = \frac{v}{\sqrt{2}} e^{i\alpha}, \quad \alpha \in [0, 2\pi).$$

The spectrum is obtained, as usual, by writing the Lagrangian (12.17) as Taylor series about the vacuum values of the fields $\phi^{vac} = \langle \phi \rangle$, $A_\mu^{vac} = 0$ and using the gauge $\mathrm{Im}\phi = 0$, i.e., $\phi = v/\sqrt{2} + h$ with real h. One finds that all fields become massive as a result of spontaneous symmetry breaking; the model contains real vector and scalar fieds with masses

$$M_V = ev \quad \text{and} \quad M_\phi = \sqrt{2\lambda}v,$$

respectively.

In the hot Universe at $T \gg v$ symmetry is restored, and both real and imaginary components of the complex field ϕ take random values at different points in space. The random phase $\alpha(\mathbf{x})$, defined formally as $\arctan[\mathrm{Re}\phi/\mathrm{Im}\phi]$, is distributed homogeneously. At critical temperature T_c, the scalar field acquires nonzero value, and the values of the phase $\alpha(\mathbf{x})$ get fixed. Generally speaking, they are different at different points. Due to the gradient term in the energy, the configuration becomes more homogeneous in the course of evolution. The phases $\alpha(\mathbf{x})$ may become equal over the horizon scale in the end, but this is not necessarily the case: the initial state may be such that there is a non-trivial winding around some contour $\mathcal{C}$. Single-valuedness of the field requires that the phase obeys

$$\Delta \alpha = \oint_C \frac{d\alpha}{d\theta} d\theta = 2\pi N, \tag{12.19}$$

where θ is the asimuthal angle in physical space and N is an integer. Configurations with $N \neq 0$ become strings in the end, and the mechanism of their production which we just described is nothing but the Kibble mechanism. Cosmic string may again be illustrated by Fig. 12.1, where the plane of the plot is now the plane orthogonal to the string, and arrows show the directions of vector $(\mathrm{Re}\phi, \mathrm{Im}\phi)$ in internal space. With this clarification, the Kibble mechanism is again illustrated by Fig. 12.2; it leads to the production of about one piece of string of length l_{cor} per volume l_{cor}^3.

Continuity of the field ϕ guarantees that strings must be either closed or infinite. The latter are open strings stretching across the entire horizon.[5] The contituity of ϕ requires also that ϕ vanishes along a line surrounded by the contour of integration C in (12.19). This means that the string configuration has large energy density there, $\mathcal{E} \sim \lambda v^4$.

At large distance from the center, the string configuration minimizes the scalar potential (12.18), but has non-trivial angular dependence of ϕ and A_μ. In the case of infinite straight string stretching along z-axis, the asymptotics in the plane (x, y) are (modulo irrelevant constant phase)

$$\phi \to \frac{v}{\sqrt{2}} e^{iN\theta}, \quad A_\mu \to \frac{N}{e} \partial_\mu \theta. \tag{12.20}$$

The winding number N is a topological invariant. The behavior of the field A_μ in (12.20) is such that asymptotically, i.e., at $x^2 + y^2 \to \infty$, there are no physical fields,

$$D_\mu \phi \to 0, \quad F_{\mu\nu} \to 0,$$

and string energy per unit length is finite. It follows from (12.20) that there is magnetic field $\mathbf{B} = \nabla \times \mathbf{A}$ inside the string core. Its flux is given by

$$\int \mathbf{B} d\mathbf{s} = \oint A_\theta d\theta = \frac{2\pi}{e} N.$$

We see that it is quantized in units of $2\pi/e$.

These defects are called Abrikosov vortices in condensed matter theory; these are magnetic flux tubes in superconductors. In particle theory, these objects are known as Abrikosov–Nielsen–Olesen strings [231, 232], or simply cosmic strings.

In gauge-Higgs theories of general type, the existence of cosmic strings is possible provided the vacuum manifold $\mathcal{M}$ is not simply connected, i.e., it contains non-contractible loops. The necessary (and often sufficient) condition for the existence of cosmic strings is thus non-trivial first homotopy group; in notations used in (12.13),

$$\pi_1(G/H) \neq 0. \tag{12.21}$$

Rotationally symmetric string Ansatz is

$$\phi = \frac{v}{\sqrt{2}} [1 - f(\rho)] e^{iN\theta}, \quad A_i = -\frac{N}{e\rho} \frac{\epsilon_{ij} x^j}{\rho} [1 - a(\rho)], \quad i, j = 1, 2, \tag{12.22}$$

where $\rho = \sqrt{(x^1)^2 + (x^2)^2}$ is the radial coordinate in the plane (x, y). Here, the rotation of the plane can be undone by the phase rotation of ϕ. Large-ρ asymptotics

---

[5] For this reason it is hard to "get rid" of strings, if they exist in the Universe.

should coincide with (12.20), which means

$$f(\rho \to \infty) \to 0, \quad a(\rho \to \infty) \to 0.$$

The fields must be non-singular at the origin, so we have the following behavior,

$$f(\rho \to 0) \to 1, \quad a(\rho \to 0) \to 1.$$

Analytical solutions of the field equations are unknown, but the solutions are straightforward to find numerically by making use of the Ansatz (12.22).

Let us estimate the string energy per unit length, i.e., the string tension. In analogy to the estimate of the monopole mass, we write

$$\mu = \frac{dE}{dz} = \int d^2x \left[D_i\phi^* D_i\phi + \frac{1}{4}F_{ij}^2 + \lambda \left(\phi^*\phi - \frac{v^2}{2} \right)^2 \right] \qquad (12.23a)$$

$$= v^2 \int d^2\xi \left[D_i\varphi^* D_i\varphi + \frac{1}{4}\mathcal{F}_{ij}^2 + \frac{\lambda}{e^2} \left(\varphi^*\varphi - \frac{1}{2} \right)^2 \right], \qquad (12.23b)$$

where the change of variables is $\phi = v\varphi$, $x = (ev)^{-1}\xi$, $A_i = v\mathcal{A}_i$. This gives the estimate for the tension,

$$\mu \sim \pi v^2, \qquad (12.24)$$

for $N \sim 1$ and $\lambda \sim e^2$. We also see that the string thickness l_s, i.e., the radius of the region where energy density is considerable, is estimated as $\xi \sim 1$, which gives $l_s \sim (ev)^{-1}$. The tension can be calculated exactly in the special case $M_\phi = M_V$ with the result

$$\mu = \pi v^2. \qquad (12.25)$$

This supports the estimate (12.24).

Problem 12.5. *Express the string tension in kg/cm for $v \sim 10^{16}$ GeV. Compare the mass of the Earth with the mass of a string encircling the Earth along equator.*

The fact that N is a conserved topological number does not tell what is the minimum energy configuration for given N at $|N| > 1$: this can be one string of winding number N or N strings of unit winding number. It turns out that the result depends on the parameters of the theory. For $M_V < M_\phi$ strings with $|N| > 1$ decay into $|N| = 1$ strings; conversely, for $M_V > M_\phi$ energy decreases if strings merge (these two cases correspond to type II and type I superconductors in condensed matter theory). The production of strings with $|N| > 1$ is suppressed in the early Universe; the phase transition creates predominantly strings with $|N| = 1$.

To understand how cosmic strings affect the space–time geometry, let us find energy–momentum tensor of a string. The general formula for the theory (12.17) is

$$T_{\mu\nu} = -\mathcal{L}g_{\mu\nu} + 2D_\mu\phi^* D_\nu\phi - F_{\mu\lambda}F_{\nu\rho}g^{\lambda\rho}.$$

We immediately find that

$$T_{00} = -T_{zz} = -\mathcal{L}.$$

Rotational symmetry in the (x, y) plane dictates the form of (ij)-components:

$$T_{ij} = T^{(0)}(\rho)\delta_{ij} + T^{(2)}(\rho)\left(n_i n_j - \frac{1}{2}\delta_{ij}\right), \tag{12.26}$$

where

$$n_i = \frac{x_i}{\rho}$$

is unit radius-vector on the plane (x, y). Note that the configuration (12.22), and, consequently, energy–momentum tensor are invariant under Lorentz boosts along the string. Clearly, this implies that motion of the entire configuration along the string direction is unphysical. The notation in (12.26) has to do with the fact that the first and second terms in the right-hand side have angular momenum squared equal to 0 and $4 = 2^2$, respectively.

Problem 12.6. *Show that*

$$L^2\left(n_i n_j - \frac{1}{2}\delta_{ij}\right) = 4\left(n_i n_j - \frac{1}{2}\delta_{ij}\right),$$

where $L = -i\partial/\partial\theta$ is angular momentum on the plane (x, y).

In what follows we will be interested in length scales much greater than the string thickness, $\rho \gg l_s \sim (ev)^{-1}$. In that case the relevant object is the energy–momentum tensor integrated over the string core,

$$\int \rho \, d\rho \, T_{\mu\nu}.$$

Let us see that the integrated trace part $T^{(0)}$ of the (ij)-components vanishes. We have explicitly

$$T^{(0)} = \frac{1}{2}T_{kk} = \frac{1}{4}F_{ik}F_{ik} - V(\phi),$$

where $V(\phi)$ is the scalar potential. Therefore, we write

$$\int \rho \, d\rho \, T^{(0)} \propto \int d^2x \left(\frac{1}{4}F_{ik}F_{ik} - V(\phi)\right). \tag{12.27}$$

To see that this integral is equal to zero, we recall that the string configuration is a minimum of the energy functional (12.23a), so this functional is stationary against all variations of the fields that vanish at infinity. Let us consider a configuration

$\tilde{\phi}(x^i) = \phi(\kappa x^i)$, $\tilde{A}_i(x^j) = \kappa A_i(\kappa x^j)$, where the right-hand sides involve the fields of the string configuration, and the parameter κ is close to 1. Upon the change of variables $x^i \to \kappa x^i$, the energy functional for the new configuration is

$$\frac{dE}{dz}[\tilde{\phi}, \tilde{A}_i] = \int d^2x \left[D_i\phi^* D_i\phi + \frac{\kappa^2}{4}F_{ij}^2 + \kappa^{-2}V(\phi) \right].$$

This expression as function of κ must have a minimum at $\kappa = 1$. This is the case only if the integral (12.27) vanishes, which is the desired result. We note that the above scaling argument goes back to Derrick and works for wide class of soliton-like solutions.

Thus, seen from far away, the energy–momentum tensor is

$$T_{\mu\nu} = \mu \cdot \text{diag}(1,0,0,-1)\delta(x)\delta(y) + T_{\mu\nu}^{(2)}, \tag{12.28}$$

where the only non-vanishing components of $T_{\mu\nu}^{(2)}$ are

$$T_{ij}^{(2)} = T^{(2)} \left(n_i n_j - \frac{1}{2}\delta_{ij} \right), \tag{12.29}$$

and μ is string tension estimated according to (12.24).

Problem 12.7. *Obtain the structure of the tensor (12.28) from the arguments based on dimensions, localization of energy, rotational symmetry, energy–momentum conservation and invariance under boosts along the string direction.*

We discuss the effects due to the first term in (12.28) later on, and here we show that the second, quadrupole term is, in fact, irrelevant. To this end, let us study linearized metric perturbations $h_{\mu\nu}$ about Minkowski metric. Let us choose harmonic gauge,

$$\partial_\mu h_\nu^\mu - \frac{1}{2}\partial_\nu h_\mu^\mu = 0. \tag{12.30}$$

Then linearized Einstein equations are (see Sec. A.9)

$$\Box h_{\mu\nu} = -16\pi G \left(T_{\mu\nu} - \frac{1}{2}\eta_{\mu\nu}T_\lambda^\lambda \right), \tag{12.31}$$

where $\Box = \partial_\lambda \partial^\lambda$ is D'Alembertian in Minkowski space–time. Since the string energy–momentum tensor is static and independent of z, the solution is independent of t and z and obeys

$$\Delta_2 h_{\mu\nu} = 16\pi G \left(T_{\mu\nu} - \frac{1}{2}\eta_{\mu\nu}T_\lambda^\lambda \right), \tag{12.32}$$

where Δ_2 is 2-dimensional Laplacian on a plane (x,y). Let us consider the effect of the term (12.29). The metric perturbation induced by this term has the same angular dependence as in (12.29),

$$h_{ij} \equiv h_{ij}^{(2)} = h^{(2)}(\rho)\left(n_i n_j - \frac{1}{2}\delta_{ij}\right),$$

and (see Problem 12.6)

$$\Delta_2 h_{ij}^{(2)} = \frac{1}{\rho}\frac{\partial}{\partial\rho}\left(\rho\frac{\partial h_{ij}^{(2)}}{\partial\rho}\right) + \frac{1}{\rho^2}\frac{\partial^2 h_{ij}^{(2)}}{\partial\theta^2} = \left(h^{(2)\prime\prime} + \frac{1}{\rho}h^{(2)\prime} - \frac{4}{\rho^2}h^{(2)}\right)\left(n_i n_j - \frac{1}{2}\delta_{ij}\right).$$

Away from the string $\Delta_2 h_{ij}^{(2)} = 0$, and the solution decays as

$$h^{(2)} \propto \frac{1}{\rho^2}.$$

(Another solution grows like ρ^2 and is unphysical.) We conclude that the quadrupole component of the metric induced by $T_{\mu\nu}^{(2)}$ rapidly decays away from the string: on dimensional grounds $h_{ij}^{(2)} \sim G\mu l_s^2/\rho^2$. This component may thus be neglected at large distances we consider, $\rho \gg l_s$.

To summarize, for studying the long-distance effects, the string energy-momentum tensor can be set equal to

$$T_{\mu\nu} = \mu \cdot \mathrm{diag}(1,0,0,-1)\delta(x)\delta(y). \tag{12.33}$$

The formula (12.33) represents the approximation of infinitely thin string; this approximation will be sufficient for our purposes.

The only non-vanishing components of the energy–momentum tensor (12.33) are the energy density T_{00} and pressure along the string T_{33}. They are equal up to sign, so we are dealing with a relativistic object which cannot be studied in the Newtonian approximation. Still, one expects that the space–time metric is nearly flat far away from the string, so that one can use linearized equation for the Newtonian potential Φ. In background Minkowski space the latter is defined by $g_{00} = 1 + 2\Phi$ and obeys the equation (see (A.119) and (A.121))

$$\Delta\Phi = 8\pi G\left(T_{00} - \frac{1}{2}\eta^{\mu\nu}T_{\mu\nu}\right) = 4\pi G(T_{00} + T_{11} + T_{22} + T_{33}). \tag{12.34}$$

The Newtonian potential is zero for the source (12.33) since the right-hand side of (12.34) vanishes. Hence, gravitational field is absent away from the string core; straight strings neither attract nor repel each other or surrounding matter! We will discuss the geometry of space in the presence of a string in Sec. 12.3.3.

We discuss here straight strings. For strings with ripples, the energy–momentum tensor can also be treated as localized along a line, provided that one considers distances larger than the length scale of ripples. However, in that case the energy density T_{00} and pressure along the string T_{33} no longer coincide up to sign; the effective energy–momentum tensor obeys $|T_{00}| > |T_{33}|$. The energy density T_{00} is somewhat larger than tension of straight string, $T_{00} > \mu$, since a distant observer sees "more string" per unit length. For similar reason $T_{33} < \mu$. It turns out that energy density and pressure are related by $T_{00} \cdot |T_{33}| = \mu^2$. Since

$|T_{00}| \neq |T_{33}|$ for strings with ripples, the right-hand side of Eq. (12.34) does not vanish: strings with ripples generate static gravitational field.

As it evolves, cosmic string sweeps a $(1+1)$-dimensional manifold, world sheet. The action for thin cosmic string equals the area of the world sheet (just like the action for a particle is the length of its world line):

$$S = -\mu \int \sqrt{-\gamma} d^2 \xi. \tag{12.35}$$

Here ξ^0 and ξ^1 are timelike and spacelike coordinates parameterizing the world sheet, $X^\mu = X^\mu(\xi)$ are space–time coordinates of a point on world sheet, and $\gamma = \det(\gamma_{\alpha\beta})$, where

$$\gamma_{\alpha\beta} = \partial_\alpha X^\mu(\xi) \partial_\beta X^\nu(\xi) g_{\mu\nu}$$

is the induced metric on world sheet. This action is known as Nambu–Goto action; it describes the dynamics of cosmic strings everywhere except for their crossing points where finiteness of string width is important.

The Nambu–Goto action can be obtained as the leading approximation to the action of moving curved cosmic string in the Abelian Higgs model (12.17). The idea of the derivation is to find approximate expression for the string configuration, insert it into the action (12.17) and integrate over coordinates transverse to the string.

Let ξ^0 and ξ^1 be timelike and spacelike coordinates on the manifold $X^\mu(\xi)$ where the scalar field ϕ takes zero value. This manifold is identified with the string world sheet. Assuming that the string curvature is small compared to its thickness, one approximates the string configuration by Eq. (12.22) generalized to nonzero string velocity. The two vectors parallel to the world sheet at a given point $X^\mu(\xi)$ are $\partial X^\mu / \partial \xi^\beta$, and one can choose two tangential spacelike vectors $e_\mu^{(\alpha)}$, $\alpha = 1, 2$, such that they are orthonormal, $e_\mu^{(\alpha)} e_\nu^{(\beta)} g^{\mu\nu} = -\delta^{\alpha\beta}$, and orthogonal to the world sheet, $e_\mu^{(\alpha)} \partial X^\mu / \partial \xi^\beta = 0$. The coordinates x^μ of each point near the world sheet can now be written as

$$x^\mu \equiv x^\mu(\zeta) = X^\mu(\xi) + \sum_{\alpha=1}^{2} e^{(\alpha)\mu}(\xi) \eta^\alpha, \quad \zeta = (\xi^0, \xi^1, \eta^1, \eta^2),$$

where we introduced two new coordinates η^α, $\alpha = 1, 2$ which in the case of straight string parameterize points on the plane orthogonal to the string. The Jacobian of the coordinate transformation from x^μ to ζ^μ is

$$\sqrt{-g} \cdot \det\left(\frac{\partial x}{\partial \zeta}\right) = \sqrt{-\det\left(g_{\mu\nu} \frac{\partial x^\mu}{\partial \zeta^\lambda} \frac{\partial x^\nu}{\partial \zeta^\rho}\right)} \approx \sqrt{-\gamma} + \cdots, \tag{12.36}$$

where we write only the terms which are not suppressed by the radius of the string curvature.

In terms of new coordinates, the string configuration is

$$\phi[x(\zeta)] = \phi^{(s)}(\eta^1, \eta^2) \quad A^\mu[x(\zeta)] = \sum_{\alpha=1}^{\alpha=2} e^{(\alpha)\mu} A_\alpha^{(s)}(\eta^1, \eta^2),$$

where (s) denotes the configuration of straight string in the plane (η_1, η_2). This is the same configuration as in (12.22) with η^1, η^2 substituted for x_1, x_2. To the leading order in the string curvature we have

$$D_\mu\phi^* D^\mu\phi \approx D_\mu\phi^{(s)*} D^\mu\phi^{(s)}, \quad F_{\mu\nu}^2 \approx F_{\mu\nu}^{(s)2}, \quad V(\phi) \approx V(\phi^{(s)}). \tag{12.37}$$

The Nambu–Goto action is now obtained by substituting the Jacobian (12.36) and expressions (12.37), (12.17) into the action of the Abelian Higgs model and integrating over the transverse coordinates η^α.

Problem 12.8. *Show that there are no corrections to the Nambu–Goto action (12.35) which are linear in the ratio of the string thickness to its radius of curvature.*

12.3.2.   Gas of cosmic strings

For a distant observer, a *closed* cosmic string of radius R looks like a particle of mass $M_s = 2\pi R\mu$. If dissipation of string energies (say, by gravitational wave emission) is negligible, their effect on expansion rate is the same as that of relativistic or non-relativistic matter, depending on their velocities. In fact, the dissipation *is* important; we consider the realistic situation in Sec. 12.3.4.

    The properties of the gas of *infinite* strings are less obvious. In what follows we consider non-interacting strings: indeed, their interaction mediated by the fields A_μ and ϕ weakens exponentially with separation,[6] while gravitational interaction is weak. To find equation of state for gas of cosmic strings, consider first the configuration of N straight strings at rest, which are parallel to z-axis and separated by distance L well exceeding the string thickness. Energy-momentum tensor of such a configuration is

$$T_{\mu\nu}^{(0)}(x, y) = \mu \cdot \mathrm{diag}(1, 0, 0, -1) \cdot \sum_i^N \delta(x - x_i)\delta(y - y_i).$$

In the limit of large number of strings the average tensor becomes

$$\langle T_{\mu\nu}^{(0)} \rangle = \frac{\int T_{\mu\nu}^{(0)}(x, y) dx dy}{\int dx dy} = \frac{\mu}{L^2} \cdot \mathrm{diag}(1, 0, 0, -1). \tag{12.38}$$

This expression is valid for static configuration. Let now the strings move along the x-axis at velocity u. Energy-momentum tensor is obtained from (12.38) by Lorentz

---

[6] We consider here Abrikosov–Nielsen–Olesen strings; the situation for global string is different.

boost,[7]

$$\langle T_{\mu\nu}\rangle^{(u_x)} = \frac{\mu}{L^2}\begin{pmatrix} \gamma^2 & \gamma^2 u & 0 & 0 \\ \gamma^2 u & \gamma^2 u^2 & 0 & 0 \\ 0 & 0 & 0 & 0 \\ 0 & 0 & 0 & -1 \end{pmatrix}, \qquad \gamma = \frac{1}{\sqrt{1-u^2}}.$$

We are going to average over the boost directions, so we omit terms linear in u. The tensor averaged over boosts along x and y is (Recall that the motion along the z-axis is unphysical for strings stretched along z-axis.)

$$\langle T_{\mu\nu}\rangle^{(u_{xy})} = \frac{\mu}{L^2}\cdot \mathrm{diag}\left(\gamma^2, \frac{\gamma^2 u^2}{2}, \frac{\gamma^2 u^2}{2}, -1\right).$$

We now repeat the procedure for string configurations parallel to x- and y-axes and average over all three directions. The resulting energy-momentum of the gas of infinite cosmic strings moving with velocity u in random directions is

$$T_{\mu\nu}^{gs} \equiv \langle T_{\mu\nu}\rangle^{(u)} = \frac{\mu}{L^2}\cdot \mathrm{diag}\left(\gamma^2, \frac{u^2\gamma^2 - 1}{3}, \frac{u^2\gamma^2 - 1}{3}, \frac{u^2\gamma^2 - 1}{3}\right). \qquad (12.39)$$

As expected, in the limit $u \to 1$ the equation of state of the string gas coincides with that of radiation, $p = \rho/3$. More relevant for the Universe is the equation of state for slow strings

$$p = -\frac{1}{3}\rho. \qquad (12.40)$$

According to the results of Sec. 3.2.4 this gives

$$\rho, p \propto a^{-2}(t). \qquad (12.41)$$

The physical reason for this behavior is that expansion affects distances between strings in two transverse directions, rather than in three directions as in the case of particles.

Problem 12.9. *Neglecting string collisions and dissipation of their energy (which is not in fact a good approximation), estimate the number of strings in the visible Universe, if they were produced in the course of phase transition at $T_c \sim 100\,GeV$. The same for $T_c \sim 10^{16}\,GeV$. Hint: Assume that strings were produced by the Kibble mechanism, so that just after the phase transition there was one piece of string of Hubble length per Hubble volume.*

The dependence (12.41) shows that strings, if exist, could become important at later stages of the cosmological expansion. Their energy density behaves in the same way as the contribution of spatial curvature to the Friedmann equation, see (4.2), and decreases slower than that of radiation or non-relativistic matter.

---

[7]This transformation accounts not only for the change of the energy–momentum of each string but also Lorentz-contraction of the distance between the strings.

Problem 12.10. *Under conditions of Problem 12.9 and assuming, within the Abelian Higgs model (12.17), that strings were produced in the course of phase transition at $T_c \sim v$, obtain the bound on the energy scale v by requiring that the string gas makes small contribution to the present energy density.*

If the late time expansion of the Universe were dominated by cosmic string gas, the scale factor would grow as

$$a(t) \propto t.$$

This shows that string domination at present epoch is inconsistent with observations: the cosmological expansion accelerates. In fact, CMB and SNe Ia data rule out string domination at a high confidence level (see (4.51) and Fig. 4.10); the cosmological data require dark energy instead. It then follows from (12.41) that strings never dominated and, if dark energy density is constant or almost constant, they will never dominate in the future.

12.3.3.   *Deficit angle*

Let us now consider other effects of infinite or very long cosmic strings. We begin with studying the spatial geometry in the presence of an infinite string. Even though gravitational potential due to cosmic string vanishes away of its core, the effect on geometry is non-trivial. To see this, we consider small perturbation $h_{\mu\nu}$ about Minkowski metric. In harmonic gauge, it obeys Eq. (12.32) where the energy–momentum tensor is given by (12.33). The only nonzero components of the combination $(T_{\mu\nu} - \frac{1}{2}\eta_{\mu\nu}T_\lambda^\lambda)$ in the right-hand side of Eq. (12.32) are (11) and (22). Thus, the nonzero components of perturbations are h_{11} and h_{22}, and they both obey

$$\Delta_2 h_{11(22)} = 16\pi G\mu\delta(x)\delta(y).$$

We conclude that in the harmonic gauge

$$h_{\mu\nu} = 4G\log\left(\frac{x^2 + y^2}{\rho_0^2}\right) \cdot \mathrm{diag}(0, 1, 1, 0), \tag{12.42}$$

where ρ_0 is a parameter of dimension of length, determined by the string thickness. With perturbation included, the metric is

$$ds^2 = dt^2 - dz^2 - \left[1 - 4G\mu\log\left(\frac{\rho^2}{\rho_0^2}\right)\right] \cdot (d\rho^2 + \rho^2 d\theta^2),$$

where we switched to cylindrical coordinates. The latter expression is valid for $G\mu \ll 1$ and $\rho \gg \rho_0$, but $4G\mu\log\left(\frac{\rho^2}{\rho_0^2}\right) \ll 1$; in this case the perturbation (12.42) is small while the details of the string core are irrelevant. It is convenient to perform coordinate transformation to the radial coordinate $\tilde{\rho}$, such that

$$d\tilde{\rho}^2 = \left(1 - 4G\mu \log\left(\frac{\rho^2}{\rho_0^2}\right)\right) d\rho^2. \tag{12.43}$$

To the first order in $G\mu$ this gives

$$\tilde{\rho} = \rho \cdot \left(1 - 4G\mu \log\frac{\rho}{\rho_0} + 4G\mu\right),$$

and

$$\rho^2 \cdot \left[1 - 4G\mu \log\left(\frac{\rho^2}{\rho_0^2}\right)\right] = \tilde{\rho}^2 \cdot (1 - 4G\mu)^2. \tag{12.44}$$

It follows from (12.43) and (12.44) that the metric of static straight string stretching along z-axis can be written as [233] (we omit tilde over ρ from now on)

$$ds^2 = dt^2 - dz^2 - d\rho^2 - (1 - 4G\mu)^2 \rho^2 d\theta^2. \tag{12.45}$$

Problem 12.11. *Find asymptotically flat form of metric for string with ripples whose energy–momentum tensor is $T_{\mu\nu} = \mathrm{diag}(\mu', 0, 0, -\mu'')\delta(x)\delta(y)$, $\mu'\mu'' = \mu^2$, $\mu' > \mu''$, to the first order in $G\mu', G\mu''$. Show that space–time is curved outside the string core.*

Metric (12.45) has conical singularity. Circles $\rho = \mathrm{const}$ have length smaller than $2\pi\rho$. Metric (12.45) becomes Minkowskian upon the coordinate transformation

$$\theta \to (1 - 4G\mu)\theta.$$

However, the polar angle θ now takes values in a narrower interval

$$0 \le \theta < 2\pi(1 - 4G\mu).$$

In this regard, one introduces the notion of deficit angle, whose value is

$$\Delta\theta = 8\pi G\mu. \tag{12.46}$$

Thus, space–time is locally flat but the geometry on (x, y)-plane is conical.[8]

The presence of deficit angle leads to a number of interesting physical phenomena. One is double image of an object behind the string. To illustrate this pictorially, let us use the coordinates in which metric is Minkowskian. Consider an observer at distance d from straight cosmic string: this is the point O in Fig. 12.3, while the string is at point S; the figure shows tangential plane to the string. The lines SA' and SA'' must be identified, and the shaded region cut out; the deficit angle is $\Delta\theta$. Light rays coming to the observer at equal angles $\alpha' = \alpha''$ to the direction OS towards the string, actually emanate from one and the same point, since the

---

[8]We obtained the expression (12.45) for $4G\mu \log\left(\frac{\rho^2}{\rho_0^2}\right) \ll 1$. It is clear, however, that (12.45) is a solution to the Einstein equations for larger ρ as well: the metric is locally flat and its Ricci tensor vanishes, $R_{\mu\nu} = 0$.

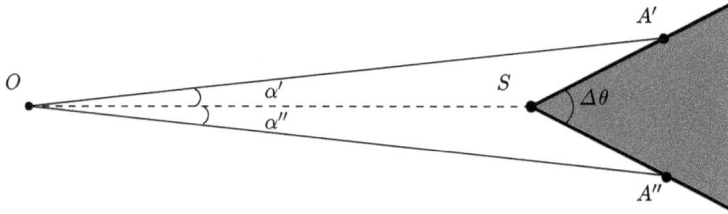

Fig. 12.3.   Light propagation from a point $A' = A''$ behind the string S to an observer O.

points A' and A'' are identified. Thus, the observer will see two images of the object placed at $A' = A''$. The angular distance between the two images, $\Delta\alpha = \alpha' + \alpha''$, is proportional to deficit angle. For small $\Delta\theta$ one has

$$\Delta\theta = \frac{l+d}{l}\Delta\alpha, \tag{12.47}$$

where l and d are the distances from the string to the source and observer, respectively. While the distance to the source is measurable (say, by determining redshift), the distance d to the string is not. Thus, the measurement of the angular distance can only give the lower bound on deficit angle,

$$\Delta\theta \geq \Delta\alpha.$$

According to (12.46), deficit angle is proportional to the string tension, so a measurement of $\Delta\alpha$ would place a bound on the energy scale v of the theory,

$$v \gtrsim \frac{M_{Pl}}{2}\sqrt{\frac{\Delta\alpha}{2\pi}}.$$

To see this lensing effect in the coordinate frame (12.45) one would cut out the shaded region in Fig. 12.3 and glue together its boundaries SA' and SA''. The resulting surface would be a cone with apex at the string position S. Light can travel from the source $A' = A''$ to the observer O along two straight lines, which pass left and right of the apex. Hence the two images of the source.

Problem 12.12. *Making use of the geodesic equation in metric (12.45) give alternative derivation of the lensing effect of cosmic string.*

We neglected the cosmological expansion in the above analysis. In general, however, the relationship between $\Delta\theta$ and $\Delta\alpha$ depends on cosmological evolution. As an example, the analog of the relation (12.47) in flat matter dominated Universe is

$$\Delta\alpha = \Delta\theta \cdot \left(1 - \frac{1 - (1+z_S)^{-1/2}}{1 - (1+z_A)^{-1/2}}\right), \tag{12.48}$$

where z_S and z_A are redshifts of string and source, respectively.

Problem 12.13. *Derive the relation (12.48). Find similar relation for the Universe dominated by cosmological constant. Generalize that formula, as well as (12.48) to string inclined to the plane of Fig. 12.3 at angle δ.*

We have considered point-like source for simplicity; in that case the observational proof that two images are due to single source would be the identity of the two spectra. The situation with extended source is more complicated, but in that case too, there exists a region behind the string such that sources in it produce two identical images.

Problem 12.14. *Find the shapes of images of a spherical object behind the string in a general case.*

Another phenomenon due to cosmic string is specific distortion of CMB anisotropy. If a string and an observer are at rest with respect to CMB, then lensing by the string leads to repetitions in the pattern of anisotropy. The source is now the surface of last scattering, and the distance to it much exceeds the distance to the string. Therefore, the angular distance between identical temperature spots is equal to deficit angle, $\Delta\theta = \Delta\alpha$.

If the string moves in the direction transverse to the line of sight, there is a new effect of systematic shift between frequencies (and hence temperature) of photons passing the string on different sides [237]. This effect is illustrated in Fig. 12.4, where the coordinate frame and notations are the same as in Fig. 12.3. Let us again consider two images A' and A'' of one and the same source. Let the string S move with velocity $u_\perp$ in direction normal to the line of sight. Then all points on the line $A'SA''$ move with the same velocity. The source A' has velocity $u_\perp$ whose projection on the line OA' points in the same direction as the photon momentum. For the source A'' this projection is opposite to photon momentum. Hence, the longitudinal Doppler effect leads to shifts of photon frequencies

$$OA': \Delta\omega = u_\perp \frac{\Delta\theta}{2}\gamma, \quad OA'': \Delta\omega = -u_\perp \frac{\Delta\theta}{2}\gamma,$$

where $\gamma = 1/\sqrt{1-u_\perp^2}$ and we have set $\Delta\alpha \approx \Delta\theta$, having in mind CMB photons. We see that photons crossing the string trajectory in front and behind the string get redshifted and blueshifted, respectively. This leads to additional CMB anisotropy.

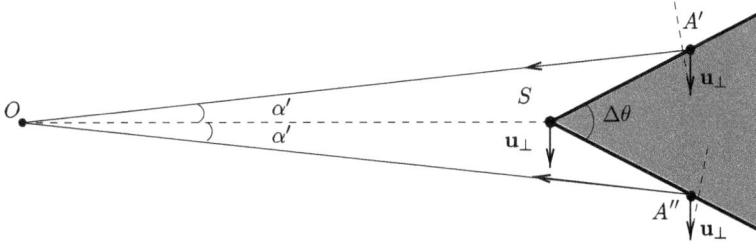

Fig. 12.4. Light propagation from source $A' = A''$ behind the string S moving with velocity $\mathbf{u_\perp}$.

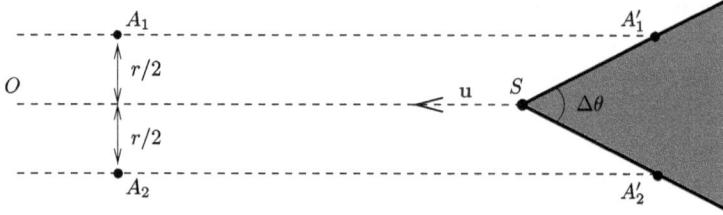

Fig. 12.5. Motion of a string S in dust-like medium.

For small $\Delta\theta$ and realistic $u_\perp$ the effect is small,

$$\frac{\delta T}{T} = \gamma u_\perp \Delta\theta = 1.7 \cdot 10^{-6} \gamma \frac{u_\perp}{10^{-1}} \frac{\mu}{(10^{16}\,\text{GeV})^2}.$$

Yet the analysis of CMB angular spectrum together with data on large scale structure rule out string networks [238] with $\mu \gtrsim (0.6 \cdot 10^{16}\,\text{GeV})^2$ and single string stretching through the visible Universe with $\mu \gtrsim (1.0 \cdot 10^{16}\,\text{GeV})^2$ [261].

It is known from observations that CMB temperature fluctuations are random and Gaussian. The small jump of temperature across the string projection onto celestial sphere is a *non-Gaussian* feature. Non-Gaussianity of this sort has not been observed, which also leads to a bound on μ [239]. This bound is of the same order as other bounds, $\mu \lesssim (0.7 \cdot 10^{16}\,\text{GeV})^2$ for an infinite straight string moving with velocity $u_\perp = 1/\sqrt{2}$.

Yet another effect due to cosmic string, moving now in dust-like medium, is the formation of overdense region (sheet-like wake) behind the string [240]. This is illustrated in Fig. 12.5, where the coordinate frame is the same as in Fig. 12.3. The string is normal to the plane of the figure. Let there be two dust particles A_1 and A_2 in this plane. Let these particles be at rest and at distance $r/2$ from the string trajectory OS. The string moves with velocity u. After the string passes between the particles, the latter move towards each other and meet behind the string. (The meeting point is $A_1' = A_2'$ in Fig. 12.5; recall that the lines SA_1' and SA_2' are identified.)

Let us calculate the velocities of the dust particles in direction towards each other. For particle A_1 this direction is orthogonal to the line SA_1'. In the string rest frame this particle moves along the line $A_1 A_1'$ with velocity u. Hence, its velocity at point A_1' in the direction to particle A_2 is

$$v_y = u \cdot \sin \frac{\Delta\theta}{2} = 4\pi G\mu u, \tag{12.49}$$

where we set $G\mu \ll 1$. Taking, as an example, $u \sim 0.1$, we find numerically

$$|\Delta v_y| = 0.8 \cdot 10^{-6} \cdot \frac{\mu}{(10^{16}\,\text{GeV})^2}.$$

We see that this velocity may be fairly high.

12.3.4. *Strings in the Universe*

Relatively large velocities of particles in wakes of moving cosmic strings lead to formation of overdense regions there. These overdensities may in principle serve as seeds for formation of structures — galaxies, clusters of galaxies, etc. The structures formed in this way would stretch along the string trajectories and hence they would be 2-dimensional. We note here that galaxy distribution indeed shows that there are 2-dimensional and also 1-dimensional structures, walls and filaments. However, detailed analysis reveals that only small fraction of matter gets into the wakes [241], so that observed voids (galaxy-poor regions) do not form.

Another structure formation mechanism in models with cosmic strings is accretion of non-relativistic matter onto string loops [242, 243]. The resulting density perturbation spectrum is almost scale invariant, in accordance with observations. However, the string mechanism of structure formation predicts CMB anisotropy angular spectrum in gross contradiction to observations [244]; see Fig. 12.6. Hence, string mechanism of structure formation cannot be dominant. The bounds on the string contribution into CMB anisotropy, and hence into structure formation, are at the level of 3%. Even stronger limits come from the search for B-mode of CMB polarization, since strings lead to the generation of tensor and, most notably, vector perturbations in metric and matter which induce the B-mode. (See the accompanyng book for the discussion of CMB polarization.)

Many properties of dynamics of cosmic strings cannot be found analytically, so one relies upon numerical simulations. These show, in particular, that just after the

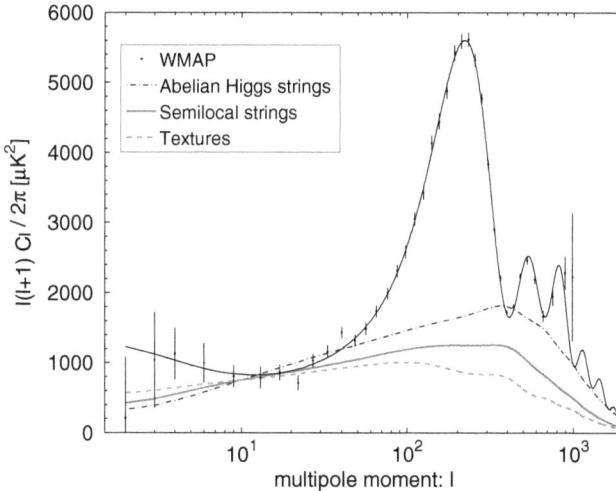

Fig. 12.6. CMB temperature angular spectra in ΛCDM model (solid line with peaks) and predictions in in models with density perturbations generated by cosmic strings and textures [244]. (See Sec. 12.5 for discussion of textures.) The absence of peaks in the latter models is due to the fact that perturbations are never superhorizon and are are generated during fairly long period of time. For this reason tha phases of oscillations in (1.23) are random.

phase transition the mass fraction of infinite strings is four times larger than that of closed strings. The initial velocities of long strings are small, but they increase in time and on average become rather large, $u \approx 0.15$. Numerical simulations have shown also that the energy density of cosmic strings decreases in time, so that it never dominates the cosmological expansion. Hence, unlike monopoles and domain walls, models with strings produced in phase transitions are still cosmologically allowed.

There are two main processes determining the evolution of cosmic strings in the Universe. One is string intersections and self-intersections, which lead to production of loops out of long strings. The second is gravitational wave emission by relatively short strings, which leads to disappearance of small loops. Let us first discuss the latter process. The power of gravitational wave emission is determined by the third time derivative of the quadrupole moment [83] Q of a string loop,

$$P_{gw} \sim \frac{1}{M_{Pl}^2} \left(\frac{d^3 Q}{dt^3} \right)^2 .$$

The quadrupole moment of a loop of radius R is estimated as $Q \sim \mu R^3$. The string equation of motion which follows from the action (12.35) shows that loops rotate and oscillate at velocities of order 1, i.e., $dR/dt \sim 1$. This gives

$$P_{gw} = C_{gw} \frac{\mu^2}{M_{Pl}^2},$$

where the constant C_{gw} is obtained by numerical simulations and turns out to be rather large, $C_{gw} \simeq 10^2$ (see, e.g., [234]). Thus, loop looses its energy at the time scale

$$t_{gw} \sim \frac{\mu R}{P_{gw}} = \frac{M_{Pl}^2 R}{C_{gw} \mu} . \tag{12.50}$$

As a result, the size of the loop becomes of order of its thickness, $R \sim v^{-1}$, and the string decays into high energy particles.[9]

Let us now turn to string intersections. Numerical analysis shows that intersection almost always leads to reconnection [235], so self-intersections give rise to decays into shorter loops. Somewhat unexpected result of numerical simulations of cosmic string dynamics in the Universe is that the evolution of the cosmic string energy density soon after the phase transition coincides with the evolution of the total energy density, $\rho \propto t^{-2}$ [236]. At any given moment of time each Hubble volume contains a dozen long strings stretching behind the horizon, many long closed strings and numerous small loops. This behavior is due to efficient production of closed strings and their subsequent decay via gravitational wave emission.

---

[9]We note that this is one of the possible mechanisms of generation of ultra-high energy cosmic rays. For $\mu = (10^{16} \text{ GeV})^2$, this mechanism works with string loops of present size of order $R \sim 1$ Mpc: it follows from (12.50) that the lifetime of these strings is of the order of the present age of the Universe.

To understand this picture at qualitative level, let us estimate the energy density of closed loops at radiation domination. The number density of loops of size R, where $R \ll H^{-1}$, decreases due to the cosmological expansion,

$$n_l(R,t) \propto a(t)^{-3} \sim t^{-3/2}.$$

In the scale-invariant regime we obtain on dimensional grounds

$$n_l(R,t) \sim \frac{1}{(Rt)^{3/2}} \sim \left(\frac{\mu}{Et}\right)^{3/2},$$

where $E \sim R\mu$ is the mass of a loop. The energy density is then

$$\rho_l(t) \sim \int_{E_{min}}^{E_{max}} E dE \frac{dn_l(R,t)}{dE} = \left(\frac{\mu}{t}\right)^{3/2} \int_{E_{min}}^{E_{max}} \frac{dE}{E^{3/2}}. \qquad (12.51)$$

This integral is saturated at lower limit, hence

$$\rho_l(t) \sim \left(\frac{\mu}{t}\right)^{3/2} \frac{1}{\sqrt{E_{min}}}, \qquad (12.52)$$

where E_{min} is the minimum possible loop energy. We see that the energy density is dominated by the lightest but numerous loops.

The minimum size of loops is determined by the requirement that the lifetime (12.50) exceeds the Hubble time, so we get

$$R_{min}(t) \sim C_{gw} \frac{\mu}{M_{Pl}^2} t, \quad E_{min}(t) \sim C_{gw} \frac{\mu^2}{M_{Pl}^2} t.$$

We finally obtain

$$\rho_l(t) \simeq \frac{\sqrt{\mu} M_{Pl}}{\sqrt{C_{gw}}} \frac{1}{t^2},$$

i.e., the energy density of strings indeed follows that of radiation, $\rho_{rad} \sim M_{Pl}^2/t^2$. The relative contribution of strings remains small, $\rho_l/\rho_{rad} \sim \sqrt{\mu/M_{Pl}^2}$. Still, it is much larger than the relative contribution of infinite strings, which is of order μ/M_{Pl}^2. This shows that it is small loops rather than long strings that may affect the spectrum of density perturbations.

The behavior $n_l(R,t) \propto (Rt)^{-3/2}$ can be seen from the following argument. Let $n_l(R,t)$ be the number density of loops of size R at time t. Loops of sizes in interval R and $R+dR$ have energy density

$$\rho_l(R,t)dR = \mu R n_l(R,t)dR.$$

Energy density of loops decreases due to the cosmological expansion, and at the same time more loops are produced by (self)intersections of long strings. Scale-invariant behavior means that the loop production is determined by a scale-invariant function which we denote by $f(R/l_H)$, where l_H is the horizon size, the length scale inherent in the Universe.

This function parameterizes the energy loss of long strings due to production of loops of sizes between R and $R + dR$ as follows,

$$\mu \frac{dR}{l_H} \cdot f(R/l_H).$$

Hence, the equation for balance of loops is

$$\frac{d\rho_l(R,t)}{dt} + 3H(t)\rho_l(R,t) = \frac{\mu}{l_H^4} f(R/l_H),$$

where $H(t) = 1/(2t)$ and $l_H = 2t$. The solution to this equation is

$$\rho_l(R,t) = \frac{1}{16} \frac{\mu}{(Rt)^{3/2}} \int_{R/t}^{\infty} \sqrt{\xi} f(\xi/2) d\xi.$$

For short loops, $R/t \to 0$, this indeed gives the evolution law (12.51).

All the discussion above assumed that the sizes of loops produced in intersections are much larger than the minimum size. Numerical simulations [245] show that the loop sizes are indeed comparable to the horizon size; at both radiation and matter domination the typical size of a newborn loop is $R(t) = \alpha t$, $\alpha \simeq 0.1$. We note that earlier simulations with worse resolution could not determine the value of α but suggested that it was small, $\alpha \ll 1$. The interpretation was that the typical size of a newborn loop was determined by the size of ripples of long strings rather than the horizon size. Were that the case, the newborn loops would decay in Hubble time, and would be irrelevant from the viewpoint of density perturbations. More sizeable would be density perturbations created by a dozen of moving infinite strings.

Finally, let us estimate the energy of gravitational waves produced by cosmic strings. With account of redshift of their frequencies, the energy density in gravity waves ρ_{gw} obeys the equation

$$\dot{\rho}_{gw} + 4H\rho_{gw} = -\frac{d\rho_l}{dt} = 2\frac{\sqrt{\mu}M_{Pl}}{\sqrt{C_{gw}}} \frac{1}{t^3}.$$

Its solution at radiation domination is

$$\rho_{gw} = \frac{1}{t^2} \int_{t_1}^{t} \left(-\frac{d\rho_l}{dt'}\right) t'^2 dt' \simeq 2\frac{\sqrt{\mu}M_{Pl}}{\sqrt{C_{gw}}} \frac{1}{t^2} \log \frac{t}{t_1}, \qquad (12.53)$$

where t_1 denotes the time when the first gravity waves are emitted, i.e., the time when the cosmic strings begin to decay. The energy density in the Universe at radiation domination is (see Sec. 3.2)

$$\rho = \frac{3M_{Pl}^2}{32\pi t^2},$$

hence the relative contribution of gravity waves is given by

$$\frac{\rho_{gw}}{\rho} \simeq \frac{64\pi}{3M_{Pl}} \frac{\sqrt{\mu}}{\sqrt{C_{gw}}} \log \frac{t}{t_1}.$$

We see that this contribution grows slowly, logarithmically, but it can be quite large. Decaying strings may be intense source of relic gravity waves.

The effect of gravity wave production enables one to place strong bounds on models with cosmic strings. In particular, pulsar timing measurements give the bound [246]

$$\mu \lesssim (2 \cdot 10^{15}\,\text{GeV})^2.$$

We note that additional sources of gravity waves are singularities on strings, kinks and cusps. These singularities emerge both as a result of the evolution of closed strings and due to (self-)intersections; see, e.g., Ref. [247]. The singularities have velocities close to the speed of light and emit intense pulses of gravity waves. The singularities may also be sources of particles of large masses and/or super-high energies. The latter aspect is of interest for cosmic ray physics.

The possibility of production of strings in the Universe is considered also in the context of superstring theory, the candidate theory unifying gravity with other forces. In superstring theory, there exist fundamental strings and other string-like objects. In simple cases their existence in the Universe is ruled out: their tension is in the range $\mu \sim (1 - 10^{-2}) \cdot M_{Pl}^2$, which is inconsistent with CMB data. Furthermore, production of very heavy strings is problematic from the viewpoint of inflationary theory. Finally, in simple versions of string theory long strings are unstable at cosmological time scale.

Nevertheless, there are versions of string theory in which strings have interesting and phenomenologically acceptable properties; see e.g. [248] and references therein. Besides fundamental strings (F-strings), another type of objects, D-strings, is of importance. This is a subclass of so called Dp-branes (where p refers to spatial dimensionality of the object).

Properties of these types of strings are quite different from those of cosmic strings. Production mechanisms of F- and D-strings are also different from the Kibble mechanism. Unlike cosmic strings, they have small probability to reconnect when intersecting. Furthermore, intersection of F- and D-strings may lead to formation of hybrid FD-strings connecting F- and D-strings. The production of loops at reconnections, and hence energy loss, is thus strongly suppressed. In some string theories, F-, D- and FD-strings form stable networks of cosmological scale. Tensions of these strings need not be extremely large, so such networks are not ruled out.

12.4. Domain Walls

Let us consider the simplest field theory model admitting domain wall solution. This is the model of one real self-interacting scalar field ϕ with the Lagrangian

$$\mathcal{L} = \frac{1}{2}\partial_\mu \phi \partial^\mu \phi - \frac{\lambda}{4}(\phi^2 - v^2)^2. \tag{12.54}$$

This Lagrangian is invariant under change of sign of the field, $\phi \to -\phi$ (symmetry group $\mathbb{Z}_2$). As the result of spontaneous symmetry breaking, the field ϕ acquires vacuum expectation value

$$\langle \phi \rangle = +v \quad \text{or} \quad \langle \phi \rangle = -v,$$

i.e., the space of vacua is disconnected and consists of two points. Phase transition in the Universe leads to domains of two types: one with positive field[10] $\langle\phi\rangle = +v$, and another with negative value $\langle\phi\rangle = -v$. The field configurations interpolating in space between different vacua $\phi = +v$ and $\phi = -v$ are known as domain walls.

The general necessary condition for the existence of domain walls is, in notations used in (12.13),

$$\pi_0(G/H) \neq 0.$$

This condition means that the vacuum manifold $\mathcal{M} = G/H$ consists of several disconnected components.

In the simplest case of static infinite domain wall stretching in flat space along the plane $z = 0$, the profile depends only on z and is the solution to the field equation

$$\frac{d^2\phi}{dz^2} - \lambda(\phi^2 - v^2)\phi = 0, \tag{12.55}$$

with boundary conditions $\phi(z \to \pm\infty) = \pm v$ (kink) or $\phi(z \to \pm\infty) = \mp v$ (antikink). These configurations are given by

$$\text{kink:} \quad \phi(z) = v\tanh\frac{z}{\Delta}, \quad \Delta^2 = \frac{2}{\lambda v^2} \tag{12.56}$$

$$\text{antikink:} \quad \phi(z) = -v\tanh\frac{z}{\Delta}. \tag{12.57}$$

Problem 12.15. *Show that kink is indeed the solution to Eq. (12.55).*

The parameter Δ has the meaning of the domain wall thickness. To see this, let us calculate the energy–momentum tensor of the domain wall. The general expression for the energy–momentum tensor is

$$T^{sc}_{\mu\nu} = \partial_\mu\phi\partial_\nu\phi - \mathcal{L}\eta_{\mu\nu},$$

and for the solution (12.56) we obtain

$$T^{DW}_{\mu\nu} = \frac{\lambda v^4}{2\cosh^4\frac{z}{4\Delta}}\,\text{diag}(1, -1, -1, 0). \tag{12.58}$$

We see that the energy–momentum tensor does not depend on x and y and is nonzero only in the region $-\Delta \lesssim z \lesssim \Delta$. Thus, Δ is indeed the wall thickness. We note in passing that domain walls produced in the Universe are not flat, generally speaking. Nevertheless, the kink approximation works well unless domain walls are strongly curved.

---

[10]The expectation value of ϕ at finite temperature does not coincide with v, but this is unimportant for what follows.

The integrals

$$\eta \equiv \int T_{00}^{DW} dz = \frac{2\sqrt{2\lambda}}{3} v^3 \quad \text{and} \quad \int T_{11}^{DW} dz = \int T_{22}^{DW} dz = -\eta$$

give the surface energy density of the wall and pressure along the wall. These are equal up to sign.[11] We note that $T_{33}^{DW} = 0$, which means that there is no tension across the wall.

Large spatial components of the energy–momentum tensor imply that domain walls are relativistic objects, which may be expected to have non-trivial properties once gravity is turned on. To illustrate this, let us make use of the linearized equation (12.34) for the Newtonian potential. With the domain wall energy–momentum tensor (12.58) we obtain

$$\Delta \Phi = -4\pi G T_{00}^{DW}.$$

Its solution in the thin wall limit, $T_{00} = \eta \delta(z)$, reads

$$\Phi = -2\pi G \eta |z|. \tag{12.59}$$

This would give gravitational field of opposite sign as compared to point-like object. Hence, if a domain wall could be at rest it would antigravitate: non-relativistic particles would be repelled by domain wall.

In fact, the notion of a static domain wall, as well as usage of the linearized Einstein equations are not valid. Hence, the picture we just presented is oversimplified. We have given it only to illustrate that gravitating domain walls have non-trivial properties.

The Kibble mechanism works for domain walls too. In general, after the phase transition the Universe consists of domains of different vacua divided by domain walls, which in turn contain closed domains of the opposite vacuum, etc. Domain walls make either open or closed non-intersecting surfaces.[12]

In vast majority of particle physics models, the existence of even a single "infinite" (horizon size) domain wall in the present Universe is ruled out. A simple way to see this is to estimate the energy of a domain wall of horizon size, $M_{DW} \sim \eta H_0^{-2}$, and compare it to the total energy inside the present horizon, $\rho_c H_0^{-3}$. The requirement that the wall energy is smaller than the total energy,

$$\eta H_0^{-2} \lesssim \rho_c H_0^{-3},$$

gives the bound on the surface energy density,

$$\eta \lesssim \rho_c H_0^{-1} \sim (10\,\text{MeV})^3. \tag{12.60}$$

---

[11] The latter property is no longer valid for curved domain walls.

[12] We consider the model (12.54) whose space of vacua consists of two points only. There can be more complicated situations (e.g., there are models with symmetry $\mathbb{Z}_n$, $n > 2$). In those cases, there are more types of domains, and hence more types of walls; these walls may intersect.

Barring the possibility of extremely small couplings, this means that the energy scale of physics responsible for domain wall must be less than $10\,\mathrm{MeV}$, which is very small from the viewpoint of extensions of the Standard Model. Reversing the argument, extensions of the Standard Model with discrete symmetries face the "domain wall problem" [249]: there must be a mechanism that forbids domain wall production in the early Universe, or a mechanism that destroys them in the course of evolution. This requirement is certainly far from trivial.

Problem 12.16. *Find equation of state for gas of non-interacting domain walls moving at non-relativistic velocities. Hint: Make use of approach similar to that of Sec. 12.3.2.*

To end this Section we note that there are other cosmological effects of domain walls besides their effect on the cosmological expansion. In particular, gravitational potential of a domain wall would affect CMB through the integrated Sachs–Wolfe effect discussed in detail in accompanying book. The dominant effect [262] comes from the variation of the gravitational (12.59) across the wall, which induces the change in photon frequency $\delta\omega/\omega = \Delta\Phi$. As a result, the CMB temperature in the direction towards the wall changes by

$$\frac{\delta T}{T} = -2\pi G\,\eta\,l\,,$$

where l is the distance from the wall to the observer. Generically, this distance is of order of the present horizon length, $l \simeq H_0^{-1}$, and by requiring that $\delta T/T < 10^{-5}$ one obtains even stronger bound as compared to (12.60):

$$\eta < (1\ \mathrm{MeV})^3.$$

12.5. *Textures

Textures are *unstable* topologically non-trivial field configurations [226, 250]. These are also produced in the course of phase transitions, and the role of topology is that their production occurs via the Kibble mechanism. Being unstable, textures do not survive up to the present epoch, and their presence in the early Universe can be detected only indirectly. One possibility is that they contribute to the generation of density perturbations that develop into structures in the Universe [251]. We note, though, that no effects due to textures have been observed so far.

Like stable topological defects, textures may exist in theories with spontaneously broken global or gauge symmetry. An example of the former type is given by a model with global symmetry $O(4)$ broken down to $O(3)$. The model contains four real scalar fields φ^a, $a = 1, \ldots, 4$, and the Lagrangian is

$$\mathcal{L} = \frac{1}{2}\partial^\mu\varphi^a\partial_\mu\varphi^a - \frac{\lambda}{4}(\varphi^a\varphi^a - v^2)^2. \tag{12.61}$$

Symmetry is broken at zero temperature, and the vacuum can be chosen as

$$\langle \varphi^a \rangle = v\delta_4^a. \tag{12.62}$$

The fields φ^1, φ^2, φ^3 are massless in this vacuum; these are Nambu–Goldstone bosons[13] corresponding to global symmetry breaking $O(4) \rightarrow O(3)$. The field h defined by $h = \varphi^4 - v$ acquires the mass m_h which we assume to be large enough.

Problem 12.17. *By calculating the quadratic action for perturbations about vacuum (12.62) show that the spectrum of the model is indeed as described above.*

The scalar potential of the model (12.61) vanishes for any fields obeying

$$\varphi^a \varphi^a = v^2. \tag{12.63}$$

This equation determines the vacuum manifold; in our case this manifold is a 3-sphere S_{vac}^3. The energy density of field configurations of the sizes greater than m_h^{-1} is not very large provided the relation (12.63) is valid at every point in space. These configurations have finite total energy as long as the gradients of φ^a vanish as $r \rightarrow \infty$, i.e., $\varphi^a(\mathbf{x})$ tends to a constant independent of angles and r. Without loss of generality we set

$$\varphi^a(r \rightarrow \infty) = v\delta_4^a.$$

From the topological viewpoint this means that our 3-dimensional space effectively has topology of sphere S_{space}^3, since all points at spatial infinity are mapped into one point in S_{vac}^3. The field configurations $\varphi^a(\mathbf{x})$ just described define, therefore, mappings from 3-sphere S_{space}^3 to 3-sphere S_{vac}^3, which may have non-trivial topology. The topological number is the number of times the sphere S_{vac}^3 is wrapped by the mapping $S_{space}^3 \rightarrow S_{vac}^3$; this number, called degree of mapping, is integer.

The simplest topologically non-trivial configuration is

$$\varphi = v \begin{pmatrix} \cos\phi \sin\theta \sin\chi \\ \sin\phi \sin\theta \sin\chi \\ \cos\theta \sin\chi \\ \cos\chi \end{pmatrix}, \tag{12.64}$$

where ϕ and θ are spherical angles in our space, while the function $\chi(r)$ obeys

$$\chi(r = 0) = 0, \quad \chi(r \rightarrow \infty) = \pi. \tag{12.65}$$

This configuration has unit topological number. This is precisely a texture of minimum topological number.

Problem 12.18. *Show that the configuration (12.64) has topological number 1.*

---

[13]The existence of massless Nambu–Goldstone bosons is a very general property of models with spontaneously broken global symmetries. It may lead to phenomenological problems; we do not discuss this aspect here.

The configuration (12.64) is unstable. Its energy is entirely due to the gradient term,

$$E = \frac{1}{2} \int d^3x \nabla \varphi^a \cdot \nabla \varphi^a. \tag{12.66}$$

Under rescaling $x \to x' = \alpha x$, energy (12.66) scales as $E \to E' = \alpha E$. Hence, this configuration is unstable against shrinking;[14] once formed, texture shrinks. When its size becomes of order m_h^{-1}, the property (12.63) need not hold any longer, and the texture unwinds and disappears. Its energy gets transmitted into energy of elementary excitations, Nambu–Goldstone bosons.

We note in passing that in static 3-dimensional space whose *geometry* is S^3, configurations similar to textures may be stable. The solution still has the form (12.64), but now χ is the third coordinate on S^3_{space}. Metric in coordinates χ, θ, ϕ is

$$d\Omega^2 = a^2 [d\chi^2 + \sin^2 \chi (d\theta^2 + \sin^2 \theta d\phi^2)]. \tag{12.67}$$

The solution (12.64) is non-trivial on entire S^3_{space}, and vector (12.64) coincides with the normal to S^3_{space}, if the latter is embedded into fictitious 4-dimensional Euclidean space.

Problem 12.19. *Show that the field configuration (12.64) is a solution to the field equations in the space with metric (12.67).*

Let us give a heuristic argument in favor of stability of the solution (12.64) in space with metric (12.67). Consider the family of field configurations of the following form:

$$\varphi = v \begin{pmatrix} \cos\phi \sin\theta \sin(\chi/\alpha) \\ \sin\phi \sin\theta \sin(\chi/\alpha) \\ \cos\theta \sin(\chi/\alpha) \\ \cos(\chi/\alpha) \end{pmatrix}, \quad \varphi = v \begin{pmatrix} 0 \\ 0 \\ 0 \\ 1 \end{pmatrix} \tag{12.68}$$

$$\text{for } 0 < \chi \leq \pi\alpha, \quad \text{for } \pi\alpha < \chi \leq \pi.$$

They have topological number 1 and describe shrinking texture as α changes from 1 to 0. The energy of the configuration (12.68) is

$$E = 4\pi \cdot a \cdot v^2 \cdot \int_0^{\pi\alpha} \left[\frac{1}{\alpha^2} + 2 \frac{\sin^2(\chi/\alpha)}{\sin^2 \chi} \right] \sin^2 \chi d\chi$$

$$= 2\pi^2 \cdot a \cdot v^2 \left[\frac{1}{2\alpha} + \alpha - \frac{\sin 2\pi\alpha}{4\pi\alpha^2} \right]. \tag{12.69}$$

Since both $\alpha = 0$ and $\alpha = 1$ are minima of energy, both limiting configurations (i.e., the original configuration with $\alpha = 1$, and shrunk configuration with $\alpha = 0$) appear classically stable; there is potential barrier between the configuration (12.64) and shrunk texture, at least along the path in configuration space defined by (12.68).

---

[14]This illustrates the fact that topology alone is not sufficient to ensure the existence of non-trivial stable configurations.

Textures are created in the course of phase transition by the Kibble mechanism. As the Universe expands, textures shrink and disappear by emitting Nambu–Goldstone bosons. The energy density in textures decreases, while the spatial regions where the field takes one and the same value expand and eventually become of the horizon size. The latter process is fast, so almost immediately after the phase transition, there is about 1 texture of size $H^{-1}(T_c)$ per Hubble volume $H^{-3}(T_c)$.

As the Universe expands, the process continues, and there remains about one texture of the horizon size per horizon volume. Since the energy density in texture is determined by the gradient energy, we have an estimate

$$\delta\rho \sim (\nabla\varphi)^2 \sim v^2 H^2(t) \sim v^2 t^{-2}.$$

The background energy density has the same dependence on time at both radiation and matter domination, so textures give perturbations of energy density at horizon crossing independent of the spatial scale of the horizon,

$$\frac{\delta\rho}{\rho} = \text{const}\frac{v^2}{M_{Pl}^2}.$$

Importantly, these perturbations exist in a wide range of spatial scales, starting from $H^{-1}(T_c)$. Hence, to zeroth approximation density perturbations have *scale-invariant* spectrum [251].

Approximate scale invariance is precisely the property of the measured spectrum of density perturbations. However, the texture mechanism of the generation of density perturbations would lead to the CMB anisotropy spectrum inconsistent with observations; see Fig. 12.6. Thus, textures as the main source of density perturbations are ruled out [244].

Problem 12.20. *Estimate the number density and energy density of Nambu–Goldstone bosons produced in the Universe within the model (12.64). Which values of v are ruled out by BBN? What is the fraction of Nambu–Goldstone bosons in the energy of relativistic matter today for $v = 10^{16}$ GeV?*

12.6. *Hybrid Topological Defects

Some models admit structures consisting of topological defects of different dimensionality. This may be the case if spontaneous symmetry breaking occurs in two or more steps. Each step leads to vacuum rearrangement, possibly new topological properties of vacuum manifold and hence new types of defects. From a topology viewpoint, a chain of phase transitions

$$G \xrightarrow{T_1} H_1 \xrightarrow{T_2} H_2 \to \cdots, \quad T_1 > T_2 > \cdots,$$

may give rise to vacuum manifolds with non-trivial homotopy groups

$$\pi_{N_1}(G/H_1) \neq 0, \quad \pi_{N_2}(G/H_2) \neq 0, \cdots$$

Examples of hybrid structures are fleece (strings with ends on domain walls), neck-laces (strings with magnetic monopoles on them), etc. As an example, necklaces are produced in two subsequent phase transitions:

$$\text{step 1}: \quad G \to G' \times U(1),$$
$$\text{step 2}: \quad G' \times U(1) \to H \times \mathbb{Z}_N.$$

The first step leads to monopole production, while at the second step these monopoles get connected by strings. The necklace case corresponds to $N = 2$, so that each monopole gets connected to two others.

The evolution of hybrid defects is, generally speaking, different from that of topological defects constituting the hybrids. Their detailed discussion is beyond the scope of this book; we only point out here that among hybrids, the most interesting are necklaces.

12.7. *Non-topological Solitons: Q-balls

Besides topological solitons discussed in previous Sections, there exist other localized field configurations, whose stability is unrelated to topology. Particularly popular are Q-balls which are stable due to the existence of conserved global charge (hence the name) and absence of massless particles carrying this charge.

12.7.1. *Two-field model*

A simple model admitting non-topological solitons [252] contains real scalar field χ and complex scalar field ϕ. The Lagrangian is

$$\mathcal{L} = \partial_\mu \phi^* \partial^\mu \phi + \frac{1}{2} \partial_\mu \chi \partial^\mu \chi - V(\chi) - h^2 \chi^2 |\phi|^2, \tag{12.70}$$

where the potential $V(\chi)$ has absolute minimum at

$$\chi = v \neq 0, \tag{12.71}$$

so that

$$V(\chi = v) = 0, \quad V(\chi = 0) = V_0 > 0.$$

The mass of the field ϕ in vacuum (12.71) is

$$m_\phi = hv.$$

We assume that the field $\delta\chi \equiv \chi - v$ is also massive.

The Lagrangian (12.70) is invariant under global $U(1)$ symmetry

$$\phi \to e^{i\alpha}\phi, \quad \phi^* \to e^{-i\alpha}\phi^*, \quad \chi \to \chi. \tag{12.72}$$

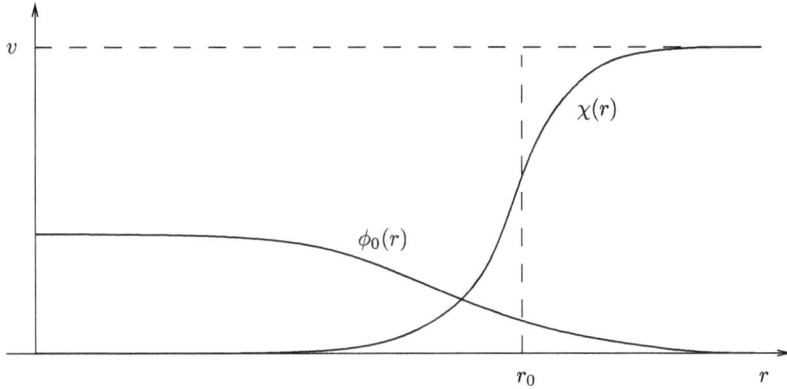

Fig. 12.7. Q-ball configuration: profile of the field $\chi(r)$ and wave function of ϕ-particle $\phi_0(r)$.

Note that vacuum (12.71) with $\phi = 0$ is invariant under this symmetry. Due to these properties, there exists conserved charge

$$Q = i \int d^3\mathbf{x}(\phi\dot{\phi}^* - \phi^*\dot{\phi}). \tag{12.73}$$

Let us find the state of minimum energy at a given value of Q. One of the states of charge Q is a collection of ϕ-particles at rest in vacuum (12.71); their number equals Q. The energy of this state is equal to $Q \cdot m_\phi$. The competing Q-ball state has the following structure. Inside a region of yet to be determined size r_0, the field χ is zero, while away from this region this field takes vacuum value (12.71); $\chi(r)$ smoothly changes from zero to v near the boundary; see Fig. 12.7. All Q particles ϕ are sitting at the lowest energy level inside the region of vanishing χ. Their mass is zero there, and the momentum is of order $1/r_0$. Hence, the energy of each of the ϕ-particles is b/r_0, where b is of order 1, and their total energy is bQ/r_0. The total Q-ball energy is thus

$$E = \frac{4}{3}\pi r_0^3 \cdot V_0 + 4\pi r_0^2 \cdot \sigma + b\frac{Q}{r_0}, \tag{12.74}$$

where the first term is the energy of the field χ inside the Q-ball (recall that $V_0 = V(\chi = 0)$), and the second one is energy in the transition region near r_0, σ being the tension of the Q-ball boundary. We will find (see (12.76)) that the Q-ball size is large for large Q, so the boundary term can be neglected, and the Q-ball energy is given by

$$E(r_0) = \frac{4}{3}\pi r_0^3 \cdot V_0 + b\frac{Q}{r_0}. \tag{12.75}$$

We now minimize this expression with respect to r_0 and find the Q-ball radius at given Q,

$$r_0(Q) = \left(\frac{bQ}{4\pi V_0}\right)^{1/4}, \tag{12.76}$$

so the Q-ball energy (mass) is

$$E[r_0(Q)] \equiv M_Q = \text{const} \cdot V_0^{1/4}Q^{3/4}, \tag{12.77}$$

with constant of order 1. We see that Q-ball energy grows with Q slower than the energy $m_\phi Q$ of free ϕ-particles in vacuum (12.71). This means that Q-ball is indeed the lowest energy state at sufficiently large Q. It is stable against decay into ϕ-particles at

$$M_Q < m_\phi Q, \tag{12.78}$$

i.e., at $Q > Q_c$ where the critical charge is of order

$$Q_c \sim \frac{V_0}{m_\phi^4}. \tag{12.79}$$

Note that for the potential $V(\chi) = \lambda \cdot (\chi^2 - v^2)^2$ the estimate (12.79) gives

$$Q_c \sim \frac{\lambda}{h^4},$$

so that the critical charge is large for $\lambda \sim h^2 \ll 1$, i.e., in the weak coupling regime without fine-tuning of parameters. If the estimate (12.79) gives formally $Q_c \lesssim 1$ (say, for $\lambda \lesssim h^4$), then the above calculation is inadequate for the critical Q-ball, and accurate analysis is required for finding Q_c. We note also that the above calculation is made under two assumptions. First, we assumed that ϕ-particles are the lightest particles carrying global $U(1)$ charge, otherwise the estimate (12.79) would contain the mass of the lightest charged particle instead of m_ϕ. Second, it is important that ϕ-particles are bosons, so they can all occupy the lowest energy level inside the Q-ball.

Problem 12.21. *Let the particles ϕ charged under $U(1)$ be fermions. Are Q-balls similar to those studied in the text stable in weakly coupled theories with $m_\phi \sim m_\chi$, where m_χ is χ-particle mass in vacuum (12.71)? Hint: Take for definiteness $V(\chi) = \lambda \cdot (\chi^2 - v^2)^2, \lambda \ll 1$.*

Let us give, for later convenience, an alternative but equivalent description of Q-balls entirely within classical field theory. Such a description is adequate at large Q. In classical field theory, one considers a classical configuration of the fields χ

and ϕ, instead of studying the state of quantum ϕ-particles. For time-independent χ and time-dependent ϕ the energy is

$$E = \int d^3\mathbf{x} \left(\dot{\phi}^* \cdot \dot{\phi} + \nabla\phi^* \cdot \nabla\phi + h^2\chi^2|\phi|^2 + \frac{1}{2}\nabla\chi \cdot \nabla\chi + V(\chi) \right).$$

To construct the Q-ball, one finds the minimum of this functional at given value of classical charge (12.73). As the ansatz for $\chi(\mathbf{x})$ one again chooses the configuration of Fig. 12.7, while for $\phi(\mathbf{x}, t)$ one writes

$$\phi(\mathbf{x}, t) = Ae^{i\omega t}f(\mathbf{x}), \tag{12.80}$$

where A is yet unknown amplitude and $f(\mathbf{x})$ is normalized by

$$\int |f(\mathbf{x})|^2 d^3\mathbf{x} = 1. \tag{12.81}$$

Making use of the latter condition, one finds the energy and charge

$$E = \omega^2 A^2 + A^2 \cdot \mathcal{E}_f + \frac{4}{3}\pi r_0^3 V_0, \tag{12.82}$$

$$Q = 2\omega A^2, \tag{12.83}$$

where we again neglected the surface term in energy of field χ and introduced the notation

$$\mathcal{E}_f = \int d^3\mathbf{x}(|\nabla f|^2 + h^2\chi^2(r)|f|^2).$$

One chooses the function $f(\mathbf{x})$ in such a way that it minimizes $\mathcal{E}_f$ under normalization condition (12.81). Thus, $f(\mathbf{x})$ obeys

$$-\Delta f + h^2\chi^2(r)f = \lambda^2 f, \tag{12.84}$$

where λ is the Lagrange multiplier. Equation (12.84) coincides with the Schrödinger equation in potential $h^2\chi^2(r)$; its lowest eigenvalue λ determines $\mathcal{E}_f$:

$$\mathcal{E}_f = \lambda^2.$$

Thus, $f(\mathbf{x})$ coincides with the wave function of the ground state of ϕ-particle in Q-ball ($f_0(r)$ in Fig. 12.7), and

$$\lambda = E_\phi = \frac{b}{r_0},$$

where b is the same coefficient as in (12.74).

Hence, the energy functional (12.82) has the form

$$E = \omega^2 A^2 + A^2 \cdot \frac{b^2}{r_0^2} + \frac{4}{3}\pi r_0^3 V_0. \tag{12.85}$$

It remains to find the minimum of this functional with respect to remaining unknowns ω, A and r_0 at given value of charge (12.83). One obtains from (12.83) that $\omega = Q/(2A^2)$. Then minimization of (12.85) with respect to A gives

$$A^2 = \frac{Qr_0}{2b},$$

and the frequency is $\omega = b/r_0$, as one should expect. The energy as a function of the only remaining parameter r_0 is given precisely by Eq. (12.75). Further analysis coincides with that given above, so we see that classical field theory approach is indeed equivalent to quantum mechanical one.

Let us discuss the simple mechanism of the cosmological production of Q-balls as dark matter candidates. Assume that the field χ has zero expectation value at high temperatures, and at some critical temperature T_c it develops nonzero expectation value χ_c due to *first order* phase transition. Let us assume further that the ϕ-particle mass is large in the new phase[15]

$$m_\phi(T_c) = h\chi_c \gg T_c. \tag{12.86}$$

We assume finally that before the phase transition, the Universe is Q-asymmetric, with the asymmetry characterized by

$$\Delta_\phi = \frac{n_\phi - n_{\bar\phi}}{s} \sim \frac{n_\phi - n_{\bar\phi}}{n_\phi + n_{\bar\phi}}. \tag{12.87}$$

(We assume that ϕ is the lightest particle carrying the global $U(1)$ charge.)

As we discussed in Sec. 10.1, first order phase transition occurs via nucleation of a few bubbles of new phase per Hubble volume; these bubbles then expand and eventually percolate. As a result, there remain islands of the old phase $\chi = 0$. These islands are precursors of Q-balls, and, indeed, entire Q-charge of the Universe concentrates in these islands. The reason is that because of (12.86), ϕ-particles cannot penetrate from regions with $\chi = 0$ to regions with $\chi = \chi_c$. Hence, the islands of old phase carry large charges Q. To make an estimate, we notice that the number of Q-balls right after the phase transition is of the order of the number of bubbles of new phase at percolation. We make use of the result (10.19) and write

$$\frac{n_Q(T_c)}{s(T_c)} \sim \sqrt{g_*}\beta^3 \frac{T_c^3}{M_{Pl}^{*3}},$$

where, in notations of Sec. 10.1

$$\beta = \frac{L^{3/2}}{u\kappa^{1/2}} \gg 1.$$

Every Q-ball collects the charge from volume of order n_Q^{-1}, so the typical Q-ball charge is

$$Q \sim n_Q^{-1}(T_c)s(T_c)\Delta_\phi. \tag{12.88}$$

---

[15]The latter assumption is in fact non-trivial. According to (10.40), (10.45) it implies, in particular, that self-coupling of ϕ is small, $\lambda \ll h^4$.

We make use of Eq. (12.77) and estimate the mass density of Q-balls in the present Universe:

$$\rho_{Q,0} = M_Q \cdot n_{Q,0} \sim V_0^{1/4} \Delta_\phi^{3/4} \beta^{3/4} g_*^{1/8} \left(\frac{T_c}{M_{Pl}} \right)^{3/4} \cdot s_0. \qquad (12.89)$$

Finally, we set $V_0^{1/4}$ and T_c roughly equal to the energy scale v, recall that $s_0 \simeq 3000 \ \mathrm{cm}^{-3}$ and set $g_* \sim 100$. We get

$$\rho_{Q,0} \sim 3 \cdot 10^{-9} \cdot \left(\frac{\beta}{10^6} \right)^{3/4} \left(\frac{\Delta_\phi}{10^{-10}} \right)^{3/4} \left(\frac{v}{1 \ \mathrm{TeV}} \right)^{7/4} \frac{\mathrm{GeV}}{\mathrm{cm}^3}.$$

We see that Q-balls are dark matter candidates, provided that the energy scale v is not very high: as an example, for $\Delta_\phi \sim 10^{-10}$ (which is similar to the baryon asymmetry) the required value $\rho_{Q,0} = \rho_{CDM} \sim 0.25\rho_c \sim 10^{-6} \ \mathrm{GeV/cm}^3$ is obtained for v in (a few)·10 TeV range. More accurate estimates [263] give for $\Delta_\phi \sim 10^{-10}$

$$v = 1 - 30 \ \mathrm{TeV}.$$

The present number density of dark matter Q-balls is

$$n_{Q,0} \simeq \frac{1}{(10^8 - 10^9 \ \mathrm{cm})^3},$$

i.e. the typical distance between Q-balls is of order of thousand kilometers. Mass and charge of typical Q-ball are in the range

$$M_Q = 10^{-3} - 10^{-7} \ \mathrm{g}, \quad Q = 10^{19} - 10^{22},$$

and the estimate for Q-ball size is

$$r_0 \simeq \frac{4\pi}{3} \frac{\Delta_\phi s_0}{\rho_{CDM}} = 2 \cdot 10^{-14} \ \mathrm{cm}, \quad \text{for } \Delta_\phi = 10^{-10}.$$

Note that the latter estimate is practically independent of the parameters of the model.

Problem 12.22. *Obtain the latter estimate.*

Problem 12.23. *In the scenario considered in the text, ϕ-particles and their antiparticles from high energy tail of momentum distribution may penetrate the new phase even if the condition (12.86) is satisfied. Let us neglect $\phi\bar{\phi}$-annihilation in the new phase (which is a reasonable approxomation, given the high energy scale of the model; see Sec. (9.2)). These particles ϕ and $\bar{\phi}$ are not sitting in Q-balls, but stay in the Universe intact until today. Find the bounds on the scenario from the requirement that the present mass density of free particles ϕ and $\bar{\phi}$ does not exceed ρ_{CDM} and thus refine the inequality (12.86).*

12.7.2.   *Models with flat directions*

Somewhat different class of Q-balls [253] exists in models with sufficiently flat scalar potentials. The simplest of these models contains single complex scalar field and has the Lagrangian

$$\mathcal{L} = \partial^\mu \phi^* \partial_\mu \phi - V(\phi^* \phi), \tag{12.90}$$

where the potential $V(\phi^* \phi)$ has absolute minimum at the origin; its other properties will be specified later on. This Lagrangian is invariant under the $U(1)$ symmetry of global phase rotations (12.72), and vacuum $\phi = 0$ does not break this symmetry. In what follows we set $V(0) = 0$, so that vacuum has zero energy.

Field equation in the model (12.90) is

$$\partial^\mu \partial_\mu \phi + \frac{\partial V}{\partial \phi^*} = 0, \tag{12.91}$$

and the total energy is given by the integral

$$E = \int [|\partial_0 \phi|^2 + |\partial_i \phi|^2 + V(\phi^* \phi)] d^3 \mathbf{x}. \tag{12.92}$$

The mass of ϕ-particle in vacuum $\phi = 0$ is

$$m = \sqrt{\frac{\partial^2 V}{\partial \phi \partial \phi^*}(0)}. \tag{12.93}$$

In analogy with (12.80), let us try to find a Q-ball solution to Eq. (12.91). This solution should carry global charge (12.73), so one expects that it oscillates (in internal space) and is spherically symmetric,

$$\phi = e^{i\omega t} f(r), \qquad \phi^* = e^{-i\omega t} f(r). \tag{12.94}$$

By substituting this ansatz into Eq. (12.91) we find

$$\frac{d^2 f}{dr^2} = -\frac{2}{r}\frac{df}{dr} - \frac{d}{df}\left(\frac{1}{2}\omega^2 f^2 - \frac{1}{2}V(f)\right). \tag{12.95}$$

Formally, this equation coincides with the equation for classical particle of unit mass with coordinate f, moving with friction in the potential

$$V_{\text{eff}}(f) = \frac{1}{2}\omega^2 f^2 - \frac{1}{2}V(f),$$

while the radial coordinate plays the role of time. We find the asymptotics of $f(r)$ at large r by requiring that energy (12.90) is finite. The energy functional for the configuration (12.94) has the form

$$E = 4\pi \int [\omega^2 f^2 + (\partial_r f)^2 + V(f)] r^2 dr. \tag{12.96}$$

Its finiteness requires that

$$f(r \to \infty) \to 0. \tag{12.97}$$

This means that the solution we search for is localized, as should be the case for Q-ball.

The behavior near $r = 0$ is found by requiring that "friction force" in Eq. (12.95) is finite. This gives

$$\left[\frac{df}{dr}(r \to 0)\right] \propto r^{1+\epsilon}, \quad \epsilon \geq 0. \tag{12.98}$$

The Q-ball profile $f(r)$ corresponds to the "particle" that starts to move at zero velocity at initial "time" $r = 0$ from some point

$$f_0 = f(r = 0) \tag{12.99}$$

and rolls down along the potential $V_{\text{eff}}(f)$ in such a way that the point $f = 0$ is reached in infinite "time", $f(r \to \infty) \to 0$.

Clearly, the existence of such a solution implies that, first, the effective potential $V_{\text{eff}}(f)$ has a minimum at $f = 0$, and second, the particle has positive energy at the start of its motion, since it experiences friction force and its energy is zero in the end. Hence, the "initial" value of the field must be such that

$$\frac{V(f_0)}{f_0^2} \equiv \omega_0^2 \leq \omega^2. \tag{12.100}$$

The region $r \lesssim r_0$ where $f(r)$ is considerably different from zero, $0 < f(r) \lesssim f_0$, is the inner region of Q-ball, and r_0 is the Q-ball size.

The charge of the Q-ball is given by

$$Q = 8\pi\omega \int_0^\infty f^2(r) r^2 dr = \frac{8\pi}{3}\omega f_0^2 r_0^3, \tag{12.101}$$

where the second equality is obtained by assuming that

$$f(r) \approx f_0 \cdot \theta(r_0 - r).$$

This is a good approximation in the thin wall regime, when the size Δr of the region where $f(r)$ changes is small compared to the size of Q-ball itself, $\Delta r \ll r_0$. We note that in the general case, the second equality in (12.101) can be considered as the definition of Q-ball size r_0.

Assuming the thin wall regime, we estimate the Q-ball energy (12.96) as

$$E \approx \frac{4\pi}{3} r_0^3 [\omega^2 f_0^2 + V(f_0)], \qquad (12.102)$$

where we neglected the contribution of the gradient term $(\nabla f)^2$. The latter comes from the Q-ball surface region, and is therefore small for large Q. The Q-ball size is obtained by minimizing the energy (12.102) with respect to r_0 at given value of the charge (12.101). To this end, we use (12.101) to express the frequency ω through r_0 and insert the result into (12.102). This gives

$$E \approx \frac{4\pi}{3} r_0^3 V(f_0) + \frac{3Q^2}{16\pi r_0^3 f_0^2}.$$

The minimum of this expression with respect to r_0 is at

$$\frac{4\pi}{3} r_0^3 = \frac{Q}{2f_0} \frac{1}{\sqrt{V(f_0)}}.$$

The energy of the Q-ball is

$$E = \frac{Q}{f_0} \sqrt{V(f_0)},$$

while the frequency ω reaches the critical value ω_0 given by (12.100). Finally, the energy of stable stationary solution must be at minimum with respect to the remaining parameter, the field value in the center f_0. This is only possible if the function $V(f)/f^2$ has a minimum at finite $f = f_0 \neq 0$,

$$\min_f \left[\frac{V(f)}{f^2} \right] = \frac{V(f_0)}{f_0^2} \neq 0, \quad f_0 \neq 0, \quad f_0 \neq \infty. \qquad (12.103)$$

It is this value that the field takes in the soliton center.

The properties that the potential $V(f)$ has global minimum at $f = 0$ and at the same time that the function $V(f)/f^2$ has minimum away from zero are quite non-trivial, especially if one recalls that $V(f)$ is actually a function of $f^2 = \phi^*\phi$. (The latter follows from invariance under global $U(1)$.) As an example, the conventional renormalizable potential $V(\phi) = m^2\phi^*\phi + \lambda(\phi^*\phi)^2$ does not have these properties. The existence of Q-balls needs more exotic scalar potentials which are sufficiently flat at least somewhere in the field space. We will briefly discuss in what models this is indeed the case, and here we proceed by assuming that the potential has the required properties.

The Q-ball stability condition is still given by (12.78), which we write as

$$\frac{Q}{f_0}\sqrt{V(f_0)} < Q \cdot m. \tag{12.104}$$

Making use of (12.93) we find that the latter condition is yet another non-trivial requirement imposed on the scalar potential,

$$\frac{2V(f_0)}{f_0^2} < \frac{d^2 V(0)}{df^2}. \tag{12.105}$$

Provided the potential has properties (12.103), (12.105) and has global minimum at the origin, there exist stable Q-balls of the type we have described.

We note that we have performed the analysis for large Q-balls, as we neglected the surface contribution to the energy. Detailed analysis shows that some models admit rather small Q-balls whose charges are not exceedingly larger than 1.

Problem 12.24. *Find approximate Q-ball solution (assuming that Q is large) in a model with gauge group $U(1)$. Show that the ratio E/Q increases with Q because of the Coulomb contribution to the energy. Show that at large charge the Q-ball is unstable against emitting charged particles from its surface. We note, though, that for sufficiently small gauge coupling, there is a range of Q-ball charges in which Q-balls are stable in models with gauge $U(1)$ symmetry.*

Let us now consider the case when the function $V(f)/f^2$ does not have a minimum at finite f. If $V(f)$ increases with f slower than f^2, then the minimum of $V(f)/f^2$ is at $f \to \infty$ and the value at minimum is zero,

$$\min_f \left[\frac{V(f)}{f^2}\right] = \left[\frac{V(f)}{f^2}\right]_{f \to \infty} = 0. \tag{12.106}$$

This is possible only in models with very flat scalar potentials. In this case our previous analysis does not go through. However, in the case of very flat potential,

$$f \to \infty : \quad V(f) \propto f^\alpha, \quad \alpha < 2,$$

the scalar potential can be neglected in Eq. (12.95) and there exists approximate solution obeying all above conditions (12.97)–(12.99),

$$f(r) = f_0 \frac{\sin \omega r}{\omega r} \cdot \theta(r_0 - r). \tag{12.107}$$

This soluton is continuous across the Q-ball boundary $r = r_0$ provided that the soliton size is related to the oscillation frequency,

$$r_0 = \frac{\pi}{\omega}. \tag{12.108}$$

This is in contrast with the case (12.103). The form of solution (12.107) shows that the thin wall approximation is not valid, as $f(r)$ varies in the entire region

$r < r_0$. Nevertheless, the solution is explicitly known, so we can proceed further. By expressing the frequency in terms of the size, we find the charge of the Q-ball,

$$Q = 4f_0^2 r_0^2, \tag{12.109}$$

and its energy (12.96)

$$E = 4\pi f_0^2 r_0 + \frac{4\pi}{3} r_0^3 bV(f_0), \tag{12.110}$$

where the parameter b is of order 1; it is determined by the shape of the scalar potential. We now have to minimize the energy with respect to parameters f_0 and r_0 at given value of the charge (12.109). By expressing r_0 through f_0 from (12.109) we find energy as function of f_0,

$$E(f_0) = 2\pi f_0 Q^{1/2} + \frac{\pi}{6} b \frac{Q^{3/2}}{f_0^3} V(f_0). \tag{12.111}$$

Its minimum occurs at

$$-\frac{b}{12} \frac{d}{df_0} \left(\frac{V(f_0)}{f_0^3} \right) = \frac{1}{Q}. \tag{12.112}$$

The left-hand side decreases as function of f_0 for potentials under discussion, so the solution $f_0(Q)$ to Eq. (12.112) increases with Q. In particular, for $V(f) \sim f^\alpha$ we have from (12.112) that

$$f_0(Q) \propto Q^{\frac{1}{4-\alpha}},$$

and the energy (12.111) behaves as

$$E(Q) \propto Q^{\frac{6-\alpha}{2(4-\alpha)}}.$$

For $\alpha < 2$ the exponent here is smaller than 1, so the Q-ball energy is smaller than energy of Q particles at least for large Q; the Q-ball is stable. We note that in the limit $\alpha \to 0$ (very flat potential at large fields), the energy behavior is

$$E \propto Q^{3/4}. \tag{12.113}$$

This is analogous to (12.77).

Q-balls exist in some realistic extensions of the Standard Model, including supersymmetric ones [254–257]. Our analysis goes through in these models, except for one important point. The analogs of our ϕ-particles are usually not the lightest particles carrying Q-charge, so the bound (12.104) gets modified. It is the mass of the lightest charged particle that enters the right-hand side of the bound (12.104) in realistic models.

The charge Q in realistic theories may be baryon number, lepton number or a combination thereof. Flatness of the scalar potential is a natural property of supersymmetric theories, which often have flat directions at the tree level. Quantum corrections lift the flat directions, but the potential remains nearly flat, since quantum corrections give weak dependence on the field, $V \propto \log |\phi|$.

In SUSY theories, Q may be the baryon number, and then the field ϕ is a combination of squark fields. The parameter m in the generalized formula (12.104) is the proton mass in that case. If Q is lepton number, then the field ϕ is a combination of sleptons, and m in (12.104) is neutrino mass. In models with potentials of the type (12.106), Q-balls are stable on cosmological time scale for large values of their charge only.

Problem 12.25. *Estimate the charges of stable B-ball and L-ball (Q being baryon and lepton number, respectively) in models with potentials of the type (12.106). For numerical estimate consider the following potential at large f,*

$$V(f) \simeq (1\,\text{TeV})^4 \cdot (f/1\,\text{TeV})^\alpha,$$

and study the cases $\alpha = 1$ and $\alpha \to 0$.

Problem 12.26. *Let us consider a model with Q-balls in which scalar particles are unstable against decay into massless fermions. Q-balls in this model are unstable, but may have long lifetime, since because of Pauli blocking, fermions evaporate from the Q-ball surface rather than from its interior. Assuming that the only parameter determining the evaporation rate per unit area is the frequency ω, find the charge for which the Q-ball lifetime exceeds the age of the Universe. Make numerical estimates for the same potentials as in the previous problem.*

Properties of large Q-balls are affected by gravity. At very large Q these objects are gravitationally unstable and collapse into black holes.

Problem 12.27. *Estimate the critical charge above which Q-balls collapse into black holes in models of the types (12.103) and (12.106). Take all dimensionful parameters of order $1\,TeV$ for numerical estimates. Compare with the results of previous two problems.*

The dominant mechanism of Q-ball production is the decay of flat directions of the scalar potential, moduli fields [256, 258]. In the limit $r_0 \to \infty$ our Q-balls degenerate into homogeneous scalar condensate filling the entire space. It can be formed in much the same way as the condensate of the Affleck–Dine field (see Sec. 11.6) but now this condensate carries certain density of $U(1)$ charge. As we will see below, the condensate is actually unstable; it decays by producing very different charge densities in different places in space. The regions with high charge density eventually evolve into Q-balls, while in regions with low charge density the charge is carried by free particles. Detailed analysis shows that this mechanism indeed efficiently produces Q-balls, so that they carry fairly large fraction of the charge initially contained in the condensate.

Let us consider this mechanism in some detail. We have seen in Sec. 11.6 that the asymmetry (here we are talking about asymmetry with respect to global $U(1)$

charge Q) is produced in narrow time interval near the epoch at which the slow roll conditions get violated,[16]

$$\frac{V'(\phi_i)}{\phi_i} \sim H^2(t_r),$$

(12.114)

where ϕ_i is the initial value of the field, and subscript r refers to the end of slow roll regime, cf. Sec. 11.6. Let small explicit breaking of $U(1)$ result in Q-asymmetry

$$\Delta_Q = \frac{n_Q}{s},$$

where n_Q is the Q-charge density. Unlike in the previous Section, the Q-charge density is due to the field ϕ itself, as this field evolves as shown in Fig. 11.17.

Let us first show that the homogeneous scalar condensate carrying Q-charge is unstable against production of inhomogeneities. To this end, we neglect the cosmological expansion (we will discuss this point later on) and write the scalar field as

$$\phi(x) = f(x) \cdot e^{i\alpha(x)}.$$

The field equations for f and α are

$$\ddot{\alpha} - \Delta\alpha + \frac{2}{f}\dot{\alpha}\dot{f} - \frac{2}{f}\boldsymbol{\nabla}\alpha \cdot \boldsymbol{\nabla}f = 0,$$

(12.115)

$$\ddot{f} - \Delta f - f\dot{\alpha}^2 + f\Delta\alpha + V'(f) = 0.$$

(12.116)

Let us begin with the situation in which the scalar field rotates along a circle in internal space,

$$f = \text{const}, \quad \alpha = \omega t, \quad \omega = \text{const},$$

(12.117)

and it follows from (12.116) that

$$\omega^2 = \frac{V'}{f}.$$

(12.118)

Note that at $t > t_r$, when slow roll conditions are violated, one has $\omega > H$. In the end, this justifies our approximation of static Universe.

Let us consider linear perturbations about the homogeneous condensate (12.117). We ask whether there exist real-valued perturbations that exponentially

---

[16]We use here the general relation (4.57) rather than (4.58) valid for power-law potentials.

grow in time. Making use of translational invariance we write

$$\delta f(x) = \delta f \cdot e^{\lambda t} \cos(\mathbf{px}),$$
$$\delta \alpha(x) = \delta \alpha \cdot e^{\lambda t} \cos(\mathbf{px}).$$

(12.119)

Equations (12.115) and (12.116) give

$$(2\lambda w) \cdot \delta f + f \cdot (\lambda^2 + \mathbf{p}^2) \cdot \delta \alpha = 0,$$
$$(\lambda^2 + \mathbf{p}^2 + V''(f) - w^2) \cdot \delta f - (2\lambda f w) \cdot \delta \alpha = 0.$$

We combine these equations to obtain the equation for eigenvalues at each $\mathbf{p}^2$,

$$\lambda^4 + \lambda^2 \cdot (2\mathbf{p}^2 + V'' + 3w^2) + \mathbf{p}^2 \cdot (\mathbf{p}^2 + V'' - w^2) = 0. \qquad (12.120)$$

It is straightforward to see that all its solutions λ^2 are real. For

$$\mathbf{p}^2 + V'' - w^2 < 0, \qquad (12.121)$$

one of the roots is positive, $\lambda^2 > 0$. This means the instability of the condensate: some modes (12.119) do increase exponentially in time. Importantly, these modes are not homogeneous in space: all values of λ are real at $\mathbf{p} = 0$.

Making use of (12.115) we find that the condition (12.121) can be satisfied if the potential is sufficiently flat,

$$V'' < \frac{V'}{f}. \qquad (12.122)$$

The latter inequality is valid for power-law potential, $V \propto f^\alpha$, at $\alpha < 2$, i.e., when the model admits Q-balls. The strongest instability (the largest λ^2) occurs for

$$|\mathbf{p}| \sim w, \qquad (12.123)$$

and the exponent is $\lambda \sim w$. This is the final justification of our static Universe approximation.

The development of instability in non-linear regime is difficult, if not impossible, to study analytically. Still, the very fact that the condition for instability (12.122) coincides with the condition for the existence of Q-balls suggests that the process ends up by Q-ball formation. Numerical simulations support this conclusion [259, 260]. The estimate (12.123) shows that Q-charge is collected into Q-ball from the region of volume

$$|\mathbf{p}|^{-3} \sim w^{-3}.$$

Both evolution of the condensate and the development of instability occur right after the end of the slow roll regime, so a crude estimate is

$$|\mathbf{p}| \sim |\lambda| \sim w \sim H(t_r),$$

where we made use of (12.114) and (12.118). Thus, of order one Q-ball is produced per Hubble volume at time t_r.

Further estimates are quite similar to our estimates in the end of Sec. 12.7.1. Let us consider for definiteness almost flat potential at large f,

$$V(f) = v^4 \left(\frac{f}{v}\right)^\alpha, \quad \alpha \ll 1,$$

where v is a parameter of dimension of mass. In this case one has the estimate (12.113), i.e., the Q-ball mass is of order

$$M_Q \sim vQ^{3/4}.$$

The results of Sec. 12.7.1 are directly translated to our model, and the parameter β is of order 1, since in our case $n_Q(t_r) \sim H^{-3}(t_r)$. We obtain (see (12.89))

$$\rho_{Q,0} = v\Delta_Q^{3/4} g_*^{1/8} \left(\frac{T_r}{M_{Pl}}\right)^{3/4} s_0.$$

The difference, though, is that the temperature T_r is determined by (12.114), so that in the limit of small α we have

$$T_r \sim v\sqrt{\frac{M_{Pl}}{f_i}}.$$

This introduces additional uncertainty into the estimate of the present Q-ball mass density related to the unknown initial value of the field ϕ. If we set $f_i \sim M_{Pl}$ as in Sec. 11.6, then the results of Sec. 12.7.1 with $\beta = 1$ are literally valid. With appropriate choice of parameters, Q-balls of our model serve as dark matter candidates.

If Q-charge is baryon number, then Q-balls may play a role in the generation of the baryon asymmetry [257]. As an example, B-balls may carry baryon number through the epoch when electroweak baryon and lepton number violating interactions are at work and tend to wash out the asymmetry.

Of particular interest are B-balls which are unstable at fairly late cosmological epoch. Their decays transmit their baryon number into quarks. This is basically a version of the Affleck–Dine mechanism. Most intriguing is the situation when Q-balls are unstable and decay into baryons and new stable particles, dark matter candidates. Such a mechanism may possibly explain approximate equality between the present energy densities of baryons and dark matter. As an example, the decay of squark condensate produces three neutralino LSP per baryon. If most of baryon number initially was in Q-balls, one obtains a simple relation between mass densities of baryons and LSP dark matter,

$$\frac{\rho_B}{\rho_{CDM}} \sim \frac{m_p}{3m_{LSP}}. \tag{12.124}$$

For realistic LSP masses, $m_{LSP} \sim 10\text{--}100\,\text{GeV}$, this is only two or three orders of magnitude smaller than the real value. We note in this regard that the neutralino density may get diluted by their annihilation near Q-ball surface. Also, the correct ratio ρ_B/ρ_{CDM} is obtained in models where LSP is a gravitino of mass of order 1 GeV.

To end this Section we note that the class of non-topological solitons is not exhausted by Q-balls. This class includes also quark nuggets, cosmic neutrino balls, soliton stars and other hypothetical objects.

Chapter 13

Color Pages

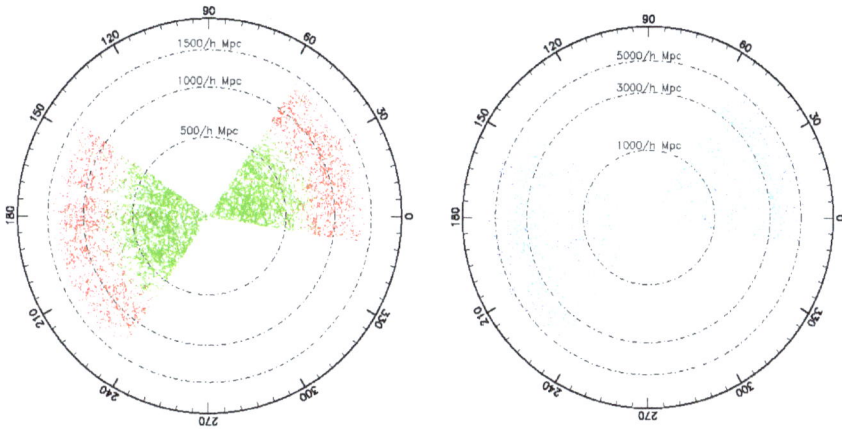

Fig. 13.1.   Spatial distribution of galaxies and quasars from the analysis of early observational data of SDSS [4]. Green dots mark all galaxies (within a given solid angle) with brightness (apparent magnitude) exceeding a certain value. Red dots show galaxies of a special type (Large Red Galaxies), which are very luminous and form fairly homogeneous population; in the comoving frame their spectrum is shifted towards the red wave-band as compared to ordinary galaxies. Turquoise and blue points show the positions of ordinary quasars. The value of parameter h is about 0.7 (see Sec. 1.2.2).

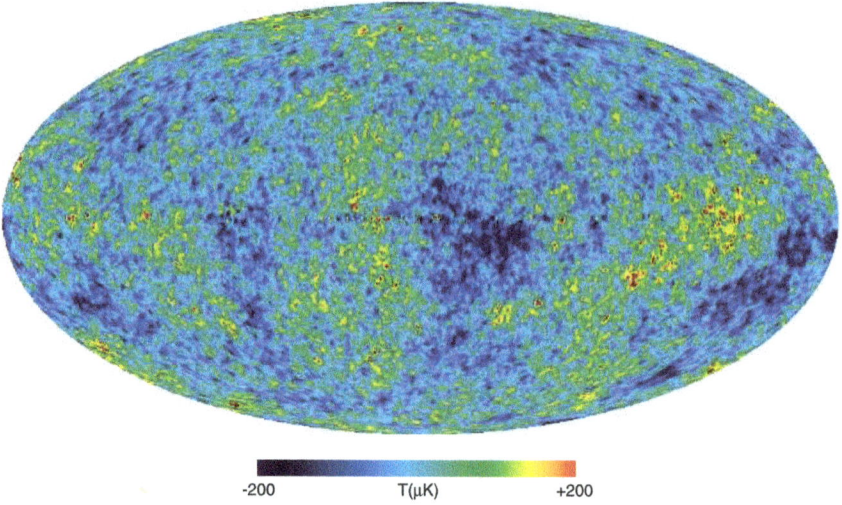

Fig. 13.2. WMAP data [9]: angular anisotropy of CMB temperature, i.e., variation of the temperature of photons coming from different directions in the sky (shown by color). The average temperature and dipole component are subtracted. The observed variation of temperature is at the level of $\delta T \sim 100\ \mu K$, i.e., $\delta T/T_0 \sim 10^{-4} - 10^{-5}$.

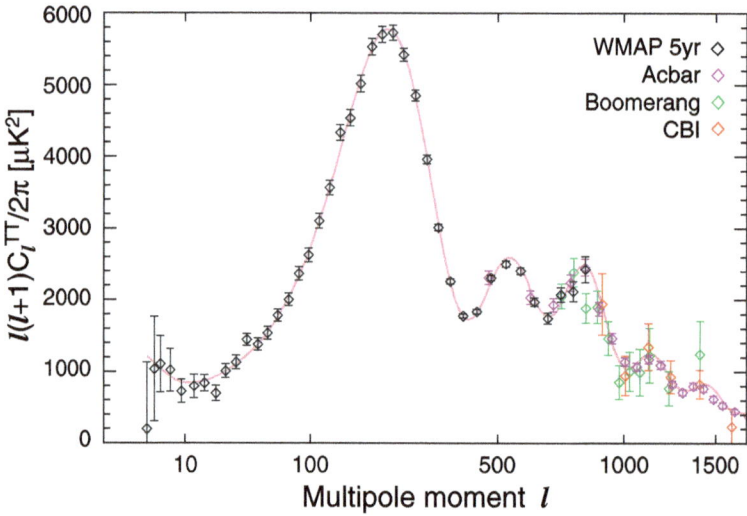

Fig. 13.3. CMB temperature anisotropy measured by various instruments [8]. The theoretical curve is the best fit of the ΛCDM model (see Chapter 4) to the WMAP data; this fit is in agreement with other experiments as well.

Fig. 13.4.   Cluster of galaxies CL0024 + 1654 [22]. Blue color in the left panel illustrates dark matter distribution; elongated blue objects in the right panel are multiple images of a galaxy behind the cluster.

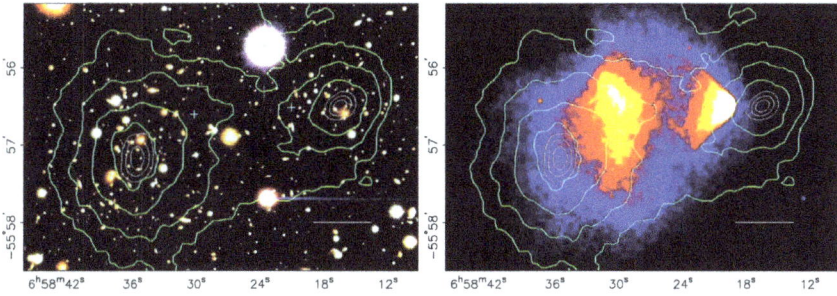

Fig. 13.5.   Observation of "Bullet cluster" 1E0657-558, two colliding clusters of galaxies [24]. Lines show gravitational equipotential surfaces, the bright regions in the right panel are regions of hot baryon gas.

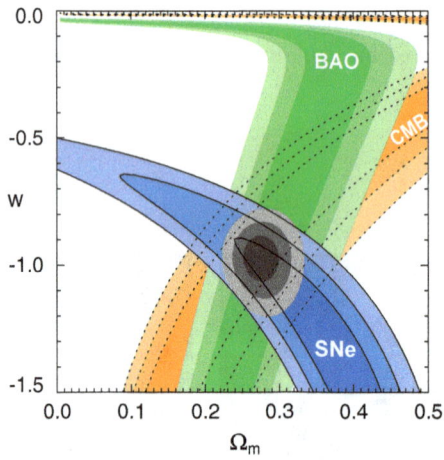

Fig. 13.6. Regions in the plane of parameters (Ω_M, w) allowed (for the flat Universe) by observations of CMB anisotropy, by large scale structures (BAO) and by SNe Ia data [67]. The intersection region corresponds to the combined analysis of all these data. Regions of smaller and larger size correspond to 68.3%, 95.4% and 99.7% confidence level, respectively.

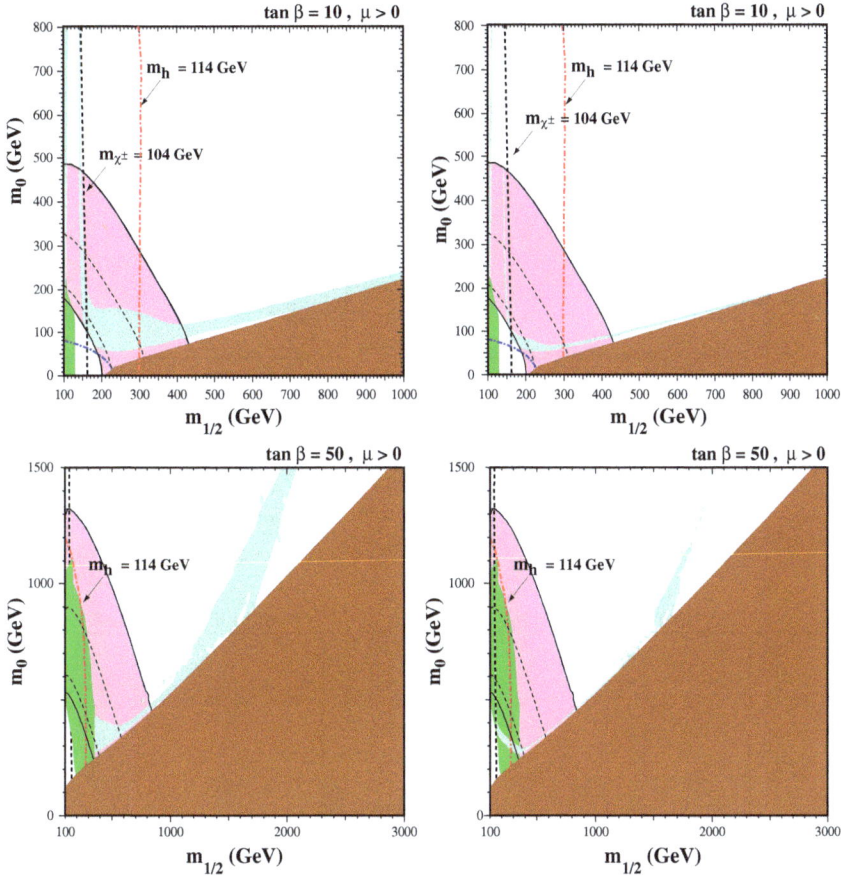

Fig. 13.7. Experimentally forbidden and cosmologically favored regions [130] in the plane $(M_{1/2}, m_0)$ in mSUGRA model with $\tan \beta = 10$ and $\tan \beta = 50$. (A is chosen to be zero at the scale $M_{GUT} \approx 10^{16}$ GeV.) Forbidden regions in each plot: the region left of the thick blue dashed line is excluded by the bound on light chargino mass $m_{\chi^\pm} > 104$ GeV, the region in the lower left corner left of blue dash-dotted line is excluded by the bound on the light slepton mass $m_{\tilde{e}} > 99$ GeV, the region left of red dash-dotted line is excluded by the bound on the lightest Higgs boson mass, the green region in the left part is excluded by the measurement of $b \to s\gamma$ decay width, the light pink strip is excluded by the measurement of the muon anomalous magnetic moment. The brown region on the right is cosmologically unfavored, since it corresponds to charged slepton LSP (mostly stau). Neutralino dark matter regions are shown in light blue; the neutralino mass density range in the left and right plots are $0.1 < \Omega_N h^2 < 0.3$ (conservative estimate) and $0.094 < \Omega_N h^2 < 0.129$, respectively. In the bulk of the experimentally allowed region, neutralinos are overproduced.

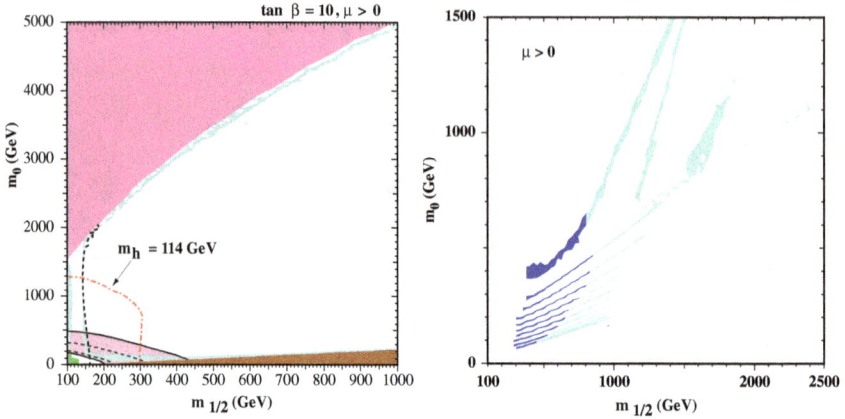

Fig. 13.8. Left: The same as in upper part of Fig. 13.7 but for larger range of masses. The pink upper left region is excluded by the requirement of electroweak symmetry breaking. Right: Cosmologically favored regions ($0.094 < \Omega_N h^2 < 0.129$) for various values of $\tan \beta = 5, 10, \ldots, 55$; lower strips correspond to smaller $\tan \beta$ [130].

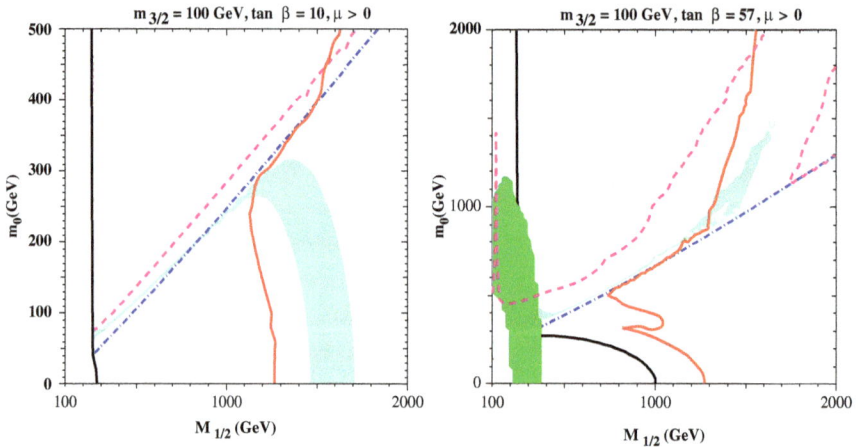

Fig. 13.9. Bounds [136] on the plane ($M_{1/2}, m_0$) in mSUGRA model with $m_{3/2} = 100$ GeV and two values of $\tan \beta$. The region where gravitino is LSP is to the right of the solid black line. The green shaded region is ruled out by $b \to s\gamma$ decays. The region to the right of the solid red line is consistent with BBN. The blue dash-dotted line divides the regions with neutralino NLSP (upper parts of figures) and the lightest slepton (lower parts). In the blue shaded region, NLSP, if stable, would have the right dark matter mass density ($0.094 \leq \Omega h^2 \leq 0.129$). The present gravitino mass density coincides with the dark matter mass density at pink dashed lines. The gravitino mass density is smaller than observed Ω_{DM} below this line (left panel) and between these lines (right panel). These reagions are cosmologically allowed in this model.

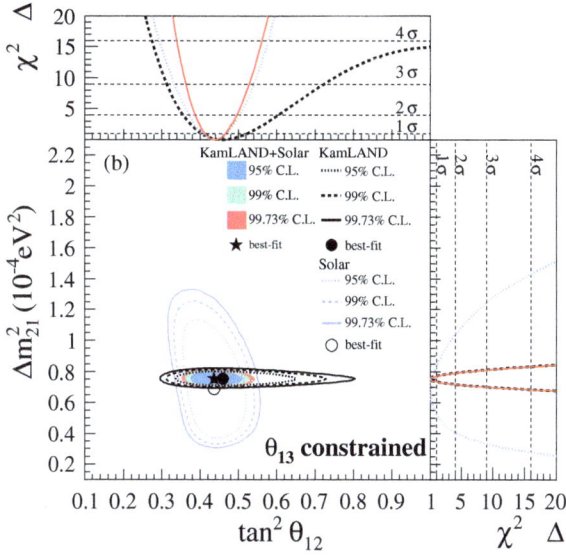

Fig. 13.10. Allowed region of parameter space for oscillations $\nu_e \leftrightarrow \tilde{\nu}$ obtained from solar neutrino experiments and KamLAND [305].

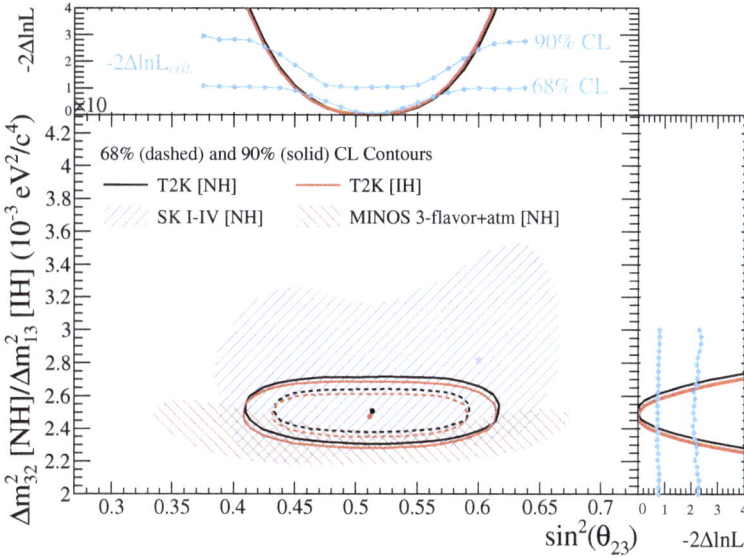

Fig. 13.11. Allowed region of parameter space for oscillations $\nu_\mu \leftrightarrow \nu_\tau$ [307]. Lines encircle areas allowed by T2K at 68% and 90% C.L. Shaded regions show the results (at 90% C.L.) of MINOS and atmospheric neutrino study at Super-K. The results are presented for normal and inverse hierarchy of neutrino masses, NH and IH, respectively. Dots show the best fit values for the two hierarchies.

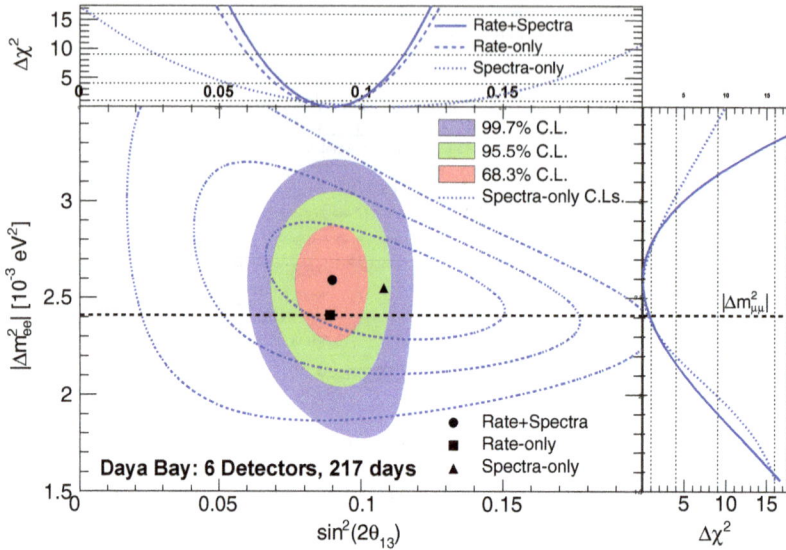

Fig. 13.12.   Allowed regions of parameter space for $\bar{\nu}_e \leftrightarrow \bar{\nu}_e$ oscillations determined by Daya Bay reactor neutrino experiment [306]. Dashed lines show the results obtained by analyzing solely the antineutrino energy spectrum. Horizontal line $\Delta m^2_{\mu\mu}$ corresponds to the measurement of the same parameter by MINOS.

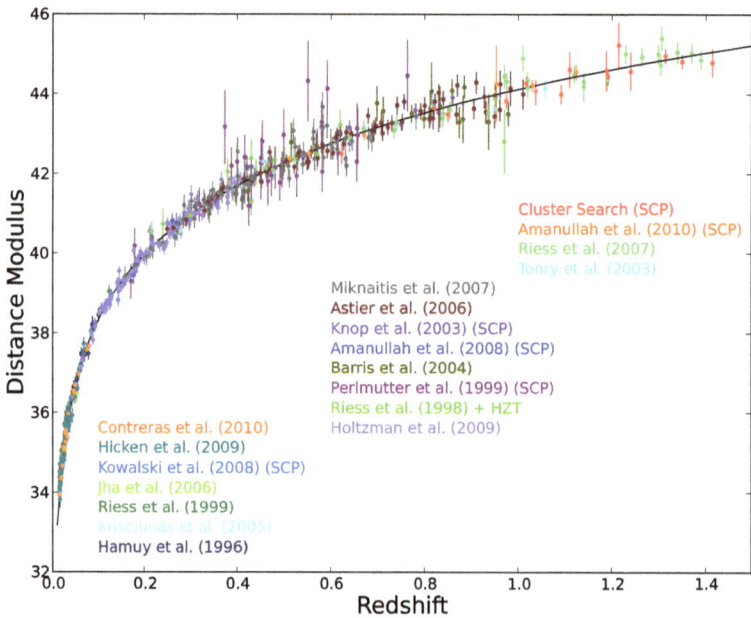

Fig. 13.13.   Hubble diagram for distant objects [75] (2011). Solid curve is the best fit for spatially flat ΛCDM model.

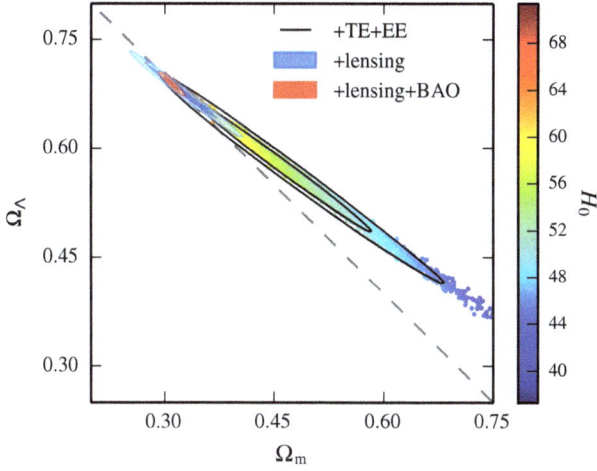

Fig. 13.14. Allowed region in the space of parameters $(\Omega_M, \Omega_\Lambda)$ from CMB anisotropy data (+TE+EE, lensing) and data on baryon acoustic oscilations (BAO), the preferable values of the present Hubble parameter H_0 are indicated in color [94]. Shown here are parameter regions allowed at 68 %–95 % C.L.

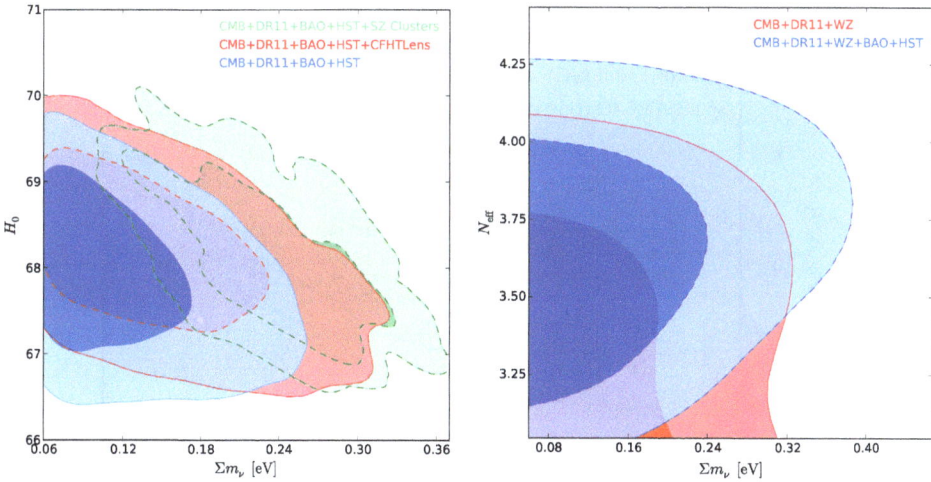

Fig. 13.15. Limits on the sum of neutrino masses obtained under various assumptions on other cosmological parameters and using different cosmological datasets [104]; H_0 (left) is the present value of the Hubble parameter, N_{eff} (right) is the effective number of neutrino species.

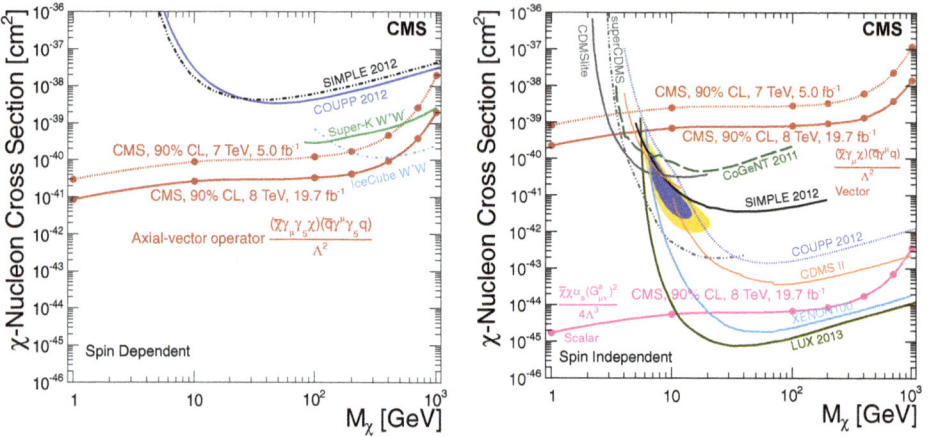

Fig. 13.16. Excluded regions in parameter plane (M_X, σ_{pX}) [162] for interactions independent of nucleon spin (left) and dependent on nucleon spin (right). Regions above the curves are excluded at 90% C.L. CMS: search at Large hadron collider (under assumption of contact interaction $\bar{X}Xf_1f_2$, see the main text); IceCube and Super-K: search for a signal from dark matter annihilation in the Sun; other data: direct search for elastic scattering of cosmic dark matter in low background detectors. Shaded region in the central part of right figure corresponds to possible signal announced by CDMS experiment.

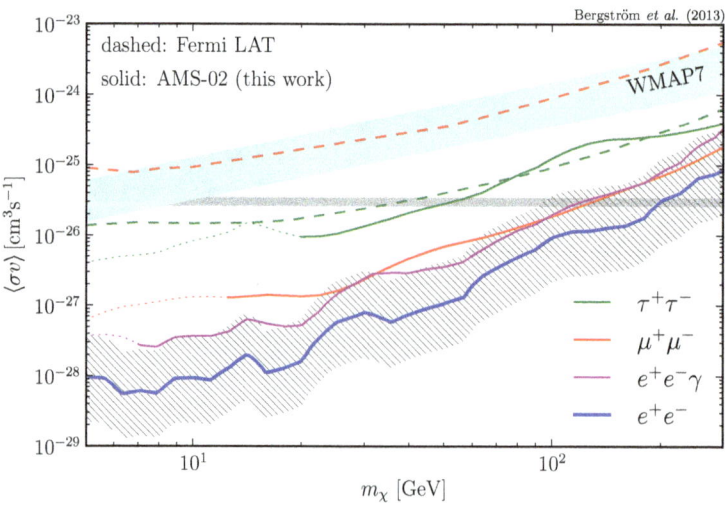

Fig. 13.17. Exclusion regions in the plane (M_X, σ_{XX}^{ann}) [318] for dark matter particles annihilating through indicated channels. The light gray line is the cross-section at freeze-out in the early Universe. Regions above the lines are ruled out at 90% C.L. AMS-02: search for antiparticles; Fermi LAT: search for photons from dark matter annihilation to $\mu^+\mu^-$ and $\tau^+\tau^-$ (upper and lower lines), WMAP7: limit from possible annihilation into charged lepton pair before recombination (upper range $\tau^+\tau^-$, lower range e^+e^-), which would affect CMB. As an example, the uncertainty for the channel e^+e^- is shown as shadow region.

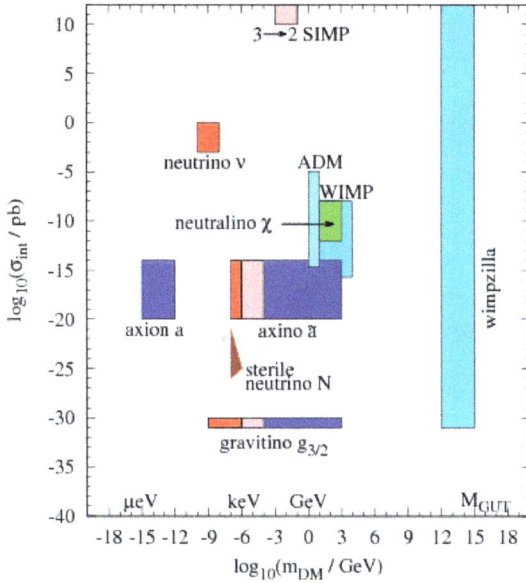

Fig. 13.18. Parameter ranges for various dark metter candidates [164]: their masses M_X and elastic scattering cross-sections off nucleon σ_{int}; $1\,\mathrm{pb} = 10^{-36}\,\mathrm{cm}^2$ (conversion cross-sections for axion and axino). Red, rose, blue and dark blue show hot, warm, cold and ultracold dark matter, respectively.

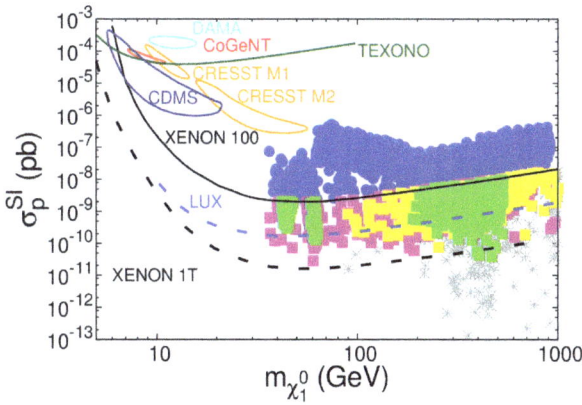

Fig. 13.19. Predictions of MSSM for spin-independent cross-section of elastic scattering of neutralino off nucleons [165]. Color regions correspond to parameters consistent with accelerator data and yielding $\Omega_N \approx 0.25$. The blue region is ruled out by direct search experiment XENON 100. The neutralino annihilation cross-section is enhanced in green regions by resonant annihilation through Z-boson or one of the Higgs bosons of MSSM, when $M_N \approx M_Z/2$ or $M_N \approx m_H/2$; see discussion below. The suppression of neutralino abundance in yellow and magenta regions is due to approximate degeneracy of neutralino and some other superpartner leading to efficient co-annihillation of the neutralino with this superpartner. Other mechansims that suppress neutralino abundance operate in gray regions; they require fine tuning of parameters too. Shown here are also the regions of possible signals in DAMA, CoGENT, CRESST and CDMS experiments and planned sensitivity of LUX and XENON 1T.

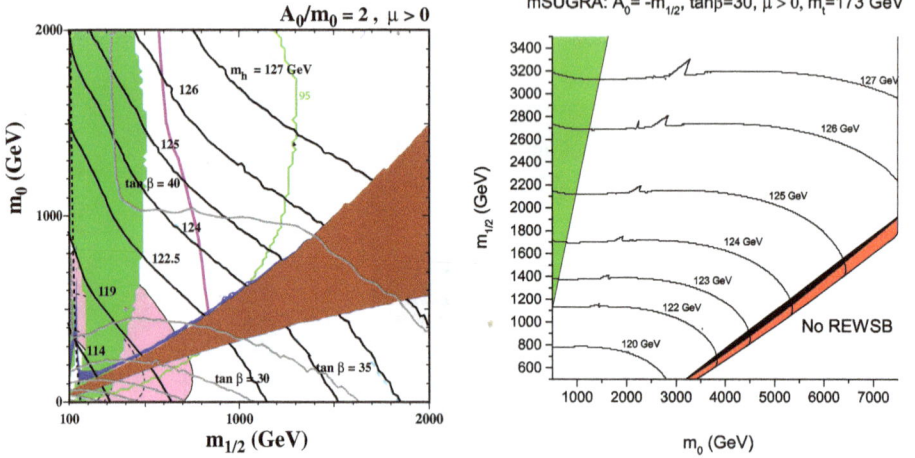

Fig. 13.20. Experimentally forbidden and cosmologically favored regions in the plane $(M_{1/2}, m_0)$ in mSUGRA model [166] with $A_0 = m_0$ and $\mu > 0$ (left) and CMSSM [167] with $A_0 = -m_{1/2}$, $\tan\beta = 30$ and $\mu > 0$ (right). Shown here are lines of constant mass of the lightest Higgs boson (realistic interval is 125–126 GeV. In the left panel, the elongated blue region is consistent with neutralino dark matter and has $m_{1/2} \approx 1300$ GeV, $m_0 \approx 850$ GeV; the region above the blue strip and brown region below are cosmologically ruled out (neuralinos are overproduced and LSP is a charged superpartner of tau-lepton, respectively); below the brown region LSP is gravitino (see disussion in Sec. 9.6.3), the region to the left of solid green line is excluded by the LHC data; gray lines show the values of $\tan\beta$. In the right panel, the cosmologically allowed region is shown in red, in the black region where $m_{1/2} \approx 1800$ GeV, $m_0 \approx 7000$ GeV neutralino abundance coincides with that of dark matter; in the green region LSP is the electrically charged superpartner of t-quark.

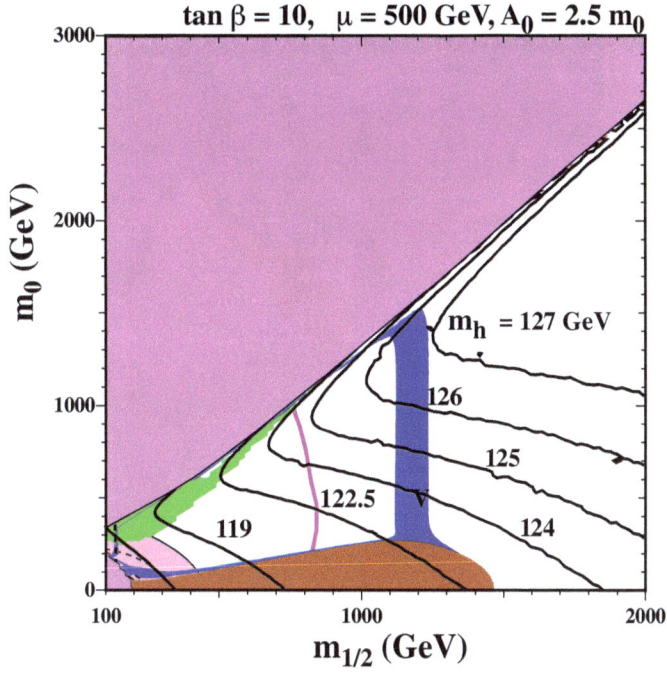

Fig. 13.21. Same as in Fig. 13.20, left, but for model [166] with an additional parameter in the Higgs sector as compared to CMSSM. The magenta region in the left upper corner is ruled out by the requirement of spontaneous breaking of the electroweak symmetry. The neutralino is dark matter particle in the blue region.

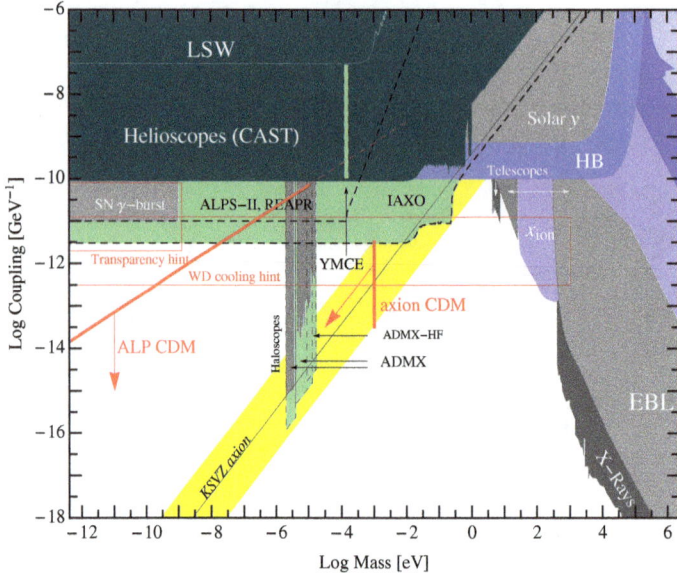

Fig. 13.22. Experimental constraints on the parameters of models with dark matter axions [170]. The inclined line "KSVZ axion" is the prediction of the KSVZ model; the yellow band around it shown predictions of the majority of axion models.

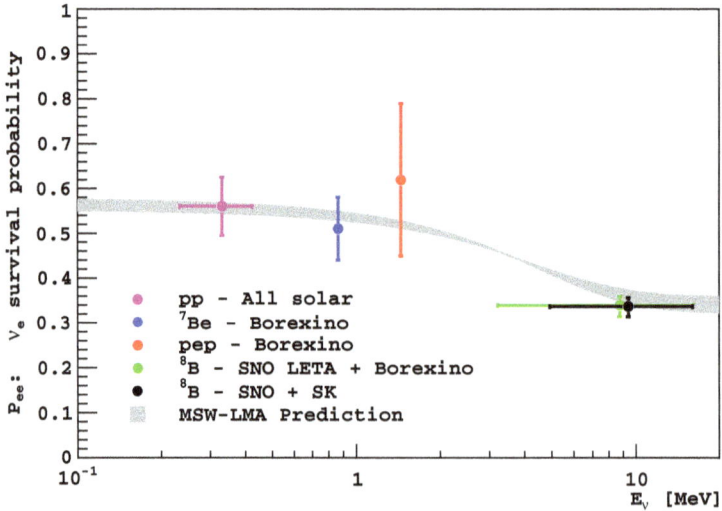

Fig. 13.23. Electron neutrino survival probability as function of energy [297].

Appendix A

Elements of General Relativity

A.1. Tensors in Curved Space–Time

In this Appendix we introduce the basic concepts of General Relativity. Our presentation does not pretend to be mathematically rigorous and comprehensive; its main purpose is to introduce the notions used in the main text and to gather in a single place a number of useful relations and formulas. A more systematic treatment of General Relativity (GR) is given in the textbooks [83, 264, 265]. Readers, interested in a more mathematically rigorous description of differential geometry, are invited to consult the book [266]. Conventions and notations used in this book are collected in Sec. A.11.

The main object of study in GR is curved 4-dimensional space (manifold) $\mathcal{M}$ describing space–time. For better understanding of abstract mathematical notions introduced below, it is sometimes helpful to think of this space as embedded into a flat enveloping space (ambient manifold) of higher dimension. We emphasize, however, that our space–time is embedded nowhere,[1] and we never rely on the possibility of such an embedding. It is worth noting here that all definitions and facts contained in this Appendix are straightforwardly generalized to (pseudo-)Riemannian spaces of arbitrary dimension.

The interval (square of the invariant distance) ds^2 between two nearby points in space–time is[2]

$$ds^2 = g_{\mu\nu}(x)dx^\mu dx^\nu, \tag{A.1}$$

where indices μ, ν take values $0, 1, 2, 3$, and metric $g_{\mu\nu}(x)$ can be considered as a 4×4 symmetric matrix. Thus, the metric is determined by ten independent functions of the coordinates $g_{\mu\nu}(x)$, $\mu \leq \nu$. Hereafter we assume that the metric has the signature $(+, -, -, -)$, i.e., that the matrix $g_{\mu\nu}(x)$ has one positive and three negative eigenvalues at each point x. Vectors dx^μ of positive, zero and negative values of ds^2 correspond to time-like, light-like (null) and space-like directions, respectively.

---

[1] We are not discussing here models with extra spatial dimensions.
[2] In what follows, unless otherwise stated, summation over repeated indices is implied.

The basic principle of GR is that all choices of the local coordinate frame are equivalent to each other. It is therefore natural to consider the functions (fields) on the manifold $\mathcal{M}$, which transform in a certain way under coordinate transformation,

$$x^\mu \to x'^\mu(x^\mu). \tag{A.2}$$

The simplest example of such an object is scalar field $\phi(x)$, whose defining property is that it transforms as

$$\phi'(x') = \phi(x).$$

Hereafter all quantities with prime belong to the new coordinate frame, while quantities without prime refer to the original one. The above relation shows that the value of the field at a given point of the manifold does not change under coordinate transformation. Another important example of an object, "well"-transforming under coordinate transformations, is a *contravariant* vector: a set of four functions $A^\mu(x)$, transforming in the same way as small increments of coordinates dx^μ, i.e.,

$$A'^\nu(x') = \frac{\partial x'^\nu}{\partial x^\mu} A^\mu(x). \tag{A.3}$$

The *covariant* vector is a set of four variables $A_\mu(x)$ which transform in the same way as the derivatives $\frac{\partial}{\partial x^\mu}$, i.e.,

$$A'_\nu(x') = \frac{\partial x^\mu}{\partial x'^\nu} A_\mu(x). \tag{A.4}$$

Using the transformation laws (A.3) and (A.4), it is easy to obtain the transformation law of the contraction $A^\mu B_\mu$,

$$
\begin{aligned}
A'^\mu(x') B'_\mu(x') &= \frac{\partial x'^\mu}{\partial x^\nu} A^\nu(x) \frac{\partial x^\lambda}{\partial x'^\mu} B_\lambda(x) \\
&= \frac{\partial x^\lambda}{\partial x^\nu} A^\nu(x) B_\lambda(x) \\
&= A^\nu(x) B_\nu(x).
\end{aligned}
\tag{A.5}
$$

We see that this contraction transforms as a scalar, i.e., its value at each point does not depend on the choice of coordinates.

From the point of view of geometry, contravariant vector $A^\mu(x)$ can be thought of as a tangent vector to the surface $\mathcal{M}$, if the latter is embedded into some enveloping space. For example, a derivative of a scalar function $\phi(x)$ along the direction defined by the tangent vector $A^\mu(x)$ has the form

$$\partial_A \phi(x) = A^\mu(x) \partial_\mu \phi(x). \tag{A.6}$$

Invariance of contraction (A.5) under coordinate transformations shows that covariant vectors $B_\mu(x)$ can be regarded as linear functionals that map, via contraction, the tangent space to real numbers.

Similarly, we can define tensor with an arbitrary number of upper and lower indices. Such an object transforms in the same way as the product of an appropriate number of covariant and contravariant vectors. For example, tensor $B^{\mu}_{\nu\lambda}$ transforms under coordinate transformations as follows,

$$B'^{\mu}_{\nu\lambda}(x') = \frac{\partial x'^{\mu}}{\partial x^{\sigma}} \frac{\partial x^{\tau}}{\partial x'^{\nu}} \frac{\partial x^{\rho}}{\partial x'^{\lambda}} B^{\sigma}_{\tau\rho}(x).$$

Directly generalizing the above result for contraction of covariant and contravariant vectors, it is straightforward to prove that by contracting an upper and a lower indices in tensor of arbitrary rank we again obtain tensor.

From the fact that the interval ds^2 defines the distance between two points, which is independent of the choice of coordinate frame, it follows that metric $g_{\mu\nu}(x)$ is a covariant tensor of the second rank, i.e., it transforms as

$$g'_{\mu\nu}(x') = \frac{\partial x^{\lambda}}{\partial x'^{\mu}} \frac{\partial x^{\rho}}{\partial x'^{\nu}} g_{\lambda\rho}(x). \tag{A.7}$$

Problem A.1. *Prove the transformation law (A.7).*

Another important example of second rank tensor is the Kronecker δ-symbol δ^{ν}_{μ} defined in an arbitrary coordinate frame as unit diagonal matrix,

$$\delta^{\nu}_{\mu} = \mathrm{diag}(1, 1, 1, 1).$$

Let us check that this definition of δ^{ν}_{μ} is consistent with the tensor transformation law. If a tensor is equal to δ^{ν}_{μ} in an original coordinate frame, then in a new coordinate frame it is equal to

$$\frac{\partial x'^{\mu}}{\partial x^{\lambda}} \frac{\partial x^{\rho}}{\partial x'^{\nu}} \delta^{\lambda}_{\rho} = \frac{\partial x'^{\mu}}{\partial x^{\lambda}} \frac{\partial x^{\lambda}}{\partial x'^{\nu}}.$$

The right-hand side here is again the Kronecker symbol δ^{μ}_{ν}, so that δ^{μ}_{ν} is a tensor indeed. Starting from the metric tensor $g_{\mu\nu}$ and the Kronecker tensor δ^{μ}_{ν} one can define a new contravariant second rank symmetric tensor $g^{\mu\nu}$ by the following equality,

$$g^{\mu\nu} g_{\nu\lambda} = \delta^{\mu}_{\lambda}. \tag{A.8}$$

In other words, the matrix $g^{\mu\nu}$ is inverse to the matrix $g_{\mu\nu}$.

Problem A.2. *Prove that $g^{\mu\nu}$ is indeed a tensor.*

Making convolutions with tensors $g_{\mu\nu}$ and $g^{\mu\nu}$, we can define the operations of raising and lowering of indices. For example,

$$A^{\nu} = g^{\nu\mu} A_{\mu}, \quad B_{\mu\nu} = g_{\mu\lambda} g_{\nu\rho} B^{\lambda\rho}.$$

If A_{μ} and $B^{\lambda\rho}$ are tensors, then A^{ν} and $B_{\mu\nu}$ are tensors too.

Another important object needed to construct the action functional in GR is the determinant of the metric tensor,

$$g \equiv \det g_{\mu\nu}.$$

In order to determine how g transforms under coordinate transformations, we write the transformation law (A.7) in the matrix form:

$$\hat{g}'(x') = \hat{J}\hat{g}(x)\hat{J}^T. \tag{A.9}$$

Hats indicate that all objects in Eq. (A.9) are matrices of size 4×4. The symbol $\hat{J}$ denotes the Jacobian matrix corresponding to the change of coordinates (A.2),

$$J^{\mu}_{\nu} = \frac{\partial x^{\mu}}{\partial x'^{\nu}},$$

while $\hat{J}^T$ is the transposed matrix. Equality (A.9) implies the following transformation law for g,

$$g'(x') = J^2 g(x), \tag{A.10}$$

where J is the Jacobian determinant of the coordinate transformation (A.2),

$$J \equiv \det\left(\frac{\partial x^{\mu}}{\partial x'^{\nu}}\right).$$

It follows from the transformation law (A.10) that the product

$$\sqrt{-g}d^4x$$

defines the invariant 4-volume element. Since the matrix $g_{\mu\nu}$ has three negative and one positive eigenvalues, the determinant g is negative. Therefore, the quantity $\sqrt{-g}$ is real.

In Minkowski space, in addition to the Kronecker delta, there is one more tensor which is invariant under Lorentz transformations. It is the Levi-Civita symbol $\epsilon^{\mu\nu\lambda\rho}$. We are reminded that $\epsilon^{\mu\nu\lambda\rho}$ is totally antisymmetric in its indices and, consequently, it is uniquely defined by the condition

$$\epsilon^{0123} = 1.$$

However, the Levi-Civita symbol is not invariant under arbitrary coordinate transformations. Indeed, if a tensor is equal to $\epsilon^{\mu\nu\lambda\rho}$ in a given coordinate frame, then in another frame it is equal to

$$\epsilon'^{\mu\nu\lambda\rho} = \frac{\partial x'^{\mu}}{\partial x^{\alpha}}\frac{\partial x'^{\nu}}{\partial x^{\beta}}\frac{\partial x'^{\lambda}}{\partial x^{\gamma}}\frac{\partial x'^{\rho}}{\partial x^{\delta}}\epsilon^{\alpha\beta\gamma\delta} = J^{-1}\epsilon^{\mu\nu\lambda\rho}. \tag{A.11}$$

The transformation law (A.11) shows that the natural generalization of the Levi-Civita symbol to the case of arbitrary curvilinear coordinates and to curved space is the Levi-Civita tensor[3]

$$E^{\mu\nu\lambda\rho} = \frac{1}{\sqrt{-g}}\epsilon^{\mu\nu\lambda\rho}.$$

This tensor is completely antisymmetric in all its indices and reduces to $\epsilon^{\mu\nu\lambda\rho}$ when metric $g_{\mu\nu}$ coincides with the metric of Minkowski space in Cartesian coordinates, $\eta_{\mu\nu} = \text{diag}(1, -1, -1, -1)$.

A.2. Covariant Derivative

In order to build the action invariant under arbitrary coordinate transformations, we have to define the covariant differentiation operator ∇_μ that converts tensors into tensors. For scalar field, it is natural to demand that this operation coincides with the usual differentiation,

$$\nabla_\mu \phi(x) \equiv \partial_\mu \phi(x). \tag{A.12}$$

It follows from the definition (A.4), that the derivative $\nabla_\mu \phi$ is covariant vector.

One cannot define the covariant derivative of vector field $A^\mu(x)$ in the same way. In order to differentiate vector field, one must learn how to subtract tangent vectors belonging to different points of the space $\mathcal{M}$. Therefore, we have to define the rule of parallel transport of vectors from one point to another.

Consider the parallel transport of contravariant vector A^μ from the point with coordinates x^μ to the point with coordinates

$$\tilde{x}^\mu = x^\mu + dx^\mu.$$

(See Fig. A.1.) Imposing the natural requirement of linearity (the sum of two vectors upon parallel transport becomes the sum of images), we observe that to the leading order in the increments of coordinates dx^μ, the image $\tilde{A}^\mu$ of the vector A^μ has the

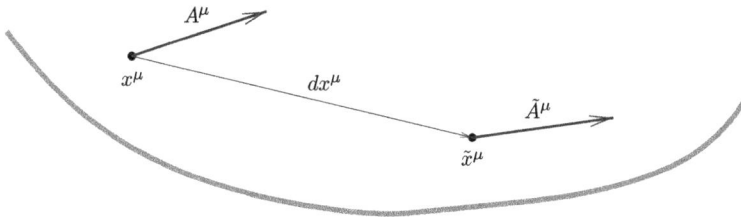

Fig. A.1. Parallel transport of a vector.

---

[3] More accurately, $E^{\mu\nu\lambda\rho}$ is pseudotensor because it transforms with "wrong" sign under the coordinate transformations which change the spatial orientation (i.e., transformations with $J < 0$).

following general form

$$\tilde{A}^\mu(\tilde{x}) = A^\mu(x) - \Gamma^\mu_{\nu\lambda}(x)A^\nu(x)dx^\lambda. \tag{A.13}$$

The quantities $\Gamma^\mu_{\nu\lambda}$ entering here are called connection coefficients. To determine the transformation law of the connection coefficients under arbitrary coordinate transformations, we perform coordinate transformation in both parts of equality (A.13) and make use of the fact that the quantities $\tilde{A}^\mu$, A^μ and dx^μ transform according to the law (A.3). The left-hand side of (A.13) transforms into

$$\tilde{A}'^\mu(\tilde{x}') = \frac{\partial x'^\mu(\tilde{x})}{\partial x^\nu}\tilde{A}^\nu(\tilde{x}) = \left(\frac{\partial x'^\mu(x)}{\partial x^\nu} + \frac{\partial^2 x'^\mu(x)}{\partial x^\nu \partial x^\lambda}dx^\lambda\right)\tilde{A}^\nu(\tilde{x})$$

$$= \frac{\partial x'^\mu(x)}{\partial x^\nu}\tilde{A}^\nu(\tilde{x}) + \frac{\partial^2 x'^\mu(x)}{\partial x^\nu \partial x^\lambda}dx^\lambda A^\nu(x), \tag{A.14}$$

where we work to the linear order in dx^μ. Upon coordinate transformation, the right-hand side of (A.13) takes the following form,

$$A'^\mu(x') - \Gamma'^\mu_{\nu\lambda}(x')A'^\nu(x')dx'^\lambda = \frac{\partial x'^\mu}{\partial x^\nu}A^\nu(x) - \Gamma'^\mu_{\nu\lambda}(x')\frac{\partial x'^\nu}{\partial x^\rho}A^\rho(x)\frac{\partial x'^\lambda}{\partial x^\sigma}dx^\sigma. \tag{A.15}$$

Equating the results of the transformations (A.14) and (A.15), convoluting both sides of the resulting equality with $\frac{\partial x^\mu}{\partial x'^\nu}$, and comparing the result with the original rule of parallel transport (A.13), we obtain the following transformation law for the connection coefficients,

$$\Gamma'^\mu_{\nu\lambda}(x') = \frac{\partial x^\rho}{\partial x'^\nu}\frac{\partial x^\sigma}{\partial x'^\lambda}\frac{\partial x'^\mu}{\partial x^\xi}\Gamma^\xi_{\rho\sigma} + \frac{\partial x'^\mu}{\partial x^\rho}\frac{\partial^2 x^\rho}{\partial x'^\nu \partial x'^\lambda}. \tag{A.16}$$

The second term on the right-hand side of (A.16) shows that connection is not tensor.

To define the covariant derivative of vector field we transport the vector $A^\mu(x)$ to the point $\tilde{x} = x + dx$, subtract the resulting vector from the value of the vector field at the point $\tilde{x}$ and write

$$A^\mu(\tilde{x}) - \tilde{A}^\mu(\tilde{x}) = \nabla_\nu A^\mu \cdot dx^\nu.$$

Using the parallel transport rule (A.13), we arrive at the following definition of the covariant derivative of vector field $A^\mu(x)$,

$$\nabla_\nu A^\mu(x) = \partial_\nu A^\mu(x) + \Gamma^\mu_{\lambda\nu}A^\lambda(x). \tag{A.17}$$

The transformation law of the connection coefficients (A.16) guarantees that $\nabla_\nu A^\mu$ is a second rank tensor with one covariant and one contravariant indices.

The parallel transport rule of covariant vector B_μ follows from the fact that the contraction $A^\mu B_\mu$ is a scalar transported in a trivial way,

$$(\tilde{A}^\mu \tilde{B}_\mu)(\tilde{x}) = (A^\mu B_\mu)(x). \tag{A.18}$$

The relation (A.18) and parallel transport law of contravariant vector (A.13) give the following parallel transport rule for covariant vector B_μ,

$$\tilde{B}_\mu(\tilde{x}) = B_\mu(x) + \Gamma^\nu_{\mu\lambda} B_\nu(x) dx^\lambda. \tag{A.19}$$

Consequently, the covariant derivative has the form

$$\nabla_\nu B_\mu(x) = \partial_\nu B_\mu(x) - \Gamma^\lambda_{\mu\nu} B_\lambda(x). \tag{A.20}$$

Now that we have defined the covariant derivatives of scalar and vectors of both types, it is not difficult to generalize these definitions to tensors of arbitrary rank. This is done using the Leibniz rule,

$$\nabla_\mu(AB) = (\nabla_\mu A)B + A\nabla_\mu B,$$

where A and B are two arbitrary tensors whose indices are not explicitly written. For example, the covariant derivative of third rank tensor with one upper and two lower indices reads

$$\nabla_\mu B^\nu_{\lambda\tau} = \partial_\mu B^\nu_{\lambda\tau} + \Gamma^\nu_{\rho\mu} B^\rho_{\lambda\tau} - \Gamma^\rho_{\lambda\mu} B^\nu_{\rho\tau} - \Gamma^\rho_{\tau\mu} B^\nu_{\lambda\rho}.$$

In principle, one could consider manifolds with arbitrary set of connection coefficients transforming according to the law (A.16). However, GR is based on the (pseudo-)Riemannian geometry,[4] in which additional conditions are imposed on the connection coefficients $\Gamma^\mu_{\nu\lambda}$. The first of these conditions is that the operation of parallel transport (or, equivalently, the operation of covariant differentiation) commutes with the operation of raising and lowering of indices. This means, in particular, that

$$g_{\mu\nu}\nabla_\lambda A^\nu = \nabla_\lambda(g_{\mu\nu} A^\nu)$$

for an arbitrary vector A^ν. The Leibniz rule implies that this is possible only if the metric tensor $g_{\mu\nu}$ is covariantly constant,

$$\nabla_\mu g_{\nu\lambda} = 0. \tag{A.21}$$

More explicitly, this condition reads

$$\partial_\mu g_{\nu\lambda} = \Gamma^\rho_{\nu\mu} g_{\rho\lambda} + \Gamma^\rho_{\lambda\mu} g_{\nu\rho}.$$

Connections satisfying the condition (A.21) are called *metric connections* (since they are consistent with metric). The second condition imposed on the connection

---

[4]Pseudo-Riemannian geometry differs from the Riemannian one by the signature of metric, which for the Riemannian geometry is Euclidean. We often do not pay attention to this terminological subtlety.

coefficients is the requirement that they are symmetric in lower indices,

$$C^\lambda_{\mu\nu} \equiv \Gamma^\lambda_{\mu\nu} - \Gamma^\lambda_{\nu\mu} = 0. \tag{A.22}$$

The transformation law of the connection (A.16) implies that $C^\lambda_{\mu\nu}$ is a tensor (called torsion tensor in the general case). Hence, the validity of (A.22) does not depend on the choice of coordinate frame. Manifold equipped with metric and torsionless metric connection is precisely what is called Riemannian manifold, and in this case the connection coefficients are called Christoffel symbols.

Equations (A.21) and (A.22) enable us to unambiguously express the Christoffel symbols in terms of the metric tensor,

$$\Gamma^\mu_{\nu\lambda} = \frac{1}{2} g^{\mu\rho} (\partial_\nu g_{\rho\lambda} + \partial_\lambda g_{\rho\nu} - \partial_\rho g_{\nu\lambda}). \tag{A.23}$$

We always assume in what follows that the equality (A.23) holds.[5]

Problem A.3. *Consider a 2-dimensional surface Σ embedded in 3-dimensional Euclidean space R^3. The space R^3 induces metric in Σ: if y^i ($i = 1, 2$) are coordinates on the surface Σ, then the square of the distance between nearby points on Σ can be written as*

$$ds^2 = g_{ij} dy^i dy^j,$$

where the metric $g_{ij}(y)$ is uniquely determined by the requirement that ds be the distance in R^3. There is an obvious definition of tangent plane at each point on the surface Σ; contravariant vectors, as discussed above, are vectors belonging to the tangent plane. Their components $A^i(y)$ in the chosen coordinate frame on Σ can, for instance, be defined by the relation

$$\partial_A \phi(y) = A^i(y) \frac{\partial \phi}{\partial y^i},$$

where $\phi(y)$ is a function on the surface Σ, and $\partial_A \phi$ is its derivative along the direction determined by the vector $\vec{A}$. Parallel transport of tangent vector along the surface Σ is naturally defined in the following manner (see Fig. A.2): first we transport the vector $\vec{A}$ from point y to point $\tilde{y}$ as a vector in R^3 (yielding the vector $\vec{A}_{||}$ in Fig. A.2), and then we project it onto the tangent plane at point $\tilde{y}$. Let the surface Σ be (locally) defined by the equations

$$x^\alpha = f^\alpha(y^1, y^2), \quad \alpha = 1, 2, 3,$$

where x^α are coordinates in R^3.

(1) Calculate the components of the metric $g_{ij}(y)$.

---

[5] In geometries more general than Riemannian, objects given by (A.23) are also called Christoffel symbols. Another term used for them is Riemannian connection.

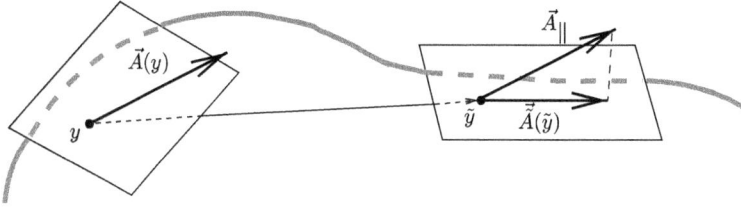

Fig. A.2.   Parallel transport of a tangent vector.

(2) *Calculate the Christoffel symbols $\Gamma_{ij}^{k}(y)$ on the surface Σ, which are associated with the above operation of parallel transport.*
(3) *Show that the properties (A.21) and (A.22) are satisfied, i.e., the geometry on the surface Σ is Riemannian.*
(4) *Propose a generalization of the parallel transport of a vector, such that the torsion tensor (A.22) is nonzero. Demonstrate this property by explicit calculation of the connection coefficients. Does the relation (A.21) remain valid?*

Problem A.4. *Derive the formula (A.23) from Eqs. (A.21) and (A.22).*

Problem A.5. *Check the following properties of the Christoffel symbols and covariant derivative:*

$$\Gamma_{\nu\mu}^{\mu} = \partial_\nu \ln \sqrt{-g}, \tag{A.24}$$

$$g^{\mu\nu}\Gamma_{\mu\nu}^{\lambda} = -\frac{1}{\sqrt{-g}}\partial_\mu(\sqrt{-g}g^{\lambda\mu}), \tag{A.25}$$

$$\nabla_\mu A^\mu = \frac{1}{\sqrt{-g}}\partial_\mu(\sqrt{-g}A^\mu), \tag{A.26}$$

for antisymmetric tensor $A^{\mu\nu}$:

$$\nabla_\mu A^{\mu\nu} = \frac{1}{\sqrt{-g}}\partial_\mu(\sqrt{-g}A^{\mu\nu}), \tag{A.27}$$

for scalar ϕ :

$$\nabla_\mu\nabla^\mu\phi = \frac{1}{\sqrt{-g}}\partial_\mu(\sqrt{-g}g^{\mu\nu}\partial_\nu\phi), \tag{A.28}$$

where $\nabla^\mu\phi \equiv g^{\mu\nu}\nabla_\nu\phi$.

The property (A.26) leads to the following generalization of the Gauss formula,

$$\int(\nabla_\nu A^\nu)\sqrt{-g}d^4x = \int \partial_\nu(\sqrt{-g}A^\nu)d^4x = \int \sqrt{-g}A^\nu d\Sigma_\nu,$$

where $d\Sigma_\nu$ is the element of surface which bounds the integration region. Together with the Leibniz rule for the covariant derivatives, this formula allows for integration by parts of invariant integrals. For example

$$\int A_\mu \nabla_\nu B^{\mu\nu} \sqrt{-g} d^4x = -\int (\nabla_\nu A_\mu) B^{\mu\nu} \sqrt{-g} d^4x + \text{surface terms}.$$

To conclude this Section, we note the following fact. By choosing a suitable coordinate frame, we can locally, at a given point, set all Christoffel symbols equal to zero; this is fully consistent with the equivalence principle because this enables us to switch off locally the gravitational field.[6] In this coordinate frame all covariant derivatives coincide with usual ones, and all first derivatives of the metric tensor vanish (by virtue of (A.21)). The transformation to such a frame at a chosen point, which we place at the origin, is

$$x^\mu \to x'^\mu = x^\mu + \frac{1}{2}\Gamma^\mu_{\nu\lambda}(0)x^\nu x^\lambda, \tag{A.29}$$

where $\Gamma^\mu_{\nu\lambda}(0)$ are the values of the Christoffel symbols in coordinates x at the origin. Using the relation (A.16), it is straightforward to see that all Christoffel symbols indeed vanish at the origin in the new frame. Note that the key role here is played by the symmetry of the Christoffel symbols in the lower indices, formula (A.22).

Since the transformation (A.29) is identity at the origin, there is still freedom in the metric tensor. This freedom can be used to reduce the metric tensor at the origin to the Minkowski tensor. To this end, one simply chooses $x^\mu = J^\mu_\nu x'^\nu$, where J^μ_ν does not depend on coordinates. In matrix notations we then have the relation (A.9). The matrix $g_{\mu\nu}$ can be cast into diagonal form by orthogonal transformation, and then converted to the Minkowski tensor by rescaling of coordinates. The resulting coordinate frame has thus the properties

$$g_{\mu\nu}(0) = \eta_{\mu\nu}, \quad \Gamma^\mu_{\nu\lambda}(0) = 0.$$

It is called the *locally-Lorentz frame*.

A.3.   Riemann Tensor

It is seen from formula (A.23) that the Christoffel symbols differ from zero if metric non-trivially depends on the coordinates x^μ. It should be understood, however, that the deviation of $\Gamma^\lambda_{\mu\nu}$ from zero does not imply that the space is not flat. Since the quantities $\Gamma^\lambda_{\mu\nu}$ do not form a tensor, they can be identically equal to zero in one coordinate frame and differ from zero in another frame.

Problem A.6. *Find the Christoffel symbols in polar coordinates on 2-dimensional plane and in spherical coordinates in 3-dimensional Euclidean space.*

---

[6] In fact, the stronger statement is valid: it is possible to make all Christoffel symbols equal to zero along any predetermined world line.

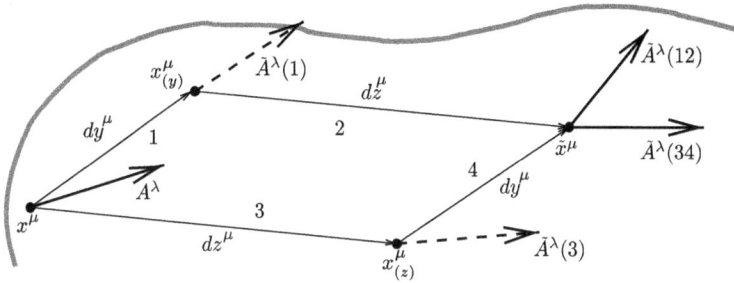

Fig. A.3.   Parallel transport of a vector along different trajectories (12) and (34).

The quantity which characterizes the geometry of manifold, rather than the choice of coordinate frame, is the Riemann tensor (curvature tensor) $R^{\mu}_{\nu\lambda\rho}$. The Riemann tensor determines how the commutator of covariant derivatives acts on tensors. For example, for an arbitrary contravariant vector A^{λ} we have

$$\nabla_{\mu}\nabla_{\nu}A^{\lambda} - \nabla_{\nu}\nabla_{\mu}A^{\lambda} = A^{\sigma}R^{\lambda}_{\sigma\mu\nu}. \tag{A.30}$$

Problem A.7. *Check that the equality* (A.30) *indeed defines the tensor* $R^{\mu}_{\nu\lambda\rho}$. *In particular, check that all terms with derivatives of* A^{λ}, *which could appear in the left-hand side of this equality, cancel out.*

Explicit expression for the Riemann tensor is

$$R^{\mu}_{\nu\lambda\rho} = \partial_{\lambda}\Gamma^{\mu}_{\nu\rho} - \partial_{\rho}\Gamma^{\mu}_{\nu\lambda} + \Gamma^{\mu}_{\sigma\lambda}\Gamma^{\sigma}_{\nu\rho} - \Gamma^{\mu}_{\sigma\rho}\Gamma^{\sigma}_{\nu\lambda}. \tag{A.31}$$

In order to better understand the geometric meaning of the Riemann tensor, let us consider parallel transport of vector A^{λ} from the point x with coordinates x^{μ} to the point $\tilde{x}$ with coordinates

$$\tilde{x}^{\mu} = x^{\mu} + dy^{\mu} + dz^{\mu},$$

where directions of vectors dy^{μ} and dz^{μ} do not coincide (see Fig. A.3). This parallel transport can be done in two different ways: (i) one first moves the vector A^{λ} along the path 1 to point y with coordinates

$$x^{\mu}_{(y)} = x^{\mu} + dy^{\mu},$$

and then proceeds along the path 2 to point $\tilde{x}$; (ii) one does the opposite, namely, first performs parallel transport along the path 3 to the point z with coordinates

$$x^{\mu}_{(z)} = x^{\mu} + dz^{\mu},$$

and then moves along the path 4 to point $\tilde{x}$. Of course, in flat space the result of parallel transport does not depend on the choice of path. In the case of curved space this is, generally speaking, incorrect. Using the parallel transport rule (A.13), one can see immediately that the result of the transport does not depend on the path to the linear order in increments of coordinates. However, in the quadratic order we obtain

$$\tilde{A}^\lambda(12) - \tilde{A}^\lambda(34) = A^\sigma R^\lambda_{\sigma\mu\nu} dz^\mu dy^\nu, \tag{A.32}$$

where $\tilde{A}^\lambda(12)$ and $\tilde{A}^\lambda(34)$ are images of the vector A^λ under parallel transport along paths (12) and (34), respectively. Thus, the tensor $R^\mu_{\nu\lambda\rho}$ determines the dependence of the parallel transport on the path along which it is done. Consequently, the Riemann tensor is indeed a non-trivial characteristic of the curvature.

Problem A.8. *Obtain equality (A.32). In particular, check that the second order terms in dx^μ, omitted in the parallel transport law (A.13), do not contribute to the difference $(\tilde{A}^\lambda(12) - \tilde{A}^\lambda(34))$ to the quadratic order.*

Problem A.9. *Using the Christoffel symbols found in Problem A.6, check by explicit calculation that all components of the Riemann tensor are zero in polar coordinates on plane and in spherical coordinates in 3-dimensional Euclidean space.*

The above analysis is valid, with minimal changes, for covariant vector A_μ. The analog of Eq. (A.30) in that case is

$$\nabla_\mu \nabla_\nu A_\lambda - \nabla_\nu \nabla_\mu A_\lambda = -A_\sigma R^\sigma_{\lambda\mu\nu}. \tag{A.33}$$

The action of the commutator of covariant derivatives

$$[\nabla_\mu, \nabla_\nu] \equiv \nabla_\mu \nabla_\nu - \nabla_\nu \nabla_\mu$$

on tensor of arbitrary rank follows from the fact that the operator $[\nabla_\mu, \nabla_\nu]$ obeys the Leibniz rule. For example,

$$[\nabla_\mu, \nabla_\nu] A^\rho_\lambda = R^\rho_{\sigma\mu\nu} A^\sigma_\lambda - R^\sigma_{\lambda\mu\nu} A^\rho_\sigma. \tag{A.34}$$

Problem A.10. *Prove the Leibnitz rule for commutatior $[\nabla_\mu, \nabla_\nu]$.*

Let us list some important properties of the Riemann tensor.

(1) Tensor

$$R_{\mu\nu\lambda\rho} \equiv g_{\mu\sigma} R^\sigma_{\nu\lambda\rho}$$

is antisymmetric in the first and second pair of indices.
(2) Tensor $R_{\mu\nu\lambda\rho}$ is symmetric under permutations of pairs of indices $(\mu\nu) \leftrightarrow (\lambda\rho)$.
(3) For any three indices, the sum of three components of the tensor $R_{\mu\nu\lambda\rho}$ with cyclic permutation of these indices is zero. For instance,

$$R_{\rho\mu\nu\lambda} + R_{\rho\lambda\mu\nu} + R_{\rho\nu\lambda\mu} = 0. \tag{A.35}$$

(4) The Bianchi identity holds:

$$\nabla_\rho R^\lambda_{\sigma\mu\nu} + \nabla_\nu R^\lambda_{\sigma\rho\mu} + \nabla_\mu R^\lambda_{\sigma\nu\rho} = 0. \tag{A.36}$$

Problem A.11. *Using the explicit expression (A.31) for the Riemann tensor, prove the properties (1) and (2).*

Problem A.12. *Using properties (1), (2), (3) determine the number of independent components of the Riemann tensor at each point in a space of dimension $D = 2, 3, 4$.*

Proving the properties (3) and (4) by using the explicit formula (A.31) would be too cumbersome. Instead, it is convenient to make use of the definition (A.30). Namely, let us write the following equality (Jacobi identity), valid for arbitrary operators,

$$[A, [B, C]] + [C, [A, B]] + [B, [C, A]] = 0.$$

Problem A.13. *Prove the Jacobi identity.*

Let us choose the operators A, B, C as covariant derivatives. Let us then act by the Jacobi identity on arbitrary scalar ϕ first,

$$[\nabla_\rho, [\nabla_\mu, \nabla_\nu]]\phi + [\nabla_\mu, [\nabla_\nu, \nabla_\rho]]\phi + [\nabla_\nu, [\nabla_\rho, \nabla_\mu]]\phi = 0. \tag{A.37}$$

Let us write for the first term

$$[\nabla_\rho, [\nabla_\mu, \nabla_\nu]]\phi = \nabla_\rho[\nabla_\mu, \nabla_\nu]\phi - [\nabla_\mu, \nabla_\nu]\nabla_\rho\phi = -[\nabla_\mu, \nabla_\nu]\nabla_\rho\phi = \partial_\sigma\phi R^\sigma_{\rho\mu\nu},$$

and similarly for the other terms. Here we first used the fact that $[\nabla_\mu, \nabla_\nu]\phi = \partial_\mu\partial_\nu\phi - \Gamma^\lambda_{\mu\nu}\partial_\lambda\phi$ is symmetric in μ, ν, and then utilized (A.33). In this way we obtain from the identity (A.37) that

$$(R^\sigma_{\rho\mu\nu} + R^\sigma_{\mu\nu\rho} + R^\sigma_{\nu\rho\mu})\partial_\sigma\phi = 0,$$

which implies (A.35) by virtue of the arbitrariness of $\partial_\sigma\phi$.

Let us now act by the Jacobi identity on arbitrary vector A^λ and get

$$[\nabla_\rho, [\nabla_\mu, \nabla_\nu]]A^\lambda + [\nabla_\mu, [\nabla_\nu, \nabla_\rho]]A^\lambda + [\nabla_\nu, [\nabla_\rho, \nabla_\mu]]A^\lambda = 0. \tag{A.38}$$

Then, using the definition (A.30), we have

$$[\nabla_\rho, [\nabla_\mu, \nabla_\nu]]A^\lambda = \nabla_\rho(R^\lambda_{\sigma\mu\nu}A^\sigma) - [\nabla_\mu, \nabla_\nu](\nabla_\rho A^\lambda). \tag{A.39}$$

Evaluating the right-hand side of Eq. (A.39) and using the Leibniz rule, we find

$$[\nabla_\rho, [\nabla_\mu, \nabla_\nu]]A^\lambda = \nabla_\rho R^\lambda_{\sigma\mu\nu}A^\sigma + R^\lambda_{\sigma\mu\nu}\nabla_\rho A^\sigma - R^\lambda_{\sigma\mu\nu}\nabla_\rho A^\sigma + R^\sigma_{\rho\mu\nu}\nabla_\sigma A^\lambda$$

$$= \nabla_\rho R^\lambda_{\sigma\mu\nu}A^\sigma + R^\sigma_{\rho\mu\nu}\nabla_\sigma A^\lambda. \tag{A.40}$$

Upon substituting the expression (A.40) and similar expressions for the other two double commutators to the Jacobi identity (A.38) we obtain, taking into account the property (3) of the Riemann tensor,

$$(\nabla_\rho R^\lambda_{\sigma\mu\nu} + \nabla_\nu R^\lambda_{\sigma\rho\mu} + \nabla_\mu R^\lambda_{\sigma\nu\rho})A^\sigma = 0.$$

Consequently, the Bianchi identity is indeed satisfied.

By contracting indices of the Riemann tensor $R_{\mu\nu\lambda\rho}$ one can construct a new tensor with smaller number of indices, still characterizing the curvature of space. The above symmetry properties of the Riemann tensor guarantee that the contraction of any two of its indices results in either zero or the following symmetric tensor of the second rank,

$$R_{\mu\nu} \equiv R^\lambda_{\mu\lambda\nu},$$

called the Ricci tensor. In many places of this book we need the explicit form of this tensor,

$$R_{\mu\nu} = \partial_\lambda \Gamma^\lambda_{\mu\nu} - \partial_\mu \Gamma^\lambda_{\lambda\nu} + \Gamma^\lambda_{\rho\lambda} \Gamma^\rho_{\mu\nu} - \Gamma^\lambda_{\rho\mu} \Gamma^\rho_{\nu\lambda}. \tag{A.41}$$

By contracting the indices of the Ricci tensor we obtain the scalar curvature,

$$R \equiv g^{\mu\nu} R_{\mu\nu}.$$

Problem A.14. *Find the metric, Christoffel symbols, Riemann and Ricci tensors and scalar curvature for 2-dimensional sphere S^2.*

Problem A.15. *Show that for an arbitrary 2-dimensional surface, the quantity $\sqrt{g}R$ is total derivative and hence the integral of the scalar curvature over the invariant volume,*

$$\frac{1}{4\pi} \int d^2 x \sqrt{g} R, \tag{A.42}$$

does not depend on the choice of metric on the surface (Gauss–Bonnet theorem). Thus, in two dimensions this integral is a characteristics of topology. Note that the scalar curvature is twice the Gaussian curvature in two dimensions. Integral (A.42) is in one-to-one correspondence with the degree of the Gauss mapping and coincides with the Euler characteristic of 2-dimensional surface. Find the value of this integral for sphere and torus.

A.4. Gravitational Field Equations

We now have at our disposal all ingredients needed for constructing General Relativity. In GR, the metric tensor is dynamical field ("gravitational field"), and the equations of GR arise as extremum conditions for the action functional. As we have already mentioned, one of the basic principles of GR is that all local coordinate frames are equivalent. This means that the equations for the gravitational field $g_{\mu\nu}$ should be written in terms of covariant quantities and their form should not depend on the choice of coordinates. To fulfill this requirement, the gravitational field action

S_{gr} must be a scalar, i.e., it must be written as an integral of a scalar Lagrangian $\mathcal{L}_{gr}$ over the invariant 4-volume,

$$S_{gr} = \int d^4x \sqrt{-g} \mathcal{L}_{gr}.$$

The simplest possibility is to take the Lagrangian equal to a constant $(-\Lambda)$ independent of metric,

$$S_\Lambda = -\Lambda \int d^4x \sqrt{-g}. \tag{A.43}$$

This term can indeed enter the action for the gravitational field and play an important role in cosmology. Since the action is dimensionless, the parameter Λ has dimension $(\text{mass})^4$. This parameter is called the cosmological constant or, for reasons that are explained in Chapter 3, vacuum energy density. However, the action (A.43) cannot be the complete gravitational field action. Indeed, S_Λ does not contain derivatives of the metric $g_{\mu\nu}$, and, consequently, upon its variation one would obtain purely algebraic equation that would not enable one to interpret $g_{\mu\nu}$ as truly dynamical field.

Another scalar quantity at our disposal is the scalar curvature R, or, in general, an arbitrary function $f(R)$. To understand which choice of the function $f(R)$ as the Lagrangian is most natural, we recall that the commonly used field equations are of the first or second order in derivatives. For the field equations to be second order at most, it is usually required that the Lagrangian does not contain second or higher derivatives. Indeed, consider field theory with the action of the form

$$S = \int d^4x \mathcal{L}(\phi, \partial\phi, \partial^2\phi, \ldots). \tag{A.44}$$

Here, the symbol ϕ denotes all fields of theory, and we omitted possible tensor indices. The variation of (A.44) under small variation of the fields, $\phi \to \phi + \delta\phi$, reads

$$\delta S = \int d^4x \left(\frac{\partial \mathcal{L}}{\partial \phi} \delta\phi + \frac{\partial \mathcal{L}}{\partial(\partial\phi)} \partial\delta\phi + \frac{\partial \mathcal{L}}{\partial(\partial^2\phi)} \partial^2\delta\phi + \cdots \right).$$

Assuming, as usual, that the variation of the fields $\delta\phi$ vanishes at infinity, and integrating by parts, we arrive at the field equation of the form

$$\frac{\partial \mathcal{L}}{\partial \phi} - \partial \frac{\partial \mathcal{L}}{\partial(\partial\phi)} + \partial^2 \frac{\partial \mathcal{L}}{\partial(\partial^2\phi)} + \cdots = 0, \tag{A.45}$$

which, generally speaking, contains the field derivatives of higher than second order.

The Riemann tensor (A.31), and, consequently, the scalar curvature R, contain the first derivatives of the Christoffel symbols $\Gamma^\lambda_{\mu\nu}$. The latter, in turn, contain the first derivatives of the metric tensor $g_{\mu\nu}$. Therefore, if the Lagrangian density $\mathcal{L}_{gr}$ depends non-trivially on the scalar curvature, the action necessarily involves the second derivatives. By the above argument, one could conclude that it would be impossible to write covariant action for the gravitational field, which would lead to second order equations. Note, however, that if the Lagrangian density $\mathcal{L}$ in Eq. (A.45) depends on second derivatives only via the terms of the form

$$f(\phi)\partial^2\phi \tag{A.46}$$

and does not contain higher derivatives, then the equations of motion do not involve higher derivatives. In fact, one can integrate the term (A.46) in the Lagrangian by parts and come to the Lagrangian which contains first derivatives only. It is straightforward to check that the action

$$S_{EH} = -\frac{1}{16\pi G}\int d^4x\sqrt{-g}R \tag{A.47}$$

depends on the second derivatives exactly in this way.

Problem A.16. *Using integration by parts, find the action equivalent to the action (A.47), which does not contain second derivatives. Is the Lagrangian in this action a scalar? Is the action itself a scalar?*

This action is called the Einstein–Hilbert action. As we will see later, the constant G here, which has the dimension M^{-2}, is Newton's gravity constant.

The full gravitational field action in GR is the sum of the terms (A.43) and (A.47),

$$S_{gr} = S_\Lambda + S_{EH}. \tag{A.48}$$

In order to obtain the gravitational field equations one has to calculate the variation of the action δS_{gr} under variation of the metric

$$g_{\mu\nu} \to g_{\mu\nu} + \delta g_{\mu\nu}.$$

Let us start with the first, simpler term S_Λ. To vary S_Λ, we make use of the following formula of linear algebra,

$$\det(M + \delta M) = \det(M)\left[1 + \text{Tr}(M^{-1}\delta M) + o(\delta M)\right], \tag{A.49}$$

where M is an arbitrary non-degenerate matrix.

Problem A.17. *Prove the formula (A.49).*

Applying the relation (A.49) to the determinant of the metric tensor, we obtain

$$\delta g = g g^{\mu\nu}\delta g_{\mu\nu}. \tag{A.50}$$

Using this result, we arrive at the following expression for the variation of S_Λ,

$$\delta S_\Lambda = -\Lambda \int d^4x\, \delta(\sqrt{-g}) = -\frac{\Lambda}{2}\int d^4x \sqrt{-g}\, g^{\mu\nu}\delta g_{\mu\nu}. \tag{A.51}$$

Let us now calculate the variation of the Einstein–Hilbert action S_{EH}. It can be written as a sum of the following three terms,

$$\delta S_{EH} = \delta S_1 + \delta S_2 + \delta S_3,$$

where

$$\delta S_1 = -\frac{1}{16\pi G}\int d^4x\, R\, \delta(\sqrt{-g}),$$

$$\delta S_2 = -\frac{1}{16\pi G}\int d^4x \sqrt{-g}\, R_{\mu\nu}\delta g^{\mu\nu},$$

and

$$\delta S_3 = -\frac{1}{16\pi G}\int d^4x \sqrt{-g}\, g^{\mu\nu}\delta R_{\mu\nu}. \tag{A.52}$$

Taking into account the relation (A.51), we immediately obtain the explicit expression for δS_1:

$$\delta S_1 = -\frac{1}{32\pi G}\int d^4x \sqrt{-g}\, R g^{\mu\nu}\delta g_{\mu\nu}. \tag{A.53}$$

In order to compute δS_2, we note that varying Eq. (A.8) we get

$$g_{\rho\lambda}\delta g^{\mu\rho} = -g^{\mu\rho}\delta g_{\rho\lambda}.$$

Contracting both sides of this equality with matrix $g^{\lambda\nu}$ we obtain

$$\delta g^{\mu\nu} = -g^{\mu\rho}\delta g_{\rho\lambda}g^{\lambda\nu}. \tag{A.54}$$

Consequently,

$$\delta S_2 = \frac{1}{16\pi G}\int d^4x \sqrt{-g}\, R^{\mu\nu}\delta g_{\mu\nu}. \tag{A.55}$$

It remains to find the variation δS_3, which at first glance looks the most complicated. To compute δS_3, we note that the transformation rules of the Christoffel symbols (A.16) imply that the *variation* $\delta\Gamma^\mu_{\nu\lambda}$ is a tensor. Then, using (A.31), we arrive at the following expression for the variation of the Riemann tensor:

$$\delta R^{\mu}_{\nu\lambda\rho} = \partial_\lambda \delta\Gamma^{\mu}_{\nu\rho} - \partial_\rho \delta\Gamma^{\mu}_{\nu\lambda} + \delta\Gamma^{\mu}_{\sigma\lambda}\Gamma^{\sigma}_{\nu\rho} + \Gamma^{\mu}_{\sigma\lambda}\delta\Gamma^{\sigma}_{\nu\rho} - \delta\Gamma^{\mu}_{\sigma\rho}\Gamma^{\sigma}_{\nu\lambda} - \Gamma^{\mu}_{\sigma\rho}\delta\Gamma^{\sigma}_{\nu\lambda}.$$

By direct calculation one can check that

$$\delta R^{\mu}_{\nu\lambda\rho} = \nabla_\lambda(\delta\Gamma^{\mu}_{\nu\rho}) - \nabla_\rho(\delta\Gamma^{\mu}_{\nu\lambda}), \tag{A.56}$$

where the covariant derivatives are taken with the unperturbed metric. From the equality (A.56) we get the following expression for the variation of the Ricci tensor

$$\delta R_{\mu\nu} = \nabla_\lambda(\delta\Gamma^{\lambda}_{\mu\nu}) - \nabla_\nu(\delta\Gamma^{\lambda}_{\mu\lambda}). \tag{A.57}$$

Substituting Eq. (A.57) into Eq. (A.52), we obtain

$$\delta S_3 = -\frac{1}{16\pi G} \int d^4x \sqrt{-g}\, g^{\mu\nu} \left[\nabla_\lambda(\delta\Gamma^{\lambda}_{\mu\nu}) - \nabla_\nu(\delta\Gamma^{\lambda}_{\mu\lambda}) \right]$$

$$= -\frac{1}{16\pi G} \int d^4x \sqrt{-g}\, \nabla_\lambda(g^{\mu\nu}\delta\Gamma^{\lambda}_{\mu\nu} - g^{\mu\lambda}\delta\Gamma^{\sigma}_{\mu\sigma}), \tag{A.58}$$

where we put the tensor $g^{\mu\nu}$ inside the covariant derivatives in the second equality and renamed the summation indices ν and λ in the second term. Finally, making use of the property (A.26), we rewrite δS_3 as an integral of the total derivative,

$$\delta S_3 = -\frac{1}{16\pi G} \int d^4x \partial_\lambda(g^{\mu\nu}\delta\Gamma^{\lambda}_{\mu\nu} - g^{\mu\lambda}\delta\Gamma^{\sigma}_{\mu\sigma}).$$

Consequently, δS_3 does not contribute to the field equations. Assembling together the variations (A.53) and (A.55), we derive the variation of the Einstein–Hilbert action,

$$\delta S_{EH} = \frac{1}{16\pi G} \int d^4x \sqrt{-g} \left(R^{\mu\nu} - \frac{1}{2}g^{\mu\nu}R \right) \delta g_{\mu\nu} \tag{A.59}$$

This formula and Eq. (A.51) give the following gravitational field equations (Einstein equations in vacuo):

$$R^{\mu\nu} - \frac{1}{2}g^{\mu\nu}R = 8\pi G\Lambda g^{\mu\nu}. \tag{A.60}$$

We see that the Einstein equations are indeed second order in derivatives. The Einstein equations are often written in the form

$$G_{\mu\nu} = 8\pi G\Lambda g_{\mu\nu},$$

where

$$G_{\mu\nu} \equiv R_{\mu\nu} - \frac{1}{2}g_{\mu\nu}R$$

is the Einstein tensor.

A.5.   Conformally Related Metrics

It is useful for some applications to have relations between the Ricci tensors and scalar curvatures of metrics, *conformally related to each other*. Suppose that there are two metrics $g_{\mu\nu}$ and $\hat{g}_{\mu\nu}$, such that

$$\hat{g}_{\mu\nu}(x) = \mathrm{e}^{2\varphi(x)} g_{\mu\nu}(x), \qquad (A.61)$$

where $\varphi(x)$ is some scalar function of coordinates. Our purpose is to express $\hat{R}_{\mu\nu}$ and $\hat{R}$ — the Ricci tensor and scalar curvature constructed from the metric $\hat{g}_{\mu\nu}$ — in terms of $R_{\mu\nu}$ and R obtained from the metric $g_{\mu\nu}$. To this end, we first find the relationship between the Christoffel symbols. The direct substitution of (A.61) in (A.23) yields

$$\hat{\Gamma}^{\mu}_{\nu\lambda} = \Gamma^{\mu}_{\nu\lambda} + \delta^{\mu}_{\lambda}\partial_{\nu}\varphi + \delta^{\mu}_{\nu}\partial_{\lambda}\varphi - g_{\nu\lambda}g^{\mu\rho}\partial_{\rho}\varphi.$$

As a result of substituting this expression in Eq. (A.41) and straightforward (but tedious) calculation we obtain

$$\hat{R}_{\mu\nu} = R_{\mu\nu} - 2\nabla_{\mu}\nabla_{\nu}\varphi - g_{\mu\nu}g^{\lambda\rho}\nabla_{\lambda}\nabla_{\rho}\varphi + 2\partial_{\mu}\varphi\partial_{\nu}\varphi - 2g_{\mu\nu}g^{\lambda\rho}\partial_{\lambda}\varphi\partial_{\rho}\varphi, \quad (A.62)$$

where the covariant derivatives are evaluated with the metric $g_{\mu\nu}$. Hence, for the scalar curvature $\hat{R} = \hat{g}^{\mu\nu}\hat{R}_{\mu\nu}$ we have

$$\hat{R} = \mathrm{e}^{-2\varphi}(R - 6g^{\mu\nu}\nabla_{\mu}\nabla_{\nu}\varphi - 6g^{\mu\nu}\partial_{\mu}\varphi\partial_{\nu}\varphi), \qquad (A.63)$$

while for the Einstein tensor we get

$$\hat{G}_{\mu\nu} \equiv \hat{R}_{\mu\nu} - \frac{1}{2}\hat{g}_{\mu\nu}\hat{R}$$

$$= G_{\mu\nu} - 2\nabla_{\mu}\nabla_{\nu}\varphi + 2\partial_{\mu}\varphi\partial_{\nu}\varphi + g_{\mu\nu}(2\nabla_{\lambda}\nabla^{\lambda}\varphi + \partial_{\lambda}\varphi\partial^{\lambda}\varphi), \quad (A.64)$$

where indices in the right-hand side are raised and lowered by the metric $g_{\mu\nu}$. Finally, for the integral entering the gravitational field action, the relationship has the form

$$\int \hat{R}\sqrt{-\hat{g}}d^{4}x = \int \mathrm{e}^{2\varphi}R\sqrt{-g}d^{4}x + 6\int \mathrm{e}^{2\varphi}g^{\mu\nu}\partial_{\mu}\varphi\partial_{\nu}\varphi\sqrt{-g}d^{4}x.$$

The latter relation is obtained with the use of (A.63) and integration by parts.

As an example of application of these formulas, we consider "non-linear gravitational theories" with actions of the form

$$S = -\int d^4x \sqrt{-g} f(R), \tag{A.65}$$

where $f(R)$ is an arbitrary function of the scalar curvature R. We are going to prove that in vacuo, these theories are dynamically equivalent to conventional gravity (General Relativity described by the Einstein–Hilbert Lagrangian) plus self-interacting scalar field.[7]

We first find the field equations obtained by varying the action (A.65). We write the variation again as the sum of three terms,

$$\delta S = \delta S_1 + \delta S_2 + \delta S_3,$$

$$\delta S_1 = -\int d^4x \, f(R) \, \delta(\sqrt{-g}),$$

$$\delta S_2 = -\int d^4x \sqrt{-g} f'(R) R_{\mu\nu} \delta g^{\mu\nu},$$

$$\delta S_3 = -\int d^4x \sqrt{-g} g^{\mu\nu} f'(R) \delta R_{\mu\nu},$$

where $f'(R) = \partial f(R)/\partial R$. The variations δS_1 and δS_2 are simple generalizations of the analogous expressions for the Einstein–Hilbert action (see (A.53) and (A.55)),

$$\delta S_1 = -\frac{1}{2} \int d^4x \sqrt{-g} f(R) g^{\mu\nu} \delta g_{\mu\nu}, \tag{A.66}$$

$$\delta S_2 = \int d^4x \sqrt{-g} f'(R) R^{\mu\nu} \delta g_{\mu\nu}. \tag{A.67}$$

To calculate δS_3 we use (A.57), where we substitute the variation of the Christoffel symbols

$$\delta \Gamma^\lambda_{\mu\nu} = \frac{1}{2} g^{\lambda\rho} (\nabla_\mu \delta g_{\nu\rho} + \nabla_\nu \delta g_{\mu\rho} - \nabla_\rho \delta g_{\mu\nu}).$$

As a result we obtain from (A.57) that

$$\delta R_{\mu\nu} = \frac{1}{2}(-\nabla_\lambda \nabla^\lambda \delta g_{\mu\nu} + \nabla^\lambda \nabla_\mu \delta g_{\lambda\nu} + \nabla^\lambda \nabla_\nu \delta g_{\lambda\mu} - \nabla_\nu \nabla_\mu \delta g^\lambda_\lambda). \tag{A.68}$$

Then the variation δS_3 reads

$$\delta S_3 = -\int d^4x \sqrt{-g} \delta g_{\mu\nu} (\nabla^\mu \nabla^\nu - g^{\mu\nu} \nabla_\lambda \nabla^\lambda) f'(R), \tag{A.69}$$

where we integrated by parts twice. In the case of the Einstein–Hilbert Lagrangian $f' = 1$, so the expression (A.69) vanishes identically.

---

[7] For the sake of convenience, in Eq. (A.65) and to the end of this Section we adopt the system of units where $16\pi G = 1$.

Finally, equating to zero the variation δS we obtain the field equations for the theory with the action (A.65):

$$\frac{1}{2}f(R)g_{\mu\nu} - f'(R)R_{\mu\nu} + (\nabla_\mu\nabla_\nu - g_{\mu\nu}\nabla^\lambda\nabla_\lambda)f'(R) = 0. \qquad (A.70)$$

Note that these equations contain fourth order derivatives.

It is convenient to introduce the new variables $\tilde{g}_{\mu\nu}$ by making a conformal transformation:

$$g_{\mu\nu} = \psi^{-1}\tilde{g}_{\mu\nu}, \quad \psi = f'(R). \qquad (A.71)$$

We will assume that $\psi > 0$, then the new metric $\tilde{g}_{\mu\nu}$ has the same signature as the metric $g_{\mu\nu}$. The relationship between the Ricci tensor and scalar curvature of these two metrics is given by Eqs. (A.62), (A.63) with $\varphi = -\frac{1}{2}\ln\psi$. We have, therefore,

$$R_{\mu\nu} = \tilde{R}_{\mu\nu} + \psi^{-1}\tilde{\nabla}_\mu\tilde{\nabla}_\nu\psi + \frac{1}{2\psi}\tilde{g}_{\mu\nu}\tilde{\nabla}_\lambda\tilde{\nabla}^\lambda\psi$$

$$- \frac{1}{2\psi^2}(\tilde{\nabla}_\nu\psi\tilde{\nabla}_\mu\psi + 2\tilde{g}_{\mu\nu}\tilde{\nabla}_\lambda\psi\tilde{\nabla}^\lambda\psi), \qquad (A.72)$$

$$R = \psi\tilde{R} + 3\tilde{\nabla}_\mu\tilde{\nabla}^\mu\psi - \frac{9}{2}\psi^{-1}\tilde{\nabla}_\lambda\psi\tilde{\nabla}^\lambda\psi, \qquad (A.73)$$

where symbols with tilde are calculated, and indices are raised and lowered with the metric $\tilde{g}_{\mu\nu}$.

Let $R_0(\psi)$ be a solution to the equation

$$f'[R_0(\psi)] - \psi = 0$$

(the analysis below can be extended to the case of multiple solutions), i.e., R_0 is the inverse function to f'. It follows from the second of Eq. (A.71) that

$$R = R_0(\psi). \qquad (A.74)$$

In terms of new variables, Eq. (A.70) takes the form

$$\tilde{R}_{\mu\nu} - \frac{1}{2}\tilde{R}\tilde{g}_{\mu\nu}$$

$$= \psi^{-2}\left\{\frac{1}{2}[f(R_0(\psi)) - \psi R_0(\psi)]\tilde{g}_{\mu\nu} + \frac{3}{2}\tilde{\nabla}_\mu\psi\tilde{\nabla}_\nu\psi - \frac{3}{4}\tilde{g}_{\mu\nu}\tilde{\nabla}_\lambda\psi\tilde{\nabla}^\lambda\psi\right\}.$$

$$(A.75)$$

We still have to account for Eq. (A.74). Its left-hand side is given by (A.73) where $\tilde{R}$ is obtained by contracting (A.75) with $\tilde{g}^{\mu\nu}$. In this way we get the equation

$$\psi\tilde{\nabla}_\lambda\tilde{\nabla}^\lambda\psi - \tilde{\nabla}_\lambda\psi\tilde{\nabla}^\lambda\psi + \frac{1}{3}\{\psi R_0(\psi) - 2f[R_0(\psi)]\} = 0. \qquad (A.76)$$

So, instead of fourth order equation (A.70), we obtain an extended system of second order equations (A.75), (A.76) in terms of the new variables $\tilde{g}_{\mu\nu}$ and ψ. This system coincides with the equations of conventional gravity interacting with the scalar field ψ. The action whose variation with respect to $\tilde{g}_{\mu\nu}$ and ψ gives Eqs. (A.75) and (A.76) is

$$S = -\int d^4x \sqrt{-\tilde{g}}\tilde{R}$$
$$+ \int d^4x \sqrt{-\tilde{g}} \left\{ \frac{3}{2} \frac{\tilde{g}^{\mu\nu}\tilde{\nabla}_\mu \psi \tilde{\nabla}_\nu \psi}{\psi^2} + \frac{R_0(\psi)}{\psi} - \frac{f[R_0(\psi)]}{\psi^2} \right\}. \qquad (A.77)$$

Problem A.18. *Obtain Eqs. (A.75) and (A.76) by varying the action (A.77).*

The kinetic term in the action (A.77) can be cast into the canonical form by the replacement $\psi = e^{\sqrt{\frac{2}{3}}\phi}$. Finally, we arrive at the action

$$S = \int d^4x \sqrt{-\tilde{g}} \left\{ -\tilde{R} + \tilde{g}^{\mu\nu}\tilde{\nabla}_\mu\phi\tilde{\nabla}_\nu\phi \right.$$
$$\left. + e^{-\sqrt{\frac{2}{3}}\phi} R_0\left(e^{\sqrt{\frac{2}{3}}\phi}\right) - e^{-2\sqrt{\frac{2}{3}}\phi} f\left[R_0\left(e^{\sqrt{\frac{2}{3}}\phi}\right)\right] \right\}.$$

It describes self-interacting scalar field ϕ in the framework of GR. As we have seen, this theory is dynamically equivalent to the theory of "non-linear" gravity with the action (A.65).

Problem A.19. *Consider scalar-tensor theory of gravity with the action*

$$S = \int d^4x \sqrt{-g} \left(-R + \frac{1}{2}\omega(\varphi)\partial_\mu\varphi\partial^\nu\varphi - V(\varphi) \right),$$

where φ is scalar field. Under what conditions on functions $\omega(\varphi)$ and $V(\varphi)$ this theory is equivalent to $f(R)$-gravity?

Note that as a result of the conformal transformation, matter fields begin to interact with the dilaton field ϕ. This results in various physical effects. In particular, for homogeneous solution $\phi = \phi(t)$ the "cosmic time" (time in the FLRW metric) differs from the "atomic time" (time determining the evolution and interaction of matter fields). In this sense, in the presence of matter fields $f(R)$-gravity is *not equivalent* to GR with scalar field.

A.6.   Interaction of Matter with Gravitational Field: Energy–Momentum Tensor

Equations (A.60) describe the dynamics of the gravitational field in vacuo. However, it is of primary interest to study gravity in the presence of matter fields which serve as sources of gravitational field. To study this general situation, we add to the action

(A.48) a new term,

$$S_m = \int d^4x \sqrt{-g} \mathcal{L}_m, \tag{A.78}$$

that describes matter and its coupling to gravity. Here the Lagrangian density $\mathcal{L}_m$ is a scalar function of the gravitational field $g_{\mu\nu}$ and matter fields. We collectively denote the latter by ψ:

$$\mathcal{L}_m = \mathcal{L}_m(\psi, g_{\mu\nu}).$$

Once we add the term (A.78), the Einstein equation (A.60) get modified as follows,

$$R_{\mu\nu} - \frac{1}{2} g_{\mu\nu} R = 8\pi G(\Lambda g_{\mu\nu} + T_{\mu\nu}). \tag{A.79}$$

Here we turned to tensors with lower indices. The symmetric tensor $T_{\mu\nu}$ is defined by the following relation,

$$\delta S_m = \frac{1}{2} \int d^4x \sqrt{-g} T_{\mu\nu} \delta g^{\mu\nu}. \tag{A.80}$$

The latter relation can be rewritten with account of (A.54),

$$\delta S_m = -\frac{1}{2} \int d^4x \sqrt{-g} T^{\mu\nu} \delta g_{\mu\nu}.$$

Equation (A.79) (with upper indices) is obtained then by using (A.59).

To understand the physical meaning of the tensor $T_{\mu\nu}$, we calculate it for two simple theories: scalar field theory and theory of electromagnetic field. The covariant action describing the real scalar field interacting with gravity reads

$$S_{sc} = \int d^4x \sqrt{-g} \mathcal{L}_{sc} = \int d^4x \sqrt{-g} \left(\frac{1}{2} g^{\mu\nu} \partial_\mu \phi \partial_\nu \phi - V(\phi) \right), \tag{A.81}$$

where the scalar potential $V(\phi)$ can be an arbitrary function of the field ϕ. Generally speaking, one can add to the action (A.81) yet another term which vanishes in flat space,

$$S_\xi = \xi \int d^4x \sqrt{-g} R U(\phi), \tag{A.82}$$

where $U(\phi)$ is an arbitrary function. We restrict ourselves to the choice $\xi = 0$. In this case, the interaction of the scalar field with gravity is called minimal.

Using the definition (A.80), we obtain the following expression for the tensor $T_{\mu\nu}$ of the scalar field,

$$T_{\mu\nu}^{sc} = \partial_\mu \phi \partial_\nu \phi - g_{\mu\nu} \mathcal{L}_{sc}. \tag{A.83}$$

Problem A.20. *Find the tensor $T_{\mu\nu}$ for free massless scalar field non-minimally coupled to gravity, $\xi \neq 0$, by choosing*

$$U(\phi) = \phi^2,$$

and $V(\phi) = 0$. At what value of the parameter ξ does the trace $g^{\mu\nu}T_{\mu\nu}$ vanish on the equations of motion? For what $V(\phi)$ does this property remain valid?

We now find the explicit form of the tensor $T_{\mu\nu}$ for the electromagnetic field. The action for the vector field A_μ coupled to gravity has the form

$$S_{em} = -\frac{1}{4} \int d^4x \sqrt{-g} F_{\mu\nu} F_{\lambda\rho} g^{\mu\lambda} g^{\nu\rho}, \tag{A.84}$$

where $F_{\mu\nu}$ is the conventional field strength tensor,

$$F_{\mu\nu} = \partial_\mu A_\nu - \partial_\nu A_\mu. \tag{A.85}$$

At first glance, the ordinary derivatives in the definition (A.85) should be replaced by covariant derivatives in curved space. But it is straightforward to check that for symmetric connection, the terms with the connection cancel out due to antisymmetrization in μ and ν, so that $\nabla_\mu A_\nu - \nabla_\nu A_\mu = \partial_\mu A_\nu - \partial_\nu A_\mu$, and $F_{\mu\nu}$ is a tensor. Using the action (A.84), we obtain the following expression for the tensor $T_{\mu\nu}$ of electromagnetic field,

$$T_{\mu\nu}^{em} = -F_{\mu\lambda} F_{\nu\rho} g^{\lambda\rho} + \frac{1}{4} g_{\mu\nu} F_{\lambda\rho} F^{\lambda\rho}. \tag{A.86}$$

Now we note that in Minkowski space, the tensor $T_{\mu\nu}$ coincides with the energy–momentum tensor for both scalar and electromagnetic field. For scalar field, $T_{\mu\nu}^{sc}$ is exactly equal to the Noether energy-momentum tensor, while for the electromagnetic field, $T_{\mu\nu}^{em}$ differs on the equations of motion from the Noether tensor by the divergence of an antisymmetric tensor.

Problem A.21. *Prove these statements.*

In particular, the (00)-components of these tensors in Minkowski space are

$$T_{00}^{sc} = \frac{1}{2}(\partial_0 \phi)^2 + \frac{1}{2}(\partial_i \phi)^2 + V(\phi)$$

and

$$T_{00}^{em} = \frac{1}{2} F_{0i}^2 + \frac{1}{4} F_{ij}^2 \equiv \frac{1}{2}(\mathbf{E}^2 + \mathbf{H}^2).$$

These are indeed the energy densities of scalar and electromagnetic fields.

Generally, the tensor $T_{\mu\nu}$ defined by Eq. (A.80) is called metric energy–momentum tensor. We emphasize that it is always symmetric. Below we prove that in Minkowski space and on equations of motion, it is always equal to the Noether energy–momentum tensor modulo the divergence of an antisymmetric tensor.

The energy–momentum tensor is conserved in flat space,

$$\partial_\mu T^{\mu\nu} = 0. \tag{A.87}$$

This leads to the conservation of energy and momentum. It is natural to assume that the generalization of the conservation law (A.87) to curved space is the covariant conservation equation

$$\nabla_\mu T^{\mu\nu} = 0. \tag{A.88}$$

To see that this is indeed the case, we take the divergence of both sides of the Einstein equations (A.79),

$$\nabla^\mu \left(R_{\mu\nu} - \frac{1}{2} g_{\mu\nu} R \right) = 8\pi G \nabla^\mu T_{\mu\nu}. \tag{A.89}$$

Let us prove that the left-hand side of Eq. (A.89) is *identically* zero. To this end, we first contract the indices λ and μ in the Bianchi identity (A.36). As a result, we obtain the following identity,

$$\nabla_\rho R_{\sigma\nu} - \nabla_\nu R_{\sigma\rho} + \nabla_\lambda R^\lambda_{\sigma\nu\rho} = 0.$$

Then we contract it with $g^{\sigma\rho}$ and arrive at

$$0 = \nabla_\rho R^\rho_\nu - \nabla_\nu R + \nabla^\lambda R_{\lambda\nu} = 2\nabla^\mu \left(R_{\mu\nu} - \frac{1}{2} g_{\mu\nu} R \right).$$

Thus we obtained the identity

$$\nabla^\mu \left(R_{\mu\nu} - \frac{1}{2} g_{\mu\nu} R \right) = 0,$$

which implies that the covariant conservation law of energy–momentum tensor (A.88) is a necessary condition for the consistency of the Einstein equations.

On the other hand, the energy–momentum tensor is entirely determined by the action of matter fields. Therefore, to check the consistency of the whole system of field equations, one needs to prove that the conservation law (A.88) is a consequence of matter field equations. The proof makes use of the invariance of the action under coordinate transformations. We begin with evaluating the first variation of the metric $g^{\mu\nu}$ under small coordinate transformation

$$x'^\mu = x^\mu + \xi^\mu. \tag{A.90}$$

Substituting Eq. (A.90) to the general formula (A.7), we get

$$g'^{\mu\nu}(x') = (\delta^\mu_\lambda + \partial_\lambda \xi^\mu)(\delta^\nu_\rho + \partial_\rho \xi^\nu) g^{\lambda\rho}(x)$$
$$= g^{\mu\nu}(x) + g^{\nu\lambda} \partial_\lambda \xi^\mu + g^{\mu\lambda} \partial_\lambda \xi^\nu, \tag{A.91}$$

where we neglected terms of the second order in ξ^μ. Expanding the left-hand side of Eq. (A.91) we write

$$g'^{\mu\nu}(x') = g'^{\mu\nu}(x) + \partial_\lambda g'^{\mu\nu}(x)\xi^\lambda + \mathcal{O}(\xi^2) = g'^{\mu\nu}(x) + \partial_\lambda g^{\mu\nu}(x)\xi^\lambda + \mathcal{O}(\xi^2).$$

In this way we obtain the following relationship between the functions $g^{\mu\nu}(x)$ and $g'^{\mu\nu}(x)$ taken at points with the same coordinates in the old and new coordinate frames,

$$g'^{\mu\nu}(x) = g^{\mu\nu}(x) - \partial_\lambda g^{\mu\nu}(x)\xi^\lambda + g^{\nu\lambda}\partial_\lambda \xi^\mu + g^{\mu\lambda}\partial_\lambda \xi^\nu. \tag{A.92}$$

Problem A.22. *Check by explicit calculation that the relation* (A.92) *can be written in the following covariant form:*

$$g'^{\mu\nu} = g^{\mu\nu} + \nabla^\mu \xi^\nu + \nabla^\nu \xi^\mu. \tag{A.93}$$

Now, the invariance of the matter action under coordinate transformations means that the variation of this action is equal to zero when simultaneously metric $g^{\mu\nu}(x)$ is transformed as in (A.93) and matter fields are transformed accordingly, $\psi'(x) = \psi(x) + \delta\psi_\xi(x)$. The form of $\delta\psi_\xi$ is dictated by the transformation properties of matter fields under the coordinate transformation (A.90). For instance, for scalar field ϕ

$$\delta\phi_\xi = -\xi^\mu \partial_\mu \phi.$$

So, in the general case we have

$$\delta_\xi S_m = \frac{1}{2} \int d^4x \sqrt{-g} T_{\mu\nu}(\nabla^\mu \xi^\nu + \nabla^\nu \xi^\mu) + \int d^4x \sqrt{-g} \frac{\delta \mathcal{L}_m}{\delta\psi} \delta\psi_\xi = 0. \tag{A.94}$$

The identity (A.94) is valid regardless of the field equations. Now we assume in addition that the matter field equations are satisfied. Then the second term on the left-hand side of Eq. (A.94) vanishes. Therefore, we proved that the matter field equations guarantee the validity of equality

$$\int d^4x \sqrt{-g} T_{\mu\nu}(\nabla^\mu \xi^\nu + \nabla^\nu \xi^\mu) = 0.$$

Since the vector ξ^μ is arbitrary and $T_{\mu\nu}$ is symmetric, we arrive, upon integration by parts, at the covariant conservation law (A.88), as required.

We now use the identity (A.94) to prove that in flat space, the metric energy–momentum tensor $T_{\mu\nu}$ coincides on the equations of motion with the Noether tensor $T_{\mu\nu}$ modulo total derivative. In flat space, the identity (A.94) takes the form

$$\int d^4x T_{\mu\nu} \partial^\mu \xi^\nu + \int d^4x \frac{\delta \mathcal{L}_m}{\delta\psi} \delta\psi_\xi = 0, \tag{A.95}$$

where we again used the symmetry of the tensor $T_{\mu\nu}$. In Minkowski space, the action is invariant under the variations of matter fields $\delta\psi_\xi$ corresponding to shifts (A.90) with constant ξ^μ. Consequently, the second term in (A.95) can be written as

$$\int d^4x \frac{\delta \mathcal{L}_m}{\delta \psi} \delta \psi_\xi = -\int d^4x \tau_{\mu\nu} \partial^\mu \xi^\nu, \tag{A.96}$$

where $\tau_{\mu\nu}$ coincides on the equations of motion with the conserved Noether energy–momentum tensor.

Problem A.23. *Modifying the proof of the Noether theorem, show the validity of the relation (A.96), where $\tau_{\mu\nu}$ is equal on the equations of motion to the Noether energy–momentum tensor.*

Integrating by parts Eq. (A.95), we see that

$$\partial^\mu (T_{\mu\nu} - \tau_{\mu\nu}) = 0.$$

This is the identity valid irrespective of the equations of motion. The identity can only hold if the difference $(T_{\mu\nu} - \tau_{\mu\nu})$ is the divergence of an antisymmetric tensor,

$$T_{\mu\nu} - \tau_{\mu\nu} = \partial^\lambda A_{\mu\nu\lambda}, \quad \text{where } A_{\mu\nu\lambda} = -A_{\lambda\nu\mu}, \tag{A.97}$$

as promised.

The total derivative of the form (A.97) does not contribute to the total energy–momentum 4-vector

$$P^\nu \equiv \int d^3x \, T^{0\nu}.$$

Indeed, because of the antisymmetry property of $A_{\mu\nu\lambda}$, its contribution is of the form

$$\int d^3x \, \partial_i A_i^{0\nu}.$$

This integral is zero for fields vanishing at spatial infinity. Hence, metric and Noether energy–momentum tensors are equivalent in flat space in the sense that they give equal values of energy and momentum.

Problem A.24. *Consider tensor of the form*

$$\Theta_{\mu\nu} = (\eta_{\mu\nu}\partial^2 - \partial_\mu\partial_\nu)f,$$

where f is an arbitrary function. Clearly, this tensor is identically conserved in flat space. Express it in the form $\Theta_{\mu\nu} = \partial^\lambda A_{\mu\nu\lambda}$, where $A_{\mu\nu\lambda} = -A_{\lambda\nu\mu}$.

Problem A.25. *Check explicitly that the metric energy–momentum tensor of scalar field coupled non-minimally to gravity, see (A.81) and (A.82), differs in flat space from the Noether energy–momentum tensor by the total derivative of the type (A.97) for arbitrary $U(\phi)$ and ξ.*

To conclude our discussion of energy–momentum tensor in GR, we make the following observation. In flat space, the differential conservation law (A.87) implies the existence of four conserved quantities, the components of the energy–momentum 4-vector. However, Eq. (A.88), in general, does not imply the existence of four integrals of motion in curved space, which could be interpreted as the energy and momentum of the system. In this regard, the concepts of energy and momentum, generally speaking, are not defined in GR. For spatially localized gravitating systems one can define the energy and momentum by making use of the asymptotics of the gravitational field far away from the system, but in general such a construction is impossible. In particular, talking about the total mass of the Universe does not make sense.

A.7.   Particle Motion in Gravitational Field

As a digression from the discussion of the properties of the Einstein equations, let us study the motion of point particles in external gravitational field. The action for point particle in GR has the same form as in special relativity,

$$S_p = -m \int ds. \tag{A.98}$$

In GR, the interval along the world line of a particle involves the space–time metric,

$$ds = \sqrt{dx^\mu dx^\nu g_{\mu\nu}} = \sqrt{\frac{dx^\mu}{d\tau}\frac{dx^\nu}{d\tau}g_{\mu\nu}}d\tau,$$

where we introduced the affine parameter τ along the world line, and $g_{\mu\nu} = g_{\mu\nu}[x(\tau)]$. In terms of the affine parameter, the action (A.98) is

$$S_p = -m \int \sqrt{\dot{x}^\mu \dot{x}^\nu g_{\mu\nu}(x)}d\tau, \tag{A.99}$$

where dot denotes differentiation with respect to τ. Equation of motion obtained by varying the action (A.99) is

$$-\frac{d}{d\tau}\left(\frac{g_{\mu\nu}\dot{x}^\nu}{\sqrt{\dot{x}^\alpha \dot{x}_\alpha}}\right) + \frac{1}{2}\frac{\dot{x}^\lambda \dot{x}^\nu \partial_\mu g_{\nu\lambda}}{\sqrt{\dot{x}^\alpha \dot{x}_\alpha}} = 0. \tag{A.100}$$

Problem A.26. *Derive Eq. (A.100).*

The affine parameter τ can be chosen in such a way that the 4-velocity vector

$$u^\mu = \frac{dx^\mu}{d\tau} \tag{A.101}$$

has unit length at each point of the world line,

$$g_{\mu\nu}u^\mu u^\nu = 1. \tag{A.102}$$

This choice identifies the affine parameter with the proper time of the particle, since Eq. (A.102) is equivalent to

$$ds = d\tau.$$

With this choice of the world line parameterization, equation of motion (A.100) becomes

$$-\frac{d}{ds}(g_{\mu\nu}u^\nu) + \frac{1}{2}\partial_\mu g_{\nu\lambda}u^\lambda u^\nu = 0. \tag{A.103}$$

Contracting this equation with $g^{\mu\rho}$, we obtain

$$-\frac{du^\rho}{ds} - g^{\mu\rho}\left(\frac{dg_{\mu\nu}}{ds} - \frac{1}{2}\partial_\mu g_{\nu\lambda}u^\lambda\right)u^\nu = 0. \tag{A.104}$$

Since $g_{\mu\nu} = g_{\mu\nu}[x(s)]$, the definition of the 4-velocity gives

$$\frac{dg_{\mu\nu}}{ds} = \partial_\lambda g_{\mu\nu}u^\lambda.$$

Substituting this expression into Eq. (A.104) and using the formula (A.23) for the Christoffel symbols, we finally arrive at the following form of the equation of motion,

$$\frac{du^\nu}{ds} + \Gamma^\nu_{\mu\lambda}u^\mu u^\lambda = 0. \tag{A.105}$$

Multiplying this equation by ds and noticing that

$$dx^\lambda = u^\lambda ds$$

along the particle world line, we rewrite Eq. (A.105) as

$$du^\nu + \Gamma^\nu_{\mu\lambda}u^\mu dx^\lambda = 0.$$

We recall now the parallel transport law of contravariant vectors (A.13) and observe that the geometric meaning of Eq. (A.105) is that under parallel transport along the world line the normalized tangential vectors $u^\mu[x(\tau)]$ transform into each other. Curves which satisfy this property are geodesics (shortest paths), and Eq. (A.105) is the geodesic equation.

The action (A.98) does not make sense for massless particles, $m = 0$. To find the world lines of these particles (e.g., rays of light) one directly uses the geodesic equation

$$\frac{du^\nu}{d\tau} + \Gamma^\nu_{\mu\lambda}u^\mu u^\lambda = 0, \tag{A.106}$$

where τ is now an arbitrary parameter along the world line, and 4-velocity u^μ is still defined by Eq. (A.101). In the massless case, the geodesic must be light-like (null),

$$ds^2 = 0,$$

or, in differential form,

$$g_{\mu\nu}\dot{x}^\mu\dot{x}^\nu \equiv g_{\mu\nu}u^\mu u^\nu = 0. \tag{A.107}$$

Problem A.27. *Show that Eq. (A.106) is consistent with the requirement (A.107).*

Problem A.28. *Check that the equation of motion of massive point particle as well as the equation for light-like geodesic can be obtained from the action*

$$S_\eta = -\frac{1}{2}\int d\tau\, \eta \left[\eta^{-2}\dot{x}^\mu\dot{x}^\nu g_{\mu\nu}(x) + m^2\right], \tag{A.108}$$

where $\eta(\tau)$ is a new auxiliary dynamical variable ("Dynamical" here means that one of the equations of motion is obtained by varying the action with respect to η.) This new variable transforms under the change of parameterization as

$$\eta'\left[\tau'(\tau)\right] = \eta(\tau)\left[\frac{\partial\tau'(\tau)}{\partial\tau}\right]^{-1}.$$

Note that $\eta^2(\tau)$ can be considered as internal metric on the world line, while the action (A.108) can be viewed as the action of four fields $x^\mu(\tau)$ in 1-dimensional space with dynamical metric.

Equation (A.106) can be obtained by considering wave equation in curved space–time. Let us discuss photons for definiteness. The Maxwell equation that follows from action (A.84) reads

$$\nabla_\mu F^{\mu\nu} = 0.$$

If the wavelength and wave period are small compared to spatial and temporal scales characteristic of metric, one neglects terms involving the Ricci tensor. One chooses the gauge $\nabla_\mu A^\mu = 0$ and writes the field equation as follows:

$$\nabla^\mu\nabla_\mu A^\nu = 0. \tag{A.109}$$

Solutions to this equation have the form

$$A^\mu = a^\mu e^{iS},$$

where $a^\mu(x)$ is a slowly varying amplitude, and $S(x)$ is a large phase called eikonal (see, e.g. the book [267]). Locally (e.g., near $x^\mu = 0$) this phase can be written as $S = \text{const} + P_\mu x^\mu$. This is a standard expression for plane wave, and P_0 and P_i are photon frequency and wave vector (momentum). Thus, the local 4-momentum is

$$P_\mu = \partial_\mu S.$$

If metric varies slowly, then P_μ varies slowly too, even though P_μ itself is large.

The latter observation shows that the leading term in Eq. (A.109) involves square of the derivative of S. Therefore, to the leading order this equation is reduced to

$$P_\mu P^\mu = 0. \tag{A.110}$$

Since P_μ is gradient of S, it obeys

$$\nabla_\mu P_\nu - \nabla_\nu P_\mu = 0.$$

One multiplies the latter identity by P^μ and makes use of (A.110) to get $P^\mu \nabla_\nu P_\mu = 0$. In this way one arrives at equation

$$P^\mu \nabla_\mu P^\nu \equiv P^\mu \partial_\mu P^\nu + \Gamma^\nu_{\mu\lambda} P^\mu P^\lambda = 0. \tag{A.111}$$

There is a set of world lines $x^\mu(\tau)$ associated with the vector field $P^\mu(x)$: vectors P^μ are tangent vectors to these lines. These are precisely photon world lines. The parameter τ along a worlds line can always be chosen in such a way that $P^\mu = dx^\mu/d\tau$. Since $P^\mu \partial_\mu P^\nu(x) = (d/d\tau)P^\nu(x(\tau))$, Eq. (A.111) can be written as

$$\frac{dP^\nu}{d\tau} + \Gamma^\nu_{\mu\lambda} P^\mu P^\lambda = 0.$$

This is precisely the geodesic equation (A.106). As a byproduct, we have shown that 4-velocity $u^\mu = dx^\mu/d\tau$ is nothing but the photon 4-momentum.

To see that the parameter τ is not arbitrary, and is rather determined by the geodesic equation, let us parameterize the geodesic line by time x^0 instead of τ. Then

$$P^i = P^0 \frac{dx^i}{dx^0}, \qquad \frac{dP^\mu}{d\tau} = \frac{dP^\mu}{dx^0} P^0,$$

and the temporal component of the geodesic equation (A.106) becomes

$$\frac{dP^0}{dx^0} + \Gamma^0_{\mu\nu} v^\mu v^\nu P^0 = 0, \tag{A.112}$$

where $v^\mu = (1, dx^i/dx^0)$. The spatial components of the geodesic equation become the system of equations for dx^i/dx^0. Once its solution $x^i(x^0)$ is known, one solves Eq. (A.112) and finds $P^\mu(x^0)$. Then the parameter τ can be found, e.g., by solving $dx^0/d\tau = P^0(x^0)$.

A.8. Newtonian Limit in General Relativity

Let us now discuss how the main object of the Newtonian gravity, the gravitational potential, arises in General Relativity and how the Newtonian gravity law follows from GR. To this end, we study the motion of a particle in weak static gravitational field, i.e., in space with metric

$$g_{\mu\nu} = \eta_{\mu\nu} + h_{\mu\nu}(\mathbf{x}), \tag{A.113}$$

where $\eta_{\mu\nu}$ is the Minkowski metric, and all components of $h_{\mu\nu}(\mathbf{x})$ are small,

$$h_{\mu\nu}(\mathbf{x}) \ll 1. \tag{A.114}$$

Furthermore, we consider non-relativistic particles,

$$v^i \equiv \frac{dx^i}{dt} \ll 1.$$

Let us write the explicit form of various components of the geodesic equation (A.105) to the linear order in the velocity v^i and gravitational field $h_{\mu\nu}$. We first note that in the linear order, the proper time of a particle ds is related to the coordinate time dt by

$$ds = \left(1 + \frac{h_{00}}{2}\right) dt. \tag{A.115}$$

Problem A.29. *Find in the general case the relation between coordinate time and proper time of a particle moving with the coordinate 3-velocity $v^i = \frac{dx^i}{dt}$. Show that in the linear order this relation indeed reduces to (A.115).*

Hence, the 4-velocity components u^μ are related to the metric and physical velocity v^i as follows,

$$u^0 \equiv \frac{dt}{ds} \approx 1 - \frac{h_{00}}{2},$$

$$u^i \equiv \frac{dx^i}{ds} \approx v^i.$$

It is straightforward to check that in the linear order, the zeroth component of the geodesic equation is satisfied identically. Indeed, the first term $\frac{d^2 t}{ds^2}$ vanishes, in view of (A.115), because the metric is static. The second term $\Gamma^\mu_{\nu\lambda} u^\nu u^\lambda$ contains small factor from the very beginning, since all Christoffel symbols vanish for the unperturbed metric $\eta_{\mu\nu}$. Furthermore, the component Γ^0_{00} is equal to zero for static metric, hence the second term must involve the velocity u^i at least once. So, all terms in the zeroth component of the geodesic equation are at least of the second order.

Spatial components of the geodesic equation in the linear approximation read

$$\frac{dv^i}{dt} + \Gamma^i_{00} = 0,$$

where we again took into account that the second term is inherently small due to the presence of the Christoffel symbols, so that the contributions involving velocities u^i drop out. Recalling the explicit expression (A.23) for the Christoffel symbols, we arrive at the following equation describing the motion of non-relativistic particles in weak static gravitational field,

$$\frac{dv^i}{dt} = -\partial_i \Phi, \tag{A.116}$$

where we introduced new function $\Phi(\mathbf{x})$ defined by

$$g_{00} = 1 + 2\Phi.$$

Equation (A.116) coincides with the equation of non-relativistic mechanics describing the motion of a particle in an external potential $\Phi(\mathbf{x})$, so the field $\Phi(\mathbf{x})$ is identified, in weak field limit, with the Newtonian gravitational potential. Note that, as follows from the above analysis, the motion of non-relativistic particles in weak static fields depends only on 00-component of metric.

To cross-check the interpretation of the field $\Phi(\mathbf{x})$ as the Newtonian potential, let us see that the Poisson equation

$$\Delta\Phi = 4\pi G\rho \tag{A.117}$$

indeed follows from the Einstein equations in the Newtonian limit. Here $\Delta \equiv (\partial_i)^2$ is the Laplace operator. At the same time we will identify the constant G, entering the Einstein–Hilbert action, with Newton's gravity constant.

To this end, let us find, starting from the Einstein equations (A.79), the metric produced by a static distribution of non-relativistic matter with energy density $\rho(\mathbf{x})$. Before doing that, it is convenient to rewrite the Einstein equations in the equivalent form. Taking the trace of both sides of the Einstein equations, we obtain

$$R = -8\pi G(4\Lambda + T), \tag{A.118}$$

where

$$T \equiv g^{\mu\nu}T_{\mu\nu}$$

is the trace of energy–momentum tensor. Substituting (A.118) back into the Einstein equations, we arrive at the following equivalent equations,

$$R_{\mu\nu} = 8\pi G\left(T_{\mu\nu} - \frac{1}{2}g_{\mu\nu}T - g_{\mu\nu}\Lambda\right). \tag{A.119}$$

This form of the Einstein equations is sometimes more convenient in practical calculations than the original one, because, as a rule, the curvature tensor $R_{\mu\nu}$ has more cumbersome structure than the energy–momentum tensor $T_{\mu\nu}$.

Coming back to the problem of calculating the gravitational field, we assume that the cosmological constant is absent, i.e., $\Lambda = 0$, and that both energy density $\rho(\mathbf{x})$ and all its spatial derivatives are small. The gravitational field produced by such a body is weak, i.e., the metric has the form (A.113). The only nonzero component of the energy–momentum tensor for static distribution of non-relativistic matter is

$$T_{00} = \rho(\mathbf{x}). \tag{A.120}$$

Consider now the 00-component of Eq. (A.119). For weak gravitational field, we neglect the quadratic terms in the expression (A.41) for the Ricci tensor. Furthermore, the second term in the expression for R_{00} vanishes for static metric. Consequently, the left-hand side of the 00-component of Eq. (A.119) takes the form

$$R_{00} = \partial_\lambda\Gamma^\lambda_{00} = \frac{1}{2}\Delta g_{00}, \tag{A.121}$$

where the latter equality is again obtained for weak and static field. Substituting this expression and the explicit form (A.120) of the energy–momentum tensor in Eq. (A.119) we obtain (at $\Lambda = 0$) Eq. (A.117). Hence, Φ is indeed the gravitational potential and G is Newton's gravity constant.

A.9. Linearized Einstein Equations about Minkowski Background

Let us generalize Eq. (A.117) to the case of weak but otherwise arbitrary gravitational field about Minkowski background. In this case the metric has the form (cf. (A.113))

$$g_{\mu\nu}(x) = \eta_{\mu\nu} + h_{\mu\nu}(x),$$

where $|h_{\mu\nu}(x)| \ll 1$, and perturbations $h_{\mu\nu}(x)$ can depend on both time and spatial coordinates. We make use of the Einstein equations in the form (A.119), and set $\Lambda = 0$, so that Minkowski space is the solution for $T_{\mu\nu} = 0$. In fact, the computation of the Ricci tensor to the linear order in $h_{\mu\nu}$ has already been performed: we can use the formula (A.68), considering it as the expression for the deviation of the Ricci tensor from its zero value in Minkowski space. Thus, we make the replacement $\delta g_{\mu\nu} \to h_{\mu\nu}$ in (A.68), replace the covariant derivatives with ordinary ones and raise and lower indices by Minkowski metric. As a result, we obtain the linearized equation (A.119) in the following form,

$$\left(-\partial_\lambda \partial^\lambda h_{\mu\nu} + \partial^\lambda \partial_\mu h_{\lambda\nu} + \partial^\lambda \partial_\nu h_{\lambda\mu} - \partial_\mu \partial_\nu h_\lambda^\lambda\right)$$
$$= 16\pi G \left(T_{\mu\nu} - \frac{1}{2}\eta_{\mu\nu}T_\lambda^\lambda\right), \tag{A.122}$$

where $T_{\mu\nu}$ is assumed to be small.

Equation (A.122) is invariant under gauge transformations

$$h_{\mu\nu} \to h_{\mu\nu} + \partial_\mu \xi_\nu + \partial_\nu \xi_\mu,$$
$$T_{\mu\nu} \to T_{\mu\nu}, \tag{A.123}$$

where $\xi_\mu(x)$ are small gauge parameters. The transformation (A.123) is nothing but the linearized transformation (A.92); since $T_{\mu\nu}$ is small, it does not change, in the linear order, under the small coordinate transformation (A.90). It is often convenient to take advantage of this gauge freedom and impose the harmonic gauge

$$\partial_\mu h_\nu^\mu - \frac{1}{2}\partial_\nu h_\lambda^\lambda = 0.$$

In this gauge, the linearized Einstein equations take particularly simple form

$$\Box h_{\mu\nu} = -16\pi G \left(T_{\mu\nu} - \frac{1}{2}\eta_{\mu\nu}T_\lambda^\lambda\right),$$

where $\Box \equiv \partial_\lambda \partial^\lambda$ is the D'Alembertian in Minkowski space. Clearly, this equation describes massless gravitational field sourced by the energy–momentum tensor.

A.10. Macroscopic Energy–Momentum Tensor

To find solutions to the Einstein equations describing the expanding Universe filled with matter (e.g., relativistic plasma or "dust"), we need the appropriate expression for the energy–momentum tensor. It is sufficient for our purposes to treat matter as ideal fluid and make use of the hydrodynamic approximation to the energy–momentum tensor. To obtain its explicit expression in curved space–time, we first consider the case of flat space. Isotropic fluid without internal rotation has the following energy–momentum tensor in its own rest frame,

$$T^{\mu\nu} = \begin{pmatrix} \rho & 0 & 0 & 0 \\ 0 & p & 0 & 0 \\ 0 & 0 & p & 0 \\ 0 & 0 & 0 & p \end{pmatrix}, \tag{A.124}$$

where ρ and p are energy density and pressure. Let us first generalize this expression to moving fluid. We note that in the rest frame, the 4-velocity vector is

$$u^\mu = (1, 0, 0, 0).$$

Therefore, we define a tensor object by

$$(p + \rho)u^\mu u^\nu - p\eta^{\mu\nu}. \tag{A.125}$$

Clearly it coincides with the energy–momentum tensor (A.124) in the rest frame. Since both energy–momentum tensor and the object (A.125) transform according to the tensor law, they coincide in all reference frames. The simplest way to generalize the expression (A.125) to curved space–time is to replace Minkowski metric $\eta^{\mu\nu}$ by the general space–time metric $g^{\mu\nu}$. Indeed, as we discussed above, there exists locally Lorentz reference frame at every given point of space–time. In this frame, the metric tensor at that point coincides with Minkowski tensor, and the matter energy–momentum tensor has the form (A.125). Performing the coordinate transformation to arbitrary frame, we arrive at the following final expression for the energy–momentum tensor,

$$T^{\mu\nu} = (p + \rho)u^\mu u^\nu - pg^{\mu\nu}. \tag{A.126}$$

It is worth noting that, generally speaking, expression (A.126) is valid only in the case of weak gravitational field. In strong field, there may appear additional terms depending on the curvature tensor.

In general, density ρ, pressure p and 4-velocity u^μ are arbitrary functions of time and spatial coordinates, with the restrictions that

$$u^\mu u_\mu = 1 \tag{A.127}$$

and

$$\nabla_\mu T^{\mu\nu} = 0. \tag{A.128}$$

Equality (A.127) is a direct consequence of the definition of 4-velocity, $u^\mu = dx^\mu/ds$, while equality (A.128) is the covariant conservation law for the energy-momentum tensor.

Problem A.30. *Write various components of the conservation law (A.128) explicitly in the case of flat space. Show that in the non-relativistic limit (i.e., for $|\mathbf{v}| \ll 1, p \ll \rho$) the resulting equations coincide with the hydrodynamic continuity equation and the Euler equation.*

To conclude this Section, we note that in the linearized theory with $\Lambda = 0$, Eqs. (A.119) and (A.121) give the following equation for the Newtonian potential in the presence of static source,

$$\Delta\Phi = 4\pi G(T_{00} + T_{ii}). \tag{A.129}$$

(Summation over i is assumed.) For the energy–momentum tensor of the form (A.124) we have

$$\Delta\Phi = 4\pi G(\rho + 3p). \tag{A.130}$$

In this sense, the source of the gravitational field in GR is not energy but rather the combination $(\rho+3p)$. In particular, an object made of hypothetical matter with $\rho+3p < 0$ would repel rather than attract non-relativistic particles. (In other words, it would "antigravitate".) The homogeneous isotropic Universe filled with such a matter would undergo accelerated expansion; see Sec. 3.2.4.

A.11.   Notations and Conventions

Indices $\mu, \nu, \ldots$ refer to space–time and take values 0, 1, 2, 3. The summation over repeated indices is assumed.

Indices $i, j, \ldots$ refer to space, $i, j = 1, 2, 3$. Spatial vectors are marked with bold-face font. The summation over repeated lower spatial indices is assumed, e.g., $a_i b_i = \mathbf{ab}$, $a_i a_i = a^2$.

The signature of metric is $(+, -, -, -)$.

The Riemann tensor is defined by

$$[\nabla_\mu, \nabla_\nu]A^\lambda = A^\sigma R^\lambda_{\sigma\mu\nu}.$$

Its explicit expression is given in (A.31). The Ricci tensor is equal to

$$R_{\mu\nu} = \partial_\lambda \Gamma^\lambda_{\mu\nu} - \partial_\mu \Gamma^\lambda_{\lambda\nu} + \Gamma^\lambda_{\rho\lambda}\Gamma^\rho_{\mu\nu} - \Gamma^\lambda_{\rho\mu}\Gamma^\rho_{\nu\lambda}.$$

Minkowski metric is denoted by $\eta_{\mu\nu} = \mathrm{diag}(1, -1, -1, -1)$. The metric with small perturbations about the spatially flat Friedmann–Lemaître–Robertson–Walker solution reads

$$ds^2 = a^2(\eta)(\eta_{\mu\nu} + h_{\mu\nu})dx^\mu dx^\nu,$$

where $x^0 = \eta$ is conformal time. In other words

$$g_{\mu\nu} = a^2(\eta)(\eta_{\mu\nu} + h_{\mu\nu}).$$

Indices of $h_{\mu\nu}$ are raised and lowered by the Minkowski metric.

Our sign convention is that

$$e^{-i\omega t}, \quad \omega > 0$$

is a negative-frequency function.

Appendix B

Standard Model of Particle Physics

In this Appendix we introduce the main elements of the Standard Model of particle physics. Of course, our presentation cannot be exhaustive; we do not consider numerous concrete phenomena in the world of elementary particles. Our purpose is to briefly describe those aspects which are used in the main text.

B.1. Field Content and Lagrangian

The Standard Model is in excellent agreement with all known to date experimental data (with the exception of neutrino oscillations, see Appendix C), obtained in low-energy physics, high-precision measurements and high-energy physics [1]. The basis of the Standard Model is quantum field theory; see, e.g., books [268–271].

The Standard Model describes the following particles thought to be elementary to date:

(a) gauge bosons: photon, gluon, $W^{\pm}$-bosons, Z-boson;
(b) quarks: u, d, s, c, b and t;
(c) leptons: electrically charged (electron e, muon μ and τ-lepton) and neutral (electron neutrino ν_e, muon neutrino ν_μ and τ-neutrino ν_τ);
(d) neutral Higgs boson h.

Particles of types "a" and "d" are bosons, particles of types "b" and "c" are fermions. Fields describing particles of type "a" are gauge fields. They are vectors under the Lorentz group and serve as mediators of gauge interactions. In particle physics, fields of types "b" and "c" are often called matter fields; we will avoid this terminology. They are spinors under the Lorentz group and participate in gauge and Yukawa interactions.[1] The field describing the Higgs boson is scalar; it participates in the Yukawa interactions and self-interactions. In addition, the Englert–Brout–Higgs field plays a special role: its vacuum expectation value gives masses to all massive Standard Model particles.

---

[1] Non-Abelian gauge fields also carry charge of their own gauge group and participate in gauge interactions.

The Standard Model gauge group is $SU(3)_c \times SU(2)_W \times U(1)_Y$. It describes strong interactions (color group $SU(3)_c$ with gauge coupling g_s) and electroweak interactions ($SU(2)_W \times U(1)_Y$ with gauge couplings g and g', respectively). The electroweak gauge group is in the Englert–Brout–Higgs phase, with the electromagnetic group $U(1)_{em}$ left unbroken. Accordingly, $W^\pm$- and Z-bosons are massive, and photon remains massless. The Standard Model fields form complete multiplets with respect to these gauge groups, i.e., they belong to certain representations of these groups.

Gauge fields are in the adjoint representations of their groups: there are eight gluon fields G_μ^a ($a = 1, \ldots, 8$, according to the number of generators of $SU(3)_c$), three gauge fields V_μ^i of $SU(2)_W$ ($i = 1, 2, 3$ corresponding to the generators of $SU(2)_W$) and one field B_μ of $U(1)_Y$. As a result of the Englert–Brout–Higgs mechanism, three combinations of the fields V_μ^i and B_μ are massive and describe $W^\pm$- and Z-bosons,

$$W_\mu^\pm = \frac{1}{\sqrt{2}}(V_\mu^1 \mp iV_\mu^2), \tag{B.1}$$

$$Z_\mu = \frac{1}{\sqrt{g^2 + g'^2}}(gV_\mu^3 - g'B_\mu). \tag{B.2}$$

The fourth combination,

$$A_\mu = \frac{1}{\sqrt{g^2 + g'^2}}(g'V_\mu^3 + gB_\mu), \tag{B.3}$$

remains massless and describes the photon. The relation between the fields Z_μ and A_μ and original gauge fields is also written in the form

$$Z_\mu = \cos\theta_W \cdot V_\mu^3 - \sin\theta_W \cdot B_\mu,$$
$$A_\mu = \cos\theta_W \cdot B_\mu + \sin\theta_W \cdot V_\mu^3,$$

where θ_W is the weak mixing angle,

$$\tan\theta_W = \frac{g'}{g}.$$

The experimental value of $\sin\theta_W$ is[2]

$$\sin\theta_W = 0.481.$$

Table B.1 contains dimensions of representations of vector fields and their Abelian charges. Note that in this Table as well as in a number of subsequent

---

[2]Here and in what follows, if not stated otherwise, we omit important details related to radiative corrections.

Table B.1. Dimensions of representations and charges of gauge and Englert–Brout–Higgs fields; symbol 0* indicates that fields B_μ and A_μ are gauge fields of groups $U(1)_Y$ and $U(1)_{em}$, respectively.

Field\Group	$SU(3)_c$	$SU(2)_w$	$U(1)_Y$	$U(1)_{em}$
G_μ	8	1	0	0
V_μ	1	3	0	
B_μ	1	1	0*	
$W_\mu^\pm$	1			± 1
Z_μ	1			0
A_μ	1			0*
H	1	2	1	

formulas, we use matrix notations

$$G_\mu \equiv \sum_{a=1}^{8} G_\mu^a \frac{\lambda^a}{2},$$

$$V_\mu \equiv \sum_{i=1}^{3} V_\mu^i \frac{\tau^i}{2},$$

where λ^a are the Gell-Mann matrices, and τ^i are the Pauli matrices ($\lambda^a/2$ and $\tau^i/2$ are generators of $SU(3)_c$ and $SU(2)_w$, respectively).

Fermion fields "b" and "c" form three generations of quarks and leptons,

$$\text{I}: \ u, \ d, \ \nu_e, \ e$$

$$\text{II}: \ c, \ s, \ \nu_\mu, \ \mu$$

$$\text{III}: \ t, \ b, \ \nu_\tau, \ \tau.$$

Particles within a single generation are discriminated by gauge interactions (they have different gauge quantum numbers), while the three "partner" particles of different generations (e.g., u-, c- and t-quarks or electron, muon and τ-lepton) have the same gauge quantum numbers but different masses and Yukawa couplings to the Higgs boson.

One way to describe fermionic fields in $(3+1)$-dimensional Minkowski space[3] is to introduce left and right two-component (Weyl) spinors χ_L and χ_R. These spinors transform independently under the *proper* Lorentz group.[4]

---

[3] We leave without discussion the description of fermion fields in curved space–time. It is considered in the accompanying book.

[4] More precisely, they transform according to fundamental (χ_L) and antifundamental (χ_R^T) representations of group $SL(2, \mathbb{C})$, which is the universal covering group for the proper Lorentz group $SO(3, 1)$. The covering is two-fold; this property is responsible, in particular, for the fact that not fermion fields themselves, but their bilinear combinations are physical observables.

Of the two-component spinors one can construct Lorentz scalars, vectors and tensors. In particular, one can show that bilinear combinations

$$\chi_L^T i\sigma_2 \chi_L, \quad \chi_R^T i\sigma_2 \chi_R,$$

are scalars, and

$$\chi_R^T \bar\sigma^\mu \chi_L, \quad \chi_L^T \sigma^\mu \chi_R$$

are vectors. Here

$$\sigma^\mu = (\mathbb{1}, \boldsymbol{\sigma}), \quad \bar\sigma^\mu = (\mathbb{1}, -\boldsymbol{\sigma}),$$

where $\boldsymbol{\sigma}$ are the conventional Pauli matrices acting on the Lorentz indices.

Problem B.1. *Check the validity of the above statements. Hint: Make use of the equivalence of the fundamental and antifundamental representations of $SU(2)$; find the transformation law for spinors χ_L and χ_R under the Lorentz boosts and 3-dimensional rotations.*

The *full* Lorentz group, besides proper transformations (boosts and rotations), also contains reflection of space (parity transformation) P and inversion of time T. Two-component spinor χ_L (or χ_R) does not transform to itself under spatial reflection. The relevant representation of the full Lorentz group is in terms of 4-component Dirac spinor ψ. The Dirac spinor includes two 2-component Weyl spinors χ_L and χ_R,

$$\psi = \begin{pmatrix} \chi_L \\ \chi_R \end{pmatrix}.$$

The free Dirac field obeys the Dirac equation

$$i\gamma^\mu \partial_\mu \psi = m\psi,$$

where m is fermion mass and γ^μ are four 4×4 Dirac matrices obeying anticommutation relations

$$\{\gamma^\mu, \gamma^\nu\} = \eta^{\mu\nu}.$$

In the chiral (Weyl) representation, the Dirac matrices are

$$\gamma^\mu = \begin{pmatrix} 0 & \sigma^\mu \\ \bar\sigma^\mu & 0 \end{pmatrix}.$$

In this representation, the Dirac equation is written in the matrix form,

$$\begin{pmatrix} 0 & i\sigma^\mu \partial_\mu \\ i\bar\sigma^\mu \partial_\mu & 0 \end{pmatrix} \begin{pmatrix} \chi_L \\ \chi_R \end{pmatrix} = m \begin{pmatrix} \chi_L \\ \chi_R \end{pmatrix}$$

Note that in the massless case, $m = 0$, the Dirac equation splits into two separate equations for each of the components χ_L, χ_R; massless solutions χ_L and χ_R are eigenfunctions of the helicity operator $\frac{\mathbf{P}\boldsymbol{\sigma}}{|\mathbf{P}|}$ with eigenvalues -1 and $+1$, respectively.[5] Therefore, in the massless case the minimum option is to introduce one 2-component spinor χ_L, so that the theory contains only particles of left helicity and antiparticles of right helicity. This is precisely the way neutrino is described in the Standard Model. Of course, parity is broken in such a situation.

To describe the Standard Model interactions in terms of the Dirac 4-component spinors, it is necessary to extract the components χ_L and χ_R. This is done by using projection operators

$$P_\mp = \frac{1 \mp \gamma^5}{2}, \quad \gamma^5 \equiv i\gamma^0\gamma^1\gamma^2\gamma^3.$$

In the chiral representation

$$\gamma^5 = \begin{pmatrix} -1 & 0 \\ 0 & 1 \end{pmatrix}.$$

In what follows we use the notations

$$\psi_L \equiv P_-\psi = \frac{1-\gamma_5}{2}\psi, \quad \psi_R \equiv P_+\psi = \frac{1+\gamma_5}{2}\psi. \tag{B.4}$$

In the chiral representation of the Dirac matrices

$$\psi_L = \begin{pmatrix} \chi_L \\ 0 \end{pmatrix}, \quad \psi_R = \begin{pmatrix} 0 \\ \chi_R \end{pmatrix}.$$

For some applications a useful observation is that $\chi_R^c \equiv i\sigma_2\chi_R^*$ is *left* spinor.

Problem B.2. *Prove the last statement above.*

If not otherwise stated, we use 4-component spinors in what follows. The most commonly used Lorentz structures bilinear in fermion fields are

$$\begin{array}{llll} \bar{\psi}\psi, & \text{scalar}, & \bar{\psi}\gamma^\mu\psi, & \text{vector}, \\ \bar{\psi}\gamma^5\psi, & \text{pseudoscalar}, & \bar{\psi}\gamma^5\gamma^\mu\psi, & \text{pseudovector}, \end{array} \tag{B.5}$$

where

$$\bar{\psi} \equiv \psi^\dagger\gamma^0$$

is the Dirac conjugate spinor.

Problem B.3. *Check the validity of the Lorentz assignment* (B.5). *Express these structures in terms of the Weyl fermions.*

---

[5] In both massless and massive cases, helicity is the projection of spin onto the direction of motion. However, helicity is Lorentz invariant only for massless fermions.

Within the Standard Model, neutrinos have only left components, in contrast to quarks and charged leptons. With respect to strong interactions, both left and right quarks form fundamental representations (triplets), so that from the standpoint of strong interactions the separation of quarks into left and right components is not required. On the other hand, right quarks and right charged leptons are singlets under $SU(2)_W$, while the left fermions form doublets

$$Q_1 = \begin{pmatrix} u \\ d \end{pmatrix}_L , \quad Q_2 = \begin{pmatrix} c \\ s \end{pmatrix}_L , \quad Q_3 = \begin{pmatrix} t \\ b \end{pmatrix}_L ,$$

$$L_1 = \begin{pmatrix} \nu_e \\ e \end{pmatrix}_L , \quad L_2 = \begin{pmatrix} \nu_\mu \\ \mu \end{pmatrix}_L , \quad L_3 = \begin{pmatrix} \nu_\tau \\ \tau \end{pmatrix}_L . \tag{B.6}$$

Similarly, the convenient notation for the right fermions is

$$U_n = u_R, \ c_R, \ t_R ; \quad n = 1, 2, 3 ;$$
$$D_n = d_R, \ s_R, \ b_R ; \tag{B.7}$$
$$E_n = e_R, \ \mu_R, \ \tau_R.$$

Dimensions of the fermion representations and Abelian charges are given in Table B.2.

The scalar Englert–Brout–Higgs field H is singlet under the group of strong interactions, doublet under $SU(2)_W$ and carries $U(1)_Y$ charge $+1$. These properties are reflected in Table B.1.

Table B.2.   Dimensions of representations and charges of fermions of the first generation; fermions of the second and third generations have exactly the same quantum numbers.

Field\Group	$SU(3)_c$	$SU(2)_W$	$U(1)_Y$	$U(1)_{em}$
$L \equiv \begin{pmatrix} \nu_e \\ e \end{pmatrix}_L$	1	2	-1	$\begin{pmatrix} 0 \\ -1 \end{pmatrix}$
$E \equiv e_R$	1	1	-2	-1
$Q \equiv \begin{pmatrix} u \\ d \end{pmatrix}_L$	3	2	$+1/3$	$\begin{pmatrix} +2/3 \\ -1/3 \end{pmatrix}$
$U \equiv u_R$	3	1	$+4/3$	$+2/3$
$D \equiv d_R$	3	1	$-2/3$	$-1/3$

In terms of the fields explicitly covariant under the gauge group $SU(3)_c \times SU(2)_w \times U(1)_Y$ the Standard Model Lagrangian reads

$$\mathcal{L}_{SM} = -\frac{1}{2}\text{Tr}G_{\mu\nu}G^{\mu\nu} - \frac{1}{2}\text{Tr}V_{\mu\nu}V^{\mu\nu} - \frac{1}{4}B_{\mu\nu}B^{\mu\nu}$$

$$+ i\bar{L}_n \mathcal{D}^\mu \gamma_\mu L_n + i\bar{E}_n \mathcal{D}^\mu \gamma_\mu E_n + i\bar{Q}_n \mathcal{D}^\mu \gamma_\mu Q_n + i\bar{U}_n \mathcal{D}^\mu \gamma_\mu U_n + i\bar{D}_n \mathcal{D}^\mu \gamma_\mu D_n$$

$$- (Y^l_{mn}\bar{L}_m H E_n + Y^d_{mn}\bar{Q}_m H D_n + Y^u_{mn}\bar{Q}_m \tilde{H} U_n + h.c.)$$

$$+ \mathcal{D}_\mu H^\dagger \mathcal{D}^\mu H - \lambda \left(H^\dagger H - \frac{v^2}{2}\right)^2. \qquad (B.8)$$

Here the first line contains gauge fields, whose strength tensors are

$$B_{\mu\nu} \equiv \partial_\mu B_\nu - \partial_\nu B_\mu,$$

$$V_{\mu\nu} \equiv \partial_\mu V_\nu - \partial_\nu V_\mu - ig[V_\mu, V_\nu],$$

$$G_{\mu\nu} \equiv \partial_\mu G_\nu - \partial_\nu G_\mu - ig_s[G_\mu, G_\nu],$$

where the square brackets denote commutators; for instance, $[V_\mu, V_\nu] \equiv V_\mu V_\nu - V_\nu V_\mu$. The field B_μ is real and fields V_μ, G_μ are Hermitean. In terms of real fields G^a_μ and V^i_μ

$$\text{Tr}G_{\mu\nu}G^{\mu\nu} = \frac{1}{2}G^a_{\mu\nu}G^{a\,\mu\nu}, \quad \text{Tr}V_{\mu\nu}V^{\mu\nu} = \frac{1}{2}V^i_{\mu\nu}V^{i\,\mu\nu},$$

with

$$G^a_{\mu\nu} = \partial_\mu G^a_\nu - \partial_\nu G^a_\mu + g_s f^{abc}G^b_\mu G^c_\nu, \quad V^i_{\mu\nu} = \partial_\mu V^i_\nu - \partial_\nu V^i_\mu + g\epsilon^{ijk}V^j_\mu V^k_\nu, \qquad (B.9)$$

where f^{abc} and ϵ^{ijk} are the structure constants of $SU(3)$ and $SU(2)$, respectively. (ϵ^{ijk} is a completely antisymmetric symbol.) The last terms in (B.9) are responsible for gluon self-interactions and for interactions between $W^\pm$-, Z-bosons and photons.

The second line in (B.8) includes free Lagrangians of fermions and fermion couplings to gauge fields. Covariant derivatives entering there are uniquely determined by gauge group representations of fermions and their $U(1)_Y$-charges: for fermion f in representations T_s and T_w of $SU(3)_c$ and $SU(2)_w$

$$\mathcal{D}_\mu f \equiv \left(\partial_\mu - ig_s T^a_s G^a_\mu - igT^i_w V^i_\mu - ig'\frac{Y_f}{2}B_\mu\right)f,$$

where T^a_s and T^i_w are generators of $SU(3)_c$ and $SU(2)_w$ and Y_f is $U(1)_Y$-charge of this fermion. For quarks $T^a_s = \lambda^a/2$, while for leptons $T^a_s = 0$. For left doublets (B.6) we have $T^i_w = \tau^i/2$ and for right singlets $T^i_w = 0$. Note that summation over generations is assumed in the second line of (B.8). In terms of the fields used in (B.8), gauge interactions are diagonal in generations.

The third line in (B.8) describes the Yukawa interactions of fermions with the Englert–Brout–Higgs field H; *h.c.* means the Hermitean conjugation and summation over repeated indices m, n labeling generations is assumed. Yukawa coupling matrices Y^l_{mn}, Y^d_{mn} and Y^u_{mn} are complex and not diagonal in generations. Below we briefly discuss the consequences of this property. We emphasize that the Lagrangian (B.8) does not describe neutrino oscillations: the relevant Yukawa terms are absent. We consider in Appendix C extensions of the Standard Model capable of describing the phenomenon of neutrino oscillations.

Let us make a remark concerning the third line in (B.8). Like the complete Lagrangian of the Standard Model, this line is invariant under $SU(3)_c \times SU(2)_W \times U(1)_Y$. To see this explicitly, we begin with the first term in the third line. Left lepton and Englert–Brout–Higgs field are doublets under $SU(2)_W$, while the right lepton is singlet. Therefore, the first term has $SU(2)_W$-structure $(L^\dagger H)E$; it is $SU(2)_W$-singlet. According to Table B.2, the total $U(1)_Y$-charge of the fields in this term is zero, which means $U(1)_Y$-invariance. The situation with the second Yukawa term is similar. The last term includes

$$\tilde{H}_\alpha \equiv i\tau_{\alpha\beta}^2 H^{*\,\beta} = \epsilon_{\alpha\beta} H^{*\,\beta}, \qquad (B.10)$$

where $\alpha = 1, 2$ and $\epsilon_{\alpha\beta}$ is the antisymmetric symbol. The field $\tilde{H}$ is in the fundamental representation[6] of $SU(2)_W$. Therefore, the third Yukawa term is $SU(2)_W$-singlet. It is also invariant under $U(1)_Y$, this is the reason for using $\tilde{H}$ there rather than the field H itself.

It is important to emphasize that the second and third lines in (B.8) give the most general gauge invariant renormalizable Lagrangian for the Standard Model fermions. In particular, *explicit* mass terms of fermions, which would also have the Lorentz structure $\overline{(f_L)} \cdot f_R + h.c.$, are forbidden by the invariance under $SU(2)_W \times U(1)_Y$.

The last line in Eq. (B.8) is the Lagrangian of the Englert–Brout–Higgs field itself. According to Table B.1, the covariant derivative of the scalar field is

$$\mathcal{D}_\mu H = \left(\partial_\mu - ig\frac{\tau^i}{2}V_\mu^i - i\frac{g'}{2}B_\mu \right) H.$$

The scalar potential of the theory — the last term in (B.8) — has minimum at nonzero value of the Englert–Brout–Higgs field such that

$$H^\dagger H = \frac{v^2}{2}.$$

Using gauge invariance, one can show that the Englert–Brout–Higgs vacuum and the perturbations of the scalar field about this vacuum, without loss of generality,

---

[6] H^* is in anti-fundamental representation. The relation (B.10) gives isomorphism between the anti-fundamental and fundamental representations. Such an isomorphism exists for $SU(2)$ but not for $SU(N)$ with $N > 2$.

can be written in the form (unitary gauge)

$$H(x) = \begin{pmatrix} 0 \\ \frac{v}{\sqrt{2}} + \frac{h(x)}{\sqrt{2}} \end{pmatrix}. \tag{B.11}$$

Thus, there is only one physical scalar excitation about the Englert–Brout–Higgs vacuum. This is the Higgs boson, described by the field h.

The Englert–Brout–Higgs vacuum breaks the gauge symmetry $SU(2)_W \times U(1)_Y$ down to $U(1)_{em}$. Conversion from explicitly $SU(2)_W \times U(1)_Y$-invariant fields to physical vector fields in this vacuum is accomplished by the change of variables (B.1), (B.3). Indeed, upon this change, free gradient terms of the fields $W_\mu^\pm$, Z_μ and A_μ maintain canonical form

$$-\frac{1}{4}\left[\sum_{i=1}^{3}(\partial_\mu V_\nu^i - \partial_\nu V_\mu^i)^2 + (\partial_\mu B_\nu - \partial_\nu B_\mu)^2\right]$$

$$= -\frac{1}{2}(\partial_\mu W_\nu^+ - \partial_\nu W_\mu^+)(\partial^\mu W^{-\,\nu} - \partial^\nu W^{-\,\mu}) - \frac{1}{4}Z_{\mu\nu}Z^{\mu\nu} - \frac{1}{4}F_{\mu\nu}F^{\mu\nu},$$

where

$$F_{\mu\nu} = \partial_\mu A_\nu - \partial_\nu A_\mu, \quad Z_{\mu\nu} = \partial_\mu Z_\nu - \partial_\nu Z_\mu. \tag{B.12}$$

The vacuum expectation value in (B.11) leads to mass terms in quadratic Lagrangian about the Englert–Brout–Higgs vacuum,

$$\mathcal{D}_\mu H^\dagger \mathcal{D}^\mu H \longrightarrow \frac{g^2 v^2}{4} W_\mu^+ W^{\mu-} + \frac{(g^2 + g'^2)v^2}{8} Z_\mu Z^\mu.$$

This is precisely what the Englert–Brout–Higgs mechanism is about. The change of variables (B.1), (B.3) is chosen in such a way that these mass terms are diagonal. Thus, the masses of $W^\pm$- and Z-bosons are

$$M_W = \frac{gv}{2}, \quad M_Z = \frac{v\sqrt{g^2 + g'^2}}{2} = \frac{M_W}{\cos\theta_W}.$$

The Yukawa couplings are responsible for the fact that quarks and charged leptons are massive as well. The masses of fermions are given by

$$m_f = \frac{y_f}{\sqrt{2}} v,$$

where y_f are eigenvalues of the matrix of Yukawa couplings entering (B.8). The only fermions that remain massless are neutrinos. (The latter property is in fact a shortcoming of the Standard Model.)

The Lagrangian of the Standard Model in terms of physical fields is obtained by substituting (B.1), (B.3) and (B.11) into (B.8) and performing the transformation of fermion fields from the original basis in which gauge interactions are diagonal, to the basis where diagonal structure is possessed by the fermion mass matrix (and Yukawa terms). As we discuss in Sec. B.3, the latter transformation gives rise to the Cabibbo–Kobayashi–Maskawa (CKM) matrix V_{mn} describing quark mixing in weak interactions. Note that we often use the same notation for fermions in the two bases.

It is convenient to present the full Lagrangian of the Standard Model written in terms of physical fields as a sum of several terms,

$$\mathcal{L}_{SM} = \mathcal{L}_{QCD} + \mathcal{L}_{lept}^{free} + \mathcal{L}_{f,em} + \mathcal{L}_{f,weak} + \mathcal{L}_Y + \mathcal{L}_V + \mathcal{L}_H + \mathcal{L}_{HV}^{int}. \qquad (B.13)$$

Here

$$\mathcal{L}_{QCD} = -\frac{1}{4}G_{\mu\nu}^a G^{a\ \mu\nu} + \sum_{quarks} \bar{q}\left(i\gamma^\mu\partial_\mu - m_q - ig_s\frac{\lambda^a}{2}G_\mu^a\right)q$$

is the Lagrangian containing quarks, gluons and their interactions with each other; summation runs over all types of quarks. This is the Lagrangian of the theory of strong interactions, quantum chromodynamics (QCD). The second term in (B.13) is the free Lagrangian of leptons

$$\mathcal{L}_{lept}^{free} = \sum_n \bar{l}_n(i\gamma^\mu\partial_\mu - m_{l_n})l_n + \sum_n \bar{\nu}_n i\gamma^\mu\partial_\mu\nu_n.$$

Here n runs over generations, $l_n = e, \mu, \tau$. The third and the fourth terms describe electromagnetic and weak interactions of quarks and leptons, respectively,

$$\mathcal{L}_{f,em} = e\sum_f q_f \bar{f}\gamma^\mu A_\mu f,$$

where

$$e = g\sin\theta_W = \frac{gg'}{\sqrt{g^2 + g'^2}} \qquad (B.14)$$

is the proton electric charge, so that eq_f is the electric charge of fermion f;

$$\mathcal{L}_{f,weak} = \frac{g}{2\sqrt{2}}\sum_n(\bar{\nu}_n\gamma^\mu(1 - \gamma^5)W_\mu^+ l_n + h.c.)$$

$$+ \frac{g}{2\sqrt{2}}\sum_{m,n}(\bar{u}_m\gamma^\mu(1 - \gamma^5)W_\mu^+ V_{mn}d_n + h.c.)$$

$$+ \frac{g}{2\cos\theta_W}\sum_f \bar{f}\gamma^\mu(t_3^f(1 - \gamma^5) - 2q_f\sin^2\theta_W)f Z_\mu. \qquad (B.15)$$

Here the sum over f means summation over all quarks and leptons, t_3^f is the weak isospin equal to $+1/2$ for up quarks u, c, t and neutrinos and $-1/2$ for down quarks d, s, b and charged leptons. The first two terms in $\mathcal{L}_{f,weak}$ are couplings of leptons and quarks to W-bosons (charged currents), the third term is the coupling to Z-boson (neutral currents). Note that the emission and absorption of W-boson changes the type (flavor) of fermion (and for quarks, generally speaking, the generation number as well, thanks to the non-diagonal CKM matrix V_{mn}), while the interaction with Z-boson does not change flavor.

Problem B.4. *Check that the following relation holds for all fermions,*

$$q_f = \frac{Y_f}{2} + t_3^f.$$

The term $\mathcal{L}_Y$ in the Lagrangian (B.13) describes the Yukawa interactions of fermions with the Higgs boson,

$$\mathcal{L}_Y = -\sum_f \frac{y_f}{\sqrt{2}} \bar{f} f h = -\sum_f \frac{m_f}{v} \bar{f} f h,$$

where summation runs over all fermion species except for neutrinos. The Yukawa couplings are proportional to the fermion masses; this is of course a reflection of the fact that all fermions obtain masses due to interactions with the Englert–Brout–Higgs field.

The term $\mathcal{L}_V$ in (B.13) contains the free Lagrangians of photons, $W^\pm$- and Z-bosons and their interaction with each other,

$$\begin{aligned}
\mathcal{L}_V = &-\frac{1}{4} F_{\mu\nu} F^{\mu\nu} - \frac{1}{4} Z_{\mu\nu} Z^{\mu\nu} + \frac{M_Z^2}{2} Z_\mu Z^\mu \\
&- \frac{1}{2} |W_{\mu\nu}^-|^2 + M_W^2 |W_\mu^-|^2 + \frac{g^2}{4} (W_\mu^- W_\nu^+ - W_\mu^+ W_\nu^-)^2 \\
&- \frac{ig}{2} (F^{\mu\nu} \sin\theta_W + Z^{\mu\nu} \cos\theta_W)(W_\mu^- W_\nu^+ - W_\mu^+ W_\nu^-),
\end{aligned} \tag{B.16}$$

where $F_{\mu\nu}$ and $Z_{\mu\nu}$ are defined by Eq. (B.12), and

$$W_{\mu\nu}^- \equiv (\partial_\mu + ieA_\mu + ig\cos\theta_W Z_\mu)W_\nu^- - (\mu \leftrightarrow \nu). \tag{B.17}$$

The relation between the electromagnetic coupling e and gauge couplings g, g' is again given by (B.14). Note that the Lagrangian (B.16), as well as the full Lagrangian (B.13), is invariant under the unbroken gauge symmetry $U(1)_{em}$, and, according to (B.17), $W^\pm$-bosons carry electric charge $\pm e$.

The term $\mathcal{L}_H$ describes the Englert–Brout–Higgs sector of the Standard Model,

$$\mathcal{L}_H = \frac{1}{2} \partial_\mu h \partial^\mu h - \frac{1}{2} m_h^2 h^2 - \lambda v h^3 - \frac{\lambda}{4} h^4,$$

where

$$m_h = \sqrt{2\lambda} v$$

is the Higgs boson mass.

Finally, the term $\mathcal{L}_{HV}^{int}$ stands for the Higgs boson coupling to the massive vector bosons,

$$\mathcal{L}_{HV}^{int} = \frac{g^2}{2} vh|W_\mu^-|^2 + \frac{g^2 + g'^2}{4} vhZ_\mu Z^\mu + \frac{g^2}{4} h^2 |W_\mu^-|^2 + \frac{g^2 + g'^2}{8} h^2 Z_\mu Z^\mu.$$

To date, all particles of the Standard Model have been observed experimentally. The values of the Standard Model parameters are[7] [1]:

$$m_e = 0.511 \text{ MeV}, \quad m_u = 1.8 - 3.0 \text{ MeV}, \quad m_d = 4.5 - 5.3 \text{ MeV},$$
$$m_\mu = 105.7 \text{ MeV}, \quad m_c = 1.25 - 1.30 \text{ GeV}, \quad m_s = 0.09 - 0.10 \text{ GeV},$$
$$m_\tau = 1.78 \text{ GeV}, \quad m_t = 172.0 - 174.4 \text{ GeV}, \quad m_b = 4.15 - 4.21 \text{ GeV},$$
$$M_Z = 91.2 \text{ GeV}, \quad M_W = 80.4 \text{ GeV}, \quad m_h = 125.1 \text{ GeV},$$

$$v = 246.22 \text{ GeV}, \quad \alpha \equiv \frac{e^2}{4\pi} = \frac{1}{137}, \quad \sin^2 \theta_W = 0.231,$$
$$\alpha_s(M_Z) = 0.118.$$

Uncertainties in quark masses (except for t-quark) are predominantly theoretical; they are due to the fact that quarks do not exist in free state.

B.2.   Global Symmetries

In addition to gauge symmetries, there are global Abelian symmetries in the Standard Model: the Lagrangian (B.13) is invariant under simultaneous phase rotations of all quark fields,

$$q \to e^{i\beta/3} q, \quad \bar{q} \to e^{-i\beta/3} \bar{q} \tag{B.18}$$

and independently under phase rotations of lepton fields of each generation,

$$(\nu_e, e) \to e^{i\beta_e}(\nu_e, e), \quad (\bar{\nu}_e, \bar{e}) \to e^{-i\beta_e}(\bar{\nu}_e, \bar{e}), \tag{B.19}$$
$$(\nu_\mu, \mu) \to e^{i\beta_\mu}(\nu_\mu, \mu), \quad (\bar{\nu}_\mu, \bar{\mu}) \to e^{-i\beta_\mu}(\bar{\nu}_\mu, \bar{\mu}), \tag{B.20}$$
$$(\nu_\tau, \tau) \to e^{i\beta_\tau}(\nu_\tau, \tau), \quad (\bar{\nu}_\tau, \bar{\tau}) \to e^{-i\beta_\tau}(\bar{\nu}_\tau, \bar{\tau}). \tag{B.21}$$

Here β, β_e, β_μ and β_τ are independent parameters of the transformations. The conserved quantum number associated with the symmetry (B.18) is baryon number[8]

$$B = \frac{1}{3}(N_q - N_{\bar{q}}),$$

---

[7] We omit the details related to the dependence of these parameters on the renormalization scale, except for the case of α_s.

[8] Baryon number is conserved in perturbation theory only. Nonperturbative effects violate baryon and lepton numbers, but these effects are very small under normal conditions (but not in the early Universe, see Chapter 11).

where N_q and $N_{\bar{q}}$ are total numbers of quarks and antiquarks of all types, respectively. The baryonic charge of quarks by definition is equal to $1/3$, and the charge of antiquarks is $(-1/3)$. With this assignment, the total baryonic charge of proton, which consists of two u-quarks and one d-quark, equals 1. Hence, in terms of numbers of baryons and antibaryons we have

$$B = \sum(N_B - N_{\bar{B}}),$$

where the summation runs over all types of baryons. A clear manifestation that baryon number is conserved in Nature with high precision is proton stability: proton is the lightest particle carrying baryon number, so it must be absolutely stable if the baryon number conservation is exact. Proton decay has not yet been discovered, and the experimental limit on the lifetime is [1]

$$\tau_p > 10^{31} - 10^{34} \text{ years,}$$

depending on the decay mode.

Symmetries (B.19)–(B.21) are associated with three independently conserved lepton numbers (electron, muon and tau)

$$L_e = (N_e + N_{\nu_e}) - (N_{e^+} + N_{\bar{\nu}_e}), \tag{B.22}$$

$$L_\mu = (N_\mu + N_{\nu_\mu}) - (N_{\mu^+} + N_{\bar{\nu}_\mu}), \tag{B.23}$$

$$L_\tau = (N_\tau + N_{\nu_\tau}) - (N_{\tau^+} + N_{\bar{\nu}_\tau}), \tag{B.24}$$

where N_e, N_{ν_e}, N_{e^+}, $N_{\bar{\nu}_e}$ are numbers of electrons, electron neutrinos, positrons and electron antineutrinos, respectively, and similarly for other generations.

A manifestation of the lepton quantum number conservation is the absence of processes violating lepton numbers, but otherwise allowed. An example of such a process is the decay

$$\mu \to e\gamma.$$

In this process, electron and muon numbers would be violated. The experimental bound on its branching ratio is [1]

$$\text{Br}(\mu \to e\gamma) < 5.7 \cdot 10^{-13}.$$

This is one of the best results showing the conservation of the muon and electron numbers. Note that the observed neutrino oscillations show that lepton numbers are actually violated in Nature; see Appendix C.

In many models generalizing the Standard Model, baryon and/or lepton numbers are violated, which should lead to new physical phenomena, such as proton decay. The search for these phenomena is an important challenge for low-energy particle physics experiments. At present, besides neutrino oscillations (see Appendix C), no experimental evidence for violation of baryon or lepton numbers has been found.

B.3. *C-*, *P-*, *T*-Transformations

Let us briefly consider discrete transformations: time reversal,

$$T\text{-transformation:}\quad (x^0,\mathbf{x}) \xrightarrow{\ \mathrm{T}\ } (-x^0,\mathbf{x}),$$

spatial reflection (parity transformation),

$$P\text{-transformation:}\quad (x^0,\mathbf{x}) \xrightarrow{\ \mathrm{P}\ } (x^0,-\mathbf{x}),$$

and charge conjugation,

$$C\text{-transformation:}\quad f(x) \xrightarrow{\ \mathrm{C}\ } f^c(x).$$

P- and *T*-transformations, together with the proper Lorentz group, form the full Lorentz group.

From the standpoint of scattering processes, time reversal means the replacement of initial state by final state and vice versa; spatial reflection implies the inversion of spatial momenta of all particles, and charge conjugation interchanges particles and antiparticles. In quantum field theory, the *CPT*-theorem is valid, which states that physical processes must be invariant under the joint action of all three transformations. One of its consequences is the equality of masses and total decay widths of particle and its antiparticle.

Lorentzian scalars, vectors and tensors[9] can be even and odd under *P*-transformations. In the latter case they are called pseudoscalars, axial vectors and pseudotensors. For instance, scalar (parity-even) and pseudoscalar (parity-odd) transform under *P*-transformation as

$$\phi(x^0,\mathbf{x}) \xrightarrow{\ \mathrm{P}\ } \phi'(x^0,\mathbf{x}) = \phi(x^0,-\mathbf{x}),$$

and

$$\phi(x^0,\mathbf{x}) \xrightarrow{\ \mathrm{P}\ } \phi'(x^0,\mathbf{x}) = -\phi(x^0,-\mathbf{x}),$$

respectively. *P*-transformation of vector (parity-even) reads

$$V_\nu(x^0,\mathbf{x}) \xrightarrow{\ \mathrm{P}\ } V'_\nu(x^0,\mathbf{x}) = \delta^0_\nu V_0(x^0,-\mathbf{x}) - \delta^i_\nu V_i(x^0,-\mathbf{x}),$$

while the *P*-transformation of axial vector is

$$A_\nu(x^0,\mathbf{x}) \xrightarrow{\ \mathrm{P}\ } A'_\nu(x^0,\mathbf{x}) = -\delta^0_\nu A_0(x^0,-\mathbf{x}) + \delta^i_\nu A_i(x^0,-\mathbf{x}),$$

Several spinor bilinear combinations of definite parity are given in (B.5).

It is clear from the form of the Lagrangian (B.13), (B.15), that weak interactions violate parity: weak bosons couple to both vector currents $\bar\psi_m\gamma^\mu\psi_n$, and axial currents $\bar\psi_m\gamma^\mu\gamma^5\psi_n$. Moreover, weak interactions violate *CP*. The source of *CP*-violation is a complex parameter in the Cabibbo–Kobayashi–Maskawa matrix. On the contrary, strong and electromagnetic interactions violate neither *P* nor *C*.

---

[9] We do not discuss here the properties of spinors under parity.

B.4. Quark Mixing

To understand how the Cabibbo–Kobayashi–Maskawa (CKM) matrix emerges, let us consider the transformation from the gauge basis of fermions to the mass basis. The Lagrangian (B.8) is written in terms of fields in gauge basis: all gauge interactions have diagonal form. As a result of spontaneous electroweak symmetry breaking, the Yukawa terms in the Lagrangian (B.8) in the unitary gauge (B.11) give rise to fermion mass terms,

$$\mathcal{L}_m = -\frac{v}{\sqrt{2}}Y^l_{mn}\bar{e}_{Lm}e_{Rn} - \frac{v}{\sqrt{2}}Y^d_{mn}\bar{d}_{Lm}d_{Rn} - \frac{v}{\sqrt{2}}Y^u_{mn}\bar{u}_{Lm}u_{Rn} + h.c., \qquad (\text{B.25})$$

as well as Yukawa couplings to the Higgs boson,

$$\mathcal{L}_Y = -\frac{h}{\sqrt{2}}Y^l_{mn}\bar{e}_{Lm}e_{Rn} - \frac{h}{\sqrt{2}}Y^d_{mn}\bar{d}_{Lm}d_{Rn} - \frac{h}{\sqrt{2}}Y^u_{mn}\bar{u}_{Lm}u_{Rn} + h.c. \qquad (\text{B.26})$$

The gauge interactions with gluons, photons and Z-bosons are still diagonal in fermions, while interactions with $W^\pm$-bosons are diagonal in generations,

$$\mathcal{L}_W = \frac{g}{\sqrt{2}}\bar{\nu}_n\gamma^\mu W^-_\mu l_{Ln} + \frac{g}{\sqrt{2}}\bar{u}_{Ln}\gamma^\mu W^-_\mu d_{Ln} + h.c.$$

The Yukawa couplings can be written in the form

$$Y^l_{mn} = U^{l_L}_{mp}Y^l_p(U^{l_R})^{-1}_{pn}, \quad Y^d_{mn} = U^{d_L}_{mp}Y^d_p(U^{d_R})^{-1}_{pn}, \quad Y^u_{mn} = U^{u_L}_{mp}Y^u_p(U^{u_R})^{-1}_{pn},$$

where constants Y^l_p, Y^d_p and Y^u_p are real, and $U^{e_L},\ldots,U^{u_R}$ are unitary matrices; summation over repeated index p is assumed.

Problem B.5. *Prove the last statement.*

It is precisely these matrices that transform the quark fields into the mass basis,

$$l_{Rm} = U^{l_R}_{mn}\tilde{l}_{Rn}, \quad d_{Rm} = U^{d_R}_{mn}\tilde{d}_{Rn}, \quad u_{Rm} = U^{u_R}_{mn}\tilde{u}_{Rn},$$
$$l_{Lm} = U^{l_L}_{mn}\tilde{l}_{Ln}, \quad d_{Lm} = U^{d_L}_{mn}\tilde{d}_{Ln}, \quad u_{Lm} = U^{u_L}_{mn}\tilde{u}_{Ln}.$$

Indeed, in terms of fields with tilde, the terms (B.25) and (B.26) are diagonal; for example

$$\frac{v}{\sqrt{2}}Y^l_{mn}\bar{l}_{Lm}l_{Rn} = \sum_p \frac{v}{\sqrt{2}}Y^l_p\bar{\tilde{l}}_{Lp}\tilde{l}_{Rp}.$$

Upon this transformation, kinetic terms of the fermion fields and also gauge interactions with gluons, photons and Z-boson remain diagonal, while the interaction with the $W^\pm$-bosons contains mixing matrices of generations.

Within the Standard Model, mixing in the lepton sector is unphysical: it can be eliminated by redefining the neutrino fields,

$$\nu_m \to \tilde{\nu}_m: \quad \tilde{\nu}_m = (U^{l_L})^{-1}_{mn} \nu_n.$$

Since all gauge interactions of neutrinos are proportional to unit matrix (in the space of generations), this redefinition does not give rise to mixing in the neutrino sector. The reason for the absence of mixing in the lepton sector of the Standard Model is that neutrinos are massless there. In reality, neutrinos are massive, and there is mixing between them. This is discussed in Appendix C.

In the quark sector the situation is different, and the transformation to the mass basis leads to mixing between quarks of different generations in the interaction vertices with $W^\pm$-bosons,

$$\mathcal{L}_W = \frac{g}{\sqrt{2}} \bar{\tilde{u}}_{L_m} \gamma^\mu W^+_\mu V_{mn} \tilde{d}_{L_n} + \frac{g}{\sqrt{2}} \bar{\tilde{d}}_{L_n} \gamma^\mu W^-_\mu V^*_{mn} \tilde{u}_{L_m}, \tag{B.27}$$

where $V_{ij} \equiv (U^{u_L})^{-1}_{ik} U^{d_L}_{kj}$ is the unitary 3×3 Cabibbo–Kobayashi–Maskawa mixing matrix.

A 3×3 unitary matrix of general form is specified by 9 real parameters (the dimension of the group of unitary matrices $U(3)$). In terms of the orthogonal subgroup $SO(3) \subset U(3)$ these parameters can be divided into 3 rotation angles and 6 phases. However, five of the six phases are unphysical. They can be eliminated from (B.27) by redefining the fermion fields $f_n \to f_n e^{i\beta_n}$. This phase rotation does not lead to complex coefficients in other parts of the Lagrangian (B.13), since fermion fields enter all its terms, except for (B.27), in combinations explicitly invariant under phase rotations, such as $\bar{f}_n \gamma_\mu f_n$. One remaining phase in the CKM matrix is the source of *CP*-violation in weak interactions (see below).

In the standard parameterization [1] in which the matrix V reads

$$V = \begin{pmatrix} c_{12}c_{13} & s_{12}c_{13} & s_{13}e^{-i\delta_{13}} \\ -s_{12}c_{23} - c_{12}s_{23}s_{13}e^{i\delta_{13}} & c_{12}c_{23} - s_{12}s_{23}s_{13}e^{i\delta_{13}} & s_{23}c_{13} \\ s_{12}s_{23} - c_{12}c_{23}s_{13}e^{i\delta_{13}} & -c_{12}s_{23} - s_{12}c_{23}s_{13}e^{i\delta_{13}} & c_{23}c_{13} \end{pmatrix}, \tag{B.28}$$

where $c_{ij} = \cos\theta_{ij}$, $s_{ij} = \sin\theta_{ij}$, the parameters θ_{ij} are mixing angles, and δ_{13} is the *CP*-violating phase. We show in the end of this Section that the matrix V can indeed be chosen in this form. The values of the parameters have been determined with good accuracy from numerous experiments [1]:

$$s_{12} = 0{,}2254, \quad s_{13} = 0{,}0035,$$
$$s_{23} = 0{,}04118, \quad \delta_{13} = 69° \pm 5°.$$

The absolute values of the elements of the quark mixing matrix are [1]

$$|V| = \begin{pmatrix} 0.9743 & 0.225 & 0.0036 \\ 0.226 & 0.9734 & 0.041 \\ 0.0087 & 0.041 & 0.9991 \end{pmatrix}, \tag{B.29}$$

with experimental errors of the order of the last digit. Note that *CP*-violating phase of the CKM matrix is large. Nevertheless, the resulting *CP*-violating effects in the Standard Model are strongly suppressed due to the smallness of mixing angles, which is clearly seen in the standard parameterization (B.28). Note also that there is strong hierarchy between the entries in (B.29): the diagonal elements are close to 1, while off-diagonal elements obey $V_{13}, V_{31} \ll V_{23}, V_{32} \ll V_{12}, V_{21} \ll 1$.

The general fact that complex entries in V_{mn} give rise to *CP*-violation, follows from the transformation law of fields under *CP*,

$$\psi_L \xrightarrow{P} \eta\gamma^0\psi_L, \quad \bar{\psi}_L \xrightarrow{P} \eta^{-1}\bar{\psi}_L\gamma_0, \tag{B.30}$$

$$\psi_L \xrightarrow{C} C\bar{\psi}_L^T, \quad \bar{\psi}_L \xrightarrow{C} \psi_L^T(C^{-1})^T, \tag{B.31}$$

where η is the sign factor ($\eta^2 = \pm 1$), and C is the charge conjugation matrix such that $\bar{\psi}\gamma^\mu\psi \xrightarrow{C} -\bar{\psi}\gamma^\mu\psi$. For the bilinear combinations of spinors which form the charged current coupled to W-boson (see (B.27)), the above formulas yield a simple transformation law (we no longer write tilde over fields in the mass basis),

$$\bar{u}_{Lm}\gamma^0 V_{mn}d_{Ln} \xrightarrow{CP} -\bar{d}_{Ln}\gamma^0 V_{mn}u_{Lm}, \quad \bar{u}_{Lm}\gamma^i V_{mn}d_{Ln} \xrightarrow{CP} \bar{d}_{Ln}\gamma^i V_{mn}u_{Lm}.$$

Taking into account the transformation law of W-bosons under *CP*-conjugation

$$W_0^\pm \xrightarrow{CP} -W_0^\mp, \quad W_i^\pm \xrightarrow{CP} W_i^\mp,$$

we finally obtain that under *CP*-transformation, the interaction (B.27) converts into

$$\mathcal{L}_W^{CP} = \frac{g}{\sqrt{2}}\bar{d}_{Ln}\gamma^\mu W_\mu^- V_{mn}u_{Lm} + \frac{g}{\sqrt{2}}\bar{u}_{Lm}\gamma^\mu W_\mu^+ V_{mn}^* d_{Ln}. \tag{B.32}$$

Comparison of (B.32) with (B.27) shows that the coupling (B.27) violates *CP* if the entries of mixing matrix V_{mn} are complex-valued.

This result reflects the general property of field theory: the initial Lagrangian transforms under *CP* into the Lagrangian with complex-conjugate couplings. If there are unremovable phases in the set of couplings, then *CP* is broken.

As an illustration, let us estimate the difference of partial widths of decays

$$t \to W^+b \quad \text{and} \quad \bar{t} \to W^-\bar{b}.$$

These widths would be equal if *CP* were exact symmetry of the Standard Model. (*CPT* ensures that the *total* widths of particles and antiparticles are the same, but

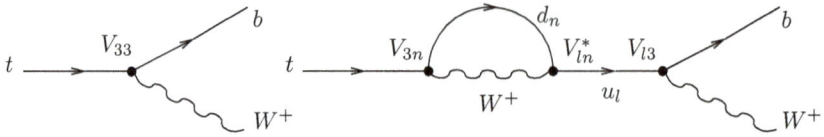

Fig. B.1.   Tree-level and one-loop diagrams giving the main contribution to the difference of partial widths of $t \to W^+b$ and $\bar{t} \to W^-\bar{b}$ in the Standard Model. The vertices are proportional to the elements of the CKM matrix which are explicitly indicated.

does not forbid different *partial* widths.) At the tree level, the partial widths are identical, and the main contribution to their difference comes from the interference of the tree and one-loop diagrams. For the decay $t \to W^+b$, these diagrams are shown in Fig. B.1.

Similar diagrams contribute to the decay $\bar{t} \to W^-\bar{b}$.

If one denotes by $V_{33}M_{tree}$ the tree-level contribution to the amplitude of the decay $t \to W^+b$, then the one-loop term is

$$M_{tree} \cdot \frac{g^2}{16\pi^2} \sum_{n,l=1}^{3} V_{3n} V_{ln}^* V_{l3} (A_{nl} + i\pi B_{nl}),$$

where A_{nl}, B_{nl} are real functions depending on masses of W-boson, t- and b-quarks, and also on masses of virtual quarks in the loop. In what follows we need only the function B_{nl}, which is conveniently written in the form

$$B_{nl} = -\frac{1}{1 - m_{u_l}^2/m_t^2} \int_0^1 x\, dx\, \tilde{C}_n \,,$$

$$\tilde{C}_n = \theta \left[-x^2 + \left(1 + \frac{m_{d_n}^2}{m_t^2} - \frac{M_W^2}{m_t^2} \right) x - \frac{m_{d_n}^2}{m_t^2} \right],$$

where $\theta(z)$ is the usual step-function. The imaginary part B_{nl} of the diagrams is related to the kinematics of virtual processes, rather than complex-valuedness of couplings: it arises from the integration over virtual momenta near mass shell. Since this is a kinematic effect, it is the same for decays of t and $\bar{t}$. Hence, the one-loop diagram for $\bar{t} \to W^-\bar{b}$ gives

$$M_{tree}^* \cdot \frac{g^2}{16\pi^2} \sum_{n,l=1}^{3} V_{3n}^* V_{ln} V_{l3}^* (A_{nl} + i\pi B_{nl}).$$

The decay widths are determined by the absolute values of the amplitudes squared,

$$|M(t \to W^+b)|^2 = |M_{tree}|^2 \tag{B.33}$$

$$\times \left\{ 1 + \frac{g^2}{16\pi^2} \left(V_{33}^* \sum_{l,n=1}^{3} V_{3n} V_{ln}^* V_{l3} (A_{nl} + i\pi B_{nl}) + h.c. \right) + \mathcal{O}(g^4) \right\}, \tag{B.34}$$

$$|M(\bar{t} \to W^-\bar{b})|^2 = |M_{tree}|^2 \tag{B.35}$$

$$\times \left\{ 1 + \frac{g^2}{16\pi^2} \left(V_{33} \sum_{l,n=1}^{3} V_{3n}^* V_{ln} V_{l3}^* (A_{nl} + i\pi B_{nl}) + h.c. \right) + \mathcal{O}(g^4) \right\}. \tag{B.36}$$

In this way, we finally obtain to the leading order in g^2:

$$\Delta_{CP} = \frac{\Gamma(t \to W^+b) - \Gamma(\bar{t} \to W^-\bar{b})}{\Gamma(t \to W^+b) + \Gamma(\bar{t} \to W^-\bar{b})} = -\frac{g^2}{8\pi} \sum_{l,n} \mathrm{Im}(V_{33}^* V_{3n} V_{ln}^* V_{l3}) B_{nl}.$$

Numerically, the relative difference of widths is tiny,

$$\Delta_{CP} = 1.7 \cdot 10^{-7}.$$

Therefore, this particular effect is unlikely to be observed experimentally in fore-seeable future. We note in this regard that the phase in the CKM matrix leads to a number of other *CP*-violating effects (notably, in processes involving neutral kaons and B-mesons), some of which have been observed in experiments. We note also that the above mechanism responsible for the difference of partial widths of parti-cles and antiparticles is quite generic. In various theories generalizing the Standard Model, this mechanism is used to generate the baryon asymmetry (see Chapter 11).

Let us show that one can make phase rotations of the quark fields in such a way that the mixing matrix is cast into the form (B.28). The relevant transformations are

$$\begin{aligned} \tilde{d}_{Ln} &\to e^{i\beta_n} \tilde{d}_{Ln}, \quad n = 1, 2, 3, \\ \tilde{u}_{Ln} &\to e^{i\gamma_n} \tilde{u}_{Ln}, \quad n = 1, 2, 3, \end{aligned} \tag{B.37}$$

where β_n and γ_n are arbitrary real numbers. The only effect of these rotations is that the matrix V is multiplied on the right by the diagonal matrix $\mathrm{diag}(e^{i\beta_1}, e^{i\beta_2}, e^{i\beta_3})$ and on the left by another diagonal matrix $\mathrm{diag}(e^{-i\gamma_1}, e^{-i\gamma_2}, e^{-i\gamma_3})$.
 Let us begin with the two-flavor case, i.e., 2×2 matrix V. An arbitrary unitary matrix 2×2 can be written as follows:

$$V = e^{i\psi} e^{i\tau_3\theta_3} e^{i(\tau_1 \sin \chi + \tau_2 \cos \chi)\varphi},$$

where τ_i are Pauli matrices. (A general prescription is to write a product of exponentials of all generators of the unitary group with arbitrary coefficients, where the generators may or may not combine into linear combinations in the exponents.) The first two factors here are diagonal, and they can be removed by the rotation (B.37). The remaining matrix is

$$e^{i(\tau_1 \sin \chi + \tau_2 \cos \chi)\varphi} = \begin{pmatrix} c & se^{i\chi} \\ -se^{-i\chi} & c \end{pmatrix},$$

where $s = \sin \varphi$, $c = \cos \varphi$. It can be written in two equivalent forms:

$$\begin{pmatrix} e^{i\chi} & 0 \\ 0 & 1 \end{pmatrix} \cdot O \cdot \begin{pmatrix} e^{-i\chi} & 0 \\ 0 & 1 \end{pmatrix} \quad \text{and} \quad \begin{pmatrix} 1 & 0 \\ 0 & e^{-i\chi} \end{pmatrix} \cdot O \cdot \begin{pmatrix} 1 & 0 \\ 0 & e^{i\chi} \end{pmatrix}, \tag{B.38}$$

where

$$O = \begin{pmatrix} c & s \\ -s & c \end{pmatrix} \tag{B.39}$$

is a real matrix. The first and last factors in (B.38) can again be removed by the quark phase rotations. Thus, the mixing matrix in the two-flavor case can be chosen real and given by (B.39).

We now turn to the three-flavor case. The matrix V can be written as

$$V = e^{i\psi} e^{i\lambda_3 \theta_3} e^{i\lambda_8 \theta_8} U_{45} U_{67} U_{12}, \tag{B.40}$$

where

$$U_{12} = e^{i(\lambda_1 \sin \chi + \lambda_2 \cos \chi)\theta_{12}}, \quad U_{45} = e^{i(\lambda_4 \sin \zeta + \lambda_5 \cos \zeta)\theta_{23}}, \quad U_{67} = e^{i(\lambda_6 \sin \xi + \lambda_7 \cos \xi)\theta_{13}}.$$

Here λ_a, $a = 1, \ldots, 8$ are Gell-Mann matrices. The matrices λ_3 and λ_8 are diagonal, so the first three factors in (B.40) are again irrelevant. The remaining Gell-Mann matrices are

$$\lambda_{1,2} = \begin{pmatrix} \tau_{1,2} & 0 \\ 0 & 1 \end{pmatrix}, \quad \lambda_{4,5} = \begin{pmatrix} 1 & 0 \\ 0 & \tau_{1,2} \end{pmatrix},$$

$$\lambda_6 = \begin{pmatrix} 0 & 0 & 1 \\ 0 & 0 & 0 \\ 1 & 0 & 0 \end{pmatrix}, \quad \lambda_7 = \begin{pmatrix} 0 & 0 & -i \\ 0 & 0 & 0 \\ i & 0 & 0 \end{pmatrix}.$$

In analogy to the two-flavor case, we write

$$U_{12} = \mathrm{diag}(1, e^{-i\chi}, 1) \cdot O_{12} \cdot \mathrm{diag}(1, e^{i\chi}, 1), \tag{B.41a}$$

$$U_{45} = \mathrm{diag}(1, 1, e^{-i\zeta}) \cdot O_{45} \cdot \mathrm{diag}(1, 1, e^{i\zeta}), \tag{B.41b}$$

$$U_{67} = \mathrm{diag}(e^{i\xi}, 1, 1) \cdot O_{67} \cdot \mathrm{diag}(e^{-i\xi}, 1, 1), \tag{B.41c}$$

where

$$O_{12} = \begin{pmatrix} c_{12} & s_{12} & 0 \\ -s_{12} & c_{12} & 0 \\ 0 & 0 & 1 \end{pmatrix}, \quad O_{45} = \begin{pmatrix} 1 & 0 & 0 \\ 0 & c_{23} & s_{23} \\ 0 & -s_{23} & c_{23} \end{pmatrix}, \quad O_{67} = \begin{pmatrix} c_{13} & 0 & s_{13} \\ 0 & 1 & 0 \\ -s_{13} & 0 & c_{13} \end{pmatrix},$$

and notations coincide with those used in (B.28). The last diagonal factor in (B.41a) and first factor in (B.41b) can be removed by the rotations of quarks, so equivalent form of V is

$$V = O_{45} \cdot \mathrm{diag}(e^{i\xi}, 1, e^{i\zeta}) \cdot O_{67} \mathrm{diag}(e^{-i\xi}, e^{-i\chi}, 1) \cdot O_{12}.$$

We perform additional rotations of quark fields which multiply the matrix V on the left and right by diagonal matrices which commute with O_{45} and O_{12}, respectively,

$$\mathrm{diag}(e^{i\alpha}, 1, 1) \cdot V \cdot \mathrm{diag}(e^{i\gamma}, e^{i\gamma}, e^{i\beta})$$
$$= O_{45} \cdot \mathrm{diag}(e^{i(\xi+\alpha)}, 1, e^{i\zeta}) \cdot O_{67} \cdot \mathrm{diag}(e^{i(\gamma-\xi)}, e^{i(\gamma-\chi)}, e^{i\beta}) \cdot O_{12}.$$

We find by direct computation that

$$\text{diag}(e^{i(\xi+\alpha)}, 1, e^{i\zeta}) \cdot O_{67} \cdot \text{diag}(e^{i(\gamma-\xi)}, e^{i(\gamma-\chi)}, e^{i\beta})$$

$$= \begin{pmatrix} c_{13}e^{i(\alpha+\gamma)} & 0 & s_{13}e^{i(\xi+\alpha+\beta)} \\ 0 & e^{i(\gamma-\chi)} & 0 \\ -s_{13}e^{i(\gamma+\zeta-\xi)} & 0 & c_{13}e^{i(\beta+\zeta)} \end{pmatrix}.$$

We now choose $\alpha = -\gamma = -\chi$, $\beta = -\zeta$ and find that the only remaining phase factor multiplies s_{13}, and the phase is $\delta \equiv \xi + \alpha + \beta = -(\gamma + \zeta - \xi)$. Thus, the CKM matrix can be written as

$$V = \begin{pmatrix} 1 & 0 & 0 \\ 0 & c_{23} & s_{23} \\ 0 & -s_{23} & c_{23} \end{pmatrix} \begin{pmatrix} c_{13} & 0 & s_{13}e^{i\delta} \\ 0 & 1 & 0 \\ -s_{13}e^{-i\delta} & 0 & c_{13} \end{pmatrix} \begin{pmatrix} c_{12} & s_{12} & 0 \\ -s_{12} & c_{12} & 0 \\ 0 & 0 & 1 \end{pmatrix}, \qquad (B.42)$$

which coincides with (B.28).

B.5. Effective Fermi Theory

Processes at low energies, $E \ll M_W$, are well described by the effective Fermi theory. The Lagrangian of this theory is obtained from (B.13) by integrating out massive vector fields.

In the Standard Model, the interaction of massive vector bosons with fermions has the form (see (B.15))

$$\mathcal{L}_{f,weak} = \frac{g}{2\cos\theta_W} J_\mu^{NC} Z^\mu + \frac{g}{2\sqrt{2}} (J_\mu^{CC} W^{\mu,-} + h.c.),$$

where the neutral J_μ^{NC} and charged J_μ^{CC} currents are

$$J_\mu^{NC} = \sum_f \bar{f}\gamma_\mu (t_3^f(1-\gamma^5) - 2q_f \sin^2\theta_W)f \qquad (B.43)$$

$$J_\mu^{CC} = \sum_m \bar{\nu}_m \gamma_\mu (1-\gamma^5) l_m + \sum_{m,n} \bar{u}_m \gamma_\mu (1-\gamma^5) V_{mn} d_n. \qquad (B.44)$$

At $E \ll M_Z$ there are only fermions (leptons and quarks, except for heavy t-quark) in the initial and final states. At these energies, the main contribution to the weak amplitudes comes from single exchange of virtual massive vector boson. In the language of Feynman diagrams "integrating out" means shrinking the propagators of W- and Z-bosons to a point (see Fig. B.2). As a result, the Standard Model diagrams of the type shown in Fig. B.3(a) transform into diagrams shown in Fig. B.3(b). The latter correspond to the Fermi theory of effective four-fermion interaction. Thus,

Fig. B.2.   Point-like approximation to the propagators of massive vector bosons.

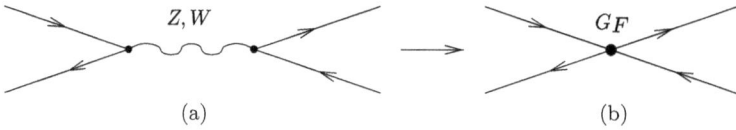

Fig. B.3.   The relation between diagrams of the Standard Model and of the Fermi theory.

integrating out the Z-boson gives rise to the interaction

$$\mathcal{L}_N = -\frac{G_F}{\sqrt{2}} J_\mu^{NC} J^{NC\ \mu},$$

while integrating out $W^\pm$-bosons yields

$$\mathcal{L}_C = -\frac{G_F}{\sqrt{2}} J_\mu^{CC} J^{CC\ \dagger\ \mu}.$$

The effective coupling G_F, Fermi constant, has dimension m^{-2}; it determines the strength of the four-fermion interaction. This coupling is related to the fundamental parameters of the Standard Model as follows (see Fig. B.2)

$$G_F \equiv \frac{g^2}{4\sqrt{2}M_W^2}.$$

Its numerical value is

$$G_F = 1.17 \cdot 10^{-5}\ \text{GeV}^{-2}.$$

Note that similar procedure of integrating out the Higgs boson also leads to effective four-fermion interaction of the form $\bar{\psi}_m\psi_m \cdot \bar{\psi}_n\psi_n$. However, the effective couplings here are proportional to the corresponding fermion Yukawa couplings, which are small compared to the weak gauge coupling.[10] Therefore, the impact of this effective interaction on processes at low energies can be neglected.

B.6.   Peculiarities of Strong Interactions

Although mediators of strong interactions, gluons, are massless, strong interactions are also described by effective theory at low energies (more accurately, at small momentum transfers Q). In contrast to the Fermi theory, the reason is that QCD is

---

[10]An exception is the t-quark Yukawa coupling, but t-quark does not participate in low energy processes discussed here.

in strong coupling regime at energies of order $\Lambda_{QCD} \simeq 200$ MeV, and even somewhat higher.

The energy-dependent ("running") strong gauge coupling $\alpha_s(Q) \equiv g_s^2(Q)/(4\pi)$ increases as energy decreases, and becomes large at $Q \sim \Lambda_{QCD}$. (At the one-loop level it tends to infinity as $Q \to \Lambda_{QCD}$.) Thus, at energy $E \sim \Lambda_{QCD}$, not only the perturbation series in gauge coupling blows up, but the very description in terms of quarks and gluons loses any sense.

This feature of the theory is fully consistent with the fact that quarks and gluons are not observed in free state. They are bound inside colorless particles — hadrons (mesons and baryons) — and begin to play an independent role only in processes with characteristic momentum transfer higher than Λ_{QCD}. This phenomenon got the name *confinement* of quarks and gluons. It is clear that the characteristic size of light hadrons — regions where u-, d- and s-quarks and gluons are confined — is precisely Λ_{QCD}^{-1}.

At low energies, interactions of the lightest hadrons (protons, neutrons, pions, kaons, etc.) between themselves and with leptons and photons are well described in the framework of the *chiral perturbation theory*. To describe heavier hadrons and strong processes in the intermediate energy region $E \gtrsim \Lambda_{QCD}$, various phenomenological approaches are used. A number of variables characterizing QCD in the strong coupling regime can be calculated from the first principles by putting the theory on the lattice and calculating the functional integral numerically by Monte Carlo method. From the point of view of cosmology, of special interest are lattice calculations of QCD transition temperature which give $T_{QCD} \simeq 170$ MeV [272].

B.7. The Effective Number of Degrees of Freedom in the Standard Model

Using the particle spectrum of the Standard Model and accounting for peculiarities of strong interactions, it is straightforward to estimate the effective number of relativistic degrees of freedom g_* as a function of temperature of the primordial plasma. The simplest way is to use step-function approximation to the temperature evolution of $g_*(T)$ near particle thresholds and QCD transition. The result is shown in Fig. B.4.

At temperature $T < 100$ MeV, only photons, electrons and neutrinos are relativistic, so at 1 MeV $\lesssim T \lesssim 100$ MeV the effective number of relativistic degrees of freedom is

$$g_*(T \lesssim 100 \text{ MeV}) = 2_\gamma + \frac{7}{8}(4_e + 3 \cdot 2_\nu) = \frac{43}{4} = 10.75.$$

At $T \gtrsim 100$ MeV, additional contribution comes from muon. Above the temperature of the QCD transition $T_{QCD} \simeq 170$ MeV, the plasma contains light quarks (u, d, s) and gluons. Lattice calculations show that their interactions do not affect dramatically the thermodynamic quantities such as energy density or pressure. Therefore,

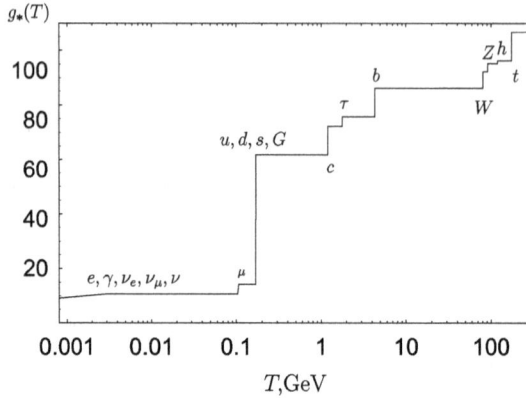

Fig. B.4. The effective number of degrees of freedom in the primordial plasma as a function of temperature. Only Standard Model particles are taken into account.

at $T \sim T_{QCD}$ the effective number of relativistic degrees of freedom changes by

$$\Delta g_*^{(QCD)} \simeq 8 \cdot 2 + \frac{7}{8} \cdot 3 \cdot 3 \cdot 4 = 47.5.$$

The first term here comes from gluons (massless vectors in eight color states), the second one is due to u-, d-, s-quarks and antiquarks, each in three color states. Steps at higher temperatures are associated with heavier particles (see the SM spectrum given in the end of Sec. B.1), as shown in Fig. B.4.

At $T > 200$ GeV the effective number of degrees of freedom, in the framework of the Standard Model with one Englert–Brout–Higgs doublet, is

$$g_*(T \gtrsim 200)$$

$$= 2_\gamma + 2 \cdot 3_W + 3_Z + 1_h + 8^{(c)} \cdot 2_G + \frac{7}{8} (3 \cdot 4_e + 3 \cdot 2_\nu + 6 \cdot 3^{(c)} \cdot 4_q) \qquad (B.45)$$

$$= 106.75.$$

Here subscripts indicate types of particles, superscript (c) refers to the number of color states, the last factors are the numbers of spin states. Note that in the above calculation, three polarizations of W- and Z-bosons and one degree of freedom of the Higgs boson are taken into account. This is appropriate for the Standard Model in the Englert–Brout–Higgs phase. In the phase of unbroken electroweak symmetry, W- and Z-bosons are massless and have 2 polarizations each, while the Higgs field — complex doublet — describes 4-scalar particles. The number of degrees of freedom is the same in these two phases, so the result (B.45) is valid in any case.

Appendix C

Neutrino Oscillations

Neutrino oscillations — transitions that change neutrino flavor — is a unique to date direct evidence for the incompleteness of the Standard Model of particle physics, obtained in laboratory experiments rather than from cosmology or astrophysical observations. Neutrino oscillations occur if neutrinos are massive and there is mixing between lepton generations analogous to quark mixing considered in Sec. B.4. Within the Standard Model, the Lagrangian cannot contain renormalizable gauge-invariant terms which would lead to neutrino masses. In this sense, neutrino oscillations are phenomenon beyond the Standard Model.

Historically, the first data pointed at neutrino oscillations were obtained in measurements of solar and atmospheric neutrino fluxes. These discoveries have been confirmed afterwards by experiments with neutrinos from nuclear reactors and accelerators.

C.1. Oscillations and Mixing

In this Section we discuss in general terms the mechanism leading to neutrino oscillations. Here we set aside the aspects related to the fact that neutrino has spin 1/2. These aspects are discussed in Sec. C.4.

C.1.1. *Vacuum oscillations*

In extensions of the Standard Model allowing for nonzero neutrino masses, there can occur oscillations between neutrinos of different types (flavors). Namely, neutrinos are produced in weak processes in full compliance with the Standard Model.[1] As an example, neutron beta-decay $n \to p\, e^+\, \nu_e$ produces electron neutrino, muon neutrino is created together with muon in the decay $\pi^+ \to \mu^+\, \nu_\mu$, while hadronic decays of τ^--lepton yield τ-neutrino. However, in the basis ν_e, ν_μ, ν_τ, the Hamiltonian describing neutrino free propagation is *non-diagonal*, which leads to oscillations.

---

[1]In this Appendix we are interested in processes at not too high energies, for which this is certainly the case.

We call this flavor (or gauge) basis, since gauge interactions of leptons are flavor-diagonal in this basis.

The situation here is quite analogous to quark mixing. In the basis where the gauge interactions of quarks are diagonal (gauge basis), the quark mass matrix is non-diagonal. Vice versa, in the basis where quark mass matrix is diagonal (mass basis) gauge couplings are non-diagonal. In the case of quarks it is convenient to work exclusively in the mass basis. When studying neutrino oscillations, one uses both gauge and mass basis.

We assume here that there are three types of neutrinos in Nature. Since the creation and detection of neutrino occur via weak interactions, it is the states of the flavor basis $|\nu_e\rangle$, $|\nu_\mu\rangle$ and $|\nu_\tau\rangle$ that are observable. The basis vectors of the flavor basis $|\nu_\alpha\rangle$, $\alpha = e, \mu, \tau$ are related to the basis vectors of the mass basis $|\nu_i\rangle$, $i = 1, 2, 3$ by unitary transformation traditionally written in the form

$$|\nu_i\rangle = U_{\alpha i}|\nu_\alpha\rangle. \tag{C.1}$$

(Summation over repeated indices is assumed.) Here the unitary matrix $U_{\alpha i}$ is the mixing matrix in the neutrino sector, dubbed Pontecorvo–Maki–Nakagawa–Sakata (PMNS) matrix. The states $|\nu_i\rangle$ are eigenstates of the free Hamiltonian, i.e., they have definite masses m_i. The inverse transformation from the mass basis to the flavor basis reads

$$|\nu_\alpha\rangle = (U^\dagger)_{i\alpha}|\nu_i\rangle \equiv U^*_{\alpha i}|\nu_i\rangle. \tag{C.2}$$

With the definition (C.1), the neutrino mass matrix in the flavor basis has simple form,

$$M_{\alpha\beta} = \langle \nu_\alpha | M | \nu_\beta \rangle = (U M^{(m)} U^\dagger)_{\alpha\beta}, \tag{C.3}$$

where $M^{(m)}$ is diagonal mass matrix in the mass basis,

$$M^{(m)}_{ij} = m_i \delta_{ij}. \tag{C.4}$$

The neutrino free evolution in the rest frame is determined by eigenvalues of the mass matrix,

$$|\nu_j(t)\rangle = e^{-im_j t}|\nu_j(0)\rangle. \tag{C.5}$$

Suppose that at time $t = 0$ there is a pure flavor state — for instance, electron neutrino $|\nu_e\rangle$ produced in the decay of (possibly virtual) W^+. Then at time t, other components of the state vector in the flavor basis also become nonzero. This implies a non-vanishing probability of detecting muon or τ-neutrino at that time.

For practical applications, one needs to calculate the transition probability $\nu_\alpha \rightarrow \nu_\beta$ in the laboratory frame at a distance L from the source of neutrino ν_α. Then the formula (C.5) for neutrino evolution in the mass basis has to be generalized,

$$|\nu_j(t, L)\rangle = e^{-i(E_j t - p_j L)}|\nu_j(0)\rangle,$$

where p_j and E_j are neutrino momentum and energy. Realistically, neutrinos are ultra-relativistic. Let us assume that neutrino energy is fixed,[2] and write in the ultra-relativistic case $p_j = \sqrt{E^2 - m_j^2} = E - m_j^2/2E$. Omitting phase factor common to all neutrino species, we find that the evolution of states in the mass basis in terms of the traveled distance is given by

$$|\nu_j(L)\rangle = e^{-i\frac{m_j^2}{2E}L}|\nu_j(0)\rangle.$$

Note that this evolution corresponds to the effective Hamiltonian

$$H_{eff} = \frac{M^2}{2E}, \tag{C.6}$$

where M is the neutrino mass matrix, which takes the forms (C.3) and (C.4) in the mass and flavor basis, respectively. As follows from (C.1), the transition amplitude of neutrino ν_α to neutrino ν_β is equal to

$$A(\alpha \to \beta) = \sum_j \langle \nu_\beta | \nu_j(L)\rangle \langle \nu_j(0)|\nu_\alpha\rangle = \sum_j \langle \nu_\beta | \nu_j \rangle e^{-i\frac{m_j^2}{2E}L} \langle \nu_j | \nu_\alpha \rangle$$

$$= \sum_j U_{\beta j} e^{-i\frac{m_j^2}{2E}L} U_{\alpha j}^*. \tag{C.7}$$

This formula enables one to calculate the probability of transition between two states of the flavor basis after traveling the distance L:

$$P(\nu_\alpha \to \nu_\beta) = |A(\alpha \to \beta)|^2$$

$$= \delta_{\alpha\beta} - 4 \sum_{j>i} \mathrm{Re}[U_{\alpha j}^* U_{\beta j} U_{\alpha i} U_{\beta i}^*] \sin^2\left(\frac{\Delta m_{ji}^2}{4E}L\right)$$

$$+ 2 \sum_{j>i} \mathrm{Im}[U_{\alpha j}^* U_{\beta j} U_{\alpha i} U_{\beta i}^*] \sin\left(\frac{\Delta m_{ji}^2}{2E}L\right) \tag{C.8}$$

where

$$\Delta m_{ji}^2 \equiv m_j^2 - m_i^2.$$

---

[2]There is a discussion in literature of delicate issues like "are neutrino states eigenstates of the energy operator $\hat{P}_0$ or the momentum operator $\hat{P}$?" The answers to questions of this sort are important for correct description of the oscillations of not too fast neutrino, as well as for studying the applicability limits of the oscillation picture. (It is obviously not valid at large distances from the source where neutrinos of different masses come in substantially different times.) We do not enter this discussion here.

The expression (C.8) describes oscillations with amplitude determined by the neutrino mixing matrix and the oscillation lengths depending on the difference of neutrino masses squared and energy. Note that in realistic situations the oscillation pattern may be washed out if the source has large spatial size and/or averaging is performed over a certain interval of neutrino energies.

It is clear that along with the oscillations of neutrinos there should be oscillations of antineutrinos. The latter have indeed been observed experimentally. Antineutrino oscillations are also described within the above formalism. CPT-theorem (see Sec. B.3) gives the following relation between the probabilities of neutrino and antineutrino transitions,

$$P(\nu_\alpha \to \nu_\beta) = P(\bar{\nu}_\beta \to \bar{\nu}_\alpha). \tag{C.9}$$

The transition probability of $\nu_\beta \to \nu_\alpha$ is equal to the transition probability of $\nu_\alpha \to \nu_\beta$ calculated with complex conjugate neutrino mixing matrix (see (C.8)). Hence, the relation (C.9) leads to equality

$$P(\bar{\nu}_\alpha \to \bar{\nu}_\beta ; U) = P(\nu_\alpha \to \nu_\beta ; U^*).$$

This equality implies that neutrino and antineutrino oscillation probabilities may be different only if the matrix U is complex (see Eq. (C.8)). This would mean CP-violation in lepton sector. Note that non-trivial CP-phase is possible only if the number of neutrino species exceeds two: in the case of two types of neutrinos, the mixing matrix U can be made real by redefining the fields (see below).

An important example is the case of oscillations between two types of neutrinos. In this case 2×2 unitary matrix $U_{\alpha i}, i, \alpha = 1, 2$ is determined by 4 real parameters. Namely, we have shown in the end of Section B.4 that any unitary matrix 2×2 can be written in the form $U = D_1 O D_2$, where $D_{1,2} = \mathrm{diag}(e^{i\phi_{1,2}}, e^{i\chi_{1,2}})$ are diagonal matrices of phase rotations (one of the phases can be set equal to zero), and O is a real orthogonal matrix. The formula (C.8) shows that matrices $D_{1,2}$ are irrelevant for oscillations, so the transition from flavor to mass basis is effectively described by the matrix

$$U_{\alpha,i} = \begin{pmatrix} \cos\theta & \sin\theta \\ -\sin\theta & \cos\theta \end{pmatrix}, \tag{C.10}$$

which depends on one parameter, mixing angle.

In the case of two-neutrino oscillations, the formula (C.8) gets simplified,

$$P(\nu_\alpha \to \nu_\beta) = \delta_{\alpha\beta} + (-1)^{\delta_{\alpha\beta}} \sin^2 2\theta \sin^2\left(\frac{\Delta m^2}{4E} L\right). \tag{C.11}$$

In other words, the probability of transition of neutrino ν_α to another type of neutrino ν_β is equal to

$$P(\nu_\alpha \to \nu_{\beta\neq\alpha}) = \sin^2 2\theta \cdot \sin^2\left(\frac{\Delta m^2}{4E} L\right), \tag{C.12}$$

while the survival probability of neutrino ν_α is

$$P(\nu_\alpha \to \nu_\alpha) = 1 - P(\nu_\alpha \to \nu_{\beta \neq \alpha}) = 1 - \sin^2 2\theta \cdot \sin^2 \left(\frac{\Delta m^2}{4E} L \right). \tag{C.13}$$

The mixing angle θ determines the oscillation amplitude, $A = \sin^2 2\theta$. The oscillation length is

$$L_{osc} = \frac{4\pi E}{\Delta m^2} = (2.5 \text{ km}) \cdot \frac{E}{\text{GeV}} \frac{\text{eV}^2}{\Delta m^2}. \tag{C.14}$$

At this distance the neutrino ν_α returns to its original state, while the maxima of the oscillation probability are at distances $L_k = L_{osc}(1/2 + k)$, $k = 0, 1, 2, \ldots$.

C.1.2.   Three-neutrino oscillations in special cases

In the three-flavor case the mixing matrix can be cast into the form (see end of Sec. B.4)

$$U^{PMNS} = D_1 U D_2, \tag{C.15}$$

where $D_{1,2}$ are again diagonal matrices of phase rotations, $D_1 = \text{diag}(e^{i\alpha_1}, e^{i\alpha_2}, e^{i\alpha_3})$, $D_2 = \text{diag}(e^{i\beta_1}, e^{i\beta_2}, e^{i\beta_3})$, and matrix U has the same form as the matrix V in (B.28), (B.42):

$$U = \begin{pmatrix} c_{12} & s_{12} & 0 \\ -s_{12} & c_{12} & 0 \\ 0 & 0 & 1 \end{pmatrix} \begin{pmatrix} c_{13} & 0 & s_{13}e^{i\delta} \\ 0 & 1 & 0 \\ -s_{13}e^{-i\delta} & 0 & c_{13} \end{pmatrix} \begin{pmatrix} 1 & 0 & 0 \\ 0 & c_{23} & s_{23} \\ 0 & -s_{23} & c_{23} \end{pmatrix}$$

$$= \begin{pmatrix} c_{12}c_{13} & s_{12}c_{13} & s_{13}e^{-i\delta} \\ -s_{12}c_{23} - c_{12}s_{23}s_{13}e^{i\delta} & c_{12}c_{23} - s_{12}s_{23}s_{13}e^{i\delta} & s_{23}c_{13} \\ s_{12}s_{23} - c_{12}c_{23}s_{13}e^{i\delta} & -c_{12}s_{23} - s_{12}c_{23}s_{13}e^{i\delta} & c_{23}c_{13} \end{pmatrix}. \tag{C.16}$$

Some phases in $D_{1,2}$ are physical (see Sec. C.4), but just like in the two-flavor case, the formula (C.8) shows that these phases are irrelevant for oscillations. Thus, in the three-flavor case the oscillations are described by three mixing angles and one phase which enter (C.16). With slight abuse of terminology, we call the matrix U of the type (C.16) as PMNS matrix.

As we discuss later, there is a hierarchy between the differences of neutrino masses squared,

$$|\Delta m_{31}^2| \gg \Delta m_{21}^2. \tag{C.17}$$

This implies, in particular, that

$$|\Delta m_{32}^2| = |\Delta m_{31}^2 - \Delta m_{21}^2| \approx |\Delta m_{31}^2| \gg \Delta m_{21}^2.$$

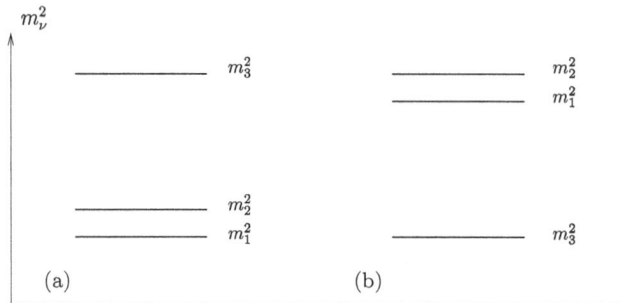

Fig. C.1. Normal (a) and inverted (b) hierarchies of neutrino masses.

Hereafter we stick to the following convention concerning the numbering of the mass eigenstates: the mass m_3 differs significantly from m_1 and m_2; the masses m_1 and m_2 are close to each other, and

$$m_2 > m_1.$$

The property (C.17) suggests either normal or inverted mass hierarchy, as illustrated in Fig. C.1. Let us show that due to this property, the formulas describing the oscillations between the three types of neutrinos, are similar in two special cases to those of two-neutrino oscillations. These cases are, in fact, of great interest, since the first of them often occurs for accelerator and reactor neutrinos, while the second one is relevant for solar neutrinos.

Let us begin with the case when the energy E and the distance between the neutrino production and detection are such that

$$\frac{\Delta m_{21}^2}{2E} L \ll 1. \tag{C.18}$$

Then the first oscillating term in (C.8) is expressed as

$$\sum_{j>i} U_{\alpha j}^* U_{\beta j} U_{\alpha i} U_{\beta i}^* \cdot \sin^2 \left(\frac{\Delta m_{ji}^2}{4E} L \right)$$

$$= U_{\alpha 3}^* U_{\beta 3} (U_{\alpha 1} U_{\beta 1}^* + U_{\alpha 2} U_{\beta 2}^*) \cdot \sin^2 \left(\frac{\Delta m_{31}^2}{4E} L \right). \tag{C.19}$$

Let us now take into account the unitarity condition

$$(UU^\dagger)_{\alpha\beta} \equiv \sum_i U_{\alpha i} U_{\beta i}^* = \delta_{\alpha\beta} \tag{C.20}$$

and write the expression (C.19) in the form

$$U_{\alpha 3}^* U_{\beta 3} (\delta_{\alpha\beta} - U_{\alpha 3} U_{\beta 3}^*) \cdot \sin^2 \left(\frac{\Delta m_{31}^2}{4E} L \right). \tag{C.21}$$

The latter formula is valid for all α and β and contains only $|U_{\alpha 3}|^2$ and $|U_{\beta 3}|^2$. Similar calculation shows that the last term in (C.8) is equal to zero. These results imply, in particular, that CP-violating effects are strongly suppressed in the regime (C.18). (They vanish in the limit $\frac{\Delta m_{21}^2}{2E}L \to 0$.) The regime (C.18) is indeed realized, as a rule, in terrestrial experiments; because of this fact, and also because of the smallness of the angle θ_{13} (see below), the observation of CP-violation in neutrino oscillations is extremely difficult.

Equations (C.8) and (C.21) give for the survival probability

$$P(\nu_\alpha \to \nu_\alpha) = 1 - \sin^2 2\theta_{\mathit{eff}} \sin^2\left(\frac{\Delta m_{31}^2}{4E}L\right), \tag{C.22}$$

where by definition

$$\sin^2\theta_{\mathit{eff}} = |U_{\alpha 3}|^2.$$

The formula (C.22) is analogous to the expression (C.13) valid for two-neutrino oscillations. The probability of appearance of neutrino of type $\beta \neq \alpha$ is

$$P(\nu_\alpha \to \nu_\beta) = 4|U_{\beta 3}|^2|U_{\alpha 3}|^2 \sin^2\left(\frac{\Delta m_{31}^2}{4E}L\right) \tag{C.23a}$$

$$= \frac{|U_{\beta 3}|^2}{\sum_{\beta' \neq \alpha}|U_{\beta' 3}|^2} \sin^2(2\theta_{\mathit{eff}}) \sin^2\left(\frac{\Delta m_{31}^2}{4E}L\right). \tag{C.23b}$$

It differs from (C.12) by the first factor which accounts for admixture of *two* states with $\beta \neq \alpha$ in the mass eigenstate ν_3.

We now turn to the second special case. It occurs when the neutrino production region is sufficiently large and/or neutrino have fairly large energy spread, so that the oscillations with phases proportional to Δm_{31}^2 and Δm_{32}^2 get averaged,

$$\left\langle \sin^2\left(\frac{\Delta m_{31}^2}{4E}L\right)\right\rangle = \frac{1}{2}, \quad \left\langle \sin\left(\frac{\Delta m_{31}^2}{2E}L\right)\right\rangle = 0.$$

The quantity of interest for applications is the survival probability $P(\nu_\alpha \to \nu_\alpha)$. Again using the unitarity condition (C.20), it can be represented as

$$P(\nu_\alpha \to \nu_\alpha) = 1 - 2|U_{\alpha 3}|^2(1 - |U_{\alpha 3}|^2) - 4|U_{\alpha 2}|^2|U_{\alpha 1}|^2 \sin^2\left(\frac{\Delta m_{21}^2}{4E}L\right).$$

We introduce the mixing angle θ'_{eff} by the relations

$$\cos^2\theta'_{\mathit{eff}} = \frac{|U_{\alpha 1}|^2}{|U_{\alpha 1}|^2 + |U_{\alpha 2}|^2} \equiv \frac{|U_{\alpha 1}|^2}{1 - |U_{\alpha 3}|^2}, \quad \sin^2\theta'_{\mathit{eff}} = \frac{|U_{\alpha 2}|^2}{|U_{\alpha 1}|^2 + |U_{\alpha 2}|^2}.$$

As a result, we obtain finally

$$P(\nu_\alpha \to \nu_\alpha) = |U_{\alpha 3}|^4 + (1 - |U_{\alpha 3}|^2)^2\left[1 - \sin^2 2\theta'_{\mathit{eff}} \sin^2\left(\frac{\Delta m_{21}^2}{4E}L\right)\right]. \tag{C.24}$$

In the case of small admixture of neutrino ν_α in the mass eigenstate ν_3, i.e., when $|U_{\alpha 3}|^2 \ll 1$, the latter formula also transforms into (C.13).

C.1.3. *Mikheev–Smirnov–Wolfenstein effect*

The formulas presented so far are valid for *vacuum* neutrino oscillations, and, more generally, when the influence of medium where neutrinos propagate is negligible. However, matter does affect neutrino properties in some situations. The corresponding phenomenon is called Mikheev–Smirnov–Wolfenstein effect [273, 274] (MSW); it is due to coherent forward neutrino scattering off electrons present in matter.

MSW effect is accounted for by introducing an effective term in the Hamiltonian describing neutrino propagation. Recall that at relatively low energies, charged current interactions are described by the Fermi theory whose Lagrangian in the lepton sector is

$$\mathcal{L}^{CC} = -\frac{G_F}{\sqrt{2}}\bar{\nu}_e\gamma^\mu(1-\gamma^5)e \cdot \bar{e}\gamma_\mu(1-\gamma^5)\nu_e = -2\sqrt{2}G_F\bar{\nu}_e\gamma^\mu e \cdot \bar{e}\gamma_\mu\nu_e \qquad (C.25)$$

(see Sec. B.5), where in the last equality we made use of the fact that neutrinos are left fermions, so that $\frac{1}{2}(1-\gamma^5)\nu_e = \nu_e$. For matter with electron number density n_e we have

$$\langle\langle \bar{e}_k\gamma^0_{kl}e_l \rangle\rangle = \langle\langle e^\dagger e \rangle\rangle = n_e, \qquad (C.26)$$

where double brackets denote matter average and k, l are spinor indices. Assuming that electric currents are negligible (this is certainly the case for non-relativistic matter), we write

$$\langle\langle \bar{e}_k\gamma^i_{kl}e_l \rangle\rangle = 0. \qquad (C.27)$$

Given that the operators $\bar{e}_k$ and e_l anticommute, we obtain from (C.26) and (C.27)

$$\langle\langle e_k\bar{e}_l \rangle\rangle = -\frac{1}{4}\gamma^0_{kl} \cdot n_e$$

(in the representation where γ^0 is symmetric). Averaging the Lagrangian (C.25) over matter, we obtain the contribution to the effective Lagrangian describing the propagation of electron neutrino,

$$\mathcal{L}_{\text{eff}} = -2\sqrt{2}G_F\bar{\nu}_e\gamma^\mu\langle\langle e\bar{e}\rangle\rangle\gamma^\mu\nu_e = 2\sqrt{2}G_Fn_e\frac{1}{4}\bar{\nu}_e\gamma^\mu\gamma^0\gamma_\mu\nu_e$$
$$= -\sqrt{2}G_Fn_e\bar{\nu}_e\gamma^0\nu_e.$$

Hence, we conclude that matter effect is accounted for by the following substitution in the Dirac operator $i\gamma^\mu\partial_\mu$,

$$i\gamma^0\partial_0 \to i\gamma^0\partial_0 - \sqrt{2}G_Fn_e\gamma^0,$$

i.e., the operator $i\partial_0$ is replaced by

$$i\partial_0 - V,$$

where

$$V = \sqrt{2}G_F n_e \tag{C.28}$$

is the matter contribution to the effective Hamiltonian. We emphasize that this contribution exists for electron neutrino only, as muons and τ-leptons are absent in matter.

The last statement is not entirely correct. The effective four-fermion Lagrangian has the neutral current terms. The relevant structure is (see Sec. B.5)

$$\sum_\alpha \bar{e}\gamma^\mu e \cdot \bar{\nu}_\alpha \gamma_\mu \nu_\alpha,$$

where summation runs over all neutrino types. These terms lead to new contribution to the neutrino effective Hamiltonian which has the form analogous to (C.28). However, this contribution is now the same for all types of neutrinos. The latter property means that in the basis $|\nu_\alpha\rangle$ as well as in any other basis, this contribution is proportional to the unit matrix. Hence, it does not affect neutrino oscillations, leading only to additional time-dependent overall phase in the state vector. There is no need to consider this contribution in what follows.

Thus, the effective Hamiltonian describing neutrino propagation in matter is different from (C.6),

$$H_{\text{eff}}(L) = \frac{M^2}{2E} + \hat{V}(L). \tag{C.29}$$

The effective potential operator $\hat{V}$ has the only nonzero matrix element in the flavor basis,

$$\hat{V}(L)_{\alpha\beta} = V(L)\delta_{e\alpha}\delta_{e\beta},$$

where $V(L) = \sqrt{2}G_F n_e(L)$ and $n_e(L)$ is the electron density at distance L from neutrino source.

The matter effect on neutrino propagation leads to a number of important and interesting phenomena. One of them is discussed in Sec. C.2.1. Here we make one simple observation. Namely, even in the two-neutrino case, the oscillation probabilities for neutrino and antineutrino are different in matter. The physical reason is that there are electrons in matter and no positrons. Hence, the presence of matter explicitly violates *CP*. At more formal level, the *CP*-transformation converts the electron density operator $\hat{n}_e$ into $(-\hat{n}_e)$, so the matter contribution in the effective Hamiltonian for antineutrino differs by sign from (C.28). This fact is responsible, in particular, for additional difficulty in search for *CP*-violation in neutrino oscillations: neutrino beams produced by accelerators will pass through the Earth, and only then will be detected, so one will have to discriminate between the "true" *CP*-violation (arising due to the phase in the PMNS matrix) and the matter effect.

C.2. Experimental Discoveries

C.2.1. *Solar neutrinos and KamLAND*

At the Earth, the major contribution to the cosmic neutrino flux comes from thermonuclear reactions in the center of the Sun, which are the source of solar energy. We list here the main reactions leading to the neutrino emission by the Sun,

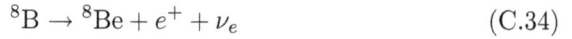

$$p + p \to {}^2\text{H} + e^+ + \nu_e \tag{C.30}$$

$$p + e + p \to {}^2\text{H} + \nu_e \tag{C.31}$$

$${}^3\text{He} + p \to {}^4\text{He} + e^+ + \nu_e \tag{C.32}$$

$${}^7\text{Be} + e^- \to {}^7\text{Li} + \nu_e \tag{C.33}$$

$${}^8\text{B} \to {}^8\text{Be} + e^+ + \nu_e \tag{C.34}$$

$${}^{13}\text{N} \to {}^{13}\text{C} + e^+ + \nu_e$$

$${}^{15}\text{O} \to {}^{15}\text{N} + e^+ + \nu_e.$$

Only *electron neutrinos* are produced in these reactions. The energies of these neutrinos range from zero to tens MeV; the energy spectrum of solar neutrinos is shown in Fig. C.2. Neutrino flux and spectrum are reliably calculated within the Standard Solar Model [275, 276] (SSM).

Low-energy neutrinos interact extremely weakly with matter; they pass through the Sun and Earth with virtually no absorption or scattering. Despite the large neutrino flux, their detection is very difficult. The number of events per unit mass of a detector is small, so one has to install massive detectors (tens of tons to several tens of kilotons, depending on the type of detector), collect statistics for many years, and reduce the background by using radioactively pure materials, placing detectors deep underground (where the cosmic ray flux is substantially reduced), etc.

Historically, the first experiment designed to measure the solar neutrino flux was built in the Homestake mine [277] (USA). It lasted for almost 30 years. Solar neutrinos were captured in the reaction

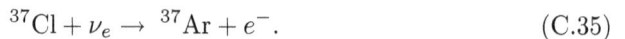

$${}^{37}\text{Cl} + \nu_e \to {}^{37}\text{Ar} + e^-. \tag{C.35}$$

A handful of ${}^{37}\text{Ar}$ atoms were chemically extracted on regular basis from the 615 tons target, and then their number was determined by counting the decays of radioactive ${}^{37}\text{Ar}$. Experiments of this type are called radiochemical. They measure the integrated neutrino flux, weighted by the energy-dependent capture cross-section. The Homestake experiment was sensitive mainly to the boron neutrinos produced in reaction (C.34), with substantial contribution expected from reaction

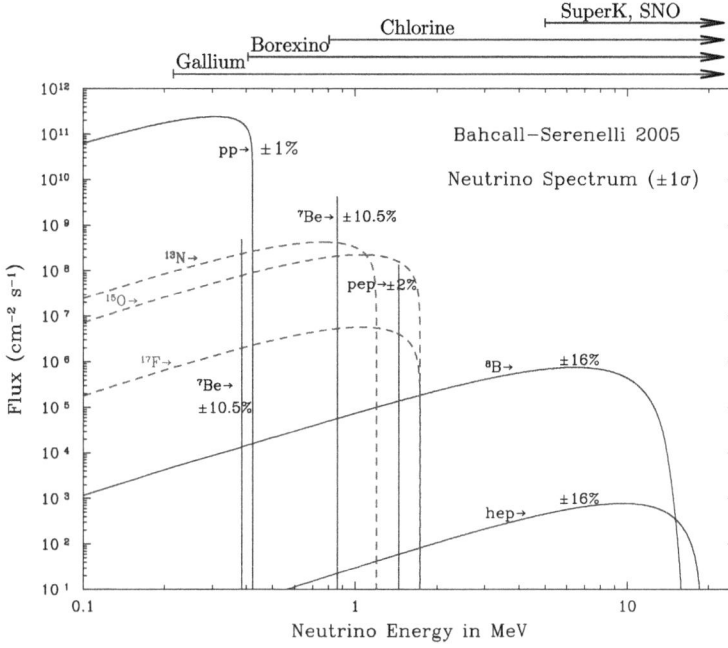

Fig. C.2. Solar neutrino spectrum at the Earth [275] in the absence of oscillations. Contributions of various thermonuclear reactions are shown with estimates of accuracy of calculations. Neutrino flux from continuum is given in units $cm^{-2}s^{-1}MeV^{-1}$. Energy ranges accessible for various solar neutrino detectors are indicated in the upper part.

(C.33) and other reactions (but not (C.30)). The measured integrated flux of electron neutrinos Φ^{Cl} turned out to be smaller than the calculated SSM flux Φ^{Cl}_{SSM},

$$\frac{\Phi^{Cl}}{\Phi^{Cl}_{SSM}} = 0.34 \pm 0.05. \tag{C.36}$$

This result was the first indication that on the way from the center of the Sun to the Earth, electron neutrino transforms into neutrinos of other types, which do not participate in the reaction (C.35).

Boron neutrino flux in the high-energy part of the spectrum was then measured by the Kamiokande detector [278] (Kamioka mine, Japan[3]), neutrino energies $E_{\nu_e} > 7$ MeV, and later by Super-K [279] neutrino energies $E_{\nu_e} > 5.5$ MeV and $E_{\nu_e} > 5.0$ MeV at different stages of the experiments. These detectors used water as detector material; target masses were about 1 kiloton and 22.5 kilotons for

---

[3] Hereafter we indicate only the geographical position of the detector. The experiments are performed by collaborations of scientists from various countries; the lists of collaboration members can be found in the original literature.

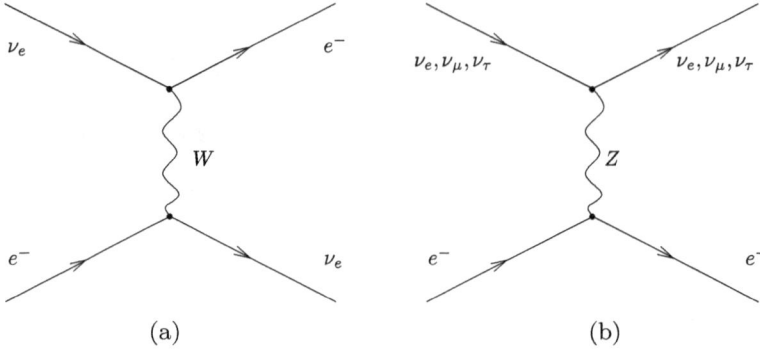

Fig. C.3.   Neutrino–electron elastic scattering via exchange of W-boson (a) and Z-boson (b).

Kamiokande and Super-K, respectively. Neutrino participates in the elastic scattering reaction

$$\nu + e^- \to \nu + e^-, \tag{C.37}$$

which results in the production of relativistic electron, whose Cherenkov radiation was detected. The measured flux of solar neutrinos was again lower than the SSM prediction (Super-K data),

$$\frac{\Phi^{S-K}}{\Phi^{S-K}_{SSM}} = 0.41 \pm 0.07. \tag{C.38}$$

Note that the error here (and to a lesser extent, in (C.36)) is mainly due to SSM uncertainties; the flux Φ^{S-K} itself is measured with much better precision.

Elastic scattering off electron is experienced by both electron neutrino and ν_μ, ν_τ (see details in Appendix B, in particular, the Lagrangian (B.15) and discussion in Sec. B.5). $\nu_e e^-$ scattering occurs through the exchange of W-boson (charged currents, Fig. C.3(a)) and Z-boson (neutral currents, Fig. C.3(b)). On the other hand, only Z-boson exchange contributes to $\nu_\mu e$ and $\nu_\tau e$ scattering (Fig. C.3(b)). The Z-boson exchange results in smaller scattering amplitude than that due to W-boson, so the effective neutrino flux detected in the elastic scattering reaction (C.37) is proportional to

$$\Phi^{\nu e}_{eff} \propto \Phi_{\nu_e} + 0.15(\Phi_{\nu_\mu} + \Phi_{\nu_\tau}). \tag{C.39}$$

This property is important for interpreting the results obtained with the detector SNO, which we discuss later on.

The next solar neutrino experiments were radiochemical; they used the reaction

$$^{71}\text{Ga} + \nu_e \to \,^{71}\text{Ge} + e^-,$$

followed by chemical extraction of ^{71}Ge atoms and counting of their radioactive decays. These are the SAGE experiment [280] (Baksan Neutrino Observatory,

Russia, 60 tons of gallium) and GALLEX/GNO [281] (Gran Sasso Laboratory, Italy, 30 tons of gallium). Unlike in other experiments, the largest contribution to the measured integrated flux comes from neutrinos produced in reaction (C.30), though sizeable contributions are due to neutrinos from the reactions (C.33), (C.34) and others. The integrated flux measured in the gallium experiments (duration of measurements exceeded 10 years) is also significantly lower than the SSM prediction (note the consistency between the two results obtained, in fact, using somewhat different techniques),

$$\frac{\Phi^{Ga}}{\Phi^{Ga}_{SSM}} = 0.54 \pm 0.06 \quad \text{SAGE},$$

$$\frac{\Phi^{Ga}}{\Phi^{Ga}_{SSM}} = 0.56 \pm 0.06 \quad \text{GALLEX/GNO}.$$

The key role of these data was that they eliminated hypothetical possibility that the observed deficit of boron neutrinos results from some error in SSM, i.e., it has astrophysical nature. Indeed, in contrast to (C.34), reaction (C.30), most relevant for gallium experiments, directly determines energy production in the Sun, so the p-p neutrino flux can be deduced in practically model-independent way from the well-measured luminosity of the Sun (barring very exotic possibilities). Hence, after gallium experiments, neutrino flavor transition became the only explanation of the solar neutrino deficit.

Final argument that directly established the fact of transitions of ν_e to ν_μ and ν_τ on their way from the center of the Sun, came from the SNO detector [282] (Sudbury Neutrino Observatory, Canada). This detector used 1 thousand tons of heavy water as the detector material.[4] Neutrinos were detected in the reaction of elastic scattering (C.37) as well as in the reactions

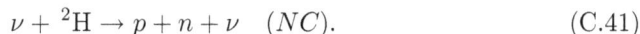

$$\nu_e + {}^2\text{H} \to p + p + e^- \quad (CC), \tag{C.40}$$

$$\nu + {}^2\text{H} \to p + n + \nu \quad (NC). \tag{C.41}$$

Like Kamiokande and Super-K, the SNO detector was sensitive to boron neutrinos of energies $E_\nu \gtrsim 5$ MeV. In reaction (C.37), the neutrino flux combination (C.39) was measured; the result was in agreement with Super-K (though it had larger statistical uncertainty). On the other hand, the charged-current reaction (C.40) exists for *electron* neutrino only, so the *electron* neutrino flux Φ_{ν_e} was obtained by measuring its rate. Finally, the reaction (C.41) is purely neutral-current, so its rate is determined by

$$\Phi_{NC} = \Phi_{\nu_e} + \Phi_{\nu_\mu} + \Phi_{\nu_\tau}. \tag{C.42}$$

---

[4]At later stages 2 tons of salt were added to increase the sensitivity to neutral currents.

Neutrino fluxes measured in reactions (C.40), (C.41) as compared to the SSM predictions are

$$\frac{\Phi_{\nu_e}^{SNO}}{\Phi_{\nu_e,\,SSM}} = 0.30 \pm 0.05, \tag{C.43}$$

$$\frac{\Phi_{NC}^{SNO}}{\Phi_{NC,\,SSM}} = 0.87 \pm 0.19. \tag{C.44}$$

The result (C.44) shows that the Standard Solar Model predicts the emitted flux of boron neutrinos correctly, while from (C.43) it follows directly that about 2/3 of them are converted from ν_e to ν_μ and ν_τ when traveling to the Earth. It is also important that the results (C.43) and (C.44) are consistent with (C.38), taking into account (C.39).

In fact, the agreement between the experimental data is even better than it might seem from (C.38), (C.43) and (C.44). As we have already noted, large contribution to errors in (C.38), (C.43), (C.44) is due to the uncertainty in the SSM calculation. By themselves, the experimental data have an error of less than 10%; this is the precision they agree among themselves. We emphasize that irrespective of SSM, the *three* measured combinations of fluxes, Φ_{ν_e}, Φ^{NC} and $\Phi_{eff}^{\nu e}$ have *two* independent parameters Φ_{ν_e} and $\Phi_{\nu_\mu} + \Phi_{\nu_\tau}$.

The fundamental result of electron neutrino oscillations has been confirmed by KamLAND experiment [283] (Kamioka mine, Japan). KamLAND detector contains 1 thousand tons of liquid scintillator and detects antineutrinos produced in nuclear reactions at Japanese nuclear power plants. Distances to them range from 70 km to 250 km and more, so that the effective baseline is about 180 km, in contrast to earlier reactor experiments with much shorter baseline. KamLAND observed the deficit of electron antineutrinos as compared to the value calculated under no-oscillation hypothesis,

$$\frac{\Phi^{KamLAND}}{\Phi_{no\,osc}} = 0.66 \pm 0.06.$$

Thus, electron antineutrino of energies $E \simeq 3$–6 MeV (the range relevant for Kam-LAND) experiences transition into other types already at distance of about 100 km.

Finally, the flux of monoenergetic neutrinos born in reaction (C.33) was measured in Borexino experiment [284] (Gran Sasso). This experiment makes use of 280 tonns of liquid scintillator. The measured ratio of neutrino flux with energy 862 keV to SSM prediction is [296]

$$\frac{\Phi^{Borexino}}{\Phi^{Borexino}} = 0.62 \pm 0.05,$$

which again implies electron neutrino transition into other neutrino flavors. In fact, Borexino is capable of measuring solar neutrinos in a wide range of energies, from *pp* to boron. Thus, different regimes of solar neutrino transitions (see below) are probed in one and the same experiment.

To describe the solar neutrino data and KamLAND results at the present level of experimental accuracy, it is sufficient to use the two-neutrino picture of oscillations between electron neutrino ν_e and some linear combination $\tilde{\nu}$ of muon neutrino and τ-neutrino. This is the second special case studied in Sec. C.1.2: the relevant mass squared difference $\Delta m_{sol}^2 \equiv \Delta m_{21}^2$ is the smallest one, and the PMNS matrix indeed has $|U_{e3}|^2 \ll 1$; see Sec. C.2.3. In the two-neutrino picture, the data are described by the following parameters of oscillations (precise ranges are given in Sec. C.3)

$$\Delta m_{sol}^2 \simeq 10^{-4}\ \mathrm{eV}^2, \tag{C.45}$$

$$\theta_{sol} \simeq 35°. \tag{C.46}$$

Notably, of great importance to solar neutrinos is the MSW effect. In the two-neutrino approximation, the effective Hamiltonian in the flavor basis $(\nu_e, \tilde{\nu})$ has the form

$$H(L) = \frac{\Delta m_{sol}^2}{4E} \begin{pmatrix} -\cos 2\theta_{sol} & \sin 2\theta_{sol} \\ \sin 2\theta_{sol} & \cos 2\theta_{sol} \end{pmatrix} + V(L) \begin{pmatrix} 1 & 0 \\ 0 & 0 \end{pmatrix}, \tag{C.47}$$

where θ_{sol} is the vacuum mixing angle, so that the mass eigenstates are

$$|\nu_2\rangle = |\nu_e\rangle \sin\theta_{sol} + |\tilde{\nu}\rangle \cos\theta_{sol}, \quad |\nu_1\rangle = |\nu_e\rangle \cos\theta_{sol} - |\tilde{\nu}\rangle \sin\theta_{sol}.$$

Recall that in vacuum, by definition, the heavier state is $|\nu_2\rangle$.

The electron density is $n_e = 6 \cdot 10^{25}$ cm^{-3} in the center of the Sun and decreases with the distance from the center. Hence, the estimate for the maximum value of the potential V is

$$V_{max} = V(L=0) \simeq 8 \cdot 10^{-12}\ \mathrm{eV}.$$

This implies that the relation

$$\frac{\Delta m_{sol}^2}{4E} \sim V_{max}$$

holds at $E \sim 3$ MeV. For a neutrino of significantly lower energies (e.g., pp-neutrino) the matter effects are negligible, and one can use (C.24) with $|U_{e3}|^2 \ll 1$. At $E \gtrsim 3$ MeV, on the contrary, the effect of solar matter is important. Two-neutrino mixing in the latter situation is characterized by effective mixing angle θ_M related to the vacuum angle θ_{sol} as follows:

$$\sin^2 2\theta_M = \frac{\sin^2 2\theta_{sol}}{\sin^2 2\theta_{sol} + (2V(L)E/\Delta m_{sol}^2 - \cos 2\theta_{sol})^2}.$$

As an example, consider neutrino produced in ^{8}B decay. Its characteristic energy is $4-10$ MeV. In this case, the matter term V dominates in the center of the Sun over the neutrino mass term in the Hamiltonian. Let $|\nu_i(L)\rangle$ be eigenvectors of the matrix (C.47) at given distance L from the solar center, and $|\nu_2(L)\rangle$ refers to larger

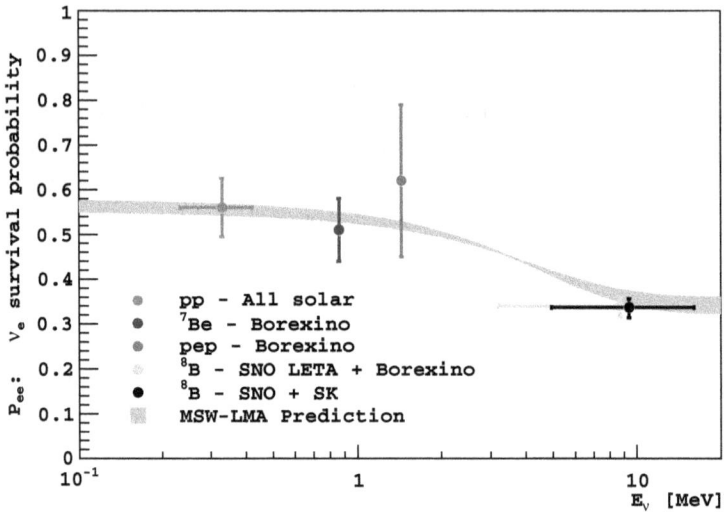

Fig. C.4.   Electron neutrino survival probability as function of energy [297]. See Fig. 13.23 for color version.

eigenvalue. Since $V > 0$, see (C.28), $|\nu_2\rangle$ coincides with $|\nu_e\rangle$ in the center. Thus, the decay of ^8B produces the eigenstate $|\nu_2\rangle$.

Let us consider further evolution of this state, making use of the adiabatic approximation. Recall that in this approximation, the quantum system is always stuck at one and the same energy level. In our case, this means that the neutrino is always in the state $|\nu_2(L)\rangle$. So, on the solar surface it is in the state which coincides with the state $|\nu_2\rangle$ in vacuum. This is still mass eigenstate, therefore it does not oscillate during the further propagation in vacuum.[5] Thus, the probability of detection of *electron neutrino* at the Earth is

$$P(\nu_e \to \nu_e) = |\langle\nu_e|\nu_2\rangle|^2 = \sin^2\theta_{sol}. \qquad (\text{C.48})$$

Equation (C.46) then implies that $P(\nu_e \to \nu_e) < 0.5$. We emphasize that the observed fact that the measured flux of boron ν_e is smaller than half of the predicted one (see (C.43)) is direct evidence for the MSW effect: in the case of vacuum oscillations, electron neutrino survival probability averaged over energies cannot be less than 50%, once two-neutrino approximation is valid; see Eq. (C.11). The solar neutrino transition data in the energy range 0.3–10 MeV, shown in Fig. C.4, nicely illustrate the transition from vacuum oscillation regime to transitions dominated by dense solar matter.

We note one feature evident from the above analysis. Consider an unrealistic case of small vacuum mixing, $|\sin\theta_{sol}| \ll 1$. In that case, the survival probability

---

[5] Note that in the adiabatic regime, neutrino *oscillations* never occur: the neutrino is constantly in the eigenstate $|\nu_2\rangle$ of the local effective Hamiltonian.

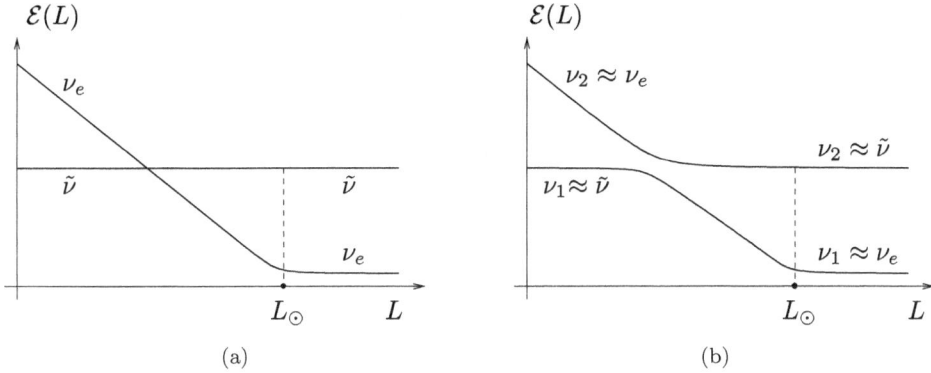

Fig. C.5.   The evolution of levels of neutrino Hamiltonian with distance from the center of the Sun. (a) No mixing, (b) Small but non-vanishing mixing.

(C.48) is low. In the case of vacuum oscillations the situation is opposite: Eq. (C.11) gives $P_{\nu_e \to \nu_e} = 1 - \mathcal{O}(\sin^2 2\theta_{sol})$. Here we have an example of the Mikheev–Smirnov resonance that enhances neutrino transition in matter. The resonance picture is as follows. In the absence of mixing, vectors $|\nu_e\rangle$ and $|\tilde{\nu}\rangle$ would be the eigenvectors of the operator (C.47) with $\sin 2\theta_{sol} = 0$. In the center of the Sun and in vacuum, $|\nu_e\rangle$ corresponds to larger and smaller eigenvalue, respectively. (The assumption that the lighter neutrino is mainly $|\nu_e\rangle$ in vacuum is important here.) The evolution of levels with L would have the form shown in Fig. C.5(a). If small mixing is switched on, the levels no longer intersect, as we know from quantum mechanics, and the evolution of the levels becomes as shown in Fig. C.5(b). The heavier neutrino in the center of the Sun almost coincides with electron neutrino, while it is almost $\tilde{\nu}$ in vacuum. In the adiabatic evolution, the transitions from level to level do not occur; this explains the low survival probability of electron neutrino for small but finite $\sin \theta_{sol}$.

Problem C.1. *Show that for small $\sin \theta_{sol}$, the eigenvalues of the Hamiltonian (C.47) indeed evolve with L as shown in Fig. C.5(b).*

Problem C.2. *Consider the (unrealistic) model of the Sun, in which the density of free electrons $n_e(L)$ changes linearly with L from its value in the center, $6 \cdot 10^{25}$ cm^{-3}, to zero at the surface (at $L_\odot = 7 \cdot 10^5$ km). In what region of parameters Δm_{sol}^2 and $\sin \theta_{sol}$ the evolution of the neutrino state with L is adiabatic? Consider separately the cases of weak and strong mixing, $|\sin \theta_{sol}| \ll 1$ and $|\sin \theta_{sol}| \sim 1$.*

Problem C.3. *Find an analog of Eq. (C.48) in the case of oscillations between three types of neutrinos, assuming $|U_{e3}|^2 \ll 1$.*

To end this discussion, we note that the MSW-effect can lead to a number of other features in neutrino experiments, such as "day-night" effect, the difference in the measured flux of solar neutrinos at night (when neutrinos pass through the Earth) and daytime.

C.2.2. *Atmospheric neutrinos, K2K and MINOS*

Oscillations of *muon* neutrino were discovered in experiments of another class. They first showed up in measurements of atmospheric neutrino flux with Kamiokande [285] and Super-K [286, 287].

Our Galaxy is filled with cosmic rays — charged particles (protons and nuclei) propagating in space. Their interaction with the Earth atmosphere gives rise to secondary particles. As a result, large number of particles are produced, among which the dominant component is the lightest hadrons, pions (with rather small admixture of kaons). Charged pions $\pi^\pm$ do not reach the Earth surface and decay in the atmosphere, producing muons and muon (anti)neutrinos,

$$\pi^+ \to \mu^+ \nu_\mu, \quad \pi^- \to \mu^- \bar\nu_\mu. \tag{C.49}$$

If the energy of the primary particle is not too high, muons, in turn, also decay, again giving rise to neutrinos:

$$\mu^+ \to e^+ \nu_e \bar\nu_\mu, \quad \mu^- \to e^- \bar\nu_e \nu_\mu. \tag{C.50}$$

Neutrinos produced in reactions like (C.49), (C.50), are called atmospheric; neutrino energies relevant for oscillations range from hundreds MeV to tens GeV.

Problem C.4. *At what energies of primary particle most muons reach the Earth surface? Assume that the average multiplicity (number of particles produced in a collision) is 10 to 500 at energies of tens GeV to hundreds EeV. Hints: Muons practically do not interact with the atmosphere; hadron (including pion) mean free path with respect to inelastic scattering is about 10% of atmospheric depth.*

The cosmic ray flux is isotropic, so the flux of atmospheric neutrinos in the absence of oscillations should be isotropic as well.[6] However, the observed flux of muon neutrinos and antineutrinos actually depends on the zenith angle (right panel of Fig. C.6). This means that muon neutrinos coming from above and flying just a few kilometers from the production point to the detector, have no time to oscillate; on the other hand, neutrinos coming from below pass through the entire Earth, and have time to partly transform into other types of neutrinos. At the same time, the effect of oscillations on the electron neutrino flux is small (left panel of Fig. C.6). The muon neutrino deficit, together with very small excess of electron neutrinos, suggest that there is $\nu_\mu - \nu_\tau$ oscillation. This result is confirmed by the entire set

---

[6]For neutrinos with energies of several GeV and above, the isotropy gets lost: the flux has a peak in the horizontal direction due to the fact that horizontal muons travel longer in the atmosphere and thus have more time to decay. This phenomenon is straightforwardly accounted for.

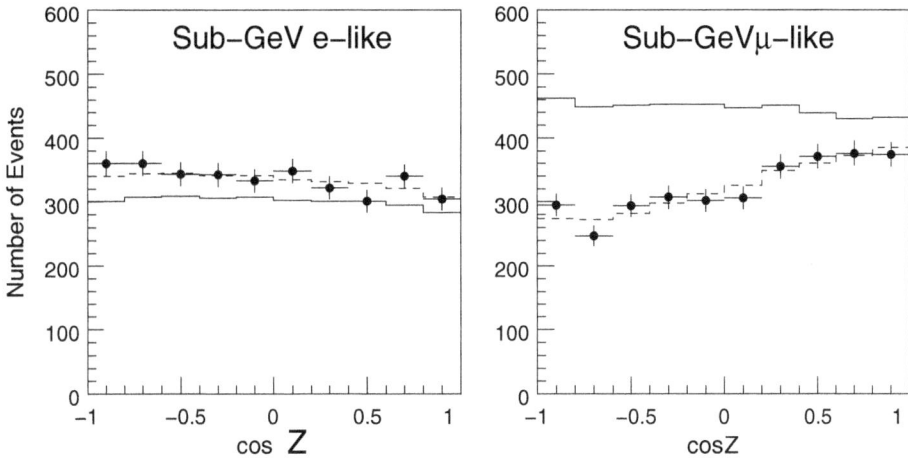

Fig. C.6. Dependence of neutrino fluxes at energies below 1 GeV on zenith angle [287]: the left and right panels refer to electron and muon neutrinos, respectively. Solid lines show the prediction in the absence of oscillations, while dotted lines are for oscillations with parameters obtained by fitting the data.

of data on atmospheric neutrinos, including the measurements of absolute fluxes of ν_e, $\bar{\nu}_e$ and ν_μ, $\bar{\nu}_\mu$, neutrino fluxes at energies above 1 GeV, etc.

Muon neutrino oscillations were confirmed by the K2K experiment [288]. Here the source of muon neutrinos are pions produced by proton beam from the accelerator of KEK Laboratory in Japan and they decay according to (C.49). These neutrinos are detected by Super-K. The distance from production to detection is 250 km (the distance between the KEK Laboratory and Kamioka mine), and the neutrino energy is 0.5–3 GeV. K2K experiment found disappearance of muon neutrinos: their flux at Super-K is smaller than the flux mesured by the "near" neutrino detector located directly at the KEK Laboratory. The results of K2K experiment are in good agreement with atmospheric neutrino data.

Study of muon neutrino oscillations is the purpose of two other experiments with accelerator neutrinos. In one of them, muon neutrinos and antineutrinos produced at Fermilab accelerator (Batavia, USA) are detected by MINOS detector (Minnesota, USA) at distance 735 km. Another experiment T2K makes use of muon neutrinos produced at JPARC accelerator complex (Tokai, Japan) and detects them by Super-K at distance 295 km. A peculiarity of T2K is that the axis of the neutrino beam, wich spreads over fairly large solid angle, is off the direction towards the detector by angle 2.5°. This leads to narrower energy distribution of neutrinos at Super-K at the expence of somwehat lower statistics. The results of MINOS [298] and T2K [299] on muon neutrino disappearance are in agreement and, together with Super-K and K2K data, give better determination of the oscillation parameters.

The results of experiments briefly reviewed in this Section are also well described within the two-neutrino oscillation picture.[7] The simplest and most plausible possibility is the oscillations of ν_μ to ν_τ. (Otherwise new types of neutrino have to be introduced.) In that case, the oscillation parameters are estimated as follows (more precise estimates are given in Sec. C.3)

$$\Delta m^2_{atm} \simeq (2-3) \cdot 10^{-3} \text{ eV}^2, \tag{C.51}$$

$$\theta_{atm} \simeq 45^\circ. \tag{C.52}$$

Note that $\theta_{atm} = 45^\circ$ corresponds to maximum mixing: the parameter $\sin^2 2\theta_{atm}$ entering (C.11) is equal to 1. Matter effects are of little importance for accelerator neutrinos, so in that case we are dealing with vacuum oscillations.

Problem C.5. *Using the parameters of ν_μ-ν_τ system, determine the energy at which neutrinos oscillate substantially when crossing the Earth. Do parent muons have enough time to decay in the atmosphere?*

Problem C.6. *Estimate what part of ν_μ disappears (transforms into ν_τ) in the K2K experiment.*

C.2.3. Accelerator and reactor neutrinos: $|U_{e3}|$

The parameter U_{e3} determines the admixture of electron neutrino ν_e in the mass eigenstate ν_3. It gives rise to two effects, showing up in experiments with accelerator neutrinos and reactor neutrinos, respectively. The first effect is the appearance of electron neutrinos in muon neutrino beam, and the second is the disappearance of electron antineutrinos at fairly short distance. The Earth matter effect is not very important in both cases, and, furthermore, the inequality (C.18) holds. Thus, we are dealing with the first of special cases considered in Section C.1.2. The appearance of ν_e in the muon neutrino beam is approximately described by the formula (C.23a) with $\alpha = \mu$, $\beta = e$, while the electron neutrino survival probability is given by (see (C.22))

$$P(\bar\nu_e \to \bar\nu_e) = 1 - 4|U_{e3}|^2\left(1 - |U_{e3}|^2\right)\sin^2\left(\frac{\Delta m^2_{31}}{4E}L\right).$$

Clearly, both effects would be absent if $U_{e3} = 0$.

The first evidence for nonzero U_{e3} was obtained at the accelerator neutrino experiments T2K [300] and MINOS [301]. They observed the excess of electron neutrinos in the beams of mostly muon neutrinos (with small admixture of electron neutrinos), although confidence level of this excess was not sufficient to claim the

---

[7]The two-neutrino oscillation approach, generally speaking, is *not* applicable to atmospheric neutrinos. In this case, almost complete absence of any distortion of electron neutrino flux is accidental: it is due to the interplay between nearly maximum mixing (C.52) and the relation between muon and electron neutrino fluxes produced in the atmosphere, $\Phi_{\nu_e}/\Phi_{\nu_\mu} \simeq 1/2$.

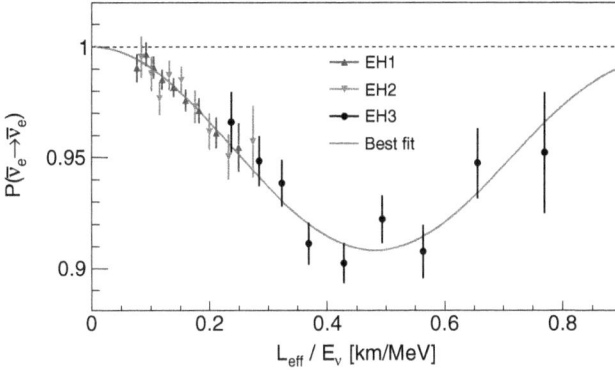

Fig. C.7. Survival probability of electron antineutrino in Daya Bay experiment. EH1 and EH2 denote near detectors, and EH3 stands for far detectors. The solid line is the theoretical expectation for best fit value of U_{e3}. The oscillatory behavior is observed due to the capability of measuring antineutrino energy.

discovery. The disappearance of electron antineutrinos was first observed, also at a confidence level insufficient for claiming discovery, in a reactor antineutrino experiment Double CHOOZ in France [302]. The disapearance effect was firmly established in reactor antineutrino experiments Daya Bay [303] in China and RENO [304] in South Korea. Antineutrinos in Daya Bay experiment are produced by several industrial reactors and detected both by near detectors (at $500-600$ meters from reactors) and far detectors at distance $1.6-1.7$ km. Antineutrino energy is 2–3 MeV, and for mass square difference Δm^2_{atm} it follows from (C.14) and (C.51) that the disappearance probability is small at near detector sites and is close to maximum at far sites. This is precisely what is observed at Daya Bay; see Fig. C.7.

Similar results are obtained by RENO experiment which employs near and far detectors 294 m and 1.4 km away from reactor, respectively. These two experiments obtain (precise estimates are given in Sec. C.3)

$$|U_{e3}|^2 \simeq 0.025.$$

Thus, the only unknown parameter of the PMNS marix (C.16) is the *CP*-phase δ.

C.3. Oscillation Parameters

Figure C.8 (see also Fig. 13.10 on color pages) shows the regions in the parameter space[8] $\tan^2 \theta_{sol}$ and Δm^2 allowed by solar neutrino experiments and KamLAND. It is seen that all data are consistent with each other in the region (C.45), (C.46).

---

[8]Sometimes the data are parameterized by $\sin^2 2\theta$ rather than $\tan^2 \theta$. In the case of vacuum oscillations, the former is more natural, see Eq. (C.11). As can be seen from (C.48), $\sin^2 2\theta$ is not an appropriate parameter in cases where matter effects are significant. Indeed, $\sin 2\theta$ does not change under the replacement $\theta \to (\pi/2 - \theta)$, while the probability of oscillations in matter does.

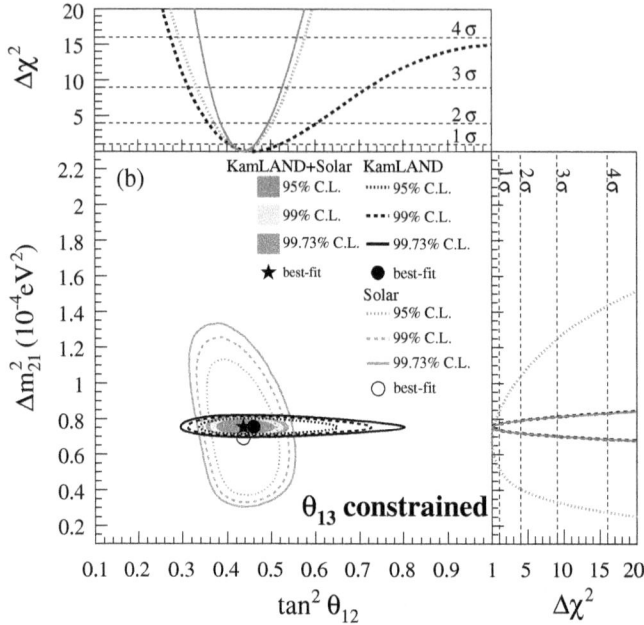

Fig. C.8. Allowed region of parameter space for oscillations $\nu_e \leftrightarrow \tilde{\nu}$ obtained from solar neutrino experiments and KamLAND [305]. (See Fig. 13.10 for color version.)

Figure C.9 shows similar data on atmospheric neutrinos, T2K and MINOS. Here the allowed region is around (C.51), (C.52). The results of reactor neutrino experiment Daya Bay are presented in Fig. C.10.

We mention here anomalies in several neutrino experiments, which may possibly point towards the existence of the third mass squared $\Delta m^2 \gg \Delta m^2_{atm}$. These are accelerator experiments LSND [308] and MiniBooNE [309], gallium experiments with artificial neutrino sources [310] and to some extent reactor neutrino experiments [311–313]. If these anomalies are confirmed, this would mean that there exists the fourth neutrino species, which is light and sterile against the Standard Model gauge interactions. For the time being, the results on the anomalies have too large statistical and systematic uncertainties, so the situation is quite unclear.

We limit ourselves with three neutrino flavors and recall that the PMNS matrix responsible for oscillations has the form (C.16). The results shown in Figs. C.8 and C.9 correspond to the following absolute values of the mixing matrix elements (3σ C.L.)

$$|U_{\alpha i}| = \begin{pmatrix} 0.795\text{--}0.846 & 0.513\text{--}0.585 & 0.126\text{--}0.178 \\ 0.205\text{--}0.543 & 0.416\text{--}0.730 & 0.579\text{--}0.808 \\ 0.215\text{--}0.548 & 0.409\text{--}0.725 & 0.567\text{--}0.800 \end{pmatrix}.$$

Current allowed intervals of the mixing angles are (3σ C.L.) [314]:

$$0.273 \leq \sin^2 \theta_{12} \leq 0.354, \tag{C.53}$$

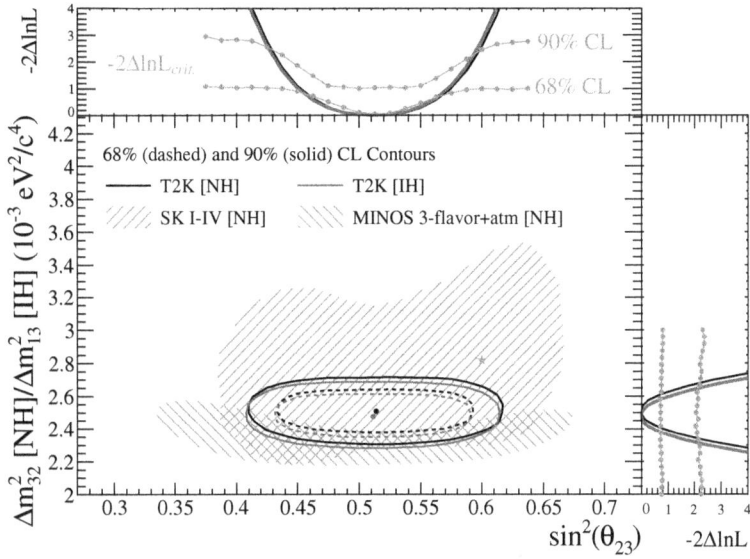

Fig. C.9. **(Color version on color pages.)** Allowed region of parameter space for oscillations $\nu_\mu \leftrightarrow \nu_\tau$ [307]. Lines encircle areas allowed by T2K at 68% and 90% C.L. Shaded regions show the results (at 90% C.L.) of MINOS and atmospheric neutrino study at Super-K. The results are presented for normal and inverse hierarchy of neutrino masses, NH and IH, respectively. Dots show the best fit values for the two hierarchies.

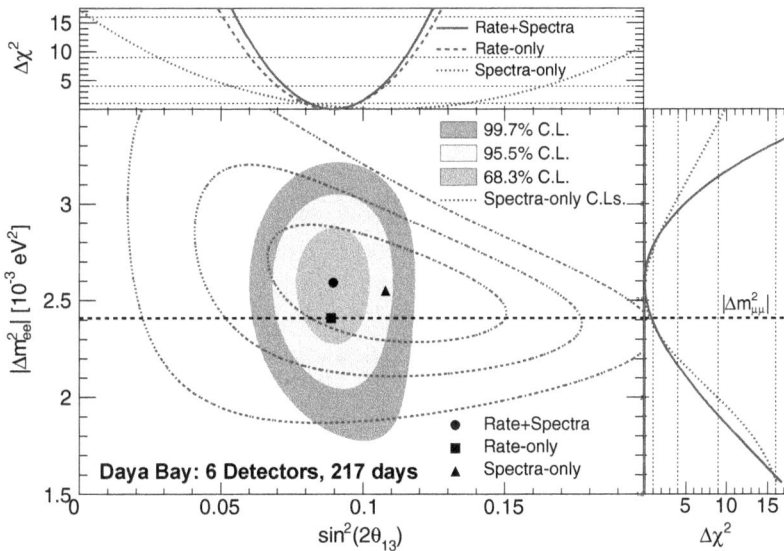

Fig. C.10. **(Color version on color pages.)** Allowed regions of parameter space for $\bar{\nu}_e \leftrightarrow \bar{\nu}_e$ oscillations determined by the Daya Bay reactor neutrino experiment [306]. Dashed lines show the results obtained by analyzing solely the antineutrino energy spectrum. Horizontal line $\Delta m^2_{\mu\mu}$ corresponds to the measurement of the same parameter by MINOS.

$$0.0181 \leq \sin^2 \theta_{13} \leq 0.0327, \tag{C.54}$$

$$0.341 \leq \sin^2 \theta_{23} \leq 0.670. \tag{C.55}$$

Note that most of the mixing matrix elements are of the same order, $U_{\alpha j} \sim 0.5$. This "anarchy" in the neutrino mixing matrix distinguishes it from its quark counterpart, the CKM matrix (B.29) exhibiting noticeable hierarchy between its elements.

The results shown in Figs. C.8 and C.9 also determine the differences of masses squared. Combined analysis of existing data yields [314]

$$7.04 \cdot 10^{-5} \text{ eV}^2 \leq \Delta m_{21}^2 \equiv \Delta m_{sol}^2 \leq 8.12 \cdot 10^{-5} \text{ eV}^2,$$
$$2.27 \cdot 10^{-3} \text{ eV}^2 \leq |\Delta m_{32}^2| \equiv \Delta m_{atm}^2 \leq 2.69 \cdot 10^{-3} \text{ eV}^2 \tag{C.56}$$

at 3σ C.L. As we discussed above, these values can be obtained with two different hierarchies in neutrino masses; see Fig. C.1.

Which of these two hierarchies exists in Nature is not yet known. At the same time, Eq. (C.56) gives lower limits on neutrino masses: at least one of them must be

$$m \geq m_{atm} \equiv \sqrt{\Delta m_{atm}^2} \simeq 0.05 \text{ eV}, \tag{C.57}$$

while another is not less than

$$m_{sol} \equiv \sqrt{\Delta m_{sol}^2} \simeq 0.009 \text{ eV}. \tag{C.58}$$

Minimal possibility is that

$$m_1 \ll m_{sol}, \quad m_2 = m_{sol}, \quad m_3 = m_{atm}, \tag{C.59}$$

(normal hierarchy without degeneracy), but other options are also widely discussed, including the case of rather heavy neutrinos almost degenerate in masses, $m_1, m_2, m_3 \gg m_{atm}$.

C.4. Dirac and Majorana Masses. Sterile Neutrinos

There are two different types of fermion masses in $3 + 1$ space–time dimensions: Majorana and Dirac. The two Lorentz-invariant mass terms in the Lagrangian for fermion f are

$$\mathcal{L}_f^M = -\frac{m_M}{2} \bar{f}_L^c f_L + h.c., \tag{C.60}$$

$$\mathcal{L}_f^D = -m_D \bar{f}_R f_L + h.c., \tag{C.61}$$

where f^c is the charge-conjugate fermion field.

Recall that 4-component Dirac spinor can be expressed in terms of 2-component Weyl spinors χ_L, ξ_R (see Sec. B.1 of Appendix B). In the Weyl basis of the Dirac matrices

$$f = \begin{pmatrix} \chi_L \\ \xi_R \end{pmatrix}.$$

Hence, the mass terms (C.60) read

$$\mathcal{L}_f^M = -\frac{m_M}{2} \chi_L^T i\sigma_2 \chi_L + h.c.,$$

$$\mathcal{L}_f^D = -m_D \xi_R^\dagger \chi_L + h.c.$$

The Dirac mass is only possible if the theory contains both left and right fermion components, while the left component (or right component) is sufficient by itself for the Majorana mass. Electrically charged fermions can only have Dirac masses, otherwise the charge would not be conserved.

The Standard Model contains only left components of neutrinos, so its minimal generalization with massive neutrino and without additional fields involves the Majorana mass term,

$$\mathcal{L}_\nu^M = -\frac{m_{\alpha\beta}}{2} \bar{\nu}_{L\alpha}^c \nu_{L\beta} + h.c., \tag{C.62}$$

where we used the flavor basis for neutrino fields, and summation over flavor indices is assumed. The mass matrix of light neutrinos $m_{\alpha\beta}$ is symmetric, and it can be diagonalized by transformation

$$m = U^{PMNS} m^{diag} (U^{PMNS})^T,$$

where U^{PMNS} is a unitary matrix and m^{diag} is a diagonal real matrix. U^{PMNS} is precisely the PMNS matrix in the case of Majorana neutrinos.

Problem C.7. *Show that the expression $\bar{\nu}_{L\alpha}^c \nu_{L\beta}$ is symmetric in α and β. Hint: note that fermion fields anticommute.*

Problem C.8. *Show that the flavor basis in related to mass basis by $\nu_i = U_{\alpha i}^{PMNS} \nu_\alpha$, which is consistent with (C.1).*

Let us come back to the issue of the physical significance of phases in diagonal matrices D_1 and D_2 which enter the general form (C.15) for the PMNS matrix. In the first place, one can extract from D_2 a factor proportional to the unit matrix; this factor commutes with the matrix U and can be absorbed into D_1. Therefore, without loss of generality one sets $D_2 = \text{diag}(e^{i\delta_1}, e^{i\delta_2}, 1)$. Now, the neutrino kinetic terms are invariant under independent phase rotations of each of the neutrino fields, $\nu_{L\alpha} \to e^{i\beta_\alpha} \nu_{L\alpha}$; the interactions with W- and Z-bosons are also invariant under these rotations supplemented by the phase rotations of charged fermion fields. Making use of these rotations one can make the matrix D_1 equal to the unit matrix. Thus, the complete PMNS matrix has the form $U^{PMNS} = U D_2$, i.e.,

$$U^{PMNS} = \begin{pmatrix} 1 & 0 & 0 \\ 0 & c_{23} & s_{23} \\ 0 & -s_{23} & c_{23} \end{pmatrix} \begin{pmatrix} c_{13} & 0 & s_{13}e^{i\delta} \\ 0 & 1 & 0 \\ -s_{13}e^{-i\delta} & 0 & c_{13} \end{pmatrix}$$

$$\times \begin{pmatrix} c_{11} & s_{12} & 0 \\ -s_{12} & c_{12} & 0 \\ 0 & 0 & 1 \end{pmatrix} \begin{pmatrix} e^{i\delta_1/2} & 0 & 0 \\ 0 & e^{i\delta_2/2} & 0 \\ 0 & 0 & 1 \end{pmatrix}.$$

We emphasize that the last factor (i.e., phases δ_1, δ_2) is physical for Majorana masses only; in the Dirac case one can get rid of this factor in the same way as in Sec. B.4.

Since the Majorana mass term mixes field with its charge-conjugate, the notions of particle and antiparticle are not quite adequate for Majorana neutrino. This means, in particular, that there is no conserved lepton number: the Majorana mass violates lepton number explicitly. Indeed, the expression (C.62) is not invariant under phase rotations $\nu \to e^{i\alpha}\nu$, $\bar{\nu} \to e^{-i\alpha}\bar{\nu}$. In the case of ultra-relativistic Majorana neutrino, the eigenstates of the Hamiltonian are states with left and right helicities, and up to corrections suppressed by the ratio m/E the left-helicity and right-helicity states coincide with neutrino and antineutrino states of massless neutrino theory, respectively.

Problem C.9. *Prove the above statement. To do this, obtain the analog of the Dirac equation in the case of the Majorana mass, find its solution in terms of creation and annihilation operators and compare it with the solution to the Weyl equation for massless left fermions.*

Problem C.10. *Show that in the ultra-relativistic case, the off-diagonal Majorana mass (C.62) gives rise to oscillations between states of one and the same helicity (i.e., left-helicity state oscillates into left-helicity state of different flavor, and not into right-helicity state, and vice versa).*

The result of the latter problem implies that oscillations $\nu_\alpha \leftrightarrow \nu_\beta$, $\bar{\nu}_\alpha \leftrightarrow \bar{\nu}_\beta$, but not $\nu_\alpha \leftrightarrow \bar{\nu}_\beta$, are possible in the case of the Majorana mass term. Here neutrino and antineutrino are understood as left- and right-helicity states, respectively. With this convention, all results of massless theory concerning neutrino interactions remain valid: for example, it is antineutrino (right-handed state) that is predominantly produced in the neutron decay, while the admixture of left-handed state (neutrino) is suppressed by powers of m_ν/E_ν.

Mass term (C.62) cannot be obtained from any $SU(3)_c \times SU(2)_W \times U(1)_Y$-invariant renormalizable interaction. Giving up renormalizability, one writes down the interaction of the form

$$\mathcal{L}_{int} = \sum_{\alpha,\beta} \frac{\xi^{\alpha\beta}}{\Lambda_\nu} \bar{L}^c_\alpha \tilde{H}^* \cdot \tilde{H}^\dagger L_\beta + h.c., \tag{C.63}$$

where we introduced dimensionless couplings $\xi^{\alpha\beta}$ (indices $\alpha, \beta = 1, 2, 3$ label generations); Λ_ν is the energy scale of a theory that generalizes the Standard Model at high energies and leads to non-renormalizable interaction (C.63) at low energies; the field $\tilde{H}$ is related to H by (B.10). The Englert–Brout–Higgs field H acquires vacuum expectation value (see Appendix B), and the coupling (C.63) gives rise to the mass term

$$\frac{v^2}{2\Lambda_\nu} \xi^{\alpha\beta} \bar{\nu}^c_\alpha \nu_\beta + h.c.,$$

which coincides with (C.62). Note that the term (C.63) is of the lowest possible order in Λ_ν^{-1} among all terms capable of giving masses to neutrinos; this is the reason we picked it up.

For $\xi^{\alpha\beta} \sim 1$, the scale of new interaction must be of order $\Lambda_\nu \sim 10^{15}$ GeV to yield neutrino masses of order 10^{-2} eV. This scale is close to the Grand Unification scale in supersymmetric theories.

Non-renormalizable effective interaction (C.63) can arise from renormalizable interaction involving new heavy fields, just like the Fermi four-fermion interaction emerges upon integrating out massive vector bosons of the Standard Model. The smallness of neutrino masses as compared to the masses of other Standard Model fermions requires strong hierarchy between the known Yukawa couplings and the new couplings, $y_{SM}^2 \gg \xi^{\alpha\beta}$, and/or between the electroweak scale and the mass scale of new heavy fields. In specific models, these hierarchies may have one or another natural explanation.

One possibility here is the so-called *see-saw* mechanism. To begin with, let us consider this mechanism in the case of one type of conventional neutrino ν. This neutrino is a component of the left lepton doublet L of the Standard Model. Let there be another left fermion N_L, which is a singlet under the Standard Model gauge group $SU(3)_c \times SU(2)_W \times U(1)_Y$. (Equivalently, one introduces right fermion N_L^c.) In contrast to the known fields of the Standard Model, the field N_L can have Majorana mass M unrelated to the vacuum expectation value of the Standard Model Englert–Brout–Higgs field. It is remarkable that the gauge invariance of the Standard Model allows the Yukawa interaction which couples N_L and ν to the Standard Model Englert–Brout–Higgs field H. So, the renormalizable Lagrangian for the fields N_L and L includes the terms

$$\mathcal{L} = -\frac{M}{2} \bar{N}_L^c N_L - y \bar{N}_L^c \tilde{H}^\dagger L + h.c., \tag{C.64}$$

where y is the Yukawa coupling. As a result of electroweak symmetry breaking, the field $\tilde{H}^\dagger$ obtains vacuum expectation value $(v/\sqrt{2}, 0)$, leading to the mass terms

$$\mathcal{L}_m = -\frac{M}{2} \bar{N}^c_L N_L - y\frac{v}{\sqrt{2}} \bar{N}^c_L \nu + h.c.. \tag{C.65}$$

Combining left fermions N_L and ν into the column

$$\psi = \begin{pmatrix} \nu \\ N_L \end{pmatrix}, \tag{C.66}$$

we find that the mass term (C.65) can be written as

$$\mathcal{L}_m = -\frac{1}{2} \bar{\psi}^c m \psi + h.c.,$$

where the matrix m is[9]

$$\begin{pmatrix} 0 & m_D \\ m_D & M \end{pmatrix} \tag{C.67}$$

and

$$m_D = \frac{yv}{\sqrt{2}}.$$

At $M \gg m_D$, the eigenvalues of the mass matrix (C.67) are (the sign of fermion mass is irrelevant)

$$m_\nu = -\frac{m_D^2}{M} = -\frac{y^2 v^2}{2M}, \tag{C.68}$$

$$m_N = M,$$

(modulo corrections suppressed by m_D/M) where the smaller eigenvalue (C.68) corresponds to the eigenvector

$$\begin{pmatrix} 1 \\ -\frac{m_D}{M} \end{pmatrix}$$

(again up to small corrections). It is seen from (C.66) that the main component of this eigenvector is the ordinary neutrino ν. Thus, as a result of this mechanism, neutrino acquires Majorana mass m_ν, which is naturally small for $M \gg m_D$. This is precisely the see-saw mechanism.[10] Note that at $y = 10^{-6} - 1$ (the range of known

---

[9]If instead of left field N_L, one uses the right field $N^c_{\tilde{L}}$, then the second term in (C.65) has the form of the Dirac mass term in which $N^c_{\tilde{L}}$ serves as the right component. Hence the notation m_D.
[10]The scenario we discuss here [203] is known as type I see-saw. There are two other types. Type II see-saw [206] employs a new scalar field which is a triplet under the electroweak $SU(2)$ and has nonzero weak hypercharge; small neutrino masses emerge due to the hierarchy between the vacuum expectation values of this field and Englert–Brout–Higgs doublet. To construct type III see-saw [315], one introduces fermionic triplets under the electroweak group, and the small neutrino masses are obtained in a way similar to type I see-saw.

Yukawa couplings of the Standard Model), the mass $m_\nu \sim 10^{-2}$ eV is obtained at

$$M \sim 10^3 - 10^{15} \text{ GeV},$$

i.e., the condition $M \gg m_D$ is satisfied indeed.

It is worth noting that the result (C.68) can be obtained by integrating out the heavy field N_L. To this end, let us write the equation obtained by varying the action with respect to N_L. With the gradient term $i\bar{N}_L \gamma^\mu \partial_\mu N_L$ in the Lagrangian and mass terms (C.65), we obtain

$$-i\partial_\mu \bar{N}_L \gamma^\mu + M\bar{N}_L^c + \frac{yv}{\sqrt{2}} \bar{\nu}^c = 0.$$

When the momentum and energy are small compared to M, the first term on the left-hand side is negligible, and the field N_L is algebraically expressed through the field ν,

$$N_L = -\frac{yv}{\sqrt{2}M} \nu.$$

We substitute this expression back into the original Lagrangian and thereby obtain the effective Lagrangian for the field ν. The correction to the kinetic term is negligible, and the main effect is the mass term

$$\mathcal{L}_{m_\nu} = \frac{y^2 v^2}{4M} \bar{\nu}^c \nu + h.c.$$

We see that the Majorana neutrino mass is indeed given by (C.68) up to sign omitted there. We also note that starting from the original Lagrangian (C.64) and integrating out the heavy field N_L, we obtain the effective Lagrangian of the form (C.63) with $\Lambda_\nu = M$ and $\xi = y^2$.

We now turn to the realistic case of three types of neutrinos. In this case, it is natural to introduce three fields N_α, $\alpha = 1, 2, 3$ (subscript L is omitted) and generalize the Lagrangian (C.64) as follows,

$$\mathcal{L} = -\frac{1}{2} M_{\alpha\beta} \bar{N}_\alpha^c N_\beta - y_{\alpha\beta} \bar{N}_\alpha^c \tilde{H}^\dagger L_\beta + h.c.$$

Here $M_{\alpha\beta}$ and $y_{\alpha\beta}$ are 3×3 matrices (generally speaking, complex) and the matrix $M_{\alpha\beta}$ is symmetric. One can always choose the basis of fields N_α in such a way that the matrix $M_{\alpha\beta}$ is real and diagonal,

$$M = diag(M_1, M_2, M_3).$$

In this basis, the field N_α describes heavy sterile neutrino of a certain mass. The easiest way to find the effective mass term for light neutrinos is to integrate out heavy fields N_α, as outlined above. As a result, we obtain the Majorana mass term for light neutrinos in the form (C.62) with the mass matrix

$$m = -m_D M^{-1} m_D^T, \tag{C.69}$$

where

$$m_{D\alpha\beta} = \frac{y_{\alpha\beta} v}{\sqrt{2}}.$$

In general, the light neutrino masses and the parameters of the PMNS matrix non-trivially depend on the elements of both the diagonal matrix M and matrix of Yukawa couplings $y_{\alpha\beta}$.

Let us now discuss the possibility that the known neutrinos have the Dirac masses. In this case one adds new light fields $\nu_{R\alpha}$, which are the right components of neutrinos. Then the Dirac mass term reads

$$\mathcal{L}_\nu^D = -m_{\alpha\beta}\bar{\nu}_{R\alpha}\nu_{L\beta} + h.c., \tag{C.70}$$

where the flavor basis is used again. These right components must be neutral (*sterile*) with respect to the Standard Model gauge group, otherwise they would give contribution, for example, to the total width of Z-boson, whlich is measured with high precision and is consistent with the Standard Model prediction.

Since the Dirac mass is invariant under charge conjugation, the notion of lepton number makes sense in this theory: the mass term (C.70), like all other terms of the Standard Model Lagrangian, are invariant under the transformation

$$\nu_\alpha \to e^{i\xi}\nu_\alpha, \quad \bar{\nu}_\alpha \to e^{-i\xi}\bar{\nu}_\alpha.$$

Were the matrix $m_{\alpha\beta}$ diagonal, there would exist conserved lepton numbers for each lepton flavor separately. The observed neutrino oscillations show that the mass matrix $m_{\alpha\beta}$ is not diagonal, hence they reveal violation of individual lepton numbers.

The mass terms (C.70) can emerge, for example, due to renormalizable Yukawa interaction (cf. (C.64))

$$\mathcal{L} = -\sum_{\alpha,\beta} y_{\alpha\beta}\bar{L}_\alpha \tilde{H}\nu_{R\beta} + h.c. \tag{C.71}$$

The Yukawa couplings $y_{\alpha\beta}$ have to be extremely small. In a number of the Standard Model extensions (for example, in supersymmetric theories and GUTs), the smallness of the Yukawa couplings is obtained naturally due to the presence of an intermediate energy scale at which the effective interaction (C.71) emerges. As a result, the Yukawa couplings are suppressed by the ratio (or its power) of the intermediate and the gravitational scales. An illustration is the interaction involving a new scalar S, which is a singlet under the Standard Model gauge group,

$$\mathcal{L} = -\frac{S}{M_{Pl}} \cdot \sum_{\alpha,\beta} Y_{\alpha\beta} \bar{L}_\alpha \tilde{H} \nu_{R\beta} + h.c.,$$

where the dimensionless couplings $Y_{\alpha\beta}$ can be of order unity. If the field S acquires nonzero vacuum expectation value $v \ll \langle S \rangle \ll M_{Pl}$, then the effective renormalizable interaction (C.71) appears at lower energies, with Yukawa couplings of order $\langle S \rangle / M_{Pl} \ll 1$.

To conclude this Section, we note that one cannot exclude the possibility that there are both Majorana and Dirac neutrino mass terms in the Lagrangian, and that both types of masses are important for describing neutrino properties. This possibility, however, does not look very natural, since the mechanisms that lead to the two different types of mass terms, generally speaking, are different. Hence it is hard to expect that these mechanisms lead to mass parameters which are equal within an order of magnitude or so.

C.5. Search for Neutrino Masses

The present direct experimental limits on neutrino masses are [1]:

$$m_{\nu_e} < 2 \text{ eV}, \qquad (C.72)$$

$$m_{\nu_\mu} < 0.19 \text{ MeV}, \qquad (C.73)$$

$$m_{\nu_\tau} < 18.2 \text{ MeV}. \qquad (C.74)$$

These limits are valid regardless of the type of neutrino mass. In the Majorana case, the constraint on the combination of neutrino masses relevant for neutrinoless double-β decay of nuclei (for details see, e.g., Ref. [1]) is stronger:

$$m_\nu < 0.3 \text{ eV}.$$

For comparison, the current limit on the sum of neutrino masses, following from the measurements of CMB anisotropy and studies of structures in the Universe, is at the level

$$\sum_i m_{\nu_i} < 0.2 - 0.5 \text{ eV},$$

depending on which cosmological parameters are fixed from other observations. We discuss the cosmological aspects of massive neutrinos in the accompanying book.

It is expected that the sensitivity of direct laboratory experiments to the mass of the electron neutrino will soon reach 0.2–0.02 eV (depending on the type of mass). Accuracy of cosmological estimates for the sum of the neutrino masses is also improving.

Appendix D

Quantum Field Theory at Finite Temperature

In this Appendix we briefly describe the method of calculation of various quantities (free energy, effective potential, static Green's functions) in quantum field theory at finite temperature. We consider predominantly the most interesting for cosmology case of zero chemical potentials, although the overall approach allows for an appropriate generalization.

We begin with a general comment. It is sometimes useful to treat quantum field theory as quantum mechanics of large but finite number of degrees of freedom. Indeed, field theory can be regularized by introducing spatial lattice of small but finite spacing (ultraviolet regularization) and considering the system in a 3-dimensional box of finite, albeit large size (infrared regularization). For our purposes, time is conveniently treated as a continuous variable.[1] Then fields $\phi(\mathbf{x}, t)$ are functions of the lattice sites[2] and time, $\phi(\mathbf{x}, t) \to \phi(\mathbf{x_n}, t)$, where $\mathbf{x_n}$ are coordinates of the lattice site, labeled by a discrete index $\mathbf{n} = (n_1, n_2, n_3)$. With this regularization, the number of dynamical coordinates $\phi(\mathbf{x_n}, t)$ is large but finite. In this way field theory reduces to quantum mechanics.

We use this approach to obtain formal results.[3] Namely, we develop finite temperature techniques in quantum mechanics, and then merely extend it to quantum field theory.

D.1. Bosonic Fields: Euclidean Time and Periodic Boundary Conditions

Let us consider quantum-mechanical system with dynamical coordinates $q = (q^{(1)}, q^{(2)}, \ldots, q^{(N)})$. Here we assume that q are bosonic coordinates, as usual in quantum mechanics. Let this system be at temperature T. As is known from

---

[1] In lattice numerical simulations, time is also discretized. This would be inconvenient for us.
[2] Gauge fields are naturally considered as living on links of the lattice, rather than on sites. This is insignificant for us.
[3] We are not going to discuss subtle points concerning removal of ultraviolet and infrared regularizations. In brief, no new ultraviolet divergencies or infrared pathologies appear at finite temperature as compared to zero temperature theory.

statistical physics, in thermal equilibrium the expectation values of operators at fixed time are given by

$$\langle \hat{O} \rangle_T = \frac{\mathrm{Tr}(e^{-\beta \hat{H}} \hat{O})}{\mathrm{Tr}(e^{-\beta \hat{H}})}, \tag{D.1}$$

where the operator $\hat{H}$ is the Hamiltonian of the system, the parameter β is

$$\beta = \frac{1}{T},$$

and trace is taken over all states of the system. Free energy F is determined by

$$e^{-\beta F} = \mathrm{Tr}(e^{-\beta \hat{H}}). \tag{D.2}$$

Our first goal is to find a convenient representation for the right-hand side of this equality.

Consider a system with one degree of freedom q and the Hamiltonian

$$\hat{H} = \frac{\hat{p}^2}{2} + V(\hat{q}). \tag{D.3}$$

As a complete set of states in (D.3) we choose eigenstates of the operator $\hat{q}$, i.e., we work in the coordinate representation. Then

$$e^{-\beta \hat{H}} = \int dq \langle q | e^{-\beta \hat{H}} | q \rangle. \tag{D.4}$$

Here and in what follows we omit the numerical factor in front of the integral, which leads only to overall shift of the free energy, $F \to F + \mathrm{const}$. We are interested in the expectation values (D.1) in which this pre-factor cancels out, and also in differences of free energies of different phases, so the shift cancels out as well.

Let us obtain a representation for the right-hand side of (D.4) in the form of functional integral, i.e., path integral in our case (for details see, e.g., Ref. [271]). We write

$$\langle q | e^{-\beta \hat{H}} | q \rangle = \langle q | \prod_i e^{-\Delta \tau_i \hat{H}} | q \rangle$$

$$= \langle q | (1 - \Delta \tau_1 \cdot \hat{H})(1 - \Delta \tau_2 \cdot \hat{H}) \cdots (1 - \Delta \tau_n \cdot \hat{H}) | q \rangle,$$

where we divided the interval of length β into n small segments of lengths $\Delta \tau_1, \ldots, \Delta \tau_n$; we take the limit $n \to \infty$, $\Delta \tau_i \to 0$ in the end. We insert the unit operator between each bracket and write

$$\int dq \langle q|e^{-\beta \hat{H}}|q\rangle = \int \prod_{k=0}^{n} dq_k \; \delta(q_0 - q_n) \cdot \langle q_0|(1 - \Delta\tau_1 \cdot \hat{H})|q_1\rangle$$
$$\times \langle q_1|(1 - \Delta\tau_2 \cdot \hat{H})|q_2\rangle \cdots \langle q_{n-1}|(1 - \Delta\tau_n \cdot \hat{H})|q_n\rangle. \tag{D.5}$$

Now, we make use of the relations

$$\langle q'|V(\hat{q})|q\rangle = V(q)\delta(q' - q) = \int \frac{dp}{2\pi} V(q)e^{ip(q-q')},$$

$$\langle q'|\frac{\hat{p}^2}{2}|q\rangle = \int \frac{dp}{2\pi}\frac{p^2}{2}e^{ip(q-q')}.$$

Omitting the numerical coefficient, we have for each factor in (D.5)

$$\langle q_{k-1}|(1 - \Delta\tau_k \cdot \hat{H})|q_k\rangle = \int dp_k e^{ip_k(q_k - q_{k-1})}e^{-\left[\frac{p_k^2}{2} + V(q_k)\right]\Delta\tau_k}, \tag{D.6}$$

where we again wrote

$$1 - \left(\frac{p_k^2}{2} + V(q_k)\right)\Delta\tau_k = e^{-\left[\frac{p_k^2}{2} + V(q_k)\right]\Delta\tau_k}.$$

The integral over dp_k in (D.6) is Gaussian and can be calculated, as usual, by shifting $p_k \rightarrow p_k - i\dot{q}_k$, where

$$\dot{q}_k = \frac{q_{k-1} - q_k}{\Delta\tau_k}.$$

This gives

$$\langle q_{k-1}|(1 - \Delta\tau_k \cdot \hat{H})|q_k\rangle = e^{-\left[\frac{\dot{q}_k^2}{2} + V(q_k)\right]\Delta\tau_k}.$$

Substituting this expression in (D.5) we obtain in the limit $n \rightarrow \infty$, $\Delta\tau_i \rightarrow 0$ the path integral representation for the free energy,

$$e^{-\beta F} = \int_{q(\beta)=q(0)} \mathcal{D}q \; e^{-S_E^{(\beta)}[q(\tau)]}, \tag{D.7}$$

where

$$S_E^{(\beta)} = \int_0^\beta d\tau \left[\frac{\dot{q}^2}{2} + V(q)\right], \tag{D.8}$$

and $\dot{q} = dq/dt$.

Let us explain the notation introduced here. S_E is the *Euclidean* action of the system with the Hamiltonian (D.1). It is obtained from the original action

$$S = \int dt \left[\frac{1}{2} \left(\frac{dq}{dt} \right)^2 - V(q) \right]$$

by formal replacement

$$t = -i\tau, \tag{D.9}$$

and then considering τ as real. The replacement (D.9) turns S into iS_E, so that

$$e^{iS} \to e^{-S_E}. \tag{D.10}$$

Further, in accordance with (D.8) the theory is considered in a finite interval of Euclidean time τ, whose length equals $\beta \equiv T^{-1}$. Finally, the functional integral (D.7) is taken over paths *periodic* in τ with period β.

The representation (D.7) of the free energy is intuitively clear. The operator $e^{-\beta\hat{H}}$ can be regarded as an evolution operator $e^{-i\hat{H}t_\beta}$ in the imaginary (Euclidean) time interval $t_\beta = -i\beta$. In accordance with this, the matrix element

$$\langle q_f | e^{-\beta\hat{H}} | q_i \rangle$$

is path integral over trajectories in Euclidean time, which start at $q = q_i$ and end at $q = q_f$. It is clear from (D.4) that the relevant trajectories are *periodic*, $q_i = q_f = q$, without any other conditions imposed on them.

The above derivation is directly generalized to quantum mechanics of multiple degrees of freedom and, in accordance with what is said in the beginning of this Appendix, to quantum theory of any *bosonic* fields. Making use of collective notation ϕ for all of the bosonic fields, we write the representation for the free energy in a form similar to (D.7),

$$e^{-\beta F} = \int \mathcal{D}\phi(\mathbf{x}, t) \cdot e^{-S_E^{(\beta)}[\phi(\mathbf{x},t)]},$$

where the integration is performed over field configurations *periodic*[4] in Euclidean time τ with period β, the Euclidean action has the form

$$S_E^{(\beta)} = \int_0^\beta d\tau \int d^3\mathbf{x} \mathcal{L}_E(\phi, \dot{\phi})$$

and is obtained from the original action by formally replacing $\tau \to -i\tau$, $iS \to -S_E$, as in (D.9), (D.10). In other words, the Euclidean Lagrangian $\mathcal{L}_E$ in the case of gauge theories with scalar fields is obtained from the original Lagrangian by replacing the

---

[4]In the case of non-Abelian gauge theories, the configurations must be periodic up to "large" (topologically non-trivial) gauge transformations; see Ref. [156]. This subtlety will be insignificant for us.

Minkowski metric with the Euclidean metric and by changing the signs in front of the scalar potential and in front of the Lagrangian of gauge fields. Schematically,

$$\mathcal{L}_E = \frac{1}{4} F^a_{\mu\nu} F^a_{\mu\nu} + D_\mu \phi^\dagger D_\mu \phi + V(\phi), \tag{D.11}$$

where summation over 4-dimensional indices μ, ν is performed with the Euclidean metric.

Problem D.1. *Show that the outlined procedure for obtaining the Euclidean action indeed leads to the expression* (D.11), *if the original Lagrangian in Minkowski space has the form*

$$\mathcal{L} = -\frac{1}{4} \eta^{\mu\nu} \eta^{\lambda\rho} F^a_{\mu\lambda} F^a_{\nu\rho} + \eta^{\mu\nu} D_\mu \phi^\dagger D_\nu \phi - V(\phi),$$

where $F^a_{\mu\nu}$ is the gauge field strength, ϕ denotes collectively all scalar fields which transform according to some (generally speaking, reducible and complex) representation of the gauge group. Hint: First, impose the gauge condition $A^a_0 = 0$, and then restore gauge invariance in the Euclidean formulation.

D.2. Fermionic Fields: Antiperiodic Boundary Conditions

In the case of fermions, the functional integral representation for the free energy has to be derived anew. We restrict ourselves to the case of the action quadratic in fermionic fields, although the result will be valid for the general case as well. More precisely, we consider theories where fermionic part of the Lagrangian in Minkowski space has the form

$$\mathcal{L} = i\bar{\psi}\gamma^\mu \partial_\mu \psi - \bar{\psi} M \psi, \tag{D.12}$$

where M accounts for fermion mass and interaction with bosonic fields (for example, in electrodynamics, $M = m - e\gamma^\mu A_\mu$). There can be several fermionic fields; generalization to this case is straightforward. Bosonic fields are considered external and fixed for the time being.

Given that $\bar{\psi} \equiv \psi^\dagger \gamma^0$, we write the Lagrangian (D.12) in the form

$$\mathcal{L} = i\psi^\dagger \partial_0 \psi - H, \tag{D.13}$$

where

$$H = -i\psi^\dagger \gamma^0 \gamma^i \partial_i \psi + \psi^\dagger \gamma^0 M \psi. \tag{D.14}$$

As is seen from (D.13), $p_\psi = i\psi^\dagger$ is the generalized momentum conjugate to the generalized coordinate ψ, and H is the Hamiltonian of the theory.

In contrast to bosonic fields, fermionic fields obey *anti*commutation relations; at equal times

$$\{\psi(\mathbf{x}, t), \psi(\mathbf{x}', t)\} = \{\psi^\dagger(\mathbf{x}, t), \psi^\dagger(\mathbf{x}', t)\} = 0,$$
$$\{\psi(\mathbf{x}, t), \psi^\dagger(\mathbf{x}', t)\} = \delta(\mathbf{x} - \mathbf{x}').$$

The latter equality is equivalent to the canonical relation $\{\psi(\mathbf{x}, t), p_\psi(\mathbf{x}', t)\} = i\delta(\mathbf{x} - \mathbf{x}')$. If spatial lattice and finite spatial box are introduced, we come to quantum mechanics of operators obeying (in the Schrödinger representation) anticommutation relations

$$\{\hat{\psi}_m, \hat{\psi}_n\} = \{\hat{\psi}_m^\dagger, \hat{\psi}_n^\dagger\} = 0, \quad \{\hat{\psi}_m, \hat{\psi}_n^\dagger\} = \delta_{mn},$$

while the discretization of (D.14) leads to the Hamiltonian of the type

$$\hat{H} = \hat{\psi}_m^\dagger h_{mn} \hat{\psi}_n.$$

Our purpose is to find the functional integral representation for $\mathrm{Tr}(e^{-\beta \hat{H}})$ in this theory.

Let us consider a theory with one fermion operator $\hat{\psi}$ and its conjugate $\hat{\psi}^\dagger$. They satisfy the following relations

$$\{\hat{\psi}, \hat{\psi}\} = \{\hat{\psi}^\dagger, \hat{\psi}^\dagger\} = 0, \quad \{\hat{\psi}, \hat{\psi}^\dagger\} = 1.$$

These coincide with the anticommutation relations for the fermion creation and annihilation operators. For definiteness, we assume that ψ is the creation operator. Then the space of states of the system has two basis vectors $|0\rangle$, $|1\rangle$, such that

$$\hat{\psi}^\dagger|0\rangle = 0, \quad \hat{\psi}|0\rangle = |1\rangle,$$
$$\hat{\psi}^\dagger|1\rangle = |0\rangle, \quad \hat{\psi}|1\rangle = 0.$$

It is convenient to consider this space of states as space of functions $\Psi(\psi)$ of the anticommuting (Grassmannian) variable ψ, whose fundamental property is nilpotency,

$$\psi \cdot \psi = 0. \tag{D.15}$$

Let us associate the vector $|0\rangle$ with the unit function, $\Psi(\psi) = 1$, and the vector $|1\rangle$ with the function $\Psi_1(\psi) = \psi$. Then the linear space with two basis vectors $|0\rangle$ and $|1\rangle$ is equivalent to the space of functions of the form

$$\Psi(\psi) = \alpha + \beta\psi,$$

where α and β are complex numbers. In fact, all functions $\Psi(\psi)$ are of this type. This is easy to see by writing the Taylor expansion in ψ and using (D.15). The operators $\hat{\psi}$ and $\hat{\psi}^\dagger$ act in this space as follows:

$$\hat{\psi}\Psi(\psi) = \psi\Psi(\psi), \quad \hat{\psi}^\dagger\Psi(\psi) = \frac{\partial}{\partial\psi}\Psi(\psi).$$

It is useful to present these formulas in an integral form. We introduce the Berezin integral; by definition,

$$\int d\psi = 0, \quad \int d\psi \cdot \psi = 1.$$

This definition is sufficient for evaluating the integral of any function $\Psi(\psi)$. It is straightforward to check by direct substitution that the following relations hold:

$$\Psi(\psi) = \int d\tilde{\psi}d\tilde{\psi}^\dagger e^{-\tilde{\psi}^\dagger(\psi-\tilde{\psi})}\Psi(\tilde{\psi}), \tag{D.16}$$

$$\hat{\psi}\Psi(\psi) = \int d\tilde{\psi}d\tilde{\psi}^\dagger e^{-\tilde{\psi}^\dagger(\psi-\tilde{\psi})}\tilde{\psi}\Psi(\tilde{\psi}), \tag{D.17}$$

$$\hat{\psi}^\dagger\Psi(\psi) = \int d\tilde{\psi}d\tilde{\psi}^\dagger e^{-\tilde{\psi}^\dagger(\psi-\tilde{\psi})}\tilde{\psi}^\dagger\Psi(\tilde{\psi}), \tag{D.18}$$

$$\hat{\psi}^\dagger\hat{\psi}\Psi(\psi) = \int d\tilde{\psi}d\tilde{\psi}^\dagger e^{-\tilde{\psi}^\dagger(\psi-\tilde{\psi})}\tilde{\psi}^\dagger\tilde{\psi}\Psi(\tilde{\psi}), \tag{D.19}$$

where all variables and differentials ψ, $\tilde{\psi}$, $\tilde{\psi}^\dagger$, $d\tilde{\psi}$, $d\tilde{\psi}^\dagger$ are treated as anticommuting.

Now we are ready to write the functional integral for the quantity

$$(e^{-\beta\hat{H}}\Psi)(\psi),$$

where the Hamiltonian is of the form $\hat{H} = c\hat{\psi}^\dagger\hat{\psi}$. We proceed in analogy to the bosonic case and write

$$e^{-\beta\hat{H}}\Psi = (1 - \hat{H}\Delta\tau_1)\cdots(1 - \hat{H}\Delta\tau_n)\cdot\Psi.$$

Using the formulas (D.16) and (D.19) we obtain

$$(e^{-\beta\hat{H}}\Psi)(\psi)$$
$$= \int\prod_{k=1}^{n}d\psi_k d\psi_k^\dagger e^{-\psi_1^\dagger(\psi-\psi_1)-H(\psi_1)\Delta\tau_1}\dots e^{-\psi_n^\dagger(\psi_{n-1}-\psi_n)-H(\psi_n)\Delta\tau_n}\Psi(\psi_n).$$

It is useful to note that, as in the bosonic case, $\psi_{k-1} - \psi_k = \dot{\psi}(\tau_k)\cdot\Delta\tau_k$ at small $\Delta\tau_k$. Hence, in the limit $n\to\infty$, $\Delta\tau_i\to 0$ we obtain the functional integral representation,

$$(e^{-\beta\hat{H}}\Psi)(\psi) = \int\mathcal{D}\psi\mathcal{D}\psi^\dagger e^{-S_E^{(\beta)}}\Psi(\psi_i), \tag{D.20}$$

where

$$S_E^{(\beta)} = \int_0^\beta\left[\psi^\dagger\frac{\partial\psi}{\partial\tau} + H(\psi^\dagger\psi)\right]d\tau.$$

Note that the functional integral in (D.20) includes integration over ψ_i and $\psi_i^\dagger$ at "initial time" $\tau = 0$ (with $\psi_i = \psi(\tau = 0)$), but does not include the integration over ψ and $\psi^\dagger$ at "final time" $\tau = \beta$. Just as in the bosonic case, the Euclidean action S_E is obtained from the action in real time

$$S = \int dt(i\psi^\dagger \partial_t \psi - H)$$

via the formal substitution $t \to -i\tau$, $iS \to -S_E$.

It remains to find the boundary conditions leading to $\mathrm{Tr}(e^{-\beta \hat{H}})$. Let us write

$$(e^{-\beta \hat{H}}\Psi)(\psi) = \int d\psi_i U(\psi, \psi_i)\Psi(\psi_i), \qquad (D.21)$$

where

$$U(\psi, \psi_i) = \int \mathcal{D}'\psi \mathcal{D}\psi^\dagger e^{-S_E^{(\beta)}},$$

and prime means that the integration is not performed over the initial "value" $\psi(\tau = 0) = \psi_i$. (This integration is left for (D.21).) The general expression for the function of two Grassmannian variables reads

$$U(\psi, \psi_i) = u_0 + u_1 \psi + u_{-1}\psi_i + u_2 \psi \psi_i.$$

Then we obtain

$$\int d\psi_i U(\psi, \psi_i) \cdot 1 = u_{-1} - u_2 \psi, \qquad \int d\psi_i U(\psi, \psi_i) \cdot \psi_i = u_0 - u_1 \psi.$$

In the operator language this means

$$e^{-\beta \hat{H}}|0\rangle = u_{-1}|0\rangle - u_2|1\rangle, \qquad e^{-\beta \hat{H}}|1\rangle = u_0|0\rangle - u_1|1\rangle.$$

Consequently,

$$\mathrm{Tr}(e^{-\beta \hat{H}}) = u_{-1} - u_1 = \int d\psi_i U(-\psi_i, \psi_i).$$

So, we finally get

$$\mathrm{Tr}(e^{-\beta \hat{H}}) = \int_{\psi(\beta)=-\psi(0)} \mathcal{D}\psi \mathcal{D}\psi^\dagger e^{-S_E^{(\beta)}},$$

i.e., the integration is performed over Grassmannian trajectories with *antiperiodic* boundary conditions on $\psi(\tau)$ in the interval $(0, \beta)$. The variable $\psi^\dagger(\tau)$ can also be considered antiperiodic: in the interval $(0, \beta)$ any function $\psi^\dagger(\tau)$ is represented as a sum of periodic and antiperiodic functions, and the periodic part does not contribute to $S_E^{(\beta)}$, since it is convoluted with antiperiodic $\psi(\tau)$.

The above derivation is fully applicable to the systems of many fermionic degrees of freedom, and, consequently, to the theory of fermionic fields. Here an important role is played by the relations (D.17) and (D.18) which we have not used so far. Although our derivation is given for the case where bosonic fields are external, it is easy to understand that this is actually not a limitation: in a theory with both bosonic and fermionic fields, the integral over fermions can be considered as internal (there bosonic fields are fixed), and then the integral over bosonic fields is evaluated. Thus, the free energy is given by the integral

$$
e^{-\beta F} \equiv Z = \int \mathcal{D}\phi \mathcal{D}\psi^\dagger \mathcal{D}\psi e^{-S_E^{(\beta)}}, \tag{D.22}
$$

where $S_E^{(\beta)}$ is the Euclidean action in the interval $(0, \beta)$, and bosonic fields ϕ (fermionic fields $\psi, \psi^\dagger$) satisfy periodic (antiperiodic) boundary conditions there.

To conclude this Section, we note that the formalism can be generalized to the case of nonzero chemical potential. In general, the chemical potential is introduced when the medium has nonzero density of conserved (at given temperature) quantum number. In the cosmological context, the baryon and lepton numbers are of the greatest interest. The corresponding operators have a structure like

$$
Q = \int d^3 \mathbf{x} \bar\psi \gamma^0 \psi.
$$

The nonzero average density $n = \bar\psi \gamma^0 \psi$ is taken into account by the additional term $(-\mu Q)$ in the effective Hamiltonian, i.e.,

$$
H_{eff} = H - \mu Q, \tag{D.23}
$$

where μ stands for the chemical potential. Within the formalism considered, this leads to the following change of the Euclidean action,

$$
S_E^{(\beta)} \to S_E^{(\beta)} - \mu \int_0^\beta d\tau \int d^3 \mathbf{x} \bar\psi \gamma^0 \psi. \tag{D.24}
$$

The partition function is then again given by (D.22). In this case, the quantity $F(T, \mu)$ is called the *Grand potential* or *Landau potential*.

D.3. Perturbation Theory

The approach described in Secs. D.1 and D.2 is useful for calculating the free energy, the effective potential $V_{eff}(T, \phi)$ introduced in Chapter 10, as well as the static Green's functions. The latter characterize the response of the system to time-independent external probes. For example, suppose that static source $J(\mathbf{x})$ is introduced into a theory of quantum field $\hat\phi$. This implies the following modification of the Hamiltonian

$$H \rightarrow H - \int J(\mathbf{x})\hat{\phi}(\mathbf{x})d^3\mathbf{x} \equiv H_J,$$

where $\hat{\phi}(\mathbf{x})$ is the Schrödinger field. In the presence of this source, the partition function

$$Z_J = \mathrm{e}^{-\beta F_J} = \mathrm{Tr}(\mathrm{e}^{-\beta H_J}),$$

is represented in the form of functional integral (D.22), and its expansion in J has the static Green's functions as coefficients,

$$G(\mathbf{x_1}, \ldots, \mathbf{x_n}) = Z^{-1} \int \mathcal{D}\phi\, \mathrm{e}^{-S^{(\beta)}[\phi]}$$

$$\times \frac{1}{\beta} \int_0^\beta d\tau_1 \phi(\mathbf{x_1}, \tau_1) \times \cdots \times \frac{1}{\beta} \int_0^\beta d\tau_n \phi(\mathbf{x_n}, \tau_n). \quad \text{(D.25)}$$

(The normalization by partition function Z without the source and factors β^{-1} are introduced for convenience.); the subscript E in the notation of the Euclidean action here and below is omitted.

A simple example is the field expectation value in the presence of the static source,

$$\langle\phi(\mathbf{x})\rangle_J = \frac{\mathrm{Tr}(\mathrm{e}^{-\beta H_J}\phi(\mathbf{x}))}{\mathrm{Tr}(\mathrm{e}^{-\beta H_J})}.$$

To the leading order in J it is equal to (assuming $\langle\phi\rangle_{J=0} = 0$)

$$\langle\phi(\mathbf{x})\rangle_J = \int G(\mathbf{x}, \mathbf{y}) J(\mathbf{y}) d\mathbf{y}.$$

The difference between $G(\mathbf{x}, \mathbf{y})$ and the free propagator at zero temperature corresponds to the modification of the Coulomb or Yukawa law in the presence of the medium.

Note that the static correlation functions (D.25) do not represent all interesting classes of Green's functions. Computational technique for correlators at different times (for instance, the Keldysh method) is quite complicated and not required here.

To generalize (D.25) we consider the Euclidean Green's functions

$$G(\mathbf{x_1}, \tau_1; \ldots; \mathbf{x_n}, \tau_n) = Z^{-1} \int \mathcal{D}\phi\, \mathrm{e}^{-S^{(\beta)}[\Phi]} \phi(\mathbf{x_1}, \tau_1) \ldots \phi(\mathbf{x_n}, \tau_n), \quad \text{(D.26)}$$

where ϕ denotes collectively all fields in the theory, and the integration is performed over the bosonic and fermionic fields, periodic and antiperiodic in the interval $[0, \beta]$, respectively. The normalization factor Z is given by a similar integral; see (D.22).

Proceeding from the representation (D.26), it is straightforward to construct the diagram technique for perturbative calculations, similar to the Feynman technique in theories at zero temperature. As usual, we first consider free theories with sources. In the scalar and fermion cases, the expressions for the quadratic actions are

$$S_\varphi^{(\beta)} = \int_0^\beta d\tau \int d^3\mathbf{x} \left[\frac{1}{2} \partial_\mu \varphi \partial_\mu \varphi + \frac{m^2}{2} \varphi^2 - J_\varphi \varphi \right], \tag{D.27}$$

$$S_\psi^{(\beta)} = \int_0^\beta d\tau \int d^3\mathbf{x} \left[\bar\psi \gamma^\mu \partial_\mu \psi + m\bar\psi\psi - \bar J_\psi \psi - \bar\psi J_\psi \right]. \tag{D.28}$$

Here $x^0 \equiv \tau$, summation over indices is performed with the Euclidean metric, and the Euclidean γ-matrices are Hermitean and obey the rule $\{\gamma^\mu, \gamma^\nu\} = \delta^{\mu\nu}$.

Problem D.2. *Check that the Euclidean action of the free Dirac field with a source has the form* (D.28).

As the field φ is periodic in τ with period β, so is the source $J_\varphi(\mathbf{x}, \tau)$. And vice versa, J_ψ and $\bar J_\psi$ are antiperiodic.

The functional integral (D.22) with the quadratic action and source (D.27) is Gaussian and is calculated by shifting $\varphi(\mathbf{x}, \tau) \to \varphi(\mathbf{x}, \tau) + \varphi_c(\mathbf{x}, \tau)$, where the function $\varphi_c(\mathbf{x}, \tau)$ is the solution to the equation

$$-\partial_\mu \partial_\mu \varphi_c + m^2 \varphi_c = J_\varphi. \tag{D.29}$$

According to our prescription, φ_c has to be periodic in τ with period β. One writes

$$\varphi_c(\mathbf{x}, \tau) = \int_0^\beta d\tau' \int d^3\mathbf{x}' D(\mathbf{x}, \tau; \mathbf{x}', \tau') J_\varphi(\mathbf{x}', \tau'),$$

where D is the free propagator at finite temperature. Given the periodicity of J_φ, it is straightforward to see that Eq. (D.29) and periodic boundary conditions are obeyed if the free propagator has the form

$$D(\mathbf{x}, \tau; \mathbf{x}', \tau') = \frac{1}{(2\pi)^3 \beta} \sum_{n \in \mathbb{Z}} \int d^3\mathbf{p} \, \frac{e^{i\mathbf{p}(\mathbf{x}-\mathbf{x}') + i\omega_n(\tau-\tau')}}{\mathbf{p}^2 + \omega_n^2 + m^2},$$

where

$$\omega_n = \frac{2\pi n}{\beta}, \quad n = 0, \pm 1, \pm 2, \ldots \tag{D.30}$$

are the Matsubara frequencies of bosonic fields. In contrast to the field theory at zero temperature, the frequencies form a *discrete* set.

The free propagator for the vector field is constructed similarly and is also a sum over frequencies (D.30).

In the case of fermionic field, the analog of Eq. (D.29) is

$$\gamma^\mu \partial_\mu \psi_c + m\psi_c = J_\psi.$$

Both $J_\psi(\mathbf{x}, \tau)$ and $\psi_c(\mathbf{x}, \tau)$ are antiperiodic in τ with period β. The solution is

$$\psi_c(\mathbf{x}, \tau) = \int_0^\beta d\tau' \int d^3\mathbf{x}' S(\mathbf{x}, \tau; \mathbf{x}', \tau') J_\psi(\mathbf{x}', \tau'),$$

where the free propagator is given by

$$S(\mathbf{x}, \tau; \mathbf{x}', \tau')$$

$$= \frac{1}{(2\pi)^3 \beta} \sum_{n' = \pm\frac{1}{2}, \pm\frac{3}{2}, \dots} \int d^3\mathbf{p} \, \frac{-i\gamma^0 \omega_{n'} - i\gamma\mathbf{p} + m}{\mathbf{p}^2 + \omega_{n'}^2 + m^2} e^{i\mathbf{p}(\mathbf{x}-\mathbf{x}') + i\omega_{n'}(\tau-\tau')}.$$

$$(D.31)$$

Here

$$\omega_{n'} = \frac{2\pi n'}{\beta}, \quad n' = \pm\frac{1}{2}, \pm\frac{3}{2}, \dots \tag{D.32}$$

are the Matsubara frequencies for fermions. The fact that n' runs over half-integer values is obviously due to the antiperiodicity of fermionic fields.

Further development of the diagram technique proceeds along the same lines as in the (Euclidean) field theory at zero temperature. The expressions for interaction vertices in theories at $T = 0$ and $T \neq 0$ coincide. Since integration over $d\tau$ in the action runs from 0 to β, instead of the δ-function of energy conservation the following factor appears at every vertex,

$$\beta\delta\left(\sum\omega\right), \tag{D.33}$$

where $\sum\omega$ is a sum over the Matsubara frequencies of all lines considered as incoming (frequencies (D.30) and (D.32) for bosonic and fermionic lines, respectively), and the function $\delta(\sum\omega)$ equals 1 if $\sum\omega = 0$ and zero in all other cases.

Note that turning on chemical potential, according to (D.24), leads to the replacement $\partial_0 \to \partial_0 - \mu$ in the action (D.28). The corresponding change in the free fermion propagator (D.31) is the replacement

$$\omega_{n'} \to \omega_{n'} + i\mu$$

in the pre-exponential factor of the integrand in (D.31), while the Matsubara frequencies remain intact in the exponential factor $\exp[i\omega_{n'}(\tau - \tau')]$.

D.4. One-Loop Effective Potential

As the first example of the use of the technique described in Secs. D.1 and D.2, let us re-derive the expression (10.24) for the first temperature-dependent correction

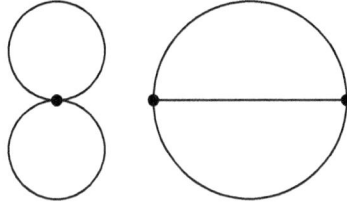

Fig. D.1. Some diagrams contributing to the first order correction to the effective potential.

to the effective potential. Our aim is to calculate the free energy as function of the homogeneous background scalar field ϕ, neglecting interactions between particles in the medium. Our starting point is the formula (D.22) for the free energy. In our approximation, the action $S^{(\beta)}$ is quadratic in quantum fields, and the background field ϕ enters into it only through the particle masses. The integral (D.22) is factorized into a product of integrals over different fields, so the free energy has the structure (10.22) indeed.

Note that in perturbation theory, the higher-order corrections to the effective potential are given by diagrams without external lines, where masses and vertices depend on the background field ϕ. The simplest of these diagrams are schematically shown in Fig. D.1. These diagrams begin with two loops. Therefore, within the described formalism, the zeroth order is naturally called *one-loop approximation.*

Turning back to the one-loop approximation, consider, for example, the contribution of the scalar field, whose action is given by formula (D.27) with $J_\varphi = 0$. The integral over φ of the type (D.22) is Gaussian and equal to

$$\int \mathcal{D}\varphi e^{-S^{(\beta)}[\varphi]} = \left[\mathrm{Det}(-\partial_\mu\partial_\mu + m^2)\right]^{-1/2},$$

where $m^2 = m^2(\phi)$, and the determinant can be understood as the product of eigenvalues of the operator $(-\partial_\mu\partial_\mu + m^2)$ with boundary conditions of periodicity in τ with period β. If the system is put into a spatial box of large size L, then the eigenvalues are

$$\lambda_{n,n_1,n_2,n_3} = \mathbf{p}^2 + w_n^2 + m^2,$$

where

$$\mathbf{p} = \left(\frac{2\pi n_1}{L}, \frac{2\pi n_2}{L}, \frac{2\pi n_3}{L}\right), \quad n_1, n_2, n_3 \in \mathbb{Z}, \tag{D.34}$$

and w_n are Matsubara frequencies (D.30). Therefore, the contribution to the free energy reads

$$F_\varphi = \sum_n \sum_{n_1,n_2,n_3} \frac{1}{2\beta} \log\left[\frac{\mathbf{p}^2 + w_n^2 + m^2}{\Lambda^2}\right],$$

where the parameter Λ makes the ratio dimensionless. This parameter leads only to an overall shift in the free energy, and hence it is insignificant. In the limit of large L

$$\sum_{n_1,n_2,n_3} \to L^3 \int \frac{d^3\mathbf{p}}{(2\pi)^3},$$

so that the free energy is indeed proportional to the volume, and the contribution to the effective potential is

$$f_\varphi = \frac{1}{2\beta} \sum_n \int \frac{d^3\mathbf{p}}{(2\pi)^3} \log\left[\frac{\mathbf{p}^2 + w_n^2 + m^2}{\Lambda^2}\right]. \qquad (D.35)$$

This expression is divergent at large $\mathbf{p}$ and/or n. We assume that it is regularized (see also Sec. 10.4) and proceed to work with it as if it was finite.[5]

To calculate the sum over all integer n, we note that it can be represented in the form

$$\sum_{n=0,\pm1,\dots} u(n) = \frac{1}{2i} \oint \cot(\pi z) u(z) dz, \qquad (D.36)$$

where integration is performed along closed contour in the complex plane encircling the real axis counterclockwise; see Fig. D.2(a). To prove the formula (D.36), it is sufficient to note that $\cot \pi z$ has poles at integer $z = 0, \pm1, \dots$ with residues equal

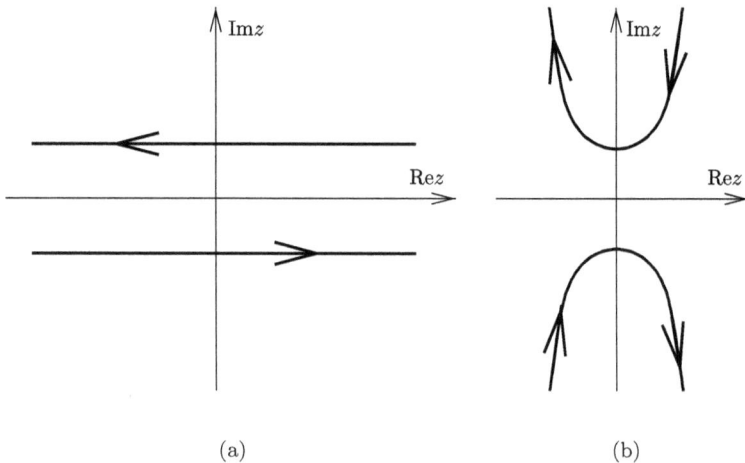

(a)   (b)

Fig. D.2.   (a) The contour of integration in Eq. (D.36); (b) Deformed contour.

---

[5]Particularly convenient for this general treatment is Pauli–Villars regularization; see, e.g., the book [268].

to π^{-1}. In our case

$$u(z) = \log\left(\frac{\mathbf{p}^2 + m^2 + (2\pi/\beta)^2}{\Lambda^2}\right) . \tag{D.37}$$

This function has branch cuts at positive and negative semi-axes starting at

$$z = \pm i\frac{\beta}{2\pi}\sqrt{\mathbf{p}^2 + m^2} ,$$

so the contour of integration in (D.36) can be deformed as shown in Fig. D.2(b). On the positive imaginary semi-axis, the jump across the cut is

$$u(i\zeta - \epsilon) - u(i\zeta + \epsilon) = -2\pi i .$$

We recall that $\cot(i\pi\zeta) = -i\coth(\pi\zeta)$ and find the contribution of the cut along the positive semi-axis:

$$f_\varphi^{positive} = -\frac{1}{2\beta}\int\frac{d^3\mathbf{p}}{(2\pi)^3}\int_{\frac{\beta}{2\pi}\sqrt{\mathbf{p}^2+m^2}}^\infty d\zeta\,\pi\coth(\pi\zeta) .$$

The same contribution comes from the cut along the negative imaginary semi-axis. We perform the integration over ζ and discard the contribution from the upper limit of integration and finite terms, which do not depend on parameter m. (Recall that we pretend that the integral is regularized.) Thus,

$$f_\varphi = \frac{1}{\beta}\int\frac{d^3\mathbf{p}}{(2\pi)^3}\log\left[2\sinh\left(\frac{\beta}{2}\sqrt{\mathbf{p}^2+m^2}\right)\right] .$$

This expression can be represented as

$$f_\varphi = f_\varphi(T=0) + f_\varphi^{(T)} ,$$

where

$$f_\varphi(T=0) = \frac{1}{2}\int\frac{d^3\mathbf{p}}{(2\pi)^3}\sqrt{\mathbf{p}^2 + m^2} \tag{D.38}$$

does not depend on temperature, while the temperature-dependent part is

$$f_\varphi^{(T)} = \frac{1}{\beta}\log\left(1 - e^{-\beta\sqrt{\mathbf{p}^2+m^2}}\right) . \tag{D.39}$$

The zero-temperature contribution (D.38) to the effective potential is simply the sum of zero-point energies of oscillators of the field φ,

$$f_\varphi(T=0) = \frac{1}{L^3}\sum_{n_1,n_2,n_3}\frac{1}{2}\sqrt{\mathbf{p}^2 + m^2}, \tag{D.40}$$

where for clarity we returned back to the theory in finite spatial volume; the momentum $\mathbf{p}$ is given by (D.34). The effects of this contribution are discussed in Sec. 10.4.

The contribution (D.39), relevant at finite temperatures, is, in fact, the free energy of ideal gas of bosons [78]. To see explicitly that it is equal to ($-$ pressure), we write Eq. (D.39) as the integral over $p = |\mathbf{p}|$ and integrate by parts:

$$
\begin{aligned}
f_\varphi^{(T)} &= \frac{1}{\beta} \int_0^\infty \frac{p^2\,dp}{2\pi^2} \log\left(1 - \mathrm{e}^{-\beta\sqrt{p^2+m^2}}\right) \\
&= -\frac{1}{6\pi^2} \int_0^\infty p^3\,dp \frac{\partial}{\partial p} \log\left(1 - \mathrm{e}^{-\beta\sqrt{p^2+m^2}}\right) \\
&= -\frac{1}{6\pi^2} \int_0^\infty \frac{p^4\,dp}{\sqrt{p^2+m^2}} \frac{1}{\mathrm{e}^{\beta\sqrt{p^2+m^2}} - 1},
\end{aligned}
$$

which is precisely the boson integral (10.24) with $g_i = 1$. (We consider one real scalar field φ.)

The calculation of the fermionic contribution to the one-loop thermal effective potential proceeds in a similar way. For the theory with the action (D.28) and $J_{\bar\psi} = \bar{J}_\psi = 0$, the functional integral (D.22) equals

$$
\int \mathcal{D}\bar\psi \mathcal{D}\psi\, \mathrm{e}^{-S_\psi^{(\beta)}} = \mathrm{Det}\left[\gamma^\mu \partial_\mu + m(\phi)\right],
$$

where the eigenfunctions of the Euclidean Dirac operator must be antiperiodic in τ with period β. For fixed 3-momentum $\mathbf{p}$ and Matsubara frequency (D.32), there are two doubly degenerate eigenvalues of the Dirac operator,

$$
\lambda_\pm = m \pm i\sqrt{\mathbf{p}^2 + \omega_{n'}^2}.
$$

As a result, for each momentum we have the factor $(\lambda_+\lambda_-)^2$ in the determinant, and instead of (D.35) we obtain

$$
f_\psi = -\frac{2}{\beta} \int \frac{d^3\mathbf{p}}{(2\pi)^3} \sum_{n'=\pm\frac{1}{2},\pm\frac{3}{2},\dots} \ln\left[\frac{\mathbf{p}^2 + \omega_{n'}^2 + m^2}{\Lambda^2}\right]. \tag{D.41}
$$

We emphasize that the difference in sign as compared to (D.35) is due to the fact that we are dealing with fermions. When calculating f_ψ, we encounter the sum over half-integer n', which can be written as

$$
\sum_{n'=\pm\frac{1}{2},\pm\frac{3}{2},\dots} u(n') = -\frac{1}{2i} \oint \tan(\pi z) u(z)\,dz, \tag{D.42}
$$

where the integration contour is the same as shown in Fig. D.2(a), and $u(z)$ is still given by formula (D.37). Further calculation basically repeats the calculation for the scalar field. The contribution of fermions also has zero-temperature and finite-temperature parts. The former,

$$f_\psi(T = 0) = -2 \int \frac{d^3\mathbf{p}}{(2\pi)^3} \sqrt{\mathbf{p}^2 + m^2}, \tag{D.43}$$

can be interpreted as the contribution of the Dirac sea (negative energy states, $\omega = -\sqrt{\mathbf{p}^2 + m^2}$, doubly degenerate at each $\mathbf{p}$). The temperature-dependent term

$$f_\psi = -\frac{4}{\beta} \int \frac{d^3\mathbf{p}}{(2\pi)^3} \log\left(1 + e^{-\beta\sqrt{\mathbf{p}^2+m^2}}\right)$$

is the free energy of ideal gas of fermions [78] and coincides, upon integration by parts, with the fermion integral (10.24), given that the total number of spin states of fermion and antifermion is $g = 4$.

Thus, in the framework of the formalism presented in this Appendix, the difference between the Bose and Fermi statistics manifests itself, in particular, in the difference between the Matsubara frequencies (D.30) and (D.32). In Sec. 10.3 we discuss the importance of this difference for the infrared properties of the theory at high temperatures.

Problem D.3. *In the one-loop approximation, find the Grand potential and fermion number density of fermionic matter at chemical potential μ and temperature T. Consider the limiting cases $T \gg \mu \gg m$ and $T \ll \mu$. Hint: Make use of the property*

$$\frac{\partial F(\mu, T)}{\partial \mu} = -\langle Q \rangle_{T,\mu}.$$

This property follows from (D.23).

D.5. Debye Screening

As the second example, let us consider the one-loop contribution $\Pi_{\mu\nu}$ to the photon self-energy in quantum electrodynamics at finite temperature and zero chemical potential; see Fig. D.3. As usual, it modifies the photon propagator

$$\mathcal{D}_{\mu\nu} \to \left[\mathcal{D}_{\mu\nu}^{-1} + \Pi_{\mu\nu}\right]^{-1},$$

where $\Pi_{\mu\nu}$ is also called the photon polarization operator. Let us consider the static propagator; see (D.25). So, we are interested in the self-energy at zero Matsubara

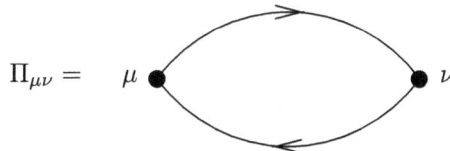

Fig. D.3.   One-loop photon self-energy.

frequency,

$$\Pi_{\mu\nu}(\mathbf{p}) = \Pi_{\mu\nu}(\mathbf{p}, \omega_n = 0).$$

The sought-for contribution modifies the static Maxwell equations, which in the medium obtain the following form (in momentum space)

$$\mathbf{p}^2 A_0 + \Pi_{00} A_0 + \Pi_{0i} A_i = j_0,$$

$$\mathbf{p}^2 A_i - p_i \mathbf{p} \mathbf{A} + \Pi_{i0} A_0 + \Pi_{ik} A_k = j_i,$$

where $j_\mu(\mathbf{p})$ are time-independent charge and current densities.

Before doing the calculation, we note that due to gauge invariance of electrodynamics, which holds in the presence of matter as well, the self-energy $\Pi_{\mu\nu}(\mathbf{p}, \omega_n)$ has to be transverse,

$$p^\mu \Pi_{\mu\nu} = 0,$$

where $p^\mu = (\omega_n, \mathbf{p})$. Lorentz-invariance is explicitly broken by the presence of matter, but the symmetry with respect to spatial rotations remains intact. Therefore, the general structure of the self-energy is

$$\Pi_{00} = \Pi^{(E)},$$

$$\Pi_{i0} = -\frac{p_i p_0}{\mathbf{p}^2} \Pi^{(E)},$$

$$\Pi_{ij} = \frac{p_i p_j p_0^2}{\mathbf{p}^4} \Pi^{(E)} + \left(\delta_{ij} - \frac{p_i p_j}{\mathbf{p}^2} \right) \Pi^{(M)},$$

where "electric" and "magnetic" terms, $\Pi^{(E)}$ and $\Pi^{(M)}$, depend on $\mathbf{p}^2$ and $p_0 \equiv \omega_n$. In the static limit $p_0 \equiv \omega_n = 0$, only Π_{00} and transverse part of Π_{ij} are nonzero, thus the modified Maxwell equations take the following form,

$$(\mathbf{p}^2 + \Pi^{(E)}) A_0 = j_0, \tag{D.44}$$

$$(\mathbf{p}^2 + \Pi^{(M)}) \left(\delta_{ik} - \frac{p_i p_k}{\mathbf{p}^2} \right) A_i = j_k. \tag{D.45}$$

We will be interested in the field behavior at large distances. Hence, in the end we take the limit $\mathbf{p}^2 \to 0$. The order of limits is important here: one should first set $p_0 \equiv \omega_n = 0$, and then take the limit of small $\mathbf{p}^2$.

The interaction of fermions with electromagnetic field is introduced, as usual, by the replacement $\partial_\mu \to \partial_\mu - ieA_\mu$ in the action (D.28), so the diagram in Fig. D.3 gives

$$\Pi^{\mu\nu}(p, p') = -e^2 \int \text{tr} \left[\gamma^\mu S(x,y) \gamma^\nu S(y,x) \right] e^{ip_\lambda^{(i)} x^\lambda} e^{-ip_\rho^{(f)} y^\rho} d^4x d^4y,$$

where $p^\mu = (\omega_{n_i}, \mathbf{p}^{(i)})$, $p'^\mu = (\omega_{n_f}, \mathbf{p}^{(f)})$ are momenta of incoming and outgoing photons, integration over x^0 and y^0 is performed in the interval $(0, \beta)$, and the

Fig. D.4. The interpretation of self-energy in terms of photon rescattering in medium.

fermion propagator is given by (D.31). Extracting the δ-function of energy and momentum conservation (The former is understood in the sense of (D.33).) and setting $\omega_{n_i} = \omega_{n_f} = 0$, $\mathbf{p}^{(i)} = \mathbf{p}^{(f)} = \mathbf{p}$, we obtain

$$\Pi^{\mu\nu}(\mathbf{p}) = \frac{e^2}{(2\pi)^3\beta} \int d^3q \sum_{n'} \frac{\mathrm{Tr}\left[\gamma^\mu(-i\hat{q} + m)\gamma^\nu(-i(\hat{q} + \hat{p}) + m)\right]}{(q^2 + m^2)((q + p)^2 + m^2)},$$

where $\hat{q} = \gamma^\mu q_\mu$, $q_0 = \frac{2\pi n'}{\beta}$, photon momentum is equal to $p^\mu = (0, \mathbf{p})$, the sum is evaluated over half-integer n', and momentum squared is understood in the Euclidean sense. We again perform summation over the Matsubara frequencies by making use of (D.42). Taking the limit of small photon momentum, we arrive at

$$\Pi_{\mu\nu}(\mathbf{p}\to 0, \omega = 0) = \frac{2e^2}{(2\pi)^3} \int d^3q \oint \frac{dq_0}{2\pi i} \tan\left(\frac{\beta}{2}q_0\right) \cdot \frac{2q_\mu q_\nu - \delta_{\mu\nu}(q_0^2 + \mathbf{q}^2 + m^2)}{(q_0^2 + \mathbf{q}^2 + m^2)^2}. \quad (D.46)$$

Poles of the integrand are at $q_0 = \pm i\sqrt{\mathbf{q}^2 + m^2}$; they show that in fact we are dealing with the photon forward scattering off fermions and antifermions existing in the medium; this is schematically illustrated in Fig. D.4, where crosses denote particles in the medium. Here we encounter the situation where nominally one-loop calculation corresponds to tree-level diagrams for scattering off the medium, while the interaction between the particles of the medium is not taken into account. This situation is analogous in some sense to that in Sec. D.4. (Formally one-loop calculation of the effective potential corresponds to the approximation of non-interacting particles in the medium.)

Integrating over dq_0 in (D.46) and omitting temperature-independent terms, we obtain for the 00-component

$$\Pi_{00}(\mathbf{p}\to 0, \omega = 0) \equiv \Pi^{(E)} = \frac{e^2}{\pi^2} \int_0^\infty \frac{dq}{\omega_q} \cdot \frac{\omega_q^2 + q^2}{e^{\frac{\omega_q}{T}} + 1}, \quad (D.47)$$

where $\omega_q = \sqrt{q^2 + m^2}$. (When performing this calculation, it is convenient to make use of the integration by parts analogous to that used in (D.42).) At the same time, the spatial components Π_{ij} are equal to zero in the limit of small photon momentum,

$$\Pi^{(M)}(\mathbf{p}\to 0, \omega = 0) = 0.$$

Recalling (D.44) and (D.45), we see that our results imply that electric field gets screened in the medium, while magnetic field does not. (Of course, the latter property has been demonstrated within one-loop approximation only.) Indeed, in the coordinate representation, the solution to Eq. (D.44), in the case of point charge q placed at the origin, has the following form at large distances,

$$A_0(\mathbf{x}) = q \int \frac{d^3\mathbf{p}}{(2\pi)^3} \frac{e^{i\mathbf{p}\mathbf{x}}}{\mathbf{p}^2 + m_D^2} = \frac{q}{4\pi} \frac{e^{-m_D|\mathbf{x}|}}{|\mathbf{x}|},$$

where $m_D^2 = \Pi^{(E)}(\mathbf{p} \to 0, \omega = 0)$ is the Debye mass squared. The exponential fall-off here precisely means screening of electric field. For the magnetic field this phenomenon of *exponential* fall-off at large distances is absent.

Note that at $T \ll m$, the Debye mass is exponentially small (this is because we consider medium at zero chemical potential; the density of fermions is thus exponentially small at low temperature), while in the opposite limit

$$m_D = \frac{e}{\pi}T, \quad T \gg m, \tag{D.48}$$

i.e., the Debye screening radius $r_D = m_D^{-1}$ decreases with temperature.

At the end of this Section, we mention that the Debye screening arises also when the medium contains charged bosons rather than fermions; the contribution of bosons to the Debye mass squared is of the same order of magnitude as that of fermions of the same mass and electric charge.

Finally, the Debye screening occurs at nonzero net particle densities (nonzero chemical potentials), too. The temperature in that case may be low and even zero, the Debye radius is then determined by the density of charged particles. A particularly interesting example here is electron–proton medium.

Problem D.4. *Find the Debye radius in an electrically neutral electron–proton medium at given electron number density and temperature T in two cases: $m_p \gg T \gg m_e$ and $m_e \gg T \gg \Delta$, where Δ is the binding energy of electron in hydrogen atom ($\Delta = 13.6\,eV$). Hint: Perform the calculation at fixed chemical potentials of electrons and protons; in order to find their relation to the electron number density use the results of Problem D.3.*

Bibliography

[1] C. Amsler *et al.* (Particle Data Group), *Phys. Lett. B* **667**, 1 (2008) and 2009 partial update for the 2010 edition, http://pdg.lbl.gov/

[2] http://www.sdss.org/

[3] http://magnum.anu.edu.au/~TDFgg/

[4] http:www.sdss.org/dr1/algorithms/edrpaper.html

[5] O. Le Fevre *et al.* [The VVDS Team Collaboration], "VVDS: early results on LSS distribution to z 1.5", arXiv:astro-ph/0402203.

[6] D. N. Spergel *et al.*, *Astrophys. J. Suppl.* **148**, 175 (2003).

[7] E. Komatsu *et al.* [WMAP Collaboration], *Astrophys. J. Suppl.* **180**, 330 (2009) [arXiv:0803.0547 [astro-ph]].

[8] M. R. Nolta *et al.* [WMAP Collaboration], *Astrophys. J. Suppl.* **180**, 296 (2009). Figure 1.6 (and 13.3) is reproduced by permission of the American Astronomical Society.

[9] G. Hinshaw *et al.* [WMAP Collaboration], *Astrophys. J. Suppl.* **180**, 225 (2009). Figure 1.5 (and 13.2) is reproduced by permission of the American Astronomical Society.

[10] W. L. Freedman *et al.*, *Astrophys. J.* **553**, 47 (2001). Figure 1.2 is reproduced by permission of the American Astronomical Society.

[11] N. J. Cornish, D. N. Spergel, G. D. Starkman and E. Komatsu, *Phys. Rev. Lett.* **92**, 201302 (2004).

[12] B. M. S. Hansen *et al.*, *Astrophys. J.* **574**, L155 (2002) [arXiv:astro-ph/0205087].

[13] N. Dauphas, *Nature* **435**, 1203 (2005).

[14] A. Frebel *et al.*, *Astrophys. J.* **660**, L117 (2007).

[15] R. Gratton, A. Bragaglia, E. Carretta, G. Clementini, S. Desidera, F. Grundahl and S. Lucatello, *Astron. Astrophys.* **408**, 529 (2003) [arXiv:astro-ph/0307016].

[16] K. Hagiwara *et al.* (Particle Data Group), *Phys. Rev. D* **66**, 010001 (2002), http://pdg.lbl.gov/2002/cmb_temp_00.ps

[17] D. J. Fixsen, *Astrophys. J.* **707**, 916 (2009).

[18] C. L. Bennett *et al.*, *Astrophys. J. Suppl.* **148**, 1 (2003) [arXiv:astro-ph/0302207].

[19] R. A. Sunyaev and Ya. B. Zeldovich, *Astrophys. Space Sci.* **7**, 3 (1970); R. A. Sunyaev and Ya. B. Zeldovich, *Comm. Astrophys. Space Phys.* **4**, 173 (1972).

[20] K. Greisen, *Phys. Rev. Lett.* **16**, 748 (1966).

[21] G. T. Zatsepin and V. A. Kuzmin, *JETP Lett.* **4**, 78 (1966) [*Pisma Zh. Eksp. Teor. Fiz.* **4**, 114 (1966)].

[22] http://www.spacetelescope.org/news/html/heic0309.html, Figure 1.7 (and 13.4), left panel: credit J.-P. Kneib. http://antwrp.gsfc.nasa.gov/apod/ap9806 14.html, Figure 1.7 (and 13.4), right panel: credit W. N. Colley, E. Turner (Princeton), J. A. Tyson, Hubble Space Telescope, NASA.

[23] A. Vikhlinin, A. Kravtsov, W. Forman, C. Jones, M. Markevitch, S. S. Murray and L. Van Speybroeck, *Astrophys. J.* **640**, 691 (2006) [arXiv:astro-ph/0507092].

[24] D. Clowe, M. Bradac, A. H. Gonzalez, M. Markevitch, S. W. Randall, C. Jones and D. Zaritsky, *Astrophys. J.* **648**, L109 (2006) [arXiv:astro-ph/0608407]. Figure 1.8 (and 13.5) is reproduced by permission of the American Astronomical Society.

[25] K. G. Begeman, A. H. Broeils and R. H. Sanders, *Mon. Not. Roy. Astron. Soc.* **249**, 523 (1991).

[26] C. Alcock *et al.* [MACHO Collaboration], *Astrophys. J.* **542**, 281 (2000) [arXiv:astro-ph/0001272].

[27] P. Tisserand *et al.* [EROS-2 Collaboration], *Astron. Astrophys.* **469**, 387 (2007) [arXiv:astro-ph/0607207].

[28] A. G. Riess *et al.* [Supernova Search Team Collaboration], *Astron. J.* **116**, 1009 (1998) [arXiv:astro-ph/9805201].

[29] S. Perlmutter *et al.* [Supernova Cosmology Project Collaboration], *Astrophys. J.* **517**, 565 (1999) [arXiv:astro-ph/9812133].

[30] D. A. Kirzhnits, *JETP Lett.* **15**, 529 (1972) [*Pisma Zh. Eksp. Teor. Fiz.* **15**, 745 (1972)].

[31] D. A. Kirzhnits and A. D. Linde, *Phys. Lett. B* **42**, 471 (1972).

[32] L. Dolan and R. Jackiw, *Phys. Rev. D* **9**, 3320 (1974).

[33] S. Weinberg, *Phys. Rev. D* **9**, 3357 (1974).

[34] A. D. Sakharov, *Pisma Zh. Eksp. Teor. Fiz.* **5**, 32 (1967) [*JETP Lett.* **5**, 24 (1967 SOPUA,34,392–393.1991 UFNAA,161,61–64.1991)].

[35] V. A. Kuzmin, *Pisma Zh. Eksp. Teor. Fiz.* **12**, 335 (1970).

[36] Stephen W. Hawking and G. F. R. Ellis, *The Large Scale Structure of Space-Time*, Cambridge University Press, 1973.

[37] A. Y. Kamenshchik, U. Moschella and V. Pasquier, *Phys. Lett. B* **511**, 265 (2001) [arXiv:gr-qc/0103004].

[38] C. L. Bennett *et al.* [WMAP Collaboration], *Astrophys. J. Suppl.* **208**, 20 (2013) [arXiv:1212.5225 [astro-ph.CO]].

[39] P. A. R. Ade *et al.* [Planck Collaboration], [arXiv:1303.5062 [astro-ph.CO]].

[40] A. A. Starobinsky, *JETP Lett.* **30**, 682 (1979) [*Pisma Zh. Eksp. Teor. Fiz.* **30**, 719 (1979)].

[41] A. A. Starobinsky, *Phys. Lett. B* **91**, 99 (1980).

[42] A. H. Guth, *Phys. Rev. D* **23**, 347 (1981).

[43] A. D. Linde, *Phys. Lett. B* **108**, 389 (1982).

[44] A. Albrecht and P. J. Steinhardt, *Phys. Rev. Lett.* **48**, 1220 (1982).

[45] V. F. Mukhanov and G. V. Chibisov, *JETP Lett.* **33**, 532 (1981) [*Pisma Zh. Eksp. Teor. Fiz.* **33**, 549 (1981)].

[46] S. W. Hawking, *Phys. Lett. B* **115**, 295 (1982).

[47] A. A. Starobinsky, *Phys. Lett. B* **117**, 175 (1982).

[48] A. H. Guth and S. Y. Pi, *Phys. Rev. Lett.* **49**, 1110 (1982).

[49] J. M. Bardeen, P. J. Steinhardt and M. S. Turner, *Phys. Rev. D* **28**, 679 (1983).

[50] E. R. Harrison, *Phys. Rev. D* **1**, 2726 (1970).

[51] Y. B. Zeldovich, *Mon. Not. Roy. Astron. Soc.* **160**, 1P (1972).

[52] P. J. E. Peebles and J. T. Yu, *Astrophys. J.* **162**, 815 (1970). doi:10.1086/150713

[53] V. A. Rubakov, M. V. Sazhin and A. V. Veryaskin, *Phys. Lett. B* **115**, 189 (1982).

[54] R. Fabbri and M. D. Pollock, *Phys. Lett.* B **125**, 445 (1983).

[55] L. F. Abbott and M. B. Wise, *Nucl. Phys.* B **244**, 541 (1984).

[56] A. A. Starobinsky, *Sov. Astron. Lett.* **11**, 133 (1985).

[57] M. J. Rees, *Astrophys. J.* **153**, L1 (1968);
M. M. Basko and A. G. Polnarev, *Mon. Not. Roy. Astron. Soc.* **191**, 207 (1980);
J. Negroponte and J. Silk, *Phys. Rev. Lett.* **44**, 1433 (1980);
N. Kaiser, *Mon. Not. Roy. Astron. Soc.* **202**, 1169 (1983).

[58] A. G. Polnarev, *Astron. Zh.* **62**, 1041 (1985) [*Sov. Astron.* **29**, 607 (1985)].

[59] M. V. Sazhin and N. Benites, *Astro. Lett. Commun.* **32**, 105 (1995).

[60] R. G. Crittenden, D. Coulson and N. G. Turok, *Phys. Rev.* D **52**, 5402 (1995).

[61] M. Kamionkowski, A. Kosowsky and A. Stebbins, *Phys. Rev. Lett.* **78**, 2058 (1997) [arXiv:astro-ph/9609132].

[62] U. Seljak and M. Zaldarriaga, *Phys. Rev. Lett.* **78**, 2054 (1997) [arXiv:astro-ph/9609169].

[63] P. Creminelli, A. Nicolis and E. Trincherini, *JCAP* **1011**, 021 (2010) [arXiv:1007.0027 [hep-th]].

[64] A. V. Filippenko and A. G. Riess, *AIP Conf. Proc.* **540**, 227 (2000) [arXiv:astro-ph/0008057]. Figure 4.6 is reproduced with permission. Copyright 2000, American Institute of Physics.

[65] A. G. Riess *et al.* [Supernova Search Team Collaboration], *Astrophys. J.* **607**, 665 (2004). Figure 4.7 is reproduced by permission of the American Astronomical Society.

[66] P. Astier *et al.*, *Astron. Astrophys.* **447**, 31 (2006) [arXiv:astro-ph/0510447].

[67] M. Kowalski *et al.* [Supernova Cosmology Project Collaboration], *Astrophys. J.* **686**, 749 (2008) [arXiv:0804.4142 [astro-ph]]. Figures 4.9, 4.10 and 6.4 (and 13.6) are reproduced by permission of the American Astronomical Society.

[68] L. H. Ford, *Phys. Rev.* D **35**, 2339 (1987).

[69] C. Wetterich, *Nucl. Phys.* B **302**, 668 (1988).

[70] B. Ratra and P. J. E. Peebles, *Phys. Rev.* D **37**, 3406 (1988).

[71] E. J. Copeland, A. R. Liddle and D. Wands, *Phys. Rev.* D **57**, 4686 (1998) [arXiv:gr-qc/9711068].

[72] P. G. Ferreira and M. Joyce, *Phys. Rev.* D **58**, 023503 (1998) [arXiv:astro-ph/9711102].

[73] I. Zlatev, L. M. Wang and P. J. Steinhardt, *Phys. Rev. Lett.* **82**, 896 (1999) [arXiv:astro-ph/9807002]; P. J. Steinhardt, L. M. Wang and I. Zlatev, *Phys. Rev.* D **59**, 123504 (1999) [arXiv:astro-ph/9812313].

[74] N. G. Busca, T. Delubac, J. Rich, S. Bailey, A. Font-Ribera, D. Kirkby, J. M. Le Goff and M. M. Pieri *et al.*, *Astron. Astrophys.* **552**, A96 (2013) [arXiv:1211.2616 [astro-ph.CO]].

[75] N. Suzuki, D. Rubin, C. Lidman, G. Aldering, R. Amanullah, K. Barbary, L. F. Barrientos, J. Botyanszki *et al.*, *Astrophys. J.* **746**, 85 (2012) [arXiv:1105.3470 [astro-ph.CO]].

[76] D. J. Eisenstein *et al.* [SDSS Collaboration], *Astrophys. J.* **633**, 560 (2005) [astro-ph/0501171].

[77] L. Anderson, E. Aubourg, S. Bailey, D. Bizyaev, M. Blanton, A. S. Bolton, J. Brinkmann, J. R. Brownstein *et al.*, *Mon. Not. Roy. Astron. Soc.* **427** (2013) 4, 3435 [arXiv:1203.6594 [astro-ph.CO]].

[78] L. D. Landau, E. M. Lifshitz, *Course of Theoretical Physics* in 10 volumes; Volume V, L. D. Landau, E. M. Lifshitz, *Statistical Physics, Part 1* (3rd Edn.), Reed Educational and Professional Publishing Ltd, 1980.

[79] Ya. B. Zeldovich, V. G. Kurt and R. A. Sunyaev, *Zh. E. T. F.* **55**, 278 (1968).

[80] P. J. Peebles, *Astrophys. J.* **153**, 1 (1968).

[81] L. D. Landau and E. M. Lifshitz, *Course of Theoretical Physics* in 10 volumes; Volume IV, V. B. Berestetskii, E. M. Lifshitz and L. P. Pitaevskii, *Quantum Electrodynamics* (2nd Edn.) Reed Educational and Professional Publishing Ltd, 1982.

[82] J. A. Rubino-Martin, J. Chluba and R. A. Sunyaev, arXiv:0711.0594 [astro-ph].

[83] L. D. Landau and E. M. Lifshitz, *Course of Theoretical Physics* in 10 volumes; Volume II, L. D. Landau and E. M. Lifshitz, *The Classical Theory of Fields* (4th Edn.), Reed Educational and Professional Publishing Ltd, 1975.

[84] S. P. Goldman, *Phys. Rev. A* **40**, 1185 (1989).

[85] P. Naselsky, D. Novikov and I. Novikov, *The Physics of Cosmic Microwave Background,* Cambridge University Press, 2006.

[86] A. H. Jaffe *et al.* [Boomerang Collaboration], *Phys. Rev. Lett.* **86**, 3475 (2001). Figure 6.8 is reproduced with permission. Copyright (2001) by the American Physical Society.

[87] J. Chluba and R. A. Sunyaev, *Astron. Astroph.* **458**, (2006) L29. [arXiv:astro-ph/0608120].

[88] R. A. Sunyaev and R. Khatri, "Unavoidable CMB spectral features and blackbody photosphere of our Universe", [arXiv:1302.6553 [astro-ph.CO]].

[89] J. M. Jauch and F. Rohrlich, "The Theory of Photons and Electrons", Springer, N.-Y., 1976.

[90] A. P. Lightman, *Astrophys. J.* **244**, 392 (1981).

[91] W. Hu and J. Silk, *Phys. Rev. D* **48**, 485 (1993).

[92] D. J. Fixsen, E. S. Cheng, J. M. Gales, J. C. Mather, R. A. Shafer and E. L. Wright, *Astrophys. J.* **473**, 576 (1996) [astro-ph/9605054].

[93] W. Hu and J. Silk, *Phys. Rev. Lett.* **70**, 2661 (1993).

[94] P. A. R. Ade *et al.* [Planck Collaboration], [arXiv:1502.01589 [astro-ph.CO]].

[95] C. Lunardini and A. Y. Smirnov, *Phys. Rev. D* **64**, 073006 (2001) [arXiv:hep-ph/0012056].

[96] A. D. Dolgov, S. H. Hansen, S. Pastor, S. T. Petcov, G. G. Raffelt and D. V. Semikoz, *Nucl. Phys. B* **632**, 363 (2002) [arXiv:hep-ph/0201287].

[97] A. D. Dolgov, *Phys. Rept.* **370**, 333 (2002) [arXiv:hep-ph/0202122].

[98] S. S. Gershtein and Y. B. Zeldovich, *JETP Lett.* **4**, 120 (1966) [*Pisma Zh. Eksp. Teor. Fiz.* **4**, 174 (1966)].

[99] J. Lesgourgues and S. Pastor, *Phys. Rept.* **429**, 307 (2006) [arXiv:astro-ph/0603494].

[100] S. Dodelson and L. M. Widrow, *Phys. Rev. Lett.* **72**, 17 (1994) [arXiv:hep-ph/9303287].

[101] A. D. Dolgov and S. H. Hansen, *Astropart. Phys.* **16**, 339 (2002) [arXiv:hep-ph/0009083].

[102] T. Asaka, M. Laine and M. Shaposhnikov, *JHEP* **0701**, 091 (2007) [arXiv:hep-ph/0612182].

[103] D. Notzold and G. Raffelt, *Nucl. Phys. B* **307**, 924 (1988).

[104] E. Giusarma, E. Di Valentino, M. Lattanzi, A. Melchiorri and O. Mena, "Relic Neutrinos, thermal axions and cosmology in early 2014", [arXiv:1403.4852 [astro-ph.CO]].

[105] M. Laine and M. Shaposhnikov, *JCAP* **0806**, 031 (2008) [arXiv:0804.4543 [hep-ph]].

[106] A. Boyarsky, A. Neronov, O. Ruchayskiy and M. Shaposhnikov, *JETP Lett.* **83**, 133 (2006).

[107] X. -D. Shi and G. M. Fuller, *Phys. Rev. Lett.* **82**, 2832 (1999).

[108] E. K. Akhmedov, [hep-ph/0001264].

[109] A. Anisimov, [hep-ph/0612024].

[110] A. Anisimov and P. Di Bari, *Phys. Rev. D* **80**, 073017 (2009) doi:10.1103/PhysRevD. 80.073017 [arXiv:0812.5085 [hep-ph]].

[111] L. D. Landau and E. M. Lifshitz, *Course of Theoretical Physics* in 10 volumes; Volume III, L. D. Landau, E. M. Lifshitz, *Quantum Mechanics (Nonrelativistic Theory)* (3rd Edn.), Reed Educational and Professional Publishing Ltd, 1977.

[112] J. Bernstein, L. S. Brown and G. Feinberg, *Rev. Mod. Phys.* **61**, 25 (1989).

[113] S. Sarkar, *Rept. Prog. Phys.* **59**, 1493 (1996) [arXiv:hep-ph/9602260].

[114] F. Iocco, G. Mangano, G. Miele, O. Pisanti and P. D. Serpico, *Phys. Rept.* **472**, 1 (2009) [arXiv:0809.0631 [astro-ph]].

[115] M. S. Smith, L. H. Kawano and R. A. Malaney, *Astrophys. J.* Suppl. **85**, 219 (1993); R. H. Cyburt, *Phys. Rev. D* **70**, 023505 (2004) [arXiv:astro-ph/0401091]; P. D. Serpico *et al.*, *JCAP* **0412**, 010 (2004) [arXiv:astro-ph/0408076].

[116] M. Kusakabe, T. Kajino and G. J. Mathews, *Phys. Rev. D* **74**, 023526 (2006) [arXiv:astro-ph/0605255].

[117] M. Kawasaki, K. Kohri and T. Moroi, *Phys. Rev. D* **71**, 083502 (2005) [arXiv:astro-ph/0408426]; K. Jedamzik, *Phys. Rev. D* **74**, 103509 (2006) [arXiv:hep-ph/0604251].

[118] P. A. R. Ade *et al.* [Planck Collaboration], arXiv:1303.5076 [astro-ph.CO].

[119] G. Steigman and K. M. Nollett, "Light WIMPs, Equivalent Neutrinos, BBN, and the CMB", [arXiv:1401.5488 [astro-ph.CO]].

[120] B. Moore, *Astrophys. J.* **413**, L93 (1993) [arXiv:astro-ph/9306004].

[121] B. J. Carr and M. Sakellariadou, *Astrophys. J.* **516**, 195 (1999).

[122] J. Angle *et al.*, *Phys. Rev. Lett.* **101**, 091301 (2008) [arXiv:0805.2939 [astro-ph]].

[123] T. K. Hemmick *et al.*, *Phys. Rev. D* **41**, 2074 (1990).

[124] A. Kudo and M. Yamaguchi, *Phys. Lett. B* **516**, 151 (2001) [arXiv:hep-ph/0103272].

[125] B. W. Lee and S. Weinberg, *Phys. Rev. Lett.* **39**, 165 (1977); M. I. Vysotsky, A. D. Dolgov and Y. B. Zeldovich, *JETP Lett.* **26**, 188 (1977) [*Pisma Zh. Eksp. Teor. Fiz.* **26**, 200 (1977)]; P. Hut, *Phys. Lett. B* **69**, 85 (1977); K. Sato and M. Kobayashi, *Prog. Theor. Phys.* **58**, 1775 (1977).

[126] K. Y. Choi and L. Roszkowski, *AIP Conf. Proc.* **805**, 30 (2006) [arXiv:hep-ph/0511003]. Figure 9.5 is reproduced with permission. Copyright 2006, American Institute of Physics.

[127] Yu. A. Golfand and E. P. Likhtman, *JETP Lett.* **13**, 323 (1971) [*Pisma Zh. Eksp. Teor. Fiz.* **13**, 452 (1971)].

[128] A. Bottino, N. Fornengo and S. Scopel, *Phys. Rev. D* **67**, 063519 (2003) [arXiv:hep-ph/0212379].

[129] H. P. Nilles, *Phys. Rept.* **110**, 1 (1984).

[130] K. A. Olive, "TASI lectures on astroparticle physics", [arXiv:astro-ph/0503065].

[131] U. Amaldi, W. de Boer and H. Furstenau, *Phys. Lett. B* **260**, 447 (1991).

[132] J. R. Ellis, K. A. Olive, Y. Santoso and V. C. Spanos, *Phys. Lett. B* **565**, 176 (2003) [arXiv:hep-ph/0303043]. Figure 9.11, left panel (and 13.8, left panel) is reproduced with permission from Elsevier, Copyright (2003).

[133] G. F. Giudice and R. Rattazzi, *Phys. Rept.* **322**, 419 (1999) [arXiv:hep-ph/9801271].

[134] S. L. Dubovsky, D. S. Gorbunov and S. V. Troitsky, *Phys. Usp.* **42**, 623 (1999) [*Usp. Fiz. Nauk* **169**, 705 (1999)] [arXiv:hep-ph/9905466].

[135] D. Gorbunov, A. Khmelnitsky and V. Rubakov, *JCAP* **0810**, 041 (2008) [arXiv:0808.3910 [hep-ph]].

[136] J. R. Ellis, K. A. Olive and E. Vangioni, *Phys. Lett. B* **619**, 30 (2005) [arXiv:astro-ph/0503023]. Figure 9.13 (and 13.9) is reproduced with permission from Elsevier, Copyright (2005).

[137] A. de Gouvea, T. Moroi and H. Murayama, *Phys. Rev. D* **56**, 1281 (1997) [arXiv:hep-ph/9701244]. Figure 9.14 is reproduced with permission. Copyright (2008) by the American Physical Society.

[138] G. 't Hooft, *Phys. Rev. Lett.* **37**, 8 (1976).

[139] C. G. Callan, R. F. Dashen and D. J. Gross, *Phys. Lett. B* **63**, 334 (1976).

[140] R. Jackiw and C. Rebbi, *Phys. Rev. Lett.* **37**, 172 (1976).

[141] R. J. Crewther, P. Di Vecchia, G. Veneziano and E. Witten, *Phys. Lett. B* **88**, 123 (1979) [*Erratum-ibid. B* **91**, 487 (1980)].

[142] J. E. Kim and G. Carosi, "Axions and the Strong *CP* Problem", [arXiv:0807.3125 [hep-ph]].

[143] R. D. Peccei and H. R. Quinn, *Phys. Rev. Lett.* **38**, 1440 (1977).

[144] S. Weinberg, *Phys. Rev. Lett.* **40**, 223 (1978).

[145] F. Wilczek, *Phys. Rev. Lett.* **40**, 279 (1978).

[146] M. Dine, W. Fischler and M. Srednicki, *Phys. Lett. B* **104**, 199 (1981).

[147] A. R. Zhitnitsky, *Sov. J. Nucl. Phys.* **31**, 260 (1980) [*Yad. Fiz.* **31**, 497 (1980)].

[148] J. E. Kim, *Phys. Rev. Lett.* **43**, 103 (1979).

[149] M. A. Shifman, A. I. Vainshtein and V. I. Zakharov, *Nucl. Phys. B* **166**, 493 (1980).

[150] A. Vilenkin and A. E. Everett, *Phys. Rev. Lett.* **48**, 1867 (1982).

[151] R. A. Battye and E. P. S. Shellard, "Axion string cosmology and its controversies", [arXiv:astro-ph/9909231].

[152] J. Preskill, M. B. Wise and F. Wilczek, *Phys. Lett. B* **120**, 127 (1983).

[153] L. F. Abbott and P. Sikivie, *Phys. Lett. B* **120**, 133 (1983).

[154] M. Dine and W. Fischler, *Phys. Lett. B* **120**, 137 (1983).

[155] G. G. Raffelt, *J. Phys. A* **40**, 6607 (2007) [arXiv:hep-ph/0611118].

[156] D. J. Gross, R. D. Pisarski and L. G. Yaffe, *Rev. Mod. Phys.* **53**, 43 (1981).

[157] E. W. Kolb, D. J. H. Chung and A. Riotto, "WIMPzillas!", [arXiv:hep-ph/9810361].

[158] S. G. Mamaev, V. M. Mostepanenko and A. A. Starobinsky, *Zh. Eksp. Teor. Fiz.* **70**, 1577 (1976).

[159] C. Bonati, M. D'Elia, M. Mariti, G. Martinelli, M. Mesiti, F. Negro, F. Sanfilippo and G. Villadoro, *JHEP* **1603**, 155 (2016), doi:10.1007/JHEP03(2016)155 [arXiv:1512.06746 [hep-lat]].

[160] G. Grilli di Cortona, E. Hardy, J. Pardo Vega and G. Villadoro, *JHEP* **1601**, 034 (2016), doi:10.1007/JHEP01(2016)034 [arXiv:1511.02867 [hep-ph]].

[161] M. Kaplinghat, R. E. Keeley, T. Linden and H. B. Yu, *Phys. Rev. Lett.* **113**, 021302 (2014) [arXiv:1311.6524 [astro-ph.CO]].

[162] V. Khachatryan *et al.* [CMS Collaboration], "Search for dark matter, extra dimensions, and unparticles in monojet events in proton-proton collisions at sqrt(s) = 8 TeV", [arXiv:1408.3583 [hep-ex]].

[163] A. D. Avrorin *et al.* [Baikal Collaboration], *Astroparticle Physics* (2015), pp. 12–20 [arXiv:1405.3551 [astro-ph.HE]].

[164] H. Baer, K. Y. Choi, J. E. Kim and L. Roszkowski, "Non-thermal dark matter: supersymmetric axions and other candidates", [arXiv:1407.0017 [hep-ph]].

[165] T. Han, Z. Liu and A. Natarajan, *JHEP* **1311**, 008 (2013) [arXiv:1303.3040 [hep-ph]].

[166] O. Buchmueller, M. J. Dolan, J. Ellis, T. Hahn, S. Heinemeyer, W. Hollik, J. Marrouche, K. A. Olive *et al.*, *Eur. Phys. J. C* **74**, 2809 (2014) [arXiv:1312.5233 [hep-ph]].

[167] V. E. Mayes, *Int. J. Mod. Phys. A* **28**, 1350061 (2013) [arXiv:1302.4394 [hep-ph]].

[168] F. S. Queiroz and K. Sinha, *Phys. Lett. B* **735**, 69 (2014) [arXiv:1404.1400 [hep-ph]].

[169] S. M. Barr and D. Seckel, *Phys. Rev. D* **46**, 539 (1992).

[170] A. Ringwald, *J. Phys. Conf. Ser.* **485**, 012013 (2014). [arXiv:1209.2299 [hep-ph]].

[171] L. Visinelli and P. Gondolo, *Phys. Rev. D* **80**, 035024 (2009) [arXiv:0903.4377 [astro-ph.CO]].

[172] E. Witten, *Phys. Rev. D* **30**, 272 (1984).

[173] K. Kajantie, M. Laine, K. Rummukainen and M. E. Shaposhnikov, *Phys. Rev. Lett.* **77**, 2887 (1996) [arXiv:hep-ph/9605288]; F. Karsch, T. Neuhaus, A. Patkos and J. Rank, *Nucl. Phys. B* **474**, 217 (1996) [arXiv:hep-lat/9603004]; F. Karsch, T. Neuhaus, A. Patkos and J. Rank, *Nucl. Phys. Proc. Suppl.* **53**, 623 (1997) [arXiv:hep-lat/9608087].

[174] S. R. Coleman and F. De Luccia, *Phys. Rev. D* **21**, 3305 (1980).

[175] V. A. Rubakov, *Classical Theory of Gauge Fields*, Princeton University Press, 2002.

[176] I. Y. Kobzarev, L. B. Okun and M. B. Voloshin, *Sov. J. Nucl. Phys.* **20**, 644 (1975) [*Yad. Fiz.* **20**, 1229 (1974)].

[177] S. R. Coleman, *Phys. Rev. D* **15**, 2929 (1977) [*Erratum-ibid. D* **16**, 1248 (1977)].

[178] C. G. Callan and S. R. Coleman, *Phys. Rev. D* **16**, 1762 (1977).

[179] M. N. Chernodub, F. V. Gubarev, E. M. Ilgenfritz and A. Schiller, *Phys. Lett. B* **443**, 244 (1998) [arXiv:hep-lat/9807016].

[180] E. H. Fradkin and S. H. Shenker, *Phys. Rev. D* **19**, 3682 (1979).

[181] T. Banks and E. Rabinovici, *Nucl. Phys. B* **160**, 349 (1979).

[182] A. D. Linde, *Phys. Lett. B* **96**, 289 (1980).

[183] D. Y. Grigoriev and V. A. Rubakov, *Nucl. Phys. B* **299**, 67 (1988); D. Y. Grigoriev, V. A. Rubakov and M. E. Shaposhnikov, *Phys. Lett. B* **216**, 172 (1989).

[184] J. Ambjorn, T. Askgaard, H. Porter and M. E. Shaposhnikov, *Phys. Lett. B* **244**, 479 (1990); J. Ambjorn, T. Askgaard, H. Porter and M. E. Shaposhnikov, *Nucl. Phys. B* **353**, 346 (1991).

[185] A. H. Guth and E. J. Weinberg, *Phys. Rev. D* **23**, 876 (1981).

[186] M. Garny and T. Konstandin, *JHEP* **1207**, 189 (2012) [arXiv:1205.3392 [hep-ph]].

[187] G. W. Anderson and L. J. Hall, *Phys. Rev. D* **45**, 2685 (1992); J. R. Espinosa, T. Konstandin and F. Riva, *Nucl. Phys. B* **854**, 592 (2012) [arXiv:1107.5441 [hep-ph]].

[188] L. N. Lipatov, *Sov. Phys. JETP* **45**, 216 (1977) [*Zh. Eksp. Teor. Fiz.* **72**, 411 (1977)].

[189] S. R. Coleman and E. J. Weinberg, *Phys. Rev. D* **7**, 1888 (1973).

[190] D. Buttazzo, G. Degrassi, P. P. Giardino, G. F. Giudice, F. Sala, A. Salvio and A. Strumia, *JHEP* **1312**, 089 (2013) [arXiv:1307.3536].

[191] D. Bodeker, G. D. Moore and K. Rummukainen, *Nucl. Phys. Proc. Suppl.* **83**, 583 (2000) [arXiv:hep-lat/9909054]; G. D. Moore, "Do we understand the sphaleron rate?", [arXiv:hep-ph/0009161].

[192] A. Dolgov and J. Silk, *Phys. Rev. D* **47**, 4244 (1993); M. Y. Khlopov, S. G. Rubin and A. S. Sakharov, *Phys. Rev. D* **62**, 083505 (2000) [arXiv:hep-ph/0003285]; A. D. Dolgov, M. Kawasaki and N. Kevlishvili, *Nucl. Phys. B* **807**, 229 (2009) [arXiv:0806.2986 [hep-ph]].

[193] M. Y. Khlopov, S. G. Rubin and A. S. Sakharov, "Antimatter regions in the baryon-dominated universe", arXiv:hep-ph/0210012; C. Bambi and A. D. Dolgov, *Nucl. Phys. B* **784**, 132 (2007) [arXiv:astro-ph/0702350]; A. D. Dolgov, "Cosmic antimatter: models and phenomenology", [arXiv:1002.2940].

[194] V. A. Kuzmin, V. A. Rubakov and M. E. Shaposhnikov, *Phys. Lett. B* **155**, 36 (1985).

[195] E. W. Kolb and M. S. Turner, *The Early Universe*, Addison-Wesley, Redwood City, 1990; *Front. Phys.* **69**.

[196] F. R. Klinkhamer and N. S. Manton, *Phys. Rev. D* **30**, 2212 (1984).

[197] A. A. Belavin, A. M. Polyakov, A. S. Schwartz and Yu. S. Tyupkin, *Phys. Lett.* B **59**, 85 (1975).

[198] P. Arnold, D. Son and L. G. Yaffe, *Phys. Rev. D* **55**, 6264 (1997) [arXiv:hep-ph/9609481].

[199] D. Bodeker, G. D. Moore and K. Rummukainen, *Nucl. Phys. Proc.* Suppl. **83**, 583 (2000) [arXiv:hep-lat/9909054]; G. D. Moore, "Do we understand the sphaleron rate?", [arXiv:hep-ph/0009161].

[200] S. Y. Khlebnikov and M. E. Shaposhnikov, *Nucl. Phys. B* **308**, 885 (1988); K. Kajantie, M. Laine, K. Rummukainen and M. E. Shaposhnikov, *Nucl. Phys.* B **458**, 90 (1996).

[201] H. Georgi and S. L. Glashow, *Phys. Rev. Lett.* **32**, 438 (1974).

[202] H. Georgi, H. R. Quinn and S. Weinberg, *Phys. Rev. Lett.* **33**, 451 (1974).

[203] P. Minkowski, *Phys. Lett. B* **67**, 421 (1977).

[204] T. Yanagida, "Horizontal gauge symmetry and masses of neutrinos", In *Proceedings of the Workshop on the Baryon Number of the Universe and Unified Theories, Tsukuba, Japan, 1979*, preprint KEK-79-18-95.

[205] M. Gell-Mann, P. Ramond and R. Slansky, "Complex Spinors And Unified Theories", In *Supergravity*, eds. P. van Nieuwenhuizen and D. Z. Freedman, North Holland Publ. Co., 1979.

[206] R. N. Mohapatra and G. Senjanovic, *Phys. Rev. D* **23**, 165 (1981).

[207] M. Fukugita and T. Yanagida, *Phys. Lett. B* **174**, 45 (1986).

[208] W. Buchmuller, P. Di Bari and M. Plumacher, *Annals Phys.* **315**, 305 (2005) [arXiv:hep-ph/0401240]; W. Buchmuller, P. Di Bari and M. Plumacher, *Nucl. Phys.* B **665**, 445 (2003) [arXiv:hep-ph/0302092].

[209] A. I. Bochkarev and M. E. Shaposhnikov, *Mod. Phys. Lett. A* **2**, 417 (1987).

[210] M. S. Carena, M. Quiros and C. E. M. Wagner, *Phys. Lett. B* **380**, 81 (1996) [arXiv:hep-ph/9603420].

[211] A. G. Cohen, D. B. Kaplan and A. E. Nelson, *Ann. Rev. Nucl. Part. Sci.* **43**, 27 (1993) [arXiv:hep-ph/9302210].

[212] I. Affleck and M. Dine, *Nucl. Phys. B* **249**, 361 (1985).

[213] M. Dine, L. Randall and S. D. Thomas, *Nucl. Phys. B* **458**, 291 (1996) [arXiv:hep-ph/9507453].

[214] G. Engelhard, Y. Grossman, E. Nardi and Y. Nir, *Phys. Rev. Lett.* **99**, 081802 (2007) doi:10.1103/PhysRevLett.99.081802 [hep-ph/0612187].

[215] A. G. Cohen and D. B. Kaplan, *Nucl. Phys. B* **308**, 913 (1988).

[216] G. Lambiase, S. Mohanty and A. R. Prasanna, *Int. J. Mod. Phys. D* **22**, 1330030 (2013) [arXiv:1310.8459 [hep-ph]].

[217] H. Davoudiasl, D. E. Morrissey, K. Sigurdson and S. Tulin, *Phys. Rev. Lett.* **105**, 211304 (2010) [arXiv:1008.2399 [hep-ph]].

[218] H. Davoudiasl and R. N. Mohapatra, *New J. Phys.* **14**, 0950511 (2012) [arXiv:1203.1247 [hep-ph]].

[219] E. K. Akhmedov, V. A. Rubakov and A. Y. Smirnov, *Phys. Rev. Lett.* **81**, 1359 (1998) [hep-ph/9803255].

[220] T. Asaka and M. Shaposhnikov, *Phys. Lett. B* **620**, 17 (2005) [hep-ph/0505013].

[221] K. Krizka, A. Kumar and D. E. Morrissey, *Phys. Rev. D* **87**, 9, 095016 (2013) [arXiv:1212.4856 [hep-ph]].

[222] M. Joyce, T. Prokopec and N. Turok, *Phys. Rev. Lett.* **75**, 1695 (1995) [Erratum-ibid. **75** (1995) 3375] [hep-ph/9408339].

[223] M. Joyce, T. Prokopec and N. Turok, *Phys. Rev. D* **53**, 2930 (1996) [hep-ph/9410281].

[224] M. Joyce, T. Prokopec and N. Turok, *Phys. Rev. D* **53**, 2958 (1996) [hep-ph/9410282].

[225] T. Konstandin, *Phys. Usp.* **56**, 747 (2013) [arXiv:1302.6713 [hep-ph]].

[226] T. W. B. Kibble, *J. Phys. A* **9**, 1387 (1976).

[227] G. 't Hooft, *Nucl. Phys. B* **79**, 276 (1974).

[228] A. M. Polyakov, *JETP Lett.* **20**, 194 (1974) [*Pisma Zh. Eksp. Teor. Fiz.* **20**, 430 (1974)].

[229] Y. B. Zeldovich and M. Y. Khlopov, *Phys. Lett. B* **79**, 239 (1978).

[230] J. Preskill, *Phys. Rev. Lett.* **43**, 1365 (1979).

[231] A. A. Abrikosov, *Sov. Phys. JETP* **5**, 1174 (1957) [*Zh. Eksp. Teor. Fiz.* **32**, 1442 (1957)].

[232] H. B. Nielsen and P. Olesen, *Nucl. Phys. B* **61**, 45 (1973).

[233] S. Deser, R. Jackiw and G. 't Hooft, *Annals Phys.* **152**, 220 (1984); J. R. I. Gott, *Astrophys. J.* **288**, 422 (1985).

[234] B. Allen and E. P. S. Shellard, *Phys. Rev. D* **45**, 1898 (1992).

[235] E. P. S. Shellard, *Nucl. Phys. B* **283**, 624 (1987).

[236] B. Allen and E. P. S. Shellard, *Phys. Rev. Lett.* **64**, 119 (1990).

[237] N. Kaiser and A. Stebbins, *Nature* **310**, 391 (1984).

[238] M. Wyman, L. Pogosian and I. Wasserman, *Phys. Rev. D* **72**, 023513 (2005) [*Erratum-ibid. D* **73**, 089905 (2006)] [arXiv:astro-ph/0503364].

[239] E. Jeong and G. F. Smoot, *Astrophys. J.* **624**, 21 (2005) [arXiv:astro-ph/0406432].

[240] J. Silk and A. Vilenkin, *Phys. Rev. Lett.* **53**, 1700 (1984).

[241] A. Stebbins, S. Veeraraghavan, R. H. Brandenberger, J. Silk and N. Turok, *Astrophys. J.* **322**, 1 (1987).

[242] Y. B. Zeldovich, *Mon. Not. Roy. Astron. Soc.* **192**, 663 (1980).

[243] A. Vilenkin, *Phys. Rev. Lett.* **46**, 1169 (1981) [*Erratum-ibid.* **46**, 1496 (1981)].

[244] J. Urrestilla, N. Bevis, M. Hindmarsh, M. Kunz and A. R. Liddle, *JCAP* **0807**, 010 (2008) [arXiv:0711.1842 [astro-ph]]. Figure 12.6 is reproduced with permission from Institute of Physics.

[245] K. D. Olum and V. Vanchurin, *Phys. Rev. D* **75**, 063521 (2007) [arXiv:astro-ph/0610419].

[246] F. A. Jenet *et al.*, *Astrophys. J.* **653**, 1571 (2006) [arXiv:astro-ph/0609013].

[247] A. Vilenkin and E. P. S. Shellard, *Cosmic Strings and Other Topological Defects*, Cambridge University Press, 1994.

[248] E. J. Copeland and T. W. B. Kibble, "Cosmic strings and superstrings", [arXiv:0911.1345 [hep-th]].

[249] Y. B. Zeldovich, I. Y. Kobzarev and L. B. Okun, *Zh. Eksp. Teor. Fiz.* **67**, 3 (1974) [*Sov. Phys. JETP* **40**, 1 (1974)].

[250] R. L. Davis, *Phys. Rev. D* **35**, 3705 (1987).

[251] N. Turok, *Phys. Rev. Lett.* **63**, 2625 (1989).

[252] R. Friedberg, T. D. Lee and A. Sirlin, *Phys. Rev. D* **13**, 2739 (1976).

[253] S. R. Coleman, *Nucl. Phys. B* **262**, 263 (1985) [*Erratum-ibid. B* **269**, 744 (1986)].

[254] A. Kusenko, *Phys. Lett. B* **405**, 108 (1997) [arXiv:hep-ph/9704273].

[255] G. R. Dvali, A. Kusenko and M. E. Shaposhnikov, *Phys. Lett. B* **417**, 99 (1998) [arXiv:hep-ph/9707423].

[256] A. Kusenko and M. E. Shaposhnikov, *Phys. Lett. B* **418**, 46 (1998) [arXiv:hep-ph/9709492].

[257] K. Enqvist and J. McDonald, *Phys. Lett. B* **425**, 309 (1998) [arXiv:hep-ph/9711514].

[258] K. M. Lee, *Phys. Rev. D* **50**, 5333 (1994) [arXiv:hep-ph/9404293].

[259] S. Kasuya and M. Kawasaki, *Phys. Rev. D* **61**, 041301 (2000) [arXiv:hep-ph/9909509]; S. Kasuya and M. Kawasaki, *Phys. Rev. D* **62**, 023512 (2000) [arXiv:hep-ph/0002285].

[260] K. Enqvist, A. Jokinen, T. Multamaki and I. Vilja, *Phys. Rev. D* **63**, 083501 (2001) [arXiv:hep-ph/0011134]; T. Multamaki and I. Vilja, *Phys. Lett. B* **535**, 170 (2002) [arXiv:hep-ph/0203195].

[261] O. S. Sazhina, D. Scognamiglio and M. Sazhin, "Observational constraints on the types of cosmic strings", [arXiv:1312.6106 [astro-ph.CO]].

[262] A. Friedland, H. Murayama and M. Perelstein, *Phys. Rev. D* **67**, 043519 (2003) [astro-ph/0205520].

[263] E. Krylov, A. Levin and V. Rubakov, "Cosmological phase transition, baryon asymmetry and dark matter Q-balls", [arXiv:1301.0354 [hep-ph]].

[264] S. Weinberg, *Gravitation and Cosmology: Principles and Applications of the General Theory of Relativity*, John Wiley & Sons, Inc, 1972.

[265] C. W. Misner, K. S. Thorne and J. A. Wheeler, *Gravitation*, Freeman, 1973.

[266] B. A. Dubrovin, S. P. Novikov and A. T. Fomenko, *Modern Geometry. Methods and Applications*, (Parts I, II) Springer-Verlag, New York, Inc., 1984, 1985.

[267] L. D. Landau and E. M. Lifshitz, *Course of Theoretical Physics* in 10 volumes; Volume I, L. D. Landau and E. M. Lifshitz, *Mechanics* (3rd Edn.), Reed Educational and Professional Publishing Ltd, 1977.

[268] N. N. Bogolyubov and D. V. Shirkov, *Introduction to the Theory of Quantized Fields*, (3rd Edn.), Wiley, New York, 1980.

[269] C. Itzykson and J. B. Zuber, *Quantum Field Theory*, McGraw-Hill, New York, 1980.

[270] M. E. Peskin and D. V. Schroeder, *An Introduction to Quantum Field Theory*, Addison-Wesley Publishing Company, 1995.

[271] S. Weinberg, *The Quantum Theory of Fields* in 3 volumes, Cambridge University Press, 1996.

[272] C. Bernard *et al.* [MILC Collaboration], *Phys. Rev. D* **71**, 034504 (2005) [arXiv:hep-lat/0405029].

[273] L. Wolfenstein, *Phys. Rev. D* **17**, 2369 (1978).

[274] S. P. Mikheev and A. Y. Smirnov, *Sov. J. Nucl. Phys.* **42**, 913 (1985) [*Yad. Fiz.* **42**, 1441 (1985)].

[275] J. N. Bahcall, A. M. Serenelli and S. Basu, *Astrophys. J.* **621**, L85 (2005) [arXiv:astro-ph/0412440]. Figure C.2 is reproduced by permission of the American Astronomical Society.

[276] S. Turck-Chieze *et al.*, *Phys. Rev. Lett.* **93**, 211102 (2004) [arXiv:astro-ph/0407176].

[277] R. J. Davis, D. S. Harmer and K. C. Hoffman, *Phys. Rev. Lett.* **20**, 1205 (1968); B. T. Cleveland *et al.*, *Astrophys. J.* **496**, 505 (1998).

[278] K. S. Hirata *et al.* [Kamiokande-II Collaboration], *Phys. Rev. Lett.* **63**, 16 (1989); Y. Fukuda *et al.* [Kamiokande Collaboration], *Phys. Rev. Lett.* **77**, 1683 (1996).

[279] J. Hosaka *et al.* [Super-Kamiokande Collaboration], *Phys. Rev. D* **73**, 112001 (2006) [arXiv:hep-ex/0508053].

[280] A. I. Abazov *et al.*, *Phys. Rev. Lett.* **67**, 3332 (1991); J. N. Abdurashitov *et al.* [SAGE Collaboration], *Phys. Rev. C* **80**, 015807 (2009) [arXiv:0901.2200 [nucl-ex]].

[281] P. Anselmann *et al.* [GALLEX Collaboration], *Phys. Lett. B* **285**, 376 (1992); M. Altmann *et al.* [GNO Collaboration], *Phys. Lett. B* **616**, 174 (2005) [arXiv:hep-ex/0504037].

[282] Q. R. Ahmad *et al.* [SNO Collaboration], *Phys. Rev. Lett.* **89**, 011301 (2002) [arXiv:nucl-ex/0204008]; B. Aharmim *et al.* [SNO Collaboration], *Phys. Rev. C* **72**, 055502 (2005) [arXiv:nucl-ex/0502021]; R. G. H. Robertson [SNO Collaboration], *J. Phys. Conf. Ser.* **136**, 022002 (2008).

[283] K. Eguchi *et al.* [KamLAND Collaboration], *Phys. Rev. Lett.* **90**, 021802 (2003) [arXiv:hep-ex/0212021]; S. Abe *et al.* [KamLAND Collaboration], *Phys. Rev. Lett.* **100**, 221803 (2008) [arXiv:0801.4589 [hep-ex]].

[284] C. Arpesella *et al.* [Borexino Collaboration], *Phys. Lett. B* **658**, 101 (2008) [arXiv:0708.2251 [astro-ph]].

[285] K. S. Hirata *et al.* [Kamiokande-II Collaboration], *Phys. Lett. B* **205**, 416 (1988); K. S. Hirata *et al.* [Kamiokande-II Collaboration], *Phys. Lett. B* **280**, 146 (1992); Y. Fukuda *et al.* [Kamiokande Collaboration], *Phys. Lett. B* **335**, 237 (1994).

[286] Y. Fukuda *et al.* [Super-Kamiokande Collaboration], *Phys. Rev. Lett.* **81**, 1562 (1998) [arXiv:hep-ex/9807003].

[287] Y. Ashie *et al.* [Super-Kamiokande Collaboration], *Phys. Rev. D* **71**, 112005 (2005) [arXiv:hep-ex/0501064]. Figure C.6 is reproduced with permission. Copyright (2005) by the American Physical Society.

[288] M. H. Ahn *et al.* [K2K Collaboration], *Phys. Rev. Lett.* **90**, 041801 (2003) [arXiv:hep-ex/0212007]; M. H. Ahn *et al.* [K2K Collaboration], *Phys. Rev. D* **74**, 072003 (2006) [arXiv:hep-ex/0606032].

[289] D. G. Michael *et al.* [MINOS Collaboration], *Phys. Rev. Lett.* **97**, 191801 (2006) [arXiv:hep-ex/0607088].

[290] M. Apollonio *et al.* [CHOOZ Collaboration], *Eur. Phys. J. C* **27**, 331 (2003) [arXiv:hep-ex/0301017].

[291] H. Murayama, `http://hitoshi.berkeley.edu/neutrino/`

[292] P. Adamson *et al.* [MINOS Collaboration], *Phys. Rev. Lett.* **101**, 131802 (2008) [arXiv:0806.2237 [hep-ex]]. Figure C.9 is reproduced with permission. Copyright (2008) by the American Physical Society.

[293] C. Athanassopoulos *et al.* [LSND Collaboration], *Phys. Rev. Lett.* **81**, 1774 (1998) [arXiv:nucl-ex/9709006].

[294] A. A. Aguilar-Arevalo *et al.* [MiniBooNE Collaboration], *Phys. Rev. Lett.* **98**, 231801 (2007) [arXiv:0704.1500 [hep-ex]].

[295] A. A. Aguilar-Arevalo *et al.* [MiniBooNE Collaboration], *Phys. Rev. Lett.* **102**, 101802 (2009) [arXiv:0812.2243 [hep-ex]].

[296] L. Ludhova *et al.* [BOREXINO Collaboration], [arXiv:1205.2989 [hep-ex]].

[297] G. Bellini *et al.* [Borexino Collaboration], *Phys. Rev. Lett.* **108**, 051302 (2012).

[298] P. Adamson *et al.* [MINOS Collaboration], *Phys. Rev. Lett.* **108**, 191802 (2012) [arXiv:1202.2772 [hep-ex]].

[299] K. Abe *et al.* [T2K Collaboration], *Phys. Rev. D* **85**, 031103 (2012) [arXiv:1201.1386 [hep-ex]].

[300] K. Abe *et al.* [T2K Collaboration], *Phys. Rev. Lett.* **107**, 041801 (2011) [arXiv:1106.2822 [hep-ex]]; K. Abe *et al.* [T2K Collaboration], [arXiv:1304.0841 [hep-ex]].

[301] P. Adamson *et al.* [MINOS Collaboration], *Phys. Rev. Lett.* **107**, 181802 (2011) [arXiv:1108.0015 [hep-ex]]; P. Adamson *et al.* [MINOS Collaboration], [arXiv:1301.4581 [hep-ex]].

[302] Y. Abe *et al.* [DOUBLE-CHOOZ Collaboration], *Phys. Rev. Lett.* **108**, 130801 (2012) [arXiv:1112.6353 [hep-ex]].

[303] F. P. An *et al.* [DAYA-BAY Collaboration], *Phys. Rev. Lett.* **108**, 171803 (2012) [arXiv:1203.1669 [hep-ex]].

[304] J. K. Ahn *et al.* [RENO Collaboration], *Phys. Rev. Lett.* **108**, 191802 (2012) [arXiv:1204.0626 [hep-ex]].

[305] A. Gando *et al.* [KamLAND Collaboration], *Phys. Rev. D* **88**, 3, 033001 (2013) http://arxiv.org/abs/arXiv:1303.4667.

[306] F. P. An *et al.* [Daya Bay Collaboration], *Phys. Rev. Lett.* **112**, 061801 (2014) [arXiv:1310.6732].

[307] K. Abe *et al.* [T2K Collaboration], "Precise measurement of the neutrino mixing parameter θ_{23} from muon neutrino disappearance in an off-axis beam", [arXiv:1403.1532 [hep-ex]].

[308] A. Aguilar-Arevalo *et al.* [LSND Collaboration], *Phys. Rev. D* **64**, 112007 (2001) [arXiv: hep-ex/0104049].

[309] A. A. Aguilar-Arevalo *et al.* [MiniBooNE Collaboration], *Phys. Rev. Lett.* **102**, 101802 (2009) [arXiv:0812.2243 [hep-ex]]; A. A. Aguilar-Arevalo *et al.* [MiniBooNE Collaboration], [arXiv:1207.4809 [hep-ex]].

[310] J. N. Abdurashitov, V. N. Gavrin, S. V. Girin, V. V. Gorbachev, P. P. Gurkina, T. V. Ibragimova, A. V. Kalikhov and N. G. Khairnasov *et al.*, *Phys. Rev. C* **73**, 045805 (2006) [nucl-ex/0512041].

[311] G. Mention, M. Fechner, T. Lasserre, T. A. Mueller, D. Lhuillier, M. Cribier and A. Letourneau, *Phys. Rev. D* **83**, 073006 (2011) [arXiv:1101.2755].

[312] P. Huber, *Phys. Rev. C* **84**, 024617 (2011) [*Erratum-ibid. C* **85**, 029901 (2012)] [arXiv:1106.0687].

[313] M. Praszalowicz and T. Stebel, *JHEP* **1303**, (2013) 090 [arXiv:1211.5305 [hep-ph]].

[314] M. C. Gonzalez-Garcia, M. Maltoni, J. Salvado and T. Schwetz, *JHEP* **1212**, 123 (2012) [arXiv:1209.3023 [hep-ph]].

[315] R. Foot, H. Lew, X. G. He and G. C. Joshi, *Z. Phys. C* **44**, 441 (1989).

[316] F. L. Bezrukov and M. Shaposhnikov, *Phys. Lett. B* **659**, 703 (2008) doi:10.1016/j.physletb.2007.11.072 [arXiv:0710.3755 [hep-th]].

[317] N. K. Nielsen, Theories", *Nucl. Phys. B* **101**, 173 (1975). doi:10.1016/0550-3213(75)90301-6; R. Fukuda and T. Kugo, *Phys. Rev. D* **13**, 3469 (1976). doi:10.1103/PhysRevD.13.3469.

[318] L. Bergstrom, T. Bringmann, I. Cholis, D. Hooper and C. Weniger, *Phys. Rev. Lett.* **111**, 171101 (2013). doi:10.1103/PhysRevLett.111.171101 [arXiv:1306.3983 [astr-ph.HE]].

Books and Reviews

We give here an (incomplete) list of books and reviews where various aspects of the Hot Big Bang theory and related issues are discussed.

Books

Ya. B. Zeldovich and I. D. Novikov, *The Structure and Evolution of the Universe (Relativistic Astrophysics, Volume 2)*, University of Chicago Press, 1983.

A. D. Linde, *Particle Physics and Inflationary Cosmology*, Harwood, Chur, 1990.

E. W. Kolb and M. S. Turner, *The Early Universe*, Addison–Wesley, Redwood City, 1990, Frontiers in physics, 69.

A. D. Dolgov, M. V. Sazhin and Ya. B. Zeldovich, *Basics of Modern Cosmology*, Ed. Frontieres, Gif-sur-Yvette, 1991.

P. J. E. Peebles, *Principles of Physical Cosmology*, Princeton University Press, 1993.

A. Vilenkin and E. P. S. Shellard, *Cosmic Strings and Other Topological Defects*, Cambridge University Press, 1994.

J. A. Peacock, *Cosmological Physics*, Cambridge University Press, 1999.

S. Dodelson, *Modern Cosmology*, Academic Press, Amsterdam, 2003.

V. Mukhanov, *Physical Foundations of Cosmology*, Cambridge University Press, 2005.

S. Weinberg, *Cosmology*, Oxford University Press, 2008.

General Reviews

A. D. Dolgov and Y. B. Zeldovich, Cosmology and elementary particles, *Rev. Mod. Phys.* **53**, 1 (1981).

R. H. Brandenberger, Particle physics aspects of modern cosmology [arXiv:hep-ph/9701276].

M. S. Turner and J. A. Tyson, Cosmology at the millennium, *Rev. Mod. Phys.* **71**, S145 (1999) [arXiv:astro-ph/9901113].

W. L. Freedman and M. S. Turner, Measuring and understanding the Universe, *Rev. Mod. Phys.* **75**, 1433 (2003) [arXiv:astro-ph/0308418].

V. Rubakov, Introduction to cosmology, PoS **RTN2005**, 003 (2005).

Lectures at Summer Schools on High Energy Physics

J. A. Peacock, Cosmology and particle physics, Proc. 1998 European School of High-Energy Physics, St. Andrews, Scotland, 23 Aug–5 Sep 1998.

M. Shaposhnikov, Cosmology and astrophysics, Proc. 2000 European School of High-Energy Physics, Caramulo, Portugal, 20 Aug–2 Sep 2000.

V. A. Rubakov, Cosmology and astrophysics, Proc. 2001 European School of High-Energy Physics, Beatenberg, Switzerland, 2001.

I. I. Tkachev, Astroparticle physics, Proc. 2003 European School on High-Energy Physics, Tsakhkadzor, Armenia, 24 Aug–6 Sep 2003 [arXiv:hep-ph/0405168].

Reviews on Topics Covered in Separate Chapters

If necessary, relevant Sections are indicated in parenthesis.

Chapter 4

S. Weinberg, The cosmological constant problem, *Rev. Mod. Phys.* **61**, 1 (1989).

V. Sahni and A. A. Starobinsky, The case for a positive cosmological Lambda-term, *Int. J. Mod. Phys. D* **9**, 373 (2000) [arXiv:astro-ph/9904398].

S. Weinberg, The cosmological constant problems, arXiv:astro-ph/0005265.

A. D. Chernin, Cosmic vacuum, *Phys. Usp.* **44**, 1099 (2001) [Usp. Fiz. Nauk **44**, 1153 (2001)].

T. Padmanabhan, Cosmological constant: The weight of the vacuum, *Phys. Rept.* **380**, 235 (2003) [arXiv:hep-th/0212290].

P. J. E. Peebles and B. Ratra, The cosmological constant and dark energy, *Rev. Mod. Phys.* **75**, 559 (2003) [arXiv:astro-ph/0207347].

V. Sahni, Dark matter and dark energy, *Lect. Notes Phys.* **653**, 141 (2004) [arXiv:astro-ph/0403324].

E. J. Copeland, M. Sami and S Tsujikawa, Dynamics of dark energy, *Int. J. Mod. Phys. D* **15**, 1753 (2006) [arXiv:hep-th/0603057].

V. Sahni and A. Starobinsky, Reconstructing dark energy, *Int. J. Mod. Phys. D* **15**, 2105 (2006) [arXiv:astro-ph/0610026].

Chapter 6

R. A. Sunyaev and J. Chluba, Signals from the epoch of cosmological recombination, *Astron. Nachr.* **330**, 657 (2009) [arXiv:0908.0435 [astro-ph.CO]].

Chapter 7

A. D. Dolgov, Cosmological implications of neutrinos, *Surveys High Energy Phys.* **17**, 91 (2002) [arXiv:hep-ph/0208222].

A. D. Dolgov, Neutrinos in cosmology, *Phys. Rept.* **370**, 333 (2002) [arXiv:hep-ph/0202122].

J. Lesgourgues and S. Pastor, Massive neutrinos and cosmology, *Phys. Rept.* **429**, 307 (2006) [arXiv:astro-ph/0603494].

M. Drewes *et al.*, A White Paper on keV Sterile Neutrino Dark Matter, *White Paper* [arXiv:1602.04816 [hep-ph]].

Chapter 8

A. Merchant Boesgaard and G. Steigman, Big Bang Nucleosynthesis: Theories and observations, *Ann. Rev. Astron. Astrophys.* **23**, 319 (1985).

S. Sarkar, Big Bang nucleosynthesis and physics beyond the standard model, *Rept. Prog. Phys.* **59**, 1493 (1996) [arXiv:hep-ph/9602260].

K. A. Olive, G. Steigman and T. P. Walker, Primordial nucleosynthesis: Theory and observations, *Phys. Rept.* **333**, 389 (2000) [arXiv:astro-ph/9905320].

F. Iocco, G. Mangano, G. Miele, O. Pisanti and P. D. Serpico, Primordial nucleosynthesis: From precision cosmology to fundamental physics, *Phys. Rept.* **472**, 1 (2009) [arXiv:0809.0631 [astro-ph]].

Chapter 9

J. R. Primack, D. Seckel and B. Sadoulet, Detection of cosmic dark matter, *Ann. Rev. Nucl. Part. Sci.* **38**, 751 (1988).

P. F. Smith and J. D. Lewin, Dark matter detection, *Phys. Rept.* **187**, 203 (1990).

A. Bottino and N. Fornengo, Dark matter and its particle candidates, [arXiv:hep-ph/9904469].

K. A. Olive, Dark matter, [arXiv:astro-ph/0301505].

G. Bertone, D. Hooper and J. Silk, Particle dark matter: Evidence, candidates and constraints, *Phys. Rept.* **405**, 279 (2005) [arXiv:hep-ph/0404175].

G. Jungman, M. Kamionkowski and K. Griest, Supersymmetric dark matter, *Phys. Rept.* **267**, 195 (1996) [arXiv:hep-ph/9506380] (Section 9.6).

H. P. Nilles, Supersymmetry, supergravity and particle physics, *Phys. Rept.* **110**, 1 (1984) (Sec. 9.6).

G. F. Giudice and R. Rattazzi, Theories with gauge-mediated supersymmetry breaking, *Phys. Rept.* **322**, 419 (1999) [arXiv:hep-ph/9801271] (Sec. 9.6).

S. L. Dubovsky, D. S. Gorbunov and S. V. Troitsky, Gauge mechanism of mediation of supersymmetry breaking, *Phys. Usp.* **42**, 623 (1999) [*Usp. Fiz. Nauk* **169**, 705 (1999)] [arXiv:hep-ph/9905466] (Sec. 9.6).

D. I. Kazakov, Beyond the standard model (in search of supersymmetry) [arXiv:hep-ph/0012288] (Sec. 9.6).

M. I. Vysotsky and R. B. Nevzorov, Selected problems of supersymmetry phenomenology, *Phys. Usp.* **44**, 919 (2001) [*Usp. Fiz. Nauk* **44**, 939 (2001)] (Sec. 9.6).

J. E. Kim, Light pseudoscalars, particle physics and cosmology, *Phys. Rept.* **150**, 1 (1987) (Sec. 9.7.1).

M. S. Turner, Windows on the axion, *Phys. Rept.* **197**, 67 (1990) (Sec. 9.7.1).

J. E. Kim and G. Carosi, Axions and the strong *CP* problem [arXiv:0807.3125 [hep-ph]] (Sec. 9.7.1).

Chapter 10

V. A. Rubakov and M. E. Shaposhnikov, Electroweak baryon number non-conservation in the early universe and in high-energy collisions, *Usp. Fiz. Nauk* **166**, 493 (1996) [*Phys. Usp.* **39**, 461 (1996)] [arXiv:hep-ph/9603208].

Chapter 11

E. W. Kolb and M. S. Turner, Grand unified theories and the origin of the baryon asymmetry, *Ann. Rev. Nucl. Part. Sci.* **33**, 645 (1983).

A. D. Dolgov, NonGUT baryogenesis, *Phys. Rept.* **222**, 309 (1992).

A. D. Dolgov, Baryogenesis, 30 years after [arXiv:hep-ph/9707419].

A. Riotto and M. Trodden, Recent progress in baryogenesis, *Ann. Rev. Nucl. Part. Sci.* **49**, 35 (1999) [arXiv:hep-ph/9901362].

V. A. Rubakov and M. E. Shaposhnikov, Electroweak baryon number non-conservation in the early universe and in high-energy collisions, *Usp. Fiz. Nauk* **166**, 493 (1996) [*Phys. Usp.* **39**, 461 (1996)] [arXiv:hep-ph/9603208].

W. Buchmuller, P. Di Bari and M. Plumacher, Leptogenesis for pedestrians, *Annals Phys.* **315**, 305 (2005) [arXiv:hep-ph/0401240] (Sec. 11.4).

W. Buchmuller, R. D. Peccei and T. Yanagida, Leptogenesis as the origin of matter, *Ann. Rev. Nucl. Part. Sci.* **55**, 311 (2005) [arXiv:hep-ph/0502169] (Sec. 11.4).

A. Strumia and F. Vissani, Neutrino masses and mixings and..., [arXiv:hep-ph/0606054] (Sec. 11.4).

S. Davidson, E. Nardi and Y. Nir, Leptogenesis, *Phys. Rept.* **466**, 105 (2008) [arXiv: 0802.2962 [hep-ph]] (Sec. 11.4).

A. G. Cohen, D. B. Kaplan and A. E. Nelson, Progress in electroweak baryogenesis, *Ann. Rev. Nucl. Part. Sci.* **43**, 27 (1993) [arXiv:hep-ph/9302210] (Sec. 11.5).

M. Trodden, Electroweak baryogenesis, *Rev. Mod. Phys.* **71**, 1463 (1999) [arXiv:hep-ph/9803479] (Sec. 11.5).

K. Enqvist and A. Mazumdar, Cosmological consequences of MSSM flat directions, *Phys. Rept.* **380**, 99 (2003) [arXiv:hep-ph/0209244] (Sec. 11.6).

Chapter 12

A. Vilenkin, Cosmic strings and domain walls, *Phys. Rept.* **121**, 263 (1985).

M. B. Hindmarsh and T. W. B. Kibble, Cosmic strings, *Rept. Prog. Phys.* **58**, 477 (1995) [arXiv:hep-ph/9411342].

K. Enqvist and A. Mazumdar, Cosmological consequences of MSSM flat directions, *Phys. Rept.* **380**, 99 (2003) [arXiv:hep-ph/0209244] (Sec. 12.7).

Appendix C

S. S. Gershtein, E. P. Kuznetsov and V. A. Ryabov, The nature of neutrino mass and the phenomenon of neutrino oscillations, *Phys. Usp.* **40**, 773 (1997) [*Usp. Fiz. Nauk* **167**, 811 (1997)].

R. N. Mohapatra, ICTP lectures on theoretical aspects of neutrino masses and mixings, [arXiv:hep-ph/0211252].

S. M. Bilenky, Neutrino masses, mixing and oscillations, *Phys. Usp.* **46**, 1137 (2003) [*Usp. Fiz. Nauk* **46**, 1171 (2003)].

W. M. Alberico and S. M. Bilenky, Neutrino oscillations, masses and mixing, *Phys. Part. Nucl.* **35**, 297 (2004) [*Fiz. Elem. Chast. Atom. Yadra* **35**, 545 (2004)] [arXiv:hep-ph/0306239].

A. de Gouvea, 2004 TASI lectures on neutrino physics, [arXiv:hep-ph/0411274].

S. F. King, Neutrino mass models, *Rept. Prog. Phys.* **67**, 107 (2004) [arXiv:hep-ph/0310204].

G. Altarelli and F. Feruglio, Models of neutrino masses and mixings, *New J. Phys.* **6**, 106 (2004) [arXiv:hep-ph/0405048].

R. N. Mohapatra and A. Y. Smirnov, Neutrino mass and new physics, *Ann. Rev. Nucl. Part. Sci.* **56**, 569 (2006) [arXiv:hep-ph/0603118].

A. Yu. Smirnov, Recent developments in neutrino phenomenology, In the *Proceedings of IPM School and Conference on Lepton and Hadron Physics* (IPM-LHP06), Tehran, Iran, 15–20 May 2006, p. 0003 [arXiv:hep-ph/0702061].

E. K. Akhmedov, Neutrino physics, Proceedings: *Summer School in Particle Physics: Trieste*, Italy, June 21-July 9, 1999, [arXiv:hep-ph/0001264].

D. S. Gorbunov, Sterile neutrinos and their role in particle physics and cosmology, *Phys. Usp.* **57**, 503 (2014) [Usp. Fiz. Nauk **184**, 545 (2014)], doi:10.3367/UFNe.0184.201405i.0545.

Appendix D

E. V. Shuryak, Quantum chromodynamics and the theory of superdense matter, *Phys. Rept.* **61**, 71 (1980).

D. J. Gross, R. D. Pisarski and L. G. Yaffe, QCD and instantons at finite temperature, *Rev. Mod. Phys.* **53**, 43 (1981).

Index

www.ingramcontent.com/pod-product-compliance
Lightning Source LLC
Chambersburg PA
CBHW072005230326
41598CB00082B/6770

9 789813 209886